Methods in Enzymology

Volume 350

GUIDE TO YEAST GENETICS AND MOLECULAR
AND CELL BIOLOGY

Part B

METHODS IN ENZYMOLOGY

EDITORS-IN-CHIEF

John N. Abelson Melvin I. Simon

DIVISION OF BIOLOGY
CALIFORNIA INSTITUTE OF TECHNOLOGY
PASADENA, CALIFORNIA

FOUNDING EDITORS

Sidney P. Colowick and Nathan O. Kaplan

Methods in Enzymology

Volume 350

Guide to Yeast Genetics and Molecular and Cell Biology

Part B

EDITED BY

Christine Guthrie

DEPARTMENT OF BIOCHEMISTRY AND BIOPHYSICS
UNIVERSITY OF CALIFORNIA, SAN FRANCISCO
SAN FRANCISCO, CALIFORNIA

Gerald R. Fink

WHITEHEAD INSTITUTE FOR BIOMEDICAL RESEARCH
MASSACHUSETTS INSTITUTE OF TECHNOLOGY
CAMBRIDGE, MASSACHUSETTS

ACADEMIC PRESS

An imprint of Elsevier Science

Amsterdam Boston London New York Oxford Paris
San Diego San Francisco Singapore Sydney Tokyo

Front cover photograph: Mex67-GFP localization in *Saccharomyces cerevisiae.* Photograph courtesy of Anne de Bruyn Kops.

This book is printed on acid-free paper. 

Academic Press
An imprint of Elsevier Science.
525 B Street, Suite 1900, San Diego, California 92101-4495, USA
http://www.academicpress.com

Academic Press
84 Theobalds Road, London WC1X 8RR, UK
http://www.academicpress.com

International Standard Book Number: 0-12-182253-2 (case)
International Standard Book Number: 0-12-310671-0 (cb)

PRINTED IN THE UNITED STATES OF AMERICA
02 03 04 05 06 07 SB 9 8 7 6 5 4 3 2 1

Table of Contents

Section I. Basic Techniques

Section II. Making Mutants

Section III. Genomics

Section IV. Proteomics

Contributors to Volume 350

Article numbers are in parentheses following the names of contributors.
Affiliations listed are current.

REY ANDRADA (19), *Department of Genetics, Stanford University, Stanford, California 94305*

MANUEL ARES, JR. (22), *Department of Molecular, Cell, and Developmental Biology, and Center for Molecular Biology of RNA, University of California, Santa Cruz, California 95064*

NURJANA BACHMAN (13), *Department of Molecular Biology and Genetics, Johns Hopkins University School of Medicine, Baltimore, Maryland 21205*

RAMA BALAKRISHNAN (19), *Department of Genetics, Stanford University, Stanford, California 94305*

CATHERINE A. BALL (19), *Department of Genetics, Stanford University, Stanford, California 94305*

MUNIRA A. BASRAI (24), *Genetics Branch, National Cancer Institute, Bethesda, Maryland 20889*

MATTHEW C. BIERY (13), *Rosetta Inpharmatics, Kirkland, Washington 98034*

GAIL BINKLEY (19), *Department of Genetics, Stanford University, Stanford, California 94305*

ANDERS BLOMBERG (32), *Department of Cell and Molecular Biology, Lundberg Laboratory, University of Göteborg, 41390 Göteborg, Sweden*

JEF D. BOEKE (13), *Department of Molecular Biology and Genetics, Johns Hopkins University School of Medicine, Baltimore, Maryland 21205*

NATHALIE BONNEFOY (5), *Center for Molecular Genetics, Laboratoire propre du CNRS associé à l'Université Pierre et Marie Curie, 91198 Gif-sur-Yvette Cedex, France*

DAVID BOTSTEIN (19), *Department of Genetics, Stanford University, Stanford, California 94305*

KAY BRINKMANN (14), *MediGene AG, D-82152 Martinsreid/Munich, Germany*

JOAN BROOKS (20), *Proteome Division, Incyte Genomics, Beverly, Massachusetts 01915*

IRENE CASTAÑO (11), *Department of Molecular Biology and Genetics, Johns Hopkins University School of Medicine, Baltimore, Maryland 21205*

JOHN CHANT (7), *Department of Molecular and Cellular Biology, Harvard University, Cambridge, Massachusetts 02138*

TRACY CHEN (7), *Department of Molecular and Cellular Biology, Harvard University, Cambridge, Massachusetts 02138*

J. MICHAEL CHERRY (19), *Department of Genetics, Stanford University, Stanford, California 94305*

KAREN R. CHRISTIE (19), *Department of Genetics, Stanford University, Stanford, California 94305*

BRENDAN CORMACK (11), *Department of Molecular Biology and Genetics, Johns Hopkins University School of Medicine, Baltimore, Maryland 21205*

MARIA C. COSTANZO (20), *Proteome Division, Incyte Genomics, Beverly, Massachusetts 01915*

NANCY L. CRAIG (13), *Department of Molecular Biology and Genetics, Johns Hopkins University School of Medicine, Howard Hughes Medical Institute, Baltimore, Maryland 21205*

CSILLA CSANK (20), *Proteome Division, Incyte Genomics, Beverly, Massachusetts 01915*

KARA DOLINSKI (19), *Department of Genetics, Stanford University, Stanford, California 94305*

STAN DONG (19), *Department of Genetics, Stanford University, Stanford, California 94305*

SELINA S. DWIGHT (19), *Department of Genetics, Stanford University, Stanford, California 94305*

HIDEKI ENDOH (30), *Yamanouchi Pharmaceuticals, Tsukuba, Japan*

STANLEY FIELDS (28, 31), *Departments of Genome Sciences and Medicine, Howard Hughes Medical Institute, University of Washington, Seattle, Washington 98195*

DIANNA G. FISK (19), *Department of Genetics, Stanford University, Stanford, California 94305*

THOMAS D. FOX (5), *Department of Molecular Biology and Genetics, Cornell University, Ithaca, New York 14853*

MICHELINE FROMONT-RACINE (29), *Department of Biotechnology, Macromolecular Interactions Genetics Unit, Institut Pasteur, CNRS URA 2171, 75724 Paris-Cedex 15, France*

ATSUSHI FUJITA (7), *Neuroscience Research Institute, National Institute of Advanced Industrial Science and Technology, 1-1-1 Higashi Tsukuba 305-8566, Japan*

MARTIN FUNK (14), *MediGene AG, D-82152 Martinsried/Munich, Germany*

JAMES GARRELS (20), *Proteome Division, Incyte Genomics, Beverly, Massachusetts 01915*

AUDREY P. GASCH (23), *Department of Genome Sciences, Lawrence Berkeley National Laboratory, Berkeley, California 94720*

BILYANA GEORGIEVA (16), *Department of Genetics and Development, Columbia University College of Physicians and Surgeons, New York, New York 10032*

JOSEPH F. GERA (28), *Departments of Hematology and Oncology, University of California/Veterans Administration Medical Center, Los Angeles, California 90073*

R. DANIEL GIETZ (4), *Department of Biochemistry and Medical Genetics, University of Manitoba, Winnipeg, Manitoba, Canada R3M 0T5*

DANIEL E. GOTTSCHLING (9), *Division of Basic Sciences, Fred Hutchinson Cancer Research Center, Seattle, Washington 98109*

LESLIE GRATE (22), *Department of Computer Science and Engineering, Baskin School of Engineering and Center for Molecular Biology of RNA, University of California, Santa Cruz, California 95064*

ELIZABETH J. GRAYHACK (31), *Department of Biochemistry and Biophysics, University of Rochester School of Medicine, Rochester, New York 14642*

JAMES E. HABER (8), *Rosenstiel Center and Department of Biology, Brandeis University, Waltham, Massachusetts 02454*

MIDORI HARRIS (19), *European Molecular Biology Laboratory, Wellcome Trust Genome Campus, European Bioinformatics Institute, Cambridge CB10 1SD, United Kingdom*

TONY R. HAZBUN (28), *Departments of Genome Sciences and Medicine, Howard Hughes Medical Institute, University of Washington, Seattle, Washington 98195*

JOHANNES H. HEGEMANN (17). *Institute for Microbiology, Heinrich-Heine-Universitaet Duesseldorf, D-40225 Duesseldorf, Germany*

THOMAS HENKEL (14), *MediGene AG, D-82152 Martinsried/Munich, Germany*

PHILIP HIETER (18, 24), *Department of Medical Genetics, The Centre for Molecular Medicine and Therapeutics, University of British Columbia, Vancouver, British Columbia, Canada V5Z 4H4*

JODI HIRSCHMAN (20), *Proteome Division, Incyte Genomics, Beverly, Massachusetts 01915*

PETER HODGES (20), *Proteome Division, Incyte Genomics, Beverly, Massachusetts 01915*

CHRISTINE E. HORAK (26), *Department of Molecular, Cellular, and Developmental Biology, Yale University, New Haven, Connecticut 06520*

LAURIE ISSEL-TARVER (19), *Department of Genetics, Stanford University, Stanford, California 94305*

YVES JACOB (30), *Lyssavirus Laboratory, Institut Pasteur, 75015 Paris, France*

MARK JOHNSTON (17), *Department of Genetics, Washington University School of Medicine, St. Louis, Missouri 63110*

JANICE E. KRANZ (20), *Proteome Division, Incyte Genomics, Beverly, Massachusetts 01915*

ANUJ KUMAR (12), *Department of Molecular, Cellular, and Developmental Biology, Yale University, New Haven, Connecticut 06520*

CHRISTOPHER W. LAWRENCE (10), *Department of Biochemistry and Biophysics, University of Rochester School of Medicine and Dentistry, Rochester, New York 14642*

PIERRE LEGRAIN (29), *Hybrigenics, 75014 Paris, France*

FRAN LEWITTER (21), *Whitehead Institute for Biomedical Research, Cambridge, Massachusetts 02142*

HAO LI (27), *Department of Biochemistry and Biophysics, University of California, San Francisco, California 94143*

MICHAEL LISBY (15), *Department of Genetics and Development, Columbia University College of Physicians and Surgeons, New York, New York 10032*

HAOPING LIU (3), *Department of Biological Chemistry, University of California, Irvine, California 92697*

MARK S. LONGTINE (25), *Department of Biochemistry and Molecular Biology, Oklahoma State University, Stillwater, Oklahoma 74078*

MATTHEW LORD (7), *Department of Molecular and Cellular Biology, Harvard University, Cambridge, Massachusetts 02138*

MARY MANGAN (20), *Proteome Division, Incyte Genomics, Beverly, Massachusetts 01915*

MARK R. MARTZEN (31), *CEPTYR, Inc., Bothell, Washington 98021*

STEPHEN M. MCCRAITH (31), *Pacific Northwest Research Institute, Seattle, Washington 98122*

VIVIEN MEASDAY (18), *Department of Medical Genetics, The Centre for Molecular Medicine and Therapeutics, University of British Columbia, Vancouver, British Columbia, Canada V5Z 4H4*

REBECCA K. MONTAGNE (31), *Department of Biochemistry and Biophysics, University of Rochester School of Medicine, Rochester, New York 14642*

DOMINIK MUMBERG (14), *Experimental Oncology, Schering AG, D-13342 Berlin, Germany*

RAINER NIEDENTHAL (14), *Institute for Physiological Chemistry, Medizinische Hochschule Hannover, D-30625 Hannover, Germany*

KATHY E. O'NEILL (20), *Proteome Division, Incyte Genomics, Beverly, Massachusetts 01915*

MARIE E. PETRACEK (25), *Department of Biochemistry and Molecular Biology, Oklahoma State University, Stillwater, Oklahoma 74078*

ERIC M. PHIZICKY (31), *Department of Biochemistry and Biophysics, University of Rochester School of Medicine, Rochester, New York 14642*

JEAN-CHRISTOPHE RAIN (29), *Hybrigenics, 75014 Paris, France*

ELÉONORE RÉAL (30), *Lyssavirus Laboratory, Institut Pasteur, 75015 Paris, France*

ROBERT J. D. REID (15), *Department of Genetics and Development, Columbia University College of Physicians and Surgeons, New York, New York 10032*

LINDA RILES (17), *Department of Genetics, Washington University School of Medicine, St. Louis, Missouri 63110*

LAURA S. ROBERTSON (20), *Proteome Division, Incyte Genomics, Beverly, Massachusetts 01915*

VOLKER RÖNICKE (14), *MediGene AG, D-82152 Martinsried/Munich, Germany*

RODNEY ROTHSTEIN (15, 16), *Department of Genetics and Development, Columbia University College of Physicians and Surgeons, New York, New York 10032*

STEFFEN RUPP (6), *Department of Molecular Biotechnology, Fraunhofer Institute for Interfacial Engineering and Biotechnology, 70569 Stuttgart, Germany*

MARK SCHROEDER (19), *Department of Genetics, Stanford University, Stanford, California 94305*

ANAND SETHURAMAN (19), *Department of Genetics, Stanford University, Stanford, California 94305*

FRED SHERMAN (1), *Department of Biochemistry and Biophysics, University of Rochester Medical School, Rochester, New York 14642*

NEIL P. SHULL (31), *Department of Biochemistry and Biophysics, University of Rochester School of Medicine, Rochester, New York 14642*

MAREK S. SKRZYPEK (20), *Proteome Division, Incyte Genomics, Beverly, Massachusetts 01915*

MICHAEL SNYDER (12, 26), *Department of Molecular, Cellular, and Developmental Biology, Yale University, New Haven, Connecticut 06520*

SHERRY L. SPINELLI (31), *Department of Chemistry, University of Rochester, Rochester, New York 14627*

CORA STYLES (2), *105 Cypress Point, Hendersonville, North Carolina 28739*

FRANCY M. TORRES (31), *Department of Biochemistry and Biophysics, University of Rochester School of Medicine, Rochester, New York 14642*

KANE TSE (19), *Department of Genetics, Stanford University, Stanford, California 94305*

FRED VAN LEEUWEN (9), *Division of Basic Sciences, Fred Hutchinson Cancer Research Center, Seattle, Washington 98109*

CERI VAN SLYKE (31), *Department of Biochemistry and Biophysics, University of Rochester School of Medicine, Rochester, New York 14642*

MARC VIDAL (30), *Department of Genetics and Dana-Farber Cancer Institute, Harvard Medical School, Boston, Massachusetts 02115*

SUSANA VIDAN (12), *Department of Molecular, Cellular, and Developmental Biology, Yale University, New Haven, Connecticut 06520*

SYLVIE VINCENT (30), *Enanta Pharmaceuticals Inc., Cambridge, Massachusetts 02139*

ALBERTHA J. M. WALHOUT (30), *Department of Genetics and Dana-Farber Cancer Institute, Harvard Medical School, Boston, Massachusetts 02115*

SHUAI WENG (19), *Department of Genetics, Stanford University, Stanford, California 94305*

ROBIN A. WOODS (4), *Department of Biology, University of Winnipeg, Winnipeg, Manitoba, Canada R3B 2E9*

FENG XING (31), *Department of Biochemistry and Biophysics, University of Rochester School of Medicine, Rochester, New York 14642*

Preface

Volumes 350 and 351 of *Methods in Enzymology,* "Guide to Yeast Genetics and Molecular and Cell Biology," Parts B and C, reflect the enormous burst of information on *Saccharomyces cerevisiae* since publication of Part A, Volume 194. The ten years between these publications witnessed the emergence of *Saccharomyces cerevisiae* as the most technically advanced experimental organism, extending its versatility as a system to drug discovery, cancer research, and aging. As the first eukaryotic genome to be completely sequenced (April 1996), yeast provided the inaugural view of the basic functions common to all nucleated cells. The availability of the complete yeast genome sequence ($\sim$13 mbp) coupled with facile databases that are easily accessible on the internet quickly fueled the discovery of new techniques such as two hybrid analysis, transcriptional and protein arrays, and sophisticated microscopic techniques, all of which have completely changed the landscape of today's biology. Because of these neoteric advances, Volumes 350 and 351 contain chapters on proteomics and genomics that provide convenient links to reliable sites on the internet. These information-based tools extend the power intrinsic to the traditional yeast genetic system. This vibrancy is evident in the creation of a library containing a null allele for each of the 6100 yeast genes predicted to encode a protein of 100 amino acids or more. This library permits a comprehensive screen of all such genes in the genome for any loss of function phenotype without the biases of random mutant hunts.

These remarkable advances like a searchlight in a cave also reveal many unexplored areas. Some are technical; there is still no reliable method for obtaining pure yeast nuclei. And of the 6000 genes in the genome there are $\sim$35% whose function is not known. Other unexplored areas are of a more theoretical nature. Though we know much about the information coding capacity of yeast genomic DNA there are many molecules whose information content is unknown. The work on yeast prions shows that some proteins contain heritable information not coded in the DNA. How widespread is this phenomenon? Do the lipids, polysaccharides, and RNA molecules passed mitotically from mother to daughter cell or through meiosis to the progeny also exist in alternative states that influence phenotype? It is our hope that the techniques recounted in these volumes will help answer these many questions.

CHRISTINE GUTHRIE
GERALD R. FINK

METHODS IN ENZYMOLOGY

VOLUME XVII. Metabolism of Amino Acids and Amines (Parts A and B)
Edited by HERBERT TABOR AND CELIA WHITE TABOR

VOLUME XVIII. Vitamins and Coenzymes (Parts A, B, and C)
Edited by DONALD B. MCCORMICK AND LEMUEL D. WRIGHT

VOLUME XIX. Proteolytic Enzymes
Edited by GERTRUDE E. PERLMANN AND LASZLO LORAND

VOLUME XX. Nucleic Acids and Protein Synthesis (Part C)
Edited by KIVIE MOLDAVE AND LAWRENCE GROSSMAN

VOLUME XXI. Nucleic Acids (Part D)
Edited by LAWRENCE GROSSMAN AND KIVIE MOLDAVE

VOLUME XXII. Enzyme Purification and Related Techniques
Edited by WILLIAM B. JAKOBY

VOLUME XXIII. Photosynthesis (Part A)
Edited by ANTHONY SAN PIETRO

VOLUME XXIV. Photosynthesis and Nitrogen Fixation (Part B)
Edited by ANTHONY SAN PIETRO

VOLUME XXV. Enzyme Structure (Part B)
Edited by C. H. W. HIRS AND SERGE N. TIMASHEFF

VOLUME XXVI. Enzyme Structure (Part C)
Edited by C. H. W. HIRS AND SERGE N. TIMASHEFF

VOLUME XXVII. Enzyme Structure (Part D)
Edited by C. H. W. HIRS AND SERGE N. TIMASHEFF

VOLUME XXVIII. Complex Carbohydrates (Part B)
Edited by VICTOR GINSBURG

VOLUME XXIX. Nucleic Acids and Protein Synthesis (Part E)
Edited by LAWRENCE GROSSMAN AND KIVIE MOLDAVE

VOLUME XXX. Nucleic Acids and Protein Synthesis (Part F)
Edited by KIVIE MOLDAVE AND LAWRENCE GROSSMAN

VOLUME XXXI. Biomembranes (Part A)
Edited by SIDNEY FLEISCHER AND LESTER PACKER

VOLUME XXXII. Biomembranes (Part B)
Edited by SIDNEY FLEISCHER AND LESTER PACKER

VOLUME XXXIII. Cumulative Subject Index Volumes I-XXX
Edited by MARTHA G. DENNIS AND EDWARD A. DENNIS

VOLUME XXXIV. Affinity Techniques (Enzyme Purification: Part B)
Edited by WILLIAM B. JAKOBY AND MEIR WILCHEK

VOLUME XXXV. Lipids (Part B)
Edited by JOHN M. LOWENSTEIN

VOLUME XXXVI. Hormone Action (Part A: Steroid Hormones)
Edited by BERT W. O'MALLEY AND JOEL G. HARDMAN

VOLUME XXXVII. Hormone Action (Part B: Peptide Hormones)
Edited by BERT W. O'MALLEY AND JOEL G. HARDMAN

VOLUME XXXVIII. Hormone Action (Part C: Cyclic Nucleotides)
Edited by JOEL G. HARDMAN AND BERT W. O'MALLEY

VOLUME XXXIX. Hormone Action (Part D: Isolated Cells, Tissues, and Organ Systems)
Edited by JOEL G. HARDMAN AND BERT W. O'MALLEY

VOLUME XL. Hormone Action (Part E: Nuclear Structure and Function)
Edited by BERT W. O'MALLEY AND JOEL G. HARDMAN

VOLUME XLI. Carbohydrate Metabolism (Part B)
Edited by W. A. WOOD

VOLUME XLII. Carbohydrate Metabolism (Part C)
Edited by W. A. WOOD

VOLUME XLIII. Antibiotics
Edited by JOHN H. HASH

VOLUME XLIV. Immobilized Enzymes
Edited by KLAUS MOSBACH

VOLUME XLV. Proteolytic Enzymes (Part B)
Edited by LASZLO LORAND

VOLUME XLVI. Affinity Labeling
Edited by WILLIAM B. JAKOBY AND MEIR WILCHEK

VOLUME XLVII. Enzyme Structure (Part E)
Edited by C. H. W. HIRS AND SERGE N. TIMASHEFF

VOLUME XLVIII. Enzyme Structure (Part F)
Edited by C. H. W. HIRS AND SERGE N. TIMASHEFF

VOLUME XLIX. Enzyme Structure (Part G)
Edited by C. H. W. HIRS AND SERGE N. TIMASHEFF

VOLUME L. Complex Carbohydrates (Part C)
Edited by VICTOR GINSBURG

VOLUME LI. Purine and Pyrimidine Nucleotide Metabolism
Edited by PATRICIA A. HOFFEE AND MARY ELLEN JONES

VOLUME LII. Biomembranes (Part C: Biological Oxidations)
Edited by SIDNEY FLEISCHER AND LESTER PACKER

VOLUME LIII. Biomembranes (Part D: Biological Oxidations)
Edited by SIDNEY FLEISCHER AND LESTER PACKER

VOLUME LIV. Biomembranes (Part E: Biological Oxidations)
Edited by SIDNEY FLEISCHER AND LESTER PACKER

VOLUME 108. Immunochemical Techniques (Part G: Separation and Characterization of Lymphoid Cells)
Edited by GIOVANNI DI SABATO, JOHN J. LANGONE, AND HELEN VAN VUNAKIS

VOLUME 109. Hormone Action (Part I: Peptide Hormones)
Edited by LUTZ BIRNBAUMER AND BERT W. O'MALLEY

VOLUME 110. Steroids and Isoprenoids (Part A)
Edited by JOHN H. LAW AND HANS C. RILLING

VOLUME 111. Steroids and Isoprenoids (Part B)
Edited by JOHN H. LAW AND HANS C. RILLING

VOLUME 112. Drug and Enzyme Targeting (Part A)
Edited by KENNETH J. WIDDER AND RALPH GREEN

VOLUME 113. Glutamate, Glutamine, Glutathione, and Related Compounds
Edited by ALTON MEISTER

VOLUME 114. Diffraction Methods for Biological Macromolecules (Part A)
Edited by HAROLD W. WYCKOFF, C. H. W. HIRS, AND SERGE N. TIMASHEFF

VOLUME 115. Diffraction Methods for Biological Macromolecules (Part B)
Edited by HAROLD W. WYCKOFF, C. H. W. HIRS, AND SERGE N. TIMASHEFF

VOLUME 116. Immunochemical Techniques (Part H: Effectors and Mediators of Lymphoid Cell Functions)
Edited by GIOVANNI DI SABATO, JOHN J. LANGONE, AND HELEN VAN VUNAKIS

VOLUME 117. Enzyme Structure (Part J)
Edited by C. H. W. HIRS AND SERGE N. TIMASHEFF

VOLUME 118. Plant Molecular Biology
Edited by ARTHUR WEISSBACH AND HERBERT WEISSBACH

VOLUME 119. Interferons (Part C)
Edited by SIDNEY PESTKA

VOLUME 120. Cumulative Subject Index Volumes 81–94, 96–101

VOLUME 121. Immunochemical Techniques (Part I: Hybridoma Technology and Monoclonal Antibodies)
Edited by JOHN J. LANGONE AND HELEN VAN VUNAKIS

VOLUME 122. Vitamins and Coenzymes (Part G)
Edited by FRANK CHYTIL AND DONALD B. MCCORMICK

VOLUME 123. Vitamins and Coenzymes (Part H)
Edited by FRANK CHYTIL AND DONALD B. MCCORMICK

VOLUME 124. Hormone Action (Part J: Neuroendocrine Peptides)
Edited by P. MICHAEL CONN

VOLUME 125. Biomembranes (Part M: Transport in Bacteria, Mitochondria, and Chloroplasts: General Approaches and Transport Systems)
Edited by SIDNEY FLEISCHER AND BECCA FLEISCHER

VOLUME 126. Biomembranes (Part N: Transport in Bacteria, Mitochondria, and Chloroplasts: Protonmotive Force)
Edited by SIDNEY FLEISCHER AND BECCA FLEISCHER

VOLUME 127. Biomembranes (Part O: Protons and Water: Structure and Translocation)
Edited by LESTER PACKER

VOLUME 128. Plasma Lipoproteins (Part A: Preparation, Structure, and Molecular Biology)
Edited by JERE P. SEGREST AND JOHN J. ALBERS

VOLUME 129. Plasma Lipoproteins (Part B: Characterization, Cell Biology, and Metabolism)
Edited by JOHN J. ALBERS AND JERE P. SEGREST

VOLUME 130. Enzyme Structure (Part K)
Edited by C. H. W. HIRS AND SERGE N. TIMASHEFF

VOLUME 131. Enzyme Structure (Part L)
Edited by C. H. W. HIRS AND SERGE N. TIMASHEFF

VOLUME 132. Immunochemical Techniques (Part J: Phagocytosis and Cell-Mediated Cytotoxicity)
Edited by GIOVANNI DI SABATO AND JOHANNES EVERSE

VOLUME 133. Bioluminescence and Chemiluminescence (Part B)
Edited by MARLENE DELUCA AND WILLIAM D. MCELROY

VOLUME 134. Structural and Contractile Proteins (Part C: The Contractile Apparatus and the Cytoskeleton)
Edited by RICHARD B. VALLEE

VOLUME 135. Immobilized Enzymes and Cells (Part B)
Edited by KLAUS MOSBACH

VOLUME 136. Immobilized Enzymes and Cells (Part C)
Edited by KLAUS MOSBACH

VOLUME 137. Immobilized Enzymes and Cells (Part D)
Edited by KLAUS MOSBACH

VOLUME 138. Complex Carbohydrates (Part E)
Edited by VICTOR GINSBURG

VOLUME 139. Cellular Regulators (Part A: Calcium- and Calmodulin-Binding Proteins)
Edited by ANTHONY R. MEANS AND P. MICHAEL CONN

VOLUME 140. Cumulative Subject Index Volumes 102–119, 121–134

VOLUME 141. Cellular Regulators (Part B: Calcium and Lipids)
Edited by P. MICHAEL CONN AND ANTHONY R. MEANS

VOLUME 142. Metabolism of Aromatic Amino Acids and Amines
Edited by SEYMOUR KAUFMAN

VOLUME 161. Biomass (Part B: Lignin, Pectin, and Chitin)
Edited by WILLIS A. WOOD AND SCOTT T. KELLOGG

VOLUME 162. Immunochemical Techniques (Part L: Chemotaxis and Inflammation)
Edited by GIOVANNI DI SABATO

VOLUME 163. Immunochemical Techniques (Part M: Chemotaxis and Inflammation)
Edited by GIOVANNI DI SABATO

VOLUME 164. Ribosomes
Edited by HARRY F. NOLLER, JR., AND KIVIE MOLDAVE

VOLUME 165. Microbial Toxins: Tools for Enzymology
Edited by SIDNEY HARSHMAN

VOLUME 166. Branched-Chain Amino Acids
Edited by ROBERT HARRIS AND JOHN R. SOKATCH

VOLUME 167. Cyanobacteria
Edited by LESTER PACKER AND ALEXANDER N. GLAZER

VOLUME 168. Hormone Action (Part K: Neuroendocrine Peptides)
Edited by P. MICHAEL CONN

VOLUME 169. Platelets: Receptors, Adhesion, Secretion (Part A)
Edited by JACEK HAWIGER

VOLUME 170. Nucleosomes
Edited by PAUL M. WASSARMAN AND ROGER D. KORNBERG

VOLUME 171. Biomembranes (Part R: Transport Theory: Cells and Model Membranes)
Edited by SIDNEY FLEISCHER AND BECCA FLEISCHER

VOLUME 172. Biomembranes (Part S: Transport: Membrane Isolation and Characterization)
Edited by SIDNEY FLEISCHER AND BECCA FLEISCHER

VOLUME 173. Biomembranes [Part T: Cellular and Subcellular Transport: Eukaryotic (Nonepithelial) Cells]
Edited by SIDNEY FLEISCHER AND BECCA FLEISCHER

VOLUME 174. Biomembranes [Part U: Cellular and Subcellular Transport: Eukaryotic (Nonepithelial) Cells]
Edited by SIDNEY FLEISCHER AND BECCA FLEISCHER

VOLUME 175. Cumulative Subject Index Volumes 135–139, 141–167

VOLUME 176. Nuclear Magnetic Resonance (Part A: Spectral Techniques and Dynamics)
Edited by NORMAN J. OPPENHEIMER AND THOMAS L. JAMES

VOLUME 177. Nuclear Magnetic Resonance (Part B: Structure and Mechanism)
Edited by NORMAN J. OPPENHEIMER AND THOMAS L. JAMES

VOLUME 178. Antibodies, Antigens, and Molecular Mimicry
Edited by JOHN J. LANGONE

VOLUME 179. Complex Carbohydrates (Part F)
Edited by VICTOR GINSBURG

VOLUME 180. RNA Processing (Part A: General Methods)
Edited by JAMES E. DAHLBERG AND JOHN N. ABELSON

VOLUME 181. RNA Processing (Part B: Specific Methods)
Edited by JAMES E. DAHLBERG AND JOHN N. ABELSON

VOLUME 182. Guide to Protein Purification
Edited by MURRAY P. DEUTSCHER

VOLUME 183. Molecular Evolution: Computer Analysis of Protein and Nucleic Acid Sequences
Edited by RUSSELL F. DOOLITTLE

VOLUME 184. Avidin-Biotin Technology
Edited by MEIR WILCHEK AND EDWARD A. BAYER

VOLUME 185. Gene Expression Technology
Edited by DAVID V. GOEDDEL

VOLUME 186. Oxygen Radicals in Biological Systems (Part B: Oxygen Radicals and Antioxidants)
Edited by LESTER PACKER AND ALEXANDER N. GLAZER

VOLUME 187. Arachidonate Related Lipid Mediators
Edited by ROBERT C. MURPHY AND FRANK A. FITZPATRICK

VOLUME 188. Hydrocarbons and Methylotrophy
Edited by MARY E. LIDSTROM

VOLUME 189. Retinoids (Part A: Molecular and Metabolic Aspects)
Edited by LESTER PACKER

VOLUME 190. Retinoids (Part B: Cell Differentiation and Clinical Applications)
Edited by LESTER PACKER

VOLUME 191. Biomembranes (Part V: Cellular and Subcellular Transport: Epithelial Cells)
Edited by SIDNEY FLEISCHER AND BECCA FLEISCHER

VOLUME 192. Biomembranes (Part W: Cellular and Subcellular Transport: Epithelial Cells)
Edited by SIDNEY FLEISCHER AND BECCA FLEISCHER

VOLUME 193. Mass Spectrometry
Edited by JAMES A. MCCLOSKEY

VOLUME 194. Guide to Yeast Genetics and Molecular Biology
Edited by CHRISTINE GUTHRIE AND GERALD R. FINK

VOLUME 195. Adenylyl Cyclase, G Proteins, and Guanylyl Cyclase
Edited by ROGER A. JOHNSON AND JACKIE D. CORBIN

Volume 213. Carotenoids (Part A: Chemistry, Separation, Quantitation, and Antioxidation)
Edited by Lester Packer

Volume 214. Carotenoids (Part B: Metabolism, Genetics, and Biosynthesis)
Edited by Lester Packer

Volume 215. Platelets: Receptors, Adhesion, Secretion (Part B)
Edited by Jacek J. Hawiger

Volume 216. Recombinant DNA (Part G)
Edited by Ray Wu

Volume 217. Recombinant DNA (Part H)
Edited by Ray Wu

Volume 218. Recombinant DNA (Part I)
Edited by Ray Wu

Volume 219. Reconstitution of Intracellular Transport
Edited by James E. Rothman

Volume 220. Membrane Fusion Techniques (Part A)
Edited by Nejat Düzguünes

Volume 221. Membrane Fusion Techniques (Part B)
Edited by Nejat Düzgünes

Volume 222. Proteolytic Enzymes in Coagulation, Fibrinolysis, and Complement Activation (Part A: Mammalian Blood Coagulation Factors and Inhibitors)
Edited by Laszlo Lorand and Kenneth G. Mann

Volume 223. Proteolytic Enzymes in Coagulation, Fibrinolysis, and Complement Activation (Part B: Complement Activation, Fibrinolysis, and Nonmammalian Blood Coagulation Factors)
Edited by Laszlo Lorand and Kenneth G. Mann

Volume 224. Molecular Evolution: Producing the Biochemical Data
Edited by Elizabeth Anne Zimmer, Thomas J. White, Rebecca L. Cann, and Allan C. Wilson

Volume 225. Guide to Techniques in Mouse Development
Edited by Paul M. Wassarman and Melvin L. DePamphilis

Volume 226. Metallobiochemistry (Part C: Spectroscopic and Physical Methods for Probing Metal Ion Environments in Metalloenzymes and Metalloproteins)
Edited by James F. Riordan and Bert L. Vallee

Volume 227. Metallobiochemistry (Part D: Physical and Spectroscopic Methods for Probing Metal Ion Environments in Metalloproteins)
Edited by James F. Riordan and Bert L. Vallee

Volume 228. Aqueous Two-Phase Systems
Edited by Harry Walter and Göte Johansson

VOLUME 248. Proteolytic Enzymes: Aspartic and Metallo Peptidases
Edited by ALAN J. BARRETT

VOLUME 249. Enzyme Kinetics and Mechanism (Part D: Developments in Enzyme Dynamics)
Edited by DANIEL L. PURICH

VOLUME 250. Lipid Modifications of Proteins
Edited by PATRICK J. CASEY AND JANICE E. BUSS

VOLUME 251. Biothiols (Part A: Monothiols and Dithiols, Protein Thiols, and Thiyl Radicals)
Edited by LESTER PACKER

VOLUME 252. Biothiols (Part B: Glutathione and Thioredoxin; Thiols in Signal Transduction and Gene Regulation)
Edited by LESTER PACKER

VOLUME 253. Adhesion of Microbial Pathogens
Edited by RON J. DOYLE AND ITZHAK OFEK

VOLUME 254. Oncogene Techniques
Edited by PETER K. VOGT AND INDER M. VERMA

VOLUME 255. Small GTPases and Their Regulators (Part A: Ras Family)
Edited by W. E. BALCH, CHANNING J. DER, AND ALAN HALL

VOLUME 256. Small GTPases and Their Regulators (Part B: Rho Family)
Edited by W. E. BALCH, CHANNING J. DER, AND ALAN HALL

VOLUME 257. Small GTPases and Their Regulators (Part C: Proteins Involved in Transport)
Edited by W. E. BALCH, CHANNING J. DER, AND ALAN HALL

VOLUME 258. Redox-Active Amino Acids in Biology
Edited by JUDITH P. KLINMAN

VOLUME 259. Energetics of Biological Macromolecules
Edited by MICHAEL L. JOHNSON AND GARY K. ACKERS

VOLUME 260. Mitochondrial Biogenesis and Genetics (Part A)
Edited by GIUSEPPE M. ATTARDI AND ANNE CHOMYN

VOLUME 261. Nuclear Magnetic Resonance and Nucleic Acids
Edited by THOMAS L. JAMES

VOLUME 262. DNA Replication
Edited by JUDITH L. CAMPBELL

VOLUME 263. Plasma Lipoproteins (Part C: Quantitation)
Edited by WILLIAM A. BRADLEY, SANDRA H. GIANTURCO, AND JERE P. SEGREST

VOLUME 264. Mitochondrial Biogenesis and Genetics (Part B)
Edited by GIUSEPPE M. ATTARDI AND ANNE CHOMYN

VOLUME 265. Cumulative Subject Index Volumes 228, 230–262

VOLUME 303. cDNA Preparation and Display
Edited by SHERMAN M. WEISSMAN

VOLUME 304. Chromatin
Edited by PAUL M. WASSARMAN AND ALAN P. WOLFFE

VOLUME 305. Bioluminescence and Chemiluminescence (Part C)
Edited by THOMAS O. BALDWIN AND MIRIAM M. ZIEGLER

VOLUME 306. Expression of Recombinant Genes in Eukaryotic Systems
Edited by JOSEPH C. GLORIOSO AND MARTIN C. SCHMIDT

VOLUME 307. Confocal Microscopy
Edited by P. MICHAEL CONN

VOLUME 308. Enzyme Kinetics and Mechanism (Part E: Energetics of Enzyme Catalysis)
Edited by DANIEL L. PURICH AND VERN L. SCHRAMM

VOLUME 309. Amyloid, Prions, and Other Protein Aggregates
Edited by RONALD WETZEL

VOLUME 310. Biofilms
Edited by RON J. DOYLE

VOLUME 311. Sphingolipid Metabolism and Cell Signaling (Part A)
Edited by ALFRED H. MERRILL, JR., AND YUSUF A. HANNUN

VOLUME 312. Sphingolipid Metabolism and Cell Signaling (Part B)
Edited by ALFRED H. MERRILL, JR., AND YUSUF A. HANNUN

VOLUME 313. Antisense Technology (Part A: General Methods, Methods of Delivery, and RNA Studies)
Edited by M. IAN PHILLIPS

VOLUME 314. Antisense Technology (Part B: Applications)
Edited by M. IAN PHILLIPS

VOLUME 315. Vertebrate Phototransduction and the Visual Cycle (Part A)
Edited by KRZYSZTOF PALCZEWSKI

VOLUME 316. Vertebrate Phototransduction and the Visual Cycle (Part B)
Edited by KRZYSZTOF PALCZEWSKI

VOLUME 317. RNA–Ligand Interactions (Part A: Structural Biology Methods)
Edited by DANIEL W. CELANDER AND JOHN N. ABELSON

VOLUME 318. RNA–Ligand Interactions (Part B: Molecular Biology Methods)
Edited by DANIEL W. CELANDER AND JOHN N. ABELSON

VOLUME 319. Singlet Oxygen, UV-A, and Ozone
Edited by LESTER PACKER AND HELMUT SIES

VOLUME 320. Cumulative Subject Index Volumes 290–319

VOLUME 321. Numerical Computer Methods (Part C)
Edited by MICHAEL L. JOHNSON AND LUDWIG BRAND

VOLUME 322. Apoptosis
Edited by JOHN C. REED

VOLUME 323. Energetics of Biological Macromolecules (Part C)
Edited by MICHAEL L. JOHNSON AND GARY K. ACKERS

VOLUME 324. Branched-Chain Amino Acids (Part B)
Edited by ROBERT A. HARRIS AND JOHN R. SOKATCH

VOLUME 325. Regulators and Effectors of Small GTPases (Part D: Rho Family)
Edited by W. E. BALCH, CHANNING J. DER, AND ALAN HALL

VOLUME 326. Applications of Chimeric Genes and Hybrid Proteins (Part A: Gene Expression and Protein Purification)
Edited by JEREMY THORNER, SCOTT D. EMR, AND JOHN N. ABELSON

VOLUME 327. Applications of Chimeric Genes and Hybrid Proteins (Part B: Cell Biology and Physiology)
Edited by JEREMY THORNER, SCOTT D. EMR, AND JOHN N. ABELSON

VOLUME 328. Applications of Chimeric Genes and Hybrid Proteins (Part C: Protein-Protein Interactions and Genomics)
Edited by JEREMY THORNER, SCOTT D. EMR, AND JOHN N. ABELSON

VOLUME 329. Regulators and Effectors of Small GTPases (Part E: GTPases Involved in Vesicular Traffic)
Edited by W. E. BALCH, CHANNING J. DER, AND ALAN HALL

VOLUME 330. Hyperthermophilic Enzymes (Part A)
Edited by MICHAEL W. W. ADAMS AND ROBERT M. KELLY

VOLUME 331. Hyperthermophilic Enzymes (Part B)
Edited by MICHAEL W. W. ADAMS AND ROBERT M. KELLY

VOLUME 332. Regulators and Effectors of Small GTPases (Part F: Ras Family I)
Edited by W. E. BALCH, CHANNING J. DER, AND ALAN HALL

VOLUME 333. Regulators and Effectors of Small GTPases (Part G: Ras Family II)
Edited by W. E. BALCH, CHANNING J. DER, AND ALAN HALL

VOLUME 334. Hyperthermophilic Enzymes (Part C)
Edited by MICHAEL W. W. ADAMS AND ROBERT M. KELLY

VOLUME 335. Flavonoids and Other Polyphenols
Edited by LESTER PACKER

VOLUME 336. Microbial Growth in Biofilms (Part A: Developmental and Molecular Biological Aspects)
Edited by RON J. DOYLE

VOLUME 337. Microbial Growth in Biofilms (Part B: Special Environments and Physicochemical Aspects)
Edited by RON J. DOYLE

VOLUME 338. Nuclear Magnetic Resonance of Biological Macromolecules (Part A)
Edited by THOMAS L. JAMES, VOLKER DÖTSCH, AND ULI SCHMITZ

VOLUME 339. Nuclear Magnetic Resonance of Biological Macromolecules (Part B)
Edited by THOMAS L. JAMES, VOLKER DÖTSCH, AND ULI SCHMITZ

VOLUME 340. Drug–Nucleic Acid Interactions
Edited by JONATHAN B. CHAIRES AND MICHAEL J. WARING

VOLUME 341. Ribonucleases (Part A)
Edited by ALLEN W. NICHOLSON

VOLUME 342. Ribonucleases (Part B)
Edited by ALLEN W. NICHOLSON

VOLUME 343. G Protein Pathways (Part A: Receptors)
Edited by RAVI IYENGAR AND JOHN D. HILDEBRANDT

VOLUME 344. G Protein Pathways (Part B: G Proteins and Their Regulators)
Edited by RAVI IYENGAR AND JOHN D. HILDEBRANDT

VOLUME 345. G Protein Pathways (Part C: Effector Mechanisms)
Edited by RAVI IYENGAR AND JOHN D. HILDEBRANDT

VOLUME 346. Gene Therapy Methods
Edited by M. IAN PHILLIPS

VOLUME 347. Protein Sensors and Reactive Oxygen Species (Part A: Selenoproteins and Thioredoxin)
Edited by HELMUT SIES AND LESTER PACKER

VOLUME 348. Protein Sensors and Reactive Oxygen Species (Part B: Thiol Enzymes and Proteins)
Edited by HELMUT SIES AND LESTER PACKER

VOLUME 349. Superoxide Dismutase
Edited by LESTER PACKER

VOLUME 350. Guide to Yeast Genetics and Molecular and Cell Biology (Part B)
Edited by CHRISTINE GUTHRIE AND GERALD R. FINK

VOLUME 351. Guide to Yeast Genetics and Molecular and Cell Biology (Part C)
Edited by CHRISTINE GUTHRIE AND GERALD R. FINK

VOLUME 352. Redox Cell Biology and Genetics (Part A) (in preparation)
Edited by CHANDAN K. SEN AND LESTER PACKER

VOLUME 353. Redox Cell Biology and Genetics (Part B) (in preparation)
Edited by CHANDAN K. SEN AND LESTER PACKER

VOLUME 354. Enzyme Kinetics and Mechanism (Part F: Detection and Characterization of Enzyme Reaction Intermediates) (in preparation)
Edited by DANIEL L. PURICH

Section I

Basic Techniques

[1] Getting Started with Yeast

By FRED SHERMAN

Yeast as a Model System

The yeast *Saccharomyces cerevisiae* is now recognized as a model system representing a simple eukaryote whose genome can be easily manipulated. Yeast has only a slightly greater genetic complexity than bacteria and shares many of the technical advantages that permitted rapid progress in the molecular genetics of prokaryotes and their viruses. Some of the properties that make yeast particularly suitable for biological studies include rapid growth, dispersed cells, the ease of replica plating and mutant isolation, a well-defined genetic system, and most important, a highly versatile DNA transformation system.[1] Being nonpathogenic, yeast can be handled with little precautions. Large quantities of normal baker's yeast are commercially available and can provide a cheap source for biochemical studies.

Strains of *S. cerevisiae,* unlike most other microorganisms, have both a stable haploid and a stable diploid state and are viable with a large number of markers. Thus, recessive mutations are conveniently manifested in haploid strains, whereas complementation tests can be carried out with diploid strains. The development of DNA transformation has made yeast particularly accessible to gene cloning and genetic engineering techniques. Structural genes corresponding to virtually any genetic trait can be identified by complementation from plasmid libraries. Plasmids can be introduced into yeast cells either as replicating molecules or by integration into the genome. In contrast to most other organisms, integrative recombination of transforming DNA in yeast proceeds exclusively via homologous recombination. Cloned yeast sequences, accompanied by foreign sequences on plasmids, can therefore be directed at will to specific locations in the genome.

In addition, homologous recombination coupled with high levels of gene conversion in yeast has led to the development of techniques for the direct replacement of genetically engineered DNA sequences into their normal chromosome locations. Thus, normal wild-type genes, even those having no previously known mutations, can be conveniently replaced with altered and disrupted alleles. The phenotypes arising after disruption of yeast genes have contributed significantly toward understanding of the function of certain proteins *in vivo.* Many investigators have been shocked to find viable mutants with few or no detrimental phenotypes after distrupting "essential" genes. Genes can be directly replaced at high efficiencies in yeasts and other fungi, but only with difficulty in other eukaryotic organisms. Also unique to yeast, transformation can be carried out directly with short

[1] R. D. Gietz and R. A. Wooda, *Methods Enzymol.* **350,** [4], 2002 (this volume).

0076-6879/01 $35.00

single-stranded synthetic oligonucleotides, permitting the convenient productions of numerous altered forms of proteins. These techniques have been extensively exploited in the analysis of gene regulation, structure–function relationships of proteins, chromosome structure, and other general questions in cell biology.

Saccharomyces cerevisiae was the first eukaryote whose genome was completely sequenced.[2] Subsequently, yeast became one of the key organisms for genomic research,[3] including extensive use of DNA microarrays for investigating the transcriptome,[4–7a,7b,7c] as well as genome-wide analysis of gene functions by gene disruption,[8] of serial analysis of gene expression (SAGE),[9] of protein localization,[10] of 2-D protein maps,[11,12] of enzymatic activities,[13,14] and of protein–protein interactions by two-hybrid analysis.[15,16]

The overriding virtues of yeast are illustrated by the fact that mammalian genes are routinely being introduced into yeast for systematic analyses of the functions of the corresponding gene products. Many human genes related to disease have orthologs in yeast,[17] and the high conservation of metabolic and regulatory mechanisms has contributed to the widespread use of yeast as a model eukaryotic

[2] A. Goffeau, B. G. Barrell, H. Bussey, R. W. Davis, B. Dujon, H. Feldmann, F. Galibert, J. D. Hoheisel, C. Jacq, M. Johnston, E. J. Louis, H. W. Mewes, Y. Murakami, P. Philippsen, H. Tettelin, and S. G. Oliver, *Science* **274,** 546 (1996).

[3] A. Kumar and M. Snyder, *Nature Rev. Genet.* **2,** 302 (2001).

[4] R. J. Cho, M. J. Cambell, E. A. Winzeler, L. Steinmetz, A. Conway, L. Wodicka, T. G. Wolfsberg, A. E. Gabrielian, D. Landsman, D. J. Lockhart, and R. W. Davis, *Mol. Cell* **2,** 65 (1998).

[5] J. L. DeRisi, V. R. Iyer, and P. O. Brown, *Science* **278,** 680 (1997).

[6] P. O. Brown and D. Botstein, *Nature Genet.* **21,** 33 (1999).

[7a] P. Marc, F. Devaux, and C. Jacq, *Nucleic Acids Res.* **29,** e63 (2001).

[7b] http://transcriptome.ens.fr/ymgv/

[7c] A. P. Gash, *Methods Enzymol.* **350,** [23], 2002 (this volume).

[8] S. Oliver, *Nature* **379,** 597 (1996).

[9] V. E. Velculescu, L. Zhang, W. Zhou, J. Vogelstein, M. A. Basrai, D. E. Bassett, Jr., P. Hieter, B. Vogelstein, and K. W. Kinzler, *Cell* **88,** 243 (1997).

[10] P. Ross-Macdonald, A. Sheehan, C. Friddle, G. S. Roeder, and M. Snyder, *Methods Enzymol.* **303,** 512 (1999).

[11] J. I. Garrels, C. S. McLaughlin, J. R. Warner, B. Futcher, G. I. Latter, R. Kobayashi, B. Schwender, T. Volpe, D. S. Anderson, R. Mesquita-Fuentes, and W. E. Payne, *Electrophoresis* **18,** 1347 (1997).

[12] A. Blomberg, *Methods Enzymol.* **350,** [32], 2002 (this volume).

[13] M. R. Martzen, S. M. McCraith, S. L. Spinelli, F. M. Torres, S. Field, E. J. Grayhack, and E. M. Phizicky, *Science* **286,** 1153 (1999).

[14] E. M. Phizicky, M. R. Martzen, S. M. McCraith, S. L. Spinelli, F. Xeng, N. P. Shull, C. Yanslyke, R. K. Montagne, F. M. Torres, S. Fields, and E. J. Grayhack, *Methods Enzymol.* **350,** [31], 2002 (this volume).

[15] P. Uetz, L. Giot, G. Cagney, T. A. Mansfield, R. S. Judson, J. R. Knight, D. Lockshon, V. Narayan, M. Srinivasan, P. Pochart, A. Qureshi-Emili, Y. Li, B. Godwin, D. Conover, T. Kalbfleisch, G. Vijayadamodar, M. Yang, M. Johnston, S. Fields, and J. M. Rothberg, *Nature* **403,** 623 (2000).

[16] T. Ito, K. Tashiro, S. Muta, R. Ozawa, T. Chiba, M. Nishizawa, K. Yamamoto, S. Kuhara, and Y. Sakaki, *Proc. Natl Acad. Sci. U.S.A.* **97,** 1143 (2000).

[17] R. Ploger, J. Zhang, D. Bassett, R. Reeves, P. Hieter, M. Boguski, and F. Spencer, *Nucleic Acids Res.* **28,** 120 (2000).

system for diverse biological studies. Furthermore, the ability of yeast to replicate artificial circular and linear chromosomes has allowed detailed studies of telomeres, centromeres, length dependencies, and origins of replication. Mitochondrial DNA can be altered in defined ways by transformation,[18] adding to the already impressive genetic and biochemical techniques that have allowed detailed analysis of this organelle. The ease with which the genome of yeast can be manipulated is truly unprecedented for any other eukaryote. Many of these techniques are reviewed in this volume.

Information on Yeast

A general introduction to a few selected topics on yeast can be found in the book chapters "Yeast as the *E. coli* of Eucaryotic Cells" and "Recombinant DNA at Work."[19] An introduction to the genetics and molecular biology of *S. cerevisiae* is presented on the Internet in a review by Sherman.[20] Comprehensive and excellent reviews of the genetics and molecular biology of *S. cerevisiae* are contained in three volumes entitled *Molecular Biology of the Yeast Saccharomyces*.[21–23] An important source for methods used in genetics and molecular biology of yeast is contained in this volume and in other volumes in this series also edited by Guthrie and Fink.[24,24a] Overviews of numerous subjects are also covered in other sources,[21–26] including protocols applicable to yeasts[27] and introductory material.[28] *Methods in Yeast Genetics: A Cold Spring Harbor Laboratory Course Manual*[29] contains useful

[18] N. Bonnefoy and T. Fox, *Methods Enzymol.* **350,** [5], 2002 (this volume).

[19] J. D. Watson, N. H. Hopkins, J. W. Roberts, J. A. Steitz, and A. M. Weiner, "Molecular Biology of the Gene," Chapters 18 and 19. Benjamin/Cummings Pub., Menlo Park, CA, 1987.

[20] http://dbb.urmc.rochester.edu/labs/sherman_f/yeast/Index.html

[21] J. R. Broach, E. W. Jones, and J. R. Pringle (eds.), "The Molecular and Cellular Biology of the Yeast *Saccharomyces,*" Vol. 1. Genome Dynamics, Protein Synthesis, and Energetics. Cold Spring Harbor Laboratory Press, Cold Spring Harbor, NY, 1991.

[22] E. W. Jones, J. R. Pringle, and J. R. Broach (eds.), "The Molecular and Cellular Biology of the Yeast *Saccharomyces,*" Vol. 2. Gene Expression. Cold Spring Harbor Laboratory Press, Cold Spring Harbor, NY, 1992.

[23] J. R. Pringle, J. R. Broach, and E. W. Jones (eds.), "The Molecular and Cellular Biology of the Yeast *Saccharomyces,*" Vol. 3. Cell Cycle and Cell Biology. Cold Spring Harbor Laboratory Press, Cold Spring Harbor, NY, 1997.

[24] C. Guthrie and G. R. Fink (eds.), *Methods Enzymol.* **194** (1991).

[24a] C. Guthrie and G. R. Fink (eds.), *Methods Enzymol.* **351** (2002).

[25] A. J. P. Brown and M. F. Tuite (eds.), "Yeast Gene Analysis," *Methods Microbiol.* **26.** Academic Press, New York, 1998.

[26] A. E. Wheals, A. H. Rose, and J. S. Harrison (eds.), "The Yeast," Vol. 6, 2nd Ed. Yeast Genetics. Academic Press, London, 1995.

[27] S. Fields and M. Johnson (eds.), *Methods, A Companion to Methods Enzymol.* **5,** 77 (1993).

[28] G. M. Walker, "Yeast Physiology and Biotechnology." John Wiley and Son, Chichester, UK, 1998.

[29] D. Burke, D. Dawson, and T. Stearns, "Methods in Yeast Genetics: A Cold Spring Harbor Laboratory Course Manual." Cold Spring Harbor Laboratory Press, Cold Spring Harbor, NY, 2000.

elementary material and protocols and is frequently updated. The use of yeast as an instructional tool at the secondary school level has been provided by Manney and others,[30] including classroom guides and kits.[31] A more comprehensive listing of earlier reviews can be found in an article by Sherman.[32] Interesting and amusing accounts of developments in the field are covered in *The Early Days of Yeast Genetics*.[33] The journal *Yeast* publishes research articles, reviews, short communications, sequencing reports, and selective lists of current articles on all aspects of *Saccharomyces* and other yeast genera.[34] A journal, *FEMS Yeast Research,* began publication in 2001.[35] Current and frequently updated information and databases on yeast can be conveniently retrieved on the Internet through the World Wide Web, including the "*Saccharomyces* Genomic Information Resource"[36,37] and linked files containing DNA sequences, lists of genes, home pages of yeast workers, and other useful information concerning yeast. From the MIPS page[38] one can access the annotated sequence information of the genome of *Saccharomyces cerevisiae* and view the chromosomes graphically or as text, and more. The YPD page[39,40,41] contains a protein database with emphasis on the physical and functional properties of the yeast proteins.

Strains of *Saccharomyces cerevisiae*

Although genetic analyses have been undertaken with a number of taxonomically distinct varieties of yeast, extensive studies have been restricted primarily to the many freely interbreeding species of the budding yeast *Saccharomyces* and to the fission yeast *Schizosaccharomyces pombe*. Although "*Saccharomyces cerevisiae*" is commonly used to designate many of the laboratory stocks of

[30] http://www-personal.ksu.edu/~bethmont/rl2k/yeast.html

[31] https://www3.carolina.com/onlinecatalog/Templates/Default/mainscreen2frame.asp?workspace=home&

[32] F. Sherman, *Methods Enzymol.* **194,** 3 (1991).

[33] M. N. Hall and P. Linder (eds.), "The Early Days of Yeast Genetics." Cold Spring Harbor Laboratory Press, Cold Spring Harbor, NY, 1992.

[34] http://www.interscience.wiley.com/jpages/0749-503X/

[35] http://www.elsevier.nl/locate/femsyr

[36] http://genome-www.stanford.edu/Saccharomyces/

[37] L. Issel-Tarver, K. R. Christie, K. Dolinski, R. Andrada, R. Balakrishnan, C. A. Binkley, S. Dong, S. S. Dwight, D. G. Fisk, M. Harris, M. Schroede, A. Sethuraman, K. Tse, S. Weng, D. Botstein, and J. M. Cherry, *Methods Enzymol.* **350,** [19], 2002 (this volume).

[38] http://www.mips.biochem.mpg.de/

[39] P. E. Hodges, A. H. McKee, B. P. Davis, W. E. Payne, and J. I. Garrels, *Nucl. Acids. Res.* **27,** 69 (1999).

[40] C. Csank, M. C. Costango, J. Hirachmann, P. Hodges, J. E. Kranz, M. Mangan, K. E. O'Neill, L. S. Robertson, K. S. Skrzypak, J. Brooks, and J. Garrels, *Methods Enzymol.* **350,** [20], 2002 (this volume).

[41] http://www.proteome.com/YPDhome.html

Saccharomyces used throughout the world, it should be pointed out that most of these strains originated from the interbred stocks of Winge, Lindegren, and others who employed fermentation markers not only from *S. cerevisiae* but also from *S. bayanus, S. carlsbergensis, S. chevalieri, S. chodati, S. diastaticus,* etc.[42,43] Nevertheless, it is still recommended that the interbreeding laboratory stocks of *Saccharomyces* be denoted as *S. cerevisiae* in order to conveniently distinguish them from the more distantly related species of *Saccharomyces*.

Care should be taken in choosing strains for genetic and biochemical studies. Unfortunately there are no truly wild-type *Saccharomyces* strains that are commonly employed in genetic studies. Also, most domesticated strains of brewer's yeast and probably many strains of baker's yeast and true wild-type strains of *S. cerevisiae* are not genetically compatible with laboratory stocks. It is often not appreciated that many "normal" laboratory strains contain mutant characters. This condition arose because these laboratory strains were derived from pedigrees involving mutagenized strains, or strains that carry genetic markers. Many current genetic studies are carried out with one or another of the following strains or their derivatives, and these strains have different properties that can greatly influence experimental outcomes: S288C; W303; D273–10B; X2180; A364A; Σ1278B; AB972; SK1; and FL100. The haploid strain S288C (*MATα SUC2 mal mel gal2 CUP1 flo1 flo8-1 hap1*) is often used as a normal standard because the sequence of its genome has been determined, many isogenic mutant derivatives are available,[43,44] and it gives rise to well-dispersed cells. However, S288C contains a defective *HAP1* gene,[45] making it incompatible with studies of mitochondrial and related systems. Also, in contrast to Σ1278B, S288C does not form pseudohyae.[44] While true wild-type and domesticated baker's yeast give rise to less than 2% ρ^- colonies (see below), many laboratory strains produce high frequencies of ρ^- mutants. Another strain, D273–10B, has been extensively used as a typical normal yeast, especially for mitochondrial studies. One should examine the specific characters of interest before initiating a study with any strain. Also, there can be a high degree of inviability of the meiotic progeny from crosses among these "normal" strains.

Many strains containing characterized auxotrophic, temperature-sensitive, and other markers can be obtained from the Yeast Genetics Stock Culture Center of the American Type Culture Collection,[46] including an almost complete set of

[42] C. C. Lindegren, "The Yeast Cell. Its Genetics and Cytology." Educational Publishers, St. Louis, MO, 1949.

[43] R. K. Mortimer and J. R. Johnson, *Genetics* **113,** 35 (1986).

[44] A. Rogowska-Wrzesinska, P. Mose Larsen, A. Blomberg, A. Görg, P. Roepstorff, J. Norbeck, and S. J. Fey, *Comparative and Functional Genomics* **2,** 207 (2001).

[45] M. Gaisne, A.-M. Bécam, J. Verdière, and C. J. Herbert, *Curr. Genet.* **36,** 195 (1999).

[46] http://www.atcc.org/SearchCatalogs/YeastGeneticStock.cfm

deletion strains.[47] Currently this set consists of 20,382 strains representing deletants of nearly all nonessential open reading frames (ORFs) in different genetic backgrounds. Deletion strains are also availabe from EUROSCARF[48] and Research Genetics.[49] Other sources of yeast strains include the National Collection of Yeast Culture,[50] the Centraalbureau voor Schimmelcultures[51] and the Culture Collection of *Saccharomyces cerevisiae* (DGUB, Bratislava, Slovak Republic).[52] Before using strains obtained from these sources or from any investigator, it is advisable to test the strains and verify their genotypes.

The Genome of *Saccharomyces cerevisiae*

Saccharomyces cerevisiae contains a haploid set of 16 well-characterized chromosomes, ranging in size from 200 to 2200 kb. The total sequence of chromosomal DNA, constituting 12,052 kb, was released in 1996.[2] A total of 6183 open reading frames (ORFs) more than 100 amino acids long were reported, and approximately 5800 of them were predicted to correspond to actual protein-coding genes. A larger number of ORFs were predicted by considering shorter proteins. In contrast to the genomes of multicellular organisms, the yeast genome is highly compact, with genes representing 72% of the total sequence. The average size of yeast genes is 1.45 kb, or 483 codons, with a range from 40 to 4910 codons. A total of 3.8% of the ORFs contain introns. Approximately 30% of the genes already have been characterized experimentally. Of the remaining 70% with unknown function, approximately one-half either contain a motif of a characterized class of proteins or correspond to genes encoding proteins that are structurally related to functionally characterized gene products from yeast or from other organisms.

Ribosomal RNA is coded by approximately 120 copies of a single tandem array on chromosome XII. The DNA sequence revealed that yeast contains 262 tRNA genes, of which 80 have introns. In addition, chromosomes contain movable DNA elements, retrotransposons, that vary in number and position in different strains of *S. cerevisiae,* with most laboratory strains having approximately 30.

Other nucleic acid entities, presented in Table I,[53–56] also can be considered part of the yeast genome. Mitochondrial DNA encodes components of the

[47] http://www-deletion.stanford.edu/cgi-bin/deletion/search3.pl.atcc

[48] http://www.uni-frankfurt.de/fb15/mikro/euroscarf/col_index.html

[49] http://www.resgen.com/products/YEASTD.php3

[50] http://www.ncyc.co.uk/Sacchgen.html

[51] http://www2.cbs.knaw.nl/yeast/webc.asp

[52] http://www.natur.cuni.cz/fccm/federacs.htm#dgub

[53] G. F. Carle and M. Olson, *Proc. Natl. Acad. Sci. U.S.A.* **82,** 3756 (1985).

[54] J. R. Broach, *Methods Enzymol.* **101,** 307 (1983).

[55] B. Dujon, *in* "Molecular Biology of the Yeast *Saccharomyces,*" Vol. I, "Life Cycle and Inheritance" (J. N. Strathern, E. W. Jones, and J. R. Broach, eds.), p. 505. Cold Spring Harbor Laboratory Press, Cold Spring Harbor, NY, 1981.

[56] R. B. Wickner, *FASEB* **3,** 2257 (1989).

TABLE I

GENOME OF DIPLOID *Saccharomyces cerevisiae* CELL[a]

Inheritance:	Mendelian			Non-Mendelian				
Nucleic acid:	Double-stranded DNA			Double-stranded RNA				
Location:	Nucleus			Cytoplasm				
				RNA virus				
Genetic determinant	Chromosomes	2-μm plasmid	Mitochondrial DNA	L-A	M	L-BC	T	W
Relative amount (%)	85	5	10	80	10	9	0.5	0.5
Number of copies	2 sets of 16	60–100	~50 (8–130)	103	170	150	10	10
Size (kb)	13,500 (200–2200)	6.318	70–76	4.576	1.8	4.6	2.7	2.25
Deficiencies in mutants	All kinds	None	Cyto. $a \cdot a_3$, b	Killer toxin		None		
Wild type	*YFG1*$^+$	cir$^+$	ρ^+	KIL-k_1				
Mutant or variant	*yfg1-1*	cir^o	ρ^-	KIL-o				

[a] A wild-type chromosomal gene is designated as *YFG1*$^+$ (Your Favorite Gene) and the mutation as *yfg1-1*. Adapted from Refs. 53–56.

mitochondrial translational machinery and approximately 15% of the mitochondrial proteins. ρ^0 mutants completely lack mitochondrial DNA and are deficient in the respiratory polypeptides synthesized on mitochondrial ribosomes, i.e., cytochrome *b* and subunits of cytochrome oxidase and ATPase complexes. Even though ρ^0 mutants are respiratory deficient, they are viable and still retain mitochondria, although the mitochondria are morphologically abnormal.

The 2-μm circle plasmids, present in most strains of *S. cerevisiae,* apparently function solely for their own replication. Generally *cir*^{0} strains, which lack 2-μm DNA, have no observable phenotype. However, a certain chromosomal mutation, *nib1,* causes a reduction in growth of *cir*$^+$ strains, due to an abnormally high copy number 2-μm DNA.[57,58]

Similarly, almost all *S. cerevisiae* strains contain intracellular dsRNA viruses that constitute approximately 0.1% of total nucleic acid. RNA viruses include three families with dsRNA genomes, L-A, L-BC, and M. Two other families of dsRNA, T and W, replicate in yeast but so far have not been shown to be viral. M dsRNA encodes a toxin, and L-A encodes the major coat protein and components required for the viral replication and maintenance of M. The two dsRNA, M and L-A, are packaged separately with the common capsid protein encoded by L-A, resulting in virus-like particles that are transmitted cytoplasmically during vegetative growth and conjugation. L-B and L-C (collectively denoted L-BC), similar to L-A, have an RNA-dependent RNA polymerase and are present in intracellular particles. *KIL*-o mutants, lacking M dsRNA and consequently the killer toxin, are readily induced by growth at elevated temperatures and by chemical and physical agents.

Yeast also contains a 20S circular single-stranded RNA (not shown in Table I) that appears to encode an RNA-dependent RNA polymerase that acts as an independent replicon and is inherited as a non-Mendelian genetic element.

Only mutations of chromosomal genes exhibit Mendelian 2 : 2 segregation in tetrads after sporulation of heterozygous diploids; this property is dependent on the disjunction of chromosomal centromeres. In contrast, non-Mendelian inheritance is observed for the phenotypes associated with the absence or alteration of other nucleic acids described in Table I.

Genetic Nomenclature

Chromosomal Genes

The accepted genetic nomenclature for chromosomal genes of the yeast *S. cerevisiae* is illustrated in Table II, using *ARG2* as an example. Whenever possible, each gene, allele, or locus is designated by three italicized letters, e.g.,

[57] C. Holm, *Cell* **29,** 585 (1982).

[58] R. Sweeny and V. A. Zakian, *Genetics* **122,** 749 (1989).

TABLE II
GENETIC NOMENCLATURE USING *ARG2* AS AN EXAMPLE

Gene symbol	Definition
ARG$^+$	All wild-type alleles controlling arginine requirement
ARG2	Locus or dominant allele
arg2	Locus or recessive allele conferring an arginine requirement
arg2$^-$	Any *arg2* allele conferring an arginine requirement
ARG2$^+$	Wild-type allele
arg2-9	Specific allele or mutation
Arg$^+$	Strain not requiring arginine
Arg$^-$	Strain requiring arginine
Arg2p	Protein encoded by *ARG2*
Arg2 protein	Protein encoded by *ARG2*
ARG2 mRNA	mRNA transcribed from *ARG2*
arg2-Δ1	Specific complete or partial deletion of *ARG2*
ARG2::LEU2	Insertion of the functional *LEU2* gene at the *ARG2* locus; *ARG2* remains functional and dominant
arg2::LEU2	Insertion of the functional *LEU2* gene at the *ARG2* locus; *arg2* is or became nonfunctional
arg2-10::LEU2	Insertion of the functional *LEU2* gene at the *ARG2* locus, and the specified *arg2-10* allele which is nonfunctional
cyc1-arg2	Fusion between the *CYC1* and *ARG2* genes, where both are nonfunctional
P_{CYC1}-*ARG2*	Fusion between the *CYC1* promoter and *ARG2,* where the *ARG2* gene is functional

ARG, which is usually a describer, followed by a number, e.g., *ARG2*. Unlike most other systems of genetic nomenclature, dominant alleles are denoted by using uppercase italics for all letters of the gene symbol, e.g., *ARG2,* whereas lowercase letters denote the recessive allele, e.g., the auxotrophic marker *arg2*. Wild-type genes are designated with a superscript "plus" (*sup6*$^+$ or *ARG2*$^+$). Alleles are designated by a number separated from the locus number by a hyphen, e.g., *arg2-9*. The symbol Δ can denote complete or partial deletions, e.g., *arg2-Δ1*. (Do not use the symbols Δ*arg2* or *arg2*Δ for deletions.) Insertion of genes follow the bacterial nomenclature by using the symbol ::. For example, *arg2::LEU2* denotes the insertion of the *LEU2* gene at the *ARG2* locus in which *LEU2* is dominant (and functional) and *arg2* is recessive (and defective).

Phenotypes are denoted by cognate symbols in roman type and by the superscripts + and –. For example, the independence and requirement for arginine can be denoted by Arg$^+$ and Arg$^-$, respectively. Proteins encoded by *ARG2,* for example, can be denoted Arg2p, or simply Arg2 protein. However, gene symbols are generally used as adjectives for other nouns, for example, *ARG2* mRNA, *ARG2* strains. Resistance and sensitivity phenotypes are designated by superscript R and S, respectively. For example, resistance and sensitivity to canavanine sulfate are designated Can^R and Can^S, respectively.

Although most alleles can be unambiguously assigned as dominant or recessive by examining the phenotype of the heterozygous diploid crosses, dominant and recessive traits are defined only with pairs, and a single allele can be both dominant and recessive. For example, because the alleles *CYC1*$^{+}$, *cyc1-717,* and *cyc1-*Δ*1* produce, respectively, 100%, 5%, and 0% of the gene product, the *cyc1-717* allele can be considered recessive in the *cyc1-717/CYC1*$^{+}$ cross and dominant in the *CYC1-717/cyc1-*Δ*1* cross. Thus, it is less confusing to denote all mutant alleles by lowercase letters, especially when considering a series of mutations having a range of activities.

Wild-type and mutant alleles of the mating-type locus and related loci do not follow the standard rules. The two wild-type alleles of the mating-type locus are designated *MAT***a** and *MAT*α. The wild-type homothallic alleles at the *HMR* and *HML* loci are denoted *HMR***a,** *HMR*α, *HML***a,** and *HML*α. The mating phenotypes of *MAT***a** and *MAT*α cells are denoted simply **a** and α, respectively. The two letters *HO* denote the gene encoding the endonuclease required for homothallic switching.

Auxiliary gene symbols can be used to further describe the corresponding phenotypes, including the use of superscript R and S to distinguish genes conferring resistance and sensitivity, respectively. For example, the genes controlling resistance to canavanine sulfate (*can1*) and copper sulfate (*CUP1*) and their sensitive alleles could be denoted, respectively, as *can*^{R}*1, CUP*^{R}*1, CAN*^{S}*1,* and *cup*^{S}*1.*

Dominant and recessive suppressors are designated, respectively, by three uppercase or three lowercase letters, followed by a locus designation, e.g., *SUP4, SUF1, sup35, suf11.* In some instances UAA ochre suppressors and UAG amber suppressors are further designated, respectively, **o** and **a** following the locus. For example, *SUP4-***o** refers to suppressors of the *SUP4* locus that insert tyrosine residues at UAA sites; *SUP4-***a** refers to suppressors of the same *SUP4* locus that insert tyrosine residues at UAG sites. The corresponding wild-type locus that encodes the normal tyrosine tRNA and that lacks suppressor activity can be referred to as *sup4*$^{+}$. Intragenic mutations that inactivate suppressors can denoted, for example, *sup4*$^{-}$ or *sup4-***o***-1*. Frameshift suppressors are denoted as *suf* (or *SUF*), whereas metabolic suppressors are denoted with a variety of specialized symbols, such as *ssn* (suppressor of *snf1*), *srn* (suppressor of *rna1-1*), and *suh* (suppressor of *his2-1*).

Capital letters are also used to designate certain DNA segments whose locations have been determined by a combination of recombinant DNA techniques and classical mapping procedures, e.g., *RDN1,* the segment encoding ribosomal RNA.

The general form YCRXXw is used to designate genes deduced from the sequence of the yeast genome, where Y designates yeast; C (or A, B, etc.) designates the chromosome III (or I, II, etc.); R (or L) designates the right (or left) arm of the chromosome; XX designates the relative position of the start of the open-reading frame from the centromere; and w (or c) designates the Watson (or Crick) strand. For example, YCR5c denotes *CIT2,* a previously known but unmapped gene situated on the right arm of chromosome III, fifth open reading-frame from the centromere on the Crick strand.

Escherichia coli genes inserted into yeast are usually denoted by the prokaryotic nomenclature, e.g., *lacZ.*

A current list of gene symbols can be found on the Internet.[59]

Mitochondrial Genes

Special consideration should be given to the nomenclature describing mutations of mitochondrial components and function that are determined by both nuclear and mitochondrial DNA genes. Growth on media containing nonfermentable substrates (Nfs) as the sole energy and carbon source (such as glycerol or ethanol) is the most convenient operational procedure for testing mitochondrial function. Lack of growth on nonfermentable media (Nfs^- mutants), as well as other mitochondrial alterations, can be due to either nuclear or mitochondrial mutations as outlined in Table III. Nfs^- nuclear mutations are generally denoted by the symbol *pet*; however, more specific designations have been used instead of *pet* when the gene products were known, such as *cox4* and *hem1*.

The complexity of nomenclatures for mitochondrial DNA genes, outlined in Table III, is due in part to the complexity of the system, polymorphic differences of mitochondrial DNA, complementation between exon and intron mutations, the presence of intron-encoded maturases, diversed phenotypes of mutations within the same gene, and the lack of agreement among various workers. Unfortunately, the nomenclature for most mitochondrial mutations does not follow the rules outlined for nuclear mutations. Furthermore, confusion can occur between phenotypic designations, mutant isolation number, allelic designations, loci, and cistrons (complementation groups).

Non-Mendelian Determinants

Where necessary, non-Mendelian genotypes can be distinguished from chromosomal genotypes by enclosure in brackets, e.g., [*KIL*–o] *MAT***a** *trp1–1*. Although it is advisable to employ the above rules for designating non-Mendelian genes and to avoid using Greek letters, the use of well-known and generally accepted Greek symbols should be continued; thus, the original symbols ρ^+, ρ^-, Ψ^+, and Ψ^- or their transliterations, rho^+, rho^-, psi^+, and psi^-, respectively, have been retained.

In addition to the non-Mendelian determinants described in Table I (2 μm plasmid, mitochondrial genes, and RNA viruses), yeast contains prions, i.e., infectious

[59] ftp://genome-ftp.stanford.edu/pub/yeast/gene_registry/registry.genenames.tab

TABLE III
MITOCHONDRIAL GENES AND MUTATIONS[a]

Wild type	Mutation (with examples)	Mutant phenotype or gene product
Nuclear genes		
PET$^+$	*pet*$^-$	Nfs$^-$
	pet1	Unknown function
	cox4	Cytochrome *c* oxidase subunit IV
	hem1	δ-Aminolevulinate synthase
	cyc3	Cytochrome *c* heme lyase
Mitochondrial DNA		
	Gross aberrations	
ρ^+	ρ^-	Nfs$^-$
	ρ^o	ρ^- mutants lacking mitochondrial DNA
	Single-site mutations	
ρ^+	*mit*$^-$	Nfs$^-$, but capable of mitochondrial translation
[*COX1*]	[*cox1*]	Cytochrome *c* oxidase subunit I
[*COX2*]	[*cox2*]	Cytochrome *c* oxidase subunit II
[*COX3*]	[*cox3*]	Cytochrome *c* oxidase subunit III
[*COB1*]	[*cob1*] or [*box*]	Cytochrome *b*
[*ATP6*]	[*atp6*]	ATPase subunit 6
[*ATP8*]	[*atp8*]	ATPase subunit 8
[*ATP9*]	[*atp9*] or [*pho2*]	ATPase subunit 9
[*VAR1*]		Mitochondrial ribosomal subunit
ρ^+	*syn*$^-$	Nfs$^-$, deficient in mitochondrial translation
	tRNA^{Asp} or M7-37	Mitochondrial tRNA^{Asp} (CUG)
	ant^R	Resistant to inhibitors
[*ery*^{S}]	*ery*^{R} or [*rib1*]	Resistant to erythromycin, 21S rRNA
[*cap*^{S}]	*cap*^{R} or [*rib3*]	Resistant to chloramphenical, 21S rRNA
[*par*^{S}]	*par*^{R} or [*par1*]	Resistant to paromomycin, 16S rRNA
[*oli*^{S}]	*oli*^{R} or [*oli1*]	Resistant to oligomycin, ATPase subunit 9

[a] Nfs$^-$ denotes lack of growth on nonfermentable substrates.

TABLE IV
SOME NON-MENDELIAN DETERMINANTS OF YEAST[a]

Wild type	Mutant or polymorphic variant	Genetic element	Mutant phenotype
ρ^+	ρ^-	Mitochondrial DNA	Deficiency of cytochromes $a \cdot a_3, b$, and respiration
KIL-k$_1$	*KIL*-o	RNA plasmid	Sensitive to killer toxin
cir$^+$	cir^o	2-μm circle plasmid	None
Ψ^-	Ψ^+	Sup35p prion	Increased efficiency of certain suppression
[PIN$^-$]	[*PIN*$^+$]	Prion	Decreased induction of Ψ^+
[*URE3*]	[*ure3*$^-$]	Ure2p prion	Deficiency in ureidosuccinate utilization

[a] Adapted from Refs. 54–56, 60–63. Other non-Mendelian determinants have been reported [R. Sweeney and V. A. Zakian, *Genetics* **122,** 749 (1989)].

TABLE V
SIZE AND COMPOSITION OF YEAST CELLS

Characteristic	Haploid cell	Diploid cell
Volume (μm^3)	70	120
Composition (10^{-12}g)		
Wet weight	60	80
Dry weight	15	20
DNA	0.017	0.034
RNA	1.2	1.9
Protein	6	8

proteins. The nomenclature of these prions, representing alternative protein states, is presented in Table IV,[54–56,60–63] along with other non-Mendelian determinants.[58]

Growth and Size

"Normal" laboratory haploid strains have a doubling time of approximately 90 min in YPD medium (see below) and approximately 140 min in synthetic media during the exponential phase of growth. However, strains with greatly reduced growth rates in synthetic media are often encountered. Usually strains reach a maximum density of 2×10^8 cells/ml in YPD medium. Titers 10 times this value can be achieved with special conditions, such as pH control, continuous additions of balanced nutrients, filtered–sterilized media, and extreme aeration that can be delivered in fermentors.

The sizes of haploid and diploid cells vary with the phase of growth[64] and from strain to strain. Typically, diploid cells are 5×6-μm ellipsoids and haploid cells are 4-μm diameter spheroids.[65] The volumes and gross composition of yeast cells are listed in Table V. During exponential growth, haploid cultures tend to have higher numbers of cells per cluster compared to diploid cultures. Also, haploid cells have buds that appear adjacent to the previous one, whereas diploid cells have buds that appear at the opposite pole.[66]

[60] M. Aigle and F. Lacroute, *Mol. Gen. Genet.* **136,** 327 (1975).

[61] M. F. Tuite, P. M. Lund, A. B. Futcher, M. J. Dobson, B. S. Cox, and C. S. McLaughlin, *Plasmid* **8,** 103 (1982).

[62] R. B. Wickner and Y. O. Chernoff, *in* "Prion Biology and Diseases" (S. Prusiner, ed.), p. 229. Cold Spring Harbor Laboratory Press, Cold Spring Harbor, NY, 1999.

[63] I. L. Derkatch, M. E. Bradley, S. V. L. Masse, S. P. Zadorsky, G. V. Polozkov, S. G. Inge-Vechtomov, and S. W. Liebman, *EMBO J.* **19,** 1942 (2000).

[64] A. E. Wheals, *in* "The Yeast." Vol. 1, Biology of Yeasts, 2nd Ed. (A. H. Rose and J. S. Harrison, eds.), p. 283. Academic Press, New York, 1987.

[65] R. K. Mortimer, *Radiation Res.* **9,** 312 (1958).

[66] D. Freidfelder, *J. Bacteriol.* **80,** 567 (1960).

TABLE VI
COMPLEX MEDIA

Medium	Components	Composition
YPD (for routine growth)	1% Bacto-yeast extract	10 g
	2% Bacto-peptone	20 g
	2% Dextrose	20 g
	2% Bacto-agar	20 g
	Distilled water	1000 ml
YPG [containing a nonfermentable carbon source (glycerol) that does not support the growth of ρ^- or *pet* mutants]	1% Bacto-yeast extract	10 g
	1% Bacto-yeast extract	10 g
	2% Bacto-peptone	20 g
	3% (v/v) Glycerol	30 ml
	2% Bacto-agar	20 g
	Distilled water	970 ml
YPDG (used to determine the proportion of ρ^- cells; ρ^+ and ρ^- colonies, appear, respectively, large and small on this medium)	1% Bacto-yeast extract	10 g
	2% Bacto-peptone	20 g
	3% (v/v) Glycerol	30 ml
	0.1% Dextrose	1 g
	2% Bacto-agar	20 g
	Distilled water	970 ml
YPAD (used for the preparation of slants; adenine is added to inhibit the reversion of *ade1* and *ade2* mutations)[a]	1% Bacto-yeast extract	10 g
	1% Bacto-yeast extract	10 g
	1% Bacto-yeast extract	10 g
	2% Bacto-peptone	20 g
	2% Dextrose	20 g
	0.003% Adenine sulfate	40 mg
	Distilled water	1000 ml
	2% Bacto-agar	20 g

[a] The medium is dissolved in a boiling water bath and 1.5-ml portions are dispensed with an automatic pipetter into 1-dram (3-ml) vials. The caps are screwed on loosely, and the vials are autoclaved. After autoclaving, the rack is inclined so that the agar is just below the neck of the vial. The caps are tightened after 1 to 2 days.

Growth and Testing Media

For experimental purposes, yeast are usually grown at 30°C on the complete medium, YPD (Table VI), or on synthetic media, SD and SC (Tables VII[67] and VIII). For industrial or certain special purposes when large amounts of high titers are desirable, yeast can be grown in cheaper media with high aeration and pH control.[68] The ingredients of standard laboratory media are presented in Tables VII

[67] L. J. Wickerham, *U.S. Dept. Agric. Tech. Bull.* No. 1029 (1951).

[68] J. White, "Yeast Technology." J. Wiley and Sons Inc., New York, 1954.

TABLE VII
SYNTHETIC MINIMAL GLUCOSE MEDIUM (SD)[a]

Component	Composition
0.67% Bacto-yeast nitrogen base (without amino acids)	6.7 g
2% Dextrose	20 g
2% Bacto-agar	20 g
Distilled water	1000 ml
	Amount per liter
Carbon source	
Dextrose	20 g
Nitrogen source	
Ammonium sulfate	5 g
Vitamins	
Biotin	22 μg
Calcium pantothenate	400 μg
Folic acid	2 μg
Inositol	2 mg
Niacin	400 μg
p-Aminobenzoic acid	200 μg
Pyridoxine hydrochloride	400 μg
Riboflavin	200 μg
Thiamin hydrochloride	400 μg
Compounds supplying trace elements	
Boric acid	500 μg
Copper sulfate	40 μg
Potassium iodide	100 μg
Ferric chloride	200 μg
Manganese sulfate	400 μg
Sodium molybdate	200 μg
Zinc sulfate	400 μg
Salts	
Potassium phosphate monobasic	850 mg
Potassium phosphate dibasic	150 mg
Magnesium sulfate	500 mg
Sodium chloride	100 mg
Calcium chloride	100 mg

[a] This synthetic medium is based on media described by Wickerham[67] and is marketed, without dextrose, by Difco Laboratories (Detroit, MI) as "Yeast nitrogen base without amino acids."

and VIII. Synthetic media[67] are conveniently prepared with Bacto-yeast nitrogen base without amino acids (Difco Laboratories, Detroit, MI), containing the constituents presented in Table VIII. Nutritional requirements of mutants are supplied with the nutrients listed in Table VI. Growth on nonfermentable carbon sources can be tested on YPG medium (Table VI), and fermentation markers can

TABLE VIII
SYNTHETIC COMPLETE (SC) MEDIA[a]

Constituent	Final concentration (mg/liter)	Stock per 100 ml	Amount of stock (ml) for 1 liter
Adenine sulfate	20	200 mg[b]	10
Uracil	20	200 mg[b]	10
L-Tryptophan	20	1 g	2
L-Histidine hydrochloride	20	1 g	2
L-Arginine hydrochloride	20	1 g	2
L-Methionine	20	1 g	2
L-Tyrosine	30	200 mg	15
L-Leucine	30	1 g	3
L-Isoleucine	30	1 g	3
L-Lysine hydrochloride	30	1 g	3
L-Phenylalanine	50	1 g[b]	5
L-Glutamic acid	100	1 g[b]	10
L-Aspartic acid	100	1 g[b,c]	10
L-Valine	150	3 g	5
L-Threonine	200	4 g[b,c]	5
L-Serine	400	8 g	5

[a] SC contains synthetic minimal medium (SD) with various additions. It is convenient to prepare sterile stock solutions which can be stored for extensive periods. All stock solutions can be autoclaved for 15 min at 120°. The appropriate volume of the stock solutions (see below) is added to the ingredients of SD medium and sufficient distilled water is added so that the total volume is 1 liter. The threonine and aspartic acid solutions should be added separately after autoclaving. Given below are the concentrations of the stock solutions (amount per 100 ml). Some stock solutions should be stored at room temperature in order to prevent precipitation, whereas the other solutions may be refrigerated. It is best to use HCl salts of amino acids wherever applicable.

[b] Store at room temperature.

[c] Add after autoclaving the medium.

be determined with indicator media (Table IX) on which acid production induces color changes.

Saccharomyces cerevisiae strains can be sporulated at 30° on the media listed in Table X; most strains will readily sporulate on the surface of sporulation medium after replica plating fresh cultures from a YPD plate.[69]

Media for petri plates are prepared in 2-liter flasks, with each flask containing no more than 1 liter of medium, which is sufficient for approximately 40 standard

[69] R. R. Fowell, *Nature (London)* **170,** 578 (1952).

TABLE IX
INDICATOR MEDIA

Indicator medium	Components	Composition
MAL[a]	1% Bacto-yeast extract	10 g
	2% Bacto-peptone	20 g
	2% Maltose	20 g
	Bromcresol purple solution (0.4% stock)	9 ml
	2% Agar	20 g
	Distilled water	1000 ml
GAL[b]	1% Yeast extract	10 g
	2% Peptone	20 g
	2% Agar	20 g
	Bromthymol blue (4 mg/ml stock)	20 ml
	Distilled water	880 ml

[a] Maltose indicator medium (MAL) is a fermentation indicator medium used to distinguish strains which ferment or do not ferment maltose. Owing to the pH change, the maltose-fermenting strains will produce a yellow halo on a purple background. A 0.4% bromcresol purple solution is prepared by dissolving 20 mg of the indicator in 50 ml of ethanol.

[b] Galactose indicator medium (GAL) is a fermentation indicator medium used to distinguish strains which ferment or do not ferment. The galactose-fermenting strains will produce a yellow halo on a blue background. After autoclaving, add 100 ml of a filter-sterilized 20% galactose solution.

plates. Unless stated otherwise, all components are autoclaved together for 15 min at 120° and 15 pounds pressure. The plates should be allowed to dry at room temperature for 2–3 days after pouring. The plates can be stored in sealed plastic bags for more than 3 months at room temperature. The agar is omitted for liquid media.

Different types of synthetic media, especially omission media, can be prepared by mixing and grinding dry components in a ball mill.

Practical information on the preparation of media is presented in this volume by Styles.[70]

Small batches of liquid cultures can be grown in shake flasks using standard bacteriological techniques with high aeration. High aeration can be achieved by vigorously shaking cultures having liquid volumes less than 20% of the flask volume.

[70] C. Styles, *Methods Enzymol.* **350,** [2], 2002 (this volume).

TABLE X
SPORULATION MEDIA

Sporulation medium	Components	Composition
Presporulation[a]	0.8% Bacto-yeast extract	0.8 g
	0.3% Bacto-peptone	0.3 g
	10% Dextrose	10 g
	2% Bacto-agar	2 g
	Distilled water	100 ml
Sporulation[b]	1% Potassium acetate	10 g
	0.1% Bacto-yeast extract	1 g
	0.05% Dextrose	0.5 g
	2% Bacto-agar	20 g
	Distilled water	1000 ml
Minimal sporulation[c]	1% Potassium acetate	10 g
	2% Bacto-agar	20 g
	Distilled water	1000 ml

[a] Strains are grown 1 or 2 days on presporulation medium before being transferred to sporulation medium. This is only necessary for strains that do not sporulate well when incubated on sporulation medium directly.

[b] Strains will undergo several divisions on sporulation medium and then sporulate after 3 to 5 days' incubation. Sporulation of auxotrophic diploids is usually increased by adding the nutritional requirements to the sporulation medium at 25% of the levels given above for SD complete medium.

[c] Diploid cells will sporulate on minimal sporulation medium after 18–24 hr without vegetative growth. Nutritional requirements are added as needed for auxotrophic diploids as for sporulation medium described above (25% of level for SD complete medium).

Testing of Phenotypes and Gene Functions

Hampsey[71] has compiled a useful list of phenotypes that can be conveniently scored or selected, including the use media for testing a sensitivity and resistance to a large number of different chemical and physical agents. Furthermore, known mutant genes corresponding to each of the phenotypes have been tabulated.[72] An international project, designated EUROFAN 2 (European Network for the Functional Analysis of Yeast Genes Discovered by Systematic DNA Sequencing), is dedicated to the the functional analysis of all 6000 yeast ORFs by using gene

[71] M. Hampsey, *Yeast* **13,** 1099 (1997).

[72] http://www.mips.biochem.mpg.de/proj/yeast/catalogues/phenotype/index.html

disruptants.[73] The phenotypic analysis includes testing for sensitivity and resistance to toxic compounds under 300 different types of growth conditions.[74]

Strain Preservation

Yeast strains can be stored for short periods of time at 4° on YPD medium in petri dishes or in closed vials (slants). Although most strains remain viable at 4° for at least 1 year, many strains fail to survive even for a few months.

Yeast strains can be stored indefinitely in 15% (v/v) glycerol at −60° or lower temperature. (Yeast tend to die after several years if stored at temperatures above −55°.[75])

Many workers use 2-ml vials (35 × 12 mm) containing 1 ml of sterile 15% (v/v) glycerol. The strains are first grown on the surfaces of YPD plates; the yeast is then scraped up with sterile applicator sticks and suspended in the glycerol solution. The caps are tightened and the vials shaken before freezing. The yeast can be revived by transferring a small portion of the frozen sample to a YPD plate.

Micromanipulation and Micromanipulators

The separation of the four ascospores from individual asci by micromanipulation is required for meiotic genetic analyses and for the construction of strains with specific markers. In addition, micromanipulation is used to separate zygotes from mass-mating mixtures and, less routinely, for positioning of vegetative cells and spores for mating purposes and for single-cell analyses, such as used in aging studies. The relocation and transfer of ascospores, zygotes, and vegetative cells are almost exclusively carried out on agar surfaces with a fine glass microneedle mounted in the path of a microscope objective and controlled by a micromanipulator. Although specialized equipment and some experience is required to carry out these procedures, most workers can acquire proficiency within a few days of practice.

Micromanipulators used for yeast studies operate with control levers or joysticks that can translate hand movements into synchronously reduced movements of microtools.[76–78] Most of the instruments were designed so that movement of the tool in the horizontal (x and y) plane is directly related to the movement of the

[73] http://www.mips.biochem.mpg.de/proj/eurofan/

[74] K. J. Rieger, M. El-Alama, G. Stein, C. Bradshaw, P. P. Slonimski, and K. Maundrell, *Yeast* **15,** 973 (1999).

[75] A. M. Well and G. G. Stewart, *Appl. Microbiol.* **26,** 577 (1973).

[76] F. Sherman, *Methods Cell Biol.* **11,** 189 (1975).

[77] H. M. El-Badry, "Micromanipulators and Micromanipulation." Academic Press, New York, 1963.

[78] F. Sherman and J. B. Hicks, *Methods Enzymol.* **194,** 21 (1991).

TABLE XI
COMMERCIALLY AVAILABLE MICROMANIPULATORS WITH SINGLE-LEVER CONTROLS

Distributor and micromanipulator	Ref.
Carl Zeiss, Inc. (Thornwood, NY)	79
Tetrad Microscope	
Schütt Labortechnik GmbH (Göttingen, Germany)	80
Tetrad Dissection Microscope, TDM 400 E, Type I	
Tetrad Dissection Microscope, TDM 400 E, Type II	
Singer Instrument Co. Ltd. (Watchet, Somerset, U.K.)	82
Singer MSM System Series 200	
Singer MSM Manual	
Technical Products International (St. Louis, MO)	84
TPI de Fonbrune-type micromanipulator (without microscope)	
TPI de Fonbrune-type micromanipulator (with Olympus microscope)	

control handle, whereas vertical (*z* plane) tool movement is controlled by rotating a knob, located either on or near the horizontal control handle. Other designs have other combinations in which the joystick controls the *x* and *z* planes, or a screw controls the *z* plane. The main commercially available micromanipulators that have single control levers and that are commonly used for yeast studies are listed in Table XI, along with the distributors.

Transmission of hand motions to the tool with the de Fonbrune micromanipulator is based on pneumatic principles, whereas the other units rely on several ingenious mechanical principles involving direct coupling to sliding components.

The Zeiss Tetrad "Advanced Yeast Dissection Microscope" (Table XI, Fig. 1),[79] and the TDM 400 E "Tetrad Dissection System" (Table XI),[80] based on the design described by Sherman,[81] are primarily intended for dissection of asci.

The Zeiss Tetrad Microscope incorporates a modified Zeiss Axioskop fixed-stage microscope and a stable stage-mounted micromanipulator, with joystick control and adjustable *y–z* movement ranging from 0.1 to 5 mm. Stage movement incorporates click-stops at 5-mm intervals in both *x* and *y* directions, and an engraved *x* scale on the holder for an inverted petri dish facilitates the systematic relocation of spores. The stage assembly for tetrad dissection is easily removed, allowing the microscope to be used for general purposes. Also, the manipulator can be mounted on either the left- or right-hand side of the stage.

[79] Carl Zeiss, Inc., Microscopy Division, One Zeiss Drive, Thornwood, NY 10594.

[80] http://www.schuett-labortechnik.de/Produkte/Welcome/Products/Tetrad_Dissection_Microscope_T/tetrad_dissection_microscope_t.html

[81] F. Sherman, *Appl. Microbiol.* **26,** 829 (1973).

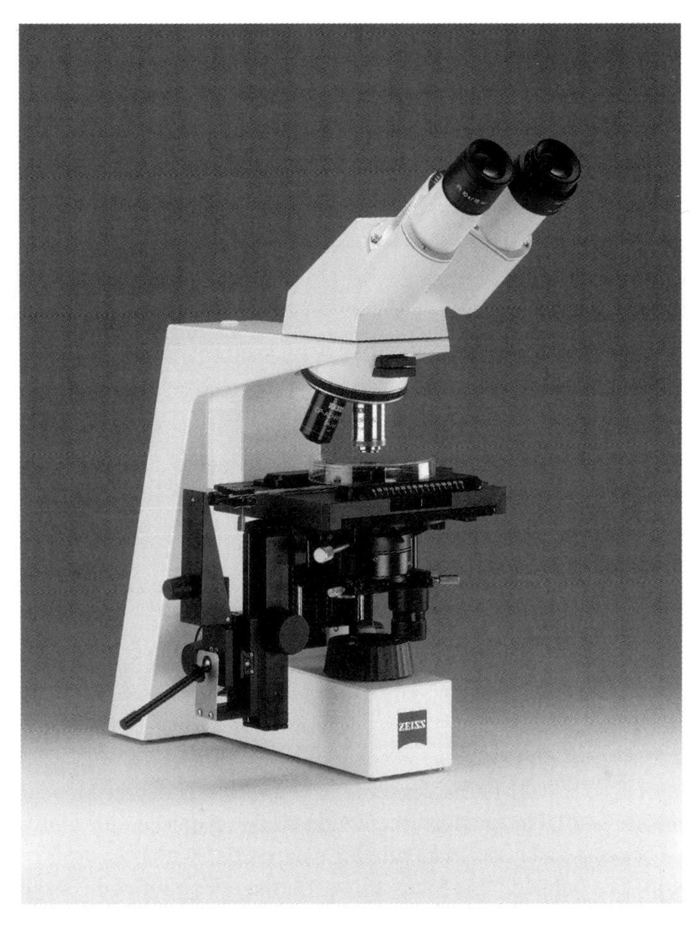

FIG. 1. The Zeiss Tetrad "Advanced Yeast Dissection Microscope," showing the micromanipulator mounted on a Zeiss Axioskop fixed-stage microscope modified for tetrad dissection.

Similarly, the TDM 400 E system incorporates the Nikon Eclipse E 400 microscope and includes a joystick micromanipulator, a holder for 100 mm petri dishes, and a calibrated stage. The microscope is complete with long-working-distance optics for viewing through the inverted dish containing spores or vegetative cells. The micromanipulator is normally mounted on the left side of the stage and both move in concert when the microscope is focused, thus eliminating the need for a fixed stage. A joystick controls the y and z motion, whereas a knurled knob is rotated for movement along the x axis. The mechanical stage has coaxial control knobs, which are tension adjustable, and the stage is indexed with click-stops every 5 mm on the x and y axes. Ten-cm petri dishes can be accommodated on the stage.

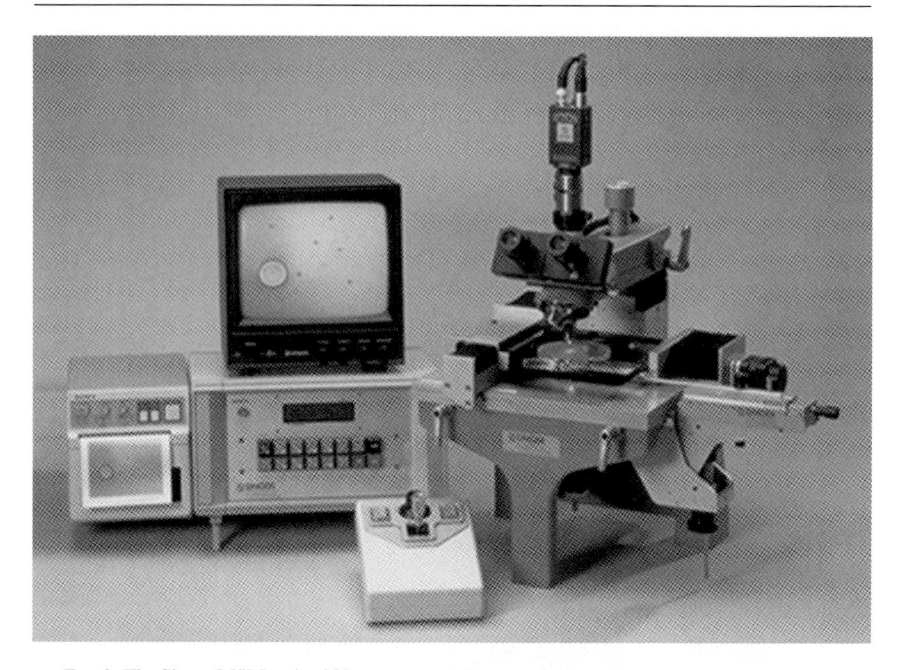

FIG. 2. The Singer MSM series 200 system, showing the microscope with a motor-driven stage, the micromanipulator, the control console, the joystick, and, and the optional accessories, a CCD CCTV camera and monitor and video printer.

The Nikon Eclipse E 400 Microscope is designed with a 25 degree binocular body, and options are available to adjust the height of the eyepieces for unusually tall operators. Focusing and micromanipulation can be done with both arms resting on the benchtop. The one-side fine focus allows for the operation of the stage and the fine focus with one hand. The TDM 400 E is offered in two versions, type I and type II, having, respectively, 150× and 200× maximum magnification. Both types come with an Abbe condenser which has been modified so that it will provide proper Köhler illumination over the extended distance to the specimen. Stand and stage are built for right-handed operation, but left-handed models are optional.

The Singer MSM System series 200 (Table XI, Fig. 2)[82] is a complete, computer-controlled workstation for micromanipulation in yeast genetics, including tetrad dissection, pedigree analysis, and cell and zygote isolation. Repetitive movements can be automated with a resolution of 4 μm, a repeatability of 2 μm, and an overall movement of 15 cm × 10 cm, using a computer-controlled motorized stage that accepts standard petri dishes. The workstation includes a integral trinocular microscope having 15× wide-field eye pieces, 4× and 20× XLWD

[82] http://www.singerinst.co.uk/system.html

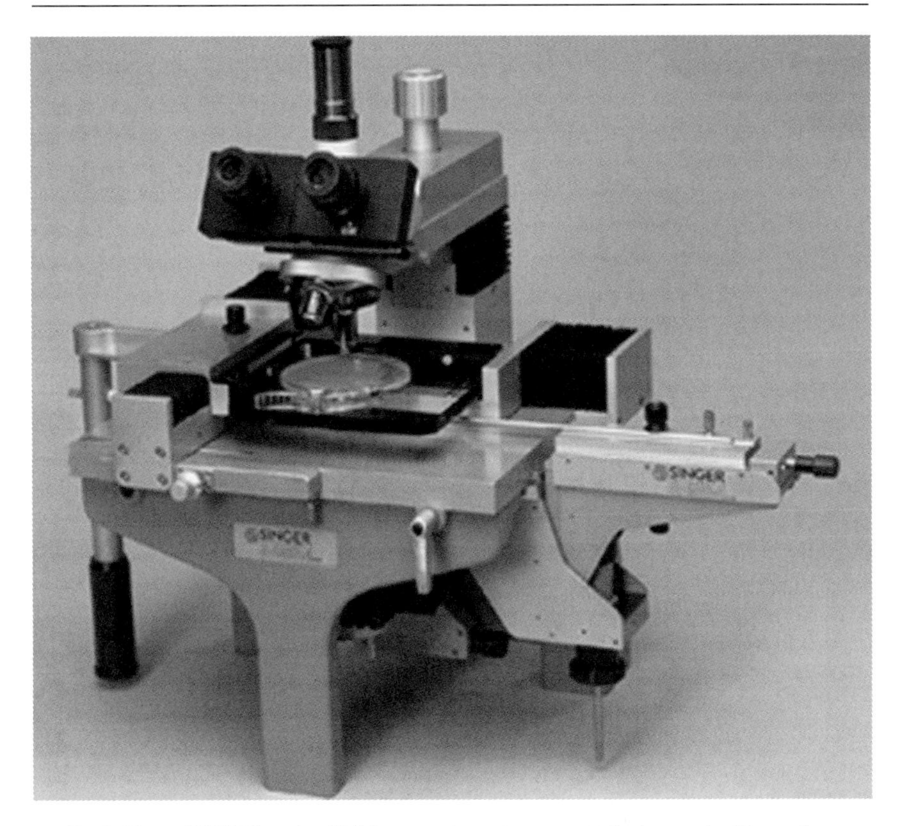

FIG. 3. Singer MSM Manual, which incorporates a stage-mounted micromanipulator and a manually operated stage and microscope. The fine focus is operated by knobs positioned on each side, underneath the stage at the bench height, whereas a pendant stage handle controls the x–y movement of the petri dish.

objectives, electronically controlled fine focusing from the joystick, and a hinged overarm to conveniently clear petri dishes. A television camera can be conveniently attached to the unit. The MSM micromanipulator can be locked by a single handle to either side of the substage for right- or left-hand operation. Horizontal movements (x and y) of the needle are controlled by a pendant joystick, whereas the vertical (z) drive is controlled by a coaxial ring. Coarse adjustments of the needle are also convenient. The Singer MSM System series 200 is supplied with a needle holder and needles.

The manually operated Singer MSM Manual (Table XI, Fig. 3)[82] includes the same microscope and stage-mounted manipulator as the Singer MSM System series 200 unit. Spring-loaded stops allow the detection of matrix grid points along the y axis, whereas an incrementing stop contols stage movement along the y axis,

FIG. 4. The de Fonbrune-type micromanipulator. The control unit (right) and the receiver (left) are interconnected by flexible tubing on opposite sides of a Leitz Laborlux II microscope. The complete assembly is on a vibration eliminator (Vibration Damping Mount, Vibrasorb, Electron Microscopy Sciences, Fort Washington, PA).

thus allowing the rapid positioning of the petri dish in a 6 mm grid, which is ideally suited for asci dissection. In addition, the *y* axis in the area of the inoculum streak is resticted, and a trim screw enables the operator to return to certain positions in the *x* axis, making the search for asci convenient.

The de Fonbrune micromanipulator, shown in Fig. 4, pneumatically transmits a fine degree of motion from a single joystick.[83,84] The micromanipulator consists of two free-standing units: (1) a joystick controlling three piston pumps that is connected by tubing to (2) three diaphragms or aneroids that actuate a lever holding the microtool. Lateral movement of the joystick controls *x* and *y* horizontal movement, whereas rotation of the joystick controls vertical *z* movement. A moveable collar on the joystick provides simple ratio adjustment control that can be varied from 1:50 to 1:2500. Thus, the movement of the microtool can be

[83] P. de Fonbrune, "Technique de Micromanipulation." Masson, Paris, 1949.

[84] http://www.techprodint.com/fsmicro.htm

adjusted to correspond to the magnification of the optical system or to increase or decrease the control sensitivity. In addition, mechanical controls on the receiver unit provide fast and coarse adjustments. The de Fonbrune micromanipulator, which is not directly attached to the stage, should be used in conjunction with various microscopes having fixed stages and tube focusing, Fine mechanical stages with graduations are essential with all micromanipulators. It is convenient to have long working distance objectives for magnifications in the range of 150–300× . Long working distances can be achieved with 10× and 15× objectives, and the appropriate magnifications with 20× or 25× eyepieces.

Because of their low cost and compactness, the Tetrad Microscopes, or Tetrad Dissection Microscopes, are the most commonly used models and are highly recommended. Although it has been our experience that the skill of asci dissection can be taught more quickly with the de Fonbrune-type micromanipulator, this and other micromanipulators not attached directly to the microscope stage require more space, and in some instances heavy base plates or vibration eliminators (see Fig. 4). The Singer MSM System series 2000, although expensive, is the ultimate apparatus for dissection of asci.

Microneedles

The separation of ascospores, zygotes, and vegetative yeast cells can be carried out with simple glass microneedles attached to any one of the micromanipulators described above. Microneedles can be made from a stock of commercial glass fibers[85,86] or glass fiber strands with polished ends,[87,88] they can be made individually from glass rods,[78] or they can be obtained from the Singer Instrument Co.[82]

Microneedles are most commonly constructed from glass fibers by a procedure that involves two steps: (1) the preparation of a stock of glass fibers, and (2) the gluing of a short segment of glass fiber perpendicular to a glass or metal mounting rod (Fig. 5). The glass mounting rod is made by first heating a 2-mm rod in a burner and pulling slowly to form a taper. When the rod has sufficient taper, the end is pulled quickly at right angles, similar to the procedure shown in Fig. 6; the end is broken so that the right angle projection is approximately 2 mm. The mounting rod should be cut with a file to approximately the size required to fit on the microscope stage, taking into account the distance from the manipulator to the center of the microscope field of view. The microneedle is attached to the micromanipulator and positioned under the microscope objective as shown in Fig. 7.

[85] K. E. Scott and R. Snow, *J. Gen. Appl. Microbiol.* **24,** 295 (1978).

[86] D. J. Eichinger and J. D. Boeke, *Yeast* **6,** 139 (1990).

[87] www.corastyles.com

[88] http://genome-www.stanford.edu/Saccharomyces/needle.html

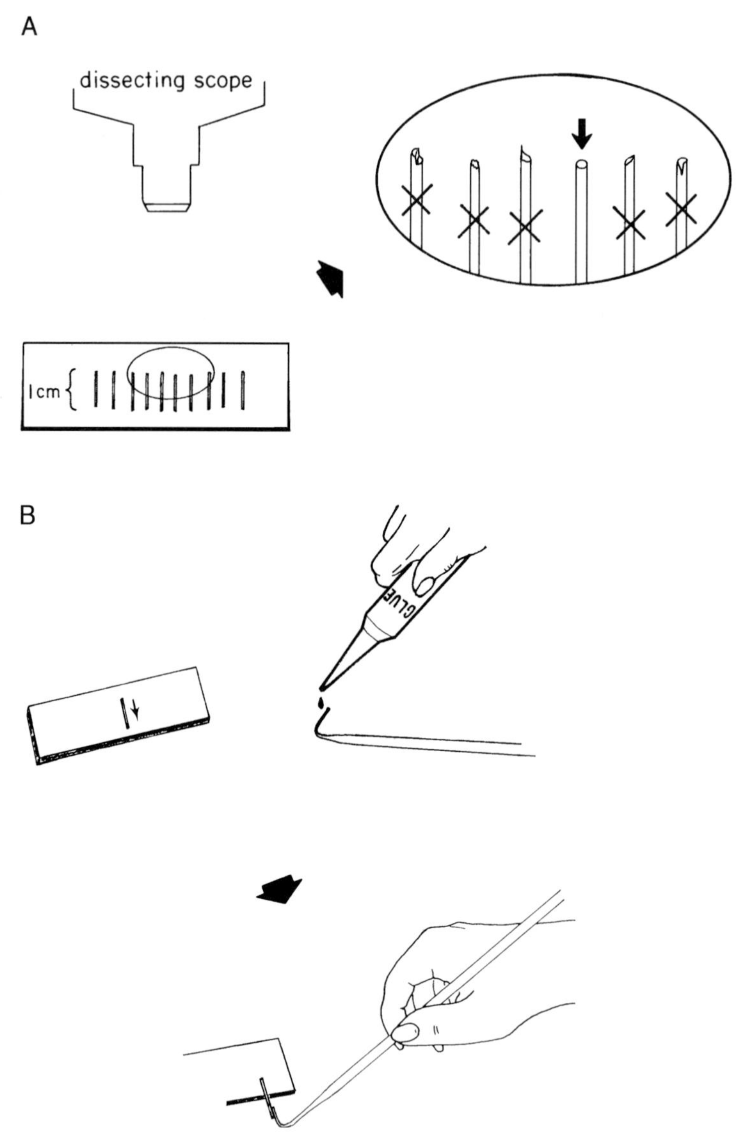

FIG. 5. Use of glass fibers for constructing microneedles. (A) A glass fiber approximately 40 μm in diameter is broken into segments approximately 1 cm long with a razor blade and examined with a microscope. (B) The segments containing flat ends are attached at a right angle to a mounting rod with cyanoacrylic Super Glue [F. Sherman and J. B. Hicks, *Methods Enzymol.* **194,** 21 (1991)].

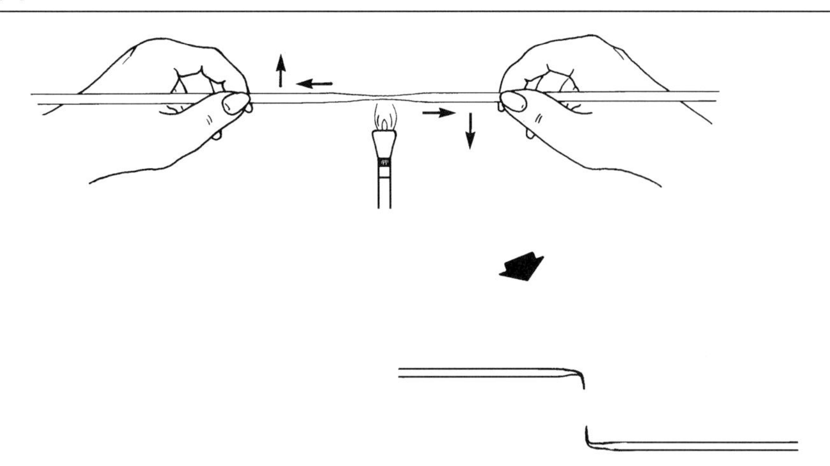

FIG. 6. Construction of microneedles. Microneedles required for the separation of ascospores can be made by first drawing out a 2-mm glass rod to a fine tip and then drawing out the end to an even finer tip at a right angle [F. Sherman and J. B. Hicks, *Methods Enzymol.* **194,** 21 (1991)].

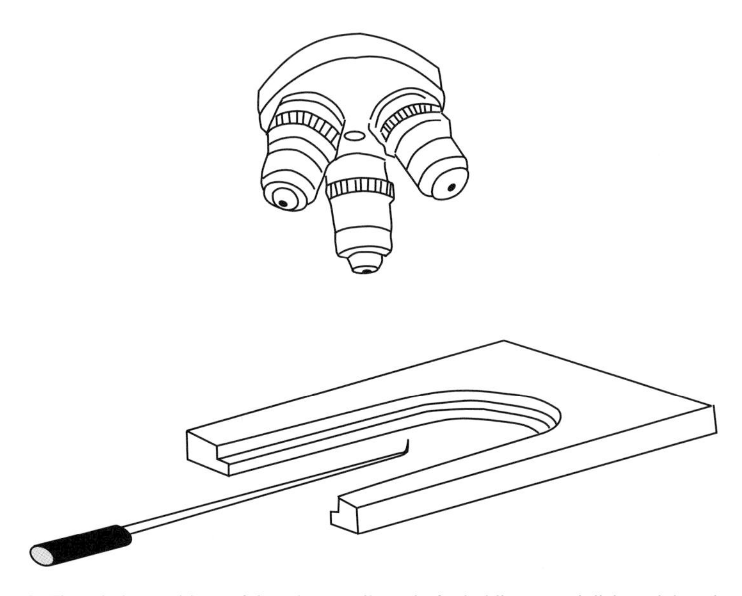

FIG. 7. The relative positions of the microneedle, a rig for holding a petri dish, and the microscope objectives [F. Sherman and J. B. Hicks, *Methods Enzymol.* **194,** 21 (1991)].

Most researchers use optical glass fibers,[86] which are commercially available (0.002 inch diameter, Edmund Scientific, 101 East Gloucester Pike, Barrington, NJ 08007) and which have a uniform size. However, glass fibers can be made by drawing thin filaments from a 2-mm glass rod.[85] The glass fibers can be broken with the fingers or cut with scissors or coverslips. The segments are placed on a microscope slide for examination under a dissecting microscope (Fig. 5). Segments of about 1 cm are usually desired for dissection on a petri dish with a standard micromanipulator. The exact length is not too important at this point, because the microneedle eventually can be cut to size with a coverslip.

The segments of glass fibers are examined under a low-power dissecting microscope to determine which of them will make a good needle, i.e., which have tips with a flat surface perpendicular to the long axis of the needle and no burrs or cracks in the tip (Fig. 5). However, a needle with minor imperfections (e.g., a half circle) sometimes will work if it has a flat working surface.

The glass-fiber segment with the best tip is moved down on the slide so that the good end is on the slide and the end to be glued is hanging off the edge. A small drop of Super Glue (cyanoacrylic) (Pacer Technology, Rancho Cucamonga, CA) is applied to the mounting rod and the glass fiber is glued to the end as shown in Fig. 5. The easiest way to apply the glue is to place a drop on a microscope slide and then to dip the whisker of the mounting rod into the drop. After contact, the glass fiber will usually come off the slide and stick to the mounting rod without any coaxing. If the glass fiber is not perpendicular to the stock, one may quickly adjust the angle before the glue sets.

Microneedles can be more conveniently prepared from commercially available glass fiber segments with polished ends.[87,88]

The preparation of individual glass microneedle from glass rods requires more skill and patience, but allows the construction of microneedle with different diameters. The individually prepared glass microneedle can be made with the small flame from the pilot light of an ordinary bunsen burner. A 2-mm-diameter glass rod is drawn out to a fine tip with the bunsen burner; by using the pilot flame, an even finer tip is drawn out at a right angle with an auxiliary piece of glass rod as illustrated in Fig. 6. The end is broken off so that the tip has a diameter of 10 to 100 μm and a length of a few millimeters. The drawn-out tip can be cut with a razor blade or broken between the surface and edge of two glass slides. It is critical that the microneedles have a flat end, which sometimes requires several attempts. The exact diameter is not critical, and various investigators have different preferences. Spores are more readily picked up and transferred with microneedles having tips of larger diameters, whereas manipulations in crowded areas having high densities of cells are more manageable with microneedles having smaller diameters. An approximately 40-μm-diameter microneedle is an acceptable compromise. Some investigators prefer larger diameters for picking up zygotes. The length of the perpendicular end should be compatible with the height of the petri dish or chamber; too short an end may result in optical distortions from the main shank of the microneedle. Longer microneedles are required for manipulations on the surfaces of petri dishes.

The needle is mounted into the micromanipulator and centered in the field. The adjustment of the needle is made most easily first at low magnification and then at higher magnification. As recommended above, asci are usually dissected at 150× or greater magnification.

Dissection of Asci[78]

Digestion of Ascus Sac

Sporulated cultures usually consist of unsporulated vegetative cells, four-spored asci, three-spored asci, etc. Dissection of asci requires the identification of four-spored asci and the relocation of each of the four ascopores to separate positions where they will form isolated spore colonies. The procedure requires the digestion of the ascus wall with Zymolyase, or another enzyme, without dissociating the four spores from the ascus.[89] (With very unusual strains that are particularly sensitive to enzyme treatments, the separation of ascospores can be carried out by rupturing the ascus wall with a microneedle.)[42,90]

Sporulated cells from the surface of sporulation medium are suspended in 50 μl of a stock solution of Zymolyase T100 (ICN, Costa Mesa, CA) (50 μg/ml in 1 *M* sorbitol), and the suspension is incubated at 30° for approximately 10 min. The exact time of incubation is strain dependent, and the progress of the digestion can be followed by removing a loopful of the digest to a glass slide and examining it under phase contrast at 400× magnification. The sample is ready for dissection when the spores in most of the asci are visible as discrete spheres, arranged in a diamond shape. Typical digested asci are seen in Fig. 8A. If a majority of the asci are still arranged in tightly packed tetrahedrons or diamond shapes in which the spores are not easily resolved, digestion is incomplete and the spores will not be easily separated by micro dissection. It is convenient to use a Zymolyase concentration that will digest the ascus wall in approximately 10 min. The digestion is terminated by placing the tube on ice and gently adding 150 μl of sterile water. Extensive treatment sometimes can decrease the viability and dissociate the clusters of four spores. The culture is suspended by gently rotating the tube; an aliquot is transferred with a wire loop to the surface of a petri plate or agar slab. It is important not to agitate the spores once they have been treated. If the treated spores are vortexed or shaken, the integrity of the ascus cannot be assured since the contents of one ascus may disperse and reassemble with the contents of another.

[89] J. R. Johnston and R. K. Mortimer, *J. Bacteriol.* **78,** 292 (1959).

[90] Ö. Winge and O. Lausten, *C. R. Trav. Lab Carlsberg, Ser. Physiol.* **22,** 99 (1937).

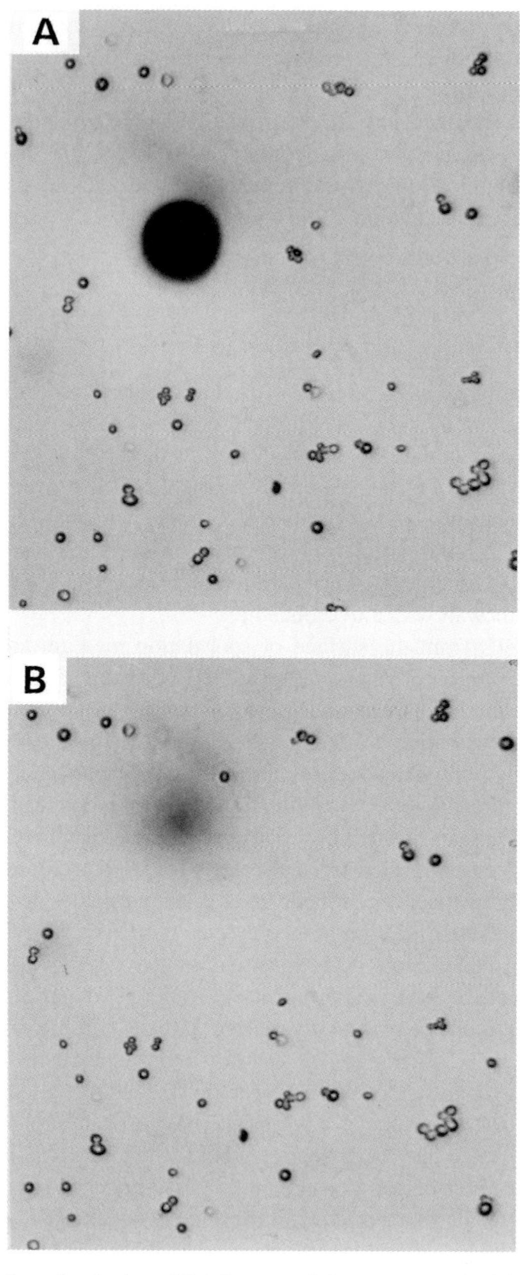

FIG. 8. A field of sporulated culture. (A) A four-spored cluster is seen at the right of the microneedle tip. (B) The cluster was picked up on the microneedle, which was lowered beneath the focal plane. The ascospores and the tip of the microneedle are, respectively, approximately 5 and 50 μm in diameter [F. Sherman and J. B. Hicks, *Methods Enzymol.* **194,** 21 (1991)].

The digestion can also be carried out with snail juice, which can be obtained commercially as Glusulase (NEN Research Products) or Suc d'*Helix pomatia* (L'Industrie Biologique Francaise, Genevilliers, France), or which can be prepared from snails, *Helix pomatia* or *Helix aspersa*.[89]

Separation of Ascospores

Micromanipulation can be implemented directly on the surfaces of ordinary petri dishes filled with nutrient medium or in special chambers on thin agar slabs. The petri dish (or chamber) is positioned so that the inoculum is in the microscope field over the microneedle. Examination of the streak should reveal the presence of the desired four-spored clusters as well as smaller clusters and vegetative cells. A typical preparation is shown in Fig. 8. A cluster of four spores is picked up on the microneedle by positioning microneedle tip next to the four-spored cluster on the surface of the agar. The microneedle is moved in a sweeping action, first touching the agar surface and then lowering the microneedle with a single motion. The absence of the four spores from the agar surface indicates that they have been transferred to the microneedle. Several attempts may be required to pick up all four ascospores. The microneedle can be considered a platform to which the spores are transferred. It is obvious from the relative sizes of the microneedle and spores (Fig. 8) that the microneedle does not "poke" the tetrad of spores to pick them up. The flat surface of the microneedle does not interact with the spores themselves, but rather with the water layer on the surface of the agar. When the microneedle approaches the surface of the agar, a meniscus forms and often a halo of refracted light can be seen around the shadow of the microneedle. At this time a column of water connects the microneedle and the agar (Fig. 9). The spores disappear from view into the meniscus. The combined sideways and downward sweeping motion is an attempt to coax the spores into the half of the meniscus that remains on the microneedle surface as it breaks away. Success in this endeavor is assayed by the disappearance of the spores from the visual field in the microscope. At the new position the process is repeated, this time with the hope that spores go from the microneedle meniscus to the surface of the agar.

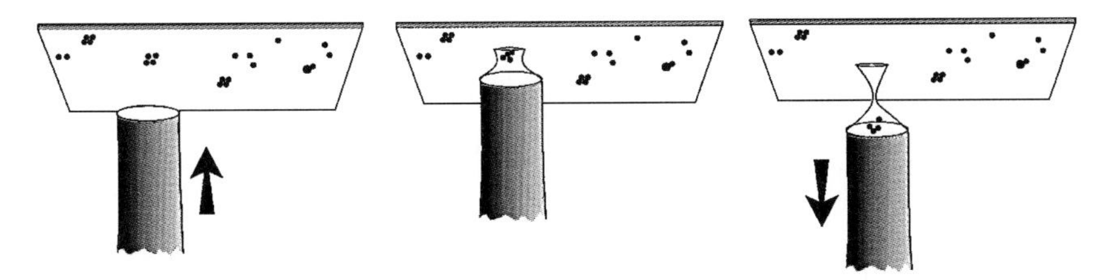

FIG. 9. The transfer of four spores from the surface of agar to the platform of a microneedle, by way of a water meniscus [F. Sherman and J. B. Hicks, *Methods Enzymol.* **194,** 21 (1991)].

Once the four spores have been transferred to the first position, it is necessary to separate at least one spore from the rest so that it can be left behind. A simple technique for achieving this goal is to move the microneedle onto the surface of the agar, forming the crisp image and halo, directly next to the cluster of spores and to vibrate the microneedle by gently tapping on the table near the microscope or on the microscope stage. The spores will often be separated by several microneedle diameters by this action. Three spores can be collected by sweeping the surface of the agar with the needle tip, and the process is repeated at the next three stops.

Note the position on the mechanical stage and place the four spores on the surface of the agar at least 5 mm from the streak. Pick up three spores and move the dish away from the streak another 5 mm (Fig. 10). Deposit the three spores and pick up two spores. Move the chamber an additional 5 mm; deposit the two spores and pick up one spore. Move the chamber 5 mm more and plant the remaining spore. Move the chamber 5 mm from the line of the four spores and select another four-spore cluster. Separate the spores as before at 5-mm intervals. Continue until a sufficient number of asci are dissected or until the entire dish is covered.

After picking up the four spores from an ascus, it is often convenient to set the stage micrometer so that each group of four spore colonies falls on cardinal points such as 15, 20, and 25. This makes it easier to keep track of progress and

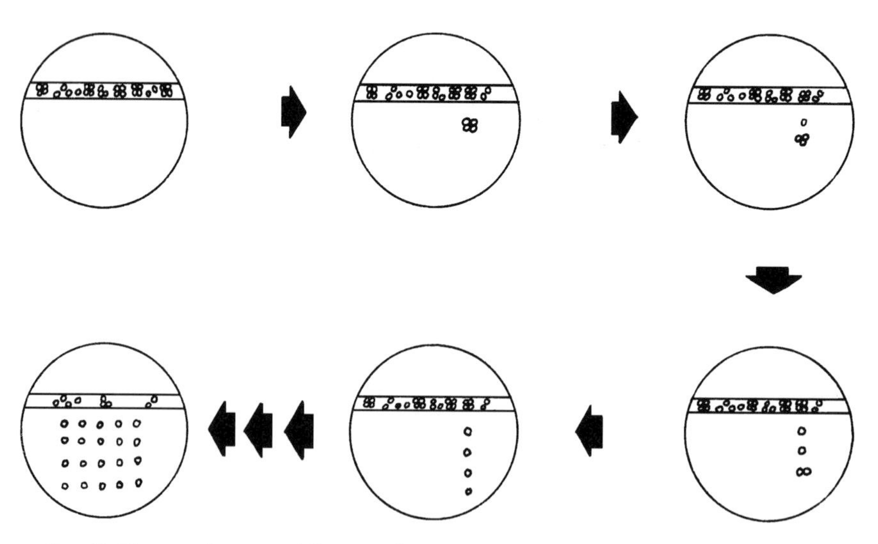

FIG. 10. The steps for sequentially separating the cluster of four ascospores approximately 5 mm apart on petri dishes [F. Sherman and J. B. Hicks, *Methods Enzymol.* **194,** 21 (1991)].

prevents the spore colonies from growing too close together. Likewise, positions on the *y* axis can be marked on the stage micrometer so that the four spore colonies from each ascus are evenly spaced. Take care not to break the microneedle when removing the dish or chamber from the stage. The thin agar slab is transferred from the chamber to the surface of a nutrient plate, which is then incubated for 3 days until the spore colonies are formed. Petri dishes containing separated spores are similarily incubated. As shown in Fig. 11, colonies derived from ascospores that were separated directly on a petri dish can be replica plated directly to media for testing nutritional requirements.

Although considerable patience is required to master ascus dissection, most workers are able to carry out this procedure after a few days of practice.

Isolation of Cells

In addition to ascus dissection, micromanipulation is occasionally required for separating zygotes from mating mixtures, for pairing vegetative cells and spores for mating, and for separating mother cells and daughter cells during vegetative growth. Zygotes usually can be picked up on microneedles, although vegetative cells usually cannot. However, vegetative cells can be separated simply by dragging them across the agar surface with microneedles. The use of microneedles is rather effective, since the cells usually follow closely in the wake of the microneedle as it is moved along the liquid surface film of the agar.

Tetrad Analysis

Meiotic analysis is the traditional method for genetically determining the order and distances between genes of organisms having well-defined genetics systems. Yeast is especially suited for meiotic mapping because the four spores in an ascus are the products of a single meiotic event, and the genetic analysis of these tetrads provides a sensitive means for determining linkage relationships of genes present in the heterozygous condition. It is also possible to map a gene relative to its centromere if known centromere-linked genes are present in the cross. Although the isolation of the four spores from an ascus is one of the more difficult techniques in yeast genetics, requiring a micromanipulator and practice, tetrad analysis is routinely carried out in most laboratories working primarily with yeast. Even though linkage relationships are no longer required for most studies, tetrad analysis is necessary for determining if a mutation corresponds to an alteration at a single locus, for constructing strains with new arrays of markers, and for investigating the interaction of genes.

There are three classes of tetrads from a hybrid which is heterozygous for two markers, $AB \times ab$: PD (parental ditype), NPD (nonparental ditype), and T

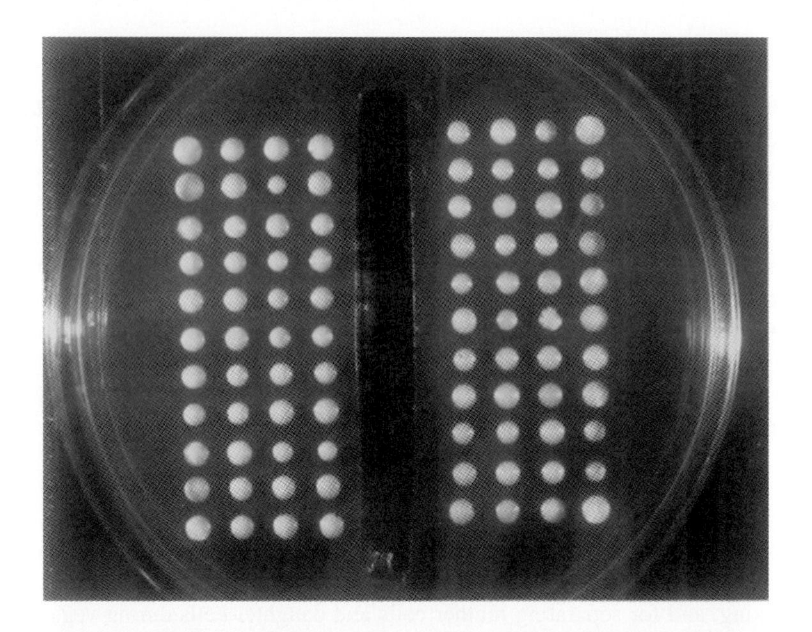

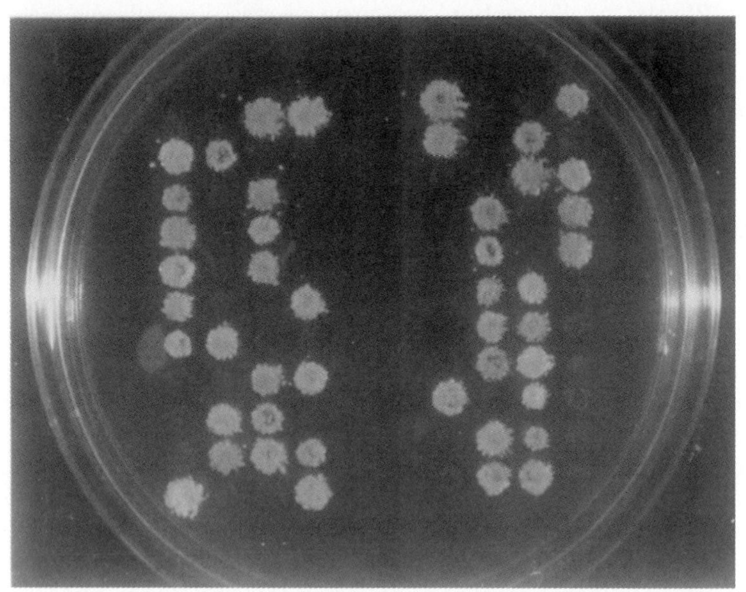

FIG. 11. Spore colonies derived from asci separated on the surface of a petri dish (*top*). The central area of the dish, containing a streak of the sporulated culture, was cut out and removed after dissection. The spore colonies were replica plated to a synthetic medium lacking a nutrient (*bottom*). The 2 : 2 segregation of a heterozygous marker is revealed by the growth pattern on the selective medium. The complete viability and uniform colony size shown are not typical of the meiotic progeny from most diploids; however, these properties can be chosen during the course of strain constructiion [F. Sherman and J. B. Hicks, *Methods Enzymol.* **194,** 21 (1991)].

(tetratype), as shown in Fig. 12. The following ratios of these tetrads can be used to deduce gene and centromere linkage:

	PD	NPD	T
	AB	*aB*	*AB*
	AB	*aB*	*Ab*
	ab	*Ab*	*ab*
	ab	*Ab*	*aB*
Random assortment	1 :	1 :	4
Linkage	>1 :	<1	
Centromere linkage	1 :	1 :	<4

There is an excess of PD to NPD asci if two genes are linked. If two genes are on different chromosomes and are linked to their respective centromeres, there is a reduction of the proportion of T asci. If two genes are on different chromosomes and at least one gene is not centromere-linked, or if two genes are widely separated on the same chromosome, there is independent assortment and the PD : NPD : T ratio is 1 : 1 : 4. The origins of different tetrad types are illustrated in Fig. 12.

The frequencies of PD, NPD, and T tetrads can be used to determine the map distance in cM (centimorgans) between two genes if there are two or fewer exchanges within the interval[91]:

$$\text{cM} = \frac{100}{2}\left[\frac{\text{T} + 6\text{NPD}}{\text{PD} + \text{NPD} + \text{T}}\right]$$

The equation for deducing map distances, cM, is accurate for distances up to approximately 35 cM. For larger distances up to approximately 75 cM, the value

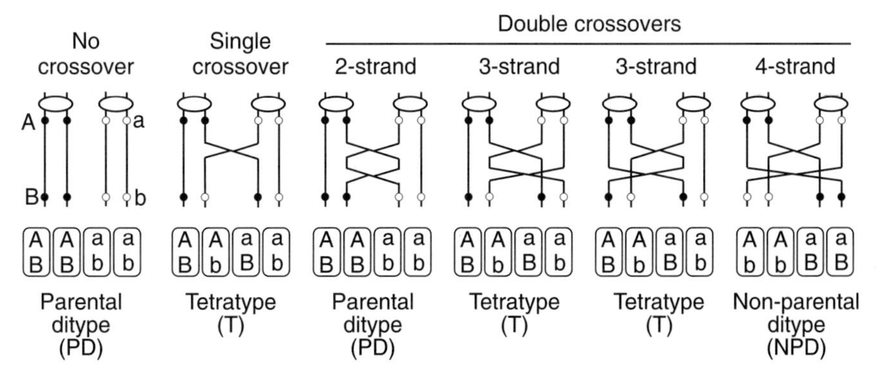

FIG. 12. Different tetrad types originating from a $AB \times ab$ heterozygous cross after either no crossover or single or double crossovers between the *A–B* interval. The fractions of the different tetrad types reveal whether A and B are linked, and the values can be used to calculate the map distance.

can be corrected by the following empirically derived equation[92]:

$$\text{cM (corrected)} = \frac{(80.7)(\text{cM}) - (0.883)(\text{cM})^2}{83.3 - \text{cM}}$$

Similarly, the distance between a marker and its centromere cM′, can be approximated from the percentage of T tetrads with a tightly linked centromere marker, such as *trp1*:

$$\text{cM}' = \frac{100}{2}\left[\frac{\text{T}}{\text{PD} + \text{NPD} + \text{T}}\right]$$

Gene Mapping

Recombinant DNA procedures have by and large replaced traditional genetic methods for determining the chromosomal positions of genes and subsequently their identification. The cloning of a DNA segment corresponding to a mutation and the sequence of the complementing fragment is a rapid method method for identifying the mutant gene. The chromosomal position can be easily determined from the database.[36]

However, there are rare occasions when a mutation cannot be identified by complementation with plasmid libraries. A new recessive mutation should be first tested by genetic complementation, involving crossing the unknown mutant to known, characterized mutants and examining the phenotype of the diploid. New mutants should be crossed to a series of known mutants having the same or similar phenotypes. Lack of complementation of two recessive mutations is almost always indicative of allelism. Meiotic analysis of the presumed homozygous diploid should reveal complete linkage and therefore identity. However, complementation of recessive mutations does not establish that they correspond to different genes. A meiotic analysis could be carried out when allelic complementation is suspected, especially when the diploid appears to have a partial mutant phenotype.

The second step in characterizing an unknown mutation should involve a meiotic analysis to determine if the mutant phenotype is controlled by a single gene. This is particularly critical when the mutant was derived from heavily mutagenized cells, such as those commonly used to obtain temperature-sensitive mutations and related defects. There have been numerous examples where temperature-sensitive growth and a particular enzyme deficiency segregated independently from each other, indicating mutations of two separate genes.

The mutant haploid strain should be crossed to a strain carrying at least one centromere-linked marker, such as *trp1*. Thus, a meiotic analysis would reveal both single-gene segregation and centromere linkage. The diploid should then be

[91] D. D. Perkins, *Genetics* **34,** 607 (1949).

[92] C. Ma and R. K. Mortimer, *Mol. Cell Biol.* **3,** 1886 (1983).

sporulated, the asci dissected, and the haploid segregants tested according to the methods outlined above. A 2 : 2 segregation of the mutant phenotype is indicative of a single-gene mutation. Less than two-thirds second division segregation (less than two-thirds tetratype asci relative to *trp1*) is indicative of centromere linkage of the unmapped gene. If centromere linkage is suspected, the mutant should be crossed to a set of centromere-tester strains that have markers near the centromeres of each of the 16 chromosomes.[46] The unmapped gene should exhibit linkage to one of the centromere-linked markers and should be further analyzed with additional markers on the assigned chromosome.

If the mutant gene is not centromere-linked, it is advisable to next determine on which chromosome it resides, using the 2 μm mapping or other procedures.[93]

Other Techniques for Genetic Analysis

In addition to the major techniques used for genetic analysis that are covered above in this chapter and elsewhere in this and the previous[24] volume, there are other simple procedures worthy of mention.

Replica Plating

Testing of strains on numerous media can be carried out by the standard procedure of replica plating with velveteen.[94] However, subtle differences in growth are better revealed by transferring diluted suspensions of cells with specially constructed spotting apparatuses. An array of inoculating rods, fastened on a metal plate, is dipped into microtiter or other compartmentalized dishes, containing yeast suspensions. Small and uniform aliquots can be repetively tranferred to different types of media.

Mating and Complementation

A few crosses can be simply carried out by mixing equal amounts of the *MAT***a** and *MAT*α strains on a YPD plate and incubating at 30° for at least 6 hr and preferably overnight. Prototrophic diploid colonies can then be selected on appropriate synthetic media if the haploid strains contain complementing auxotrophic markers. Similarily, testing mating types or other markers of meiotic progenies, requiring the selection of numerous diploid hybrids, can be carried out by replica plating, using any one of a number techniques such as cross-streaking and spotters. Prototrophic diploids also can be selected by overlaying a mixture of the two haploid strains directly on minimal plates, although the frequencies of matings may be slightly reduced. If the diploid strain cannot be selected, zygotes can be

[93] F. Sherman and L. P. Wakem, *Methods Enzymol.* **194,** 38 (1991).

[94] J. Lederberg and E. M. Lederberg, *J. Bacteriol.* **63,** 399 (1952).

isolated from the mating mixture with a micromanipulator. Zygotes, which can be identified by their characteristic thick zygotic neck, are best isolated 4–6 hr after mixing, when the mating process has just been completed[95]; diploids isolated by micromanipulation should be verified by sporulation and the lack of mating.

Formation of prototrophic diploids, indicative of complementation, is used to test *MAT***a** and *MAT*α mating types and to determine unknown markers in new mutants and meiotic segregants. Mating type tests are best carried out with *MAT***a** and *MAT*α tester strains, each containing markers not in the strains to be tested.

Complementation analysis consists of testing diploid strains that were constructed from two haploid mutants which have the same mutant phenotype, such as a specific amino acid requirement or sensitivity to UV. If the mutant character is found in the diploid and if the two mutant genes are recessive, it can be concluded that the two mutant genes are allelic, i.e., the lesions are in genes controlling the same function, or in most cases, the same polypeptide chain. In rare and special instances, a double heterozygous diploid strain may exhibit the phenotype of the recessive marker, confusing this test of complementation.

However, because of allelic (or intragenic) complementation, the growth of double heterozygous diploids does not always indicate that the two mutations are in different genes. Some cases of allelic complementation occur when the normal enzyme is composed of two or more identical subunits. The enzymes formed by allelic complementation are mutant proteins containing two different altered polypeptides, in which each of the mutant polypeptides compensates for each other's defects to produce a catalytically active protein. Allelic complementation can be pronounced when the enzyme contains separate domains carrying out different catalytic functions, such as the *HIS4A, HIS4B,* and *HIS4C* regions. Allelic complementation is frequent in yeast. For example, mutations in five of the 10 genes controlling histidine biosynthesis show extensive allelic complementation. Because of allelic complementation, frequencies of meiotic recombination are required to determine if two complementing mutants are alleles of the same gene. The frequencies of recombination are extremely low if the mutations are in the same gene, whereas the frequencies of normal meiotic segregants can be as high as 25% if the mutations are in different genes.

Complementations test are required for scoring meiotic progeny from hybrids heterozygous for two or more markers controlling the same character. For example, a $HIS3^+$ $his4^-$ X $his3^-$ $HIS4^+$ diploid will produce tetratype tetrades having the following genotypes:

$HIS3^+$ $HIS4^+$
$HIS3^+$ $his4^-$
$his3^-$ $HIS4^+$
$his3^-$ $his4^-$

Intergenic complementation tests are required to determine the segregation of the *his* alleles in the $HIS3^+$ $his4^-$, $his3^-$ $HIS4^+$, and $his3^-$ $his4^-$ segregants. These tests are carried out with *MAT***a** and *MAT*α tester strains having either $HIS3^+$ $his4^-$ or $his3^-$ $HIS4^+$ markers. Diploids homozygous for either $his3^-$ or $his4^-$ will not grow on histidine deficient medium, as indicated below:

		X $HIS3^+$ $his4^-$	X $his3^-$ $HIS4^+$
$HIS3^+$ $HIS4^+$	+	+	+
$HIS3^+ his4^-$	−	−	+
$his3^-$ $HIS4^+$	−	+	−
$his3^-$ $his4^-$	−	−	−

Random Spores

Although dissection of asci and recovery of all four ascospores is the preferred procedure for obtaining meiotic progeny, random spores can be used when isolating rare recombinants, or when analyzing a large number of crosses. Several techniques have been devised for eliminating or reducing unsporulated diploid cells from the culture. The proportion of random spores can be increased by sporulating the diploid strain on a medium containing a high concentration of potassium acetate (2%), which kills vegetative cells of many sporulated strains.[96] The sporulated culture is treated with Zymolyase, as described above, the spores are separated by a minimal level of sonication, and various dilutions are plated for single colonies. Alternatively, the spores of Zymolyase treated culture can be dispersed by vortexing with 0.5 mm glass beads. Furthermore, vegetative cells can be preferentially killed by treating a sporulated culture with an equal volume of diethyl ether.[97] Another convenient method for producing random spores relies on the selection against vegetative diploid cells that are heterozygous for can^R1, the recessive marker confirming resistance to canavanine sulfate.[98] If the $can^R1/+$ diploid is sporulated and the spores are separated and plated on canavanine medium, one-half of the haploid spores will germinate and grow, while the other half of the haploids and all of the diploids will not.

Acknowledgments

We thank Carl Zeiss, Inc. (Thornwood, NY) and Singer Instrument Co. Ltd. (Watchet, Somerset, U.K.) for photographs of micromanipulators. The writing of this chapter was supported by the U.S. Public Health Research Grant GM12702 from the National Institutes of Health.

[95] E. P. Sena, D. N. Radin, and S. Fogel, *Proc. Natl. Acad. Sci. U.S.A.* **70,** 1373 (1973).
[96] R. J. Rothstein, R. E. Esposito, and M. S. Esposito, *Genetics* **85,** 35 (1977).
[97] B. Rockmill, E. J. Lambie, and S. G. Roeder, *Methods Enzymol.* **194,** 110 (1991).
[98] F. Sherman and H. Roman, *Genetics* **48,** 255 (1963).

[2] How to Set up a Yeast Laboratory

By CORA STYLES

Introduction

The organizational strategies recommended in this chapter are biased in favor of a highly organized and cooperative laboratory community. The emphasis will be on efficient laboratory practices, pools of common supplies, and providing glasswashing and media preparation protocols for support personnel. Laboratory equipment, supplies, and experiment protocols constantly evolve and specific product recommendations soon become outdated. A helpful starting point is to see what other new researchers have bought recently to set up their yeast laboratories.

Role of Technician Manager

The work of organizing and maintaining pooled collections of chemicals, restriction enzymes, laboratory strains, antibodies, and radioactive chemicals for use by members of the laboratory community should be overseen by a long-term laboratory resident, a permanent research or technical associate. Laboratories that can retain a long-term employee in this role have the advantage of continuity in a community where others are transients. An effective laboratory manager should be perceived by laboratory members as having the full support of the principal investigator.

Orientation and Departure

A great time-saver is to develop written statements in the form of handouts that encapsulate information and state the practices to be followed by laboratory members. Three recommended handouts are the following.

Newcomer Orientation

This document should include personnel to contact in the administration, other staff to get acquainted with, and contact persons to make appointments with to receive radiation and safety training, which are necessary prerequisites to starting research work. A sample handout is shown in Fig. 1.

Laboratory Jobs

Each individual is assigned tasks and responsibilities for management, equipment care, and stocking common supplies. The manager should post current lists of laboratory job assignments. A list of job assignments is shown in Fig. 2. Ideally,

0076-6879/02 $35.00

Welcome to the Lab: Guide for Newcomers (Sample)

Preliminaries

Introduce yourself and learn the names of:

1. Office staff
2. Safety coordinator
3. Laboratory technician
4. See safety coordinator and the laboratory safety representative for training.
5. Radiation Office: Call and make appointment for training.
6. Glasswashers and media preparation staff.

Learn the location of necessary items:

1. Office supplies and Rolodex finder.
2. Orders are placed by computer requisition. Ask for instruction.
3. Reference books and common laboratory handbooks.

Common Items and Standard Practices

Distilled water: Produced by reverse osmosis through a resin bed (R/O water).
Melting agar in microwave: Get instruction on the microwave oven settings for melting agar. Loosen the bottle cap well. *Caution:* Superheated molten agar boils explosively when agitated. Handle with care.
Deionized (Milli-Q) water: The filter system is in Media Prep Room. Instructions are attached to the apparatus. There are reserve tanks of nonsterile water in carboys at the sink. Sterilized Milli-Q water in purple-cap bottles is on shelves in the hallway. These PURPLE-CAP bottles are for WATER ONLY. They are never washed. Always keep caps on bottles, even when empty. Place on sinkboards at designated sites.
Hoods: The working surface is for *experiments in progress* only. LABEL your experiment and clean up promptly. IDENTIFY all reagents used in hood. See the hood's manager for permission to store anything in the hood.

FIG. 1. A sample orientation handout for newcomers.

Laboratory agents: Animal agent, safety representative
Management and organization: Hoods and hazardous wastes, antibodies, balance room, coldroom benches, gel electrophoresis room, oligonucleotide library, curators of laboratory strains, chemicals databases, radiation materials, phages
Equipment care and maintenance: Shakers, roller drums, and water baths; centrifuges and microfuges; pH meter and test papers; gel dryers, vacuum pumps; Speed Vac, traps, and oil changes; computers; sonicator; spectrophotometer; PCR machine; electroporator; hybridizer machine; PhosphorImager plates; X-OMAT film processor; freezer -80° frost maintenance; electrophoresis equipment; microscopes; pipettors; scintillation counters and vials
Stock and preparation of common supplies and reagents: Chemicals, solvents, and ethanol; plastic goods and other disposables; films; hexamers; PCR kits, and restriction enzymes; 20× SSC; Southern base and neutralizer; TAE, TBE; protein gel buffer

FIG. 2. A list of laboratory job assignments.

each job holder will keep notes on newly accumulated information to add to the original laboratory job descriptions.

Departure Checklist

This checklist (Fig. 3) is a key handout facilitating turnover of laboratory space. The manager should give the checklist to researchers about a month before departure. The essential "tour" is a physical inspection by the manager together with the researcher of all laboratory sites occupied by the researcher. The manager verifies firsthand that all spaces are completely empty.

Management Strategies

Radioactivity

The principal investigator obtains federal and state licenses and has the ultimate responsibility for compliance. Regulations usually require that radioactive substances be stored in a separate secured place, such as a locked refrigerator–freezer. The ordering, stocking, posting of usage signup sheets, and arranging for waste pickup plus general oversight are best assigned to a single individual. One policy decision to make is whether to have a single location where all radioactive work is performed or to have individuals do radioactive work at their own benches. The former has the disadvantage that the individuals responsible for spills at a common site cannot be readily identified, and responsibility for cleaning up can become an issue.

Restriction Enzymes

The most cost-efficient method for managing restriction enzymes and other enzymes for manipulating DNA is to establish a common supply, stored alphabetically in labeled racks in the top freezer section of a household type refrigerator–freezer. To protect enzymes from inadvertent contamination, a dedicated 10-microliter pipettor and pipette tips with filters should be provided for exclusive use with the enzyme collection. For additional protection the freezer should be connected to an emergency power source.

Water

Two levels of purified water are needed in a yeast laboratory: (1) Water for preparing growth media requires that the chlorine, heavy metals, and other toxic impurities found in tap water be removed. Facilities designed for research work may have a separate plastic plumbing system which delivers water treated by reverse osmosis ("R/O"), a process which purifies tap water by passing it through a resin bed. (2) Deionized water is needed for biochemical work. Deionizing filtration

LABORATORY AND INSTITUTION DEPARTURE CHECKLIST (Sample)

People who are departing are expected to leave their desk and bench clean. It should be ready to be occupied by the next research worker. The preparation for departure and clean-up are also important so that no hazardous chemicals of unknown composition or origin remain.

ONE MONTH BEFORE LEAVING

___1.Notify the lab manager, the radiation officer and the secretary of your expected departure date. Be sure to:

- ___ Turn in your access/ID cards.
- ___ Empty out your clothing locker(s).
- ___ Turn in keys to your desk and other keys.
- ___ Turn in your telephone card.

___2. Provide Yeast Strain Curator and Bacterial Strain Curator with Xerox copies or a database file of all published strains or strains that will be published. Check the stock lists to be sure all of your strains are in the Laboratory Collection. It is helpful to provide the curator(s) with an annotated copy of published papers showing the names of all strains described.

___3. Turn over antibodies and antisera to the Lab Curator, and provide details in writing.

___4. Transmit information about your lab job(s) in writing to the next person.

___5. If you plan to continue to working with yeast, make copies of the strains that you want to take. Requests for additional strains after you leave will have to be approved by the Principal Investigator, as are all requests from outside labs.

TWO WEEKS BEFORE LEAVING

___1. Clean out your radioactive probes and check out with the lab Radiation Officer.

___2. Return all small bench hand tools to the technician.

___3. Clear desk and drawers of all contents. Return materials to office supplies.

___4. Return all borrowed books and journals.

___5. Reagents: Donate them to people who will put their own names on them. Place reagents on display for people to scavenge. You dispose of the remainders. Dispose of the contents in a NON-HAZARDOUS, SAFE MANNER, following advice from the Safety Coordinator. Rinse bottles clean and remove all adhesives and labels except washable labels.

___6. Place unwanted X-ray film in the silver recovery box in the XOMAT darkroom.

___7. Give away unused media to other lab members.

___8. Remove all your personal reagents from, the hood cold room, and elsewhere.

___9. Clear out all of your frozen strains, competent cells, and cassettes.

___10. Make arrangements if you wish to transport frozen material from the laboratory.

___11. Equipment purchased with your grant money for your research is the property of the laboratory, and cannot be removed without the Principal Investigator's consent.

___12. Chemicals at your bench which are still in the original container need to be integrated into the laboratory collection. Provide the technician with a list of the new locations.

___13. Leave your new address and phone number with the secretary. Notify the post office of your address change. Your mail will not be forwarded.

___14. Give computer-related information, especially passwords to the secretary.

___15 Check with the Principal Investigator concerning the disposition of your notebooks. There may be institutional and legal restrictions that apply.

___16. Goods may be left behind to be reclaimed later. Make arrangements with another person to take charge of your goods during the interim to make sure they are not discarded.

THE TOUR: Show the technician these cleaned areas:

Your clean and bare bench, shelves and desk with nothing left for the next person.
Your –20° shelf   The –80° freezer space   Radioactive storage areas
A check of the hoods   Your cold room shelf
Labware and equipment which will become laboratory property
Any materials and goods you are leaving for the general public
The list of new locations of your bench chemicals.

FIG. 3. A sample checklist for researchers leaving a yeast laboratory.

systems designed by Millipore Corporation (Bedford, MA)[1] work well. The choice of purification system depends on the purity of the incoming water supply and the degree of "polishing" needed. It is not necessary to remove endotoxins and pyrogens for yeast work.

Bottles of sterile, deionized water for making biochemical reagents and for mixing with sterile media components are a great convenience. To keep this water supply pristine, it is advisable to maintain a separate set of bottles solely for deionized sterile water. The glass bottles can be marked with a paint pen, and caps of an identifying color should be purchased for them. Nothing foreign should be introduced into the bottles and caps should be kept on them at all times. The bottles are never washed, only refilled and autoclaved.

Glasswash Procedures

Glassware to be recycled needs three sites: (a) a collection site for soiled labware, (b) a washing and processing site, and (c) a storage site for the cleaned and sometimes sterilized labware. Laboratory rules regarding the condition of labware that researchers place at the collection site should be formulated to shield laboratory aides from harm and to make their work easier. Agar-containing bottles and flasks should be left standing filled with water to prevent the agar residue from drying into a tough film. No agar in any form should be allowed to enter sink drains. Generally, labware used for cultures must be decontaminated with Lysol or Clorox and rinsed with tap water. It must be free of toxic chemicals and radioactivity. Broken glassware should be discarded. Recyclable glass serological pipettes need separate plastic collection containers 1/3 full of water with 10% Clorox. Used velveteens need their own collection container 1/2 full of water with a capful of Lysol. Detailed procedures for processing labware are shown in Table I.

Design for Media Preparation by Support Staff

The basic media for culturing yeast are "complete" (YPD), "minimal" (SD), and "synthetic complete" (SC). Bacteria are cultured in "LB medium" (LB) and, occasionally, in "minimal" medium (M9). The media are used as liquid cultures or combined with agar to make petri plates. For a discussion of media and formulas, see Sherman.[2] Medium is more efficiently prepared in batches by support staff. Here is a description of one system which is designed to serve about 15 users. Some media products are made for immediate use (Table II). Other products are designed to be mixed together by the researcher, usually to make agar plates (Table III). Media cannot be made too far in advance because some types of media deteriorate with time. Table IV shows a list of titles of protocols for preparing

[1] http://www.millipore.com/catalogue.nsf/docs/C1756

[2] F. Sherman, *Methods Enzymol.* **350,** [1], 2002 (this volume).

TABLE I
GLASSWASH PROCEDURES

Product and process
Cleaned sterile glass pipettes
(1) Collect soiled pipettes. Bring back to glasswash facility.
(2) Pour diluted (2%) Micro cleaning agent over pipettes to cover. Autoclave for 10 min.
(3) Separate pipettes from Micro and load pipettes into rinsing baskets. Lower baskets into siphon rinsing units and rinse in tap water followed by water (R/O).
(4) Place pipettes in drying baskets and dry in a hot oven.
(5) Sort cooled pipettes into cans and bake at high temperature for several hours.
Cleaned glass and plastic labware
(1) Collect soiled labware and deliver to glasswash facility.
(2) Hand-brush soiled labware and visually inspect. Separate by type (plastic, glass, large, small) and load glasswasher. Add minimum amount of detergent and wash.
(3) Nonsterile glass and plastic nonsterile beakers are dried in oven. Sterile labware is capped and autoclaved.
Sterile velveteens
(1) Collect used velveteens from buckets, place them in clothes washer, but do not add detergent (the liquid in the buckets already contains Lysol.). Fluff dry in dryer.
(2) Flatten and brush squares and wrap nap-side down 10 each in aluminum foil.
(3) Autoclave sterilize on the dry cycle for 35 min. Then bake to dry in a low oven (160°F).

media. See Appendix for protocols. Some of the formulas are alternatives to the ones described by Sherman.[2] Table V provides additional information about media components, naming, and sources.

Using the Media System

This system affords versatility and economy. As soon as a need for media is perceived, a researcher can quickly prepare the media he or she wants from premade ingredients. To prepare to pour plates, bottles with agar are heated in a microwave oven or in an autoclave to melt the agar. The molten agar may become superheated, making the opening of the bottles hazardous. It is advisable to let the

TABLE II
READY-TO-USE MEDIA

Container size	Quantity
500-ml media bottle	300 ml 1× LB with 2% agar
500-ml flask	100 ml LB liquid
2-liter flask	500 ml LB liquid
500-ml media bottle	300 ml 1× YPD

TABLE III
PREMADE MEDIA COMPONENTS[a]

Media bottle size	Component
1 liter	300 ml 4% (w/v) Agar
500 ml	300 ml 2× LB
500 ml	300 ml 2× YEP
500 ml	300 ml 2× Minimal
500 ml	300 ml 2× Minimal + amino acid mixes
500 ml	200–300 ml 40% (w/v) Glucose
160 ml (rectangular)	100 ml Single amino acid stock solutions

[a] The amounts and concentrations are based on an approximate final volume of 600 ml.

bottles stand for a minute before agitation. The worker then combines 300 ml of nutrient medium with a carbon source, e.g., 30 ml of 40% (w/v) glucose, and cautiously adds this mixture to the molten agar. The contents need to be thoroughly mixed. After mixing standard media for yeast, the bubbles can be safely dispersed with a short spray of ethanol. If YPD or YNB agar media should solidify in the bottle before plates can be poured, these can be remelted. Note that SC agar media cannot be remelted. The amino acid mix hydrolyzes the agar while it is remelting, and it will no longer solidify.

The site for plate pouring may be on individual benches or at a common site. In buildings with poor air quality control, plate pouring will need to be done in a sterile hood. Agar plates work best for micromanipulation and replica plating if allowed to stand for 2 to 3 days before being wrapped in a tightly closed bag. Plates standing unwrapped for more than 7 days dry up and are of little use. The shelf-life of well-wrapped media plates can be about 6 months, except for Synthetic Complete media. Light and heat lead to media deterioration.

Laboratory Databases

A technician with good organizational skills should be encouraged to develop and maintain laboratory database systems.

Laboratory Storage System

An efficient approach to codifying locations of stored items is to assign a sequence of permanent numbers to all cupboards, shelf sets, and drawers within a given room. The location of the item then is simply stated as: Room number, site number. The names of stored items may be listed on Rolodex cards or in a computer database.

TABLE IV
MEDIA PREPARATION PROTOCOLS[a]

LB liquid medium
LB 2×
LB 2% agar
2×-YT
Agar 4%
YNB 2×
Yep 2×
YPD 1×
YNB 2× + AA dropout
Glucose 40%
Histidine 100 m*M*
Adenine 30 m*M*
Uracil 20 m*M*
Lysine 100 m*M*
Leucine 100 m*M*
Tryptophan 40 m*M*
Inositol 200 m*M*
M9A 20×
SOB
LB top agar
Soft agar
T agar
Casamino acids
Sodium hydroxide 1 *N*
Amino acid powder mix, complete
Presporulation GNA
Sporulation Agar NGS
Sporulation Agar SPOR
Terrific broth
Fortified broth
T broth
SOB from capsules
Inositol-free minimal

[a] See Appendix for protocols.

Placing Orders in a Database

Regardless of differences among institutional ordering systems, it is invaluable to maintain a database within the laboratory of all orders placed. The database design (Fig. 4) should be useful to both laboratory workers and office staff. Computerized ordering has the advantage of being legible, using less paper, and providing a permanent searchable file of past orders. Records can be duplicated and edited to place repeat orders. A categories field is useful for inventorying chemicals and for creating summaries for accounting purposes. All computer databases need to

TABLE V
ADDITIONAL INFORMATION ON MEDIA

Medium	Comments
Agar	Comercially available in various degrees of purity. For yeast media, the same level of concern paid to laboratory water should also apply to agar. Neither product needs to be completely pure, but the level of purity needs to be unvarying. For yeast laboratories, we prefer Difco Bacto-agar.
YEP	Represents the ingredients, yeast extract (Difco) and Bacto-peptone (Difco). 2× YEP is made without a carbon source to allow the researcher to add one of choice, usually glucose.
2× Minimal medium	Also provided without a carbon source. We call this YNB, naming it after the sole ingredient, Yeast Nitrogen Base (Difco) (without amino acids).
2× Minimal + amino acid mixes	Refers to SC (without glucose) and is called 2× YNB + all amino acids (AA). The amino acid mixes usually have one or more of the amino acids omitted, hence the name "drop-out" medium. This so-called amino acid mix also contains adenine, uracil, and *p*-aminobenzoic acid (PABA). Dry powder mixtures of amino acids can be stored almost indefinitely. Once mixed with water, however, they have a limited shelf-life of about 1 month. The composition of Synthetic Complete medium varies as it is modified for particular strains or mutants. We have settled on a formulation that uses equal amounts by weight of each ingredient, except 1/10 that amount of *p*-aminobenzoic acid with two adjustments: (1) to support wild-type growth of leucine auxotrophs, the amount of L-leucine is doubled, and (2) if one wishes *ade1* and *ade2* mutants to turn red, the amount of adenine is reduced. A variety of SC drop-out powder mixes are commerically available.[a]
Minimal medium	Made from Difco Yeast Nitrogen Base supports weak growth of *ino*-strains. For research not involving inositol mutants the presence or absence of inositol probably can be ignored. Problems arise when wanting to distinguish inositol auxotrophs from wild type. To produce inositol-free medium,[b,c] minimal medium must be made from scratch (Table II).
LB Medium	Made from yeast extract, Bacto-tryptone, and sodium chloride and pH-adjusted with sodium hydroxide. Handling sodium hydroxide can be eliminated by purchasing LB medium as a complete powder or liquid concentrate, already pH-adjusted.[a]

[a] http://www.bio101.com/

[b] F. Sherman, G. R. Fink, and J. Hicks, "Cold Spring Harbor Laboratory Course Manual for Methods in Yeast Genetics," p. 165. Cold Spring Harbor Laboratory Press, Cold Spring Harbor, NY, 1986.

[c] C. W. Lawrence, *Methods Enzymol.* **194,** 280 (1991).

Requisition Form **New Lab**

No. [ ]

Date [ ]

User Information

Name [ ]

Phone [ ]

Room No [ ]

Purchase for [ ]

Vendor Information

Name [ ]

Number [ ]

Address [ ]

Phone [ ]

Contact [ ]

Accounting Information / Office use only

Area/Org [ ]

Total Cost [ ]

	Fund	Area/Org	Object
Line 1			
2			
3			

Rx [ ]

Date Required [ ]

Comments [ ]

Description	Type	Quant.	Unit	Code	Catalog No.	Unit Cost	Cost

Received [ ]

PO Number [ ]

Contact [ ]

Standing order? [ ]

FIG. 4. A sample record in one layout of an orders database based on FilemakerPro 4.1. Two fields have automatic functions: Record Number, a "number field," automatically enters serial numbers in increments of one, and Date, a "date field," enters the current date. Several fields have drop lists from which to choose an entry: User Name, Accounting information fields, Type, and Unit fields. Other fields are designated "Look-up" fields. When an entry is made in User Name, the application locates a second FilemakerPro 4.1 database which has User Names, User Phones, and User Room Numbers. The information found in the second database is placed in the appropriate fields of the Orders database record. The same is true for Vendor Name, which supplies the Vendor Number from a third database. A fourth database holds Descriptions and looks up Type, Quantity, Unit, Code, Catalog No., and Unit Cost. The Look-up databases need to be kept up to date. The Description field in this design needs to contain information clarifying the Quantity entry. For example, if an item is sold "10 per case," that information must be entered in the description. Under Quantity enter "1" and in Unit enter "case." The Description field is a repeating field. If one searches for an item in a repeating field, all Description fields will be searched. It is convenient to duplicate a previous order and edit it to create a new order. If the secretarial staff has access to laboratory computers via a network, paper requisitions may be eliminated altogether.

be safeguarded with backups and password protection. Password protection should allow laboratory users all activities except deleting files. That function should be reserved under password for the database manager.

Chemicals Inventory

Federal law requires research laboratories to maintain an inventory of all chemicals on the premises. Careful maintenance of a laboratory chemicals database can reduce the labor of taking a physical inventory. The database can also serve to inform workers not only whether a chemical exists in the laboratory, but also where it is located. All chemicals, even those kept at workers' benches which are still in the manufacturer's bottle, are included in the chemicals database. For keeping track of chemicals, an ideal system would be to assign an identification number to each chemical unit coming into the laboratory. The simplest way to track chemicals with precision is to assign a new identification number to each bottle. The system is amenable to barcoding, which would facilitate taking inventory. Essential information to include in the database is (1) the ID number, (2) the year ordered, (3) the volume in the original container, (4) the person who ordered it, (5) the type of chemical or the purpose for which it was bought, and (6) the storage location, such as room number, site in room, and detail within the site if appropriate, e.g., "Room 557, Chemicals Freezer, Shelf 3, Storage jar #2." From the database the manager can extract and can post alphabetized lists of chemicals stored at various locations, such as the contents of a chemical storage freezer.

The entire database, write-protected, should be available to all laboratory members. A sample layout of a chemicals database is shown in Fig. 5. The database should be managed by a laboratory manager working together with the laboratory member in charge of chemicals. The manager should monitor the Orders database regularly to be alert to incoming chemicals. Incidentally, purchasers of toxic chemicals should be made aware of the high cost of disposing of unused and expired toxic chemicals.

When chemicals are used up or a bottle discarded, the bottle identification number is reported to the database manager, who will delete it from the file. When laboratory personnel leave, one of their departure duties is to integrate their bench collection into the common collection. They must report to the chemicals database manager where each bottle in their personal collection has been relocated in detail.

Yeast and Plasmid Laboratory Strain Collections

The yeast strain collection belonging to the principal investigator will form the starting point for a permanent collection unique to his or her laboratory. Researchers in the laboratory are expected to contribute strains produced in the laboratory to the permanent collection. A well-managed, carefully documented collection will grow into a powerful asset.

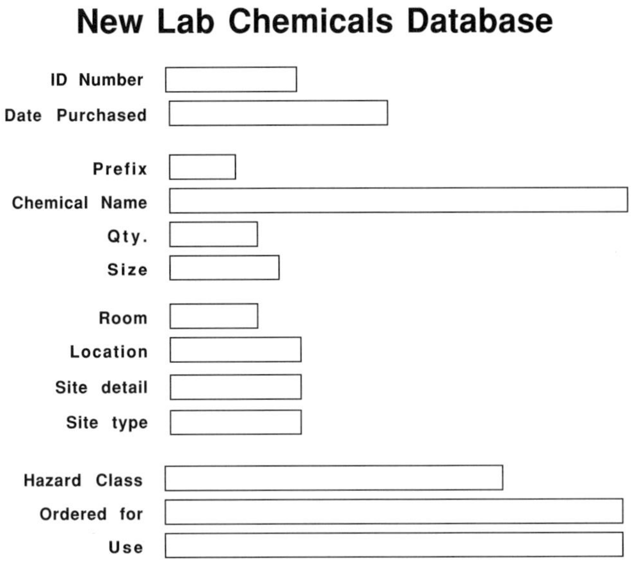

New Lab Chemicals Database

ID Number	
Date Purchased	
Prefix	
Chemical Name	
Qty.	
Size	
Room	
Location	
Site detail	
Site type	
Hazard Class	
Ordered for	
Use	

FIG. 5. A sample record in one layout of a Chemicals Database based on FilemakerPro 4.1. This Chemicals Database has additional layouts for producing special-purpose lists to post at storage sites and for laboratory members to use when relocating their bench chemicals collection.

The simplest method for organizing strains is to keep all of them in a single collection where each new entry receives the next number in a series. Strain numbers are conventionally given one or more prefixed letters, such as one's initials followed by Y for yeast. Naming all yeast strains in a single numerical sequence simplifies freezer space organization. Establish one database for yeast (*Saccharomyces cerevisiae*) and another for bacteria with plasmids.

A computerized strains database performs best as a search tool rather than as an archive of all information. Information not in the database can be organized in binders with numbered pages which the database records can reference. An example of a record in a yeast database is shown in Fig. 6a. An important feature is that mutant genes are listed individually in a repeating field. The advantage of using a repeating field is that a FIND request will search all entries within the field. This type of field also enables the database manager to produce a summary report of all mutations in the collection, which is useful for editing purposes and as a tool to maintain uniformity in naming. In other fields where entries are limited to only a small number of options, such as "Mating Type," uniformity can be ensured by attaching a drop-list to the field.

The plasmid database (Fig. 6b) is similar to the yeast database. In addition to the computer records, a hard copy of the entire database is handy.

(a) **New Lab Yeast Strains**

Lab Strain Number

Mating Type

Plasmid

Donor

Former Name

Parents

Strain Background

Comments

Genotype

Creation Date

card number

(b) **New Lab Plasmid Strains**

Lab Strain Name PJB0001

Plasmid name

Plasmid markers

Bacterial host

Insert Organism

Insert

Type

Map available ○ yes ○ no

Source

via

Reference

Comments

It is essential that researchers contribute a select number of their strains to the permanent laboratory collection. Recommended guidelines for contributions are the following: All strains received from other laboratories should be deposited promptly, together with the sender's accompanying documentation. The database record should state clearly all limits on the use and distribution of strains received. Strains received through Material Transfer Agreements need great attention to compliance, since they are covered by legal agreements. All strains that researchers produce in the laboratory which are generally useful should be added, including transformants that are hard to reproduce and strains obtained from other laboratory members, who may not have submitted them. Finally, all strains likely to appear in publications, together with the strains from which they were derived, must be submitted to the laboratory collection. Adding strains to the collection should be overseen by a single individual, usually a technician in the role of curator. Strains in the permanent frozen (−80°) collection are stored in sterilized 2-ml polyethylene vials with 1 to 1.5 ml glycerol (15% v/v glycerol/water) for yeast and 50% (v/v) glycerol for bacteria carrying plasmids.

Laboratory Equipment

Basic Small Equipment

A microwave oven needs an interior chamber height to accommodate 1-liter media bottles. When choosing roller drum racks and other tube-holding equipment, make sure the size of holes matches the choice of culture tube size (see Glass Culture tubes below).

Bench Tools

Multiprong blocks ("froggers") are useful for transferring cultures from solid medium to solid or liquid medium. For transferring liquid samples from 96-well

FIG. 6. Sample records of strains databases for yeast and for plasmids based on FilemakerPro 4.1. Both databases are password protected. The password "master" allows complete access. The password "user" allows all functions except editing, deleting, and creating, which are reserved for the Strains Curator, who knows the master password. (a) In Yeast Strains, Lab Strain Number is a "number" field which automatically assigns each new record a serial number increasing in increments of one. The number may be preceded by the Principal Investigator's initials. Card Number is another automatic numbering field. Creation Date is a "date" field, which automatically fills in the current date. Other fields are "text" fields. Mating Type and Strain Background have drop lists which limit the variety of entry notations. Genotype is a "repeating (text) field," which allows multiple fields to be included in a single search. (b) In the Plasmid Strains database, the strain name and creation date fields are like those in Yeast Strains. Marker List is a repeating field. Type has a drop list limiting the choice of entries. Both databases have additional layouts useful for summary purposes. The "Hard Copy" layout in Plasmid Strains is used to print out paper forms to store in notebook binders.

plates to solid medium or to other plates, multichannel mechanical or electronic pipettors are recommended to minimize cross-contamination of adjacent wells.

Velveteen cloth (all cotton) for replica plating can be purchased from major fabric stores. Hemming is optional. Unhemmed velveteen squares when first washed and fluff dried produce a large ball of waste threads. On subsequent washing the fringed velveteens are relatively stable.

Glass Bottles and Accessories

Media and reagent bottles should be heat-resistant glass. Glass culture tubes are sold both as disposable and reuseable. One problem with reusing glass culture tubes is their fragility. The turbulent spray in some glasswashing machines may break culture tubes and clog the water lines with shards. Glass culture tube sizes are 13 mm diameter and 16 or 18 mm diameter. One should decide early on whether to use 16- or 18-mm diameter tubes and choose other equipment, such as tube racks and roller drum tube trays, to match. Rolls of labels can be purchased with water-soluble adhesive[3] for easy removal. These can be hung on a toilet paper roll holder at convenient locations.

Microscopes and Accessories

Three leading manufacturers are Zeiss, Olympus, and Nikon. Finding a reliable, knowledgeable sales representative is as important as selecting the brand. A low-power microscope with a reflective adjustable mirror or mirror slit is convenient for viewing small yeast colonies. A higher power inverted microscope is a convenient option for viewing yeast cells and liquid well cultures. A high-power fluorescence microscope is needed to observe antibody staining. Ideally, each microscope should have a camera port (see Refs. 4 and 5). Dissection microscopes are fully discussed by Sherman.[2]

Refrigeration Units

A 20-cubic-foot household refrigerator–freezer with a top freezer cabinet is useful for a restriction enzyme collection. The lower section may be a storage site for refrigerated chemicals. As the laboratory grows, an additional small chest freezer is recommended for storing backup supplies of restriction enzymes and kits. At least one -20° freezer, such as an undercounter model, should have external contacts to permit safe storage of flammables. Full-size upright household freezers work well for storing frozen chemicals and provide shelves for researchers to store

[3] Shamrock Scientific, Bellwood, IL, 1 (800) 323-0249. Blank labels 1×2.5 in. $\times$ 1000, wash-away adhesive.

[4] S. J. Kron, *Methods Enzymol.* **351,** in preparation (2002).

[5] D. R. Rines, X. He, and P. K. Sorger, *Methods Enzymol.* **351,** in preparation (2002).

plasmids and reagents. Ultracold freezers (−80°), upright models with shelves and drawers with compartments to hold boxes, work well for storing strain collections, plasmids, and antibodies.

All freezers should be connected to the emergency power system of the building.

Labware

Standard 100-mm diameter plastic petri plates vary in size and design. Test sample plates with equipment designed to hold plates, e.g., the dissection microscope stage plate holder and plate holder accessories for an inverted microscope stage. If the plates are too big for this equipment, the recommendation is to find a competent machinist to enlarge the plate holders to fit the plates rather than buying expensive plates to fit the equipment.

Appendix: Protocols[6,7] for Preparing Media

LB Liquid Medium

Prepare in 20-liter carboy with spigot.

Ingredients	For 20 liters
LB Medium powder (BIO-101 low salt, from Q Biogene) (or [small] capsules)	500 g
Water (R/O)	20 liters

Dissolve with stir bar in 3–5 liters water. Transfer to more convenient height. Add remaining water.

Dispense as needed:

(a) 100 ml in 500-ml flasks
(b) 300 ml in 500-ml bottles
(c) 500 ml in 2-liter flasks
(d) 100 ml in 500-ml flask

Cover flasks with foil or paper towel and 2 paper cups.

Autoclave 35 min or less.

Let hot media rest 10–15 min. Place in front of fan to let cool.

Apply labels with date. Record batch with lot numbers in daily log. When no longer warm, tighten caps. Stock shelves.

[6] D. Burke, D. Dawson, and T. Stearns, "Cold Spring Harbor Laboratory Course Manual for Methods in Yeast Genetics," Appendix A—Media. Cold Spring Harbor Laboratory Press, Cold Spring Harbor, NY, 2000.

[7] J. Miller, "Experiments in Molecular Genetics," p. 431. Cold Spring Harbor Laboratory Press, Cold Spring Harbor, NY, 1972.

LB 2×

Prepare in 20-liter carboy with spigot.

Ingredients	For 10 liters
LB Medium powder (BIO-101 low salt) (or [small] capsules)	500 g
Water (R/O)	10 liters

Dissolve with stir bar in 3–5 liters water. Transfer to more convenient height. Add remaining water.
Dispense 300 ml in 500-ml bottles.
Autoclave 35 min or less.
Let hot media rest 10–15 min. Place in front of fan to let cool.
Apply labels with date. Record batch with lot numbers in daily log. When no longer warm, tighten caps. Stock shelves.

LB 2% Agar

This product is ready-made LB agar in 300 ml quantity. Convenient for melting, adding antibiotic, and pouring about 10 to 12 plates. Prepare in two 6-liter flasks.

Ingredients	For each flask
LB Medium powder or capsules (BIO-101 low salt)	75 g
Agar (Difco)	60 g
Water (R/O)	3 liters

Dissolve in small autoclave 20–25 min to melt agar. Add stir bar afterward and mix thoroughly until "lines" are gone (1 min). (Apply additional heat and stir longer if agar was not completely melted.) Keep hot so agar does not thicken and solidify.
Dispense 300 ml in 500-ml bottles. Be careful of this HOT liquid. You can wait till it cools down to a more comfortable temperature. Place the flasks on an insulated surface while cooling (paper towel).
Loosely cap bottles.
Autoclave 15 min or more.
Apply labels with date. Record batch with lot numbers in daily log. When no longer warm, tighten down caps. Stock shelves.

2×-YT

See comments for LB media. Prepare in 20-liter carboy with spigot.

Ingredients	For 10 liters
2×-YT Medium powder (BIO-101 Low Salt)	310 g
Water (R/O)	10 liters

Dissolve with stir bar in 3–5 liters water. Transfer to more convenient height. Add remaining water.
Dispense 300 ml in 500-ml bottles.
Autoclave 35 min or less.
Let hot media rest 10–15 min. Place in front of fan to let cool.
Apply labels with date. When no longer warm, tighten down caps. Stock shelves.

Agar 4%

Prepare in any number of 1-liter bottles.

Ingredients	For each bottle
Agar (Difco)	12 g
Water (R/O)	300 ml

Label 1-liter bottles with permanent marker ("4%").
Weigh 12 g—or simply fill a cut-away coffee scoop (practice weighing to get right amount) with agar.
Pour agar into 1-liter bottle with aid of large powder funnel.
Add 300 ml water with aid of large powder funnel.
Loosely cap bottles.
Autoclave 35 min
When no longer warm, tighten down caps. Stock shelves.

YNB 2×

YNB (yeast nitrogen base) is minimal medium lacking glucose. Difco's YNB without ammonium sulfate is less expensive than the same product with ammonium sulfate. Inositol is added to YNB so that the medium will support wild-type growth of inositol-requiring strains. Inositol-free medium for detecting inositol auxotrophs must be made from scratch. Prepare in two 6-liter flasks.

Ingredients	In each flask
Difco Yeast Nitrogen Base w/o AA, AS[a]	9 g
Ammonium sulfate	30 g
Inositol (200 m*M*) stock (*myo*-inositol)	6 ml
Water (R/O)	3 liters

[a] With and without amino acids (AA) ammonium sulfate (AS). Record lot number in log.

Dissolve with stir bar. Heat may be applied.
Dispense 300 ml in 500-ml bottles.
Autoclave 35 min or less.
Let hot media rest 10–15 min. Place in front of fan to let cool.
Apply labels with date. When no longer warm, tighten caps. Stock shelves.

YEP 2×

YEP is yeast extract and [Bacto] peptone with no carbon source. Prepare in two 6-liter flasks.

Ingredients	In each flask
Bacto-peptone	120 g
Yeast extract	60 g
L-Tryptophan	0.9 g
Water (R/O)	3 liters

Dissolve with stir bar. Heat may be applied.
Dispense 300 ml in 500-ml bottles.
Autoclave 35 min or less.
Let hot media rest 10–15 min. Place in front of fan to let cool.
Apply labels with date. Record lot numbers. When no longer warm, tighten down caps. Stock shelves.

YPD 1×

This product is ready-made for users to aliquot directly into culture tubes. Here the glucose is autoclaved with the media. Watch out for over-autoclaving as indicated by a dark brown color indicating that the glucose has caramelized. The medium is supplemented with additional tryptophan. Prepare in two 6-liter flasks.

Ingredients	In each flask
Bacto-peptone	60 g
Yeast extract	30 g
L-Tryptophan	0.45 g
D-Glucose (dextrose)	60 g
Water (R/O)	3 liters

Dissolve with stir bar. Heat may be applied.
Dispense 300 ml in 500-ml bottles and 100 ml in 200-ml square bottles.
Autoclave 35 min or less.
Let hot media rest 10–15 min. Place in front of fan to let cool.
Apply labels with date. When no longer warm, tighten down caps. Stock shelves.

YNB 2× + AA Drop-Out: Generic Protocol

Prepare in two 6-liter flasks.

Ingredients	In each flask
Difco Yeast Nitrogen Base w/o AA, AS[a]	9 g
Ammonium sulfate	30 g
Inositol (200 m*M*) stock	6 ml
"All Amino Acids" Powder mix drop-out	12 g
Water (R/O)	3 liters

[a] Record lot number in log.

Dissolve with stir bar. Heat may be applied.
Dispense 300 ml in 500-ml bottles.
Autoclave 35 min or less.
Apply labels with date. When no longer warm, tighten down caps. Stock shelves.

Glucose 40%

Prepare in large Nalgene beakers (must fit into the microwave).

Ingredients	Per 1 liter final volume
D-Glucose (dextrose)	400 g
Water (R/O)	780 ml (to make final volume 1 liter)

Dissolve in Nalgene beaker (see www.nalgenunc.com) by covering with Saran wrap and microwaving on high for 17 min (microwave ovens vary). Handle the hot liquid carefully. Add a stir bar *after* sugar is dissolved and mix thoroughly.
Dispense approx. 300 ml in 500-ml bottles.
Autoclave 35 min or less.
Apply labels with date. When no longer warm, tighten down caps. Stock shelves.

HIS 100 m*M*

The following applies to all six stock solutions listed here: The recommended amount to add per 600 ml is appropriate for supplementing minimal medium. It is not equivalent to the amount of amino acid in the dry powder mix used for making SC medium. Prepare in a 2-liter flask.

Ingredients	For 1 liter
L-Histidine hydrochloride (Sigma, St. Louis, MO)	20.9 g
Water (R/O)	1 liter

Dissolve in 2-liter flask with stir bar. Use heat if necessary.
Dispense 100 ml in 200-ml square bottles.
Autoclave 18 min (small autoclave).
Apply labels with date. Add instruction: "Use 1.8 ml/600 ml."
When no longer warm, tighten down caps. Stock shelves.

ADE 30 m*M*

Prepare in a 2-liter flask.

Ingredients	For 1 liter
Adenine hemisulfate (molecular weight 184.2; Sigma)	5.5 g
Water (R/O)	1 liter

Dissolve in 2-liter flask with stir bar. Use heat if necessary.
Dispense 100 ml in 200-ml square bottles.
Autoclave 18 min (small autoclave).
Apply labels with date. Add instruction: "Use 6 ml/600 ml."
When no longer warm, tighten caps. Stock shelves.

URA 20 m*M*

Prepare in a 2-liter flask.

Ingredients	For 1 liter
Uracil (Sigma)	2.24 g
Water (R/O)	1 liter

Dissolve in 2-liter flask with stir bar. Use heat if necessary.
Dispense 100 ml in 200-ml square bottles.
Autoclave 18 min (small autoclave).
Apply labels with date. Add instruction: "Use 6 ml/600 ml."
When no longer warm, tighten down caps. Stock shelves.

LYS 100 m*M*

Prepare in a 2-liter flask.

Ingredients	For 1 liter
L-Lysine (Sigma)	18.3 g
Water (R/O)	1 liter

Dissolve in 2-liter flask with stir bar. Use heat if necessary.
Dispense 100 ml in 200-ml square bottles.
Autoclave 18 min (small autoclave).
Apply labels with date. Add instruction: "Use 6 ml/600 ml."
When no longer warm, tighten down caps. Stock shelves.

LEU 100 m*M*

Prepare in a 2-liter flask.

Ingredients	For 1 liter
L-Leucine (Sigma L-8000)	13.1 g
Water (R/O)	1 liter

Dissolve in 2-liter flask with stir bar. Use heat if necessary.
Dispense 100 ml in 200-ml square bottles.
Autoclave 18 min (small autoclave).
Apply labels with date. Add instruction: "Use 10 ml/600 ml."
When no longer warm, tighten down caps. Stock shelves.

TRP 40 m*M*

Prepare in a 2-liter flask.

Ingredients	For 1 liter
L-Tryptophan (formula weight 204.2; Sigma)	8.0 g
Water (R/O)	1 liter

Dissolve in 2-liter flask with stir bar. Use heat if necessary.
Dispense 100 ml in 200-ml square bottles.
Autoclave 18 min (small autoclave).
Cool bottles in dark. Wrap in foil.
Apply labels with date. Add instruction: "Use 6 ml/600 ml."
When no longer warm, tighten down caps. Stock shelves.

Inositol 200 m*M*

Prepare in a 2-liter flask.

Ingredients	For 1 liter
myo-Inositol	36 g
Water (R/O)	1 liter

Dissolve in 2-liter flask with stir bar.
Dispense 100 ml in 200-ml square bottles.
Autoclave 18 min (small autoclave).
Apply labels with date. Add instruction: "Use 1 ml/600 ml."
When no longer warm, tighten down caps. Stock shelves.

M9A 20×

Prepare in 2-liter flask.

Ingredients	For 1 liter
Na_2HPO_4 (disodium phosphate, dibasic, anhydrous)	116 g
KH_2PO_4 (potassium phosphate, monobasic)	60 g
NaCl (sodium chloride)	10 g
NH_4Cl (ammonium chloride)	20 g
Water (R/O)	900 ml

Start with 900 ml water + stir bar in 2-liter flask. Add and dissolve each ingredient, one after the other.
Pour solution into 1000 ml graduated cylinder and add water, if needed to bring volume to 1 liter.
Dispense 100 ml each into square bottles. Cap loosely.
Autoclave 18 min (small autoclave).
Apply labels with date.
When no longer warm, tighten down caps. Stock shelves.

SOB

Prepare in two 6-liter flasks.

Ingredients	For each flask
Bacto-tryptone	60 g
Yeast extract	15 g
NaCl (sodium chloride) (10 m*M*)	1.74 g
KCl (potassium chloride) (25 m*M*)	0.57 g
NaOH 1 *N* solution (wear goggles and gloves)	4 ml
Water (R/O)	3000 ml

Dissolve with stir bar. Heat may be applied.
Dispense 300 ml each into 500 ml bottles.
Autoclave 30 min.
Remove and let cool.
Label and store.

LB Top Agar

Prepare in two 6-liter flasks.

Ingredients	For each flask
LB Medium powder or capsules (BIO-101 low salt)	75 g
Agar (Difco)	18 g
Water (R/O)	3 liters

Dissolve in small autoclave 13 min. Add stir bar afterward and mix thoroughly until "mixing lines" are gone. (Apply additional heat if needed.)
Dispense 300 ml in 500-ml bottles. Be careful of this HOT liquid.
You can wait till it cools down to a more comfortable temperature. Place the flasks on an insulated surface (paper towel).
Loosely cap bottles.
Autoclave 15 min or more.
Apply labels with date. When no longer warm, tighten caps. Stock shelves.

Soft Agar

Prepare in two 6-liter flasks.

Ingredients	For each flask
Nutrient broth powder	8 g
NaCl (sodium chloride)	15 g
Agar (Difco)	19.5 g
Water (R/O)	3 liters

Dissolve in small autoclave 13 min. Add stir bar afterward and mix thoroughly until "lines" are gone. (Apply additional heat if needed.)
Dispense 300 ml in 500-ml bottles. Be careful of this HOT liquid.
You can wait till it cools down to a more comfortable temperature. Place the flasks on an insulated surface (paper towel).
Loosely cap bottles.
Autoclave 15 min or more.
Apply labels with date. When no longer warm, tighten caps. Stock shelves.

T Agar

Prepare in two 6-liter flasks.

Ingredients	For each flask
Bacto-tryptone	30 g
NaCl (sodium chloride)	7.5 g
Agar (Difco)	19.5 g
Water (R/O)	3 liters

Dissolve in small autoclave 13 min. Add stir bar afterwards and mix thoroughly until "lines" are gone. (Apply additional heat if needed.)
Dispense 300 ml in 500-ml bottles. Be careful of this HOT liquid.
You can wait till it cools down to a more comfortable temperature. Place the flasks on an insulated surface (paper towel).
Loosely cap bottles.
Autoclave 15 min or more.
Apply labels with date. When no longer warm, tighten caps. Stock shelves.

Casamino Acids

Prepare in a 2-liter flask.

Ingredients	For 1 liter
Casamino acids	100 g
Water (R/O)	1 liter

Dissolve in 2-liter flask with stir bar.
Dispense 100 ml in 200-ml square bottles.
Autoclave 18 min (small autoclave).
Apply labels with date. Add instruction: "Use 1 ml/600 ml."
When no longer warm, tighten caps. Stock shelves.

NaOH 1 *N*

Prepare in 1-liter flask.

Ingredients	Per 100 ml final volume[a]
NaOH (sodium hydroxide)	4 g
Water (R/O)	100 ml

[a] Or a multiple of 100 ml up to 500 ml.

HAZARD: Pellets and solution are strong base. Wear goggles and gloves.
If contact, flush 15 min in water and report accident.
Add pellets to water with stir bar and mix.
Dispense 100 ml in 200-ml square bottles. Tighten caps.
Apply labels with date.

AA Powder Mix Complete

Make 10×

Amino acid amount	Amino acid amount	Amino acid amount
Ade 2 g	Gly 2 g	Ser 2 g
Ala 2 g	Leu 4 g	Thr 2 g
Arg 2 g	Ile 2 g	Trp 2 g
Asn 2 g	Lys 2 g	Tyr 2 g
Asp 2 g	His 2 g	Ura 2 g
Cys 2 g	Met 2 g	Val 2 g
Gln 2 g	Phe 2 g	PABA[a] 0.2 g
Glu 2 g	Pro 2 g	

[a] *p*-Aminobenzoic acid.

Weigh each and add to plastic jar. Check off each item as it is added.
Include several steel balls to facilitate mixing powder (or use a powder grinding mill).
Close jar and shake to mix, longer than you think necessary.

GNA (Presporulation Media)

This medium, when autoclaved all together, regularly boils over. Here parts are autoclaved separately and then combined. Prepare in 3 flasks.

Ingredients	For 3 liters
Flask 1 (4-liter)	
Yeast extract (a very fine powder; wear mask)	30 g
Agar (Difco-Bacto)	60 g
Water (R/O)	1500 ml
Flask 2 (2-liter)	
Nutrient broth (a very fine powder; wear mask)	90 g
Water (R/O)	1125 ml
Flask 3 (1-liter)	
D-Glucose (dextrose)	150 g
Water (R/O)	375 ml

No stir bar and no premixing are necessary.
Cover with foil or paper towel and cups.
Autoclave 20 min (small autoclave).
After autoclaving, let the flasks stand 5 min to cool (so they are no longer superheated). But keep the agar flask warm as in a water bath so the agar does not solidify on the bottom. Swirl the flask to make the agar uniformly distributed. Then, using *sterile technique,* pour Flask 2 and

Flask 3 into Flask 1. *Swirl* Flask 1 until mixing is complete (schlieren lines are gone). Do not worry about bubbles.

Deliver flask to plate pouring room. Place flask in the water bath and notify the requester that the medium is ready. Note the time of day.

NGS Agar (for Sporulation)

(No Glucose Sporulation.) This medium works successfully with most laboratory strains. Prepare in 6-liter flask (A) and in a 1-liter flask (B).

Ingredients	For 3 liters total
(A) For 6-liter flask	
Potassium acetate	30 g
Agar (Difco)	60 g
Water (R/O)	2500 ml
(B) For 1-liter flask	
Amino acid mix (AA – URA – TRP)	1.5 g
Uracil stock solution	7.5 ml
Tryptophan stock solution	7.5 ml
Water (R/O)	500 ml

No stir bar and no premixing are necessary.

Cover with foil or paper towel and cups.

Autoclave 35 min (small autoclave).

After autoclaving, let flasks stand 3 min to cool (so they are no longer superheated). Then pour Flask B into Flask A, without touching the sterile areas. *Swirl* by hand (wear gloves) until schlieren lines are gone (mixing is complete).

Place flask on a paper towel on a dry cart and deliver to bay of requester.

SPOR(ulation) Agar

Prepare in 6-liter flask.

Ingredients	For 3 liters
Potassium acetate	30 g
D-Glucose (dextrose)	3 g
Yeast extract	3.75 g
Agar (Difco)	60 g
Water (R/O)	3000 ml

Autoclave 35 min (small autoclave).

After autoclaving, let flask stand 3 min to cool (so it is no longer superheated). Then *swirl* by hand (wear gloves) until schlieren lines are gone (mixing is complete).

Place flask on a paper towel on a dry cart and deliver to bay of requester.

TB (Terrific Broth)

Prepare on request for individual users. Prepare in 6-liter flask.

Ingredients	For 3 liters
Bacto-tryptone	36 g
Yeast extract	72 g
Glycerol	12 ml
KH_2PO_4 (potassium phosphate, monobasic)	3.9 g
K_2HPO_4 (potassium phosphate, dibasic, anhydrous)	37.5 g
Water (R/O)	3000 ml

No stir bar and no premixing are necessary.
Cover with foil or paper towel and cups.
Autoclave 35 min (small autoclave).
Remove and let cool.
Deliver to bay of requester.

FB (Fortified Broth)

Prepare on request for individual users. Prepare in 6-liter flask.

Ingredients	For 3 liters
Bacto-tryptone	75 g
Yeast extract	22.5 g
D-Glucose (dextrose)	3 g
NaCl (sodium chloride)	18 g
2 *M* Tris-Cl, pH 7.6	75 ml
Water (R/O)	2925 ml

No stir bar and no premixing are necessary.
Cover with foil or paper towel and cups.
Autoclave 35 min (small autoclave).
Remove and let cool.
Deliver to bay of requester.

T Broth

Prepare on request for individual users. Prepare in 6-liter flask.

Ingredients	For 3 liters
Bacto-tryptone	30 g
NaCl (sodium chloride)	7.5 g
Water (R/O)	3000 ml

No stir bar, no premixing necessary.
Cover with foil or paper towel and cups.
Autoclave 35 min (small autoclave).
Remove and let cool.
Deliver to bay of requester.

SOB Capsules

To make 20 × 100-ml media bottles of 50 ml each by filter sterilization. Prepare in one 2-liter flask.

Ingredients	
SOB powder (in capsules) (BIO 101)	31 g
Water (R/O)	1000 ml

Dissolve with stir bar. Use heat sparingly.
Gather materials: 20 sterile 100-ml media bottles, one sterile1-liter media bottle, one disposable 500-ml sterilizing filter cup, one vacuum flask liquid trap.
Using sterile technique, remove the cap from the 1-liter media bottle. Open the package and screw on the sterilizing filter firmly. Insert the white filter unit tip into the vacuum trap tube.
Pour the media into the cup and turn on the vacuum. Continue adding media to the cup until it has all been filtered through. Remove the tube from the white tip and turn off the vacuum.
Arrange the 100-ml bottles in a row with the caps very loose. Remove the white tip from the filter unit. With one hand, lift the cap off the bottle and with the other hand, pour about 50 ml into the first bottle. Replace the cap and go on to the next bottle.
When finished, screw down all the caps and keep the bottles 2–3 days (one at 36°) to see if they remain clear. Then stock them on the shelves.

The filter sterilizer is disposable, and it is not reusable.

Inositol-Free Minimal Medium[4,5]

Make stock solutions/suspensions in 1-liter flasks.

Ingredients: (A) trace elements stock mix	For 1 liter
Boric acid	50 mg
Copper sulfate	4 mg
Potassium iodide	10 mg
Ferric chloride	20 mg
Manganese sulfate	40 mg
Sodium molybdate	20 mg
Zinc sulfate	40 mg
Water (R/O)	100 ml

Ingredients: (B) vitamin stock mix	For 1 liter
Biotin	2 mg
Calcium pantothenate	400 mg
Folic acid	2 mg
Niacin	400 mg
p-Aminobenzoic acid	200 mg
Pyridoxine hydrochloride	400 mg
Riboflavin	200 mg
Thiamin hydrochloride	400 mg
Water (R/O)	1000 ml

Ingredients: (C) salts and nitrogen source solution	For 1 liter
Potassium phosphate monobasic	10 g
Magnesium sulfate	5 g
Sodium chloride	1 g
Calcium chloride	1 g
Ammonium sulfate	50 g
Water (R/O)	1000 ml

Transfer media A, B, and C to bottles with tight caps. B should be stored in brown glass. A and B will be suspensions; they will not dissolve.

Add 50 ml chloroform to prevent growth of contaminants. Store in the cold.

To make medium use per liter 1 ml of trace elements, 1 ml of vitamins, 100 ml of salts plus nitrogen, and 20 g of glucose. Add 20 g of agar for solid medium.

Acknowledgment

The writing of this chapter was supported by N.I.H Grant No. GM35010-19.

[3] Constructing Yeast Libraries

By HAOPING LIU

Introduction

Having a good library is crucial for gene cloning in *Saccharomyces cerevisiae*. Different types of libraries are used for different applications. The most frequently used approach in cloning a yeast gene is complementation of a recessive mutant with a yeast genomic library on a low-copy-number yeast vector. Literally, hundreds of yeast genes have been cloned this way. Another popular approach that has efficiently identified many yeast genes is by functional cloning from overexpression libraries, based on the suppression of a recessive mutant by gene overexpression or other phenotypic consequences associated with overexpression. Overexpression libraries include high-copy-number yeast genomic libraries or cDNA libraries under the control of a strong inducible promoter. Libraries under the control of inducible promoter allows the isolation of genes that are toxic at elevated level and of low abundant genes that are not expressed to a level high enough from a high-copy-number vector. Other approaches of functional cloning in yeast include phage library screens, two-hybrid interactions, and insertional mutagenesis, some of which are described elsewhere in this volume.[1a–c]

The easiest way of obtaining a library is by mail. The last volume of this book listed several *S. cerevisiae* libraries.[1d] Since then, many new libraries have been constructed, and some of them are listed in Table I. These libraries are constructed with more recent multipurpose yeast expression vectors. Therefore, they should give higher yield in DNA preparation and be more convenient in molecular manipulation of the cloned genes. Table I also lists some *Candida albicans* genomic libraries.

Although numerous yeast genomic libraries are available, many circumstances require the construction of a new library. Methods for *de novo* construction of yeast genomic libraries and considerations involved in the process have been described in the previous volume in this series.[1d] Here I would like to describe two specific strategies that have been used successfully for the construction of several yeast libraries. One strategy should be applicable to the construction of any genomic library, and the other for any cDNA libraries.

[1a] M. Fromont-Racine, J.-C. Rain, and P. Legrain, *Methods Enzymol.* **350,** [29], 2002 (this volume).
[1b] J. F. Gesa, T. R. Hagbun, and S. Fields, *Methods Enzymol.* **350,** [28], 2002 (this volume).
[1c] A. Kumar, S. Vidan, and M. Soryder, *Methods Enzymol.* **350,** [12], 2002 (this volume).
[1d] M. D. Rose and J. R. Broach, *Methods Enzymol.* **194,** 195 (1991).

0076-6879/02 $35.00

TABLE I
YEAST LIBRARIES

Library types	Insert DNA		Vector information		Source or reference
	Strain/Genotype	Insert size	Name	Selectable marker/type	
S. cerevisiae cDNA	S288c *Mat***a** *ura3-52*		pRS316	*GAL1p/URA3/CEN*	2
S. cerevisiae genomic	Z28 *Mat***a**/*Matα mal-/mal-gl2/gal2*	8–10 kb	pCT3	*URA3/CEN*	3
S. cerevisiae genomic	S1278b *Mat***a**/*Matα*	4–8 kb	pRS316	*URA3/CEN*	4
S. cerevisiae genomic	—	~8 kb	pRS200	*TRP1/CEN*	5
S. cerevisiae genomic	—	—	pRS202	*URA3/2μ*	6
S. cerevisiae genomic	—	9–12 kb	p366	*LEU2/CEN*	ATCC 77162, Ref. 7
S. cerevisiae genomic	S1278 × S288c *Matα ura3-52 trp1 put1-54 MPR1 MPR2 AZC*^{r}	>5 kb	pYES2	*URA3/2μ*	8
S. cerevisiae GAL1-genomic	SNY243 *Mat***a** *leu2-04 ade1 ade6 circ*^{o}	4–7 kb	1YESR	*GAL1p/URA3/CEN*	ATCC 87311, Ref. 9
S. cerevisiae genomic	W303-1A	5–15 kb	pRS423	*HIS3/2μ*	10
C. albicans genomic	SC5314	5–20 kb	YEp13	*LEU2/2μ*	11
C. albicans genomic	655	5–10 kb	p1041	*CaURA3/CaARS1/2μ*	12
C. albicans genomic	B792	—	PYSK35	*LEU2/CEN*	13
C. albicans genomic	B792	—	YEp13	*LEU2/2μ*	14
C. albicans genomic	1006	>4 kb	pRS202	*URA3/2μ*	15

Construction of Yeast Genomic Libraries

Cloning Strategy

To clone a gene with dominant alleles or a gene whose activity is absent from the strains used for the construction of available yeast libraries, it is necessary to construct a new genomic library *de novo*. A major concern in constructing a genomic library is to prevent vector self-ligation, so that most of the clones in the library contain a genomic DNA fragment. This vector self-ligation can be efficiently reduced to minimal by using a cloning strategy where partially filled *Sau*3AI genomic DNA fragments are ligated to partially filled *Xho*I or *Sal*I ends of a vector, as shown in Fig. 1. Partial *Sau*3AI digestion of yeast genomic DNA generates fragments with 5′-GATC as the overhang ends. Filling the ends with dG and dA leaves only 5′-GA as the overhang ends. *Sal* I or *Xho*I restriction of vector DNA generates 5′-TCGA as the overhang ends. Filling in the ends with dT and dC leaves only 5′-TC for base pairing. Therefore, the 5′-TC overhang of the vector can only pair with the 5′-GA overhang of genomic DNA fragments. Neither the inserts nor the vector can self-ligate. This strategy has been used successfully for the construction of several yeast genomic libraries [3,4,15] and is highly recommended for anyone interested in constructing a new genomic library.

Preparing Cut Vector

Currently used yeast vectors are described elsewhere in this volume. Yeast centromeric plasmids should be used to avoid cross-suppressing clones if the goal is to clone a gene by complementing a recessive mutation in yeast or to isolate a dominant allele of a gene from a particular mutant strain. The vector should have a selectable marker usable for the transformation of the recipient strain. The vector

[2] H. Liu, J. Krizek, and A. Bretscher, *Genetics* **132,** 665 (1992).

[3] C. M. Thompson, A. J. Koleske, D. M. Chao, and R. A. Young, *Cell* **73,** 1361 (1993).

[4] H. Liu, C. A. Styles, and G. R. Fink, *Genetics* **144,** 967 (1996).

[5] P. Hieter, personal communication (1992).

[6] C. Connelly and P. Hieter, personal communication (1992).

[7] H. V. Goodson, B. L. Anderson, H. M. Warrick, L. A. Pon, and J. A. Spudich, *J. Cell Biol.* **133,** 1277 (1996).

[8] H. Takagi, M. Shichiri, M. Takemura, M. Mohri, and S. Nakamori, *J. Bacteriol.* **182,** 4249 (2000).

[9] S. W. Ramer, S. J. Elledge, and R. W. Davis, *Proc. Natl. Acad. Sci. U.S.A.* **89,** 11589 (1992).

[10] K. Thevissen, B. P. Cammue, K. Lemaire, J. Winderickx, R. C. Dickson, R. L. Lester, K. K. Ferket, F. Van Even, A. H. Parret, and W. F. Broekaert, *Proc. Natl. Acad. Sci. U.S.A.* **97,** 9531 (2000).

[11] A. M. Gillum, E. Y. Tsay, and D. R. Kirsch, *Mol. Gen. Genet.* **198,** 179 (1984).

[12] A. K. Goshorn, S. M. Grindle, and S. Scherer, *Infect. Immun.* **60,** 876 (1992).

[13] M. E. Fling, J. Kopf, A. Tamarkin, J. A. Gorman, H. A. Smith, and Y. Koltin, *Mol. Gen. Genet.* **227,** 318 (1991).

[14] A. Rosenbluh, M. Mevarech, Y. Koltin, and J. A. Gorman, *Mol. Gen. Genet.* **200,** 500 (1985).

[15] H. Liu, J. Kohler, and G. R. Fink, *Science* **266,** 1723 (1994).

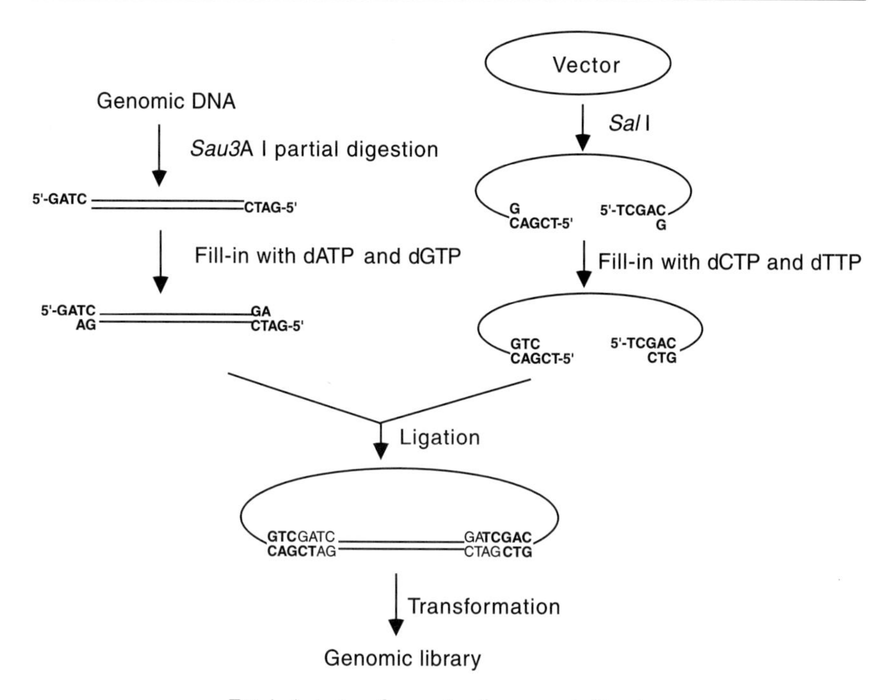

FIG. 1. A strategy for constructing genomic libraries.

must have a unique *Sal*I or *Xho*I site, which is preferably located in the middle of the polylinker to allow the efficient recovery of inserts. Once the vector is chosen, follow the following steps to prepare cut vector.

1. Digest the vector DNA with *Sal*I or *Xho*I to completion. Use overnight digestion followed by adding more of the enzyme for 2 additional hr of digestion to ensure that the cleavage is thorough. Extract DNA with phenol and precipitate with ethanol.
2. Fill in the ends with dCTP and dTTP in a 100-μl reaction.
   25 μg DNA
   10 μl of 10× Restriction buffer for *Sal*I or *Xho*I
   5 μl of 0.5 m*M* dCTP
   5 μl of 0.5 m*M* dTTP
   10–15 U Klenow fragment
   Bring the reaction to 100 μl with H_2O.
   Incubate at 30° for 30 min.
3. Ligate overnight and purify linear DNA away from the ligated circular vector. Precipitate DNA and set up a ligation reaction with T4 DNA ligase for

overnight. This will ligate any molecules whose ends are not filled in. Then, purify linear DNA away from ligated circular DNA by agarose gel electrophoresis, and extract linear DNA from gel slices with QIAEX gel extraction kit (from Qiagen, Valencia, CA). Resuspend the DNA to 100 ng/μl. This ligation step significantly reduces the background of vector self-ligation.

Preparation of Fragmented Yeast Genomic DNA

For yeast genomic DNA isolation, we have used the procedure described in the previous volume of this series.[16] An outline of a scaled-up preparation with the procedure is given here.

1. Grow 250 ml yeast culture in YPD overnight to saturation.
2. Spin down cells, resuspend in 25 ml of SE buffer (0.9 *M* sorbitol, 0.1 *M* EDTA, pH 8.0).
3. Spin down cells, resuspend in 10 ml SE buffer, add 10 μl of 2-mercapto-ethanol (2-ME).
4. Add 2.5 ml of 2 mg/ml Zymolyase 100K (ICN, Costa Mesa, CA) in SE buffer. Incubate 45 min at 37° until spheroplasts are formed.
5. Spin 5 min at 5000 rpm. Resuspend gently in 20 ml TE.
6. Add 4.5 ml TSE solution [1.2 ml 2 *M* Tris, pH 7.6, 1.2 ml 10% sodium dodecyl sulfate (SDS), 3 ml 0.5 *M* EDTA, pH 8.0], mix, and incubate for 30 min at 65°.
7. Add 4 ml of 5 *M* potassium acetate, then place on ice for at least 60 min.
8. Spin 20 min at 10,000 rpm. Transfer the supernatant and precipitate by adding at least two volumes of ethanol.
9. Spin 5 min at 8000 rpm. Rinse pellet with 70% (v/v) ethanol.
10. Resuspend pellet in 12.5 ml TE (let the pellet resuspend overnight).
11. Spin out debris. Transfer supernatant and add 65 μl of 10 mg/ml RNase A. Incubate for 30 min at 37°.
12. Add 13 ml 2-propanol, mix gently, spin, rinse pellet with 70% ethanol. Air-dry. Resuspend DNA in 1 ml TE overnight.

To obtain an optimal *Sau*3AI partial digestion of genomic DNA, we recommend using different dilutions of *Sau*3AI in the digestion reactions. For example, use 1 : 100, 1 : 50, 1 : 25, and 1 : 10 dilutions of 4U/μl *Sau*3AI (in 50% glycerol) in the following reaction:

50 μl genomic DNA of 1 μg/μl
50 μl 10× *Sau*3AI restriction buffer

[16] P. Philippsen, A. Stotz, and C. Scherf, *Methods Enzymol.* **194,** 169 (1991).

390 μl H_2O
10 μl *Sau*3AI

Incubate the reactions at room temperature. Remove 250 μl at 15 min and let the remaining reactions to incubate for another 15 min. Stop digestions by phenol/chloroform extraction. Precipitate DNA.

To determine which is the optimal reaction of partial *Sau*3AI digestion, fractionate the *Sau3*AI digested DNA on a 0.7% agarose gel. Run the rest of the appropriate reactions on an agarose gel and cut out the region that contains DNA fragments of desired size. The size of inserts is determined based on the ease of screening or selection method that will be used with the library. Use a QIAEX gel extraction kit (Qiagen) to extract DNA from gel slices. Determine the amount of DNA.

Fill in the *Sau*3AI overhangs with dATP and dGTP by Klenow, using the same reaction condition as described for the *Sal*I cut vector. Extract with phenol/chloroform twice. Precipitate DNA with ethanol. Wash and dry. Resuspend DNA to obtain a DNA concentration around 300–500 ng/μl.

Assembly of Library

We have used the following condition for ligation reactions (5 μl):

1 μl genomic DNA fragments at 300–500 ng/μl
1 μl cut vector DNA at 100 ng/μl
0.5 μl 10× T4 DNA ligase buffer
0.5 μl T4 DNA ligase, about 4 units
1.5 μl H_2O
Incubate overnight at 16°.

A control ligation reaction without the genomic DNA fragments should be set up at the same time. We recommend the use of electrocompetent *Escherichia coli* or commercially available competent cells in library transformation, because high transformation efficiency is important for boosting the size and complexity of the library.

Transformation efficiency from the reaction with genomic inserts should be at least 10–50 times higher than the control. If the control ligation gives a transformation efficiency that is too high, the filling of vector overhangs with dC and dT may have not occurred properly. If the ligation with genomic inserts does not give high transformation efficiency, the dA and dG filling of genomic fragments may be a problem. Whether a ligation reaction has proceeded properly can be examined by running the remaining ligation reaction on an agarose gel next to the cut vector

DNA. Successfully ligated DNA should show a smear above the position of the cut vector.

The quality of a genomic library is determined by its coverage of the genome and the percentage of clones carrying a genomic insert. The second standard can be assessed by randomly picking some clones to perform restriction digestions with enzymes that release the insert from the vector. The average size of inserts, the percentage of clones with inserts, and the number of colonies from initial transformation can be used to estimate the size of the library in terms of its genome coverage.

Construction of Yeast cDNA Libraries

Cloning Strategy

The currently used yeast cDNA library is constructed from mRNA expressed in *Mat***a** cells grown in YPD medium and is therefore limited to genes transcribed in this cell type and growth condition. A new cDNA library needs to be constructed if one is interested in identifying genes expressed in other cell types or growth conditions, such as meiosis or starvation and other stress conditions, with a cDNA library. The major consideration in designing a cloning strategy for the construction of a cDNA library is unidirectional ligation of cDNA inserts into the vector. The restraint in ligation orientation is essential to render the expression of cDNA inserts under the control of a promoter from the vector as well as to eliminate vector self-ligation. Directionality is obtained typically by introducing two different restriction endonuclease sites at the ends of cDNA, which is initiated by using a primer–adapter to initiate first strand synthesis. The SuperScript Plasmid System for cDNA synthesis and plasmid cloning from Invitrogen Corporation features this strategy in their design (Fig. 2), and the system has been used successfully in constructing a yeast cDNA library.[2] A *Not*I primer–adapter, which contains 15 dT and restriction sites of *Not*I and other enzymes, is used to prime the first strand cDNA synthesis at the poly(A) tail of mRNAs (Fig. 2). The product of the first and second strand reactions is blunt-ended cDNA, to which *Sal*I adapters are added. The *Sal*I adapter is duplex oligomers that are blunt-ended at one terminus and contains a 4-base overhang at the other terminus. Only the blunt-ended 5′-end of the *Sal*I adapter is phosphorylated, which eliminates self-ligation of the adapters during ligation to the cDNA (Fig. 2). The *Sal*I adapter also contains an additional *Mlu*I site which can be used, together with *Not*I, to release the cDNA insert from the vector. After the addition of *Sal*I adapters to both termini, the *Not*I terminus is exposed by *Not*I digestion. Thus, the 3′ end of the cDNA is identified with *Not*I, and the 5′ end with *Sal*I. The cDNA can then be ligated to a *Sal*I- and *Not*I-digested vector, with the *Sal*I site positioned at the terminus with an inducible promoter (Fig. 2).

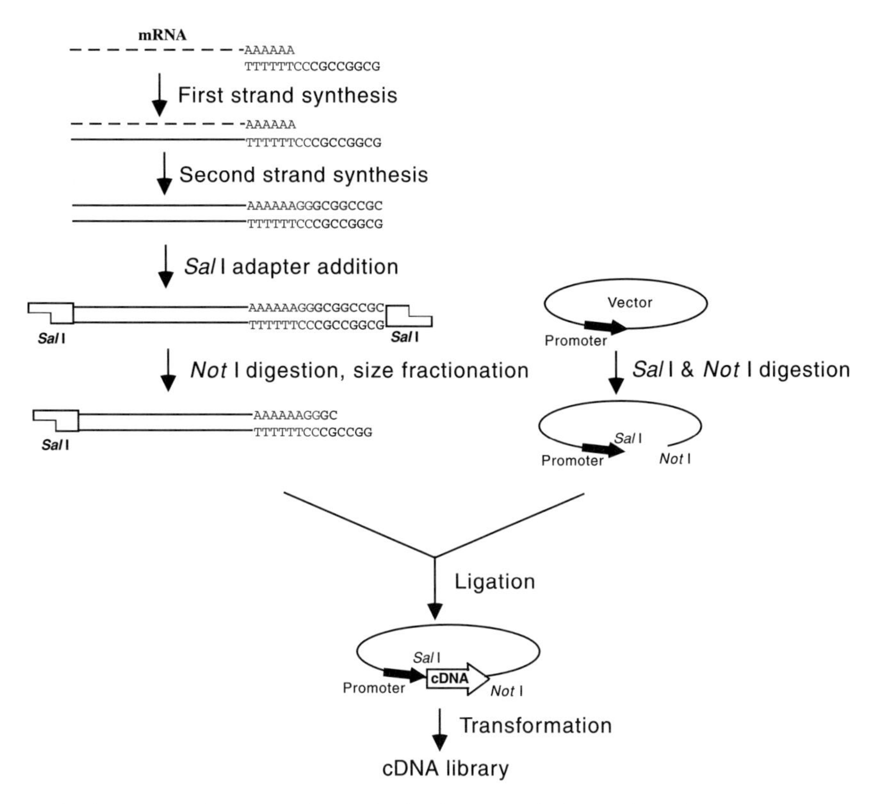

FIG. 2. A strategy for constructing cDNA libraries (adapted with permission from the SuperScript Plasmid System for cDNA synthesis and plasmid cloning of Invitrogen Corporation).

Preparation of Cut Vector

A YCp vector is usually used to construct a cDNA library. The vector must contain an inducible promoter, preferably with a high level of expression under the induction condition and completely off under a noninducing condition. If the cloning strategy in Fig. 2 will be used for this construction, the vector must also contain a unique *Not*I site in the polylinker region and one *Sal*I or *Xho*I site between the *Not*I and the inducible promoter.

To prepare the cut vector, digest the vector DNA with *Not*I to completion. Monitor the digestion on an agarose gel. Digest the *Not*I cut vector with *Sal*I and gel-purify the linear DNA. Extract the cut DNA from gel slices with a QIAEX gel extraction kit.

Because the *Not*I and *Sal*I sites are close to each other in the polylinker region of the vector, it is difficult to determine whether the second digestion is complete.

To monitor the completeness of second digestion, we have end-labeled 3% of *Not*I-digested DNA with ^{32}P and then digested the DNA with the second enzyme, *Sal*I. The extent of *Sal*I digestion can be checked by digesting with an enzyme that cuts the vector backbone. Two ^{32}P-labeled bands are expected if the second digestion has been complete. A partial *Sal*I digestion should give three ^{32}P bands. Alternatively, transformation efficiency of ligated *Sal*I-digested vector can be used to assess whether the second digestion is complete. Low transformation efficiency is expected if the second digestion has occurred successfully. As described in the previous section for preparing cut vector, ligating the *Not*I- and *Sal*I-digested vector, followed by gel purifying the linear unligated DNA, can be used to purify double digested vector DNA away from single restricted vector.

Total RNA Extraction and mRNA Purification

Construction of a good cDNA library begins with the preparation of high-quality mRNA. Generally, 5 μg of mRNA is sufficient to construct a cDNA library containing about 10^6 clones in *E. coli*. Since only 1.3 to 1.4% of a total RNA population in yeast is mRNA, it is necessary to start with at least several milligrams of total RNA. To avoid RNase contamination, it is important to use sterile disposable plasticware and diethyl pyrocarbonate (DEPC)-treated H_2O throughout RNA purification.

The hot acidic phenol method[17] works well for the extraction of total RNA from yeast. For a scaled-up preparation, about 10^{10} (500 ml culture) *S. cerevisiae* cells in mid-exponential phase are collected in 50-ml polypropylene tubes, washed once with ice-cold water. Resuspend cells in 12 ml TES (10 m*M* Tris-Cl, pH 7.5, 10 m*M* EDTA, 0.5% SDS). Add 12 ml acid phenol (unbuffered liquefied phenol from Sigma, St. Louis, MO) to the tube and vortex vigorously for 10 sec. Incubate for 30 to 60 min at 65° with occasional vortexing. Centrifuge and transfer the aqueous (top) phase to a clean 50-ml tube, add 12 ml acid phenol, vortex vigorously, and centrifuge 10 min at 3000 rpm in a swing-bucket rotor in a Sorvall table top centrifuge. Transfer aqueous phase to a new tube, add 1/10 volume of 3 *M* sodium acetate, pH 5.3, and 2 volumes of ice-cold 100% ethanol, and precipitate. Centrifuge at top speed for 30 min at 4°. Wash the RNA pellet with 70% ice-cold ethanol. Centrifuge, air-dry the RNA pellet, and resuspend the pellet in 0.5 ml H_2O. Fully dissolve RNA before measuring OD_{260}. The yield should be around 5 mg.

mRNA is isolated from total RNA by oligo(dT) cellulose column chromatography.[18] For 5 mg RNA, use 0.75 ml of oligo(dT)-cellulose in a column. About

[17] F. M. Ausubel, R. Brent, R. E. Kingston, D. Moore, J. G. Seidman, J. A. Smith, and K. Struhl (eds.), *in* "Current Protocols in Molecular Biology," Vol. 1. John Wiley & Sons, New York, 1987.

[18] J. Sambrook, E. F. Fritsch, and T. Maniatis, "Molecular Cloning: A Laboratory Manual," Vol. 1. Cold Spring Harbor Laboratory Press, Cold Spring Harbor, New York, 1989.

1/10 of the RNA is expected to be recovered from the first round of purification. Then, the oligo(dT) column is regenerated with 0.1 *M* NaCl and equilibrated. The eluted poly(A) mRNA can be loaded onto the oligo(dT) column directly without the need for precipitation for the second round of purification by adjusting the NaCl concentration of the eluted mRNA to 0.5 *M*. About 50% of the applied RNA is expected to be recovered in the elution. After three cycles of purification through the oligo(dT) column, most of the tRNA and rRNA should be removed, and the material should be highly enriched in mRNA. Quality of total RNA and purified mRNA should be evaluated by a RNA gel. Because SDS is in the loading buffer of the oligo(dT) column and can interfere with the cDNA synthesis, it is necessary to precipitate and wash the mRNA twice after the column elution to reduce the content of SDS in the mRNA.

cDNA Synthesis and Fractionation

For cDNA synthesis and fractionation, we recommend using the SuperScript Plasmid System from Invitrogen Corporation. The system contains all necessary reagents sufficient for three experiments, each converting up to 5 μg of mRNA to fractionated ready-to-ligate cDNA. The instruction manual from Invitrogen, Carlsbad, CA, is very explicit, and if followed closely, one should succeed in making a cDNA library without too much trouble. The important steps for preparing cDNA inserts are highlighted below.

First-Strand Synthesis. First strand cDNA synthesis from mRNA is achieved with SuperScript II RT (reverse transcriptase) in the following 20 μl reaction. The engineered RT enzyme SuperScript II does not contain RNase H activity, but still exhibits an improved polymerase activity.

1. Add 2 μl of *Not*I primer–adapter to 1 to 5 μg of mRNA; dilute as needed to make the final reaction 20 μl (including RT and 8-μl reagents that will be added in step 2).
2. Heat to 70° for 10 min and quick-chill on ice. Spin briefly and add:
   4 μl of 5× First strand buffer
   2 μl of 0.1 *M* DTT
   1 μl of 10 m*M* dNTP mix
   1 μl of [α-^{32}P]dCTP (1 μCi/μl)
3. Incubate at 37° for 2 min to equilibrate the temperature. Add SuperScript II RT (use 1 μl of RT for each μg of mRNA used in the reaction).
4. Incubate at 37° for 1 hr. Place the tube on ice to terminate the reaction.
5. To determine the yield of first strand synthesis, remove 2 μl from the reaction and add it to 43 μl of 20 m*M* EDTA (pH 7.5) and 5 μl of 1 μg/μl yeast tRNA. From the diluted sample, spot duplicate 10-μl aliquots onto glass-fiber filters. Dry one to determine the specific activity (SA) of the dCTP

reaction. Wash the other filter three times, 5 min each, in a beaker containing 50 ml of ice-cold 10% TCA with 1% sodium pyrophosphate, and once in 50 ml of 95% ethanol. Dry the filter to determine the yield of first strand cDNA.

The specific activity (SA) is determined by dividing the counts per min from the 10-μl aliquot from the unwashed filter with the quantity (in pmol) of the same nucleotide in the 10-μl aliquot:

$$\text{SA(cpm/pmol dCTP)} = \frac{\text{cpm}/10\,\mu\text{l}}{200\,\text{pmol dCTP}/10\,\mu\text{m}}$$

Once the SA is determined, the amount of cDNA synthesized in the first-strand reaction can be calculated from the amount of acid-precipitable radioactivity.

$$\text{Amount of cDNA}(\mu\text{g}) = \frac{(\text{acid-precipitable cpm}) \times (50\,\mu\text{l}/10\,\mu\text{l}) \times (20\,\mu\text{l}/2\,\mu\text{l}) \times (4\,\text{pmol dNTP/pmol dCTP})}{(\text{SA cpm/pmol dCTP}) \times (3030\,\text{pmol dNTP}/\mu\text{g cDNA})}$$

"3030" is the amount of nucleotide equivalent to 10 μg of single-strand DNA.

The yield of first-strand synthesis can be calculated by dividing the amount of cDNA synthesized by the amount of starting mRNA. We have had around 38% yield from 3.1 μg of yeast mRNA. A lower yield of first-strand synthesis does not necessarily indicate that a library cannot be made, since each ligation reaction only requires 10 ng of cDNA. It is critical, however, that the size distribution of the cDNA products be similar to that of the mRNA. Shorter cDNA products may indicate the danger of RNase contamination.

Second-Strand Synthesis

1. On ice, add the following reagents to the remaining 18-μl first-strand reaction:
   93 μl of H_2O
   30 μl of 5× Second strand buffer
   3 μl of 10 m*M* dNTP mix
   1 μl *E. coli* DNA ligase (10 U/μl)
   4 μl *E. coli* DNA polymerase I (10 U/μl)
   1 μl *E. coli* RNase H (2 U/μl)
   Final volume is 150 μl.
2. Incubate for 2 hr at 16°.
3. Add 2 μl (10 units) of T4 DNA polymerase, continue incubating at 16° for 5 min.
4. Place the reaction on ice, and add 10 μl of 0.5 *M* EDTA.

5. Extract with 150 μl phenol : chloroform : isoamyl alcohol (25 : 24 : 1). Precipitate the cDNA with 70 μl of 7.5-m ammonium acetate and 0.5 ml ethanol (−20°). Wash with 0.5 ml of 70% ethanol (−20°). Air-dry at 37° for 10 min.
6. To determine the quality of synthesized cDNA, run some (5 μl leftover from the aqueous phase of each phenol extraction is enough) second-strand reactions on an agarose gel and dry the gel. Visualize by exposing the gel to an X-ray film. Compare the range of cDNA smear to that of the starting mRNA to determine the quality of cDNA.

SalI Adapter Addition and NotI Digestion

1. Set up a 50-μl ligation reaction to add *Sal*I adapter to the cDNA from step 5 of second-strand synthesis.
   cDNA
   25 μl H_2O
   10 μl 5× T4 DNA Ligase buffer
   10 μl *Sal*I Adapters
   5 μl T4 DNA Ligase
2. Incubate at 16° overnight.
3. Extract with phenol : chloroform : isoamyl alcohol. Precipitate, wash, and air-dry DNA.
4. Set up a 50-μl *Not*I digestion.
5. Incubate for 2 hr at 37°.
6. Extract with phenol : chloroform : isoamyl alcohol. Precipitate, wash, and air-dry DNA.

Fractionation by Column Chromatography. To ensure the quality of your cDNA library, it is crucial to remove all residual *Sal*I adapters and the *Not*I fragments released by restriction digestion with *Not*I. This is accomplished by filtrating the cDNA through a size fractionation column provided by the SuperPlasmid System.

1. Dissolve the *Not*I digested cDNA in 100 μl TEN buffer (10 m*M* Tris-HCl pH 7.5, 0.1 m*M* EDTA, 25 m*M* NaCl).
2. Wash a cDNA fractionation column (provided in the Invitrogen SuperScript Plasmid System) with 0.8 ml TEN buffer four times; let the column drain completely for each wash.
3. Load the 100 μl cDNA to the column; collect flow-through into tube 1.
4. Add 100 μl TEN buffer to the column and collect into tube 2. Let it drain completely.

5. From the second 100 μl TEN wash, collect single drop fractions ($\sim$35 μl) into each microcentrifuge tube. Continue adding 100-μl aliquots of TEN until a total of 18 drops have been collected into tubes 3 through 20, one drop per tube.
6. Measure the volume in each tube with a fresh tip. Identify the fraction whose cumulative elution volume (sum from tube 1 to this fraction) is closest to but not exceeding 550 μl. Discard all tubes after this fraction. If fractions beyond 550 μl are used, the library will contain "empty" clones, which come from ligation of the short *Not*I–primer–adapter–*Sal*I fragment released from the *Not*I digestion.
7. Place the remaining tubes in a scintillation counter and obtain Cerenkov counts for each tube in the tritium channel without using scintillation fluid. Counts should increase after reaching 400 to 450 μl of elution volume. Calculate the amount of cDNA in each fraction by using the following equation with the SA determined for the first strand synthesis:

$$\text{Amount of ds cDNA (ng)} = \frac{(\text{Cerenkov cpm}) \times 2 \times (4\,\text{pmol dNTP/pmol dCTP}) \times (1000\,\text{ng}/\mu\text{g ds cDNA})}{\text{SA(cpm/pmol dCTP)} \times (1515\,\text{pmol dNTP}/\mu\text{g ds cDNA})}$$

8. Determine the cDNA concentration for each fraction by dividing the amount of cDNA in each fraction with the volume. If the cDNA concentration is higher than 0.71 ng/μl, 10 ng cDNA from that fraction can be used directly for ligation with the vector without precipitation. Otherwise, the cDNA has to be concentrated. To maximize the average insert size in the cDNA library, cDNA from the earliest fraction should be used for ligation. Often, the earliest fraction does not have enough cDNA and is not concentrated enough to be used directly. In this case, fractions may have to be pooled together and precipitated to obtain the 10 ng of cDNA needed for ligation.

Ligation of cDNA to Vector and Transformation

Set up ligation reactions with 10 ng of cDNA and 50 ng of *Sal*I- and *Not*I-cut vector in 20 μl reactions, as following:

14 μl of 10 ng cDNA in TEN buffer
1 μl of *Not*1–*Sal*I cut vector DNA (50 ng/μl)
4 μl of 5$\times$ T4 DNA ligase buffer
1 μl of T4 DNA ligase (1 unit/μl)
Incubate for 3 hr at room temperature.

Ligation reactions from each column fraction should be kept separately as different pools of library. A negative control ligation with the same amount of cut-vector DNA, but without cDNA, should be used. In addition, we suggest setting up a control ligation reaction using cDNA from one of the later fractions (less valuable) with the *Not*I- and *Sal*I-cut pSPORT vector DNA provided in the SuperScript Plasmid System. The transformation efficiency of this reaction will inform you whether a ligation problem is caused by cDNA or by the cut vector.

In order to generate a sufficiently large cDNA library, it is important to use the most competent *E. coli* cells in transformation. Electroporation usually offers higher transformation efficiency. With a successful ligation, we typically obtain around 10^6 transformants per ligation from electroporation of ElectroMAX DH10B cells from Invitrogen (transformation efficiency at 10^9–10^{10} transformants/μg of supercoiled plasmid). If the transformation efficiency is 100-fold lower, the library size will be on the borderline for genes of low abundance.

Transformation efficiency is also a good indication of whether the ligation reaction has worked or not. We expect that the ligation with cDNAs should give 50–100 times more transformants than the negative control without cDNAs. If not, there may be problems with ligation reactions, which can come from either cDNA or the cut vector. If the ligation reaction with the pSPORT vector gives 100-fold higher transformation efficiency than the ligation with the cut vector, the vector preparation may have a problem. If the transformation efficiencies of both ligations are low ($<10^5$ transformants/ligation, assuming that supercoiled plasmid yields $>10^9$ transformants/μg) the ligation reactions may not be proceeding properly and cDNA is likely the cause of the problem. Since synthesized cDNA has been visualized on an agarose gel and length of cDNA should not affect transformation efficiency, the first and second strand syntheses are not likely to cause the problem. On the other hand, problems in *Sal*I adapter addition and *Not*I digestion will not be easily detected during the process and will affect the ligation efficiency dramatically. To dissect which of the two steps has gone wrong, the ligation reaction can be checked by running the remaining ligation reaction on an agarose gel next to the cut vector DNA. Successfully ligated DNA should show an upward smear from the position of the cut vector. However, the same upward smear will be seen if either *Not*I or *Sal*I end of the inserts has ligated to the vector. To pinpoint the cause of a failed ligation, we have digested the ligation reaction with either *Sal*I or *Not*I and run the digestions on an agarose gel next to the ligated DNA and cut vector DNA. If *Not*I digestion, for example, can convert the upward smear to a band same size as the cut vector, the *Sal*I end must have not ligated properly. Alternative, the adapter addition step can be monitored by nondenaturing polyacrylamide gel electrophoresis during the process before the final ligation step. If 10 μl of the adapter ligation reaction is run on a 12% acrylamide gel and the DNA is visualized by ethidium bromide staining, the ligation of adapters to each other, mostly at their blunt ends but occasionally at the sticky ends, will be evidenced by a ladder of fragments of 16, 32, 60, 88 bp, etc. Although this does not verify that the *Sal*I adaptors have ligated to the cDNA, it does show that the ligation reaction is successful. To examine *Not*I digestion, 4 μl of *Not*I digestion can be run on the same 12% acrylamide gel together with *Sal*I ligation reaction. After visualizing the *Sal*I adapter ligation ladder with ethidium bromide, the gel can be dried and exposed to X-ray film to visualize a *Not*I–*Sal*I fragment of 40 bp, which should be between dimers and quadruples of the *Sal*I adapters.

Determining Quality of Library

A good cDNA library has three key characteristics that set it apart from the mediocre counterparts: (1) it should be large enough to contain genes of all genes expressed, including the ones expressed at low abundance, (2) it should contain full-length cDNA inserts, and (3) over 90% of its clones should contain cDNA inserts at the expected orientation. Either a complementation in a yeast mutant strain or a colony hybridization screen with the library could be performed to determine whether the library contains cDNAs for mRNAs of low abundance. The last two stipulations can be assessed by restriction digestion of some randomly picked clones with *Sal*I and *Not*I and by sequencing each insert with a primer from the promoter in the vector.

Acknowledgment

I thank Craig M. Thompson for sharing his protocol and expertise in making a yeast genomic library. I thank Gerald R. Fink and Anthony Bretscher for providing laboratory space and support during this work.

[4] Transformation of Yeast by Lithium Acetate/Single-Stranded Carrier DNA/Polyethylene Glycol Method

By R. DANIEL GIETZ and ROBIN A. WOODS

Introduction

The introduction of exogenous DNA into yeast by transformation has become an essential technique in molecular biology. Transformation is used to investigate the genomics and proteomics of yeast itself and also when yeast is employed as a system to study the genes and gene products of other organisms. Intact yeast cells can be transformed by a number of procedures: the lithium acetate/single-stranded carrier DNA/polyethylene glycol (LiAc/SS-Carrier DNA/PEG) method,[1–3] electroporation,[4–6] agitation with glass beads,[7] and bombardment with DNA-coated microprojectiles.[8] Cells can also be transformed after conversion to spheroplasts by treatment with Zymolyase.[9] We have reviewed the application of these techniques to *Saccharomyces cerevisiae* and other yeasts.[10] We focus here on the transformation of intact cells by the LiAc/SS-Carrier DNA/PEG method since it is the most widely applicable.

LiAc/SS-Carrier DNA/PEG Transformation

In our experience this method is the most efficient, generating yields up to 5×10^6 transformants/μg plasmid DNA/10^8 cells with many commonly used laboratory strains of yeast. This makes it particularly suitable for the assay of plasmid libraries for protein : protein interactions by the yeast two-hybrid system.[11] We have developed four protocols that can be applied to various transformation needs:

[1] R. D. Gietz, R. H. Schiestl, A. R. Willems, and R. A. Woods, *Yeast* **11,** 355 (1995).

[2] R. D. Gietz and R. A. Woods, *in* "Methods in Microbiology," Vol. 26 (A. J. P. Brown and M. F. Tuite, eds.), p. 53. Academic Press, San Diego, 1998.

[3] R. A. Woods and R. D. Gietz, *in* "Gene Transfer Methods: Introducing DNA into Living Cells and Organisms" (P. A. Norton and L. F. Steel, eds.), p. 25. Eaton Publishing, Natick, MA, 2000.

[4] D. M. Becker and L. Guarente, *Methods Enzymol.* **194,** 182 (1991).

[5] R. H. Schiestl, P. Manivasakam, R. A. Woods, and R. D. Gietz, *Methods* **5,** 79 (1993).

[6] J. R. Thompson, E. Register, J. Curotto, M. Kurtz, and R. Kelly, *Yeast* **14,** 565 (1998).

[7] M. C. Costanzo and T. D. Fox, *Genetics* **120,** 667 (1988).

[8] S. A. Johnston and M. J. DeVit, *Methods Mol. Biol.* **53,** 147 (1996).

[9] F. Spencer, G. Ketner, C. Connelly, and P. Hieter, *Methods* **5,** 161 (1993).

[10] R. D. Gietz and R. A. Woods, *Biotechniques* **30,** 816 (2001).

[11] R. D. Gietz, B. Triggs-Raine, A. Robbins, K. C. Graham, and R. A. Woods, *Mol. Cell Biochem.* **172,** 67 (1997).

0076-6879/02 $35.00

(1) the rapid transformation protocol for the introduction of a plasmid into a yeast strain; (2) the high-efficiency transformation protocol for recovering large numbers of transformants; (3) the large-scale transformation protocol for screening complex plasmid libraries such as those required for a two-hybrid screen; and (4) the microtiter plate transformation protocols[3] for the simultaneous transformation of multiple strains or multiple samples of a single strain.

In all of these protocols the number of reagents and steps has been reduced from previous versions. Growth and transformation efficiency are improved if broth cultures are grown in double-strength YPAD.[3] Cells for transformation are harvested, washed in sterile water, resuspended in Transformation Mix (PEG, LiAc, SS-carrier DNA, and plasmid DNA) without pretreatment and immediately incubated at 42°. After centrifugation and removal of the Transformation Mix the cells are resuspended in sterile water and sampled onto Synthetic Complete (SC) selection medium[12] adjusted to pH 5.6. The duration of incubation at 42° is the most important variable in the transformation reaction. For stationary phase and agar grown cells the yield of transformants with some strains is increased from $2–4 \times 10^3/\mu g$ plasmid DNA after 20 min at 42° to $>1 \times 10^6$ if the incubation is extended to 180 min. Log-phase cells show optimal transformation, 5×10^6 to $1 \times 10^7/\mu g$ plasmid DNA, after 40 to 60 min at 42°.

Reagents and Solutions

The following reagents and solutions are required for all four LiAc/SS-DNA/PEG protocols.

Lithium Acetate (1.0 M). Dissolve 5.1 g of lithium acetate dihydrate (Sigma Chemical Co. Ltd., St. Louis, MO) in 50 ml of water, sterilize by autoclaving for 15 min, and store at room temperature.

Polyethylene Glycol 3350 (50%, w/v). Dissolve 50 g of PEG 3350 (Sigma) in 30 ml of distilled, deionized water in a 150-ml beaker on a stirring hot plate. When the solution has cooled to room temperature, make the volume up to 100-ml in a 100-ml measuring cylinder and mix thoroughly by inversion. Pour the solution in to a suitable glass bottle and autoclave for 15 min. Store, securely capped, at room temperature. Evaporation of water from the solution will increase the concentration of PEG and severely reduce the yield of transformants.

Single-Stranded Carrier DNA (2.0 mg/ml). Dissolve 200 mg of salmon sperm DNA (Sigma) in 100 ml of TE buffer (10 m*M* Tris-HCl, 1 m*M* Na_2 EDTA, pH 8.0) by stirring at 4° for 1–2 hr. Store 1.0-ml samples at −20°. Denature the carrier DNA in a boiling water bath for 5 min and chill in ice/water before use. Boiled samples, stored at −20°, can be reused or reboiled three times without loss of activity.

[12] M. D. Rose, *Methods Enzymol.* **152,** 481 (1987).

Yeast Growth Media. Yeast strains are grown up on plates of YPAD agar (YPD[13] supplemented with 100 mg adenine hemisulfate per liter). The yeast cells to be transformed are usually regrown in liquid 2× YPAD medium[3] (2% Bacto-yeast extract, 4% Bacto-peptone, 4% glucose and adenine hemisulfate, 100 mg/liter). SC selection medium[12] is adjusted to pH 5.6 with 1.0 *N* NaOH and autoclaved. Commercial formulations of YPD agar (Bacto YPD agar) and broth (Bacto YPD broth) media are available from Becton Dickinson Microbiology Systems (Becton Dickinson, Sparks, MD; www.bd.com). These media should be supplemented with adenine hemisulfate as above.

Rapid Transformation Protocol

Day 1. Inoculate the yeast strain in a 2-cm^2 patch onto YPAD agar and incubate overnight at 30°. Alternatively, the yeast strain can be inoculated into 5 ml of liquid medium (2× YPAD or SC selection medium) and incubated on a shaker at 30° and 200 rpm.

Day 2

1. Heat a tube of carrier DNA in a boiling water bath for 5 min and then chill in ice/water.
2. Scrape a 50-μl portion of yeast from the YPAD plate and suspend the cells in 1 ml of sterile water in a 1.5 ml microcentrifuge tube. The suspension will contain about 5×10^8 cells. Cells grown overnight in 2× YPAD broth will reach a titer between 1 and 2×10^8/ml; the titer in SC medium will be about 5×10^7/ml. Harvest 2 ml of a YPAD culture and 5 ml of an SC culture. Note: Cells in log phase growth on agar or in liquid medium will transform with high efficiency.
3. Pellet the cells at 13,000 rpm at room temperature in a microcentrifuge for 30 sec and discard the supernatant.
4. Add the following components of the Transformation Mix (T Mix) to the cell pellet in the order listed:

Component	Volume (μl)
PEG 3350 (50%, w/v)	240
Lithium acetate 1.0 *M*	36
Boiled SS–Carrier DNA (2 mg/ml)	50
Plasmid DNA (0.1 to 1 μg) plus water	34
Total volume	360

Be sure to vortex mix the carrier DNA before pipetting it.

[13] F. Sherman, *Methods Enzymol.* **194,** 3 (1991).

5. Incubate the tube in a water bath at 42° for 40 to 60 min. Many laboratory strains will yield up to 1×10^5 transformants/μg plasmid after 60 min incubation. Extending the time at 42° to 180 min increases the yield to $>1 \times 10^6/\mu$g with some strains.
6. Microcentrifuge at 13,000 rpm at room temperature for 30 sec and remove the T Mix with a micropipettor.
7. Pipette 1.0 ml of sterile water into the tube and resuspend the cells by stirring with a micropipette tip and then vortex mixing vigorously.
8. Pipette 10- and 100-μl samples onto plates of appropriate SC selection medium, incubate at 30° for 3–4 days, and isolate transformants. The 10-μl samples should be pipetted into 100-μl puddles of sterile water.

This protocol can be used with cultures that have been stored at room temperature or in a refrigerator. The yield will be reduced with older cultures but will generally be sufficient to isolate a number of transformants of the desired genotype.

High-Efficiency Transformation Protocol

This protocol can be used to generate sufficient transformants in a single reaction to screen multiple yeast genome equivalents for plasmids that complement a specific mutation. It can also be used to transform integrating plasmids, DNA fragments and oligonucleotides[14] for yeast genome manipulation. Finally, it is used to optimize the conditions for transformation of a particular yeast strain, for example, the transformation of a plasmid library into a two-hybrid yeast strain transformed with a bait plasmid by the rapid transformation protocol. The high efficiency protocol can also be employed to transform a yeast strain simultaneously with two different plasmids, such as the two-hybrid bait and prey plasmids.

Day 1. Inoculate the yeast strain into 5 ml of liquid medium (2× YPAD or SC selection medium) and incubate overnight on a rotary shaker at 200 rpm and 30°. Place a bottle of 2× YPAD and a 250-ml culture flask in the incubator as well.

Day 2

1. Determine the titer of the yeast culture by pipetting 10 μl of cells into 1.0 ml of water in a spectrophotometer cuvette and measuring the OD at 600 nm. For many yeast strains a suspension containing 1×10^6 cells/ml will give an OD_{600} of 1.0. Alternatively, titer the culture using a hemocytometer.
2. Transfer 50 ml of the prewarmed 2× YPAD to the prewarmed culture flask and add 2.5×10^8 cells to give 5×10^6 cells/ml.
3. Incubate the flask on a rotary or reciprocating shaker at 30° and 200 rpm.
4. When the cell titer is at least 2×10^7 cells/ml, which should take about

[14] L. I. Linske-O'Connell, F. Sherman, and G. McLendon, *Biochemistry* **34,** 7094 (1995).

4 hr, harvest the cells by centrifugation at 3000*g* at room temperature for 5 min, wash the cells in 25 ml of sterile water and resuspend in 1 ml of sterile water.

5. Boil a 1.0-ml sample of carrier DNA for 5 min and chill in an ice/water bath while harvesting the cells.
6. Transfer the cell suspension to a 1.5-ml microcentrifuge tube, centrifuge for 30 sec, and discard the supernatant.
7. Add water to a final volume of 1.0 ml and vortex-mix vigorously to resuspend the cells.
8. Pipette 100-μl samples (ca. 10^8 cells) into 1.5 ml microfuge tubes, one for each transformation, centrifuge at 13,000 rpm at room temperature for 30 sec, and remove the supernatant.
9. Make up sufficient T Mix for the planned number of transformations plus one extra. Keep the T Mix in ice/water.

	Number of transformations planned		
Reagents	1	5 (6×)	10 (11×)
PEG 3350 (50%, w/v)	240 μl	1440 μl	2640 μl
Lithium acetate 1.0 *M*	36 μl	216 μl	396 μl
Boiled SS-Carrier DNA (2 mg/ml)	50 μl	300 μl	550 μl
Plasmid DNA plus water	34 μl	204 μl	374 μl
Total volume	360 μl	2160 μl	3960 μl

10. Add 360 μl of T Mix to each transformation tube and resuspend the cells by vortex mixing vigorously.
11. Incubate the tubes in a 42° water bath for 40 min.
12. Microcentrifuge at 13,000 rpm at room temperature for 30 sec and remove the T Mix with a micropipettor.
13. Pipette 1.0 ml of sterile water into each tube; stir the pellet with a micropipette tip and vortex vigorously.
14. Plate appropriate dilutions of the cell suspension onto SC selection medium. For transformation with an integrating plasmid (YIp), linear construct, or oligonucleotide, plate 200 μl onto each of five plates; for a YEp, YRp, or YCp library plasmid dilute 10 μl of the suspension into 1.0 ml of water and plate 10- and 100-μl samples onto two plates each. The 10-μl samples should be pipetted directly into 100-μl puddles of sterile water on the SC selection medium.
15. Incubate the plates at 30° for 3 to 4 days and count the number of transformants.

The transformation efficiency (transformants/1 μg plasmid/10^8 cells) can be determined by calculating the number of transformants in 1.0 ml of resuspended cells per 1.0 μg plasmid per 10^8 cells. For example, if the transformation of 1.0×10^8 cells with 100 ng plasmid resulted in 500 colonies on a plate of SC selection medium spread with 1 μl of suspension:

$$\text{Transformation efficiency} = 500 \times 1000 \text{ (plating factor)} \times 10 \text{ (plasmid factor)} \times 1 \text{ (cells/transformation} \times 10^8)$$

$$\text{Transformation efficiency} = 5 \times 10^6 \text{ transformants/1.0 } \mu\text{g plasmid/}10^8 \text{ cells}$$

Transformation efficiency declines as plasmid concentration is increased[1] but the actual yield of transformants per transformation increases. For example, 100 ng of plasmid in a transformation might give a transformation efficiency of 5×10^6 and a yield of 5×10^5 transformants, whereas with 1 μg of plasmid the transformation efficiency might be 2×10^6 and the yield 2×10^6 per transformation. In order to obtain 5×10^6 transformants it is simpler to set up two or three transformations with 1 μg of plasmid DNA, or a single 3-fold scaled up transformation, than to carry out 10 reactions with 100 ng of plasmid in each.

Large-Scale Transformation Protocol

The high efficiency transformation protocol can be scaled up 10- to 120-fold to generate the large numbers of transformants required for systems such as a two-hybrid screen.[3,11,15] It is best to scale up a transformation reaction to obtain a higher transformation yield rather than just increasing the amount of plasmid DNA. For example, for a 10-fold scale-up, use all of the cells from a 50-ml regrown culture ($\sim 1 \times 10^9$ cells) with 10 μg plasmid DNA in 3.6 ml of T Mix. The incubation at 42° should be extended to at least 60 min to allow for temperature equilibration.

Microtiter Plate Transformation

The following protocols can be used to accomplish a large number of transformation reactions in round-bottom 96-well microtiter plates. The agar plate protocol can be used to transform a plasmid into many different yeast strains and the liquid culture protocol can be used to introduce many different plasmids or constructs into a single strain. The liquid culture protocol can be used to optimize the conditions for the transformation of a specific yeast strain as suggested above. Both of these protocols require a microtiter plate centrifuge rotor, a 96-prong replicator (Fisher Scientific, Nepean, Ontario, Canada, www.fishersci.ca), 150-mm petri plates, an eight-channel micropipettor (Eppendorf or Titer Tek), and sterile troughs. We have found that an 8×8 well custom-made replicator (lacking the four corner prongs)

[15] R. A. Woods and R. D. Gietz, *in* "Methods in Molecular Biology, Vol. 177," (P. N. MacDonald, ed.), p. 85. Humana Press, Totowa, NJ.

that can be used with 100 × 15 mm regular petri dishes is often more convenient for up to 60 transformations than the full-size replicator.

Agar Plate Protocol

Day 1

1. Sterilize the replicator by dipping the prongs into a dish of 95% (v/v) ethanol and passing them through a bunsen flame.
2. Rest the replicator "prongs up" in a beaker and lower a plate of YPAD onto the prongs to make an imprint on the agar.
3. Use toothpicks or an inoculating loop to patch the yeast strains onto the imprints. Make an orientation mark on the bottom of the plate and incubate the plate overnight at 30°.

Day 2

1. Dispense 150-μl samples of sterile water into the wells of a microtiter plate.
2. Sterilize the replicator and rest it "prongs up" in a beaker.
3. Invert the YPAD plate over the replicator and align the patches of yeast with the tips of the prongs. Lower the plate onto the replicator, ensuring that all of the patches of yeast make contact, and move the plate gently in very small circles to transfer cells to the replicator. Remove the plate and inspect the prongs; use a toothpick or inoculating loop to add cells if necessary.
4. Lower the replicator into the microtiter plate wells and agitate to suspend the cells. The average number of cells/well will be $\sim 1 \times 10^7$; a second transfer will double the number. Mark the orientation of the microtiter plate.
5. Centrifuge at 3500 rpm at room temperature for 10 min using a microtiter plate rotor with an appropriate balance plate (if necessary).
6. Remove the medium by aspiration with a sterile micropipette tip attached to a vacuum line. Be careful not to touch the cell pellet with the tip. Alternatively, shake the water out of the wells into a sink. This takes practice, but is much faster than aspiration!
7. Boil carrier DNA (2 mg/μl) for 5 min and chill in ice/water.
8. Prepare T Mix minus PEG. The volumes below are for a single well and 96 wells (allowing 4 extra). Keep the T Mix minus PEG in ice/water.

Component	1 well	96 wells
Lithium acetate 1.0 *M*	15.0 μl	1.5 ml
Carrier DNA (2 mg/ml)	20.0 μl	2.0 ml
Plasmid DNA + water	15.0 μl	1.5 ml
Total volume	50.0 μl	5.0 ml

Note: We use 20 ng plasmid DNA per well; however, more can be added.

9. Pipette 50 μl T Mix minus PEG to each well. Clamp the plate on a rotary shaker and agitate at 400 rpm for 2 min to resuspend the cell pellets. The cells resuspend readily in T Mix minus PEG but not in T Mix.
10. Pipette 100 μl PEG 3350 (50% w/v) into each well. Clamp the plate on the rotary shaker at 400 rpm for 5 min to ensure that the cell suspension is homogeneous.
11. Place the microtiter plate in a ZipLoc sandwich bag or seal it with Parafilm and incubate at 42° for 3–4 hr.
12. Centrifuge the microtiter plate as before and remove the T Mix by aspiration.
13. The transformation reactions can be sampled as follows:
    (a) *Quantitative samples:* Pipette 100 μl of water into the wells. Clamp the plate on the rotary shaker at 400 rpm for 5 min to resuspend the cells. Pipette 5 μl samples into 100-μl puddles on regular plates of SC selection medium.
    (b) *Qualitative samples:* Pipette 50 μl of water into the wells. Resuspend the cells and use the sterile replicator "prongs down" to print onto plates of SC selection medium. The transfer volume is approximately 10 μl. Additional samples can be overlaid with care if required.
14. Incubate the plates at 30° for 2 to 4 days and recover transformants.

We have obtained up to 8000 transformants/well with 1×10^7 cells/well, 20 ng plasmid, and 4 hr incubation at 42°.

Liquid Culture Protocol. This protocol is used when transforming a single strain with multiple plasmids or DNA constructs. The yeast culture is grown overnight and regrown for two divisions as in the high efficiency transformation protocol. A complete microtiter plate (96 wells) with 4×10^7 cells/well will require 200 ml of regrown culture and 8 μg plasmid. The cells of the regrown culture should be harvested, washed, and resuspended in water at 4×10^8 cells/ml as described in the high efficiency transformation protocol.

Day 1. As in high-efficiency transformation protocol.

Day 2

1. Dispense 100-μl samples (4×10^7 cells) of the suspension of regrown cells into the wells of the microtiter plate. Centrifuge and remove the supernatants.
2. Continue from step 9 of the agar plate protocol with the following changes:
    (a) Increase the amount of plasmid in the T Mix minus PEG accordingly.
    (b) Incubate the plates at 42° for 60 min.
3. Sample the wells by plating or replica plating onto SC selection medium.

Frequently Asked Questions

For several years we have maintained a Web site[16] devoted to yeast transformation by the LiAc/SS-DNA/PEG protocol. The most frequently asked questions are concerned with the following:

(a) *A very low number of transformants.* Many factors can result in low numbers of transformants. (i) Check the dilution factor and plate a larger sample if necessary: we regularly plate 200-, 20-, and 2-μl samples when using a new yeast strain or plasmid. (ii) Check the integrity of the plasmid preparation prior to use. Plasmid degradation severely reduces the recovery of transformants. (iii) PEG concentration is critical for good transformation. Make fresh 50% (w/v) PEG 3350 reagent with care to ensure the proper concentration. PEG at this concentration is a viscous solution and is difficult to mix and pour. Use a plastic measuring cylinder and beaker to make the solution and carefully pour into a securely capped container. Evaporation of water from the PEG solution over time will increase the concentration and reduce the recovery of transformants. (iv) Other possible factors may include lithium acetate or temperature sensitivity of the yeast strain.

(b) A very large number of transformants. Check that the genetic markers in the yeast strain and plasmid correspond to the medium used to select for transformants. For example, if the medium selects for uracil prototrophy and the yeast strain is *URA3*, all the cells plated will grow.

(c) Purity of plasmid DNA. Plasmid DNA does not have to be extensively purified to be used for transformation by these protocols. Plasmid DNA isolated using the classic mini-preparation procedure usually transforms better than more highly purified plasmid DNA.

Optimization of Transformation for Specific Yeast Strains

Yeast strains vary tremendously in their transformation characteristics; some transform well and others poorly. The protocols described above will work well for most strains. If a strain transforms poorly it is best to obtain one with good transformation characteristics from a reputable source. When using a particular yeast strain, optimize the protocol(s) by investigating the following parameters in the order listed.

1. Duration of 42° incubation
   (a) Rapid and agar plate (microtiter plate) transformation protocols (60–300 min)
   (b) High efficiency and liquid culture (microtiter plate) protocols (20 to 80 min)

[16] R. D. Gietz, Web site: http//www.umanitoba.ca/faculties/medicine/units/biochem/gietz

2. Amount of carrier DNA (25–75 μl)
3. Amount of plasmid DNA (0.1–5 μg)
4. Number of cells (0.5–4 $\times 10^8$)
5. Lithium acetate concentration (18–54 μl)
6. PEG concentration (220–270 μl)

Several workers have reported that the addition of dimethyl sulfoxide (DMSO) and/or ethanol to the Transformation Mix increases the recovery of transformants significantly.[17–21] In our experience these additions at best double or triple the yield if other conditions, particularly the duration of incubation at 42°, have been optimized. Our findings for the strain DY2389 are summarized below:

Protocol	Additive	Min at 42°	Time of addition	Increase in yield
Rapid	DMSO, 5%	180	With T Mix	3.3×
Rapid	Ethanol, 5%	180	After 60 min at 42°	2.8×
High efficiency	DMSO, 1%	40	With T Mix	1.7×
High efficiency	Ethanol, 2.5%	40	After 10 min at 42°	1.9×

Some strains do show a marked response to additives. For example, Y190, which transforms well using the high efficiency protocol, 5×10^6 transformants/μg,[5] transforms poorly by the rapid protocol. Only 16,800 transformants/μg were recovered after incubation at 42° for 180 min; the addition of 5% DMSO to the T Mix increased the yield nearly 14-fold to 231,100 transformants/μg.

Summary

In this chapter we have provided instructions for transforming yeast by a number of variations of the LiAc/SS-DNA/PEG method for a number of different applications. The rapid transformation protocol is used when small numbers of transformants are required. The high efficiency transformation protocol is used to generate large numbers of transformants or to deliver DNA constructs or oligonucleotides into the yeast cell. The large-scale transformation protocol is primarily applicable to the analysis of complex plasmid DNA libraries, such as those required for the yeast two-hybrid system. The microtiter plate versions of the rapid and high efficiency transformation protocols can be applied to high-throughput screening technologies.

[17] V. Lauermann, *Curr. Genet.* **20,** 1 (1991).
[18] J. Hill, K. A. Ian, G. Donald, and D. E. Griffiths, *Nucleic Acids Res.* **19,** 5791 (1991).
[19] R. Soni, J. P. Carmichael, and J. A. Murray, *Curr. Genet.* **24,** 455 (1993).
[20] P. L. Bartel and S. Fields, *Methods Enzymol.* **254,** 241 (1995).
[21] G. Cagney, P. Uetz, and S. Fields, *Methods Enzymol.* **328,** 3 (2001).

[5] Genetic Transformation of *Saccharomyces cerevisiae* Mitochondria

By NATHALIE BONNEFOY and THOMAS D. FOX

Introduction

A key feature of the yeast nuclear genetic system that has made it a preeminent tool for genetic and cell biological research is the fact that DNA transformed into the nuclear chromosomes of *Saccharomyces cerevisiae* is incorporated into the genome only via homologous recombination.[1] This fact allows the researcher to add, subtract, and alter genetic information in a highly controlled fashion and essentially rewrite the yeast genome at will.[2]

In *S. cerevisiae,* and to date only in that species, similar manipulations based on homologous recombination have also been carried out on the mitochondrial genome. This article briefly summarizes some basic features of yeast mitochondrial genetics, describes current methods for delivery of DNA into the organelle, and outlines strategies employing homologous recombination that allow one to create directed mutations in mitochondrial genes and to insert new genes into mitochondrial DNA (mtDNA). These subjects have been reviewed previously and the reader is referred to several previous articles for more detailed discussions of the mitochondrial genetics underlying the transformation strategies discussed here.[3–7] General methods for yeast genetics have been compiled by Guthrie and Fink.[8]

[1] A. Hinnen, J. B. Hicks, and G. R. Fink, *Proc. Natl. Acad. Sci. U.S.A.* **75,** 1929 (1978).

[2] R. Rothstein, *Methods Enzymol.* **194,** 281 (1991).

[3] R. A. Butow, M. Henke, J. V. Moran, S. M. Belcher, and P. S. Perlman, *Methods Enzymol.* **264,** 265 (1996).

[4] B. Dujon, *in* "The Molecular Biology of the Yeast *Saccharomyces:* Life Cycle and Inheritance" (J. N. Strathern, E. W. Jones, and J. R. Broach, eds.), p. 505. Cold Spring Harbor Laboratory Press, Cold Spring Harbor, NY, 1981.

[5] T. D. Fox, L. S. Folley, J. J. Mulero, T. W. McMullin, P. E. Thorsness, L. O. Hedin, and M. C. Costanzo, *Methods Enzymol.* **194,** 149 (1991).

[6] P. S. Perlman, C. W. Birky, Jr., and R. L. Strausberg, *Methods Enzymol.* **56,** 139 (1979).

[7] L. Pon and G. Schatz, *in* "The Molecular and Cellular Biology of the Yeast *Saccharomyces:* Genome Dynamics, Protein Synthesis, and Energetics" (J. R. Broach, J. R. Pringle, and E. W. Jones, eds.), Vol. 1, p. 333. Cold Spring Harbor Laboratory Press, Cold Spring Harbor, NY, 1991.

[8] C. Guthrie and G. R. Fink (eds.), *Methods Enzymol.* **194** (1991).

0076-6879/02 $35.00

Important Features of *Saccharomyces cerevisiae* Mitochondrial Genetics

Phenotypes Associated with Mitochondrial Gene Expression

The common phenotype of mutations that affect mitochondrial genes or their expression is the inability to grow on nonfermentable carbon sources. Wild-type *S. cerevisiae* strains grow well on complete medium containing nonfermentable carbon sources, such as ethanol and glycerol [YPEG : 1% yeast extract (w/v), 2% peptone (w/v), 3% ethanol (v/v) plus 3% glycerol (v/v)]. Mutants that lack a functioning oxidative phosphorylation system cannot grow on such nonfermentable medium, but grow relatively well on medium containing fermentable carbon sources such as glucose [YPD : 1% yeast extract (w/v), 2% peptone (w/v), 2% dextrose (w/v)].

Respiratory growth of wild-type yeast can be impaired by several inhibitors of bacterial protein synthesis, and mutations conferring resistance can serve as genetic markers. Mutations in mitochondrial ribosomal RNA genes can lead to resistance to chloramphenicol,[9] erythromycin,[10] and paromomycin.[11] Mutations causing resistance to the ATP synthase inhibitor oligomycin[12,13] and the cytochrome *b* inhibitor diuron[14] also provide mitochondrial genetic markers. However, it is important to note that these drug-resistant phenotypes arise spontaneously and can only be observed on nonfermentable medium in strains that respire. Thus, they are not ideal for use as selective markers in transformation experiments.

Novel mitochondrial phenotypes have been generated by placing foreign genes into mtDNA. Phenotypes based on foreign genes can serve as mitochondrial genetic markers independently of respiratory function. One such phenotype is based on the fact that nuclear genes such as *URA3* and *TRP1* cannot be expressed when inserted into mtDNA, but will escape from mitochondria to the nucleus at high frequency, leading to detectable growth phenotypes that can be scored on petri plates.[15,16]

At least some nuclear genes can be expressed phenotypically within mitochondria if they are rewritten in the *S. cerevisiae* mitochondrial genetic code,[17] providing novel selectable markers. Expression of the synthetic gene $ARG8^m$ within mitochondria allows nuclear *arg8* mutants to grow without arginine.[18] This protein, Arg8p, is normally imported into mitochondria from the cytoplasm, but also functions when synthesized within the organelle. Thus, Arg^+ prototrophy can become

[9] B. Dijon, *Cell* **20,** 185 (1980).

[10] F. Sor and H. Fukuhara, *Nucleic Acids Res.* **10,** 6571 (1982).

[11] M. Li, A. Tzagoloff, K. Underbrink-Lyon, and N. C. Martin, *J. Biol. Chem.* **257,** 5921 (1982).

[12] B. G. Ooi, C. E. Novitski, and P. Nagley, *Eur. J. Biochem.* **152,** 709 (1985).

[13] W. Sebald, E. Wachter, and A. Tzagoloff, *Eur. J. Biochem.* **100,** 599 (1979).

[14] J. P. di Rago, X. Perea, and A. M. Colson, *FEBS Lett.* **208,** 208 (1986).

[15] P. E. Thorsness and T. D. Fox, *Nature* **346,** 376 (1990).

[16] P. E. Thorsness and T. D. Fox, *Genetics* **134,** 21 (1993).

[17] T. D. Fox, *Ann. Rev. Genet.* **21,** 67 (1987).

[18] D. F. Steele, C. A. Butler, and T. D. Fox, *Proc. Natl. Acad. Sci. U.S.A.* **93,** 5253 (1996).

a phenotype dependent on mitochondrial gene expression. This new mitochondrial marker provides a convenient way to disrupt endogenous mitochondrial genes[19] as well being a useful reporter for studying mitochondrial gene expression[18,20] and genetic instability.[21] In addition, because *ARG8*^{m} specifies a soluble protein, translational fusions to endogenous mitochondrial genes can create chimeric proteins useful for the study of targeting and membrane translocation of mitochondrial translation products.[22,23] A visible reporter phenotype based on mitochondrial gene expression can also be generated by insertion into mtDNA of a synthetic gene, *GFP*^{m}, encoding the green fluorescent protein rewritten in the yeast mitochondrial code.[23a]

Replication of mtDNA

Replication of yeast mtDNA is a complex and poorly understood process. Cells typically contain between 50 and 100 genome equivalents of mtDNA,[4] which are organized into a smaller number of nucleoid structures. The nucleoids, which are visible by fluorescence microscopy, are the genetic elements transmitted to daughter cells during cell division.[24,25] Replication of complete, or *rho*$^+$, yeast mitochondrial genomes is thought to depend on a limited number of specific sites in the chromosome[26,27] and, for unknown reasons, requires mitochondrial protein synthesis.[28]

The most frequent mutants in wild-type *S. cerevisiae* strains are the nonrespiring *rho*$^-$, or cytoplasmic petite mutants. These strains have large deletions of mtDNA that destroy the organellar gene expression machinery by deleting components of its translation system.[4] The DNA sequences retained in *rho*$^-$ mutants are typically reiterated, such that the *rho*$^-$ cell contains roughly the same amount of mtDNA as the wild type. Replication of *rho*$^-$ mtDNA differs from that of *rho*$^+$ mtDNA in that it does not require mitochondrial protein synthesis. Interestingly, the mtDNA sequences replicating in *rho*$^-$ strains can be derived from any portion of the chromosome, demonstrating that there is no clear requirement for a specific

[19] M. E. Sanchirico, T. D. Fox, and T. L. Mason, *EMBO J.* **17,** 57961 (1998).
[20] N. Bonnefoy and T. D. Fox, *Mol. Gen. Genet.* **262,** 1036 (2000).
[21] E. A. Sia, C. A. Butler, M. Dominska, P. Greenwell, T. D. Fox, and T. D. Petes, *Proc. Natl. Acad. Sci. U.S.A.* **97,** 250.
[22] S. He and T. D. Fox, *Mol. Biol. Cell* **8,** 1449 (1997).
[23] S. He and T. D. Fox, *Mol. Cell. Biol.* **19,** 6598 (1999).
[23a] J. S. Cohen and T. D. Fox, *Mitochondrion* **1,** 181 (2001).
[24] D. Lockshon, S. G. Zweifel, L. L. Freeman-Cook, H. E. Lorimer, B. J. Brewer, and W. L. Fangman, *Cell* **81,** 947 (1995).
[25] S. M. Newman, O. Zelenaya-Troitskaya, P. S. Perlman, and R. A. Butow, *Nucleic Acids Res.* **24,** 386 (1996).
[26] M. de Zamaroczy, G. Faugeron-Fonty, G. Baldacci, R. Goursot, and G. Bernardi, *Gene* **32,** 439 (1984).
[27] M. E. Schmitt and D. A. Clayton, *Curr. Opin. Genet. Dev.* **3,** 769 (1993).
[28] A. M. Myers, L. K. Pape, and A. Tzagoloff, *EMBO J.* **4,** 2087 (1985).

replication origin sequence in rho^{-} mtDNAs. This is advantageous in creating mitochondrial transformants containing defined mtDNAs, as rho^{0} yeast strains, entirely lacking mtDNA, can be transformed with bacterial plasmid DNAs that subsequently propagate as "synthetic' rho^{-} molecules.[29]

Recombination and Segregation of mtDNA

Unlike the highly differentiated situation in animals and plants, there is true equality of the sexes in yeast mating. Haploid cells mate by fusion, cytoplasms are mixed, and mitochondria of the haploid cells fuse to form an essentially continuous compartment.[30,31] Homology-dependent recombination between the parental mtDNAs occurs at a high rate[4] in the medial portion of the zygote.[22] Of great importance to the manipulation of mitochondrial DNA is the fact that rho^{-} mtDNA sequences recombine readily with complete rho^{+} genomes. Thus, if a rho^{-} mtDNA contains wild-type genetic information in a particular region, it can recombine with a rho^{+} mtDNA bearing a mutation in that region. The result is that the mating of the nonrespiring rho^{-} with the nonrespiring rho^{+} mutant yields respiring recombinants at high frequency, whose growth can be selected on nonfermentable medium.

Yeast cells containing two different kinds of mtDNA can be created by mating or by mutation. Such heteroplasmic cells rapidly give rise to homoplasmic progeny, a phenomenon known as mitotic segregation.[4] (However, exceptional cases of stable heteroplasmy in *S. cerevisiae* have been reported.[32]) *S. cerevisiae* differs in this regard from plant and animal cells, which can maintain heteroplasmic states for extended periods of growth.[33,34] In this connection it is important to note that, unlike most animal and plant cells, *S. cerevisiae* divides by a highly asymmetric budding process. New buds receive relatively few copies of mtDNA from mother cells,[35] facilitating mitotic segregation. Heteroplasmic cells generated by transformation and subsequent homologous recombination also produce pure recombinant clones.[36]

Taken together, these features of the *S. cerevisiae* mitochondrial–genetic system allow DNA introduced from outside the cell to be propagated within the organelle as a plasmid, and the plasmid-borne mitochondrial sequences to recombine homologously with complete rho^{+} mtDNA.[29]

[29] T. D. Fox, J. C. Sanford, and T. W. McMullin, *Proc. Natl. Acad. Sci. U.S.A.* **85,** 7288 (1988).

[30] R. Azpiroz and R. A. Butow, *Mol. Biol. Cell* **4,** 21 (1993).

[31] J. Nunnari, W. F. Marshall, A. Straight, A. Murray, J. W. Sedat, and P. Walter, *Mol. Biol. Cell* **8,** 1233 (1997).

[32] A. S. Lewin, R. Morimoto, and M. Rabinowitz, *Plasmid* **2,** 474 (1979).

[33] M. R. Hanson and O. Folkerts, *Int. Rev. Cytol.* **141,** 129 (1992).

[34] D. C. Wallace, *Ann. Rev. Biochem.* **61,** 1175 (1992).

[35] A. R. Zinn, J. K. Pohlman, P. S. Perlman, and R. A. Butow, *Plasmid* **17,** 248 (1987).

[36] S. A. Johnston, P. Q. Anziano, K. Shark, J. C. Sanford, and R. A. Butow, *Science* **240,** 1538 (1988).

Delivery of DNA to Mitochondrial Compartment of *rho*^{0} Cells and Detection of Mitochondrial Transformants

Overview of Transformation Procedure

Exogenous DNA can be introduced in mitochondria of yeast cells via microprojectile bombardment.[36] The standard device for microprojectile bombardment is the PDS-1000/He System, available from Bio-Rad (Hercules, CA; http://www.bio-rad.co/templates/html/64343_products.html). This instrument uses a helium shock wave in an evacuated chamber to accelerate microscopic metal particles toward a lawn of cells on a petri plate. The shock wave is generated by rupture of a membrane at high pressure and accelerates a second membrane (the macrocarrier or flying disk), carrying the metal particles, toward the plate. Some cells on the plate are penetrated by particles and survive. DNA precipitated on the particles is thus introduced into cells and is taken up readily by the nucleus. In addition, mitochondria of a small fraction of such transformants also take up DNA. The PDS-1000 functions reproducibly for transformation of *S. cerevisiae* mitochondria.

In a typical mitochondrial transformation experiment, a large number of *rho*^{0} cells are bombarded randomly by a large number of particles (Fig. 1).[17,20,37,38,51] In the first step, cells that have been hit and that survived are allowed to make colonies on the petri plates by selecting for a nuclear genetic marker that is included in the DNA precipitated on the particles. Mitochondrial transformants are identified among these colonies by genetic tests for the presence of new genetic information in the mitochondrial genome. This new information is typically a portion of the wild-type mtDNA sequence that can rescue a known mitochondrial marker mutation by recombination, after the transformants are mated to an appropriate *rho*$^+$ tester strain, resulting in recombinants with a detectable growth phenotype. The new wild-type sequence may be an unaltered region of the gene of interest or it may be another piece of wild-type mtDNA incorporated into a vector. Such marker rescue can work with as little as 50 bp of homologous sequence flanking the site of the mutation in the tester mtDNA. As shown in Fig. 1, transformation of *rho*$^+$ mutants can also be detected using this marker rescue strategy (or, as discussed in a later section, by directly selecting for a phenotypic change). Transformants can also be identified by scoring for expression of complete mitochondrial genes that function *in* trans.[3,29]

Experimental Details for Transformation and Identification of Mitochondrial Transformants

Yeast Strains. Strain background is an important factor affecting the efficiency of transformation (Fig. 1). We have obtained the best results with strains in the S288c background, in particular those derived from DBY947.[37] Strains derive

[37] N. F. Neff, J. H. Thomas, P. Grisafi, and D. Botstein, *Cell* **33,** 211 (1983).

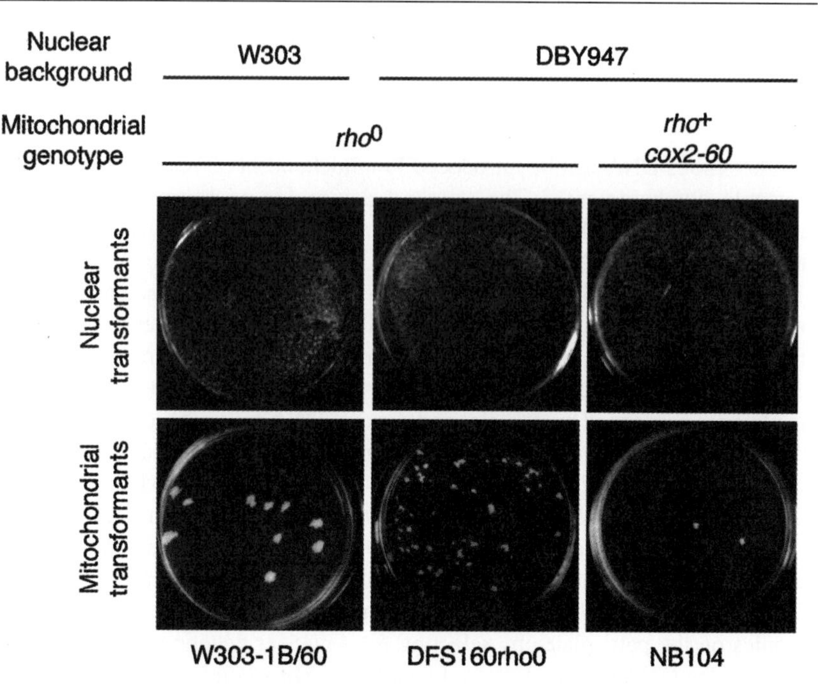

FIG. 1. Nuclear transformants and mitochondrial cotransformants obtained by bombardment of different yeast strains. The nuclear LEU2 plasmid Yep351 [J. E. Hill, A. M. Myers, T. J. Koerner, and A. Tzagoloff, *Yeast* **2,** 163 (1986)] and the COX2 plasmid pNB69 [N. Bonnefoy and T. D. Fox, *Mol. Gen. Genet.* **262,** 1036 (2000)] were precipitated together onto tungsten particles and bombarded on lawns of the rho^0 strains W303-1B/60 [*MATα, ade2-1, ura3-1, his3-11,15, trp1-1, leu2-3,112, can1-100* (rho^0)] [a rho^0 derivative of-W303-1B [B. J. Thomas and R. Rothstein, *Cell* **56,** 619 (1989)] and DFS160 (*MATα ade2-101, leu2Δ, ura3-52, arg8Δ::URA3, karl-1* (rho^0)] [T. D. Fox, *Ann. Rev. Genet.* **21,** 67 (1987)] or on lawns of the rho^+ strain NB104. W303-1B/60 was derived from W303 (ATCC200060). DFS160 was derived from DBY947 [N. F. Neff, J. H. Thomas, P. Grisafi, and D. Botstein, *Cell* **33,** 211 (1983)]. NB104 rho^+ mtDNA carries a 129-bp deletion, *cox2-60,* located around the *COX2* first codon [N. Bonnefoy and T. D. Fox, *Mol. Gen. Genet.* **262,** 1036 (2000)] and is isonuclear to DFS160. The top plates correspond to minimal medium supplemented with sorbitol and lacking leucine. Typical plates showing about 3000 nuclear transformants for each strain are shown. Nuclear transformants were crossed by replica plating to the nonrespiring tester strain (NB160), carrying a mutation of the *COX2* initiation codon [N. Bonnefoy and T. D. Fox, *Mol. Gen. Genet.* **262,** 1036 (2000)] and mitochondrial transformants (bottom plates) were detected by replica plating the mated cells onto nonfermentable medium.

from W303 (ATCC 200060)[38] give lower but satisfactory efficiencies, whereas strains in the D273-10B (ATCC 24657) background are difficult to transform. Excellent hosts for mitochondrial transformation, derived from DBY947, can be obtained from the American Type Culture Collection Manassas, VA: MCC109rh0

[38] B. J. Thomas and R. Rothstein, *Cell* **56,** 619 (1989).

[*MAT*α, *ade2-101, ura3-52, kar1-1* (*rho*^{0})][39] and MCC123rho0, which is the identical strain with *MAT*a, available as ATCC 201440 and 201442, respectively.

Preparation of Cells

(a) Grow the *rho*^{0} (or *rho*$^{+}$) strain to be bombarded for 2 to 3 days (stationary phase) at 30° with agitation in complete liquid medium (YP) containing either 2% raffinose or 2% galactose. These media may be supplemented with 0.1% glucose (to accelerate growth) and/or 100 mg/ml adenine (for Ade^{-} auxotrophs).
(b) Harvest cells and concentrate 40 to 100 times in liquid YPD medium to reach a cell density of 1 to 5 × 10^{9} cells/ml.
(c) Spread 0.1 ml of cells onto minimal glucose medium (0.67% yeast nitrogen base, 5% glucose, 100 mg/ml adenine, 3.3% agar) containing 1 *M* sorbitol and supplemented to provide the appropriated prototrophic selection.

Preparation of Microprojectiles and Precipitation of DNA. We use 0.5 μm tungsten powder obtained from Alfa-Aesar which, unfortunately, is no longer available; 0.4- and/or 0.7-μm tungsten particles available from Bio-Rad, Cat. Nos. 165-2265 and 165-2266, respectively, should be essentially equivalent. The following describes our current procedure using tungsten particles. With the exception of the sterilization step described later, the same procedure gives comparable results with gold particles (0.6 μm, Bio-Rad Cat. No. 165-2262) in our hands. A slightly different procedure for the preparation of gold particles has been described previously[3] and gives equivalent results in our hands.

(a) Sterilize up to 50 mg of tungsten particles by suspension in 1.5 ml of 70% ethanol in a microfuge tube and incubation at room temperature for 10 min. Wash the particles with 1.5 ml of sterile water and resuspend at 60 mg/ml in sterile 50% glycerol. Particles can be kept frozen for several months. Gold particles should be sterilized in 100% ethanol.[3]
(b) In a microfuge tube, mix 5 μg of plasmid carrying the nuclear marker and a nuclear replication origin with 15 to 30 μg of plasmid carrying the mitochondrial DNA of interest, in a total volume of 15–20 μl. Add and mix 100 μl of tungsten particles, 4 μl of 1 *M* spermidine-free base, and 100 μl of ice-cold 2.5 *M* $CaCl_2$. Incubate for 10 min on ice with occasional vortexing.
(c) Spin briefly and remove the supernatant. Resuspend the particles thoroughly in 200 μl of 100% ethanol, taking extreme care to fragment aggregates of particles, using the pipette tip. Repeat at least once until the particles resuspend easily.

[39] M. C. Costanzo and T. D. Fox, *Mol. Cell. Biol.* **13,** 4806 (1993).

(d) Spin briefly, remove the supernatant, and add 50–60 μl of 100% ethanol. Distribute the resulting suspension evenly at the center of six macrocarriers (flying disks) placed in their holders, allowing the ethanol to evaporate (there is no need to prewash the macrocarriers or desiccate them after coating).

Bombardment

(a) Follow the manufacturer's instructions carefully for use of the PDS-1000 apparatus. Place the rupture disk in its retaining cap and tighten using a torque wrench. Rupture disks of 1100 to 1350 psi can be used for the efficient transformation of yeast, although in our hands 1100 psi disks tend to give better results.
(b) Load the macrocarrier in its holder into the assembly system. Interestingly, we have found that simply allowing the carrier disk to fly to the surface of the petri plate, by not assembling the stopping screen, yields more transformants than if the stopping screen is employed.
(c) Place the open petri plate carrying the lawn of cells at 5 cm from the macrocarrier assembly. Shorter distances result in very high colony densities in the center of the plate with few colonies at the periphery, whereas longer distances decrease the transformation efficiency.
(d) Evacuate the vacuum chamber to a reading of 29 to 29.5 inch Hg on the gauge of the PDS-1000. We have found that failure to draw the greatest vacuum possible reduces transformation efficiency. Cell viability is not affected significantly by a prolonged stay under these vacuum conditions.
(e) Fire.
(f) Remove any fragments of the macrocarrier disk with a sterile forceps. Incubate the plate at 30° for 4 to 5 days until colonies appear (between 1000 and 10,000 per plate for S288c-related strains).

Identification of Mitochondrial Transformants

(a) During the incubation of the bombarded plates, set up a liquid YPD culture of an appropriate rho^+ mutant (mit^-) tester strain.
(b) Replica plate the transformants onto a lawn of the tester strain freshly spread on a YPD plate.
(c) Incubate at 30° for 2 days to allow mating and recombination.
(d) Print to YPEG medium (or another appropriate selection medium) to detect respiring diploids. In cases where a high number of nuclear transformants are present, it may be useful to also replicate the mated cells on medium that selects for the diploids, as comparison of the resulting plates may facilitate the identification of the desired transformant on the original bombarded plate.

(e) Pick colonies off the bombarded plate that correspond to the position of respiring recombinants. Streak these colonies on YPD and repeat the marker rescue with the tester strain as described earlier. Such subcloning and retesting must usually be done three times before pure stable synthetic *rho*$^-$ clones are obtained. (Cells usually lose the nuclear marker plasmid during these subcloning steps if no selection is applied for its maintenance.)

Strategies for Gene Replacement in *Saccharomyces cerevisiae* mtDNA

In cases where the mitochondrial gene under study encodes an active RNA molecule, such as the mitochondrial RNase P RNA, it may be possible to assay the activity of wild-type and mutant genes in the primary synthetic *rho*$^-$ transformants.[40] More commonly, however, mutations affecting protein-coding genes must be placed into *rho*$^+$ mtDNA by a double recombination event.

Integration of Altered mtDNA Sequences by Homologous Double Crossovers

The most basic method for putting a mutant version of a mitochondrial gene, or a foreign piece of DNA flanked by mtDNA sequences, into the chromosome is to first introduce the altered sequence into a *rho*^{0} strain to create a synthetic *rho*$^-$. This donor transformant (identified as described in Section III) is then (in a second step) mated with a wild-type *rho*$^+$ recipient strain. As a result of this second mating, mitochondria from the two strains fuse, and recombination between the two mtDNAs produces recombinant *rho*$^+$ strains in which the new mtDNA sequence is integrated by double crossover events. Pure recombinant strains are generated by subsequent mitotic segregation. Because mitochondrial DNA recombination and segregation are so frequent, this simple procedure typically yields the desired integrants at frequencies between 1 and 50% of clones derived from zygotes.

If one of the strains in such a cross carries the karyogamy-defective mutation *karl-1*,[41] which allows efficient mitochondrial fusion but reduces nuclear fusion greatly, haploid mitochondrial mutant cytoductants can be isolated after such a mating. This simple strategy has been used successfully with variations specific to each study in several laboratories.[16,42–47]

[40] P. Sulo, K. R. Groom, C. Wise, M. Steffen, and N. Martin, *Nucleic Acids Res.* **23,** 856 (1995).

[41] J. Conde and G. R. Fink, *Proc. Natl. Acad. Sci. U.S.A.* **73,** 3651 (1976).

[42] S. C. Boulanger, S. M. Belcher, U. Schmidt, S. D. Dib-Hajj, T. Schmidt, and P. S. Perlman, *Mol. Cell. Biol.* **115,** 4479 (1995).

[43] L. S. Folley and T. D. Fox, *Genetics* **129,** 659 (1991).

[44] R. M. Henke, R. A. Butow, and P. S. Perlman, *EMBO J.* **14,** 5094 (1995).

[45] J. J. Mulero and T. D. Fox, *Mol. Gen. Genet.* **242,** 383 (1994).

[46] H. Speno, M. R. Taheri, D. Sieburth, and C. T. Martin, *J. Biol. Chem.* **270,** 25363 (1995).

[47] T. Szczepanek and J. Lazowska, *EMBO J.* **15,** 3758 (1996).

Experimental Details for Mating and Isolation of Recombinant Cytoductants

Mating

(a) Grow cultures of the subcloned synthetic *rho*$^-$ strain and the recipient wild-type *rho*$^+$ strain overnight in liquid YPD. At least one of these two strains (usually the synthetic *rho*$^-$) must carry the *karl-1* mutation.

(b) If the synthetic *rho*$^-$ donor and the *rho*$^+$ recipient strains share nuclear markers, and therefore cannot be distinguished selectively on glucose medium, mating mixtures should contain equal numbers of cells of both strains. If nuclear auxotrophic or drug resistance markers allow selection against the synthetic *rho*$^-$ donor strain, then the mating mixture should contain a fivefold excess of donor cells.

(c) We have successfully used two different mating protocols for producing cytoductants.

1. Mix 0.5 ml of each parent (alternatively 1 ml of synthetic *rho*$^-$ and 0.2 ml of wild-type *rho*$^+$) in a microfuge tube, spin, remove the supernatant, resuspend in residual liquid, and spread the mixture onto a YPD plate. Incubate at 30° for 4 to 5 h. Check zygote formation microscopically. Scrape the mating cells from the plate and use them to inoculate fresh YPD liquid medium. Incubate at 30° with agitation for a few hours to overnight.
2. Alternatively, mix both parents, in proportions as just described, in 10 ml of liquid YPD and shake at 30° for 3 h. Spin the culture in a tube and incubate the pellet at 30° for 1 h without removing the medium. Resuspend by vortexing, transfer to a fresh flask, and incubate at 30° with agitation for at least 3 h.

Isolation of Cytoductants

(a) Dilute the culture to obtain single colonies and plate on minimal medium, selecting for the recipient nuclear genotype and against the donor nuclear genotype, if possible. Alternatively, plate on YPD medium. Densities of 50 to 200 colonies per plates should be obtained.

(b) Replica plate the colonies thus obtained to medium that will reveal the altered phenotype of the recipient strain as a result of integration of the mutant donor sequences into its mtDNA. For example, print to YPEG to identify clones that have acquired a mutation preventing respiratory growth.

(c) Mate nonrespiring candidate clones to a *rho*$^+$ tester mutant whose mitochondrial mutation is located outside of the region carried by the synthetic *rho*$^-$. The desired *rho*$^+$ recombinant cytoductants will produce respiring diploids after mating to this tester strain. (This step eliminates cytoductants that simply acquired the mtDNA of the donor strain.)

Streamlining Integration of Multiple Mutations in Short Region by Use of rho$^+$ Recipients Containing Defined Deletion

In situations where many nonfunctional mutations are to be placed in the same region of mtDNA, the strategy just described can be altered and made more efficient by using a *rho*$^+$ recipient that has a defined deletion in the region of interest.[39,48] It is, of course, often necessary to isolate such a recipient strain using the simple recombination method described previously. However, once the recipient is in hand, the nonrespiring recombinant cytoductants from crosses between nonrespiring synthetic *rho*$^-$ strains and the nonrespiring recipient can be identified by a positive marker rescue screen employing an appropriate tester strain (Fig. 2). Following mating between the synthetic *rho*$^-$ (carrying the experimental mutation "e" in Fig. 2) and the recipient, the cell population is plated on medium selecting for the recipient nuclear genotype. Recombinant cytoductants, unaltered recipient cells, and diploid cells will form colonies. However, only the recombinant cytoductants will be able to form respiring diploids when mated to a *rho*$^+$ tester strain carrying a marker mutation ("m" in Fig. 2) located in the deleted region (and distinct from the experimentally induced mutation).

Experimental Details for Identification of Nonrespiring Cytoductants by Marker Rescue

The following steps assume that one already has in hand a *rho*$^+$ recipient with a deletion mutation in the region of interest, as well as a *rho*$^-$ strain that carries wild-type information and that can recombine with the recipient deletion mutant.

1. Carry out a mating between the synthetic *rho*$^-$ donor and the *rho*$^+$ recipient containing a defined deletion to generate cytoductants as described in Section IV,B,1 (at least one of these two strains must carry the *karl-1* mutation), and spread dilutions of the cell mixture as described in Section IV. Some of these colonies will be *rho*$^+$ recombinants that have the deleted region restored and contain the desired mutation (Fig. 2A).

2. Replica plate colonies from the mating mixture onto a YPD plate bearing a freshly spread lawn of a *rho*$^+$ tester strain that has a marker mutation within the deleted region of the recipient but distinct from the new mutation to be introduced (Fig. 2B).

3. After 2 days of incubation at 30°, print the mated cells to YPEG. Identify haploid cytoductant clones on the plates from step 1 that correspond to respiring diploids in the mating of step 2.

[48] T. M. Mittelmeier and C. L. Dieckmann, *Mol. Cell. Biol.* **13,** 4203 (1993).

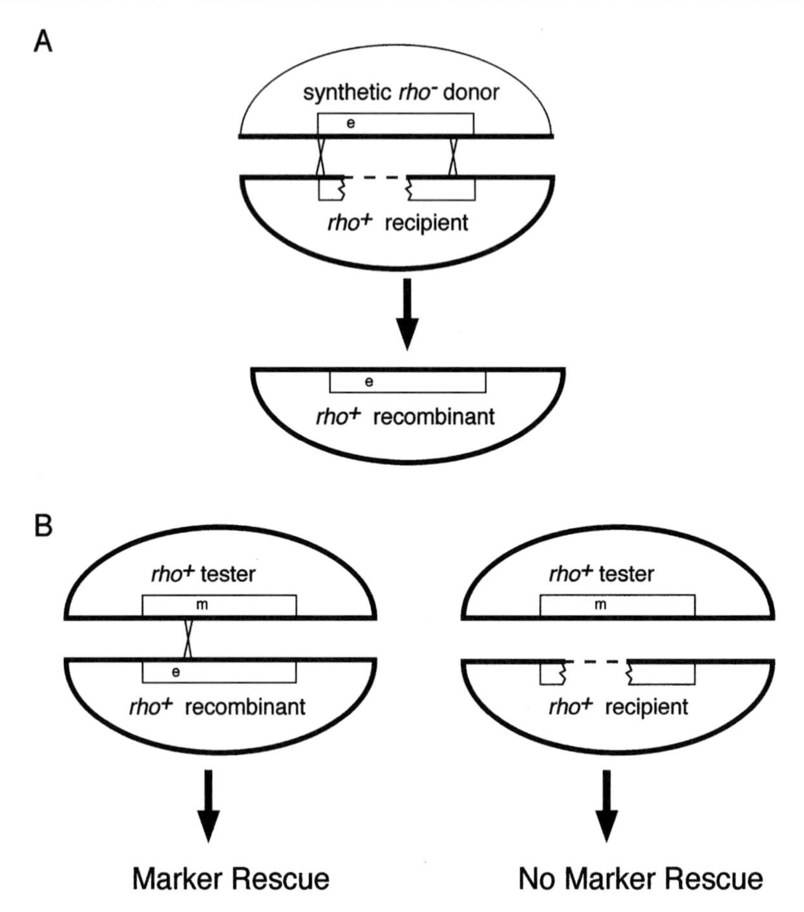

FIG. 2. Schematic diagram of recombination events that allow identification of nonrespiring recombinant cytoductants by marker rescue. Thick lines represent mtDNA sequences, and thin lines represent vector DNA. The box represents a gene under study. (A) A karyogamy defective (*karl-1*) synthetic *rho*$^-$ donor containing an experimentally induced mutation "e" is mated to a *rho*$^+$ recipient strain with a deletion in the region of interest. The desired *rho*$^+$ recombinant cytoductants are among the cells present in the mixture after mating. (B) To distinguish the desired *rho*$^+$ recombinant cytoductants from the unaltered recipient cells and other cell types present, clones derived from the mating mixture are mated to a *rho*$^+$ tester strain bearing the marker mutation "m." The desired *rho*$^+$ recombinant cytoductants can yield respiring recombinants when mated to this tester by a crossover between "e" and "m" (and a second resolving crossover anywhere else). The ability to produce respiring recombinants identifies the desired cytoductant clones. Unaltered recipient clones, and other cell types present, cannot yield such respiring recombinants.

4. Test the candidate cytoductant clones to make sure that they are *rho*$^+$ mutants by checking to see that they yield respiring diploids when mated to a *rho*$^-$ strain carrying wild-type information in the region deleted in the original recipient strain. This will eliminate cytoductants that simply contain the unrecombined mtDNA of the synthetic *rho*$^-$ donor.

Transformation of *rho*$^+$ Cells with Plasmids or Linear DNA Fragments

The first demonstration of the ability of the microprojectile bombardment to deliver DNA into yeast mitochondria depended on integration of the transforming DNA directly into a mutant *rho*$^+$ strain, converting a nonrespiring mutant into a respiring transformant.[36] In addition to selection for restoration of respiratory growth, *rho*$^+$ transformants can also be obtained with DNA sequences causing expression of *ARG8*^{m} and selection for Arg$^+$ growth. In cases where a mutated DNA sequence provides a function that can be selected for phenotypically, direct transformation of *rho*$^+$ strains bearing deletions in the region of interest can be used to integrate mutations into mtDNA.[20]

In our hands, mitochondrial transformation is 10 to 20 times more efficient during bombardment of a *rho*^{0} strain than of an isogenic *rho*$^+$ strain containing a small deletion in mtDNA (Fig. 1). This effect could be due either to physiological differences between the strains, such as the properties of the mitochondrial inner membrane, or to an advantage in establishing an incoming DNA molecule in the absence of endogenous mtDNA. Effects consistent with the latter notion have been observed in comparisons of mtDNA behavior after *rho*$^+$ × *rho*$^+$ matings as opposed to *rho*$^+$ × *rho*^{0} matings.[30]

Nevertheless, the relative inefficiency of transformation of *rho*$^+$ hosts is offset in some situations by its convenience, because *rho*$^+$ recombinants may be obtained more quickly. In addition, this strategy can be extended to transformation with linear DNA molecules obtained either from plasmid clones or by polymerase chain reaction (PCR) amplification. For example, we have found that linear DNA fragments having as little as 260 bp of homologous sequence flanking each side of a deletion mutation in a *rho*$^+$ recipient were able to yield respiring transformants at frequencies similar to those obtained with circular plasmids. The ability to use PCR-generated fragments to transform defined mtDNA deletion recipients can accelerate strain construction substantially.

Selection for transformation by DNA fragments with wild-type or near-wild-type function is straightforward by selecting first for nuclear transformants on glucose medium, as described in Section III, and then screening for the mitochondrial phenotype by replica plating (Fig. 3).[36] Interestingly, two- to fivefold fewer transformants are detected by this phenotypic selection for the integration of transforming DNA into *rho*$^+$ mtDNA than are detected by mating of the *rho*$^+$

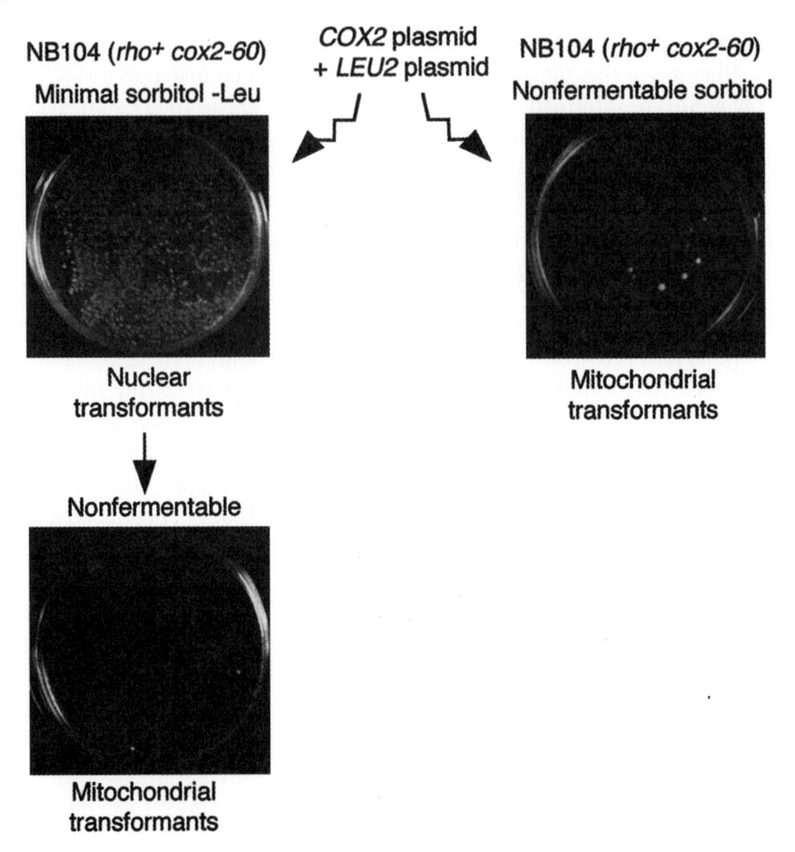

FIG. 3. Selection of rho$^+$ mitochondrial transformants directly after bombardment. The nuclear shuttle vector Yep351 (*LEU2*) and plasmid pNB69 (*COX2*) were bombarded together onto lawns of the *rho*$^+$ *cox2-60* strain NB104 (see legend of Fig. 1). The bombarded lawns had been spread either on minimal medium supplemented with sorbitol but lacking leucine (top right) or on nonfermentable YPEG medium supplemented with sorbitol and 0.1% glucose (top left). Leu$^+$ transformants were replica plated to nonfermentable medium (bottom plate) to select for mitochondrial transformants.

colonies to a tester strain and scoring for respiring recombinants (as was done in the experiment of Fig. 1). Thus it appears that only a fraction of the plasmids entering a *rho*$^+$ mitochondrion during bombardment manage to integrate and become expressed phenotypically.

Selection of respiring transformants from nonrespiring host strains generally requires a period of outgrowth on a fermentable carbon source to allow establishment of the respiratory phenotype prior to selection for it. This phenotypic lag affects selection for respiring transformants with both nuclear[49] and mitochondrial

[49] P. P. Müller and T. D. Fox, *Mol. Gen. Genet.* **195,** 275 (1984).

genes.[36] However, we have been able to successfully select respiring mitochondrial transformants in a single step by directly bombarding lawns of mutant *rho*$^+$ cells spread on YPEG plates supplemented with 0.1% glucose (to allow a brief period of outgrowth) and 1 *M* sorbitol (Fig. 3). We have also been able to select for *rho*$^+$ transformants expressing *ARG8*^{m} by directly bombarding lawns spread on appropriate minimal glucose medium (see Section III).

The detection of transformants of a *rho*$^+$ recipient that have incorporated a DNA sequence bearing a nonfunctional mutation is difficult, but possible, using essentially the same strategy as outlined in Fig. 2. If the recipient strain has a defined deletion, and integration of the transforming DNA restores the deleted sequence, then the desired integrants can be detected by their ability to rescue a known marker mutation in the deleted region.

Concluding Remarks

The methods developed for manipulation of the *S. cerevisiae* mitochondrial genome should provide a useful model for other systems. Indeed, in *Chlamydomonas,* another single-celled eukaryote in which nonreverting deletion mutations of mtDNA have been isolated, mitochondrial transformation and integration of wild-type DNA by homologous recombination have been achieved.[50] Thus, as appropriate selectable markers are developed for other species, it seems likely that their mitochondrial genomes will become amenable to *in vivo* experimental analysis.

Acknowledgments

We thank David M. MacAlpine and Ronald A. Butow for advice on the use of gold particles. N.B. was a Human Frontier Science Program Organization long-term fellow (LT22/96) during the early stages of this work and is currently supported by the Association Française contre les Myopathies. T.D.F. is supported by a grant from the U.S. National Institutes of Health (GM29362). This chapter has appeared previously[52] and is republished with permission.

[50] J. E. Boynton and N. W. Gillham, *Methods Enzymol.* **264,** 279 (1996).
[51] J. E. Hill, A. M. Myers, T. J. Koerner, and A. Tzagoloff, *Yeast* **2,** 163 (1986).
[52] N. Bonnefoy and T. D. Fox, *Methods Cell Biol.* **65,** 381 (2001).

[6] *LacZ* Assays in Yeast

By STEFFEN RUPP

Introduction

To obtain the answers to the biological questions, one needs the tools to visualize them. The use of reporter molecules in various designs (epitopes, GFP, auxotrophic/resistance markers) has been very successful in pointing to the answers to many of the questions posed. One of the most versatile reporters used in *Saccharomyces cerevisiae* is β-galactosidase, encoded by *lacZ* of *Escherichia coli*. *LacZ* assays have been used in a variety of screening procedures, the most well known being the two-hybrid system.[1] Furthermore, promoter activation screens on a genome wide level,[2] the N-end rule,[3] or dissection of complex promoters[4–8] could be worked out using *lacZ* as a reporter. β-Galactosidase can also be used as a tag to determine protein localization using immunofluorescence techniques. Especially useful for dissecting complex signal transduction pathways are reporter constructs, that give a fast and robust readout of the signals running through the pathway of interest. A variety of *lacZ* fusions exist that will help to assign the gene(s) (YFG) to one or more pathways.

This chapter is meant to give a short overview and to provide a practical guide on how to use *lacZ* assays in yeast. The first part of this chapter describes the tools available to create the *lacZ* fusion required for answering questions. In the second part some of the questions that have been asked using *lacZ* reporter constructs are described. Finally, the assays used to visualize or quantify the reporter activities are described and evaluated with regard to their best use.

When to Use *lacZ* Reporters

LacZ reporters have been designed to visualize transcriptional activity of genes using colorimetric assays. The assays are based on the expression of the *E. coli lacZ*

[1] S. Fields and O. Song, *Nature* **340,** 245 (1989).
[2] N. Burns, B. Grimwade, P. B. Ross-Macdonald, E. Y. Choi, K. Finberg, G. S. Roeder, and M. Snyder, *Genes Dev.* **8,** 1087 (1994).
[3] A. Bachmair, D. Finley, and A. Varshavsky, *Science* **234,** 179 (1986).
[4] D. W. Russell, R. Jensen, M. J. Zoller, J. Burke, B. Errede, M. Smith, and I. Herskowitz, *Mol. Cell. Biol.* **6,** 4281 (1986).
[5] L. Breeden and K. Nasmyth, *Cold Spring Harb. Symp. Quant. Biol.* **50,** 643 (1985).
[6] S. Sagee, A. Sherman, G. Shenhar, K. Robzyk, N. Ben-Doy, G. Simchen, and Y. Kassir, *Mol. Cell. Biol.* **18,** 1985 (1998). [Published erratum appears in *Mol. Cell. Biol.* **18,** 3645 (1998)].
[7] S. Rupp, E. Summers, H. J. Lo, H. Madhani, and G. Fink, *EMBO J.* **18,** 1257 (1999).
[8] M. Gagiano, D. Van Dyk, F. F. Bauer, M. G. Lambrechts, and I. S. Pretorius, *J. Bacteriol.* **181,** 6497 (1999).

0076-6879/02 $35.00

gene in yeast. The amount of β-galactosidase expressed reflects the transcriptional activity of the sequence elements cloned in front of the *lacZ* gene and can be measured using simple enzymatic assays. With this tool one can for example study important regions of a promoter by isolating these promoter fragments and cloning them in front of *lacZ*. Using this approach *lacZ* reporters have been proven extremely useful in characterizing and dissecting a wide variety of promoters.[4–8] UAS (upstream activating sequence) and URS (upstream repressing sequence) elements can be defined, as well as the potential *trans*-acting factors binding to these sequences. Direct evidence for binding of the protein to the DNA, however, has to be created by other means, e.g., bandshift experiments. *LacZ* reporters are also very useful in monitoring signaling activity. Signaling activity can be measured by using a *lacZ* fusion to a downstream gene activated transcriptionally. Multiple reporters exist that monitor the output from various signal transduction pathways a particular gene or mutant might control (e.g., *FUS1-lacZ* for output from the mating pathway,[9] *HO-lacZ* for mating type switch,[4] *FLO11-lacZ* for output through the filamentous growth pathways,[7] *IME1-lacZ* for the expression of meiosis-specific genes,[10] *GAP1-lacZ* for monitoring nitrogen metabolism,[11] *GCN4-lacZ* for monitoring general control,[12] and many more as described in TRIPLES[13]).

LacZ assays are especially valuable for all kinds of screens, of which only a few are mentioned in the following sections of this chapter. *LacZ* assays have been used to screen for regulated genes on a genome-wide level in a cost-effective manner and are an effective way to dissect complex signal transduction pathways. Thus, *lacZ* assays are a valuable tool to identify the players in the biological process of interest and define the specific experiments to be done out of a plethora of possibilities.

No doubt, measuring the actual message of the gene (using Northern blots for example) will give the most accurate information about the regulation of the gene. Effects due to mRNA stability or posttranscriptional regulation can never be accounted for using any kind of reporter. Thus, if the goal is to describe the regulation of one specific gene under various conditions, Northern blots might give more accurate and reliable results.

Tools to Generate Galactosidase Fusions

Many *lacZ* fusions already exist in the yeast community. However, if the one that best suits a particular purpose is not available, several possibilities for

[9] J. Trueheart, J. D. Boeke, and G. R. Fink, *Mol. Cell. Biol.* **7,** 2316 (1987).

[10] M. Shefer-Vaida, A. Sherman, T. Ashkenazi, K. Robzyk, and Y. Kassir, *Dev. Genet.* **16,** 219 (1995).

[11] M. Stanbrough and B. Magasanik, *J. Bacteriol.* **177,** 94 (1995).

[12] A. G. Hinnebusch, *Proc. Natl. Acad. Sci. U.S.A.* **81,** 6442 (1984).

[13] A. Kumar, K. H. Cheung, P. Ross-Macdonald, P. S. Coelho, P. Miller, and M. Snyder, *Nucleic Acids Res.* **28,** 81 (2000).

constructing *lacZ* fusions exist. If one wants to look at the regulation of a specific gene, one must place *lacZ* under the control of the entire promoter of this gene. This can be done by cloning "your favorite promoter" (YFP) into available vector systems that allow it to be hooked up to *lacZ*. Another possibility is to use PCR-mediated integration of *lacZ* into the genome, under the control of the endogenous promoter.

Two types of vector systems for the cloning of YFP to *lacZ* can be distinguished. One class of vectors only provides the galactosidase gene for fusion to YFP, requiring a core promoter (provided usually by YFP) and an in-frame ATG for β-galactosidase expression. Myers *et al.*[14] published a set of vectors, available both as high copy and as integrative plasmids. These vectors provide a polylinker in all three frames for convenient cloning and come with *URA3* or *LEU2* as selection marker (GenBank accession number for YEp353:U03500). A CEN version of one of these vectors has also been described.[6] The *lacZ* gene in these vectors lacks the first eight codons of *lacZ*.

The second class of vectors provides a functional core promoter, but is deprived of a UAS or a URS. This type of vector can be used to introduce small pieces of promoter sequences with potential UAS or URS activities. The potential UAS/URS can now be investigated in an isolated context. Most of the plasmids used for cloning of UAS elements are derived from plasmid pLG670Z.[15,16] This high copy plasmid (derived from YEp24) contains a unique *Xho*I site for cloning of any potential UAS into an UAS-less *CYC1* core promoter (for detailed analysis of the promoter see Ref. 17.) The complete sequence of this vector (pLG670Z) can be found at http://www.sacs.ucsf.edu/home/HerskowitzLab/protocols/cyc1.html. Several CEN versions have been derived from this plasmid.[8] pLG670Z also allows in frame fusions with *lacZ* using a unique *Bam*HI site, spanning the third and fourth codon of *lacZ*.

A rapid method for precise integration of *lacZ* into the genome (which works also with plasmids) can be designed using modules for PCR-based insertion.[18] The integration into the genome is mediated by primer sequences homologous to the locus where integration should occur. The second part of the primer is used to amplify the *lacZ* module from a plasmid. These modules are based on a heterologous selection marker, *kanMX,* giving a high rate of correct insertions. The *lacZ* activity now reflects the promoter activity of YFG in the context of the original chromosomal environment.

Centromeric and integrative plasmids based on an UAS-less *MEL1* core promoter using both α-galactosidase and β-galactosidase as reporter enzymes have

[14] A. M. Myers, A. Tzagoloff, D. M. Kinney, and C. J. Lusty, *Gene* **45,** 299 (1986).
[15] L. Guarente and M. Ptashne, *Proc. Natl. Acad. Sci. U.S.A.* **78,** 2199 (1981).
[16] L. Guarente, *Methods Enzymol.* **101,** 181 (1983).
[17] J. Chen, M. Ding, and D. S. Pederson, *Proc. Natl. Acad. Sci. U.S.A.* **91,** 11909 (1994).
[18] A. Wach, A. Brachat, R. Pohlmann, and P. Philippsen, *Yeast* **10,** 1793 (1994).

been reported.[19] α-Galactosidase is encoded by an endogenous yeast gene (*MEL1*) that is not transcribed in S288c or W303 backgrounds. An easy test to see if the yeast strain expresses *MEL1* is to check if it grows on melibiose as the sole carbon source. Strains lacking *MEL1* activity do not grow on melibiose and thus have no intrinsic α-galactosidase activity. Because the enzyme is cell wall associated,[20] the enzymatic activity can be measured directly on plates using living cells.[21] Although Mel1p is mainly cell wall associated, it is secreted to the media with increasing cultivation time. In stationary cells up to 50% of the activity actually is secreted.[20] Thus this reporter is especially suited for colony assays on plates, but less convenient for liquid assays (see below). β-Galactosidase from *E. coli,* however, is not secreted, making large screens directly on plates less convenient. The plasmids described by Melcher *et al.*[19] have been optimized for minimal background activities on different carbon sources. They are available from Euroscarf (http://www.rz.uni-frankfurt.de/FB/fb16/mikro/euroscarf/data/km_rep.html). Clontech (Heidelberg, Germany) also offers α-galactosidase reporter systems, designed for detection of two-hybrid interactions. A *lacZ* reporter that can also be used for the medically important yeast *Candida albicans* has been described by Uhl and Johnson.[22] The reporter plasmids described employ the β-galactosidase gene from *Streptococcus thermophilus*. The β-galactosidase gene from *S. thermophilus* has a distinct codon usage from the *E. coli* gene and thus can be efficiently expressed both in *Candida albicans* and *S. cerevisiae*.

Examples for Use of Galactosidase Reporters

Dissection of Complex Promoters

The first step in dissecting the promoter of interest usually is the cloning of the potential promoter (including the core promoter) in front of *lacZ*. This enables measurement of the activity of the entire promoter under various conditions and provides the basic tool from which to start a promoter dissection.

Several approaches have been used to dissect promoters. In *S. cerevisiae* all intergenic regions—at least for the sequenced S288c background—are known. There are only 10 intergenic regions (potential promoter sequences) that are between 3.6 kbp and 2.6 kbp,[7] whereas most of the potential promoter sequences are within the range of $\sim$1 kb. This makes a systematic PCR-based approach constructing multiple ordered deletions and an overlapping set of potential UAS regions a feasible approach for all yeast promoters. For this purpose the entire intergenic region upstream of the ATG should be amplified and introduced into

[19] K. Melcher, B. Sharma, W. V. Ding, and M. Nolden, *Gene* **247,** 53 (2000).
[20] P. S. Lazo, A. G. Ochoa, and S. Gascon, *Eur. J. Biochem.* **77,** 375 (1977).
[21] R. S. Tubb and P. L. Liljeström, *J. Inst. Brew.* **92,** 588 (1986).
[22] M. A. Uhl and A. D. Johnson, *Microbiology* **147,** 1189 (2001).

an appropriate *lacZ/MEL1* vector, e.g., YEp355.[14] This will be the basic tool with which to characterize the entire promoter and start its dissection. Sequential deletions at any point within the potential promoter sequences can be created using a primer overlap method.[23] Two primers have to be designed that consist of about 40 nucleotides, of which the first 20 nucleotides hybridize next to the 3′ end of the desired deletion and the second 20 nucleotides hybridize next to the 5′ end of the desired deletion. These two 40-mers hybridize perfectly to each other. Using these primers the PCR creates overlapping 40-bp overhangs omitting the part of the promoter to be deleted. It is important to use an enzyme that does not displace the primer hybridized to the template or adds additional nucleotides such as *Taq* (we use *Pfu,* 15 cycles, with 3 min extension per kb, ~10 ng template). The template has to be removed after PCR by digestion with *Dpn*I, a restriction endonuclease (4-cutter) that cleaves *dam* methylated DNA only (implying that the template has to be from a *dam*-positive *E. coli*) but leaves the PCR products intact. The *Dpn*I digested PCR products can be transformed directly into *E. coli,* which will ligate the overlapping overhangs *in vivo*. To facilitate the PCR we performed the mutagenesis in a smaller plasmid (Bluescript KS, Stratagene, La Jolla, CA) and cloned the promoter after mutagenesis into YEp355. Thus, using two specifically designed 40-mers can create a deletion at any point if a circular plasmid is used as a template.

Short promoter segments containing potential UAS or URS sequences are created by a primer set spread across the promoter region, including the restriction sites needed for cloning into the appropriate vector (e.g., *Xho*I sites for cloning into pLG670Z). The smaller the size of the deletions and UAS/URS segments, the more informative the analysis will be. Depending on the size of the potential promoter, a compromise will have to be found between the size of the fragments and a manageable number of constructs. Using this strategy to dissect the *FLO11* promoter—which regulates a gene essential for invasion and pseudohyphal growth in *S. cerevisiae*—200-bp deletions were inserted over 3 kb of intergenic region upstream of the ATG, resulting in 14 deletion constructs. Furthermore, 14 individual promoter elements of 400 bp length overlapping by 200 bp were cloned in front of *lacZ*. This set of promoter deletion vectors and UAS/URS vectors was transformed into wild-type and various mutant strains to detect regulatory elements under various conditions.[7] It is advisable to test several independent clones for each of the PCR-generated plasmids for reproducible results, using a filter assay (see below), and to sequence across the deletions introduced. After verification of the constructs one can proceed with an exact quantitation (see below). For larger amounts of constructs and conditions (e.g., different media, growth phases, mutated strains), the quantitative assay can be performed in a 96-well format using ELISA (enzyme-linked immunosorbent assay) readers (see below).

[23] R. Higuchi, *in* "PCR Technology" (H. Erlich, ed.). Stockton Press, New York, 1989.

Another common method for promoter dissections is the use of sequential deletions from the 3′ end as can be obtained using exonucleases or available restriction sites. By analyzing the promoter fragments generated, the different UAS/URS elements can be determined.[6,24] However, this method restricts analysis to those parts of the promoter not cut off. Convenient restriction sites for internal deletions or cloning of fragments in front of *lacZ* are not always available to avoid this limitation.

Promoter Activation Screening

Transcriptional changes in yeast are induced by environmental changes (e.g., heat shock, starvation, pheromone). To identify the signaling pathways responding to the environment defined reporters as well as randomly created *lacZ* fusions can be used. The idea of cloning regulated yeast genes from a pool of randomly created *lacZ* fusions was realized shortly after the first use of *lacZ* as a reporter gene in yeast.[25] Plasmid-based screens[26] and integrative libraries have been developed that enable the user to perform large-scale screens with the purpose to identify all the genes regulated by a specific signal in *S. cerevisiae* and to determine the subcellular locations of the fusion proteins[2,27] (http://ygac.med.yale.edu/). The integrative libraries have been constructed by mutagenizing a yeast genomic library in *E. coli* using transposons. The transposon is carrying both the *lacZ* gene and a selectable marker (*LEU2* or *URA3* versions exist) disrupting the yeast genomic DNA. This library can be integrated into the yeast genome. Within the pool of transformants, functional in-frame *lacZ* fusions are present that can be screened for regulation under the specific conditions of interest (e.g., by replica plating from rich medium to starvation media). These fusion proteins can also be used for localization of the respective protein using immunolocalization. The regulation of essential genes can be investigated in diploid cells only. These libraries also have been used as a mutagen, leaving tagged disruptions in the genome (for detailed description see Ref. 27a). Several screens using this method have been published. For example, genes regulated by α-factor as well as genes involved in cell wall architecture or pseudohyphal growth have been identified using this approach.[28–30] A large

[24] M. G. Lambrechts, I. S. Pretorius, V. S. D'Aguanno, P. Sollitti, and J. Marmur, *Gene* **146,** 137 (1994).

[25] S. W. Ruby, J. W. Szostak, and A. W. Murray, *Methods Enzymol.* **101,** 253 (1983).

[26] J. G. Coe, L. E. Murray, C. J. Kennedy, and I. W. Dawes, *Mol. Microbiol.* **6,** 75 (1992).

[27] P. Ross-Macdonald, P. S. Coelho, T. Roemer, S. Agarwal, A. Kumar, R. Jansen, K. H. Cheung, A. Sheehan, D. Symoniatis, L. Umansky, M. Heidtman, F. K. Nelson, H. Iwasaki, K. Hager, M. Gerstein, P. Miller, G. S. Roeder, and M. Snyder, *Nature* **402,** 413 (1999) [see comments].

[27a] A. Kumar, S. Vidan, and M. Snyder, *Methods Enzymol.* **350,** [12], 2002 (this volume).

[28] H. U. Mosch and G. R. Fink, *Genetics* **145,** 671 (1997).

[29] M. Lussier, A. M. White, J. Sheraton, T. di Paolo, J. Treadwell, S. B. Southard, C. I. Horenstein, J. Chen-Weiner, A. F. Ram, J. C. Kapteyn, T. W. Roemer, D. H. Vo, D. C. Bondoc, J. Hall, W. W. Zhong, A. M. Sdicu, J. Davies, F. M. Klis, P. W. Robbins, and H. Bussey, *Genetics* **147,** 435 (1997).

[30] S. Erdman, L. Lin, M. Malczynski, and M. Snyder, *J. Cell. Biol.* **140,** 461 (1998).

collection of strains with defined in-frame *lacZ* insertions and several libraries exist and are available to the public through the database TRIPLES maintained at the Yale Genome Analysis Center (http:ycmi.med.yale.edu/ygac/triples.html).[13]

Identification of Upstream Regulators in a One-Hybrid-like Manner

The different *n*-hybrid systems are described in detail elsewhere in this volume.[30a] Here I would like to discuss the use of single reporter genes in a one-hybrid-like manner. The one-hybrid system was designed to identify proteins that bind to DNA by fusion of a cDNA library to a transcription–activation domain. The bait is the reporter that consists of a set of multimerized *cis*-acting sequences that will be able to bind the *trans*-acting factor, resulting in transcription of the reporter.[31,32]

Of course this approach also works the other way around. To determine the transcriptional activity of a protein one must create a fusion of the protein to a DNA binding domain, such as *lexA* or *GAL4,* and monitor the activity of a reporter plasmid carrying the recognition sequence of the respective DNA binding domain and *lacZ* as a reporter. If the protein has a transcriptional activation domain, induction of the reporter will occur. The activation domain can be mapped further by partial deletions or allelic screens.[33,34] The large amounts of constructs generated can be conveniently tested using filter assays. If α-galactosidase is used as a reporter the assay can be performed *in situ* on plates (see below).

Galactosidase reporters are also of great use to identify proteins regulating the gene(s). This can be done by mutagenizing a reporter strain to interrupt signaling, or by transformation of a reporter strain using high-copy libraries to enhance signaling. We used this system to screen for regulators of *FLO11* (see later section "General Remarks for Screens"). For this purpose the reporter construct (e.g., *FLO11-lacZ*) should be integrated into the genome to reduce the variability due to copy number and recombination effects that occur with plasmids. For investigating the entire promoter, direct replacement of the open reading frame (ORF) with *lacZ* is the method of choice.[18] Smaller segments of promoter can be ligated into a vector providing a core promoter (e.g., *MEL1* or *CYC1*) which is integrated into the genome (e.g., at the chromosomal locus of the auxotrophic marker of the vector) to examine these segments independently. Mutagenesis or overexpression screens using this reporter strain can reveal proteins that induce or repress the promoter of YFG in a direct or indirect way. This approach is not limited to *S. cerevisiae*. Similar developmental processes in other organisms, such as hyphal development in *Candida albicans,* can be investigated in *S. cerevisiae*. Screening of *C. albicans*

[30a] J. F. Gera, T. R. Hazbun, and S. Fields, *Methods Enzymol.* **350,** [28], 2002 (this volume).
[31] M. M. Wang and R. R. Reed, *Nature* **364,** 121 (1993).
[32] J. J. Li and I. Herskowitz, *Science* **262,** 1870 (1993) [see comments].
[33] H. E. Smith, S. E. Driscoll, R. A. Sia, H. E. Yuan, and A. P. Mitchell, *Genetics* **133,** 775 (1993).
[34] S. Mandel, K. Robzyk, and Y. Kassir, *Dev. Genet.* **15,** 139 (1994).

high-copy libraries in *S. cerevisiae* carrying a *FLO11-lacZ* reporter revealed several transcription factors that are important for hyphal development in *C. albicans* (e.g., *CaTEC1, CaMCM1, EFG1, CPH1*[35,36]; S. Rupp, unpublished, 2001 (Figs. 1–3).

Protein Stability Assays

LacZ reporters can also be used to measure protein stability. Since the protein degradation rate of the fusion protein affects the amount of active enzyme present, quantification of the enzymatic activity will reveal components involved in the degradation machinery. For this purpose genetic screens have been designed to measure protein stability by fusion of proteolytically regulated proteins to β-galactosidase. For example, selection for mutations that stabilize *CLN1*-β-galactosidase fusion proteins revealed regulatory mechanisms of cyclin-controlled G_1 cell division.[37] Furthermore, the *N*-end rule has been established using β-galactosidase fusion proteins.[3]

Assays

Several types of assays can be performed measuring galactosidase activities in yeast. Assays for quantification of β-galactosidase use substrates that release soluble chromophores such as ONPG (*o*-nitrophenyl-β-D-galactopyranoside) (Sigma, St. Louis, MO; Glycosynth, Warrington, UK), whereas assays for screening procedures use precipitable water-insoluble chromophores such as X-Gal (5-bromo-4-chloro-3-indolyl-β-D-galactopyranoside) (Sigma, Glycosynth).[38] For α-galactosidase the same type of substrates exists as the respective α-galactopyranosides[39,40] (X-α-Gal can be purchased from Glycosynth; http://www.glycosynth.co.uk/). Instead of *o*-nitrophenyl-α-D-galactopyranoside (α-ONPG), *p*-nitrophenyl-α-D-galactopy-ranoside (α-PNPG) should be used for quantitative measurements. It is less expensive and has a higher extinction coefficient (ε), enhancing the sensitivity of the assay (Sigma, Glycosynth).

All assays described here can be performed both with α-galactosidase and β-galactosidase. The two enzymes differ in pH optima (for α-galactosidase pH $\sim$ 4, for β-galactosidase pH $\sim$ 7), so different buffers have to be used. I will describe the β-galactosidase assays and point out the differences when α-galactosidase is used.

[35] A. Schweizer, S. Rupp, B. N. Taylor, M. Rollinghoff, and K. Schroppel, *Mol. Microbiol.* **38,** 435 (2000).

[36] J. F. Ernst, *Microbiology* **146,** 1763 (2000).

[37] Y. Barral, S. Jentsch, and C. Mann, *Genes Dev.* **9,** 399 (1995).

[38] J. H. Miller, "Experiments in Molecular Genetics." Cold Spring Harbor Laboratory Press, Cold Spring Harbor, NY, 1972.

[39] S. Aho, A. Arffman, T. Pummi, and J. Uitto, *Anal. Biochem.* **253,** 270 (1997).

[40] M. P. Ryan, R. Jones, and R. H. Morse, *Mol. Cell. Biol.* **18,** 1774 (1998).

Qualitative Assays

Qualitative assays are fast and robust methods to get a general idea about expression levels of YFG (s) under the conditions desired. They are especially suitable for any of the screens described above. If α-galactosidase is used as a reporter, lysis of the cells is not required. The most relevant assay for this reporter is described below in "Growing Yeast on X-Gal Plates."

General Remarks for Screens

For all kinds of screening procedures it is important to have positive and negative controls for comparison. Furthermore, it is of utmost importance to carefully define the conditions used for the screen. This will determine the specificity of the screen.

Because the different assays described have different sensitivities, we select the best assay for the screen by testing all assays in a pilot screen with a few plates. The assay that shows the strongest differences between the controls in a reasonable amount of time is the one to choose. Figures 1–3 show an example of a screen looking for *Candida* inducers of *FLO11-lacZ* to identify genes relevant for hyphal development in *Candida*. In this case the filter assay gives better results (Fig. 1). In a first step the reporter strain containing an integrated *FLO11-lacZ* reporter

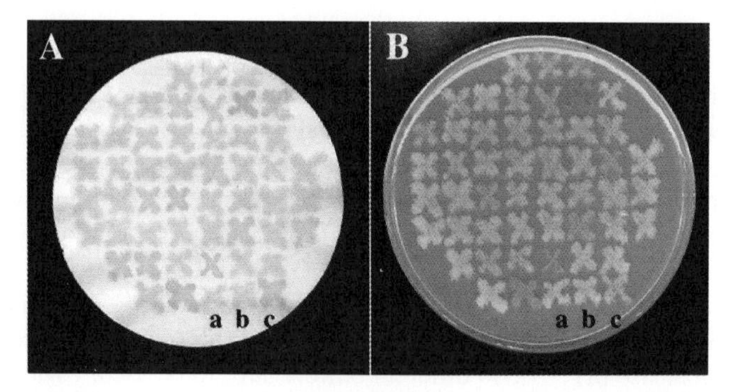

FIG. 1. Comparison of filter assay (A) and overlay assay (B) of replica plated colonies containing a reporter (*MAT* **a**/α *FLO11::lacZ/FLO11 ura3-52/ura3-52*) and *Candida albicans* library plasmids (2 μ, *URA3*). The colonies have been grown for 24 hr after replica plating at 30°. The filter assay (A) was stopped after 80 min incubation at 30° by soaking in 1 *M* Na_2CO_3 and dried. The overlay assay was incubated for 43 hr at 30° and photographed. The controls are: (a) *MAT* **a**/α *FLO11/FLO11 ura3-52/ura3-52* (no reporter), (b) *MAT* **a**/α *FLO11::lacZ/FLO11 ura3-52/ura3-52* pRS426 (WT situation), and (c) *MAT* **a**/α *FLO11::lacZ/FLO11 ura3-52/ura3-52* pRS426-FLO8 (inducer of Flo11p). Blue spots indicate high *FLO11-lacZ* activity; white-yellow indicates no activity. Comparison of both plates shows that the filter assay allows detection of about twice as many colonies with induced *FLO11-lacZ* activities (23) than does the overlay assay (10). Furthermore, the filter assay is 30 times faster than the overlay assay. Screening for *Candida* regulators of *FLO11-lacZ* thus was performed using the filter assay.

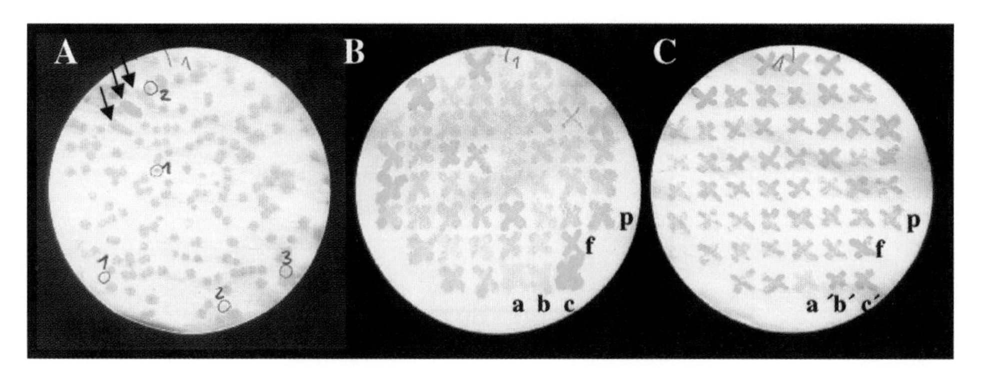

FIG. 2. Screening for *Candida* regulators of *FLO11-lacZ* using the filter assay. (A) One of 50 plates containing the reporter strain (*MAT* **a**/*α* *FLO11::lacZ/FLO11 ura3-52/ura3-52*) transformed with a genomic *C. albicans* library. Red numbers indicate putative inducers of *FLO11-lacZ;* black numbers indicate putative repressors of *FLO11-lacZ.* The controls, visible as bars (indicated by arrows), have been placed on the plate after the transformants have been grown for 2 days at 30°. Replica plating onto filters was done 24 hr later. The cells were grown on filters for another 24 hr at 30° and processed as described. (B) Colonies picked as putative inducers of *FLO11-lacZ.* (C) The same colonies cured for the library plasmid (by replica plating to 5-FOA). "False positives" show the same color in (C) as in (B). One example is marked with f. For f the induction observed is not dependent on the plasmid. Induction depending on the plasmid is visible in the patches marked with p or c. The induction observed in (B) is reduced to WT levels in (C) (compare b and c or b and p on filters (B) and (C). (a) *MAT* **a**/*α* *FLO11/FLO11 ura3-52/ura3-52* (no reporter), (b) *MAT* **a**/*α* *FLO11::lacZ/FLO11 ura3-52/ura3-52* pRS426 (WT situation), and (c) *MAT* **a**/*α* *FLO11::lacZ/FLO11 ura3-52/ura3-52* pRS426-*FLO8* (inducer of Flo11p). f, False positive; p, positive; in (C) plasmids have been cured. Blue spots indicate high *FLO11-lacZ* activity; white-yellow indicates no activity.

is transformed with a *URA3*-based *Candida* library and replica plated onto filters (Fig. 2A). The colonies that show the desired results are transferred to a second plate (Fig. 2B). To test if the result is dependent on the library plasmid, the plasmid is eliminated by replica plating onto 5-fluoroorotic acid (5-FOA). The colonies containing the plasmid and the colonies cured are retested and the "false positives" are eliminated (Fig. 2C). The library plasmids from the colonies that have been verified during this process will be isolated and retransformed into the starting strain. If the transformants reproduce the original result, the desired clone has been isolated (Fig. 3A). One of the ORFs on the plasmid will be responsible for the induction observed. After subcloning of the ORF, further analysis, e.g., by quantitation of *lacZ* activities as described below (Fig. 3B), can occur.

Yeast Filter Galactosidase Assay

This method is especially suitable for large numbers of qualitative assays and screens, e.g., promoter activation screens or screening for upstream regulators of

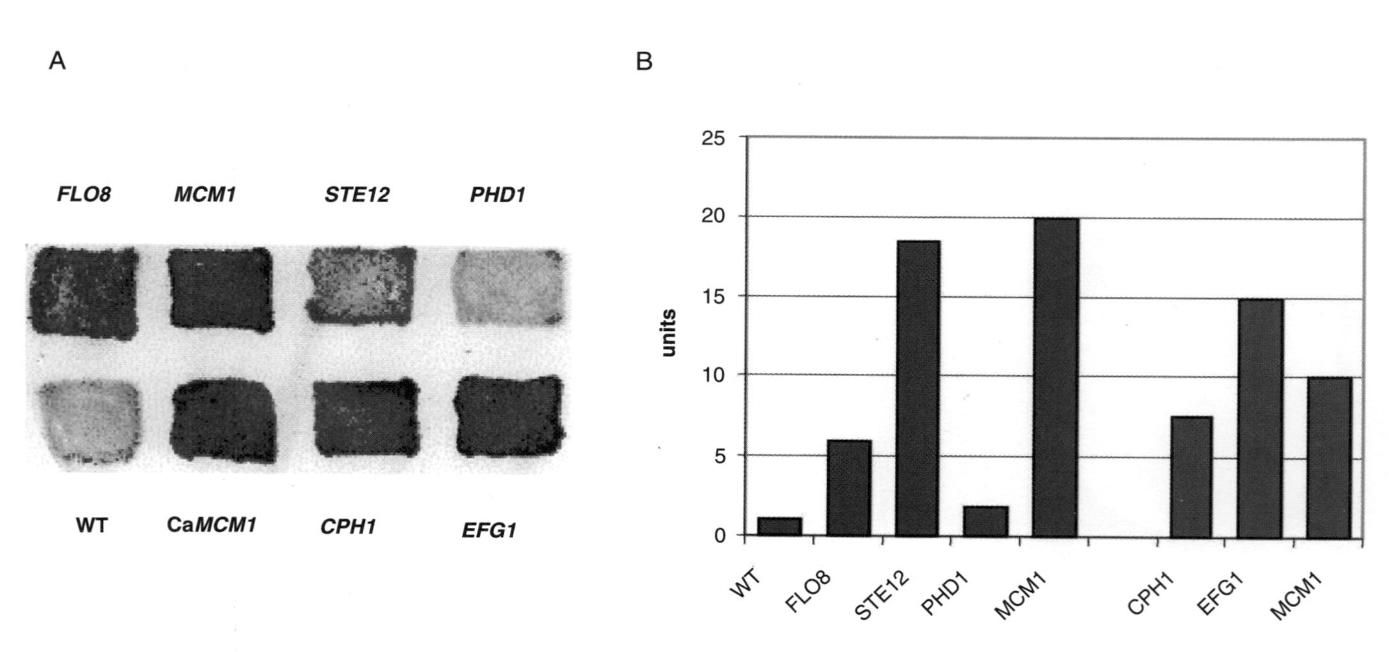

FIG. 3. Comparison of qualitative filter assay with quantitative β-galactosidase assay using crude extracts. (A) Patches of the reporter strain (*MAT* **a**/α *FLO11::lacZ/FLO11 ura3-52/ura3-52*) containing the isolated *Candida* genes identified in the screen in Fig. 2 on a 2μ plasmid and compared to the homolog genes from *S. cerevisiae,* also present on a 2μ plasmid. (B) Quantification of *FLO11-lacZ* levels in the strains shown in (A) in protein extracts using ONPG as the substrate. The numbers are normalized to WT (set to 1).

YFG. The protocol is a modification of protocols described[5,41] and modified by C. Styles (unpublished, 1990). The yeast cells are grown directly on filter paper that is placed on top of the agar plate containing the desired medium. The filter can easily be removed after the cells have been grown in order to perform the assay. The best way to transfer the yeast colonies to a filter paper is to replica plate them from a master plate to the filter-coated plate(s), representing the selection conditions of the screen. The filter paper has to be able to pick up cells from the velveteen of the replica. Thus it has to have an absorbent surface. We use Whatman #50 (VWR, West Chester, PA) or Schleicher and Schuell (Keene, NH) #576, custom cut to 8.3 cm in diameter. Autoclave the filter paper. Before being placed on top of the agar the filter papers should be marked with pencil. It is important to remove all air bubbles between the paper and the agar surface, e.g., with a sterile spreader, to ensure proper growth of the yeast colonies. After replication the plates can be incubated at the desired temperature anywhere from 2 hr to several days. After the cells have been grown, the filters are removed from the plates and the cells are lysed by freezing in liquid nitrogen for at least 1 min. To perform the enzymatic assay the frozen filters are thawed (on paper), yeast facing up, and placed on top of a second filter (we use Whatman #3 or Schleicher and Schuell #593) in a petri dish, containing 2.0 ml of the reaction buffer (Z-buffer). Again the yeast has to be facing up. The second filter is soaked with the reaction buffer to avoid spilling and blurring of colonies. Air bubbles between the filters should be avoided, to obtain even staining. Expression of β-galactosidase will result in a blue precipitate coloring the yeast cells. The reaction should be carried out at 30° or 37° and can be as fast as 20 min, but may be extended until the controls show β-galactosidase activity. Incubations longer than 1 hr should be performed in a sealed bag to avoid evaporation of the buffer. The reaction should be stopped before individual colonies are saturated (test several time points). Otherwise the differences between strains tested can become less obvious. To stop the reaction the test filters are removed, soaked with 0.5 ml 1 *M* Na_2CO_3 for at least 10 min in the lid of the petri dish, and dried.

The individual colonies showing the desired result should be recovered from a master plate. However, it is possible to isolate colonies directly from the filter, before stopping the reaction with Na_2CO_3.

Buffers and Reagents

5× Z Buffer: For 1000 ml add:

27.5 g $NaH_2PO_4 \cdot xH_2O$ (final concentration 1×: 60 m*M*)
42.5 g Na_2HPO_4 anhydrous (final concentration 1×: 40 m*M*)
3.75 g KCl (final concentration 1×: 10 m*M*)

[41] T. Durfee, K. Becherer, P. L. Chen, S. H. Yeh, A. E. Kilburn, W. H. Lee, and S. J. Elledge, *Genes Dev.* **7,** 555 (1993).

0.6 g $MgSO_4$ (final concentration 1×: 1 m*M*)
100 μl Triton X-100

pH has to be adjusted to pH 7.0 with NaOH or HCl if necessary.

Add 30 μl of a 2% X-Gal solution [20 mg/ml in dimethylformamide (DMF) or dimethyl sulfoxide (DMSO)] to 2.0 ml of 1× Z buffer just before use. The X-Gal stock solution has to be kept at −20°, protected from light. For densely packed plates and highly expressing reporters as shown in Fig. 2B,C we use twice the amount of X-Gal to obtain a better contrast of the colored yeast.

Strains mutant in *ADE2* develop a red color that interferes with the detection of X-Gal staining. To avoid this effect any test media should contain 0.6 m*M* adenine (100 mg adenine hydrochloride/liter, twice the standard amount).

To enhance the sensitivity of the filter test, a modification has been described by the Snyder Laboratory (http://ygac.med.yale.edu/mtn/protocol.stm):

Filters are lifted from the plates and placed in the lid of a 9-cm glass petri dish. This lid is then placed inside a closed 15-cm glass petri dish containing chloroform for 10 to 30 min. The minimum exposure time necessary for a particular yeast strain can be determined empirically. Filters are placed colony-side up onto X-Gal plates (120 μg/ml X-Gal, 0.1 *M* $NaPO_4$ [pH 7], and 1 m*M* $MgSO_4$ in 1.6% agar) and incubated at 30° for up to 2 days. These plates can be very thin; their use increases the signal over that obtained by simply soaking the filters in a buffered X-Gal solution.

X-Gal Overlay

This assay is based on a protocol by Barral *et al.*[37] Cells are grown on an agar plate. The assay is started by adding an overlay agar containing the reaction buffer, including X-Gal and solvents to lyse the cells. This assay is less sensitive than the filter assay. This could be due to a reduced amount of cell lysis as compared to the liquid nitrogen treatment. A direct comparison of filter assay and overlay assay is shown in Fig. 1.

Use 10–15 ml overlay agar per plate (9.0 cm diameter), composed of:

0.5 *M* KP_i pH 7.0 (phosphate buffer)
6% DMF
0.1% SDS
0.1 mg/ml X-Gal
0.5% Agarose [Seakam LE agarose (BMA, Rockland, ME) for DNA electrophoresis]

Boil the agarose in phosphate buffer until melted. Keep at 60° and add DMF, sodium dodecyl sulfate (SDS), and X-Gal (20 mg/ml in DMF). Pour at 60° fast, but—if possible—without disturbing the colonies. The overlay agar has to be poured hot; otherwise the cells are not lysed efficiently and are stained differently because of a temperature gradient on the plate. The plates are incubated until the controls show β-galactosidase activity. A word of caution has to be mentioned here. Because a significant part of the cells in a colony will survive the treatment, they might change their expression pattern after the agar is poured on top of them. Furthermore, hydrolysis of X-Gal also might occur by uptake of the substrate, which can vary in the different colonies on the plates. Thus extended incubation of the cells might lead to results not necessarily connected to the intention of the experiment, especially in large screens ("false positives"). Usually cells can be recovered from the assay plate; however, backup of the colonies using a master plate is strongly recommended.

A modification of this assay has been described,[42] enhancing the sensitivity of the assay. This protocol uses chloroform to lyse the cell prior to adding the overlay agar. Here the plates with the colonies are flooded with chloroform and incubated for 5 min in a hood (chloroform is harmful when inhaled; the plastic of petri dishes will be dissolved slowly by chloroform). After discarding the chloroform and drying of the plates an overlay agar composed of 1% agarose, 1 mg/ml X-Gal, and 100 mM KP_i, pH 7.0, is poured at 42° over the colonies. The plates are incubated at 30° until the controls show β-galactosidase activity.

Growing Yeast on X-Gal Plates

This assay is the easiest to perform because the cells are directly grown on the assay plate. For this type of assay α-galactosidase is the reporter to use, because α-galactosidase is a secreted enzyme. Furthermore, the pH optimum of the enzyme is around pH 4. Thus the enzyme is active on most unbuffered media used (the pH of SC media is ~pH 5). Yeast acidifies the media while growing, further optimizing the pH for α-galactosidase. Thus the media do not need to be buffered, avoiding interference of the pH of the plate with the experiment.

If α-galactosidase is used as a reporter, just add X-α-Gal (final concentration: 0.1 mg/ml X-Gal, stock solution 20 mg/ml in DMF) to the molten agar/media at 60° and pour the plates, or spread 100 μl of the X-α-Gal stock solution evenly onto the plate. Lysis is not required since 80% of the enzyme is cell wall associated or secreted; thus an overlay assay will not enhance the sensitivity of the assay significantly.

It is also possible to assay β-galactosidase in yeast cells grown directly on agar plates containing X-β-Gal. However, the plates have to be buffered at pH 7.0, the pH optimum of β-galactosidase. This might interfere with the desired growth conditions. Furthermore, it shows the least sensitivity of all assays presented, since β-galactosidase is not secreted. Thus the activity seen is dependent on the uptake of the substrate by the cells or on cells spontaneously lysing.

[42] H. M. Duttweiler, *Trends Genet.* **12,** 340 (1996).

X-β-Gal Plates. Buffer the agar/media to 50 m*M* KP_i pH 7.0 (final concentration), add X-Gal to a final concentration of 0.1 mg/ml (stock solution 20 mg/ml in DMF) at 60°, pour plates, and use right away.

Quantitative Assays

The quantitative *lacZ* assays described below are modified versions of the assays previously described.[16,43] Assays to determine α-galactosidase have been described as well.[19] Both assays can be performed using crude protein extracts[44] or permeabilized cells.[16] As outlined below, using crude extracts in general is more accurate, but using the permeabilized cell assay is less labor intensive. Quantitative *lacZ* assays based on the above-mentioned assays are also described in Ref. 45.

General Remarks

Accurate quantification of β-galactosidase activity requires the preparation of protein extracts from yeast cells. For this purpose the strains can be grown under any condition desired. However, a few points have to be considered. Plasmid effects could disturb the regulation of the gene especially if high copy plasmids are used, e.g., due to titration of regulators. Because of a possible background of the plasmid, always maintain the appropriate controls, e.g., the original plasmid without insert. If plasmid-based *lacZ/MEL1* reporters are used, media selecting for the plasmid are required. If that is not desired, integrate the constructs into the genome. Using standard protocols, cells are grown in rich media to exponential growth phase ($0.3 < OD_{600} < 1.0$). These cultures are easy to handle since they express low levels of proteases and yield high amounts of protein after breaking the cells. Furthermore, these cells are well defined with regard to growth phase. Different conditions sometimes require adjustments of the protocol. For example, cells grown beyond the diauxic shift, under conditions of nitrogen starvation, or cells scraped from plates usually express more proteases and are harder to break. This can be compensated for by using more cells and adding additional protease inhibitors. Most important, however, is that one can define precisely the growth phase of the cells at the point of harvesting to get reproducible results. A second method to quantify β-galactosidase is based on cell permeabilization. Using this method, the activity is normalized to the number of cells assayed (via OD). Using different strain backgrounds and different conditions might, however, give different OD values for the same cell numbers and thus lead to incorrect results. Clumpy strains cannot be used for this assay. Investigating only one strain under one condition with different constructs will give reliable results.

[43] M. Rose and D. Botstein, *Methods Enzymol.* **101,** 167 (1983).

[44] M. Rose, M. J. Casadaban, and D. Botstein, *Proc. Natl. Acad. Sci. U.S.A.* **78,** 2460 (1981).

[45] C. Kaiser, S. Michaelis, and A. Mitchell, "Methods in Yeast Genetics." Cold Spring Harbor Laboratory Press, Plainview, NY, 1994.

If α-galactosidase is used as a reporter, bear in mind that the enzyme is secreted. Thus, using crude protein extracts the results will always be lower than the activity present.

The assay should be performed at least in duplicate with at least two data points for each of the independent duplicate experiments. Disposable plastic cuvettes should be used for the Bradford assay[46] because of precipitates forming in the cuvettes. The assay can also be performed in 96-well plates using an ELISA reader for large amounts of samples (see below).

α-Galactosidase has been reported to be three times as sensitive as the comparable β-galactosidase reporter construct (due to the higher molar extinction coefficient of *p*-nitrophenol as compared to *o*-nitrophenol). Because of the low background activity described for these plasmids,[19] they are a valuable tool for detecting weak UAS activities.

Isolation of Total Protein

The standard protocol is derived from Rose and Botstein.[43]

1. Grow a 5–10 ml culture to an $OD_{600} < 1$.
2. Harvest the cells by centrifugation (2500*g*, 5 min, 4°) and keep on ice from this point on.
3. Discard the remaining media and transfer cells into 1.5-ml microfuge tubes using 0.5 ml ice-cold H_2O. Spin down (1 min, 13,000*g*), remove H_2O using an aspirator, and add 250 μl breaking buffer. The cells can be frozen at this point.
4. Break the cells by adding roughly the same amount of glass beads as the volume of cell pellet (100–200 μl equivalent). Vortex up to 10 times at top speed, 4°, for 1 min. Chill on ice for 1 min in between (a multivortexer in a coldroom can conveniently be used).
5. Clarify the extract by centrifugation at top speed for 10 min in a microfuge. Transfer supernatant to new tube. Usually the amount of protein obtained from this procedure is enough for the assays (from exponentially growing cells ~10 mg/ml protein can be expected; starved cells yield up to 10-fold less). Optional, you can extract the pellet with the glass beads again with another 250 μl of breaking buffer and pool for a more complete recovery of protein (this dilutes the sample).
6. Measure protein concentration, e.g., after Bradford.[46] For exponentially grown cells use 1 to 3 μl of the protein extract (dilute if necessary); for extracts from cells grown under limiting conditions 10–20 μl of extract might be required. To determine the protein concentration a standard curve using dilutions of bovine serum albumin (BSA) has to be used (between 1 and 20 mg/ml).

[46] M. M. Bradford, *Anal. Biochem.* **72,** 248 (1976).

Breaking Buffer

100 m*M* Tris-Cl (pH 8.0)
1 m*M* Dithiothreitol (DTT) [interferes with phenylmethylsulfonyl fluoride (PMSF)]
10% Glycerol
2 m*M* PMSF or 4-(2-Aminoethyl)benzenesulfonyl fluoride (AEBSF)

PMSF/AEBSF Stock Solution. Forty mM in isopropanol. Add 12.5 μl per 250 μl breaking buffer (final concentration 2 m*M*) just before breaking the cells.

Caution! PMSF is extremely destructive to mucous membranes of the respiratory tract, the eyes, and the skin. It may be fatal if inhaled, swallowed, or adsorbed through the skin. In case of contact, immediately flush eyes or skin with copious amounts of water and discard contaminated clothing.

PMSF and AEBSF [4-(2-aminoethyl)benzenesulfonyl fluoride hydrochloride, Perfabloc SC, Roche, Basel, Switzerland] are serine protease inhibitors. For standard assays PMSF is good enough. However, DTT reverses the inhibitory function of PMSF. DTT can be omitted from the breaking buffer if PMSF is used. PMSF is unstable at neutral and alkaline pH (half-life at pH 8.0 ~20 min), whereas AEBSF is more stable and nontoxic with comparable activity. Conditions inducing high protease levels, e.g., extended starvation, might require additional protease inhibitors (e.g., Complete protease inhibitor cocktail tablets, Roche).

Quantification of β-Galactosidase Activity

The amount of protein used for the assay has to be adjusted according to the activity of the promoter. Between 50 mg and 1000 mg total protein per assay are commonly used. The reaction can be performed in 1.5-ml microfuge tubes at 30° or 37°. For exact values the temperature during the assay has to be constant, meaning all components must be preincubated at the temperature at which the assay will be performed. Always carry a blank reaction without protein extract.

1. Add 5–100 μl of protein extract to Z-buffer, resulting in a total volume of 800 μl. Initiate the reaction by adding 200 μl of ONPG stock solution (4 mg/ml in Z-buffer) and record the time.
2. Incubate until the mixture has aquired a yellowish color.
3. Stop the reaction by adding 400 μl of 1 *M* Na_2CO_3 and record the time again.

Addition of Na_2CO_3 will inactivate the enzyme and alkalinize the solution, which is essential to obtain the deprotonated dye. The protonated dye is colorless! It is advisable to keep the time in between 30 min and 4–6 hr by modifying the amount of protein used and the temperature at which the assay is performed.

Measure the OD_{420} of the solution. The best accuracy is obtained if the reading is between 0.2 and 1.0.

5× *Z-Buffer.* For 1000 ml add:

27.5 g $NaH_2PO_4 \cdot xH_2O$ (final concentration 1×: 60 m*M*)
42.5 g Na_2HPO_4 anhydrous (final concentration 1×: 40 m*M*)
3.75 g KCl (final concentration 1×: 10 m*M*)
0.6 g $MgSO_4$ (final concentration 1×: 1 m*M*)

pH has to be adjusted to pH 7.0 with NaOH or HCl if necessary. Add 270 μl 2-mercaptoethanol per 100 ml 1× Z-buffer (50 m*M* final) just before use.

Express the specific activity of the extract according to the following formula:

$$OD_{420} \times 1.4/0.0045 \times \text{protein} \times \text{extract volume} \times \text{time}$$

OD_{420} is the optical density of the product *o*-nitrophenol at 420 nm. The factor 1.4 corrects for the reaction volume. 0.0045 is the optical density of a 1 nmol/ml solution of *o*-nitrophenol (molar extinction coefficient ε at 420 nm). Protein concentration is expressed as mg/ml. Specific activity is expressed as nmol/min/mg protein.

For quantification of α-galactosidase the same general remarks hold true. However, the assay has to be performed at pH 4.0.[19,40]

α-Galactosidase Incubation Buffer (Z-α-Buffer)

61 m*M* Citric acid
77 m*M* Na_2HPO_4

Adjust to pH 4.0 if necessary.

1. Add 3–20 μl of protein extract to 180–197 μl of Z-α-buffer.
2. Start the reaction by adding 200 μl of 7 m*M* *p*-nitrophenyl-α-D-galactopyranoside (2 mg/ml in Z-α-buffer) and incubate at 30°.
3. Remove 100 μl aliquots at intervals of 10–15 min and add to 900 μl 0.1 *M* Na_2CO_3.

p-Nitrophenol is colorless at pH 4.0. The yellow color will appear only on transfer to 0.1 *M* Na_2CO_3, after deprotonation of *p*-nitrophenol. Extinction is measured at 400 nm (OD_{400}).

Express the specific activity of the extract according to the following formula (see above):

$$OD_{400} \times 1.0/0.0182 \times \text{protein} \times \text{extract volume} \times \text{time}$$

0.0182 is the optical density of a 1 nmol/ml solution of *p*-nitrophenol (molar extinction coefficient ε at 400 nm). Protein concentration is expressed as mg/ml. Specific activity is expressed as nmol/min/mg protein.

Quantification Using ELISA Reader

Both the quantification of *o*- or *p*-nitrophenol and Bradford assay can be performed in a 96-well format using an ELISA reader. The general procedure follows the protocol as described above. Using 96-well plates that hold 300 μl per well the assay is composed of 150 μl Z-buffer and 10 μl protein extract (optimized for *FLO11-lacZ* activities).[7] Forty μl ONPG (4 mg/ml in Z-buffer) is used to start the assay; 100 μl 1 *M* Na_2CO_3 is used to stop the reaction, as soon as a yellowish color appears in most wells. Starting and stopping the reaction must be done using multichannel pipettes to avoid large time differences between the samples. To mix the samples gently pipette the mixture up and down a few times when adding ONPG or Na_2CO_3. The specific activity is calculated as described above.

Permeabilized Cell Assay

Here the assay is described for β-galactosidase only. However, this assay can be performed as well with α-galactosidase[40] if the Z-α-buffer, as described above is used and the extinction is measured at 400 nm.

1. Grow the cells as required and measure the OD_{600}. Harvest about 1×10^6–1×10^7 cells by centrifugation and discard the supernatant (~1 ml of exponentially growing cells). Keep cells and buffers on ice until starting the reaction.
2. Resuspend the cell pellet in 800 μl Z-buffer.
3. Add 60 μl of chloroform and 40 μl of 0.1% SDS and vortex at top speed for about 10 sec (~13,000*g*).
4. Preincubate the samples at 30° (approximately 5 min) and start the reaction by adding 200 μl ONPG (4 mg/ml in Z-buffer), record the time, and vortex briefly to stir up the cells.
5. Incubate the mixture until a yellowish color has developed.
6. Stop the reaction by adding 400 μl of 1 *M* Na_2CO_3 and record the time again.
7. Centrifuge for 10 min at top speed to pellet the cells and discard the pellet.
8. Measure the OD_{420} of the supernatant.

Units of β-galactosidase activity can be expressed as:

$$\text{Activity} = \text{OD}_{420}/\text{OD}_{600} \text{ of assayed culture} \times \text{volume assayed} \times \text{time}$$

In case of light scattering due to not completely cleared cell debris, correct for that using the following formula:

$$\text{Activity corrected} = \text{OD}_{420} - (\text{assay volume} \times \text{OD}_{550})/\text{OD}_{600} \text{ of assayed culture} \times \text{volume assayed} \times \text{time}$$

OD_{420} is absorbance by *o*-nitrophenol (+light scattering); OD_{600} is the optical density of the culture at the time of harvesting. OD_{550} is the light scattering by cell debris; volume assayed is the volume of culture used in the assay in ml; time is reaction time in min.

Acknowledgments

I thank Matthias Rottmann and Sonja Dieter for providing the figures and working on the protocols described in this article, and Henrike Lotz and Kai Sohn for comments on the manuscript. This work was supported in part by Grant #03121805 from the BMBF and Ru608/2-1 from the DFG.

[7] Analysis of Budding Patterns

By MATTHEW LORD, TRACY CHEN, ATSUSHI FUJITA, and JOHN CHANT

Introduction

It has long been known that *Saccharomyces* can bud and divide in spatially ordered patterns.[1] These patterns, determined by cell type and environmental conditions, serve as a paradigm for regulated morphogenetic differentiation. In the past decade or so, efforts to understand the molecular mechanisms underlying the production and regulation of these patterns have intensified. This chapter describes the most commonly employed methods for analyzing budding pattern, bud scar staining, the *ace2* colony assay, the use of pseudohyphal growth, and the microcolony method. Each has its own utility. The method chosen depends on the goal of one's experiment.

Vegetatively growing yeast can produce two patterns of bud-site selection (Fig. 1A). Haploids (**a** or α cells) exhibit the axial pattern. Diploids (**a**/α cells) exhibit the bipolar pattern. Budding pattern is determined by cell type rather than ploidy. For example, **a/a** or α/α diploids bud in the axial pattern. In the axial pattern, cells bud immediately adjacent to the previous site of division between mother and daughter (Fig. 1A).[1–3] In the bipolar pattern, cells bud from the poles of their ellipsoidal shapes.[1–3] Diploid cells choose poles in no particular order, although biases do exist. The strongest bias is that newborn daughters bud from the pole furthest from their mothers (the distal pole) (Fig. 1A).[3]

[1] D. Freifelder, *J. Bacteriol.* **80,** 567 (1960).

[2] J. B. Hicks, J. N. Strathern, and I. Herskowitz, *Genetics* **85,** 373 (1977).

[3] J. Chant and J. R. Pringle, *J. Cell Biol.* **129,** 751 (1995).

0076-6879/02 $35.00

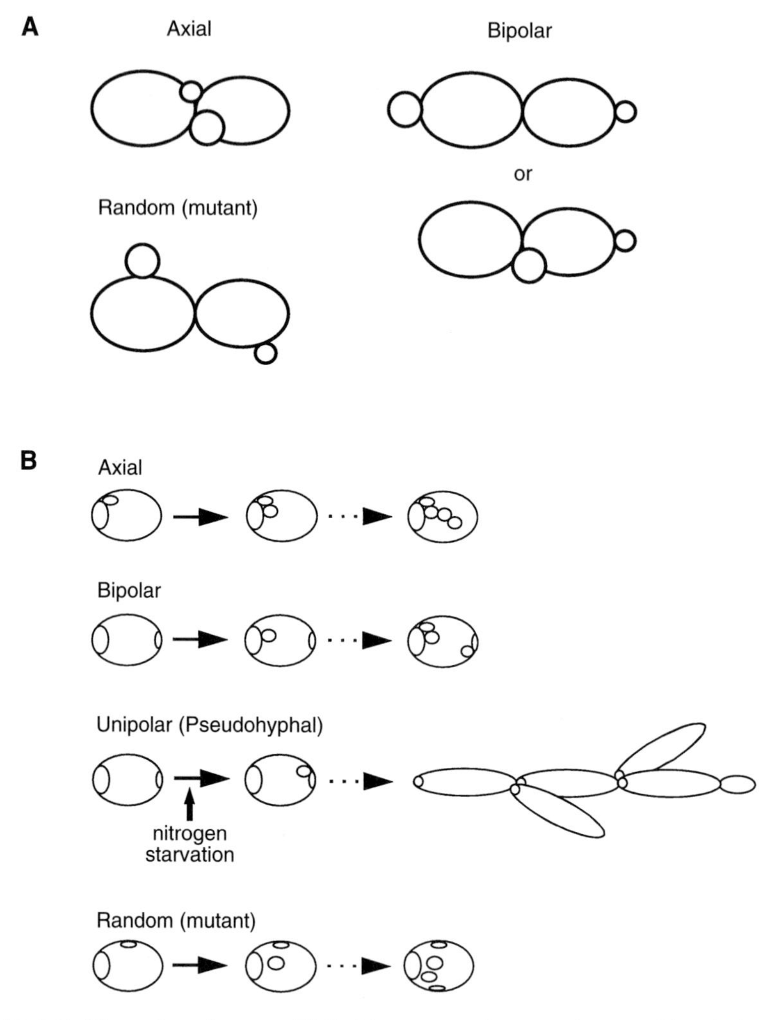

FIG. 1. Budding patterns in yeast. (A) Budding pattern of yeast at the two-cell/two-bud stage: the axial, the bipolar, and the random budding patterns. (B) Budding patterns as represented by the arrangement of division scars on the cell surface. Small rings represent bud scars. The larger ring, depicted at the left pole of each cell, represents the birth scar.

The most commonly employed method for analyzing budding pattern is the observation of patterns of division scars on the cell surface (Fig. 1B).[3] Every cell is born with a birth scar marking the former site of attachment to its mother. Each time a cell completes a round of budding, a bud scar remains on the cell surface. Because yeast possess a thick cell wall, these scars are permanent and serve as a historical record of a cell's budding pattern over many generations. These scars can also be used to assess a mother cell's age.[4] In the axial pattern, in which a cell buds adjacent to its most recent division site, bud scars form a connected chain starting at the birth scar (Fig. 1B). The path of this chain is essentially a random walk, except that cells cannot form buds on top of preexisting bud scars. A cluster, a line, or some intermediate pattern of scars can result. In the bipolar pattern, cells develop clusters of scars at their poles (Fig. 1B). The number of scars at each pole is variable because there is no set order for using one pole or the other. For a comprehensive description of these patterns, see Chant and Pringle.[3]

Much of the work surrounding the problem of budding pattern involves the use of mutants exhibiting specific alterations. Mutations in several genes (*RSR1/BUD1*, *BUD2*, or *BUD5*)[5–7] alter the axial and bipolar patterns to an undirected pattern that approaches random selection of sites (Figs. 1A,B). Mutations in other genes (*BUD3*, *BUD4*, *BUD10/AXL2/SRO4*, and *AXL1*)[6,8–10] convert haploids from budding in the axial pattern to budding in the bipolar pattern but have little to no effect on the bipolar pattern of diploids. Correspondingly, there are a number of mutations which affect the bipolar pattern without affecting the axial pattern. Mutations in *BUD8* and *BUD9* convert the bipolar pattern to unipolar patterns.[11] *BUD8* mutants choose buds exclusively from the birth scar pole. *BUD9* mutants choose buds from the distal pole only (Fig. 1B). Mutations in *RAX1* and *RAX2* convert the bipolar pattern to a mixture of random and axial.[12] A very large number of mutations simply randomize the bipolar pattern without affecting the axial pattern (*BUD7*, *BNI1*, *SPA2*, *PEA2*, and others).[11,13–15]

Calcofluor Staining of Bud Scars

Calcofluor staining is the most accurate way to analyze a cell's budding pattern in a detailed and quantitative manner. Calcofluor is a fluorescent dye that stains

[4] M. Hayashibe and S. Katohda, *J. Gen. Appl. Microbiol.* **19,** 23 (1973).
[5] A. Bender and J. R. Pringle, *Proc. Natl. Acad. Sci. U.S.A.* **86,** 9976 (1989).
[6] J. Chant and I. Herskowitz, *Cell* **65,** 1203 (1991).
[7] J. Chant, K. Corrado, J. R. Pringle, and I. Herskowitz, *Cell* **65,** 1213 (1991).
[8] A. Halme, M. Michelitch, E. L. Mitchell, and J. Chant, *Curr. Biol.* **6,** 570 (1996).
[9] T. Roemer, K. Madden, J. Chang, and M. Snyder, *Genes Dev.* **10,** 777 (1996).
[10] A. Fujita, C. Oka, Y. Arikawa, T. Katagai, A. Tonouchi, S. Kuhara, and Y. Misumi, *Nature* **372,** 567 (1994).
[11] J. E. Zahner, H. A. Harkins, and J. R. Pringle, *Mol. Cell. Biol.* **16,** 1857 (1996).
[12] T. Chen, T. Hiroko, A. Chaudhuri, F. Inose, M. Lord, S. Tanaka, J. Chant, and A. Fujita, *Science* **290,** 1975 (2000).
[13] N. Valtz and I. Herskowitz, *J. Cell Biol.* **135,** 725 (1996).
[14] M. Snyder, *J. Cell Biol.* **108,** 1419 (1989).
[15] S. Yang, K. R. Ayscough, and D. G. Drubin, *J. Cell Biol.* **136,** 111 (1997).

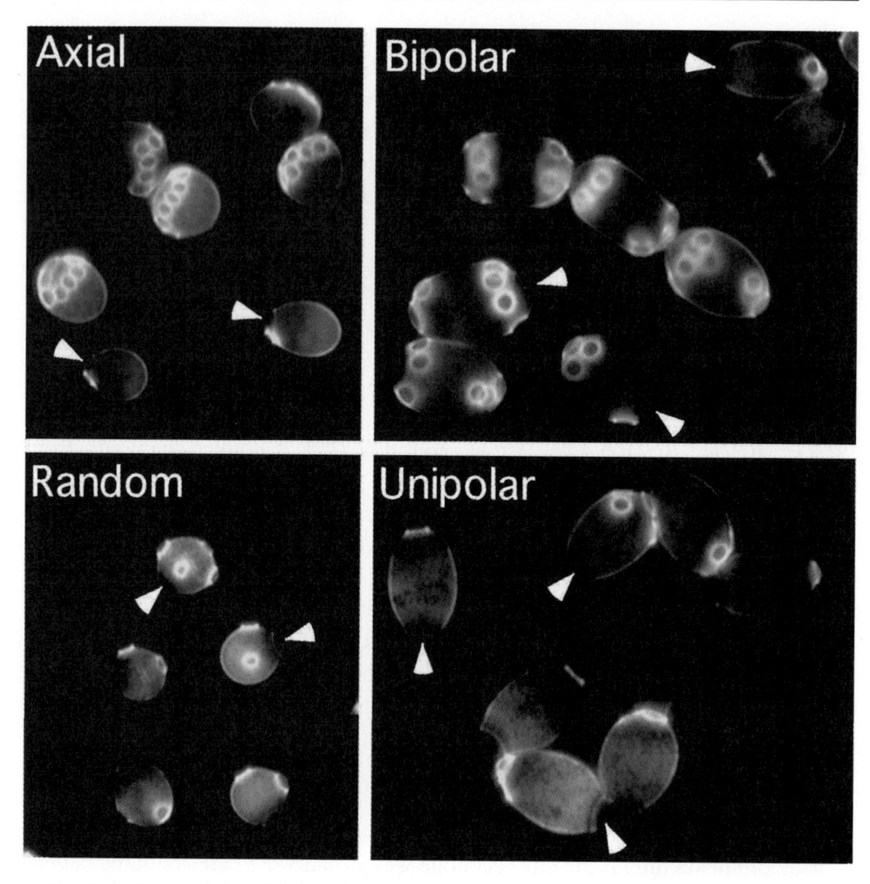

FIG. 2. Bud scar staining with Calcofluor. *Upper left:* The axial budding pattern of haploids. *Upper right:* The bipolar budding pattern of diploids. *Lower left:* The random budding pattern in *bud5*Δ haploids. *Lower right:* The unipolar budding pattern in *bud9*Δ/*bud9*Δ diploids. Examples of birth scars on cells are indicated by arrowheads.

chitin-rich structures of the *Saccharomyces cerevisiae* cell wall.[16] Examples of budding patterns seen from Calcofluor-stained cell samples are shown in Fig. 2. Bud scars stain brightly, while birth scars are considerably more subtle (as indicated by arrowheads in Fig. 2).

Standard Method

1. Grow a logarithmic culture overnight to an OD_{600} of approximately 0.2–1. Typically we dilute a saturated culture 5000-fold (1 μl in 5 ml) and grow overnight.

[16] J. R. Pringle, *Methods Enzymol.* **194,** 732 (1991).

2. Fix the cells by adding formaldehyde to the medium at a final concentration of 3.7%. Incubate the culture for a further 45–60 min at 30° or room temperature. (Formaldehyde is toxic. Please observe the recommended precautions.)

3. Pellet cells and wash twice with water (1 ml per wash).

4. Pellet cells and resuspend in Calcofluor solution (1 mg/ml in water; a 10 mg/ml stock can be kept indefinitely at −20°). We resuspend cells derived from a 5-ml culture in 0.5 ml of the Calcofluor solution.

5. Incubate the sample at room temperature for 5 min.

6. (Optional) Sonicate for 5 sec on a low setting with a microtip probe (e.g., setting 3 out of 10 on a Branson 450 Sonifier) to eliminate cell clumping, which varies from strain to strain. We strongly recommend this step whenever performing quantitation.

7. Pellet cells and wash with water. Repeat. Resuspend the final pellet in an appropriate volume of water (50–500 μl). At this stage samples can be stored for months at 4°.

8. Mount cells on a slide (typically 3–5 μl with an 18 mm × 18mm coverslip) and squash cells between the slide and coverslip as follows: lay the slide (coverslip up) on a pad of paper towels and place a small pad of paper towels on top with an appropriate weight (a heavy biochemistry textbook is useful). Press for 5–20 min. Squashing immobilizes the cells and allows observation of the cells in one focal plane. (Note that some drying of the cell suspension occurs, resulting in parts of the sample being wet with other parts having dried out. Observe the cells in the remaining droplets of moisture.)

9. Observe stained cells by epifluorescence microscopy using a DAPI filter set. Photobleaching is rarely a problem when using a standard fluorescent microscope since Calcofluor is extremely bright.

Important Considerations

(i) For accurate assessment of budding patterns, one must use cultures grown overnight in log phase since budding patterns, especially axial, lose fidelity when cells are in stationary phase.[3] If one wishes to observe cells with multiple scars, the cells should have been grown in logarithmic phase for the period during which the scars were produced. If a cell has eight scars, then this pattern was produced over roughly a 12-hour period in rich medium.

(ii) One should be sure to use extremely clean slides and coverslips—a fresh batch. Dust and other small particles hinder efficient cell pressing between slide and coverslip. Dirt also photographs readily.

(iii) Efficient flattening of cells between slide and coverslip is essential for observing the cell surface and photographing the cells.

(iv) Strain background and growth conditions can affect budding pattern. It is essential to include control samples (wild-type cells from the relevant strain background) to which one can clearly compare the phenotypes of the samples under study.

Quantitative Scoring

Some degree of quantitation with controls is often necessary as first impressions from gazing through the microscope can be misleading. Investigators have generally devised their own schemes to score budding patterns as axial, bipolar, unipolar, or random. Most schemes are very similar. Here we outline our scoring methods.[12,17] We generally score cells with 1 bud scar and cells possessing 3–4 bud scars. We count 100–200 cells per sample. Quantitative scoring is essential if one wishes to distinguish subtle differences in budding pattern.

First Bud Scar Counts. The position of the first bud scar in relation to the birth scar is scored as follows.

Proximal: If the first bud scar is immediately adjacent to the birth scar, we class this position as proximal. Cells budding axially place their first bud scar proximally 98–100% of the time. Some strains budding in the bipolar pattern place their first bud sites at the proximal pole to a limited degree (0–15%).

Distal: If the first bud scar is positioned at the opposite pole from the birth scar we class this position as distal. Distal is typical of the bipolar budding pattern.

Medial: If the first bud scar is neither adjacent to the birth scar nor within a bud scar's diameter of the distal pole, we class this position as medial. Medial is typical of the random budding pattern. When cultures are grown exponentially, cells budding in the axial or bipolar patterns essentially never bud in a medial position.

Three to Four Bud Scar Counts. Three to four bud scar counts are scored as follows.

Axial: If all three/four bud scars are connected in a chain with no spaces between scars and at least one scar is immediately adjacent to the birth scar, we class this cell as axial.

Bipolar: If all three/four bud scars are positioned at the distal pole or distributed between the proximal and distal poles, we class this cell as bipolar. Bud scars at either pole need not be touching; however, they should be either adjacent to the birth scar, touching another bud scar, or within a bud scar's diameter from the distal pole.

[17] M. Lord, M. C. Yang, M. Mischke, and J. Chant, *J. Cell Biol.* **151,** 1501 (2000).

Random: If any one or more of the three/four bud scars is situated in the cell's midsection, we class this cell as random.

Fast Method

This approach is recommended for preliminary checking of budding phenotypes. This method is not recommended for preparing cells for in-depth analysis, or for measurement of subtle budding phenotypes. Some cells will be in stationary phase, a point at which bud-site selection is not maintained with great efficiency. The axial pattern is particularly sensitive to stationary phase or slow growth.[3]

1. With a toothpick scrape an appropriate amount of cells from a patch on a freshly grown overnight plate . Resuspend the cells in 0.5 ml of 1 mg/ml-Calcofluor solution.

2. Mount cells on a slide, press, and observe (as described in instructions 8 and 9 of the Standard Method). If speed is critical, cells can be squashed by briefly pressing cells between slide and coverslip using one's thumb.

Use of Calcofluor Staining in Combination with Other Analyses

Certain dye-based methods for examining subcellular structure (e.g., DNA[18] or actin staining[19]) can be performed in conjunction with the standard bud scar staining protocol. Covisualization of proteins is best performed using green fluorescent protein (GFP) fusions. Immunofluorescence is not especially compatible with bud scar staining, as permeating the cells for antibody access requires the removal of the cell wall, which results in loss of bud scars.

This procedure relies on quick preparation of live samples and is particularly useful for GFP fusions.

1. From a freshly grown overnight patch of cells, thinly repatch some cells on a growth plate and incubate for a further 4–6 hr at 30°.

2. Take a toothpick scraping of the new cell patch and resuspend in an appropriate small volume (30–100 μl) of 0.1 mg/ml Calcofluor solution.

3. Mount 3–5 μl of the cell suspension on a slide, cover with an 18 mm × 18 mm coverslip, and press briefly using the thumb. Immediately view cells by fluorescence microscopy using DAPI (bud scar staining) and GFP filters. Generally, GFP fusions should be observed first since GFP is more prone to photobleaching than Calcofluor.

[18] J. R. Pringle, A. E. M. Adams, D. G. Drubin, and B. K. Haarer, *Methods Enzymol.* **194,** 565 (1991).
[19] A. E. M. Adams and J. R. Pringle, *Methods Enzymol.* **194,** 729 (1991).

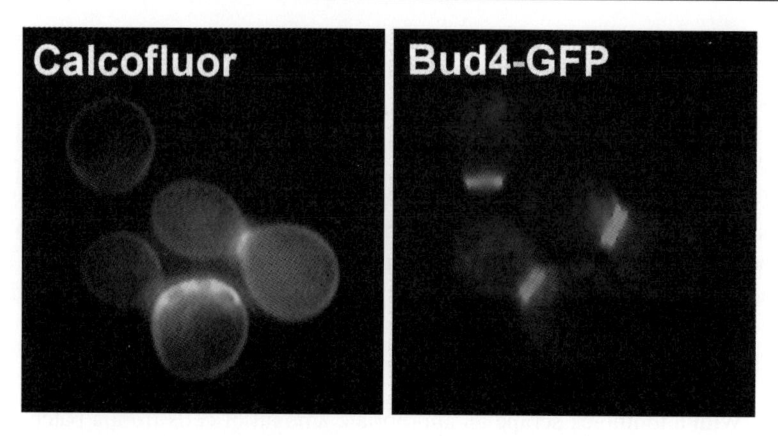

FIG. 3. Covisualization of Calcofluor-stained bud scars and Bud4-GFP.[21] *Left:* Calcofluor staining of an axial budding strain. *Right:* Bud4-GFP.

Figure 3 depicts an example of this method: covisualization of Bud4-GFP (an axial landmark factor[20]) with bud scars.

Colony Morphology-Based Analysis of Budding Patterns

ace2Δ Colony Morphology-Based Analysis

Use of colony morphology to determine the budding pattern of a cell is a facile method to screen a large population of isolates for alterations in budding pattern. The *ace2Δ* method can be used to distinguish cells budding in an axial manner from cells budding in a bipolar or random manner.

ACE2 encodes a transcription factor that is involved in activation of *CTS1*, encoding a chitinase necessary for cell separation.[22,23] *ace2Δ* mother and daughter cells cannot separate from each other resulting in attached clumps of cells (Fig. 4A). *ace2Δ* haploid cells possessing an axial budding pattern form a tight cluster, whereas *ace2Δ/ace2Δ* diploid cells possessing a bipolar budding pattern form extended chains of cells. These defects in cell separation influence colony morphology (Fig. 4B). Colonies of *ace2Δ* cells budding in an axial manner are round with a lustrous surface, an appearance very similar to that of wild-type ACE2 colonies with which yeast biologists are familiar. *ace2Δ* cells exhibiting a bipolar pattern produce colonies with a notched outline and a rough surface. *ace2Δ*

[20] S. L. Sanders and I. Herskowitz, *J. Cell Biol.* **134,** 413 (1996).

[21] M. Lord and J. Chant (unpublished observations).

[22] G. Butler and D. J. Thiele, *Mol. Cell. Biol.* **11,** 476 (1991).

[23] P. R. Dohrmann, G. Butler, K. Tamai, S. Dorland, J. R. Greene, D. J. Thiele, and D. J. Stillman, *Genes Dev.* **6,** 93 (1992).

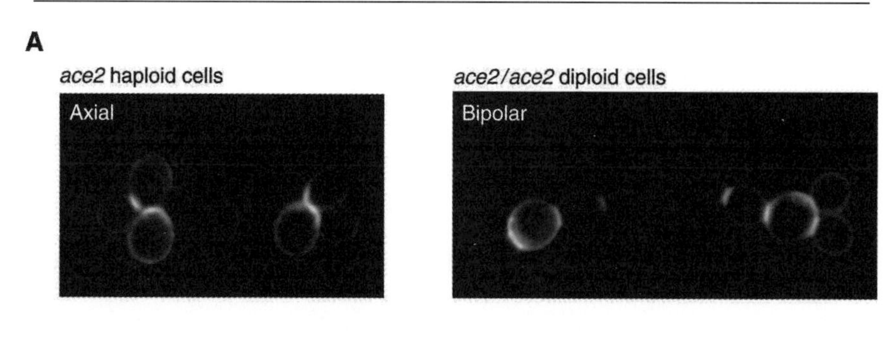

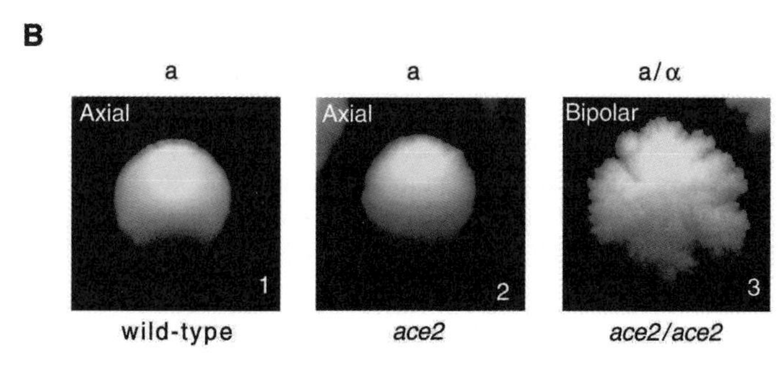

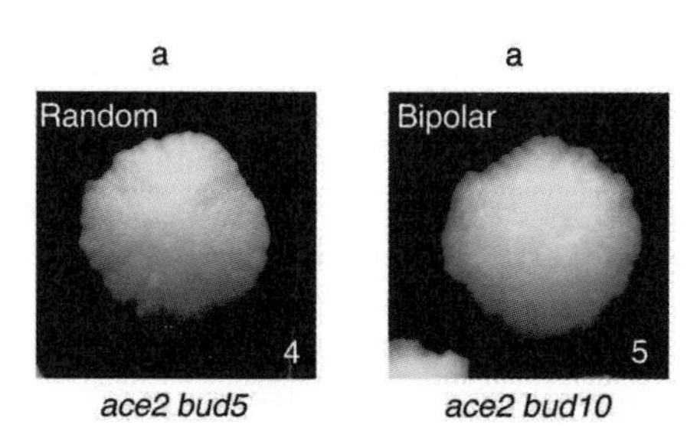

FIG. 4. The morphologies of *ace2*Δ cells and colonies. (A) *Left: ace2*Δ haploid cells exhibiting the axial budding pattern. *right: ace2*Δ/*ace2*Δ diploid cells exhibiting the bipolar budding pattern. Cells were stained with Calcofluor. (B) Colony morphologies of various strains: (1) A colony of wild-type haploid cells budding in an axial pattern; (2) a colony of *ace2*Δ haploid cells budding in an axial manner; (3) a colony of *ace2*Δ/*ace2*Δ diploid cells budding in a bipolar pattern; (4) a colony of *ace2*Δ *bud5*Δ haploid cells budding in a random pattern; (5) a colony of *ace2*Δ *bud10*Δ haploid cells budding in a bipolar pattern.

*bud5*Δ haploids budding in a random fashion produce colonies with a slightly notched outline and a surface that is a little smoother than that of a colony budding in a bipolar manner (Fig. 4B). The bipolar–random difference is very subtle; consequently, use of *ace2*Δ colony morphology to distinguish bipolar from random budding is to be approached cautiously. *ace2*Δ *bud10*Δ haploids exhibiting the bipolar budding pattern show the same colony morphology as *ace2*Δ/*ace2*Δ diploids (Fig. 4B).

Use of YPD plates is optimal for performing this assay. CSM plates can also be used. To screen large numbers of colonies, spread cells to yield 100–1000 colonies per plate. A stereoscopic microscope is helpful for the observation of colony morphology.

Analysis of budding pattern by this method provides a powerful means by which to screen for mutants altered in bud-site selection. For example, mutations in *AXL1*, an axial-specific gene, were identified in a screen for nonaxial colonies,[10] and the bipolar-specific genes, *RAX1* and *RAX2*, were discovered as partial axial revertants of an *axl1*Δ strain.[12] It should be kept in mind that mutations unrelated to budding pattern can affect colony morphology. Direct observation of budding pattern by Calcofluor staining should be relied upon as a secondary screen.

Pseudohyphal Colony Morphology-Based Analysis

Diploid cells starved of nitrogen adopt a unipolar budding pattern in which buds emerge at the distal cell pole.[24] This unipolar pattern enables diploids to form filaments of elongated cells called pseudohyphae on appropriate media (SLAD).[24] Pseudohyphal colonies form a distinctive morphology: A highly fuzzy outline of filamentous growth surrounds the colony.[24] Haploid mutants that bud in the bipolar pattern can also exhibit pseudohyphal-type colonies in appropriate strain backgrounds.[20,25] Diploids possessing a random pattern of budding are defective in pseudohyphal growth and form colonies which have bumpy outlines.[24] Unlike the *ace2*Δ assay, pseudohyphal colony morphology permits random budding colonies to be clearly distinguished from bipolar budding colonies.

Analysis of Microcolony Budding Patterns

This method utilizes direct microscopic observation of growing microcolonies on agar plates. Growing cells can be monitored over time and their budding patterns scored by analyzing the position at which two new buds emerge from a mother–daughter cell pair.[6] One sees patterns essentially as shown in Fig. 1A.

[24] C. J. Gimeno, P. O. Ljungdahl, C. A. Styles, and G. R. Fink, *Cell* **68,** 1077 (1992).
[25] H.-U. Mösch and G. R. Fink, *Genetics* **145,** 671 (1997).

Method

1. Grow up a logarithmic culture to an OD of approximately 0.2–1.
2. Take 5 μl of the cell culture and dilute it in 1 ml water.
3. Disrupt cell clumps by sonicating for 5 sec on a low setting with a microtip probe (e.g., setting 3 out of 10 on a Branson 450 Sonifier).
4. Plate a drop of the cell suspension on a YPD agar plate, allow to dry, and incubate at 30° for 2–3 hr.
5. View cells on a tetrad dissection microscope.
6. Score budding pattern at the two cell/two bud stage. Patterns should resemble those depicted in Fig. 1A.

Considering the ease by which the budding pattern of a given cell can be scored quantitatively by staining with Calcofluor, this plate-assay method is not recommended to characterize the phenotype of a single strain. However, this method provides a way to screen through mutagenized cells to detect alterations in budding patterns. Indeed, the first four genes found to play a direct role in bud-site selection (*RSR1/BUD1*, *BUD2*, *BUD3*, *BUD4*)[6] were found using this method of screening.

Acknowledgments

M. Lord was supported by a Human Frontiers Science Program Organization long-term fellowship. Work in J. Chant's laboratory is supported by the National Institutes of Health Grant GM 49782.

[8] Uses and Abuses of HO Endonuclease

By JAMES E. HABER

Introduction

The site-specific HO endonuclease is a member of a family of endonucleases implicated in the transposition of intron sequences.[1] HO evolved to catalyze the homothallic switching of mating-type (*MAT*) genes in *Saccharomyces cerevisiae,* but over the past three decades, the uses to which HO has been put go much further than simply changing the mating type of a cell. HO-induced double-strand breaks (DSBs) have been used to study and define in detail several different mechanisms of homologous recombination in both mitotic and meiotic cells, as well as to

[1] B. Dujon, *Gene* **82,** 91 (1989).

0076-6879/02 $35.00

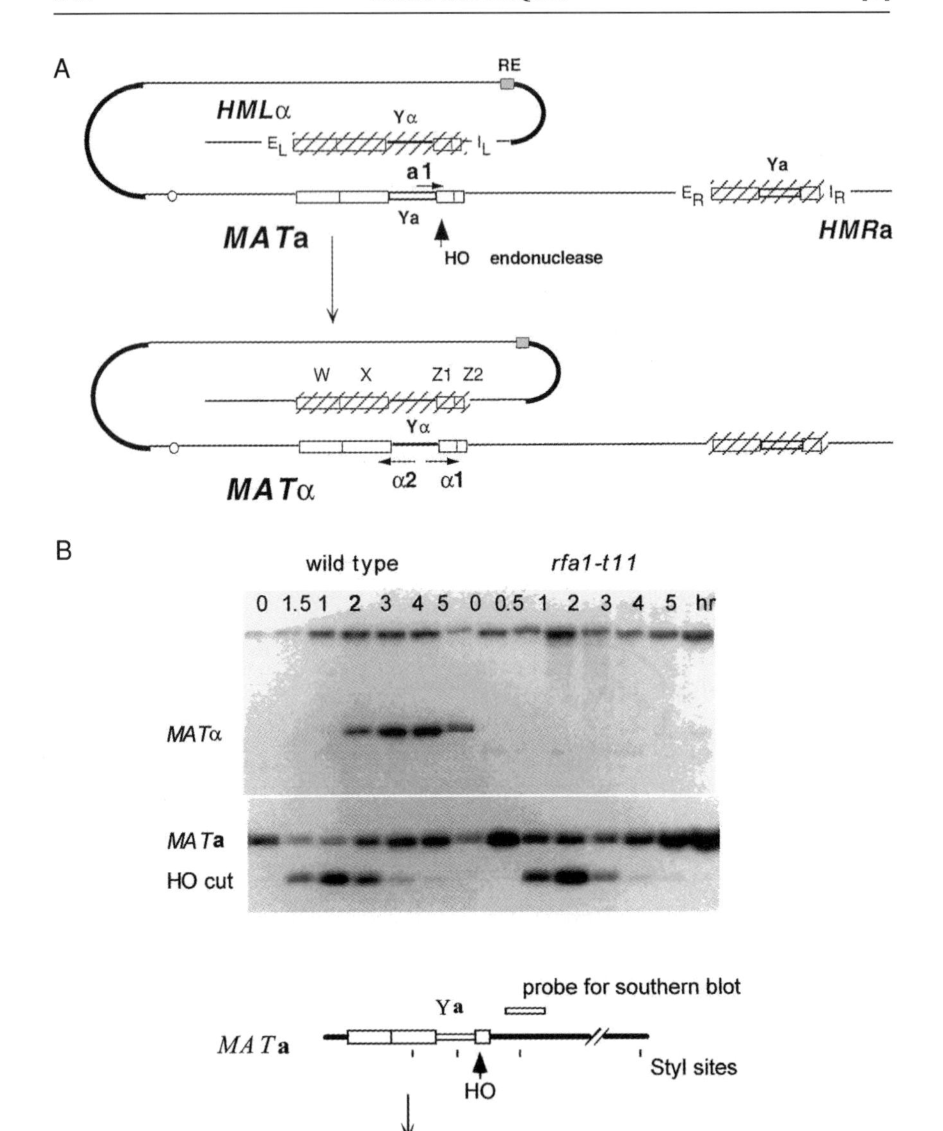

FIG. 1. Mating type gene switching in *S. cerevisiae*. (A) A DSB induced by HO endonuclease catalyzes a gene conversion event in which the Y**a** or Yα sequences at MAT can be replaced with sequences of the opposite mating type. MAT**a** encodes Mat**a**1p, while MATα encodes two proteins, Matα1p and Matα2p, that regulate aspects of sexual identity. MAT**a** switching to MATα is shown. HO cannot cleave equivalent recognition sites in HMLα or HMR**a** sequences, which serve as donors

identify a surprisingly large number of different pathways of nonhomologous recombination. HO has also been used to create defined DNA damage that elicits a DNA damage checkpoint response. Finally, HO-induced events have provided insights into chromatin structure and chromosome architecture.

In nature, the object of homothallic switching is to produce *MAT***a**/*MAT*α diploids from haploid *MAT***a** or *MAT*α cells. MAT switching has been extensively reviewed.[2–4] During this process, a double-strand break is created by HO at the *MAT* locus and the Y**a**- or Yα-specific sequences at *MAT* are replaced. This DSB is repaired by a gene conversion process, using as a donor one of two unexpressed, heterochromatic mating-type sequences at *HML*α or *HMR***a** (Fig. 1). Although both *HML* and *HMR* contain recognition sites for HO, they are not cut, because of highly positioned nucleosomes that apparently prevent access to the site.

Under its own promoter, the *HO* gene is expressed only in the G_1 phase of the cell cycle and only in cells that have previously divided (i.e., mother cells). Thus, a haploid spore divides, producing two cells of the original mating type. When the mother cell divides a second time, it and its new daughter will usually switch to the opposite mating type, while the original daughter cell and its new daughter remain unswitched. By this pattern, two cells of each mating type are in close proximity and conjugate. Once *MAT***a**/*MAT*α diploids are formed, switching is turned off by the Mat**a**1-Matα2 corepressor, whose subunits are encoded by the *MAT***a** and *MAT*α loci, respectively.

Genetics Analysis with Homothallic Strains

When homothallic diploids are sporulated, they give rise to two *MAT***a** and two *MAT*α spores that germinate and can mate with strains of the opposite mating type, but if left undisturbed, all four spores will develop into nonmating, *MAT***a**/*MAT*α colonies. However, it is easy to carry out genetic crosses with homothallic strains by sporulating them and simply mixing the asci of such a strain with *MAT***a** or

[2] J. E. Haber, *in* "Mobile DNA II" (N. Craig, R. Craigie, M. Gellert, and A. Lambowitz, eds.). ASM Press, Washington, D.C., 2002.

[3] J. E. Haber, *Annu. Rev. Genet.* **32,** 561 (1998).

[4] J. N. Strathern, *in* "Genetic Recombination" (R. Kucherlapati and G. R. Smith, eds.), p. 445. Am. Soc. Microbiol., Washington, D.C., 1988.

in the recombination process, because they are heterochromatic. Silencing of the donors depends on E and I sites adjacent to these loci and the participation of many *trans*-acting silencing factors. The selection of *HML* over *HMR* in *MAT***a** switching and the opposite donor preference in *MAT*α cells is enforced by the Recombination Enhancer (RE) sequence. (B) Southern blot analysis of *Sty*I-digested DNA showing the kinetics of *MAT* switching following induction of the galactose-inducible HO gene at time 0. Failure to complete recombination in the *rfa1-t11* mutant is also shown.

MATα cells (or another sporulated homothallic diploid) carrying complementing genetic markers. The germinating spores retain their mating type and conjugate. After growth on a YEPD plate overnight, cells can be replica plated to nutritional dropout medium that will select for the complemented diploid.

Spores of any diploid can be liberated from their asci by digestion with the enzyme preparation glusulase or more purified preparations of β1,3-glucuronidase that digests the ascus wall but not the spore wall. Glusulase and more purified enzyme preparations known as Zymolyase and Lyticase are available from many companies. Suspending asci in a 1 : 10 dilution of glusulase in water will produce free spores after about 1 hr. After digestion, sonication will produce individual spores. McCusker and Haber[5] found that it was easy to UV-irradiate spores of a homothallic strain and grow them into colonies that could be tested for nutritional or other defects. Complementation tests were simple, by mixing asci of different mutants.

Changing Mating Type of Strains

Most laboratory strains of *S. cerevisiae* are heterothallic (designated *ho*) and do not switch. In these strains, the *HO* gene has accumulated several mutations and is inactive.[6] However, these cells can be induced to switch by transforming them with a plasmid carrying the *HO* gene.[7] As with homothallic strains above, HO-mediated switching has proven to be an effective way to obtain isogenic strains of both mating types, to carry out complementation tests among a set of mutations isolated in a single haploid strain, or to make homozygous *MAT***a**/*MATα* diploids to test phenotypes in meiosis. The advantage of introducing *HO* on a plasmid is that it can then be removed simply by growing cells on nonselective medium, leaving a heterothallic strain with a stable mating type.

An even more versatile method to switch mating types has been to express the *HO* gene under the control of a galactose-inducible *GAL1-10* promoter (*GAL::HO*).[8] Versions of this construct are now available either as centromere-containing plasmids marked by *URA3*, *LEU2*, and *TRP1*[8,9] and stably integrated into the genome at the *ADE3* locus.[10] Because the *GAL1-10* promoter is repressed by glucose, cells are pregrown in liquid medium containing carbon sources such as 2% (w/v) lactate, 3% glycerol, or 2% raffinose, and then 2% galactose is added. Under these conditions HO endonuclease is expressed in all cells, mothers and daughters, and at all stages of the cell cycle.[8,11] If HO is continuously expressed,

[5] J. H. McCusker and J. E. Haber, *Genetics* **119,** 303 (1988).
[6] H. Meiron, E. Nahon, and D. Raveh, *Curr. Genet.* **28,** 367 (1995).
[7] R. Jensen, G. F. Sprague, and I. Herskowitz, *Proc. Natl. Acad. Sci. U.S.A.* **80,** 3035 (1983).
[8] R. E. Jensen and I. Herskowitz, *Cold Spring Harb. Symp. Quant. Biol.* **49,** 97 (1984).
[9] J. A. Nickoloff, J. D. Singer, M. F. Hoekstra, and F. Heffron, *J. Mol. Biol.* **207,** 527 (1989).
[10] L. L. Sandell and V. A. Zakian, *Cell* **75,** 729 (1993).
[11] B. Connolly, C. I. White, and J. E. Haber, *Mol. Cell. Biol.* **8,** 2342 (1988).

a switched *MAT* locus can be cleaved again even in the same cell cycle, impairing cell cycle progression through the DNA damage checkpoint-mediated arrest prior to mitosis. Hence, in most cases it is convenient to express *HO* only for 1 hr, after which cells are washed and grown in 2% glucose medium, where *GAL::HO* expression is repressed and the HO protein rapidly is degraded.

There is a great deal of variability reported in the efficiency of HO-mediated *MAT* switching using *GAL::HO*. Some strains apparently carry weak alleles of *gal3* or other genes that slow the response of cells to the presence of galactose. In these cases it may take several hours or longer to promote recombination in most cells. In strains used in our laboratory, virtually 100% of *MAT* sites are cleaved within 1 hr.[12] A small proportion of these *DSB*s are simply religated when HO activity is degraded, but nearly 80% of cells switch to the opposite mating type and another 10–15% recombine, but with the "wrong" donor containing the same Y**a** or Yα information that was initially at *MAT*, so that no phenotypic change results. The biased use of *HML*α by *MAT***a** and an equivalent preferential use of *HMR* by *MAT***a** is controlled by a *cis*-acting recombination enhancer (RE) that in fact controls recombination along the entire left arm of chromosome III, where *HML* resides (see below).

It is also possible to induce expression of *HO* only in the G_1 phase of the cell cycle, as it is normally expressed when the HO gene is under its own promoter. Nasmyth[13] removed the *URS1* regulatory region of the gene and replaced it by *GAL1-10*. The URS2 region still ensures that expression will be confined to G_1, but now it is also under galactose regulation. This has proven useful in demonstrating that there are in fact several different pathways by which HO-cleaved DSBs can be rejoined by Ku proteins and DNA ligase 4 (see below).

Physical Analysis of *MAT* Switching, a Model for Gene Conversion

Because galactose-induced *HO* can initiate recombination in nearly all cells at the same time, it has been possible to learn a great deal about what happens to the DNA during recombination.

1. Southern Blot Analysis of Recombination in Wild-Type and Mutant Cells

*MAT***a** and *MAT*α differ by about 650 and 750 bp in the Y**a** and Yα regions, respectively, so that it is possible to distinguish the two alleles by the size of various restriction fragments such as those generated by *Hin*dIII or *Sty*I.[11,14–16] At 30°, when *GAL::HO* is induced in *MAT***a** cells for 45 min to 1 hr, one can see first

[12] S. E. Lee, F. Pâques, J. Sylvan, and J. E. Haber, *Curr. Biol.* **9,** 767 (1999).
[13] K. Nasmyth, *EMBO J.* **6,** 243 (1987).
[14] D. Raveh, S. H. Hughes, B. K. Shafer, and J. N. Strathern, *Mol. Gen. Genet.* **220,** 33 (1989).
[15] C. I. White and J. E. Haber, *EMBO J.* **9,** 663 (1990).
[16] A. Holmes and J. E. Haber, *Methods Mol. Biol.* **113,** 403 (1999).

the appearance of two HO-cleaved *MAT***a** fragments after about 30 min (Fig. 1B). One hour later, a *MAT*α restriction fragment—the product of gene conversion using *HML*α as a donor—can be seen. At 14°, *MAT* switching takes at least 3 hr and it should be possible to identify a number of slow steps in the process.[17]

Using this system it is possible to determine whether a specific gene products are required to complete recombination. An example is shown in Fig. 1B for the *rfa1-t11* mutation of the largest subunit of the single-strand DNA protein complex, RPA.[18] Of particular interest are essential genes such as DNA polymerases and associated factors. It is possible to assess the contribution of these genes by using temperature-sensitive conditional-lethal alleles.[17] Cells are first arrested by placing them at the restrictive temperature for a given conditional mutation and then the *HO* gene is induced by adding galactose to the medium. Then, without returning cells to a permissive temperature, it is possible to use Southern blots to determine if *MAT* switching can be completed in the absence of gene function.

Procedure. Yeast can be grown overnight in 5 ml glucose-containing medium at 30°. If the *GAL::HO* gene is integrated into a chromosome, YEPD can be used; otherwise, appropriate nutritional dropout medium must be used. Cells are resuspended in YEP–lactate at a density of about 0.5 to 1×10^7 cells/ml, with 0.5–1.0 liters per 4-liter flask to ensure good aeration. When cells have undergone 1–2 doublings, remove 50 ml to serve as a control of uninduced cells, and then add 1/10 volume of 20% galactose to the remaining culture. For *MAT* switching assays, 30–60 min is enough time to induce cutting in most of the cells. After this time, remove another 50-ml aliquot and add 1/10 volume 20% glucose, to stop repress further expression of *GAL::HO*. For experiments lasting more than 5–10 hr it is best to wash the cells in YEPD to remove most of the galactose; otherwise the glucose will be preferentially metabolized and eventually cells will again induce *GAL::HO*. However, for strains that do not retain the HO cut site after recombining, or that do not contain donors (Fig. 2[3]), glucose addition is not necessary. Continue taking 50-ml aliquots as desired.

Centrifuge the samples, discard the supernatant, and resuspend cells in 400 μl of extraction buffer [100 m*M* Tris, pH 8.0, 50 m*M* EDTA, 2.0% sodium dodecyl sulfate (SDS)]. Transfer the cells to a 1.5 ml Eppendorf tube, containing 400 μl of phenol and 500 μl of glass beads. The glass bead/phenol mixture can be prepared in advance and stored at 4° in the dark. Vortex the tube vigorously for 1–2 min, with occasional inversion of the tube to keep the beads well mixed. Too much vortexing will result in shearing of the DNA. Samples can be stored on ice until the other time course samples are ready ($\geq$6 hr). Centrifuge for 10–15 min at 4° in a microcentrifuge. Carefully remove the top aqueous layer to a new microcentrifuge tube. To this new tube add 400 μl of phenol, invert, leave on ice 1–2 min, centrifuge

[17] A. Holmes and J. E. Haber, *Cell* **96,** 415 (1999).

[18] K. Umezu, N. Sugawara, C. Chen, J. E. Haber, and R. D. Kolodner, *Genetics* **148,** 989 (1998).

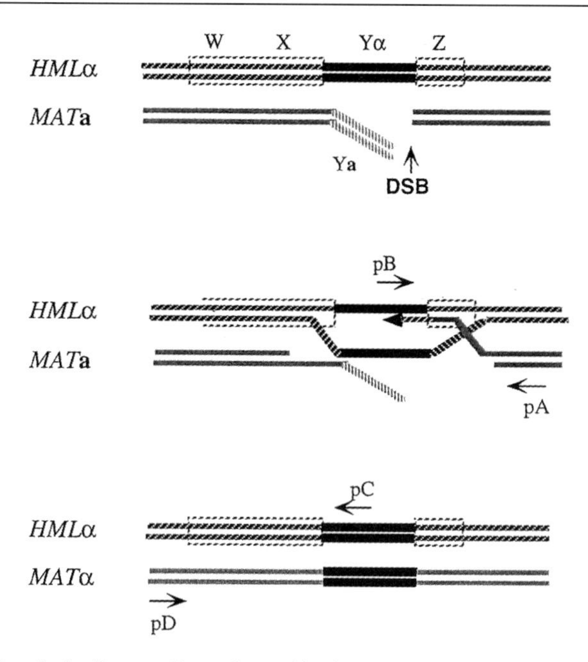

FIG. 2. PCR analysis of intermediates of recombination. Following a DSB (*top*), the end of the DSB in the Z region shared by *MAT* and *HML* invades the donor and initiates new DNA synthesis, primed from the 3′ end. (*Middle*) As soon as new DNA synthesis crosses the region homologous to primer pB, an intermediate can be amplified with the *MAT*-adjacent primer, pA. (*Bottom*) The completion of *MAT* switching can be assessed by primers pC and pD and occurs coincident with the kinetics seen by Southern blot analysis, as in Fig. 1B. [Modified with permission from J. E. Haber, *Annu. Rev. Genet.* **32,** 561 (1998).]

10–15 min as in step 6, and carefully extract the aqueous layer again. It is very important not to carry over protein from the organic/aqueous interphase. Add 50 μl 3 *M* sodium acetate, pH 5.2, and 600 μl 2-propanol. Mix. Centrifuge samples for 1 min at 4°, and discard the supernatant. Add 300 μl TE containing 10 μg of RNase A. Incubate at 37° for 30–60 min with occasional vortexing, until the pellet has dissolved. Add 30 μl of 3 *M* sodium acetate and 300 μl of isopropanol. Spin 1–5 min at 4° and discard the supernatant. Rinse with 70% ethanol, dry the pellet, and resuspend in water or TE. The DNA is then ready for restriction endonuclease digestion and Southern blotting, or for PCR analysis.

2. PCR Analysis of Intermediates of Recombination

Following induction of the DSB, DNA ends are resected by 5′ → 3′-exonucleases to produce 3′-ended single-stranded DNA. With the aid of various recombination proteins, the ssDNA end can locate homologous sequences in a donor locus and initiate strand exchange, a key step in homologous recombination. No methods have yet been presented to analyze the formation of the strand invasion intermediate *per se,* but the next step in the process is the initiation of DNA repair synthesis initiated by using the 3′ end of the invaded strand as a primer for new DNA synthesis (Fig. 2). This intermediate can be detected by using two PCR primers, one specific for Yα at the donor *HML*α and one specific for sequences distal to *MAT***a**. Initially these sequences are 200 kb apart, but once strand invasion occurs and new DNA synthesis is initiated, there is a covalent single strand of DNA that allows a novel PCR product to be amplified.[15] This intermediate appears 15 min after an HO-cleaved intermediate is found but 30 min before the recombination process is completed, as determined both by Southern blotting or by a second pair of PCR primers that detect when Yα becomes connected to *MAT*-proximal sequences (Fig. 2).

Materials. Oligonucleotide primers:

pA primer (HML): GCAGCACGGAATATGGGACT
pB primer (MAT distal): ATGTGAACCGCATGGGCAGT
pC primer (HML): AGATGAGTTTAAATCCAGCA
pD primer (MAT proximal): TGTTGTCTCACTATCTTGCC

PCR Conditions. For approximately 5 ng total purified DNA, PCR conditions are 25 cycles (each: 1.5 min, 94°; 2 min, 55°; 3 min, 72°), followed by 7 min at 72° (PTC-100 Programmable Thermal Controller, MJ Research Inc., Boston, MA).

Southern Blot and Slot-Blot Analysis of Degradation of DNA Ends

The preparation of HO-cleaved DNA ends for recombination appears to require the formation of 3′-ended ssDNA.[15] Similar intermediates have been identified in meiotic recombination, where DSBs are created by *Spo*11p exonuclease, which is region-specific but not site-specific.[19] Resection can be detected in several ways.

1. Use of Denaturing Gels for Southern Blot Analysis

If restriction enzyme-digested DNA is denatured and separated by electrophoresis in agarose gels containing 0.1 *N* NaOH, the two strands migrate separately. With HO-induced DSBs, the strand ending 5′ is resected while the other remains unaltered. Thus the 5′-ended strand is found as a smear of different-sized single strands, whereas the strand ending 3′ at the break is of a uniform size.[15]

Procedure. This denaturing gel protocol is based on that of McDonell *et al.,*[20] as modified by Maniatis *et al.*[21] Melt the agarose in 30 m*M* NaCl, 2 m*M* EDTA solution, pour into the electrophoresis tray, and let solidify. Once solidified, allow

[19] H. Sun, D. Treco, and J. W. Szostak, *Cell* **64,** 1155 (1991).
[20] M. W. McDonell, M. N. Simon, and F. W. Studier, *J. Mol. Biol.* **110,** 119 (1977).
[21] T. Maniatis, E. F. Fritsch, and J. Sambrook, "Molecular Cloning: A Laboratory Manual." Cold Spring Harbor Laboratory Press, Cold Spring Harbor, NY, 1983.

the gel to equilibrate 30 min or longer in alkaline buffer (30 m*M* NaOH, 2 m*M* EDTA, made fresh). Adding NaOH to hot agarose will cause hydrolysis of the polysaccharides in the gel, so the agarose is first melted in a neutral solution, and then allowed to become alkaline by equilibration. Because ethidium bromide does not bind to DNA at high pH, or to ssDNA very well, it is not added to the gel or alkaline buffer.

While the gel is equilibrating, the restricted DNA samples are precipitated in 0.3 *M* sodium acetate, 5 m*M* EDTA (pH 8.0), with 2 volumes of ethanol. EDTA is added to chelate Mg^{2+} so that the DNA will not precipitate in the alkaline buffer during electrophoresis. Precipitate the DNA samples on dry ice for about 10 min, and centrifuge for 10 min. Discard the supernatant, wash the pellet with 70% ethanol, and dry.

Resuspend the pellet in 10–30 μl of 6× alkaline gel loading buffer [0.3 *M* NaOH, 6 m*M* EDTA, 18% Ficoll (type 400), 0.15% bromcresol green, 0.25% xylene cyanol FF]. Denature the size markers by diluting a small volume of DNA into the loading buffer. Load the DNA onto the equilibrated gel, and cover the gel with a glass plate to keep the gel in place (gels can detach from the base plate in alkaline buffer), and to prevent the bromcresol green dye from diffusing out of the gel, which happens at high pH.

Once the DNA has migrated into the gel, carefully remove the alkaline running buffer with a pipette, leaving approximately 1 mm covering the gel. Electrophoresis should be carried out at low voltages, since alkaline gels are not buffered and thus draw more current and heat up. Run at approximately 1 volt/cm, 20–24 hr for a 0.8–1.2% gel of about 20 cm. Also, recirculate the buffer to prevent the anode from becoming too alkaline, and the cathode too acidic. After the dye has migrated halfway to two-thirds through the gel, stain the gel with 0.5 μg/ml ethidium bromide in 1× TAE or 1× TBE buffer. Because ethidium bromide binds ssDNA very poorly, stain the gel for 30–45 min, and destain for another 30 min in TAE or TBE to visualize the bands more clearly with UV light.

The DNA is transferred to a positively charged nylon membrane and hybridized according to Church and Gilbert,[22] with double-stranded DNA probes prepared by the random primer method.[23]

Analysis of $5' \rightarrow 3'$ Degradation of DSB Ends by Nucleases

1. Denaturing Gel Analysis

If $5' \rightarrow 3'$ resection proceeds past one or more restriction sites, these now single-stranded sites cannot be cleaved and thus what appear to be partial digestion products are produced.[15] If these fragments are separated by denaturing gel electrophoresis (above) and probed for sequences near the end of the DSB,

[22] G. M. Church and W. Gilbert, *Proc. Natl. Acad. Sci. U.S.A.* **81,** 1991 (1984).

[23] A. P. Feinberg and B. Vogelstein, *Anal. Biochem.* **132,** 6 (1983).

one finds a "ladder" of increasing sized single-stranded fragments ending 3′ at the DSB. These larger restriction fragments are homologous only to a single-stranded probe homologous to the strand ending 3′ at the DSB whereas a probe to the 5′-ended strand is diffuse because of the variable lengths of DNA that have been resected.

2. Slot-Blot Analysis of DNA Resection

A more quantitative and generally applicable measurement of the resection of DNA ends can be obtained by slot-blot analysis in which total DNA is fixed to a membrane and probed with strand-specific probes for sequences near the DSB. In these experiments, homologous recombination was prevented by deleting *HML* and *HMR* donor sequences, and galactose-induced HO endonuclease was continuously expressed, so that the DSB generated by *MAT* is not repaired in nearly all cells. Here, the DSB ends will be acted on by exonucleases, producing longer and longer 3′-ended ssDNA tails. By this approach it has been possible to identify mutations such as *yku70*Δ that accelerate 5′ → 3′ degradation or mutations such as *mre11*Δ or *rad50*Δ that retard the rate of degradation.[24]

Originally, methods were developed to bind nondenatured DNA to nitrocellulose membranes, so that only single-stranded DNA will be retained[16,25]; however, the present approach is to bind denatured DNA, taken at intervals after HO induction, to nylon membranes and then probe the DNA with strand-specific probes (Fig. 3).

Materials

Minifold II Slot-blot system (Schleicher and Schuell, Keene, NH).

Procedure. Dilute 1 μg of genomic DNA 1000-fold in water. Add 60 μl of 0.4 *N* NaOH into the well of a microtiter dish. Add 50 μl of the diluted DNA. Neutralize with 50 μl of 3 *M* sodium acetate/dye containing 10 μl loading dye/10 ml. Cut the nylon membrane to fit the manifold slot-blot apparatus, and assemble the dot blotter according to the manufacturers' conditions. Connect the hose of the slot-blotter to a vacuum, begin suction, and then add the samples to the wells of the apparatus. Rinse each well with 500 μl 10× SSC (still under suction). Remove the nylon filter and cross-link the DNA to the membrane with UV light. Hybridize according to Church and Gilbert at 72°,[22] with RNA strand-specific probes.[26] RNA probes are easily made using an *in vitro* transcription kit, from Promega (Madison, WI). For quantitation purposes, the signal can be normalized by reprobing the dot-blot with a probe from a region of the genome that is not undergoing specific degradation.

[24] S. E. Lee, J. K. Moore, A. Holmes, K. Umezu, R. D. Kolodner, and J. E. Haber, *Cell* **94,** 399 (1998).
[25] N. Sugawara and J. E. Haber, *Mol. Cell. Biol.* **12,** 563 (1992).
[26] D. A. Melton, P. A. Krieg, M. R. Rebagliatti, T. Maniatis, K. Zinn, and M. R. Green, *Nucleic Acids Res.* **12,** 7035 (1984).

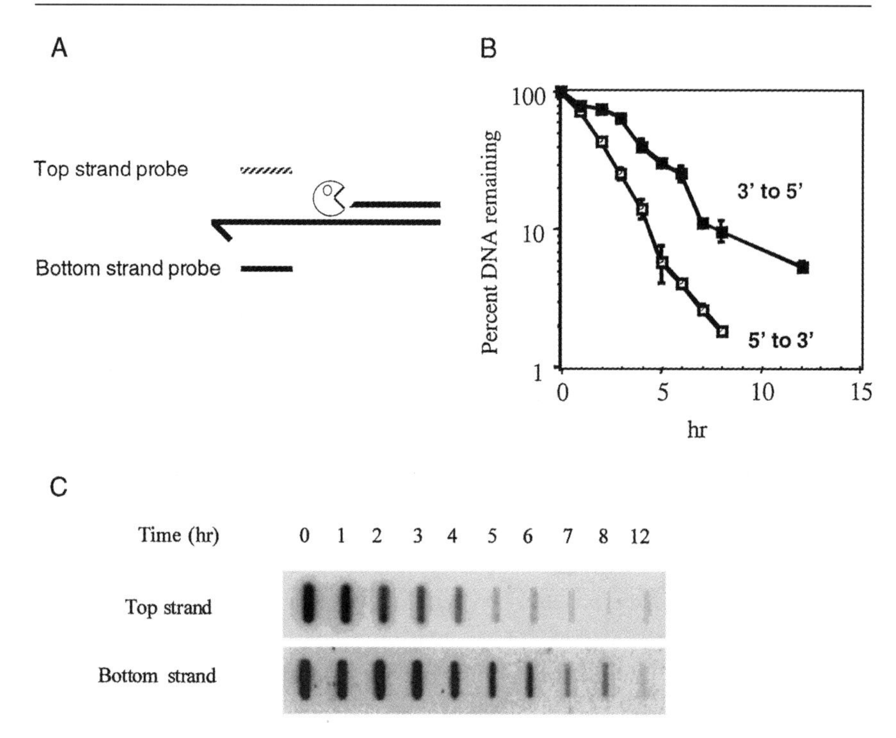

FIG. 3. Measurement of exonucleolytic digestion of the ends of an HO-induced DSB at *MAT*. In this experiment, *HML* and *HMR* have been deleted, so that nearly all the DSBs will fail to be repaired and will be resected by exonucleases for many hours. Strand-specific probes (A) are used to probe denatured DNA loaded onto a filter by a slot-blot apparatus (C). (B) The extent of DNA degradation is plotted. There is very little loss of the strand ending 3′ at the DSB for several hours whereas 5′ → 3′ degradation of the DNA is much more pronounced.

HO-Stimulated Mitotic Gene Conversion at Other Sites

It is possible to induce a DSB at sites other than the *MAT* locus simply by inserting a small HO endonuclease recognition site into another chromosomal location. The most frequently used site is a 117-bp *Bgl*II–*Hinc*II fragment from *MAT***a**[27] that is very efficiently cleaved, but smaller sites, easily obtained by PCR amplification, have also been used. The HO cleavage site (HO-cs) is degenerate, because only half of the site (in the Z1 region) is shared by *MAT***a** and *MAT*α.[28,29]

Caution: The *MAT***a** sequence found in the *Saccharomyces* genome data base is actually listed as *HMR***a** because the sequenced strain was *MAT*α and therefore

[27] R. Kostriken and F. Heffron, *Cold Spring Harb. Symp. Quant. Biol.* **49,** 89 (1984).
[28] J. A. Nickoloff, E. Y. Chen, and F. Heffron, *Proc. Natl. Acad. Sci. U.S.A.* **83,** 7831 (1986).
[29] J. A. Nickoloff, J. D. Singer, and F. Heffron, *Mol. Cell. Biol.* **10,** 1174 (1990).

MAT does not officially exist! *MAT*α1 contains the normal HO cleavage site for Yα. (The region in Z1 is underlined and the site where cleavage occurs is divided by a vertical line):

GGAATATGGGACTACTTCGCGCAACA|GTA**T**AATTTTATAA

The sequence of the Z1 region of *HMR***a** (just outside the *MAT***a**1 ORF) carries a "stuck" mutation (a T to A substitution in position Z1–11, shown below in bold) that impairs HO cleavage, although some switching still occurs.[30]

TCTTTTAGTTTCAGCTTTCCGCAACAGTAA**A**ATTTTATAAA

When this same mutation is found in *MAT*α-stk there is no detectable cleavage,[30] an indication that HO cleaves the *MAT***a** version of the cleavage site better than it cleaves *MAT*α.

Although *in vitro* a 24-bp site can be cleaved, efficient cleavage *in vivo* requires a larger site.[31] A convenient 36-bp cleavage site is the following:

AGTTTCAGCTTTCCGCAACA|GTATAATTTTATAAAC

This site works well in a plasmid context but is cut somewhat less well than a 117-bp HO-cs when inserted into *LEU2* on chromosome III.[32]

There is a substantial variation in the efficiency of HO cleavage by *GAL::HO* in different strains. This is most likely due to a combination of the efficiency of induction in different strains, the size of the HO-cs used, and the site at which the HO-cs is inserted. This has not been studied systematically.

HO-mediated gene conversion has been studied extensively at the *URA3* gene[29,33–36] and at the *LEU2* gene,[32,37,38] by inserting an cleavage site in these genes. Another popular assay system are centromeric plasmids containing direct or inverted copies of *E. coli* LacZ sequences, one of which has an HO cleavage site.[39–41]

[30] B. L. Ray, C. I. White, and J. E. Haber, *Mol. Cell. Biol.* **11,** 5372 (1991).
[31] N. Rudin and J. E. Haber, *Mol. Cell. Biol.* **8,** 3918 (1988).
[32] F. Pâques, W. Y. Leung, and J. E. Haber, *Mol. Cell. Biol.* **18,** 2045 (1998).
[33] Y. S. Weng, J. Whelden, L. Gunn, and J. A. Nickoloff, *Curr. Genet.* **29,** 335 (1996).
[34] D. B. Sweetser, H. Hough, J. F. Whelden, M. Arbuckle, and J. A. Nickoloff, *Mol. Cell. Biol.* **14,** 3863 (1994).
[35] O. Inbar, B. Liefshitz, G. Bitan, and M. Kupiec, *J. Biol. Chem.* **275,** 30833 (2000).
[36] O. Inbar and M. Kupiec, *Mol. Cell. Biol.* **19,** 4134 (1999).
[37] G. F. Richard, B. Dujon, and J. E. Haber, *Mol. Gen. Genet.* **261,** 871 (1999).
[38] G. F. Richard, G. M. Goellner, C. T. McMurray, and J. E. Haber, *EMBO J.* **19,** 2381 (2000).
[39] N. Rudin, E. Sugarman, and J. E. Haber, *Genetics* **122,** 519 (1989).
[40] M. P. Colaiácovo, F. Pâques, and J. E. Haber, *Genetics* **151,** 1409 (1999).
[41] J. Fishman-Lobell, N. Rudin, and J. E. Haber, *Mol. Cell. Biol.* **12,** 1292 (1992).

SPO13::HO-Induced DSBs in Meiosis

Meiotic recombination differs in many respects from mitotic recombination, most notably in its frequency and in the proportion of gene conversions that are accompanied by crossing-over. In meiosis, DSBs are generated by the region-specific but sequence nonspecific endonuclease, Spo11p (reviewed by Keeney[42]). To compare how the same DSB is repaired in meiotic and mitotic cells it was useful to express HO under the control of a meiosis-specific promoter. A *SPO13::HO* construct has proven very useful in this regard.[43] The results of such analysis have already shown that the special properties of meiotic recombination (high levels of crossing-over and shorter gene conversion tracts than are seen in mitotic cells) are not dependent on cleavage by Spo11p, as HO-induced events yield very similar outcomes.[44] The methods used for this study are conventional tetrad analysis and Southern blot analysis.

Alternative Mechanisms of Homologous Recombination to Repair HO-Induced DSBs

Although gene conversion events are the most common form of DSB repair, it is now evident that there are two other major types of repair events that probably occur in all eukaryotes.

1. Single-Strand Annealing

Single-strand annealing (SSA) occurs when 5′ to 3′ resection of DNA ends continues until complementary strands of flanking repeated sequences are exposed and annealing can occur (Fig. 4A). SSA is a nonconservative event in which all the sequences between the two homologous repeats are deleted. In wild-type yeast, such events may be rare, as there is little dispersed repeated DNA and in only some of these situations would a deletion be viable (i.e., an essential gene would not be lost). Such events are likely to occur in tandemly repeated ribosomal DNA (as experimentally demonstrated by an HO-induced DSB in these sequences[45]). Recombination between the approximately 300-bp delta sequences flanking a Ty element has also been demonstrated by using an HO-induced DSB.[46] SSA is efficient when there are 200–400 bp of homologous sequences flanking the break, but it can be detected even when there are as few as 30 bp.[47]

[42] S. Keeney, *Curr. Topics Dev. Biol.* **52,** 1 (2001).

[43] A. Malkova, L. Ross, D. Dawson, M. F. Hoekstra, and J. E. Haber, *Genetics* **143,** 741 (1996).

[44] A. Malkova, F. Klein, W.-Y. Leung, and J. E. Haber, *Proc. Natl. Acad. Sci. U.S.A.* **97,** 14500 (2000).

[45] B. A. Ozenberger and G. S. Roeder, *Mol. Cell. Biol.* **11,** 1222 (1991).

[46] A. Parket, O. Inbar, and M. Kupiec, *Genetics* **140,** 67 (1995).

[47] N. Sugawara, G. Ira, and J. E. Haber, *Mol. Cell. Biol.* **20,** 5300 (2000).

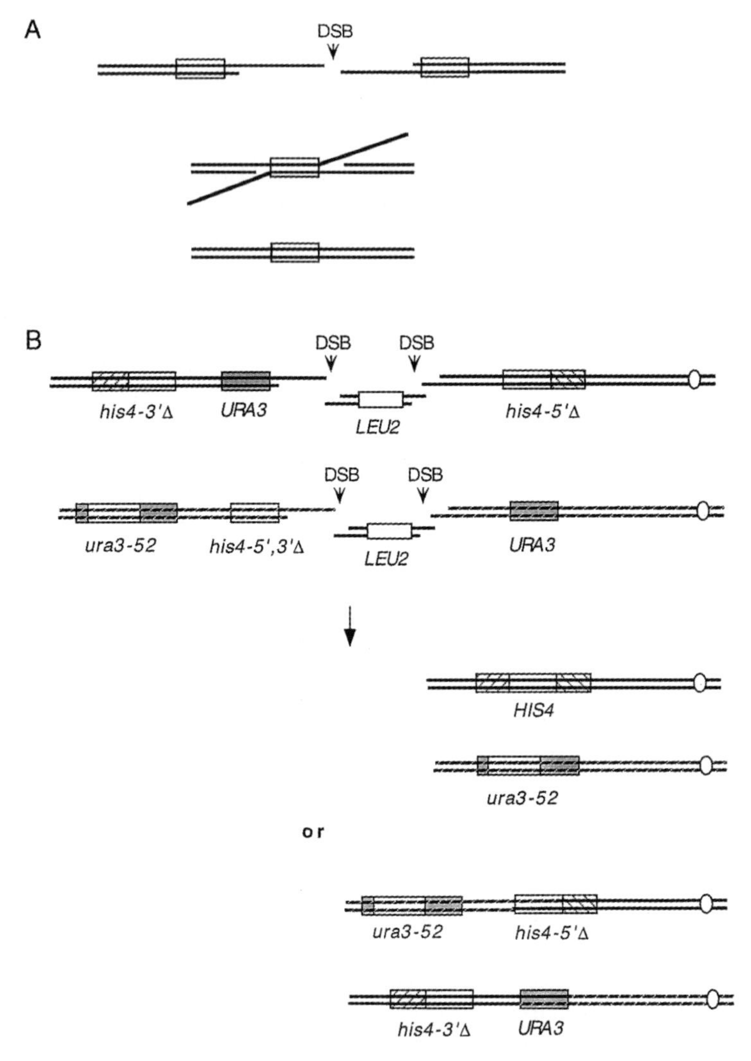

FIG. 4. Single-strand annealing. (A) $5' \rightarrow 3'$ Resection of DSB ends exposes complementary single-stranded DNA in homologous regions flanking the DSB. Annealing of these sequences and trimming of the nonhomologous 3′-ended tails creates a deletion. (B) An experiment in which DSBs were made simultaneously on two different chromosomes to ask SSA between sites on the same chromosome would be annealed more efficiently than sites on different chromosomes. In fact, the two outcomes were approximately equal.

The simplest way to study SSA is to integrate by homologous recombination a circular plasmid containing an HO cleavage site. The integration creates the flanking homologous sequences, with a cutting site in between. SSA has also been studied on centromere-containing replicating plasmids carrying direct repeats. One such plasmid, with two *E. coli* lacZ sequences one of which contains HO-cs, was used to show two important aspects of conversion vs SSA. First, gene conversions failed to produce crossings-over that would yield a reciprocal small circle; instead all deletions appeared to come from SSA.[41] Second, the kinetics of SSA were different from those of gene conversion and this could be exaggerated by increasing the distance between the direct repeats. Increasing distance required more time for resection (at about 4 kb/hr) whereas strand invasion for gene conversion appeared to be distance-independent.[41]

Perhaps surprisingly, SSA is as efficient, or more so, than a competing intrachromosomal gene conversion event, both on plasmids and on a chromosome,[39,41] and the ratio between SSA and gene conversion can be changed by increasing the distance between the repeats. We demonstrated that when a repeated sequence was placed 25 kb from the DSB, SSA required 6 hr, as expected for a 4 kb/hr resection rate.[48]

Physical analysis of the efficiency and kinetics of SSA can be carried out using Southern blots or PCR. The slot blot approach discussed above allowed Sugawara and Haber[25] to demonstrate that DSB ends were resected prior to repair of the DSB by single-strand annealing. SSA does not require the Rad51p strand exchange protein or its auxiliary proteins Rad54p, Rad55p, and Rad57p, but it needs the single-strand annealing protein Rad52p, Rad59p (a Rad52p homologs), and the single-strand DNA binding complex RPA.[25,31,47,49,50] SSA also needs a novel combination of mismatch repair and nucleotide excision repair proteins to cleave off the long ssDNA tails after annealing. This is accomplished by the Rad1p-Rad10p endonuclease and is aided especially when homologous annealing regions are short, by Msh2p and Msh3p (but no other mismatch repair proteins are needed).[49,51,52] Indeed it was the analysis of the dependence of SSA on Rad1p that led to the design of *in vitro* substrates to show that Rad1p-Rad10p was indeed a 3′-flap endonuclease.[53,54]

[48] M. B. Vaze and J. E. Haber, unpublished results, 2001.

[49] N. Sugawara, F. Paques, M. Colaiacovo, and J. E. Haber, *Proc. Natl. Acad. Sci. U.S.A.* **94,** 9214 (1997).

[50] E. L. Ivanov, N. Sugawara, L. J. Fishman, and J. E. Haber, *Genetics* **142,** 693 (1996).

[51] E. L. Ivanov and J. E. Haber, *Mol. Cell. Biol.* **15,** 2245 (1995).

[52] J. Fishman-Lobell and J. E. Haber, *Science* **258,** 480 (1992).

[53] A. E. Tomkinson, A. J. Bardwell, L. Bardwell, N. J. Tappe, and E. C. Friedberg, *Nature* **362,** 860 (1993).

[54] P. Sung, P. Reynolds, L. Prakash, and S. Prakash, *J. Biol. Chem.* **268,** 26391 (1993).

2. Break-Induced Replication (BIR)

An alternative means of repairing DSBs is by break-induced replication (BIR), also known as copy-choice or recombination-dependent DNA replication (reviewed in Refs. 55 and 56). BIR is probably of utmost importance near chromosome ends, where a DSB or degradation of a telomere region can produce a situation where only one end remains to initiate recombination. The outcome of such a process is a nonreciprocal recombination event extending to the end of a chromosome (Figs. 5A and 5B). In yeast, evidence for such one-ended events was obtained in studies of mitotic recombination,[57,58] as well as by transforming in linearized plasmids with only a single telomere.[59] When a linearized plasmid is transformed into yeast, it can initiate homologous recombination, setting up a replication fork that can proceed several hundred kilobases to a chromosome end.[60]

Using HO to Study BIR. HO endonuclease has been used to create well-defined models of this process in both mitotic and meiotic cells.[44,61,62] For example, an HO cleavage site was placed near the end of chromosome III, such that repair of the DSB could only occur in this haploid cell by new telomere formation or by a one-ended recombination event. The 70-bp centromere-proximal part of the HO cleavage site proved to be sufficient to invade and copy all the sequences from HMRa (on the opposite side of the centromere) to the end of the chromosome[61] (Fig. 5A).

BIR has also been studied in diploids in which interchromosomal homologous recombination (gene conversion) is compromised (Fig. 5C). In one series of diploids, mutations in *rad51, rad54, rad55,* and *rad57* block gene conversion but still allow BIR to occur, producing a diploid homozygous for all markers distal to the site of repair.[63] Another diploid has been devised in which there are only 46 bp of homology distal to the DSB (by truncating the chromosome that can be cleaved), so that gene conversion is very rare, but BIR is efficient.[64]

By using these strains it has become evident that the genetic requirements for maintenance of telomeres in the absence of telomerase and for a single BIR event in the middle of a chromosome are very similar.[63,65–68]

[55] J. E. Haber, *Trends Biochem. Sci.* **24,** 271 (1999).
[56] R. Rothstein, B. Michel, and S. Gangloff, *Genes Dev.* **14,** 1 (2000).
[57] M. S. Esposito, *Proc. Natl. Acad. Sci. U.S.A.* **75,** 4436 (1978).
[58] K. Voelkel-Meiman and G. S. Roeder, *Genetics* **126,** 851 (1990).
[59] B. Dunn, P. Szauter, M. L. Pardue, and J. W. Szostak, *Cell* **39,** 191 (1984).
[60] D. M. Morrow, C. Connelly, and P. Hieter, *Genetics* **147,** 371 (1997).
[61] G. Bosco and J. E. Haber, *Genetics* **150,** 1037 (1998).
[62] A. Malkova, E. L. Ivanov, and J. E. Haber, *Proc. Natl. Acad. Sci. U.S.A.* **93,** 7131 (1996).
[63] L. Signon, A. Malkova, M. Naylor, and J. E. Haber, *Mol. Cell. Biol.* **21,** 2048 (2001).
[64] M. Naylor, A. Malkova, and J. E. Haber, unpublished results, 2001.
[65] S. Teng, J. Chang, B. McCowan, and V. A. Zakian, *Mol. Cell* **6,** 947 (2000).
[66] S. C. Teng and V. A. Zakian, *Mol. Cell. Biol.* **19,** 8083 (1999).
[67] S. Le, J. K. Moore, J. E. Haber, and C. Greider, *Genetics* **152,** 143 (1999).
[68] Q. Chen, A. Ijpma, and C. W. Greider, *Mol. Cell. Biol.* **21,** 1819 (2001).

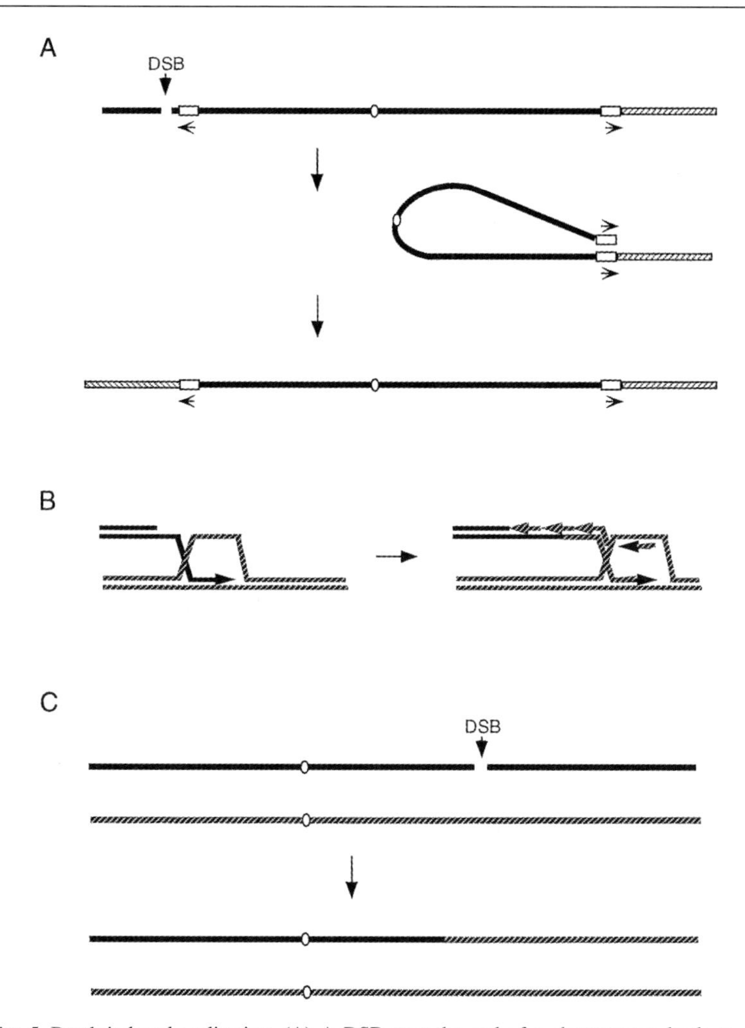

FIG. 5. Break-induced replication. (A) A DSB near the end of a chromosome leads to a one-ended search for homologous sequences. Recombination-dependent DNA replication to the end of the template then stabilizes the DSB. BIR would also occur following the degradation of a telomere until the protection that is normally afforded to an end is lost. (B) A diagram of strand invasion, leading to the formation of a unidirectional replication fork. In the version shown, both newly synthesized DNA strands are displaced from the template by the moving replication bubble. It is also possible that BIR is semiconservative, leaving each molecule with one "old" and one "new" strand. (C) BIR can also occur interchromosomally, causing all markers distal to the point of repair to become homozygous.

Analysis of *de Novo* Telomere Formation

Telomeres are elements that prevent exonucleases from degrading chromosome ends and prevent fusion between telomeres. Normally telomeres are maintained by the telomerase enzyme, an RNA-dependent DNA polymerase that can extend the G-rich strand ending 3′ (reviewed in Refs. 69 and 70). If a chromosome end is broken off, then telomerase cannot easily reconstruct a telomere, as there is no single-stranded G-rich chromosome end-sequence to which it can base-pair and copy additional nucleotides from the enzyme RNA template. However, it is possible to study both completely *de novo* telomere formation and the use of relatively short telomere sequence "seeds" to learn how now telomeres are generated.

HO endonuclease has been used to create a DSB that "lops off" the normal telomere but which can acquire a new telomere. In one series of experiments Sandell and Zakian[10] examined the rare formation of telomeres in the absence of any telomere-like sequences nearby. Kramer and Haber[71] provided a *Tetrahymena* G_4T_2 sequence (which cannot by itself act as a yeast telomere) centromere-proximal to an HO-induced DSB in a rad52 diploid strain, to provide some sort of anchor, or stabilizing role, in new telomere formation. They found that most new telomeres were added not at the G_4T_2 segment but as far as 100 bp more distal, using as little as a single G to begin templated addition. The analysis of 50 such sequences provided evidence of the RNA template sequence that was subsequently found in telomerase RNA.[72]

Diede and Gottschling[73] have taken this approach in a different direction, to learn about the cell cycle dependency and genetic requirements for telomere addition. Centromere-proximal to an HO cleavage site they placed 81 bp of authentic yeast telomere sequence. A DSB led to the rapid extension of this "seed" to the normal telomere length of about 300 bp, requiring both telomerase and lagging-strand DNA polymerase components. Telomere addition only occurs in late S or G_2/M-blocked cells. It should be noted that this process may be somewhat different from *bona fide de novo* telomere addition.

Analysis of Nonhomologous End Joining

DSBs are often repaired by illegitimate or nonhomologous recombination. Even here, the junctions formed between two ends usually contain one of a few complementary base pairs, but there is no strong bias for using sites with the

[69] B. Weiffenbach and J. E. Haber, *Mol. Cell. Biol.* **1,** 522 (1981).
[70] A. Kass-Eisler and C. W. Greider, *Trends Biochem. Sci.* **25,** 200 (2000).
[71] K. M. Kramer and J. E. Haber, *Genes Dev.* **7,** 2345 (1993).
[72] M. S. Singer and D. E. Gottschling, *Science* **266,** 404 (1994).
[73] S. J. Diede and D. E. Gottschling, *Cell* **99,** 723 (1999).

largest number of possible base pairs.[74–76] When DNA containing sequences homologous to the genome are transformed for gene targeting, end-joining is very frequently found in mammalian cells, along with mistargeting the DNA to nonhomologous locations. But even in yeast, which is reputed to be the champion organism for homologous integration of linearized DNA sequences, end-joining occurs.[77,78] However, more revealing observations have come from comparing yeast and mammalian cells that have suffered a DSB on the chromosome. Here it appears that the efficiency of end-joining and repair by homologous recombination are nearly equivalent.[12,79]

Nonhomologous end-joining (NHEJ) has been studied in several ways, including the ligation of linearized, complementary-ended, and blunt-ended plasmid DNAs transformed into yeast[75,80–82] or the repair of dicentric chromosome and plasmids that will break and then form monocentric derivatives.[76,83] And of course, one can examine repair of HO-induced DSBs, in strains where homologous recombination is prevented, either by deleting the *RAD52* gene[76] or by removing homologous donor sequences.[84]

In fact, from studies of HO-induced events, four different kinds of NHEJ events have been described[84–86] (Fig. 6). HO-cut DNA ends can be perfectly religated,[12] or else there can be different-sized deletions ranging from a few base pairs to several kilobases.[76,84] Alternatively there can be misalignment and filling in of the overhanging ends, leading to 2- and 3-bp insertions.[76,84] These last events were shown to be strongly cell cycle regulated, being high in S and/or G_2/M but nearly absent in G_1 cells. This was accomplished by using the G_1-regulated, galactose-inducible HO gene described above.[84] Finally, it was found that a DSB could be "patched up" by incorporating exogenous DNA at the DSB site. Both Ty retrotransposon DNA[85,86] and mitochondrial DNA[87,88] have been found inserted at the HO break site.

[74] D. B. Roth and J. H. Wilson, *Mol. Cell. Biol.* **6,** 4295 (1986).
[75] C. Mezard and A. Nicolas, *Mol. Cell. Biol.* **14,** 1278 (1994).
[76] K. M. Kramer, J. A. Brock, K. Bloom, J. K. Moore, and J. E. Haber, *Mol. Cell. Biol.* **14,** 1293 (1994).
[77] R. H. Schiestl, M. Dominska, and T. D. Petes, *Mol. Cell. Biol.* **13,** 2697 (1993).
[78] A. Wach, A. Brachat, R. Pohlmann, and P. Philippsen, *Yeast* **10,** 1793 (1994).
[79] M. Jasin, *Cancer Invest.* **18,** 78 (2000).
[80] G. T. Milne, S. Jin, K. B. Shannon, and D. T. Weaver, *Mol. Cell. Biol.* **16,** 4189 (1996).
[81] Y. Tsukamoto, J. Kato, and H. Ikeda, *Nucleic Acids Res.* **24,** 2067 (1996).
[82] Y. Tsukamoto, J. Kato, and H. Ikeda, *Mol. Gen. Genet.* **255,** 543 (1997).
[83] Y. Tsukamoto, J. Kato, and H. Ikeda, *Genetics* **142,** 383 (1996).
[84] J. K. Moore and J. E. Haber, *Mol. Cell. Biol.* **16,** 2164 (1996).
[85] J. K. Moore and J. E. Haber, *Nature* **383,** 644 (1996).
[86] S. C. Teng, B. Kim, and A. Gabriel, *Nature* **383,** 641 (1996).
[87] M. Ricchetti, C. Fairhead, and B. Dujon, *Nature* **402,** 96 (1999).
[88] X. Yu and A. Gabriel, *Mol. Cell* **4,** 873 (1999).

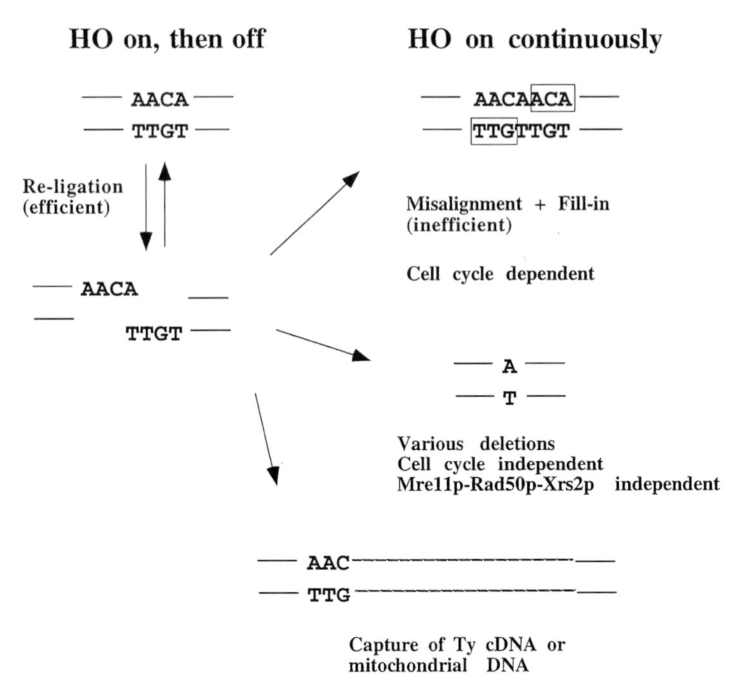

FIG. 6. Alternative nonhomologous mechanisms of DSB repair. An Ho-induced DSB can be repaired either by perfect religation or by several mutagenic processes that destroy the HO cleavage site. These include misalignment and filling in, deletions, and the capture of other DNA sequences at the site of the DSB. All of these processes depend on DNA ligase 4, Lif1p, Yku70p, and Yku80p, and all but the deletion events depend on Mre11p, Rad50p, and Xrs2p.

All of these processes have proven to require the DNA end-binding proteins Ku[80,81,89–91] and the special DNA ligase 4 and Lif1p, its associated XRCC4 homolog.[92–95] The homologs of these proteins are needed in V(D)J joining in the mammalian immune system (except that budding yeast has no obvious DNA-PKcs subunit to interact with Ku proteins). In addition, Mre11p, Rad50p, and Xrs2p have been shown to be required for all but the deletion process in yeast.[82,84] Additional components of NHEJ await discovery. At least one of these is mating-type

[89] G. T. Milne and D. T. Weaver, *Genes Dev.* **7,** 1755 (1993).
[90] S. J. Boulton and S. P. Jackson, *EMBO J.* **15,** 5093 (1996).
[91] S. J. Boulton and S. P. Jackson, *Nucleic Acids Res.* **24,** 4639 (1996).
[92] S. H. Teo and S. P. Jackson, *EMBO J.* **16,** 4788 (1997).
[93] G. Herrmann, T. Lindahl, and P. Schar, *EMBO J.* **17,** 4188 (1998).
[94] T. E. Wilson, U. Grawunder, and M. R. Lieber, *Nature* **388,** 495 (1997).
[95] W. Ramos, N. Tappe, J. Talamantez, E. C. Friedberg, and A. E. Tomkinson, *Nucleic Acids Res.* **25,** 1485 (1997).

regulated, so that NHEJ is low in diploids (where even in G_1 cells have a homolog with which to repair the DSB) and high in haploids.[12,96,97]

Procedure. A *MATα* strain, JKM179, deleted for both *HML* and *HMR* and carrying a copy of *GAL::HO* integrated at *ADE3*, is routinely used, because even 3-bp deletions and insertions at the HO cleavage site inactivate MATα1 and thus create nonmating (sterile) derivatives that are easily scored.[84] Larger deletions that remove both *MATα1* and *MATα2* become **a**-like in mating behavior. If *GAL::HO* is induced for 45 min to 1 hr and then cells are transferred to glucose medium, approximately 30% of cells survive. In a *yku70* deletion or other mutation affecting end-joining, survival is less than 0.1%. Essentially all survivors have rejoined the 4-bp 3′ overhanging ends to regenerate a normal HO cleavage site that can be recut.

If HO is continually expressed throughout the cell cycle, then religation is futile and the only survivors have in one way or the other altered the HO cleavage site. This occurs by deletions, insertions, or capture of other linear DNA fragments.

Analysis of DNA Damage Checkpoints after HO-Induced DSB

When cells fail to repair a DSB, they arrest prior to anaphase in response to a DNA damage checkpoint response mediated by Mec1p, Rad53p, Chk1p and other protein kinases (reviewed in Refs. 98 and 99). It is easy to induce a single unrepaired DSB simply by inducing HO continuously. Only 0.2% of cells repair the DSB, so the rest experience checkpoint-mediated arrest. Experiments have been done both in haploid cells,[24] where the DSB will eventually kill the cells because essential genetic information is lost, or in a *rad52* strain disomic for chromosome VII,[10,100] where the HO-cleaved chromosome cannot be repaired by homologous recombination but where the cells will always have an intact genomic complement of genes. In both cases, cells arrest for many hours with the dumbbell shape characteristic of G_2/M arrest, but then they adapt: that is, they resume growth in the absence of having repaired the DSB. Physical and genetic analysis has shown that most cells retain the broken chromosome, or at least the centromere-containing fragment, though several rounds of mitosis. Whether resection continues unabated for many cell divisions is not yet established.

The use of HO-induced DNA damage has shown many aspects of this process that were evident using more global DNA damaging agents such as UV, γ-irradiation, or methyl methane sulfonic acid (MMS). Somewhat analogous responses are seen with a temperature-sensitive *cdc13-1* mutation that causes

[96] Y. Tsukamoto, J. Kato, and H. Ikeda, *Nature* **388,** 900 (1997).

[97] S. U. Åström, S. M. Okamura, and J. Rine, *Nature* **397,** 310 (1999).

[98] N. F. Lowndes and J. R. Murguia, *Curr. Opin. Genet. Dev.* **10,** 17 (2000).

[99] M. Foiani, A. Pellicioli, M. Lopes, C. Ferrari, G. Liberi, M. Falconi, and M. Plevani, *Mutat. Res.* **451,** 187 (2000).

[100] D. P. Toczyski, D. J. Galgoczy, and L. H. Hartwell, *Cell* **90,** 1097 (1997).

degradation of telomeres and adjacent chromosome ends,[101,102] but it is not possible in that system to know exactly how much damage has been created.

A single HO-induced DSB is sufficient to elicit the checkpoint. The response to DNA damage per se (as opposed to agents that also block replication) is slow to develop, suggesting it is the long resected ssDNA that triggers the response.[103] Indeed, HO-induced MAT switching, which takes an hour or longer, does not cause the hyperphosphorylation of Rad53p kinase, whereas unrepaired DSBs begin to induce a response after an hour. At the time of adaptation, the checkpoint kinase proteins either turn over or become dephosphorylated.[103]

The use of HO has allowed the discovery of adaptation-defective mutations and the realization that yeast have very finely tuned the way they respond.[24,100] Haploid cells will adapt to one DSB, but will not adapt if there are two DSBs or if one DSB is resected twice as fast as normal. This was shown by using slot blots (discussed above) to show that *yk70*Δ strains have a faster rate of $5' \rightarrow 3'$ resection of DSB ends, whereas *mre11*Δ and *rad50*Δ slow down this degradation.[24] This same study revealed that the RPA ssDNA binding complex plays an important role in monitoring the extent of DNA damage.

Probing Chromosome Architecture with HO

Questions about the way DNA is arranged in the nucleus—the architecture of chromosomes—can be explored by recombination. HO endonuclease has proven to be an excellent tool for this purpose.

1. Exploring Territoriality of Chromosomes

Because HO is such an efficient endonuclease, it is possible to create several apparently simultaneous DSBs on different chromosomes. Haber and Leung[104] used this approach to ask whether the repair of two broken chromosomes by SSA would favor the rejoining of ends that were initially nearby on the same chromosome. They found that interchromosomal joinings (producing a pair of reciprocal translocations) were as efficient as two intrachromosomal rejoinings (Fig. 4B). The process could be followed both genetically and on Southern blots. The results suggest that DNA ends, at least when they are resected, are free to find partners anywhere in the nucleus. It will be interesting to examine similar experiments in which ends will be joined by NHEJ processes.

[101] D. Lydall and T. Weinert, *Methods Enzymol.* **283,** 410 (1997).

[102] R. Gardner, C. W. Putnam, and T. Weinert, *EMBO J.* **18,** 3173 (1999).

[103] A. Pellicioli, S. E. Lee, C. Lucca, M. Foiani, and J. E. Haber, *Mol. Cell* **7,** 293 (2001).

[104] J. E. Haber and W. Y. Leung, *Proc. Natl. Acad. Sci. U.S.A.* **93,** 13949 (1996).

2. Donor Preference: Controlling Recombination on Chromosome III

Another very different aspect of chromosome structure has also been explored. The choice of *HML* vs *HMR* ("donor preference") is a highly regulated process so that *MAT***a** recombines preferentially with *HML* whereas *MAT*α interacts mostly with *HMR*. This choice proves to be under the control of a small *cis*-acting Recombination Enhancer (Fig. 1) that appears to regulate recombination (but not transcription) along the entire left arm of chromosome III (reviewed in Refs. 2 and 105). When the cells are *MAT***a**, the entire left arm is "hot" for recombination, but in *MAT*α cells, or when the RE is deleted or mutated, the entire left arm becomes cold, or inaccessible for recombination. Thus the RE is a long-distance regulator of sequences along the chromosome. It seems that this system regulates some aspect of chromosome folding, movement, or location.

Procedure. In practice, the choice of *HML* vs *HMR* in *MAT***a** cells can be determined by replacing *HMR***a** with *HMR*α-B (*HMR*α-*Bam*HI, a single-base replacement) so that there is no ambiguity if a cell *MAT***a** strain has used *HMR***a** or not switched.[106] One need only determine what proportion of *MAT***a** switches have acquired the *Bam*HI site to know what fraction used *HMR*α-B. This can be done by examining switches *en masse,* using a Southern blot using DNA cleaved with *Bam*HI and another enzyme such as *Hin*dIII that cleaves outside of the regions of homology shared by *MAT* and the donors. The Southern blot is then probed with a sequences homologous to Yα. Although the probe hybridizes to *HML*α, *HMR*α-B, *MAT*α, and *MAT*α-B, the two last bands are the smallest and are easily analyzed by densitometry.

Similarly, *MAT*α donor selection can be measured by using *HMR***a** vs *HML***a**-*Bgl*II (a single-base substitution).[107]

Other Site-Specific Endonucleases (I-*Sce*I, I-*Ppo*I)

Some experiments in yeast have also been done with two other site-specific endonucleases of the same large family of "intron homing" endonucleases. The I-*Sce*I nuclease is normally found in mitochondria to promote proliferation of the ω intron in ribosomal DNA to genomes lacking the intron. The gene was engineered by Dujon's laboratory to have codon usage that was compatible with expression in the nucleus.[108] In all experiments done in parallel between HO and I-*Sce*I, the results are comparable.[108,37]

[105] J. E. Haber, *Trends Genet.* **14,** 317 (1998).
[106] X. Wu and J. E. Haber, *Cell* **87,** 277 (1996).
[107] X. Wu, C. Wu, and J. E. Haber, *Genetics* **147,** 399 (1997).
[108] A. Plessis, A. Perrin, J. E. Haber, and B. Dujon, *Genetics* **130,** 451 (1992).

The I-*Ppo*I nuclease comes from *Physarum* and had the interesting property of cleaving every rDNA gene once, when expressed in yeast.[109] This led to the selection of I-*Ppo*I-resistant strains that had accumulated a mutation or an I-*Ppo*I gene insert in every rDNA copy. This provides a novel way to examine the way rDNA sequences become rapidly homogenized. Once the strain is I-*Ppo*I-resistant, it could also be used for studies analogous to those carried out by HO.

Uses of HO and I-*Sce*I Endonucleases in Other Organisms

It has not escaped the attention of researchers interested in studying similar questions in higher eukaryotes that site-specific endonucleases provide a wonderful way to make defined DSBs, to initiate recombination, and to compare the results with what has been found in *S. cerevisiae*. In mammalian cells, several laboratories have used I-*Sce*I with great success to demonstrate that both gene conversion and SSA can be stimulated many orders of magnitude over background. I-*Sce*I endonuclease has been used to initiate gene conversion.[110] Also, in flies, I-*Sce*I liberation of a linearized piece of DNA from a plasmid already established in cells accomplished the elusive goal of targeting of an ends-out fragment to a homologous site in the genome,[111] a result that had not been possible by injecting "naked" DNA. The liberation of a fragment in this way was previously carried out in *S. cerevisiae* in an experiment to compare gene targeting of naked and chromatinized DNA.[112,113]

Acknowledgments

Allyson Holmes graciously provided descriptions of many methods, and Neal Sugawara and Sang Eun Lee provided autoradiograms. Work in the Haber laboratory has been supported by grants from NIH, NSF, and DOE.

[109] D. E. Muscarella and V. M. Vogt, *Mol. Cell. Biol.* **13,** 1023 (1993).
[110] Y. Bellaiche, V. Mogila, and N. Perrimon, *Genetics* **152,** 1037 (1999).
[111] Y. S. Rong and K. G. Golic, *Science* **288,** 2013 (2000).
[112] W. Leung, A. Malkova, and J. E. Haber, *Proc. Natl. Acad. Sci. U.S.A.* **94,** 6851 (1997).
[113] M. T. Negritto, X. Wu, T. Kuo, S. Chu, and A. M. Bailis, *Mol. Cell. Biol.* **17,** 278 (1997).

[9] Assays for Gene Silencing in Yeast

By FRED VAN LEEUWEN and DANIEL E. GOTTSCHLING

Introduction

Silencing in *Saccharomyces cerevisiae* is a chromatin-mediated repression of genes located within specific chromosomal domains. These domains include telomeres, the silent mating type loci *HMR* and *HML,* and the ribosomal RNA gene array (rDNA or *RDN1* locus).[1–5] Nucleosomes are the fundamental building blocks of silent chromatin, along with special structural proteins and enzymatic activities that interact with the nucleosomes to create a repressive structure. While gene repression is the typical "readout" of silent chromatin, its repressive nature also inhibits recombination[6,7] and access of enzymes that cleave or modify the DNA within.[8,9] In general, silent chromatin shuts down regions of the genome, making it inaccessible, but in one instance silent chromatin is the preferred site for activity: the retrotransposon, Ty-*5*, is preferentially targeted to silent chromatin.[10]

Each time chromosomal DNA replicates, silent chromatin must be recreated by an orchestrated series of assembly processes and events. Mutations affecting DNA replication, cell cycle progression, DNA repair, chromatin assembly, histone modification, telomere length, protein turnover, or cellular metabolism modulate silencing, demonstrating that many cellular pathways are connected to the process. A loss of silencing, particularly at the silent mating loci, also has its consequences. Absence of silencing in a haploid cell triggers a cascade of events that cause the cell to act as if it were a diploid. As a result of its intimate ties to overall cellular physiology and processes, and the fact that silent chromatin is found in all eukaryotes, silencing has become a well-studied aspect of yeast biology.

This chapter focuses on genetic and cellular assays to monitor gene silencing in yeast. In addition, strains and vectors most commonly used in these assays are described with their advantages and limitations. Finally, an overview is given of assays that can facilitate identification of new silencing components.

[1] M. Grunstein, *Cell* **93,** 325 (1998).
[2] A. J. Lustig, *Curr. Opin. Genet. Dev.* **8,** 233 (1998).
[3] J. M. Sherman and L. Pillus, *Trends. Genet.* **13,** 308 (1997).
[4] M. R. Gartenberg, *Curr. Opin. Microbiol.* **3,** 132 (2000).
[5] L. Guarente, *Nature Genet.* **23,** 281 (1999).
[6] S. Gottlieb and R. Easton Esposito, *Cell* **56,** 771 (1989).
[7] D. E. Gottschling, *Curr. Biol.* **10,** R708 (2000).
[8] J. Singh and A. J. Klar, *Genes Dev.* **6,** 186 (1992).
[9] D. E. Gottschling, *Proc. Natl. Acad. Sci. U.S.A.* **89,** 4062 (1992).
[10] S. Zou, N. Ke, J. M. Kim, and D. F. Voytas, *Genes Dev.* **10,** 634 (1996).

0076-6879/02 $35.00

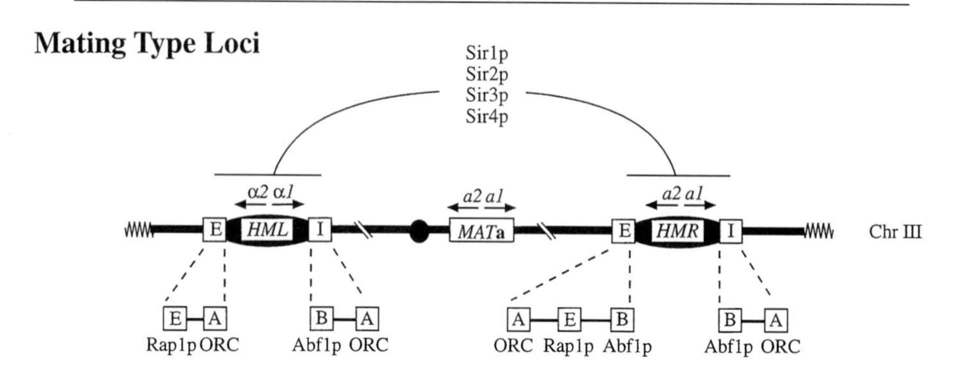

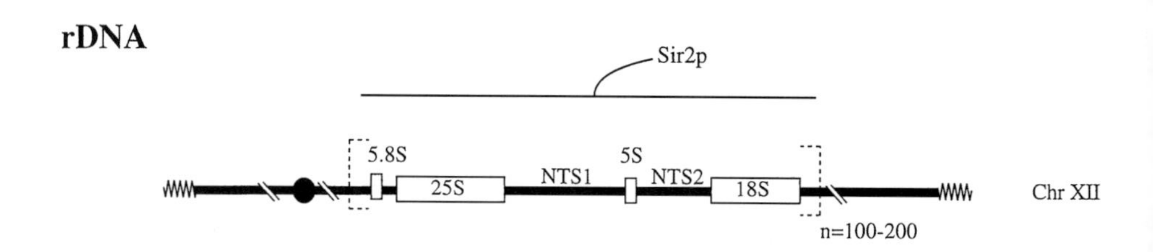

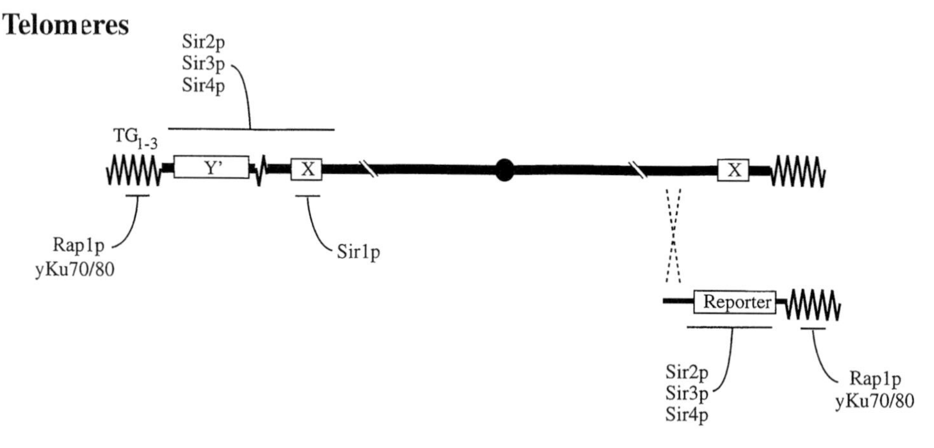

FIG. 1. The silenced chromosomal domains in yeast: the silent mating-type loci, the ribosomal RNA gene repeats (rDNA), and telomeres. The major structural proteins involved in silencing are indicated for each domain (see text). Silencers that are essential (E) or important (I) for repression mediate silencing at HML and HMR [S. Loo and J. Rine, *Ann. Rev. Cell Dev. Biol.* **11,** 519 (1995); I. Herskowitz, J. Rine, and J. Strathern, *in* "The Molecular and Cellular Biology of the Yeast *Saccharomyces cerevisiae* Gene Expression" (E. W. Jones, J. R. Pringle, and J. R. Broach, eds.), Vol. 2, p. 583. Cold Spring Harbor Laboratory Press, Cold Spring Harbor, NY, 1992]. The silencers contain binding sites for two or three of the following factors: ORC (A, ARS consensus sequence), Rap1p (E), and Abf1p (B).

Fundamentals of Silencing

The three known types of silent domains in yeast share several silencing factors, yet each has distinctive properties as well (Fig. 1).[3] Silencing at telomeres and the silent mating-type loci requires the three Sir proteins (silent information regulator) Sir2, Sir3, and Sir4. The Sir proteins interact with each other, with the amino-terminal tails of histone H3 and H4, and with other proteins.[1] Sir2 also has enzymatic activity: it is a deacetylase, most likely acting on the histone tails.[7,11] At telomeres, recruitment of the Sir proteins is dependent on the telomere DNA binding proteins Rap1 and yKu70/80. Recruitment of the Sir complex at the silent mating type loci is dependent on Rap1, Abf1, Sir1, and ORC proteins.[12] Silencing at the rDNA is somewhat more distinct in that it requires Sir2 and other factors such as Net1, but neither Sir3 nor Sir4.[13,14] In general, genes inserted near telomeres and within the rDNA are not silenced as thoroughly as the mating type genes are. This is due to some of the aforementioned differences, as well as the fact that the endogenous genes within *HML* and *HMR* have weak promoters (see below). This incomplete silencing at telomeres is manifested as a heritable, but reversible repression of telomeric genes (also known as telomeric position effect or TPE[15]; see Fig. 2). This variegated expression represents a competition between activation of the telomeric gene and assembly of silent chromatin.

Factors That Affect Silencing

As the result of detailed studies on silencing, a few fundamental rules for modulating the level of silencing at the various loci are known.

[11] L. Guarente, *Genes Dev.* **14,** 1021 (2000).

[12] S. Loo and J. Rine, *Ann. Rev. Cell Dev. Biol.* **11,** 519 (1995).

[13] J. S. Smith and J. D. Boeke, *Genes Dev.* **11,** 241 (1997).

[14] M. M. Cockell and S. M. Gasser, *Curr. Biol.* **9,** R575 (1999).

[15] D. E. Gottschling, O. M. Aparicio, B. L. Billington, and V. A. Zakian, *Cell* **63,** 751 (1990).

The rDNA is organized in 100–200 tandem transcription units of 9.1 kb. Silencing occurs throughout the repeat; reporter genes are usually inserted within the nontranscribed spacer (NTS) 2 and/or the 18S coding region [J. S. Smith and J. D. Boeke, *Genes Dev.* **11,** 241 (1997); J. S. Smith, E. Caputo, and J. D. Boeke, *Mol. Cell. Biol.* **19,** 3184 (1999)]. Native telomeres are composed of terminal TG_{1-3} repeats and subtelomeric elements. Many telomeres contain one or more middle repetitive Y' element and all telomeres have a small core X element [F. E. Pryde and E. J. Louis, *Biochemistry* **62,** 1232 (1997)]. The constructs most frequently used to insert reporter genes at telomeres contain TG_{1-3} repeats immediately downstream of the reporter gene. This is done by a single crossover and results in deletion of all native subtelomeric elements [D. E. Gottschling, O. M. Aparicio, B. L. Billington and V. A. Zakian, *Cell* **63,** 751 (1990)]. Sir1p makes a small contribution to silencing at X elements but does not play a role in silencing at nonnative telomeres [F. E. Pryde and E. J. Louis, *EMBO J.* **18,** 2538 (1999)].

SIR Proteins

Some of the Sir proteins (i.e., Sir2 and Sir3) are limiting in the cell. Consequently, silencing is sensitive to the exact dosage of Sir proteins and increased silencing at one locus can result in reduced silencing at another locus.[16,17] For example, overexpression of *SIR3* enhances telomeric silencing,[18] whereas overexpression of *SIR4* reduces telomeric silencing.[19] Furthermore, while deletion of *SIR4* abolishes telomeric, *HML*, and *HMR* silencing,[20] silencing at the rDNA locus is increased because more Sir2 is available to bind in the nucleolus.[16]

Chromatin and Replication

The histones are critical to silencing and the tails of histone H3 and H4 interact with the Sir complex.[21] Mutations in histone acetyltransferases or histone deacetylases can affect silencing by changing the acetylated state of the histone tails.[1,2] There is also an important link between histone deposition and DNA replication. Chromatin assembly factors CAF-I and RCAF have both been shown to play a role in telomeric, rDNA, and mating-type silencing.[22–24] Silencing in yeast also seems to be linked to DNA replication and cell cycle progression.[12,25] Mutants that slow S-phase progression or mutations in replication proteins can restore silencing at mutant *HMR* loci.[26,27]

Chromosome Dynamics

As mentioned above, silencing at telomeres is weaker than silencing at the mating-type loci. The strength of silencing at the telomeres is dependent not only on the Sir proteins and the histone tails but also on telomere length, subnuclear location (telomeres are usually located at the nuclear periphery[28]), and the presence

[16] J. S. Smith, C. B. Brachmann, L. Pillus, and J. D. Boeke, *Genetics* **149,** 1205 (1998).

[17] L. Maillet, C. Boscheron, M. Gotta, S. Marcand, E. Gilson, and S. M. Gasser, *Genes Dev.* **10,** 1796 (1996).

[18] H. Renauld, O. M. Aparicio, P. D. Zierath, B. L. Billington, S. K. Chhablani, and D. E. Gottschling, *Genes Dev.* **7,** 1133 (1993).

[19] M. Cockell, F. Palladino, T. Laroche, G. Kyrion, C. Liu, A. J. Lustig, and S. M. Gasser, *J. Cell Biol.* **129,** 909 (1995).

[20] O. M. Aparicio, B. L. Billington, and D. E. Gottschling, *Cell* **66,** 1279 (1991).

[21] M. Grunstein, *Nature* **389,** 349 (1997).

[22] P. D. Kaufman, R. Kobayashi, and B. Stillman, *Genes Dev.* **11,** 345 (1997).

[23] S. Enomoto and J. Berman, *Genes. Dev.* **12,** 219 (1998).

[24] J. K. Tyler, C. R. Adams, S. R. Chen, R. Kobayashi, R. T. Kamakaka, and J. T. Kadonaga, *Nature* **402,** 555 (1999).

[25] A. M. Miller and K. A. Nasmyth, *Nature* **312,** 247 (1984).

[26] H. Laman, D. Balderes, and D. Shore, *Mol. Cell. Biol.* **15,** 3608 (1995).

[27] A. E. Ehrenhofer-Murray, R. T. Kamakaka, and J. Rine, *Genetics* **153,** 1171 (1999).

[28] M. Gotta, T. Laroche, A. Formenton, L. Maillet, H. Scherthan, and S. M. Gasser, *J. Cell Biol.* **134,** 1349 (1996).

of telomere-binding proteins that interact with Sir proteins.[3] Creation of double-strand (ds) DNA breaks in the chromosome results in relocalization of yKu and Sir3p from telomeres to the site of DNA damage, and a reduction in telomeric silencing.[29,30]

General Metabolism

Sir2p is a deacetylase that likely acts on histone tails *in vivo* and is essential for silencing at telomeres, rDNA, and the silent mating-type loci.[7,11,31–33] Sir2p requires nicotinamide adenine dinucleotide (NAD^+) as a cofactor for deacetylase activity. Reduction of the cellular levels of NAD^+ by genetic or environmental changes reduces silencing at telomeres and the rDNA.[31,34] In addition, carbon sources in media can affect silencing. Silencing is best in glucose, reduced somewhat in galactose, and reduced further in the presence of nonfermentable carbon sources (e.g., glycerol, lactate, ethanol) (Ref. 35 and D.E.G., unpublished). We have also noticed that variations in batches of media constituents (e.g., amino acids, nucleosides, and agar) can affect the outcome of a silencing assay (D.E.G., unpublished). Therefore, it is imperative that strains being tested for silencing always be assayed on the same batch of plates, under the same conditions, and preferably at the same time.

Promoter Strength and Silencing Readout

Whether a gene is silenced efficiently depends not only on its location but also on the strength of the gene's promoter. A gene with a weak promoter is silenced more efficiently than a gene with a strong promoter.[36] For example, whereas the *URA3* and *ADE2* genes are significantly repressed at telomeres,[15] the *HIS3* gene is not.[18] However, silencing of the *HIS3* gene can be enhanced by using a gene with a defective promoter[37] or by deletion of *GCN4*,[18] the transactivator of *HIS3*. Similarly, silencing of the *URA3* gene can be dramatically enhanced by deletion of its transcriptional activator, *PPR1*.[36] As indicated in Table I, a variety of promoter–reporter combinations have been created which take advantage of this fact, to make the silencing assays more sensitive.

[29] S. G Martin, T. Laroche, N. Suka, M. Grunstein, and S. M. Gasser, *Cell* **97,** 621 (1999).

[30] K. D. Mills, D. A. Sinclair, and L. Guarente, *Cell* **97,** 609 (1999).

[31] J. S. Smith, C. B. Brachmann, I. Celic, M. A. Kenna, S. Muhammad, V. J. Starai, J. L. Avalos, J. C. Escalante-Semerena, C. Grubmeyer, C. Wolberger, and J. D Boeke, *Proc. Natl. Acad. Sci. U.S.A.* **97,** 6658 (2000).

[32] S. Imai, C. M. Armstrong, M. Kaeberlein, and L. Guarente, *Nature* **403,** 795 (2000).

[33] J. Landry, A. Sutton, S. T. Tafrov, R. C. Heller, J. Stebbins, L. Pillus, and R. Sternglanz, *Proc. Natl. Acad. Sci. U.S.A.* **97,** 5807 (2000).

[34] S. J. Lin, P. A. Defossez, and L. Guarente, *Science* **289,** 2126 (2000).

[35] G. J. Shei and J. R. Broach, *Mol. Cell. Biol.* **15,** 3496 (1995).

[36] O. M. Aparicio and D. E. Gottschling, *Genes Dev.* **8,** 1133 (1994).

[37] B. D. Bourns, M. K. Alexander, A. M. Smith, and V. A. Zakian, *Mol. Cell. Biol.* **18,** 5600 (1998).

TABLE I
SILENCING REPORTER CONSTRUCTS

Location	Reporter[a]	Construct/PCR template	Note	Ref.[b]
Telomere	*URA3-VIIL*	pVIIL-URA3-TEL (pADH4UCA4-IV)		(1)
	(URA3-)HIS3-VIIL	pYAHISTEL	Promoter defective *HIS3* gene	(2)
	URA3-ADE2-VIIL	pADADE2(+)		(1)
	URA3-TRP1-VIIL	pADATRP(+)	Minimal *TRP1* promoter	(1)
	URA3-VR	pVR-URA3-TEL		(1)
	ADE2-VR	pHR10-6		(3)
*HMR***a**	*URA3*	pVZ1+HMRa::URA3/ pΔmat::URA3		(4, 5)
	ADE2	LSD270/LSD271	Flanked by wild-type or mutant E silencer	(6)
	P_{URA3}-*ADE2*	HMR::URA3pr-ADE2		(7)
	TRP1	HMR**a**::TRP1	Flanked by wild-type or mutant E silencer	(8, 9)
	P_{ADH}-*GFP*	pPP46		(10)
	P_{LEU2}-*lacZ*	LEU2″LacZ		(11)
*HML*α	*URA3*	pGJ8, pMB36, pMB37		(12)
	ADE2	pYXB19		(13)
	P_{URA3}-*ADE2*	pHMLα::URA3p-ADE2		(14)
	P_{LEU2}-*lacZ*	LEU2″LacZ		(11, 15)
rDNA	*mURA3-HIS3*	pJSS51-9 PCR	*URA3* expressed by minimal *TRP1* promoter	(16)
	MET15	pJSS41-8 PCR		(16)
	ADE2-CAN1	pCAR1		(17)
	P_{URA3}-*ADE2*	pDS40		(18, 19)

[a] VIIL and VR indicate integrated at the telomere on the left arm of chromosome VII and right arm of chromosome V, respectively. P_{YFG1}-*YFG2* indicates a gene fusion (promoter swap) between the *YFG1* promoter and *YFG2*.

[b] Key to references: (1) D. E. Gottschling, O. M. Aparicio, B. L. Billington, and V. A. Zakian, *Cell* **63,** 751 (1990); (2) B. D. Bourns, M. K. Alexander, A. M. Smith, and V. A. Zakian, *Mol. Cell. Biol.* **18,** 5600 (1998); (3) M. S. Singer and D. E. Gottschling, *Science* **266,** 404 (1994); (4) M. S. Singer, A. Kahana, A. J. Wolf, L. L. Meisinger, S. E. Peterson, C. Goggin, M. Mahowald, and D. E. Gottschling, *Genetics* **150,** 613 (1998); (5) T. H. Cheng, Y. C. Li, and M. R. Gartenberg, *Proc. Natl. Acad. Sci. U.S.A.* **95,** 5521 (1998); (6) L. Sussel, D. Vannier, and D. Shore, *Mol. Cell. Biol.* **13,** 3919 (1993); (7) D. H. Rivier, J. L. Ekena, and J. Rine, *Genetics* **151,** 521 (1999); (8) L. Sussel and D. Shore, *Proc. Natl. Acad. Sci. U.S.A.* **88,** 7749 (1991); (9) S. Enomoto and J. Berman, *Genes. Dev.* **12,** 219 (1998); (10) P. U. Park, P. A. Defossez, and L. Guarente, *Mol. Cell. Biol.* **19,** 3848 (1999); (11) C. Boscheron, L. Maillet, S. Marcand, M. Tsai-Pflugfelder, S. M. Gasser, and E. Gilson, *EMBO J.* **15,** 2184 (1996); (12) X. Bi, M. Braunstein, G. J. Shei, and J. R. Broach, *Proc. Natl. Acad. Sci. U.S.A.* **96,** 11934 (1999); (13) X. Bi and J. R. Broach, *Genes. Dev.* **13,** 1089 (1999); (14) T. H. Cheng and M. R. Gartenberg, *Genes. Dev.* **14,** 452 (2000); (15) L. Maillet, C. Boscheron, M. Gotta, S. Marcand, E. Gilson, and S. M. Gasser, *Genes. Dev.* **10,** 1796 (1996); (16) J. S. Smith and J. D. Boeke, *Genes Dev.* **11,** 241 (1997); (17) C. F. Fritze, K. Verschuren, R. Strich, and R. Easton Esposito, *EMBO J.* **16,** 6495 (1997); (18) D. A. Sinclair and L. Guarente, *Cell* **91,** 1033 (1997); (19) S. Imai, C. M. Armstrong, M. Kaeberlein, and L. Guarente, *Nature* **403,** 795 (2000).

Strain Variability

Given the number of pathways that affect silencing, it is perhaps not surprising that allelic differences between strain backgrounds can influence the degree of silencing at a locus. As shown in Fig. 3B, a telomeric URA3 gene is usually silenced better in strains derived from the YPH[38] or W303 backgrounds[39] than in BY strains[40] (see below for more details). Furthermore, temperature can affect the level of silencing. Telomeric silencing in the YPH background is greatest at 37° and somewhat reduced at 23°.[41]

Mating-Type Silencing Assays

In wild-type haploid cells the mating-type genes are exclusively expressed from the MAT locus. The genes present in the MAT locus determine the mating type of the cell, either **a** or α (see Fig. 1). In addition to the MAT locus, **a** and α information is present in the silenced *HMR***a** and *HML*α loci.[42] If silencing is disrupted, both **a** and α information become expressed and the haploid acts as if it is a diploid.[12,43] These cells do not respond to mating pheromone and do not mate with cells of the opposite mating type. Both of these phenotypes can be used to measure silencing at the mating-type loci.

Quantitative Mating

The background and principles of the yeast-mating assays have been described in detail.[44] Here we outline a protocol to quantify mating as a measure of silencing at the HM loci (adapted from Ref. 44).

1. Grow cultures of the experimental strain and of the appropriate mating-type tester strain (opposite mating-type to experimental strain) to a density of about 1×10^7 cells/ml in YEPD.
2. Mix 2×10^6 cells of the experimental strain with 1×10^7 cells of the tester strain and collect the cells on a sterile 0.45 μm pore, 25 mm nitrocellulose filter disk via vacuum filtration in a filter funnel (e.g., Pall Gelman, Newport Richey, FL; Millipore, Bedford, MA).

[38] R. S. Sikorski and P. Hieter, *Genetics* **122,** 19 (1989).
[39] B. J. Thomas and R. Rothstein, *Cell* **56,** 619 (1989).
[40] C. B. Brachmann, A. Davies, G. J. Cost, E. Caputo, J. Li, P. Hieter, and J. D. Boeke, *Yeast* **14,** 115 (1998).
[41] O. M. Aparicio, Ph.D. Thesis, University of Chicago, Chicago, IL, 1993.
[42] I. Herskowitz, *Microbiol. Rev.* **52,** 536 (1988).
[43] S. G. Holmes, M. Braunstein, and J. R. Broach, *in* "Epigenetic Mechanisms of Gene Regulation" (V. E. A. Russo, R. A. Martiensen, and A. D. Riggs, eds.), p. 467. Cold Spring Harbor Laboratory Press, Cold Spring Harbor, NY, 1996.
[44] G. F. Sprague, Jr., *Methods Enzymol.* **194,** 77 (1991).

3. Place the filters on the surface of a YEPD plate (cell side up) and incubate at 30° for 5 hr.

4. Resuspend the cells on each filter in 1 ml water, sonicate, or vortex vigorously, for 5–10 seconds to disrupt cell clumps, and make a dilution series in water. Plate the cells from the mating mix onto different selective media to select for **a**/α diploids or for the experimental haploid parent strain.

5. The mating efficiency is calculated as the number of colonies of **a**/α diploids divided by the sum of **a**/α diploid plus experimental strain haploid colonies. The defect in a silencing mutant can be expressed as the fraction of the mating efficiency of wild-type cells.

Notes

1. Strains PT1 (*MAT***a**, *ile*, *hom3*) and PT2 (*MAT*α, *ile*, *hom3*) (Fred Cross, Rockefeller University, NY) are universal tester strains that can be used for mating assays with all experimental strains that are auxotrophic for at least one of the commonly used markers. Mating between PT1 or PT2 and the experimental strain will result in a prototrophic diploid that can be selected for on synthetic minimal media (Y-min) plates.[45] One liter of Y-min contains 1.2 g yeast nitrogen base (without amino acids and ammonium sulfate), 5 g $(NH_4)_2SO_4$, 10 g succinic acid, 6 g NaOH, 20 g agar, and 20 g glucose (add after autoclaving). Before adding the agar and autoclaving, make sure all the components are in solution and mixed well.

2. In addition to the mating reaction, it is recommended to collect cells of each strain on separate filters and plate them out separately on the two media as well, in order to control for reversion and/or contamination.

3. The advantage of the quantitative mating assay is that it can be done with many different strains and does not require the integration of reporter genes. However, assays using reporter genes (e.g., *hmr::TRP1*, *hmr::P*$_{URA3}$*-ADE2*) have been shown to be more sensitive to low levels of *HMR* derepression than the mating assays (see below).

α-Factor Confrontation Assay

Before the yeast cells physically interact and start the mating reaction, **a** and α haploid cells first communicate with each other by secretion of the cell type-specific pheromones **a**-factor and α-factor, respectively.[44] Whereas diploid cells are insensitive to mating factor, haploid cells respond to the mating pheromone of the opposite mating type by arresting in G_1 and forming pear-shaped mating projections called shmoos. This prepares the haploids for the cell and nuclear fusion

[45] D. Burke, D. Dawson, and T. Stearns (eds.), "Methods in Yeast Genetics: A Cold Spring Harbor Laboratory Course Manual, 2000 Edition." Cold Spring Harbor Laboratory Press, Cold Spring Harbor, NY, 2000.

events. When the *HM* loci become derepressed, haploid cells do not arrest growth or form mating projections upon mating pheromone confrontation. Experiments with mating pheromone are routinely done with α-factor and not with **a**-factor, because α-factor is a peptide of 13 unmodified amino acids that can be easily synthesized and readily diffuses in agar plates, whereas biologically active **a**-factor is a modified hydrophobic peptide (see Ref. 46).

It has been shown that α-factor confrontation can provide a more sensitive readout for small disturbances in *HM* silencing than quantitative mating assays.[23] Basically, there are two types of α-factor response assays (adapted from Ref. 23).

*A. Measuring Fraction of **a**-Cells That Responds to α-Factor*

1. Resuspend exponentially growing *MAT***a** cells in liquid YEPD medium containing 500 ng/ml α-factor and incubate for 3 hr (α-factor is stored as a 2–5 mg/ml stock in methanol at -20°). If the strains grow slowly, it may take more than 3 hr before all the cells finish their cell cycle and arrest.

2. Using a microscope, determine the proportion of divided cells (cells with one or more buds) and the proportion of arrested cells (unbudded cells with or without mating projections). In wild-type strains usually ~70–100% of the cells will arrest in G_1.

B. Measuring Ability of Cells to Sustain Arrested State on Prolonged α-Factor Treatment

1. Place 5 μl of α-factor (200 μg/ml) on a 5-mm diameter region of a solid media YEPD plate.

2. Shortly after the 5 μl has dried, streak the *MAT***a** strain across this region. Immediately identify areas of well-separated cells under a microscope and mark the locations by puncturing the agar.

3. Incubate the cells at 23° and monitor cell shape over time.

4. After overnight incubation the vast majority ($\geq$95%) of the wild-type cells arrest as individual shmooing cells. Mutants defective in maintenance of *HM* silencing during the cell cycle (as shown for *cac1* mutants[23]) initially arrest and form shmoos but will escape the arrest, start to divide, and form clusters (and eventually small colonies) of shmooing cells.

Reporter Genes Used in Silencing Assays

Silencing at telomeres and the rDNA is usually measured by placing reporter genes into the domain of interest. Also at the mating-type loci, exogenous reporter genes are frequently used because they can provide a more sensitive or more

[46] L. L. Breeden, *Methods Enzymol.* **283,** 332 (1997).

easily monitored silencing assay than quantitative mating or α-factor confrontation. Some reporters can be used to measure silencing by colony formation on positive- or negative-selection plates (*URA3*, *TRP1*, *HIS3*, *CAN1*, P_{URA3}-*ADE2*), while other markers provide convenient colony color assays (*ADE2*, *MET15*). The color markers have the added benefit that variegated expression within a colony, or relative levels of expression between colonies, can be readily detected. In addition, LacZ has been used to measure expression levels in a population of cells[47] and green fluorescent protein (GFP) has been used to measure silencing at a mutant *HMR* locus in live cells.[48] The reporter genes and plasmids used to integrate them in the genome are listed in Table I. Usually, silencing assays are used to test the effect of a genetic or physiological perturbation on chromatin-mediated silencing. However, reporter genes with nearby LexA or Gal4 binding sites have also been used to directly test whether recruitment of a specific protein can create a silenced domain.[49,50]

Colony Color Assays

A. ADE2. The color assay used for the *ADE2* reporter is a simple assay that does not require any staining or special plating. Normally, colonies expressing *ADE2* are white, while those not expressing it are red because of the accumulation of a red pigment.[51,52] Since the *ADE2* gene at a telomere stochastically switches between transcriptional states, a single colony is usually composed of red and white sectors (Fig. 2A).[15] When a cell in a colony switches, the new transcriptional state is typically inherited by the daughter cells and will propagate for many generations. The degree of sectoring is dependent on the strength of silencing and the heritability of the transcriptional states. Because of the stochastic nature of TPE, there is usually some variation in the sectoring pattern between different colonies on the same plate (Fig. 2A). Note that in general, cells that accumulate the red pigment do not grow as well as white cells,[52] which affects the relative size of the red and white sectors in a colony. It has been reported that lowering adenine and increasing glucose levels in the media reduces this disparity.[53]

Several factors influence the degree of red pigment formation. (1) Temperature affects color formation. Because pigment develops better at lower temperatures it usually helps to store the plates at 4° for several days after the colonies have grown

[47] C. Boscheron, L. Maillet, S. Marcand, M. Tsai-Pflugfelder, S. M. Gasser, and E. Gilson, *EMBO J.* **15,** 2184 (1996).

[48] P. U. Park, P. A. Defossez, and L. Guarente, *Mol. Cell. Biol.* **19,** 3848 (1999).

[49] A. J. Lustig, C. Liu, C. Zhang, and J. P. Hanish, *Mol. Cell. Biol.* **16,** 2483 (1996).

[50] C. A. Fox, A. E. Ehrenhofer-Murray, S. Loo, and J. Rine, *Science* **276,** 1547 (1997).

[51] H. Roman, *Cold Spring Harbor Symp. Quant. Biol.* **21,** 175 (1956).

[52] S. Ugolini and C. V. Bruschi, *Curr. Genet.* **30,** 485 (1996).

[53] K. W. Runge and V. A. Zakian, *Mol. Cell. Biol.* **9,** 1488 (1989).

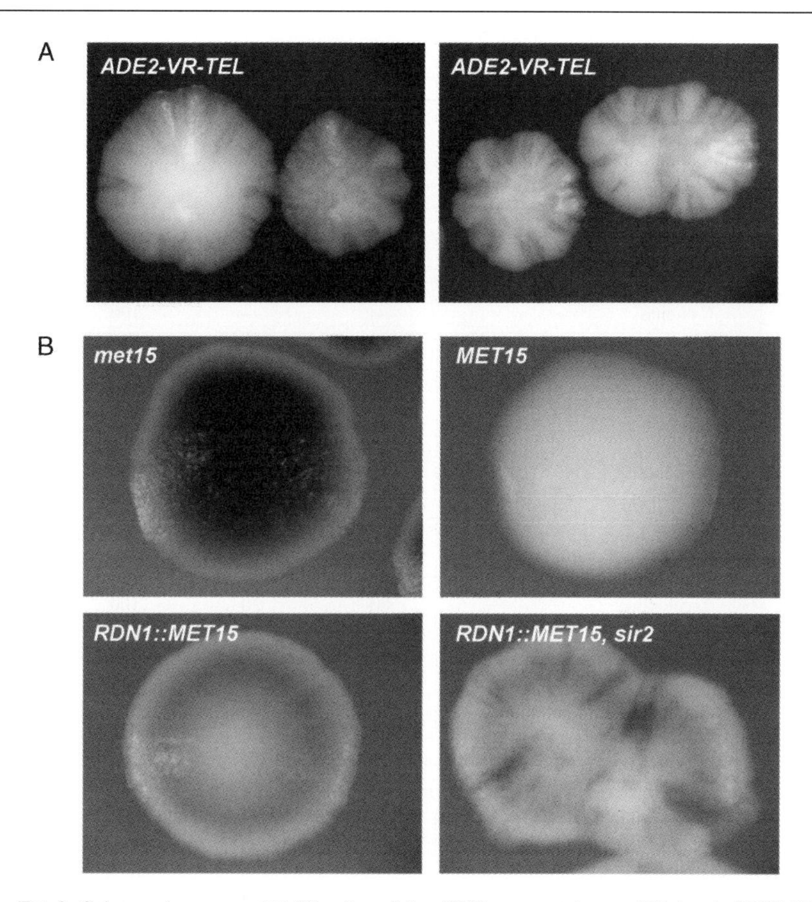

FIG. 2. Colony color assays. (A) Silencing of the *ADE2* gene at telomere VR (strain UCC3503, see Table V) results in colonies composed of red sectors (*ADE2* silenced) and white sectors (*ADE2* transcribed). The degree of sectoring depends on the strength of silencing and the heritability of the transcriptional states. (B) Silencing of the *MET15* gene at the rDNA results in brown/tan colonies when grown on MLA media (see text), whereas *MET15* and *met15* strains are white and dark brown, respectively. Deletion of *SIR2* results in reduced silencing (light tan colony) and elevated levels of loss of the *MET15* reporter gene by recombination, indicated by dark brown sectors. [Strains shown are BY4714 (*MET15*), BY4705a (*met15*), JSS237 (*RDN1::MET15*), JSS218 (*RDN1::MET15, sir2*); J. S. Smith and J. D. Boeke, *Genes Dev.* **11,** 241 (1997); C. B. Brachmann A. Davies, G. J. Cost, E. Caputo, J. Li, P. Hieter, and J. D. Boeke, *Yeast* **14,** 115 (1998)].

to the desired size. (2) Color formation depends on the type of media used. This is probably in part caused by the difference in concentration of amino acids and purines (such as adenine). In our laboratory we have found that color develops well on standard YEPD plates. On synthetic plates we have found that color develops very well on synthetic YC media, but is much fainter on synthetic HC media.

TABLE II
SYNTHETIC MEDIA

Constituent[a]	Concentration (mg/liter) HC[b]	SC[b]	YC[c]
Adenine	20	20	10
Arginine	20	20	100
Aspartic acid	100	100	50
Cysteine	—	—	100
Glutamic acid	100	100	—
Histidine	20	20	50
Isoleucine	80	30	50
Leucine	80	30	100
Lysine	120	30	100
Methionine	20	20	50
Phenylalanine	50	50	50
Proline	—	—	50
Serine	400	400	50
Threonine	200	200	100
Tryptophan	80	20	100
Tyrosine	60	30	50
Uracil	35	20	100
Valine	150	150	50
Bacto-yeast nitrogen base	1450	1700	1700
Ammonium sulfate	5000	5000	5000

[a] All compounds are from Sigma (St. Louis, MO), except Bacto-yeast nitrogen base (Difco, Detroit, MI). Sources through Sigma occasionally change, which can affect growth characteristics in silencing assays.

[b] See D. Burke, D. Dawson, and T. Stearns (eds.), "Methods in Yeast Genetics: A Cold Spring Harbor Laboratory Course Manual, 2000 Edition." Cold Spring Harbor Laboratory Press, Cold Spring Harbor, NY, 2000.

[c] Modified from ATCC (Manassas, VA) Culture Medium 1984 YC, M. DuBois and D. Gottschling, unpublished.

Details of the composition of synthetic media are shown in Table II. (3) The exact type and intensity of the color is also dependent on the strain background. Strains derived from W303–1a[39] develop a very intense red color; those from the BY series[40] (originally derived from S288C) also develop the color well, but with a brown hue. In strains derived from the YPH series[38] (S288C) the color is less intense and takes longer to develop (e.g., UCC3505, UCC4563).[54]

[54] M. S. Singer, A. Kahana, A. J. Wolf, L. L. Meisinger, S. E. Peterson, C. Goggin, M. Mahowald, and D. E. Gottschling, *Genetics* **150,** 613 (1998).

B. MET15. On media containing Pb^{2+} (MLA, see below) *met15* mutant strains form dark brown to black colonies, due to PbS precipitating on the surface of the cells. In contrast, *MET15* strains are white on MLA media.[55] When the *MET15* gene is inserted within the rDNA array in a *met15* strain, an intermediate tan color is observed (Fig. 2B). The intensity of the tan color can vary between independent isolates, which likely reflect where *MET15* integrated within the 100–200 tandem copies of the rDNA array (see Fig. 1).[13] Unlike the *ADE2* gene at telomeres, the *MET15* gene within the rDNA does not result in a variegated colony, suggesting that the *MET15* gene does not switch back and forth between active and fully repressed transcriptional states, but is constantly repressed at an intermediate level.[13] While black sectors can occasionally be seen in a colony, these represent loss of *MET15* as a result of recombination between rDNA repeats (Fig. 2B).[13] Silencing and recombination at the rDNA seem to be intimately linked: loss of silencing at the rDNA by deletion of *SIR2* results in increased recombination at the rDNA,[6,7] and consequently in higher loss rates of the reporter gene (see Fig. 2B).

The *MET15* color assay is done on solid modified lead acetate medium (MLA). One liter of MLA contains 3 g peptone, 5 g yeast extract, 200 mg $(NH_4)_2SO_4$, 20 g agar, 40 g glucose, and 1 g $Pb(NO_3)_2$. Add lead nitrate (stock solution of 10 mg/ml water, filter sterilized) and glucose to medium after autoclaving and cooling of the medium to 55°. MLA medium should fill at least half of the volume of a standard petri dish; thin plates do not support complete growth or color development. Growth on MLA plates takes about twice as long as compared to YEPD or complete synthetic media. Similar to the *ade2* color development, the brown color intensifies over time and is enhanced by incubation at 4° for one or more days. Extended incubation on MLA plates results in wild-type strains becoming yellow.

Markers Used for Selection

A. URA3. The *URA3* gene is the most versatile of the silencing assay reporters. There are both positive and negative selections for *URA3* expression. Its gene product is required for growth in media lacking uracil. The chemical 5-fluoroorotic acid (5-FOA) is used in the negative selection of URA3 expression; 5-FOA is converted into a toxic product by the *URA3* gene product.[56] When URA3 is silenced, cells are resistant to 5-FOA. In addition to positive and negative selection, 6-azauracil (6-AU) can be used to measure changes in the level of *URA3* expression.[54,57] 6-AU is a competitive inhibitor of the *URA3* enzyme; the higher the expression of *URA3*, the greater the resistance of the strain to 6-AU. By combining positive and negative selection schemes, small changes in silencing can be measured over a broad range of expression levels (Fig. 3A).

[55] G. J. Cost and J. D. Boeke, *Yeast* **12,** 939 (1996).

[56] J. D. Boeke, J. Trueheart, G. Natsoulis, and G. R. Fink, *Methods Enzymol.* **154,** 164 (1987).

[57] G. Loison, R. Losson, and F. Lacroute, *Curr. Genet.* **2,** 39 (1980).

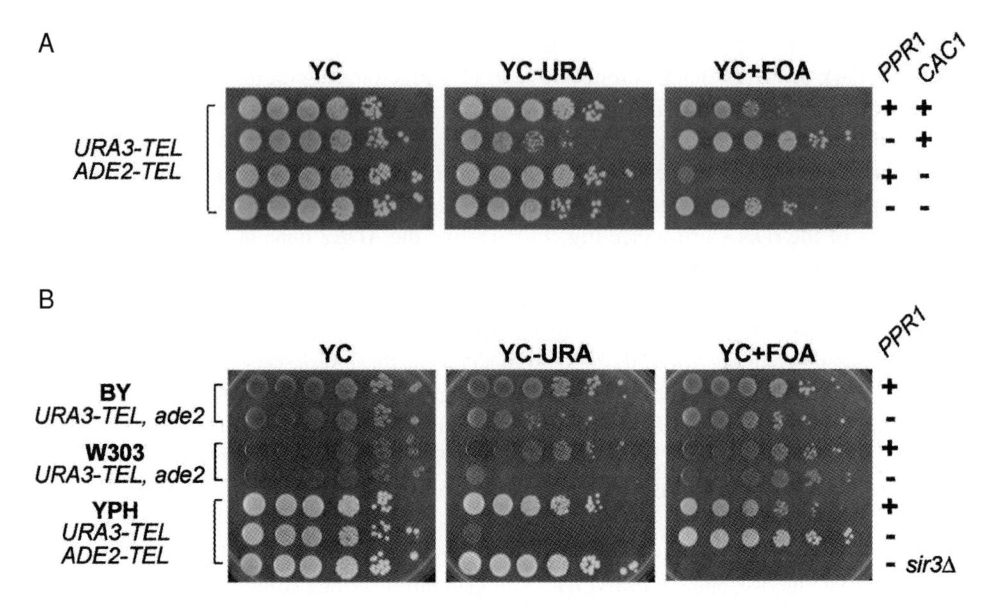

FIG. 3. Spot tests to measure silencing of a telomeric *URA3* gene. Tenfold serial dilutions of strains containing the *URA3* gene integrated at telomere VIIL. (A) The combination of positive selection (YC-URA) and negative selection (YC+5-FOA) allows the detection of differences in silencing of the *URA3* gene over a wide range of expression levels. *PPR1* is the transactivator of *URA3*; *CAC1* is a component of the chromatin assembly complex CAF-1, which when mutated partially cripples silencing (see text). The appearance of tiny colonies indicates that the repressed or active transcriptional state is not inherited in a stable fashion in the growing colony. Strains are derived from BY4705 (see Fig. 2). (B) The degree of silencing of *URA3-TEL-VIIL* depends on the strength of the promoter (absence or presence of the *PPR1* transactivator, see text) and strain background. In the absence of *PPR1*, silencing of *URA3* is very tight in YPH and W303 derivatives, whereas silencing in BY derivatives is weaker. Deletion of *SIR3* completely abolishes silencing, even in the absence of *PPR1*. Strains are derived from BY, W303-1a, or YPH as indicated (see text for details).

The level of *URA3* silencing can be increased by the deletion of *PPR1*.[36] The transcriptional activator of the *URA3* gene, *PPR1*, is important for induced expression of *URA3*. When medium lacks uracil, it induces *URA3* expression sufficiently to overcome silencing, which explains why cells can grow on 5-FOA (no *URA3* expression) and also grow when switched to media lacking uracil. When *PPR1* is mutated, silencing of the *URA3* gene is increased (Fig. 3) because the *URA3* promoter is effectively crippled and cannot respond to a lack of uracil. Thus, *PPR1* deletion can modulate the sensitivity range of the silencing assay.

We find that silencing of a telomeric *URA3* gene is usually greater in YPH[38] or W303[39] strains than it is in the BY background.[40] In YPH strains with *URA3* at the VIIL telomere, deletion of *PPR1* results in an almost total lack of growth on media without uracil and full resistance to 5-FOA (see Fig. 3B). A related way to make *URA3* more sensitive has been to replace its promoter with a minimal *TRP1* promoter (*mURA3*).[13] Strains such as these provide a powerful and sensitive genetic tool to identify perturbations in silencing.

Silencing of *URA3* at telomeres decreases with distance from the telomere[18] and is affected by proximity of subtelomeric sequences.[58,59] Thus, the most commonly used telomeric *URA3* reporters are located very close to the telomere (see Fig. 1).

B. TRP1. Silencing of *TRP1* can be measured by examining growth in the absence of tryptophan. However, a significant growth defect has been detected at *HMR***a** and near telomeres only when a *TRP1* gene without its upstream activating sequence (UAS) is used.[15,60]

C. HIS3. The normal *HIS3* gene is poorly silenced at the rDNA locus and can therefore be used on tandem gene constructs to select for integration of other reporter genes.[13] *HIS3* silencing at telomeres is very slight but can be improved by using a promoter defective version of the reporter[37] or by deleting its transcriptional activator, *GCN4* (see above).[18] Typically, measuring silencing of *HIS3* by growth on media lacking histidine is not a very sensitive assay. However, as described above for *URA3*, the histidine analog 3-amino-1,2,4-triazole (3-AT) is a competitive inhibitor of the *HIS3* gene product.[61,62] The higher the expression of *HIS3*, the greater the resistance of the strain to 3-AT. The concentration at which 3-AT inhibits growth varies with strain background, location of the *HIS3* gene, and strength of the promoter.[63]

D. ADE2. While silencing of the *ADE2* gene is usually monitored by a colony color assay[15] (see above), it can also be examined in media lacking adenine. However, analogous to *URA3* and other reporter genes, *ADE2* expression is induced in the absence of adenine. Thus, successful use of *ADE2* as a silencing reporter has been accomplished by swapping its promoter with the *URA3* promoter (which is not induced in the absence of adenine). It has been reported that the chimeric *ADE2* gene is more sensitive to measure silencing at *HMR* than quantitative mating assays.[64,65]

E. CAN1. The *CAN1* gene has been used to measure silencing at the rDNA by negative selection.[66] Canavanine is an analog of arginine and both are imported into

[58] F. E. Pryde and E. J. Louis, *EMBO J.* **18,** 2538 (1999).

[59] G. Fourel, E. Revardel, C. E. Koering, and E. Gilson, *EMBO J.* **18,** 2522 (1999).

[60] L. Sussel and D. Shore, *Proc. Natl. Acad. Sci. U.S.A.* **88,** 7749 (1991).

[61] G. M. Kishore and D. M. Shah, *Annu. Rev. Biochem.* **57,** 627 (1988).

[62] C. Alexandre, D. A. Grueneberg, and M. Z. Gilman, *Methods* **5,** 147 (1993).

[63] D. de Bruin, Z. Zaman, R. A. Liberatore, and M. Ptashne, *Nature* **409,** 109 (2001).

[64] D. H. Rivier, J. L. Ekena, and J. Rine, *Genetics* **151,** 521 (1999).

[65] T. H. Cheng and M. R. Gartenberg, *Genes Dev.* **14,** 452 (2000).

[66] C. F. Fritze, K. Verschuren, R. Strich, and R. Easton Esposito, *EMBO J.* **16,** 6495 (1997).

the cell by the yeast *CAN1* permease. Cells that are mutant for *CAN1* are resistant to canavanine, whereas cells that express *CAN1* are sensitive to canavanine.

Spot Tests

A fast and simple way to assay for silencing is to plate the cells on selective solid media in serial dilutions. This quantitative assay measures the fraction of cells in a population that express the reporter gene at levels sufficient to permit colony growth on dropout media or to make the cells sensitive to toxic compounds such as 5-FOA (see Fig. 3 and above). Typically, 6–8 strains are tested on a standard 10-cm plate.

Spot Test Protocol

1. Add 90 μl of sterile H_2O to each well of a 6×6 grid in a 96-well microtiter plate.
2. For each strain, pick a medium-size freshly grown colony (or the equivalent of that; approximately 10^6–10^7cells) with a sterile pipette tip and resuspend it in the first well of each respective row.
3. Carefully resuspend the cells and then make a 10-fold serial dilution by transferring 10 μl to the next column of wells and so on.
4. Apply 7–15 μl of each cell suspension (resuspended well with a pipette) onto the selection plates to measure silencing and onto nonselective plates to control for cell number and plating efficiency. Start at the column with the highest dilution and end with the undiluted cell suspension.
5. Let the spots dry, then incubate the plates at the appropriate temperature for approximately 2–4 days. Document colony growth at several time points (e.g., 2 days and 3 days) via photography. Although bigger colonies are easier to count and to document, small differences in colony number or colony size are sometimes more obvious at earlier time points.
6. The level of silencing can be calculated by determining the fraction of cells that can form colonies on (negative) selection media relative to the plating control.

Notes

1. The use of a multichannel pipette can be very helpful for making the dilutions and for applying the cells onto the plates.
2. The total volume of cell suspension that can be applied onto a plate depends on the spacing between the spots and the type of media. The surface tension of rich media (YEPD) is lower than that of synthetic media. We usually apply 7–10 μl onto YEPD and 8–15 μl onto synthetic plates.
3. Because silencing can be affected by the exact growth conditions, type of media, or temperature (see above), it is important that the mutant and control strains

TABLE III
DRUGS USED FOR NEGATIVE SELECTION OR COMPETITIVE INHIBITION

Drug[a]	Target gene product	Concentration	Stock[b]	Synthetic media
5-FOA	*URA3*	1 g/liter	Powder	Low uracil[c]
6-AU	*URA3*	3–20 mg/liter	5 mg/ml	Lacking uracil
3-AT	*HIS3*	5–100 m*M*	2.5 *M*	Lacking histidine
Canavanine	*CAN1*	60 mg/liter	20 mg/ml	Lacking arginine

[a] All compounds are from Sigma, except 5-FOA (Toronto Research Chemicals, North York, ON, Canada).

[b] Stocks are made in sterile water, heated to 50° to dissolve if necessary, and filter sterilized. Compounds are added to the media after autoclaving just prior to pouring the plates. The drugs do not work in rich media.

[c] For YC+5-FOA media use 20 mg/liter uracil. For HC or SC media use uracil concentration indicated in Table II.

be tested in the same experiment on the same batch of plates. To get consistent results, it is best to use fresh colonies and pregrow all the strains at the same time on plates of the same batch. To control for cell number and plating efficiency it is recommended to use nonselective media plates that resemble the selection plates most closely, e.g., if dropout media are used for positive selection, use nonselective synthetic media as a control, not YEPD.

4. Media recipes are given in Tables II and III.

5. Because reporter genes stochastically switch between transcriptional states, variation can occur between individual colonies. To control for the natural variation, measurements should be done on at least three colonies.

6. If an inducible promoter drives the expression of one of the silencing components, it is important to pregrow the strains on the solid medium that will be used for the spot test. This will allow for the induction/shutoff of the induced gene and for the subsequent silencing/derepression of the reporter gene. Full establishment of a silencing phenotype can take several cell generations.

7. The interpretation of a spot test can sometimes be complicated by variation in colony size (e.g., see Fig. 3A). However, colony size can also provide information about the heritability of silencing in the growing colony.[67] Small colony size can occur for several reasons. It may indicate that all the cells grow more slowly because they express the reporter at very low levels. It could also indicate that the reporter is initially expressed, but that the active state is not maintained in all the descendents of the cell in the colony. On negative selection plates, small colony size may indicate that cells in the colony switch from the repressed transcriptional state, which allows growth, to the activated state, which does not allow growth. Whereas in the *ADE2* color assay these cells would be represented as white sectors

[67] E. K. Monson, D. de Bruin, and V. A. Zakian, *Proc. Natl. Acad. Sci. U.S.A.* **94,** 13081 (1997).

in a red colony (Fig. 2A), under negative selection such as 5-FOA these subpopulations will die and not contribute to the colony. Maintenance or heritability of repressed states can be further studied at a single-cell level by micromanipulation (see below). Unusually large colonies typically represent spontaneous mutations that have a silencing defect.

Analysis of Individual Cells and Their Offspring by Micromanipulation

Silencing can be determined by several criteria. It can be measured as the average expression of a gene in a population of cells (e.g., by Northern or Western blot, β-Gal assays, or colony color), or as the fraction of cells that express or silence a gene (e.g., by spot test, colony sectoring, quantitative mating, or α-factor response). Silencing can also be monitored at a single-cell level by micromanipulation to determine if, and how, the transcriptional state of a cell is inherited by its descendents. By monitoring the heritability of α-factor sensitivity (or resistance) via pedigree analysis, it has been shown that *SIR1* and the continuous presence of the silencers are important for the maintenance of *HM* repression.[68,69] Micromanipulation has also been used to examine the heritability of and the rate of switching between transcriptional states of a telomeric *URA3* gene.[67] Sherman and Hicks have previously described micromanipulation in great detail.[70] Ready-to-use optical fiber dissecting needles for yeast micromanipulation are available from Cora Styles, Arlington, MA (25 μm; cstyles@tiac.net).

Overview of Recent Genetic Screens at Silent Loci

Over the past few years there have been a number of genetic screens/selections, using the assays described above, to identify components and processes involved in silencing at *HML*, *HMR*, telomeres, and the rDNA. In Table IV we present some examples of how the assays were used in specific screens. In some screens the readout was disruption of silencing; in others it was enhancement of silencing or suppression of weak silencing defects. In some cases *HML* or *HMR* was "sensitized" by deletion of one of the silencers and/or silencer elements (A, E, or B; see Fig. 1), by replacing the wild-type *HMR-E* silencer with a weaker synthetic silencer (ss), and/or by using silencing defective alleles of *SIR1* or *RAP1*. The silencing assays employed in these mutant hunts included mating, growth on selective media, and colony-color tests.

[68] L. Pillus and J. Rine, *Cell* **59,** 637 (1989).

[69] S. G. Holmes and J. R. Broach, *Genes Dev.* **10,** 1021 (1996).

[70] F. Sherman and J. Hicks, *Methods Enzymol.* **194,** 21 (1991).

TABLE IV
ASSAYS USED FOR MUTANT HUNTS

Locus	Reporter[a] (relevant genotype)	Screen	Desired phenotype	Ref.[c]
HMR/HML	*HMR***a***-e*** (*HMR***a***ΔAB*) (*MATα*)	Repression of *HMR***a** (suppress silencing defect of the crippled silencer)	Restore mating	(1, 2)
	pHMRα-E (*rap1-10, mat***a***1, ste14*)	*HMRα* derepression (results in switch from default **a**-type to α-type)[b]	Mating to **a**-type strain	(3)
	pHMRssa (*MATα, HMLα, HMRα*)	*HMR***a** derepression	Lack of mating	(3)
	HMRΔA::TRP1 (or *ADE2*) (*rap1^S*)	Suppress silencing defect caused by *rap1* mutation and weakened silencer	Growth on −tryptophan (or red colonies)	(4, 5)
	HMRssα (all **a** genes deleted)	*HMRα* gene derepression (results in switch from default **a**-type to α-type)[b]	Mating to **a**-type strain	(6)
	*HMR-ss***a** (*sir1-101*)	Suppression of silencing defect caused by hypomorphic *sir1* allele and weakened synthetic silencer.	Restore mating	(7)
	HMRss(A::Gal4), (*sir1Δ, GAL4-sir1-201/202*)	Suppression of silencing defect of *SIR1* mutations	Restore mating	(8)
	hml::TRP1	*TRP1* derepression	Growth on −tryptophan	(9)
	sir1Δ	Enhanced *sir1* mating defect	Lack of mating	(10, 11)
Telomere	*ADE2-VR-TEL* *URA3-VIIL-TEL*	*ADE2* and *URA3* derepression	Increased white sectoring and better growth on −uracil	(12)
rDNA	*RDN1::MET15* *RDN1::URA3*	*MET15* and *URA3* derepression	Lighter brown colonies and better growth on −uracil	(13)
	RDN1::MET15 *RDN1::URA3*	Increased *MET15* and *URA3* repression	Darker colonies and less growth on −uracil	(13)

[a] For details about the loci and reporter gene constructs, see Table I and Fig. 1.

[b] When no mating type gene is expressed, a haploid cell behaves like an **a**-type cell. Derepression of *HMRα* results in a phenotypic switch of the mating type to α.

[c] Key to references: (1) A. Axelrod and J. Rine, *Mol. Cell. Biol.* **11,** 1080 (1991); (2) E. Y. Xu, S. Kim, K. Replogle, J. Rine, and D. H. Rivier, *Genetics* **153,** 13 (1999); (3) M. Foss, F. J. McNally, P. Laurenson, and J. Rine, *Science* **262,** 1838 (1993); (4) L. Sussel, D. Vannier, and D. Shore, *Mol. Cell. Biol.* **13,** 3919 (1993); (5) L. Sussel, D. Vannier, and D. Shore, *Genetics* **141,** 873 (1995); (6) R. T. Kamakaka and J. Rine, *Genetics* **149,** 903 (1998); (7) P. C. Hollenhorst, M. E. Bose, M. R. Mielke, U. Muller, and C. A. Fox, *Genetics* **154,** 1533 (2000); (8) N. Dhillon and R. T. Kamakaka, *Mol. Cell* **6,** 769 (2000); (9) S. Le, C. Davis, J. B. Konopka, and R. Sternglanz, *Yeast* **13,** 1029 (1997); (10) C. Reifsnyder, J. Lowell, A. Clarke, and L. Pillus, *Nat. Genet.* **16,** 109 (1997); (11) E. M. Stone, C. Reifsnyder, M. McVey, B. Gazo, and L. Pillus, *Genetics* **155,** 509 (2000); (12) M. S. Singer and D. E. Gottschling, *Science* **266,** 404 (1994); (13) J. S. Smith, E. Caputo, and J. D. Boeke, *Mol. Cell. Biol.* **19,** 3184 (1999).

TABLE V
SILENCING STRAINS

Strain	Relevant genotype[a]	Strain background	Ref.[b]
UCC3505/UCC4563	*URA3-TEL-VIIL (ppr1), ADE2-TEL-VR*	YPH	(1, 2)
UCC3503/UCC4562	*URA3-TEL-VIIL (PPR1), ADE2-TEL-VR*	YPH	(2)
JS306/JS311	*RDN1(NTS)::Ty1-MET15, RDN1(18S)::mURA3/HIS3*	BY	(3)
PPY143	*hmr*Δ*2::ADH1-GFP, RDN1::ADE2*	YPH	(4)
YJB1633	*hmr*Δ*A::TRP1, URA3-TEL-VIIL*	W303	(5)
GCY23	*RDN1::MET15, hmr*Δ*A::TRP1, URA3-TEL-VIIL*	W303	(6)
GCY40	*RDN1::mURA3-HIS3, hmr*Δ*A::TRP1, ADE2-TEL-VIIL*	W303	(6)
CCFY100	*RDN1::ADE2-CAN1, hmr*Δ*E::TRP1, URA3-TEL-VR*	W303	(7)

[a] For details about the reporter genes see Table I and Fig. 1.

[b] Key to references: (1) M. S. Singer and D. E. Gottschling, *Science* **266,** 404 (1994); (2) M. S. Singer, A. Kahana, A. J. Wolf, L. L. Meisinger, S. E. Peterson, C. Goggin, M. Mahowald, and D. E. Gottschling, *Genetics* **150,** 613 (1998); (3) J. S. Smith, E. Caputo, and J. D. Boeke, *Mol. Cell. Biol.* **19,** 3184 (1999); (4) P. U. Park, P. A. Defossez, and L. Guarente, *Mol. Cell. Biol.* **19,** 3848 (1999); (5) S. Enomoto, S. D. Johnston, and J. Berman, *Genetics* **155,** 523 (2000); (6) G. Cuperus, R. Shafaatian, and D. Shore, *EMBO J.* **19,** 2641 (2000); (7) N. Roy and K. W. Runge, *Curr. Biol.* **10,** 111 (2000).

The variety of silencing assays described above also makes it possible to develop stringent screens that help to quickly separate relevant mutants from those that may be "less interesting." For instance, by using multiple reporter genes, mutants that are reporter gene specific can be avoided. For telomeric silencing a combination of *URA3* (weakened by deletion of *PPR1*) and *ADE2* was used[54,71]; for rDNA silencing m*URA3* and *MET15* were used.[72] In addition, strains that have a different reporter at each of the three silenced loci (*HM*, rDNA, and telomeres) can be used to screen for mutants that affect all loci or a specific subset of them. Some of the strains with useful silencing reporter combinations are listed in Table V.

Further Characterization of Silencing Mutants

If genetic or cellular assays indicate that a gene has an effect on silencing, then a better understanding of how it functions may be obtained by analyzing it in the context of other assays used to study silencing. Chromatin immunoprecipitation is perhaps one of the most informative assays to characterize silent loci. Using specific antibodies, the location along the genome of silencing components (i.e., Sir2p,

[71] M. S. Singer and D. E. Gottschling, *Science* **266,** 404 (1994).

[72] J. S. Smith, E. Caputo, and J. D. Boeke, *Mol. Cell. Biol.* **19,** 3184 (1999).

Sir3p, Sir4p) and of particular modified forms of histones can be ascertained.[73–76] New candidate genes can also be analyzed for their presence at a locus as well as how mutations affect the location of known silencing proteins. Such analysis can provide insights into the assembly of silent chromatin. Several useful protocols for chromatin immunoprecipitation are available.[77,78]

There is also a cell biological component to silencing that can be informative. Immunofluorescence and *in situ* hybridization methods have been used to study the subnuclear location of telomeres and silencing proteins.[79,80] In wild-type cells telomeres and telomeric silencing proteins form clusters and localize to the nuclear periphery and some silencing proteins colocalize with the nucleolus. However, in silencing mutants the location of silencing proteins can be dramatically altered.[17,28,81,82]

A third level of characterization is how nucleosome positioning and occupancy is affected in a silent locus. In some instances, nucleosome position is different than in nonsilenced loci.[83,84] This information can help explain why a gene is silenced. The mapping of nucleosome and transcription factor position is done with nucleases (DNase I or micrococcal nuclease) and expression of exogenous methyltransferases.[85–87] When used in combination these assays can provide a detailed view of changes in chromatin structure.

Conclusions

We hope that our description of silencing assays will be useful for developing a deeper understanding of silencing in *S. cerevisiae,* serve as a model for examining

[73] M. Braunstein, A. B. Rose, S. G. Holmes, C. D. Allis, and J. R. Broach, *Genes Dev.* **7,** 592 (1993).
[74] S. Strahl-Bolsinger, A. Hecht, K. Luo, and M. Grunstein, *Genes Dev.* **11,** 83 (1997).
[75] A. Hecht, S. Strahl-Bolsinger, and M. Grunstein, *Nature* **383,** 92 (1996).
[76] M. Braunstein, R. E. Sobel, C. D. Allis, B. M. Turner, and J. R. Broach, *Mol. Cell. Biol.* **16,** 4349 (1996).
[77] A. Hecht and M. Grunstein, *Methods Enzymol.* **304,** 399 (1999).
[78] P. B. Meluh and J. R. Broach, *Methods Enzymol.* **304,** 414 (1999).
[79] M. Gotta, T. Laroche, and S. M. Gasser, *Methods Enzymol.* **304,** 663 (1999).
[80] M. Cockell and S. M. Gasser, *Curr. Opin. Genet. Dev.* **9,** 199 (1999).
[81] B. K. Kennedy, M. Gotta, D. A. Sinclair, K. Mills, D. S. McNabb, M. Murthy, S. M. Pak, T. Laroche, S. M. Gasser, and L. Guarente, *Cell* **89,** 381 (1997).
[82] M. Gotta, S. Strahl-Bolsinger, H. Renauld, T. Laroche, B. K. Kennedy, M. Grunstein, and S. M. Gasser, *EMBO J.* **16,** 3243 (1997).
[83] K. Weiss and R. T. Simpson, *Mol. Cell. Biol.* **18,** 5392 (1998).
[84] A. Ravindra, K. Weiss, and R. T. Simpson, *Mol. Cell. Biol.* **19,** 7944 (1999).
[85] P. D. Gregory and W. Horz, *Methods Enzymol.* **304,** 365 (1999).
[86] M. P. Ryan, G. A. Stafford, L. Yu, K. B. Cummings, and R. H. Morse, *Methods Enzymol.* **304,** 376 (1999).
[87] M. P. Kladde, M. Xu, and R. T. Simpson, *Methods Enzymol.* **304,** 431 (1999).

silencing in other systems, and provide a reference for yeast researchers who find their studies lead them into the realm of silencing.

Acknowledgments

We thank M. DuBois and A. Stellwagen for media recipes and critical reading of the manuscript. F.v.L. is supported by fellowships from EMBO (ALTF 178-1998) and the Dutch Cancer Society (NKB/KWF). D.G. appreciates support from the National Institutes of Health (GM43893).

Section II

Making Mutants

[10] Classical Mutagenesis Techniques

By CHRISTOPHER W. LAWRENCE

Introduction

Generating mutants, to identify new genes and to study their properties, is the starting point for much of molecular biology. Forward mutations and metabolic suppressors obtained by reversion can provide powerful insights into the functions and relationships of normal gene products. Similarly, mutations and intragenic revertants provide the raw material for the analysis of gene product structure–function relationships. Site-specific mutagenesis and other methods based on recombinant DNA techniques are increasingly used for these purposes, and they are clearly the methods of choice where particular changes in specified genes or genetic sites are needed. Nevertheless, classical methods, in which cells are treated with mutagens, are likely to remain the chief means for inducing mutations in many circumstances because they require no prior knowledge of gene or product and are generally applicable: the user need only specify an appropriate alteration in phenotype. However, unless selection for the desired strain is possible, hunting for mutants can be extremely laborious and analyzing the material obtained even more so. Good planning, efficient mutagenesis, careful choice of strain, and effective mutant detection usually pay off in time and labor.

Choice of Mutagen and Dose

The best mutagens for most purposes are those that induce high frequencies of base-pair substitutions and little lethality. The widely used alkylating agents *N*-methyl-*N*′-nitro-*N*-nitrosoguanidine (MNNG) and ethylmethane sulfonate (EMS) fulfill these criteria but are highly specific in their action: they almost exclusively produce transitions at G-C sites.[1] For most purposes, such as forward mutagenesis, this specificity is unlikely to pose any problem, though it may be a disadvantage when, as in some kinds of reversion, particular kinds of mutations at specified sites are needed. Germicidal ultraviolet light (254 nm UV) is also a fairly efficient mutagen, and it has the advantage of producing a greater range of substitutions[2,3]: most occur in runs of pyrimidines, particularly T-T pairs, and include both transitions and transversions. UV also induces a significant frequency of frameshift mutations, almost exclusively of the single nucleotide deletion variety. One or other of the chemical mutagens together with UV are likely to satisfy

[1] S. E. Kohalmi and B. A. Kunz, *J. Mol. Biol.* **204,** 561 (1988).

[2] B. A. Kunz, M. K. Pierce, J. R. A. Mis, and C. N. Giroux, *Mutagenesis* **2,** 445 (1987).

[3] G. S. F. Lee, E. A. Savage, R. G. Ritzel, and R. C. von Borstel, *Mol. Gen. Genet.* **214,** 396 (1988).

0076-6879/02 $35.00

most experimental needs, and it may be an advantage to induce different samples of mutants with each of these agents (for an example, see Ref. 4). Although base-pair substitution mutagens have most general utility, the alkylating acridine mustard ICR-170 (2-methoxy-6-chloro-9-[3-(ethyl-2-chloroethyl)aminopropylamino] acridine dihydrochloride can be used when +1 frameshift mutations are required.[5–7] The majority of mutations induced by this agent are single G insertions in runs of two or more G's, with preference for runs of three or more G's.[7]

Choosing an optimal dose usually requires balancing the competing needs for a high mutation frequency, reasonably high survival, and avoidance of multiple mutations. The highest proportion of mutants per *treated* cell is usually found at doses giving 10 to 50% survival. The highest fraction of mutations per *surviving* cell most commonly requires a larger dose, but mutation frequencies often decline at very high doses. In any case, it is desirable to avoid doses that kill more than 95% of cells, because they may select multicell clusters or atypically resistant variants, which occur spontaneously in all cell populations. In addition, multiple mutants become more common and may interfere with analysis. An indication of the effectiveness of the mutagenic treatment in the particular strain used can be obtained by measuring the frequency of canavanine-resistant mutants that it induces.

Growth Conditions after Mutagen Treatment

After being treated with mutagens, cell cultures should be allowed to grow for several generations under nonselective or permissive conditions, to enhance the production and expression of mutations. With some mutagens, such as EMS, mutations are thought to occur principally during S-phase replication, and unrepaired damage can continue to produce mutations in successive generations. However, with others, such as UV, most mutations probably occur during G_1 excision repair synthesis. Growth is also required to promote dilution and turnover of gene products, or the synthesis of new ones, to allow full expression of the mutant or revertant phenotype. In addition, cells may require time to recover from mutagen damage, which can cause some cells to stop growing temporarily or to grow more slowly. Full recovery from mutagen damage is particularly important when mutagen enrichment procedures are used.

Various ways of accomplishing outgrowth of mutagenized cultures can be chosen, depending on experimental needs. Plating dilutions of treated cells on solid medium, to get colonies for screening, has the advantage that each induced mutation identified is of independent origin. Many of the desired mutations may occur as sectors in otherwise normal colonies, however, and therefore be hard to

[4] R. Sitcheran, R. Emter, A. Kralli, and K. R. Yamamoto, *Genetics* **156,** 963 (2000).
[5] D. J. Brusick, *Mutat. Res.* **10,** 11 (1970).
[6] M. R. Culbertson, L. Charnas, M. T. Johnson, and G. R. Fink, *Genetics* **86,** 745 (1977).
[7] L. Mathison and M. R. Culbertson, *Mol. Cell. Biol.* **5,** 2247 (1985).

detect by some screening procedures. Outgrowth in liquid medium is convenient and allows segregation of pure mutant clones, but different mutant isolates may represent repeat copies of the same, rather than independent, mutational events. If outgrowth in liquid medium is needed, independent mutations can be isolated by dividing the mutagenized culture before outgrowth and taking a single mutant from each subculture. However, more than one mutant can be taken if they are shown to be different by sequence analysis. When selective methods allow a large number of cells to be spread on each plate, as in the selection of prototrophs from an auxotrophic strain, outgrowth can be achieved by adding small amounts of the required nutrilite to the medium. In experiments of this kind it is usually advisable to spread no more than about 10^7 cells on each plate, since the efficiency with which revertants are recovered drops greatly at higher cell densities. Finally, it should be noted that all mutagens increase the frequency of rho^0 petites, some, like ICR-170, to very high levels. It may therefore be useful to grow mutagenized cultures in medium containing a nonfermentable carbon source, such as glycerol, to avoid recovering such strains.

Choice of Strain

With many experimental species, it is customary to isolate mutations in a designated wild-type strain or genetic background, but there is no such wild type of *Saccharomyces cerevisiae* in general laboratory use. However, many mutations have been isolated in the haploid strains A364A and S288C, both of which are obtainable from the Yeast Genetics Stock Center (Berkeley, CA). In addition, a pair of strains of opposite mating type, X2180-1A and X2180-1B, that are both isogenic with S288C are also available. These are useful for mutant "cleanup" (see below).

Although these strains are sometimes useful, for many purposes it will be necessary to select a strain tailored to meet the investigator's specific experimental needs, and it is particularly important to examine the strain chosen carefully, to ensure its suitability. In addition to building into it any particular mutations that may be required for enrichment, selection, mutant detection, or analytical methods, it is prudent to check that the parental strain performs satisfactorily with respect to mating, transformation, and, when crossed to other strains, sporulation, since some laboratory strains perform poorly in these respects. Nonflocculent strains that give single-budded cells directly or after brief sonication are also highly desirable. Further, haploid yeast strains may carry additional copies of one or more chromosomes, and such aneuploidy may underlie the not-uncommon failure to recover mutations at one locus, even though similar mutations at other loci are found readily. When exhaustive mutagenesis studies are planned, it may therefore be desirable to use a variety of unrelated strains. Alternatively, the presence of aneuploidy in the parental strain can be investigated genetically, by crossing it with strains carrying recessive markers and analyzing the tetrads. Aneuploidy is unlikely if 2 : 2 segregation for these markers is observed. Pulsed gel electrophoresis may also be used to detect aneuploidy.[8,9]

Mutant Enrichment Procedures

Although mutants can sometimes be selected, they more often can be isolated only by screening individual clones from mutagenized cell populations, a highly laborious process. Enrichment procedures, which increase the proportion of mutants, can sometimes be used to reduce this labor. Various procedures of this kind have been proposed,[10–13] but most depend on the same principle: the use of conditions which temporarily prevent mutant, but not nonmutant, growth and which promote the selective killing of growing cells. The method using inositol starvation[12] to achieve selective killing is convenient and has been widely used. For good enrichment with mutagenized cultures, the cells must be grown nonselectively for several generations, to allow mutant expression and to promote recovery of damaged, but nonmutant, cells. To ensure the independent origin of the mutants eventually isolated, such outgrowth can be done on solid medium. Alternatively, a single mutant can be isolated from each of a series of liquid cultures.

Mutant "Cleanup"

Since strains treated with powerful mutagens often contain mutations in more than one gene, the mutant phenotype initially observed may be a misleading compound of several individual phenotypes. It is therefore usually helpful to "clean up" the strain by placing the mutation of interest in a nonmutagenized genetic background. This can be done by repeated backcrossing to an untreated isogenic strain or, if the locus in question's identity is known, by cloning the mutant gene by PCR (polymerase chain reaction) and transferring it to an untreated strain by gene replacement.

Safety

Powerful mutagens are powerful carcinogens: their use and disposal require care. Chemical mutagens should be handled only in a hood, using protective

[8] M. V. Olsen, *in* "The Molecular Biology of the Yeast *Saccharomyces,* Genome Dynamics, Protein Synthesis, and Energetics" (J. R. Broach, J. R. Pringle, and E. W. Jones, eds.), p. 1. Cold Spring Harbor Laboratory Press, Cold Spring Harbor, NY, 1991.

[9] A. J. Link and M. V. Olsen, *Genetics,* 681 (1991).

[10] R. Snow, *Nature* (*London*) **211,** 206 (1966).

[11] B. S. Littlewood, *in* "Methods in Cell Biology" (D. M. Prescott, ed.), Vol. 11, p. 273. Academic Press, New York, 1975.

[12] S. A. Henry, T. F. Donahue, and M. R. Culbertson, *Mol. Gen. Genet.* **143,** 5 (1975).

[13] M. T. McCammon and L. W. Parks, *Mol. Gen. Genet.* **186,** 295 (1982).

clothing, gloves and eye protection. MNNG decomposes to release volatile diazomethane, a powerful carcinogen, and EMS is itself volatile. Handle open bottles only in a hood in good working condition (i.e., air flow at face height of 150 fpm) with the window closed as much as possible, and avoid inhaling the volatile materials. Keep a freshly made supply of 10% (w/v) sodium thiosulfate on hand, to deal with accidental spills. Treatments with MNNG and EMS can be stopped, and the mutagens destroyed, by making the cell suspension 5% in sodium thiosulfate, using a filter-sterilized stock solution of this reagent. ICR-170, removed from cells by centrifugation, can be destroyed by making the solution 0.1 *M* in sodium hydroxide. Supernatants of inactivated chemical mutagens are often toxic, so should be handled with care; consult your institutional hazardous waste unit with respect to their disposal. Germicidal UV (principally 254 nm UV) is particularly damaging to the eyes, but it can also cause sunburn and skin cancer, and it is important to avoid all exposure of the skin to the radiation by wearing a UV-opaque face mask, opaque gauntlet gloves, and protective clothing if exposure is anticipated. UV tubes should be housed in a wood or metal structure, with screened ventilation louvers, painted matte black. Samples being irradiated can be observed through 6-mm-thick Lucite, which effectively blocks scattered UV.

Methods

MNNG and EMS Mutagenesis

1. Inoculate 10 ml (or other appropriate volume) of liquid YPD medium with a freshly subcloned sample of the yeast strain to give approximately 1×10^6 cells/ml (just detectably turbid). Incubate overnight at 30° with vigorous shaking. In the morning, the culture should contain about 2×10^8 cells/ml.

2. Wash 2.5 ml of the overnight culture twice in 50 m*M* potassium phosphate buffer, pH 7.0, and resuspend in 10 ml of this buffer. Cell concentration should be $\sim 5 \times 10^7$ cells/ml. Check cell concentration with a hemocytometer, and adjust if necessary. Observation of the cells on the hemocytometer slide will also indicate the presence of cell clumping. If sonication is needed to disperse clumped cells, chill cell suspension in ice and sonicate for 15 sec. A second cycle of chilling and sonication can be given, but further cycles are unlikely to be of benefit. As mentioned above under "Choice of Strain," it is important to select a strain that gives single-budded cells directly, with brief sonication.

3. (a) For MNNG mutagenesis, add 40 μl of a solution of MNNG in acetone (10 mg/ml) to 10 ml of cells in a screw-cap glass tube, tighten the cap, and mix well. Carry out all operations in a hood, wear gloves and a laboratory coat, and avoid inhaling volatile substances. Incubate in the tightly capped tube at 30° without shaking for 60 min. Add 40 μl of acetone without MNNG to an identical cell suspension to serve as control and for the determination of cell survival. The MNNG solution is made by dispensing (in a hood, with window lowered as much

as possible) approximately 10 mg of MNNG into a capped, preweighed glass vial, followed by reweighing and the addition of a sufficient volume of acetone to bring the concentration to 10 mg/ml. Transfer MNNG from bottle to vial over a tray, to catch accidental spills. Do not attempt to weigh in a hood, since the air flow interferes with this process. MNNG is light sensitive, and the vial should therefore be wrapped in aluminum foil or otherwise darkened, and the mutagen handled in subdued light.

(b) For EMS mutagenesis, add 300 μl of EMS to 10 ml of cells in a screw-cap glass tube, tighten the cap well, and vortex vigorously: EMS is poorly miscible in the buffer. Incubate for 30 min at 30° with shaking. Carry out all operations in a hood, wear gloves and a laboratory coat, and avoid inhaling volatile substances. Most commercial samples of EMS contain contaminants that increase its toxicity but not mutagenicity: redistilled EMS is a significantly better mutagen. Set up an identical cell suspension without EMS to serve as control.

4. Stop MNNG and EMS mutagenesis in the cell suspensions by adding, in a hood, an equal volume of a freshly made 10% (w/v) filter-sterilized solution of sodium thiosulfate, mixing well, collecting the cells by centrifugation, and washing them twice with sterile water. Dispose of the supernatants carefully, as recommended by your institutional toxic waste facility. Treat control cells in the same manner.

5. Incubate the cells in liquid medium or on plates as appropriate for the particular experimental needs. The mutagen doses suggested above kill 50 to 90% of the cells of most strains, but it is usually desirable to check the survival level of the particular strain used by plating suitable dilutions of both treated and untreated cells. A dose–response determination, if one is needed, is best carried out by exposing cells to varying concentrations of mutagen for a fixed time, since some chemical mutagens undergo destruction in solution.

ICR-170 Mutagenesis

1. Inoculate 10 ml of liquid YPD medium with a freshly subcloned sample of the yeast strain to give approximately 1×10^6 cells/ml (just detectably turbid). Incubate overnight at 30° with vigorous shaking. Accurately determine the cell concentration with a hemocytometer.

2. Inoculate 40 ml of liquid YPD with a sufficient volume of the overnight to give a final cell concentration of 1×10^4 cell/ml, incubate at 30° with vigorous shaking, and collect the cells by centrifugation when the culture reaches $1–3 \times 10^7$ cells/ml (mid log phase). Wash the cells twice with sterile water, resuspend the pellet in 20 ml of 0.1 *M* potassium phosphate buffer (pH 7.0), and shake at 30° for 12 hr.

3. Collect the starved cells by centrifugation and resuspend the pellet in sterile water to a concentration of 2×10^6 cells/ml. Add 1 ml of an aqueous solution

containing 0.25 mg/ml of ICR-170 to 9 ml of the cell suspension, and treat for 60 min. Carry out exposure to the mutagen under red light in a dark room to avoid photodynamic effects, which principally kill cells rather than mutating them. Collect cells by centrifugation, and wash twice with water. Add 1 ml of water to 9 ml cells and handle identically to serve as control. To destroy the mutagen in supernatants, make them 0.1 *M* in sodium hydroxide. Dispose of the treated supernatants as advised by your institutional toxic waste facility.

4. Incubate the cells in liquid medium or on plates as appropriate for the particular experimental needs. The mutagen dose suggested above should result in 5–10% survival, but it is usually desirable to check survival in the particular strain used. Highest mutation frequencies are found by starving mid-log-phase cells, though the extent of their advantage over stationary-phase cells varies with strain.[5]

UV Mutagenesis

1. Inoculate 10 ml of liquid YPD medium with a freshly subcloned sample of the yeast strain to give approximately 1×10^6 cells/ml (just detectably turbid). Incubate overnight at 30° with vigorous shaking. In the morning, the culture should contain about 2×10^8 cells/ml.

2. Wash cells twice in sterile water, sonicate if necessary, and irradiate them either on plates or in suspension, according to need. Turn on UV tubes at least 10–15 min before use, to allow them to come to a constant temperature. To irradiate on plates, spread 200 μl of an appropriate dilution of the cell suspension on each plate, allow the liquid to be absorbed, and expose them, with lids removed, to 50 J/m^2 UV (or for an empirically determined time). Carry out the irradiation under illumination from "gold" fluorescent lights (e.g., F40GO, Philips Lighting Co., Somerset, New Jersey), or very low light, and incubate the plates in the dark for at least 24 hr, to avoid photoreactivation. To irradiate suspensions, 30–50 ml of washed cells in 0.9% (w/v) KCl is placed in a standard 9-cm petri dish, stirred vigorously and continuously with a magnetic mixer, and exposed to UV with petri dish lid removed. Depending on the cell concentration, most suspensions significantly absorb and scatter UV, and suspensions therefore usually need to be exposed to higher UV fluences than cells on plates, to achieve the same level of killing and mutagenesis. As a rough rule of thumb, relative to plates, suspensions of 10^6 cells/ml or less should be exposed to equal UV fluences, suspensions of 10^7 cells/ml exposed to 1.5-fold higher fluences, and suspensions of 10^8 cells/ml to 10-fold higher fluences. However, strains vary in this respect and an empirical check on killing is desirable. High cell concentrations are best given long UV exposures (5–10 min) at low fluence rates, but even so the results are often less reproducible. If very long exposures are required, evaporation can be minimized by covering the cell suspension with polyethylene film (e.g., a single layer cut

from a thin food bag), secured around the dish with a rubber band. Most films reduce the UV fluence by only 10–15%, but check the transmittance of a sample in a spectrophotometer. Protect cells from photoreactivation as before.

A convenient source of UV can be made by enclosing G8T5 germicidal UV tubes in a box containing ventilation holes screened with matte-black painted panels to prevent the escape of direct or scattered radiation. Ventilation is needed because the relative output of the tube at 254 nm, the major effective wavelength, depends on tube temperature, which optimally should be about 30°. A tube-to-sample distance of at least 50 cm is needed to give uniform radiation. Tightly woven metal mesh makes an excellent neutral filter to reduce fluence rates if this is necessary.

A fluence of 50 J/m^2 kills about 50% of the cells in many strains, but it is often easier to determine exposure time empirically. Many commercial UV meters underread fluence rates by as much as a factor of 2, because they are calibrated against collimated beams; depending on the particular arrangement of tubes and kind of container used, cells are usually exposed to a wider arc of radiation than a collimated beam. If needed, fluence rates for specific circumstances can be determined by potassium ferrioxalate actinometry.[14]

Checking Effectiveness of Mutagen Treatment

Some indication of the effectiveness of a given mutagen treatment in the particular strain chosen can be obtained by measuring the frequency of canavanine-resistant mutants that it induces. Resistance to high concentrations of canavanine, an arginine analog, results from mutations that inactivate the arginine permease, which prevents canavanine uptake. Strains auxotrophic for arginine cannot therefore be used, except those carrying *arg6* or *arg8* mutations, in which the arginine requirement can be met by substituting ornithine.

Measurement of the frequency of canavanine resistant mutants is conveniently estimated by the top agar method.[15] Spread $\sim 5 \times 10^6$ treated cells on synthetic dextrose complete medium lacking arginine (SC-ARG), and suitable dilutions of the cell suspension to measure survival. Also plate untreated cells cells to measure spontaneous mutation frequency and plating efficiency. As soon as the liquid is absorbed, overlay each plate with 10ml of SC-ARG top agar cooled to 40° to immobilize the cells. Incubate for 2–3 hr at 30° to allow for mutant expression, then overlay plates containing 5×10^6 cells with 10ml of SC-ARG top agar containing 200 μg/ml canavanine sufate. Overlay plates for estimating survival with 10 ml

[14] J. Jagger, "Introduction to Research in Ultraviolet Photobiology," p. 137. Prentice Hall, Englewood Cliffs, NJ, 1967.

[15] J. F. Lemontt and S. V. Lair, *Mutat. Res.* **93,** 339 (1982).

SC-ARG top agar without canavanine. Direct plating of mutagen treated cells on SC-ARG + canavanine can kill a variable proportion of cells, because of the initial existence of functional permease; pools of the preferentially incorporated arginine may not always be adequate to offset canavanine toxicity. Measurement of the induced frequency of canavanine resistant mutants in some standard strain, such as S288C, provides a benchmark against which to determine results from other strains. Synthetic dextrose complete medium contains, per liter, 1.67 g Difco (Detroit, MI) yeast nitrogen base (without amino acids and ammonium sulfate), 5 g ammonium sulfate, 20 gm dextrose, and such nutrilites as the strain requires (see below for concentrations), solidified with 2% (plates) or 0.75% (overlay) agar.

Mutant Enrichment by Inositol Starvation

Selective killing of growing cells by starvation for inositol, and hence enrichment of mutants unable to grow in the particular conditions used, was first described for *Neurospora*[16] and has since been developed for yeast.[12] A necessary prerequisite is the presence of one or more *ino* mutations in the parent strain. Initial studies[12] used an *ino1-13 ino4-8* double mutant, but a single mutant containing a *ino1* deletion/disruption[17] is likely to be more convenient.

1. Mutagenize cells by one of the methods described above. Inoculate the culture to be treated with a carefully subcloned strain: enrichment procedures indiscriminately increase the frequency of all slow-growing cells, such as mitochondrial petites which can occur at high frequency in old cultures. If independent mutations at any given locus are needed, distribute aliquots of the mutagenized cells to different tubes before outgrowth, and carry out the procedure on each in parallel.
2. Allow the treated cells to recover from the mutagen damage and to express mutations by resuspending them in YPD or other appropriate medium at a concentration of about 5×10^5 cells/ml, grow the culture to no more than 10^7 cells/ml, and collect the cells by centrifugation: it is important to harvest exponential-phase cells.
3. Wash the cells in prewarmed prestarvation medium, resuspend in this medium at a concentration of between 1×10^4 and 1×10^6 cells/ml, and incubate for 3–4 hr under conditions that will stop the growth of the mutants for which enrichment is desired. If histidine auxotrophs are sought, for example, prestarvation medium is synthetic complete medium that contains inositol but not histidine. For temperature-sensitive mutations, prestarvation medium is any complete medium containing inositol, but the incubation is carried out at 35°.

[16] H. E. Lester and S. R. Gross, *Science* **139,** 572 (1959).

[17] M. Dean-Johnson and S. A. Henry, *J. Biol. Chem.* **264,** 1274 (1989).

4. Wash cells twice in prewarmed starvation medium and resuspend at a concentration of no more than 5×10^6 cells/ml. Starvation medium is prestarvation medium lacking inositol. Incubate for 24 hr at 35° to enrich for temperature-sensitive mutants; otherwise incubate at 30°. At 30°, cells begin to die after 5–6 hr, and eventually only about 0.1–1% should remain viable.

5. Plate cells on any suitable permissive medium and incubate under permissive conditions to obtain well-separated colonies. Rich medium containing a nonfermentable carbon source (e.g., YPG) can be used at this stage if selection against petites is desirable.

6. Screen surviving clones for the desired mutant.

A second cycle of enrichment can be tried, but is usually not required when cells are mutagenized. Solid medium can be used in place of liquid for the starvation phase of the procedure, and surviving cells recovered from these plates by velveteen replication.[12] When this is done, the plating density needs to be adjusted to account not only for inositol-less death, but also for the fact that only about 10% cells are transferred by velveteen.

Inositol Starvation Medium per Liter

Ammonium sulfate	5 g
Potassium phosphate, monobasic	1 g
Magnesium sulfate	0.5 g
Sodium chloride	0.1 g
Calcium chloride	0.1 g
Boric acid	500 μg
Copper sulfate	40 μg
Potassium iodide	100 μg
Ferric chloride	200 μg
Manganese sulfate	400 μg
Sodium molybdate	200 μg
Zinc sulfate	400 μg
Biotin	2 μg
Calcium pantothenate	400 μg
Folic acid	2 μg
Niacin	400 μg
p-Aminobenzoic acid	200 μg
Pyridoxine hydrochloride	400 μg
Riboflavin	200 μg
Thiamin hydrochloride	400 μg
Dextrose	20 g

To prevent the selection of auxotrophs, the following nutrilites can be added to the inositol starvation medium: adenine sulfate, uracil, L-arginine hydrochloride,

L-histidine hydrochloride, L-methionine, L-tryptophan, each at 20 mg/liter; L-isoleucine, L-leucine, L-lysine hydrochloride, L-tyrosine, each at 30 mg/liter; L-phenylalanine, 50 mg/liter; L-valine, 150 mg/liter; L-aspartic acid, L-glutamic acid, each at 100 mg/liter; L-homoserine, L-threonine, each at 200 mg/liter; L-serine, 375 mg/liter. The last five nutrilites can often be omitted from the medium, since auxotrophs with these requirements are rare.

Synthetic prestarvation medium is inositol starvation medium plus 2000 μg of inositol. Starvation medium can also be made using Difco vitamin-free yeast nitrogen base (16.9 g/liter), but this provides 1% dextrose and also histidine, methionine, and tryptophan. YPD medium is 1% yeast extract, 2% peptone, and 2% dextrose. YPG medium is similar, but with 2% glycerol (v/v) replacing the dextrose.

[11] Introduction of Point Mutations into Cloned Genes

By BRENDAN CORMACK and IRENE CASTAÑO

Introduction

In the 10 years since the publication of *Guide to Yeast Genetics and Molecular Biology,*[1] the availability of the complete yeast genomic sequence and more recently of a set of haploid and diploid strains deleted for each of the yeast open reading frames (ORFs) has dramatically affected the practice of classical yeast genetics. Researchers are likely to choose, when possible, to use the collection of ORF deletion strains as the default for genetic screens and selections. There are, of course, limitations to the utility of the *Saccharomyces cerevisiae* deletion collection: global methods of mutagenesis will still be required for genetic analysis of strains with backgrounds different than the type strains used for the *S. cerevisiae* knockout project[2,3] (see chapters 12 and 13). Equally, the genetic analysis of essential genes still relies largely on the generation of point mutations in those genes (for an alternative approach, exploiting repressible promoters and ubiquitin destabilization of the gene product, see a recent review by Varshavsky.[5] This chapter first discusses briefly the principle of the plasmid shuffle technique which is critical to the generation and analysis of point mutants in essential genes. Second, two approaches to *in vitro* generalized mutagenesis of cloned yeast genes are discussed.

[1] C. Guthrie and G. R. Fink (eds.), *Methods Enzymol.* **194** (1991).

[2] N. Backman, M. Biery, J. D. Boeke, and N. L. Craig, *Methods Enzymol.* **350,** [13], 2002 (this volume).

[3] A. Kumar, S. Vidam, and M. Snyder, *Methods Enzymol.* **350,** [12], 2002 (this volume).

[4] Deleted in proof.

[5] A. Varshavsky, *Methods Enzymol.* **327,** 578 (2000).

0076-6879/02 $35.00

Lastly, several methods for the introduction of mutations into a clonable DNA fragment are laid out. Where appropriate, methods for introduction of the mutagenized fragment into the full-length yeast gene are also described. There are many available techniques for introducing sequence changes into cloned DNAs, and one focus of this chapter will be on matching a particular mutagenesis technique with the nature of the mutant being made or desired.

Plasmid Shuffling

The analysis of essential genes in yeast poses particular problems because the wild-type copy of the gene must be inactivated, generating a cell with an inviable phenotype. The method of plasmid shuffling (Fig. 1) exploits a plasmid-borne wild-type copy of the gene of interest (*YFG*) as a way of complementing a chromosomal deletion (or other null allele) of the gene. In the first step of the method, a heterozygous diploid, with one wild-type and one deleted copy of the *YFG* gene, is transformed with a plasmid carrying a wild-type copy of the *YFG* gene. This allows recovery, after sporulation, of haploid strains in which the chromosomal allele is deleted and a wild-type copy of the gene is carried on a plasmid. Mutagenized copies of the gene (*YFG*) are then introduced on a second plasmid and the plasmid carrying wild-type copy of the gene "shuffled" out of the strain generally by counterselection. This results in a strain carrying a single mutagenized copy of the *YFG* gene covering a chromosomal deletion and permits assessment of any recessive phenotype due to the mutated gene. Because the haploid strain used for the plasmid shuffle transiently maintains two different yeast vectors, it needs to be marked with a minimum of two markers (e.g., *ura3*, *trp1*, *his3*, *leu2*).

Loss of the plasmid carrying the wild-type copy of the *YFG* gene occurs at a reasonably high frequency. Yeast plasmids, even those carrying centromeric sequences, are relatively unstable and are lost at a frequency of approximately 10^{-2}/generation. The loss of the plasmid can be selected for if the plasmid carrying the wild-type copy of the *YFG* gene also carries a second gene that can be selected against by growth in the presence of certain selectively toxic compounds. The most commonly used of these toxic compounds is 5-fluoroorotic acid or 5-FOA. 5-FOA is toxic to *URA3* cells but has no effect on *ura3* cells. Therefore, if the wild-type copy of *YFG* is carried on a centromeric plasmid also carrying the *URA3* gene (abbreviated YCp-*URA3*), then growth in the presence of 5-FOA selects for loss of the *URA3* plasmid and concomitantly loss of the wild-type copy of the *YFG* gene. There are other less commonly used counterselection schemes which exploit the selective toxicity of α-aminoadipate for *LYS2* cells, canavanine for *CAN1* cells, and cyclohexamide for *CYH2* cells. *URA3* and *LYS2* have the advantage that *URA3* cells or *LYS2* cells (carrying the YCp-*URA3* or YCp-*LYS2* plasmids) can be selected

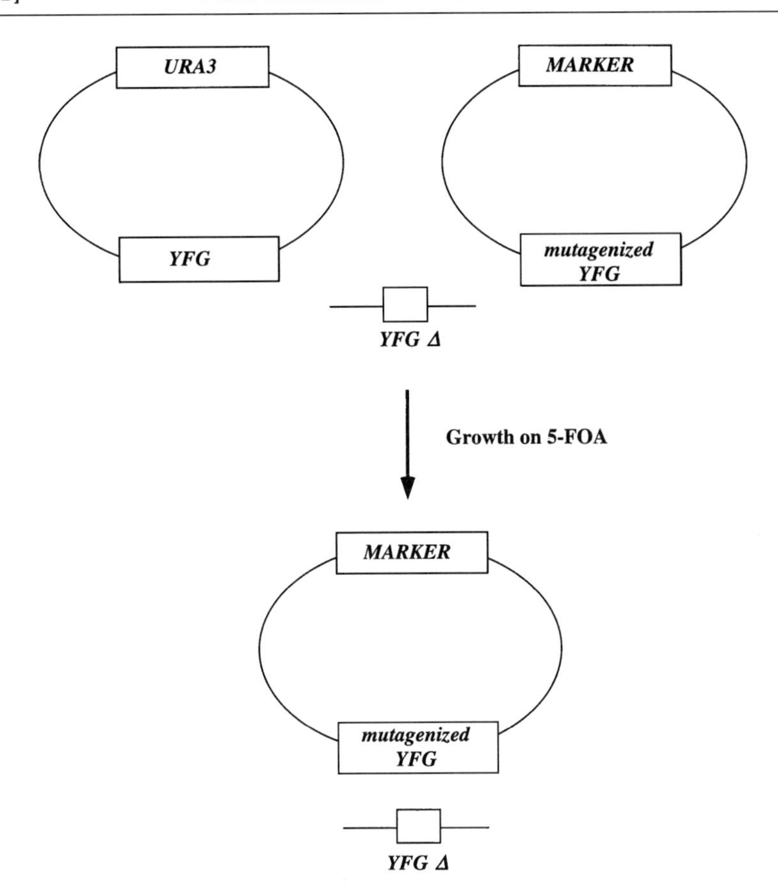

FIG. 1. The plasmid shuffle. The starting strain carries two plasmids and a deletion of the chromosomal *YFG* gene. The wild-type copy of the *YFG* is carried on a YCp (or YEp) plasmid marked with *URA3*. The mutagenized copy of *YFG* is carried on a differently marked YCp plasmid. After nonselective growth in the presence of uracil and replica plating onto 5-FOA containing plates, the YCp-*URA3* plasmid is lost; at this point, the mutagenized copy of *YFG* is the only copy of *YFG* in the cell and the phenotype of any mutation can be easily assessed.

by growth in the absence of uracil or lysine, respectively. For ease of discussion, it is assumed for the rest of this chapter that *URA3* is the counterselectable marker being used.

Once the wild-type copy of the *YFG* gene has been shuffled out of the cell, the phenotype of the mutated copy of the gene carried on the second plasmid can be assessed. Such mutations can be introduced into the *YFG* gene by the various approaches given in the next section. After replica plating onto the counterselective medium, null alleles that cannot complement the chromosomal deletion will have a

5-FOA-sensitive phenotype, since cells that have lost the YCp-*URA3-YFG* plasmid (and would therefore normally be resistant to 5-FOA) die for lack of a functional copy of *YFG*. Conditional mutants, such as temperature- or cold-sensitive mutants, are identified by replica plating onto 5-FOA plates under both restrictive and permissive conditions and identifying mutants able to grow only under the permissive condition. For such a screen, it is important to have a relatively small number of colonies (approximately 200–300) on each plate since otherwise identifying those few that do not grow can be difficult. This requires titering the transformation to determine the number of yeast cells to be plated that will yield the correct number of transformants per plate. To titer the transformation, frozen competent cells that are of consistent competency from tube to tube can be used. Alternatively, yeast cells transformed by the lithium acetate method can be stored for several days in 0.1 *M* lithium acetate at 4° with no significant change in transformation efficiency. The transformed cells are plated on media selective for the mutagenized plasmid and incubated at the permissive temperature. At this stage, selection for the *URA3* plasmid carrying the unmutagenized copy of the gene is not maintained. The result is that without selection, many cells in a given transformant colony will lose the YCp-*URA3* plasmid. This will permit, after counterselection on 5-FOA, the phenotype resulting from the mutagenized gene to be assessed directly by replica plating the transformation plates.

In the following sections, mutagenesis strategies suited to introduction of specific desired mutations into a given gene or alternatively, construction of a library of point mutants in a given gene will be discussed.

Principles of Mutagenesis

Broadly speaking, point mutations can be introduced into a cloned yeast gene by one of two methods: general mutagenesis and oligonucleotide-directed mutagenesis. Which method an investigator chooses depends largely on available information about the gene's function and on the nature of the mutations wanted. Oligonucleotide-directed mutagenesis schemes have the enormous advantage of allowing the investigator to choose the exact nature of the nucleotide or amino acid change made to the gene. As well, for generation of a library of point mutants, use of degenerate oligonucleotides allows the targeting of mutations to a specific region of the protein that might be of particular interest. Unfortunately, the region of the protein that can efficiently be mutagenized in a single library is limited (by the size of synthetic oligonucleotides) to approximately 100 nucleotides.

By contrast, for generalized mutagenesis schemes, there is effectively no limit on the size of gene or region that can be mutagenized. Notwithstanding the ease and flexibility of the generalized mutagenesis schemes described below, there is for all general mutagenesis methods available a significant bias in the types of mutations generated. This bias in classes of mutations generated cannot be compensated for by increasing the rate of mutagenesis, since this has the concomitant and generally undesirable effect of increasing the number of mutations per gene.

How to weigh these advantages and disadvantages depends largely on the nature of the mutations desired. If the investigator has only limited knowledge of the protein's functional domains and is interested, therefore, in making mutants across the entire gene, a generalized method is appropriate—in particular if a loss or partial loss of function allele is desired. In the simplest class of site-directed mutagenesis in which targeted changes to a small number of amino acids are desired, the investigator would want to limit mutants to the corresponding region of the gene and would use an oligonucleotide-directed mutagenesis method. But oligonucleotide-directed methods can also be used to tremendous advantage to generate libraries that can be screened for mutants of interest. If, for example, the investigator aims to identify specialized mutant classes from a library of mutants—for example, mutations altering substrate specificity or selectively eliminating one function for a multifunctional protein—then generalized mutagenesis techniques might yield an insufficient diversity of mutations; oligonucleotide-directed methods which yield a superior range of nucleotide substitutions (and therefore amino acid substitutions) might be more appropriate.

Lastly, whether a defined mutation or a library of mutations is being made, the mutations present on a PCR fragment or in an oligonucleotide must be introduced into the full-length gene. We outline three basic approaches. One fast and reproducible method to incorporate mutations in an oligonucleotide into the full-length gene is the QuikChange method (Stratagene, La Jolla, CA). Because of its ease and speed, this method is optimally suited to most site-directed mutagenesis projects. When constructing a library of mutants made by degenerate PCR of a portion of a gene, recombinational gap repair in *S. cerevisiae* itself can be used to efficiently clone the mutagenized region into the full-length gene. Lastly, mutations can be cloned by directly exploiting natural restriction sites or sites introduced at the 5′ end of the mutagenizing oligonucleotide. This last approach is the most reproducibly efficient method and we use it preferentially when efficiency of cloning is of prime importance as in, for example, construction of a library of mutants using a doped oligonucleotide.

Generalized *in Vitro* Mutagenesis

There are multiple methods for more or less randomly introducing mutations into cloned DNAs, including the use of UV, nitrous acid, and other chemical mutagens as well as passage of a plasmid carrying the gene of interest through mutator strains of *Escherichia coli*. However, because of their relative ease, reproducibility, and the mutation spectrum generated, two methods stand out: hydroxylamine mutagenesis and degenerate PCR.

HA Mutagenesis

Under very acidic conditions, hydroxylamine, acting as a hydroxyl donor, reacts with double-stranded DNA templates, resulting in the deamination of cytosine to uracil. Under milder acidic conditions (pH 6.0), the stable intermediate in the deamination of cytosine, hydroxyaminocytosine, is generated. This can base pair with adenosine resulting in CG-TA transition mutations. The relevant pK_a of hydroxylamine is 6.5. Mutagenesis will happen at pH 7.0, but is more efficient when carried out at pH 6.0. The following protocol (adapted from Sikorski and Boeke[6]) gave us excellent results.

1. Make a solution of 1 *M* hydroxylamine with 15 m*M* sodium pyrophosphate, 100 m*M* sodium chloride, 2 m*M* EDTA, and adjust the pH to 6.0 with HCl.
2. Add 10 μg of the target plasmid to 500 μl of the hydroxylamine solution.
3. Allow the reaction to proceed at 65°. Remove aliquots at different time points of incubation (0–5 hr) and transfer to ice.
4. Remove hydroxylamine by gel filtration over a Sephadex G-50 spin column.
5. Transform 1/100 of each mutagenized sample into *E. coli* to estimate degree of mutagenesis. This can be estimated by determining the null mutation rate for the yeast marker gene carried on the yeast–*E. coli* shuttle vector plasmid being mutagenized. Strain KC8 (available from Clontech, Palo Alto, CA) is marked with bacterial mutations that are complemented by the yeast *LEU2*, *HIS3*, *TRP1*, and *URA3* genes. After selecting for the antibiotic resistance marker in the transformation, replica plate to minimal M9 media lacking the appropriate amino acid or nucleotide. The number of transformants failing to grow on the minimal media gives the percent null alleles for the yeast auxotrophic marker carried on the plasmid.
6. Decide which of the mutagenized samples is the most appropriate based on the estimated mutation rates for each sample and the desired mutation rate. Transform this sample into *E. coli* in order to obtain DNA for transformation into yeast. The number of transformants here determines the maximum number of mutations to be screened in yeast. Pool at least 10,000–20,000 transformants by scraping and make a large-scale plasmid preparation of this mutagenized library.
7. Expected results. In our mutagenesis of the *SPT15* gene, encoding the TATA-binding protein, carried on a YCp-*TRP* plasmid, samples mutated for 5 hr at pH 7.0 or for 50 min at pH 6.0 gave approximately the same rate of mutagenesis. Of Amp^R transformants of the bacterial strain KC8, 1% were Trp^-. From the first of these libraries, temperature-sensitive alleles were recovered at a frequency of 0.1% and null alleles at a frequency of 0.4%. Four were analyzed and were all due to single transition mutations.[7] In their analysis of the large subunit of RNA

[6] R. S. Sikorski and J. D. Boeke, *Methods Enzymol.* **194,** 302 (1991).
[7] B. P. Cormack and K. Struhl, *Cell* **69,** 685 (1992).

polymerase II, Scafe *et al.*[8] mutagenized the *RPB1* gene (approximately 5.5 kb target size) with hydroxylamine. Of transformants, 2% were null alleles, 0.8% were temperature sensitive, and 0.4% were cold sensitive. Reflecting the bias inherent in hydroxylamine mutagenesis, of 21 sequenced point mutants, 19 were C to T transitions, one was a C to G transversion, and one was a T to A transversion.[8]

Degenerate PCR Mutagenesis

An alternative to the above mutagenesis procedure which gives a wider range of nucleotide substitutions is mutagenesis by low-fidelity PCR amplification. The fidelity of *Taq* polymerase is decreased by addition of Mn^{2+} to the buffer. Increasing the dCTP and dTTP concentrations in the PCR reaction increases the spectrum of mutations obtained. The following PCR conditions are taken from Cadwell and Joyce's modification[9] of an earlier method.[10]

1. PCR Conditions. The plasmid to be mutagenized, carrying the wild-type gene, is subjected to a PCR reaction containing (100 μl final) 20 fmol template (e.g., 66 ng for a 5 kb-plasmid), 0.3 μM each primer, 50 mM KCl, 10 mM Tris-HCl (pH 8.3), 0.01% gelatin, 7 mM $MgCl_2$, 0.5 mM $MnCl_2$, 0.2 mM dATP, 0.2 mM dGTP, 1 mM dCTP, 1 mM dTTP. Mutation frequency will depend on the number of template duplications and can be changed, therefore, by varying the number of mutagenic PCR cycles (see Expected Results).

2. Cloning. Clone the mutagenized fragment into the appropriately digested YCp vector either by recombinational gap repair or by using restriction sites incorporated into the PCR primers (see "Cassette Mutagenesis," below). Transform into *E. coli* (or directly into yeast if cloning by recombinational gap repair) to make a library of mutants. It is recommended that the number of transformants be maximized at this step since this number will determine the ultimate number of mutations available to be screened in yeast.

3. Expected Results. Mutation rate can be manipulated by changes to two parameters: Mn^{2+} concentration and number of rounds of mutagenic PCR. To obtain different rates of mutagenesis, investigators may wish to titrate the Mn^{2+} concentration and carry out a constant number of PCR cycles. Because the influence of the Mn^{2+} concentration on the spectrum of mutations recovered has not been exhaustively examined, we generally fix the Mn^{2+} concentration at 0.5 mM and adjust mutation rate by altering the number of mutagenic PCR cycles. In the original report, using the conditions given above, the reported mutation rate was 0.66% per base pair after 10 duplications of the template. For applications in which a precise

[8] C. Scafe, C. Martin, M. Nonet, S. Podos, S. Okamura, and R. A. Young, *Mol. Cell. Biol.* **10,** 1270 (1990).

[9] R. C. Cadwell and G. F. Joyce, *PCR Methods Appl.* **2,** 28 (1992).

[10] D. W. Leung, E. Chen, and D. V. Goeddel, *Technique* **1,** 11 (1989).

mutation rate is desirable, it is necessary to calibrate mutation rate precisely. In such instances, it is recommended to remove about 1/8 of the sample (12.5 μl) from the mutagenic PCR every 3 template duplications (i.e., an 8-fold increase in product), dilute into 100 μl final of mutagenic PCR buffer, and amplify for another 3 template duplications. The amplified product from the remaining 87.5 μl is recovered and subjected to nonmutagenic amplification. This cycle of dilution and mutagenic reamplification is repeated as often as needed to get the desired mutation frequency. This approach keeps the number of template duplications (and hence the mutation rate) linear with the total number of amplification cycles. The number of mutagenic PCR cycles required to give a template duplication (i.e., a doubling of product) needs to be calibrated since it will often require more than one PCR cycle. This approach generates a series of PCR products (corresponding to the samples removed after every three mutagenic template duplications); each of these products has a different frequency of mutation. Sequencing clones from products amplified for different numbers of mutagenic cycles allows one to precisely estimate the mutation rate for each sample, and to choose the mutation rate best suited to the envisioned experiment.

In the original description of this methodology[9] it was reported that AT and GC base pairs were mutated at approximately equal frequency and the mutations isolated included both transitions and transversions at approximately equal frequency (the ratio transition : transversion is 0.8). The mutations generated are not completely random, however. In a later careful analysis[11] of the mutations generated using a slight modification of this method, the transition : transversion ratio was also about 1 (1.05). Among the transition mutations, both possibilities were well represented (AT to GC 64%, CG to TA 36%). However, of the four possible transversion mutations (AT to TA, AT to CG, CG to AT, and CG to GC), AT to TA accounted for 84%, CG to AT for 12%, AT to CG for 4%, CG to GC for 0%. Thus, the spectrum of mutations that can be generated even under the best conditions, is clearly limited. Notwithstanding this bias, mutagenic PCR is the best available option for generalized mutagenesis and will in most cases yield the desired mutant.

Cloning PCR-Mutagenized Fragment by Gap Repair in S. cerevisiae

Once the mutagenized PCR fragment is generated, it must be introduced into the full-length gene. If the fragment has usable restriction sites, or if the PCR primers incorporate restriction sites at each end of the PCR fragment, these may be used directly to clone the mutagenized DNA into the full-length gene (see section below: "Cassette Mutagenesis"). An efficient alternative (Fig. 2), described here, exploits *S. cerevisiae* homologous recombination to repair a gapped

[11] D. P. Bartel and J. W. Szostak, *Science* **261,** 1411 (1982) [see comment].

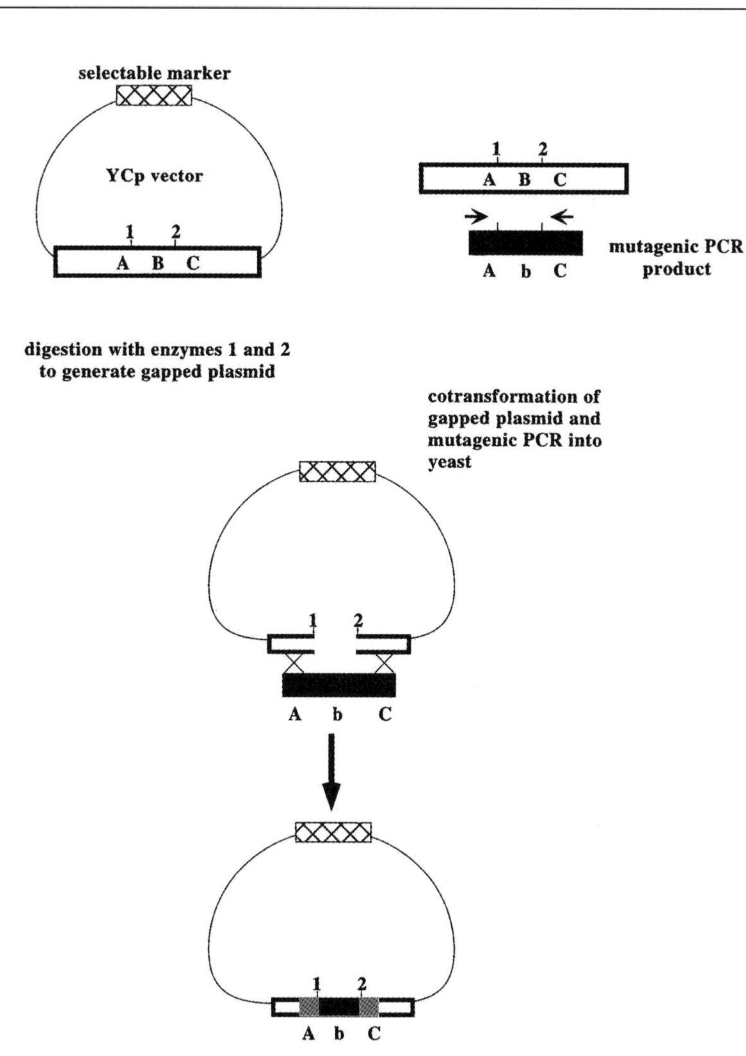

FIG. 2. Cloning by gap repair. The cloning vector carrying the gene to be mutated is digested with restriction enzymes 1 and 2, generating a gapped vector with incompatible ends 1 and 2. The PCR product to be cloned has regions of homology (A and C) with the vector at both ends. After transformation, recombinational gap repair results in repair of the gapped vector with the cotransformed PCR product, replacing the region B with the mutagenized region b. Mutations in the homologous regions A and C of the PCR product may or may not be incorporated into the recombinant plasmid depending on where the effective crossover events occur during gap repair.

plasmid using a cotransformed PCR fragment as template.[12] One great advantage of cloning by gap repair is that the cloning is done directly in yeast and analysis of the clone can often be made without the necessity of retransforming the resulting plasmid. The efficiency of the method depends on two major factors: (1) the extent of homology shared between the ends of the PCR fragment and the gapped plasmid and (2) the underlying rate of nonhomologous end joining (NHEJ) that results in recircularization of the plasmid without incorporation of the PCR fragment.

1. Generating the Gapped Vector. The vector used is a yeast–*E. coli* shuttle vector carrying the cloned gene of interest. The plasmid is first gapped by restriction enzyme digestion with two enzymes uniquely present in the gene of interest. After gapping, the plasmid can be gel purified to remove undigested or singly digested vector. The two enzymes used to gap the plasmid should be chosen such that the generated ends are not substrates for efficient religation by NHEJ. Either the plasmid can be gapped with enzymes generating blunt ends, or alternatively by two different enzymes generating noncompatible ends. Plasmids gapped with enzymes generating compatible overhanging ends are efficiently repaired by NHEJ quite independently of the PCR fragment and yield transformants at close to the same frequency as undigested supercoiled vector (for example, see Ref. 13).

2. Homology of the Vector and the Mutagenized PCR Product. Both ends of the PCR product to be cloned must share homology with the ends of the gapped plasmid. The original report[12] showed that regions of homology as little as 65 nucleotides in length were substrates for recombinational repair, although more commonly the PCR fragment carries around 200 bp of homology with the gapped plasmid. It follows, therefore, that the precise boundaries of the region to be mutagenized (by generalized PCR mutagenesis) will in some cases be dictated by the position of the restriction sites available for gapping the plasmid.

3. Transformation. Mix 0.1 to 1.0 μg of gapped plasmid with a molar excess of PCR product (often a 3–5$\times$ molar excess, although the precise molar ratio seems not to be critical). Transform this DNA into an appropriately marked yeast strain deleted, if possible, for the chromosomal locus of the gene being mutated. In the case in which the wild-type chromosomal copy of the gene has not been deleted, the gapped plasmid can also be repaired by gene conversion from the chromosome resulting in a wild type and not the desired mutant copy on the plasmid. Plate on media selective for the marker on the gapped plasmid.

As a control, transform just the gapped plasmid into the same strain. The ratio of transformants (generally around 10 : 1) resulting from the transformation with

[12] D. Muhlrad, R. Hunter, and R. Parker, *Yeast* **8,** 79 (1992).

[13] T. E. Wilson and M. R. Lieber, *J. Biol. Chem.* **274,** 23599 (1999).

and without the PCR product should indicate the efficiency with which the PCR product has been incorporated into the gapped plasmid.

4. Expected Results. In a study of the silencing gene *SIR1*, a PCR-mutagenized fragment corresponding to the C-terminal half of *SIR1* was cloned by recombinational gap repair. Mutagenesis was high with 45% of mutants being loss of function alleles. From this library, 18,000 transformants were screened for mutants specifically defective in recruitment to the HMR-E silencer, and six such mutants were found.[14]

Transformation Efficiencies. The initial publication of this method reported between 1×10^4 and 5×10^4 transformants per microgram of linearized vector.[12] This range of transformation efficiency is typical of what is reported in the literature, although higher efficiencies of cloning can be achieved. In an analysis of cis sequences required for Ty transposition (E. Bolton and J. Boeke, unpublished data, 2000), Bolton and Boeke generated a PCR-mutagenized fragment overlapping with a gapped (*Hpa*I–*Xho*I digested) 5.0 kb YCp*URA3* vector by 200 nt at each end. Using a 3 : 1 molar ratio of insert to gapped plasmid and transforming with a total of 100 ng of plasmid DNA, they recovered 3×10^6 transformants per microgram of gapped plasmid. This number of transformants was approximately 600× higher when the PCR fragment was included than when it was omitted. Of the inserts, 50% had at least one nucleotide change in the mutagenized region.

Oligonucleotide-Directed Mutagenesis

The use of oligonucleotides to introduce mutations into a gene has the significant advantage of limiting the mutations ultimately introduced into the gene to the region of the gene corresponding to the oligonucleotide; moreover, the changes introduced are determined by the sequence of the oligonucletide used in the mutagenesis. Thus, any nucleotide can be changed to any other nucleotide by synthesizing the appropriate "mutant" oligonucleotide. It is worth emphasizing here that in the previous section, mutations were introduced during the PCR reaction itself as a consequence of low polymerase fidelity. In methods in this section, by contrast, while PCR amplification might be used to generate a mutated DNA fragment, the mutations introduced are present in the synthetic oligonucleotides used as primers in the PCR reaction.

Oligonucleotides can also be used to generate a library of point mutants that can be subsequently screened for mutants of interest. Commercial suppliers of oligonucleotides can synthesize oligos in which any nucleotide position is mutated

[14] K. A. Gardner, J. Rine, and C. A. Fox, *Genetics* **151,** 31 (1999).

with a given probability simply by using an appropriate mixture of deoxynucleotide phosphoramidites during the corresponding synthetic cycle of the oligonucleotide. Practically, this means making four extra mixes of phosphoramidites in which the "wild-type" nucleotide makes up some percentage of the mix, and the three other "mutant" nucleotides make up the rest of the mix. Libraries of mutants derived from such oligonucleotides are precisely controlled in terms of frequency of nucleotide substitution at each position and therefore have excellent mutant representation.

In the following two methods, mutant oligonucleotides are incorporated into the full-length gene in fundamentally different ways. The first, QuikChange (Stratagene), permits the rapid introduction of point mutations into a plasmid. Where possible, it is our method of choice because of the speed with which mutants can be made. Where the number of independent mutant clones generated is important (as for example in the generation of a library of mutants), we use an alternative method in which the mutation is introduced into the full-length gene by direct cloning of a double-stranded (mutant) oligonucleotide or a PCR-amplified fragment that incorporates the mutagenic primer. If there are no useful natural restriction sites in the gene of interest close to the region to be mutagenized, this method can be more labor intensive than other methods of cloning mutagenized DNA fragments, since the investigator first needs to reengineer the gene to add restriction sites close the site to be mutated. In our hands, however, this method generates the largest number of independent clones ultimately in yeast and therefore is our method of choice when it is of interest to make and express complex libraries of point mutants for a given gene. When only a small number of mutants are to be analyzed, the QuikChange method outlined above is probably preferable, if only because it is likely to be faster.

QuikChange

In this method (commercialized by Stratagene), the mutation present in the synthesized oligonucleotide is introduced into the full-length gene by *Pfu* DNA polymerase-mediated linear amplification (Fig. 3). The template in the amplification reaction is the full-length wild-type gene (or an easily subclonable fragment of the wild-type gene) carried on a plasmid. The primers used for linear amplification are precisely complementary to each other and both include the mutation being introduced. The amplification with *Pfu* DNA polymerase is made using a circular plasmid carrying the wild-type gene fragment. Since the two primers are complementary, the reaction amplifies the entire plasmid, generating an *in vitro* generated product corresponding to the entire plasmid. This final product is a circular DNA with nicks corresponding to the 5′ end of each primer. The parental template DNA, which is naturally methylated, is digested with a methylation sensitive enzyme (*Dpn*I) and the result of the reaction is transformed into *E. coli*. Only the newly synthesized plasmid including the oligonucleotides with the incorporated

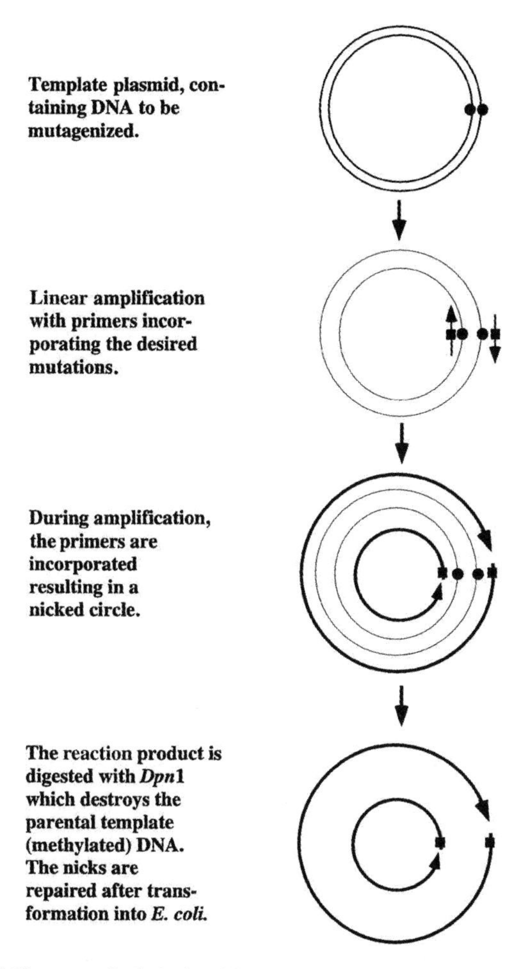

FIG. 3. The QuikChange method. A plasmid carrying the DNA to be mutagenized is subjected to linear amplification with two primers that amplify the entire plasmid. The template DNA is methylated, but the newly synthesized copies are not. The amplification reaction is digested with *Dpn*I (which does not recognize unmethylated DNA), leaving only the amplified plasmid which has incorporated the mutation(s) carried on the oligonucleotide primers. Closed circle: the wild-type nucleotide at the site to be mutated; closed squares: the mutant nucleotides incorporated in the oligonucleotides; thin lines: the parental double-stranded template plasmid; thick lines: the newly synthesized copies of the template plasmid.

mutations will be left undigested by the *Dpn*I and will be able to transform *E. coli,* resulting in efficient recovery of the mutagenized full-length gene. Since methylation is the basis for selective destruction of the unmutagenized template DNA, it follows that the template DNA must be prepared from a methylation-competent strain of *E. coli* (most standard laboratory strains are in fact in this category).

1. Design of Primers. In the QuikChange method, two primers are synthesized to introduce a given change. Both primers include the desired mutation and hybridize to the identical sequence on opposite strands. They are therefore precisely complementary to each other. The primers should be synthesized so that the T_m is at least 78°. The mutation is placed in the middle of the primer with at least 12 bp (and up to 16 bp routinely) of wild-type sequence on each side of the mutation site. We generally design the primer so that its terminal base is a G or C. The primers do not need to be phosphorylated, nor, in the majority of cases, do they need to be gel purified. Rarely, in the case of suboptimal oligonucleotide synthesis, the oligonucleotide needs to be gel purified to optimize efficiency of mutant recovery.

2. Amplification Reaction. In a small thermocycler tube, make a 50-μl reaction containing:

1 μg template DNA
250 μ*M* each dNTP
1× *Pfu* buffer
3 m*M* $MgSO_4$ (final)
30 pmol of each mutagenic primer

Heat the tube to 95° and add 1 μl of *Pfu* polymerase, mixing the reaction with a pipette tip to mix the enzyme. Cycle for 16 rounds using the following conditions:

95°	40 sec
53°	60 sec
68°	2 min per kb of template plasmid length

3. Digestion of Product with Dpn I. Following cycling, check the amplification by running on an agarose gel. If there is product (see below) remove 5 μl, add to a total of 20 μl of digestion buffer, and digest with 10 units of *Dpn*I for 1 hr. Transform 1 μl of the digested product into *E. coli.*

4. Expected Results. The success of this protocol is measured largely by the success of the amplification step. The expectation after amplification is a mix of several products: supercoiled and nicked circular template DNAs, single-stranded DNA, and the product of interest, linear DNA. The linear DNA should migrate above the supercoiled band of the template, but below the faint nicked circle band. The amplification is most efficient in our hands for plasmids with a total size less than 5 kb. Above 6 kb total size, the amplification is less efficient. Often the $MgSO_4$ concentration can be optimized by using final concentrations that vary from 2 to 3.5 m*M*. In our experience, if the amplification step is successful, more than 80% of transformants contain the desired mutation. Kits are available from Stratagene for this method; however, *Pfu* DNA polymerase can also be purchased separately and all other components are readily available.

Cassette Mutagenesis: Using Restriction Endonuclease Sites to Clone Mutant Oligonucleotide (for Construction of Single Point Mutants or Libraries of Point Mutants)

The most efficient way of introducing a mutated oligonucleotide into the full-length gene is by direct cloning of a double-stranded oligonucleotide or a PCR-amplified fragment using restriction sites incorporated into the mutagenic primer. In the QuikChange method above, oligonucleotide primers are used to amplify the entire vector containing the insert to be mutagenized. In cassette mutagenesis, the mutations, present in the oligonucleotide used for PCR, are used to amplify a DNA fragment containing restriction sites also engineered into the PCR primers. These restriction sites are then used to introduce the PCR fragment into the full-length gene.

If there are no useful natural restriction sites in the gene of interest close to the region to be mutagenized, this method can be more labor intensive than other oligonucleotide-directed mutagenesis methods, since the investigator first needs to reengineer the gene to add restriction sites close the site to be mutated. In our experience, however, this method generates the largest number of independent clones ultimately in yeast and therefore is our method of choice when it is of interest to make and express complex libraries of point mutants for a given gene. In our hands, libraries made by this method routinely consist of $1–5 \times 10^6$ independent clones. When only a small number of mutants are to be analyzed, the QuikChange method outlined above is probably preferable, if only because it is likely to be faster.

1. Restriction Site Requirements. At least one restriction site must be located close (within 100 nucleotides) to the site to be mutated. If there is a usable (i.e., unique) site occurring naturally within such a distance, then that same site can be incorporated into the mutagenic primer and used for cloning. On the other hand, if there is no useful site, then the investigator needs to examine the gene sequence for sites where mutations can be introduced that result in a restriction site but do not change the amino acid sequence in the encoded protein. These changes must first be introduced into the "wild-type" copy of the gene (by QuikChange, or indeed by cassette mutagenesis). Subsequently, these natural or engineered restriction sites can be used to efficiently clone double-stranded fragments derived from mutagenic oligonucleotides.

In the event that silent changes are in fact made to the gene, it is recommended that the investigator verify that the "wild-type" copy of the gene carrying the engineered restriction sites complements a null mutation with no obvious phenotype. This is particularly important when the restriction site is being introduced at the beginning of the gene, since translation efficiencies can be dramatically affected by nucleotide sequence just upstream of, and by codon usage just following, the start codon.

2. Direct Cloning of a Mutagenic Oligonucleotide. If there are two restriction sites flanking the site to be mutagenized, then two complementary oligonucleotides incorporating the mutation on each strand can be synthesized, allowed to anneal, and cloned directly into the appropriately digested vector.

3. PCR Amplification of Subclonable Fragment Incorporating Mutagenic Primer. More commonly, there is a single useful restriction site close to the mutation site. In such a case, PCR can be used to generate a fragment that terminates at one end with the mutagenic oligonucleotide containing the appropriate restriction site at its 5′ end. The other end of the PCR product, is of course determined by the second PCR primer (Fig. 4). This is chosen to facilitate cloning of the ultimate PCR fragment and could, for example, hybridize to a site just upstream of a naturally occurring restriction site; alternatively, the second primer could also include a 5′ restriction site corresponding to a restriction site naturally present in, or previously engineered into, the gene. After amplification, the PCR product is digested with the two restriction enzymes and cloned into the appropriately digested vector to regenerate the full-length gene. The product of this ligation is a full-length gene in which the wild-type sequence between the two restriction sites is replaced with the PCR product that incorporates the changes in the mutagenic oligonucleotide.

4. Primer Design. When using cassette mutagenesis to incorporate an oligonucleotide into a larger clonable fragment, the oligonucleotide must have four distinct regions (Fig. 5). Region 1, at the 3′ end of the oligonucleotide, is a region of perfect homology to the target gene. This is needed to ensure efficient priming from the unmutagenized template. This constant region can be as little as 16 nucleotides (nt), if it will be used to generate a clonable PCR product incorporating a simple point mutation. When using degenerate oligonucleotides to construct libraries of mutants, we generally make this constant region 25 nt in length to guarantee that mutations in the degenerate oligonucleotide just 5′ to this priming region are not biased against in the subsequent PCR step. Otherwise stated, the constant region (region 1) at the 3′ end should be long enough that region 2, the mutagenized region, makes no significant contribution to the stability of the primer–template hybrid during the PCR hybridization step. Region 2, immediately 5′ to the constant region 1, incorporates the desired point mutation or, in the case of a library of mutants, multiple mutations at a frequency chosen by the investigator (for tips on choosing the appropriate mutagenesis rate, see the section following ("Mutation Frequency and the Poisson Distribution"). Next, region 3, just 5′ to the mutagenized region includes the restriction site and can also include a further constant region, depending on the distance between the site to be mutagenized and the restriction site. Lastly, just 5′ to the restriction site (i.e., at the very 5′ end of the oligonucleotide) is region 4, the "clamp" of several nucleotides to allow efficient digestion with the restriction enzyme whose site has been incorporated in region 3 of the oligonucleotide. Many restriction enzymes will not digest restriction sites present at the very end of a DNA fragment, and so, depending on the restriction

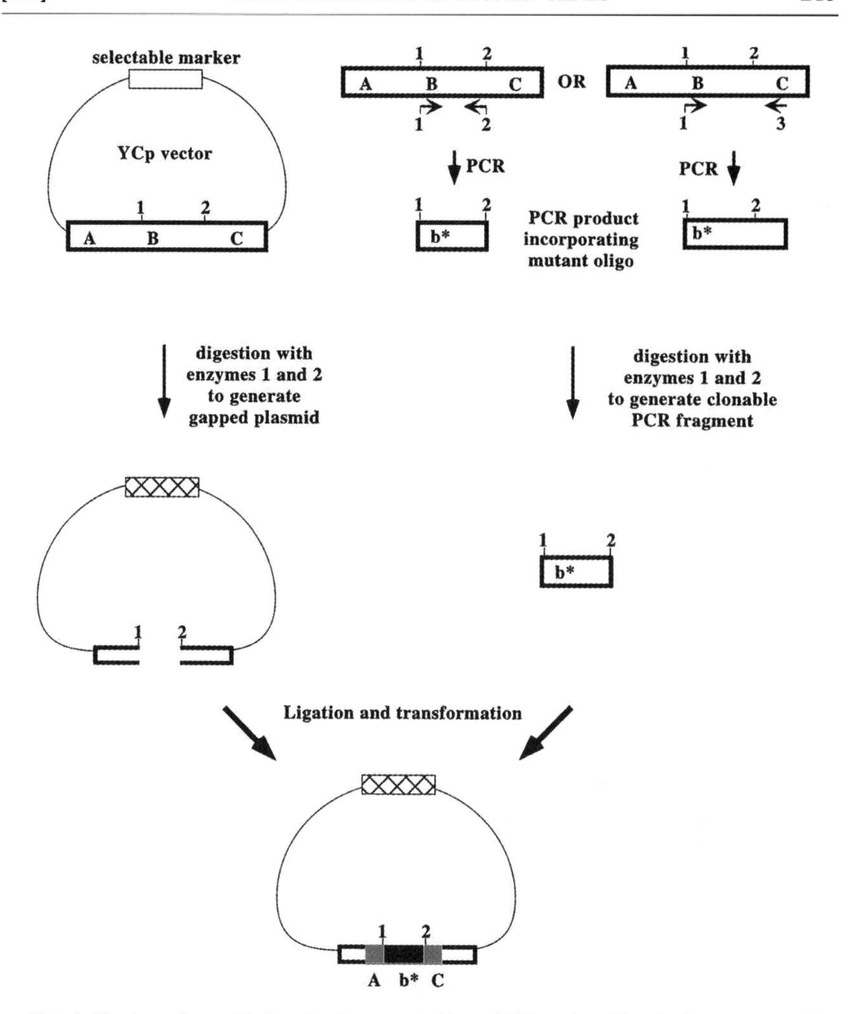

FIG. 4. Cloning using restriction sites incorporated into a PCR product. The cloning vector carrying the gene to be mutated is digested with restriction enzymes 1 and 2, generating a gapped vector with incompatible ends 1 and 2. PCR primer 1 has restriction site 1 and the mutation (b*) to be introduced into the full-length gene. The PCR product is generated with primers 1 and 2 (which include restriction sites 1 and 2) or alternatively with primers 1 and 3 (primer 1 has restriction site 1; primer 3 is downstream of restriction site 2 such that the PCR product generated will have restriction site 2 as an internal site).

enzyme, the clamp may be necessary to get efficient digestion of the PCR product. The NEB (New England Biolabs, Beverly, MA) catalog contains a table of efficiencies with which various restriction enzymes digest short oligonucleotides that contain the restriction site and variable flank sequences. This can be used as a guide to choosing the appropriate clamp nucleotides for a given enzyme.

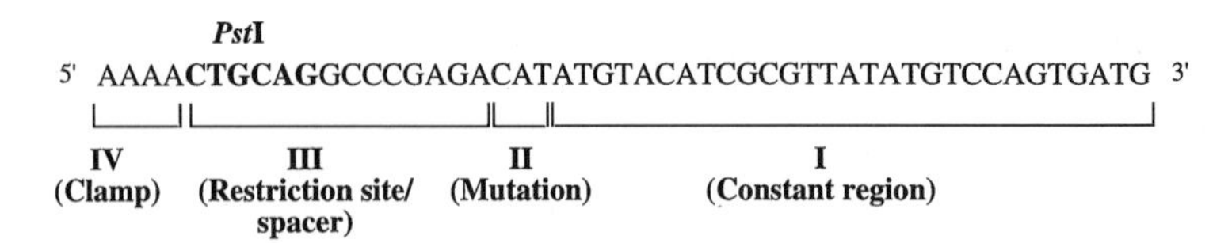

FIG. 5. Example of an oligonucleotide designed to introduce a point (CTT → CAT) mutation by PCR amplification and cassette mutagenesis. Region I: the 3′ constant region; region II: the mutant change; region III: the restriction site (a *Pst*I site) and constant spacer region between the restriction site and the mutation site; region IV: the 5′ clamp.

In our experience, the overall length of the mutagenic oligonucleotide is not an important parameter and oligonucleotides as long as 110 nt have been efficiently incorporated by PCR into a fragment.[15]

5. Expected Results. The following example details the potential power of this approach and the utility of doped oligonucleotides in generating mutant libraries with unparalleled mutant representation. In an elegant study, Strubin and Struhl[16] identified an altered specificity mutant of the yeast TATA-binding protein. They synthesized a 63-nucleotide oligonucleotide of which the core 45 nucleotides, corresponding to 15 amino acids in the DNA binding domain of TBP, were mutagenized at 8% per base pair. The oligo was flanked by *Bcl*I and *Sac*I sites used ultimately to clone the oligonucleotide into the full-length *SPT15* gene, engineered to contain corresponding *Bgl*II and *Sac*I sites. They generated a library of 2×10^6 transformants in which the average number of amino acid substitutions was 2.4 per clone. From this library they selected a rare double mutant able to support transcription from a nonconsensus TATA box. The altered specificity mutant contained two independent (both critical) transversion mutations (AT to TA and AT to CG) that would have been underrepresented in libraries derived from hydroxylamine or degenerate PCR mutagenesis. The success in this example depended on having a highly representational library of mutants and makes the point that in constructing mutant libraries where the aim is to identify rare mutant classes, doped oligonucleotides have important advantages. In order to sample even a minority of amino acid substitutions in even a small region of a protein, the bias in mutation classes must be relatively modest and the libraries must be relatively large. An example illustrates this last point. For a 10 amino acid stretch there are 7830 possible two-nucleotide changes ($30 \times 29 \times 3 \times 3$). In a library with an average of two nucleotide changes per oligonucleotide (6.6% mutagenesis) approximately 27% of the clones will have exactly two mutations; therefore a minimum of 29,000 independent clones are needed to sample all possible double nucleotide

[15] B. P. Cormack and K. Struhl, *Science* **262,** 244 (1993).
[16] M. Strubin and K. Struhl, *Cell* **68,** 721 (1992).

substitutions (and about 80% of possible single amino acid changes for the region) in an otherwise wild-type background. Similarly, in this same library, a minimum of 3.6 million clones would be needed to sample all triple nucleotide substitutions and therefore guarantee sampling of all possible single amino acid substitutions. For larger regions, therefore, it is in essence impossible to generate large enough libraries to efficiently sample even all single amino acid changes. One way around this problem is to use codon-based mutagenesis methods,[15] which permit efficient sampling of all possible amino acid substitutions in compact highly representative libraries. Codon-based mutagenesis is particularly advantageous when the aim is to identify rare mutant classes in targeted regions of a protein. For the same 10 amino acid region, there are only 320 possible codon changes (10×32). In a library with on average two codons changed per oligonucleotide, all possible single amino acid changes in an otherwise wild-type background can be sampled in a minimum of 1200 independent clones.

Lastly, it is true that for most applications (generation of null or temperature-sensitive mutants), such mutant diversity is not necessary and hydroxylamine or low-fidelity PCR mutagenesis will probably yield the desired mutant.

Mutation Frequency and Poisson Distribution

The investigator interested in making a library of point mutations needs first to decide on the frequency of mutation desired. In an oligonucleotide mutated at a given average frequency, the number of oligonucleotides with a particular number of mutational changes is given by the Poisson distribution. The equation below allows the calculation of the probability of having an oligonucleotide with exactly r mutations if the oligonucleotide has μ mutations on average.

$$P(r) = (\mu^r / r!)e^{-\mu}$$

where $P(r)$ is probability of exactly r mutations in an individual oligonucleotide; r, the exact number of mutations in an individual oligonucleotide; and μ, the average number of events per oligonucleotide. By way of example, Table I gives the calculated relative frequencies of oligonucleotides in a population having r mutations (0,1,2,3,4) when the average number of mutations (μ) ranges from 0.5 to 2.0. It will be noticed, for example, that when the average number of mutations per oligonucleotide is 1.0, approximately 1/3 are wild type, approximately 1/3 have exactly one mutation, and 1/3 have greater than one mutation. When the average number of mutations is doubled and increased to 2.0, approximately 1/3 still have one mutation, but only 13% are wild type and the rest have greater than one mutation. So, while maximizing the percent of the library made up of mutants with a single nucleotide change, the investigator has the flexibility of biasing the library to more or fewer mutations. That decision could be based, for example,

TABLE I
RELATIVE FREQUENCIES OF OLIGONUCLEOTIDES IN POPULATION HAVING r MUTATIONS

	μ (average number of mutations)				
r	0.5	0.75	1.0	1.5	2.0
0	0.607	0.472	0.368	0.223	0.135
1	0.303	0.354	0.368	0.335	0.270
2	0.076	0.133	0.184	0.251	0.270
3	0.012	0.033	0.061	0.125	0.180
4	0.002	0.006	0.016	0.047	0.09

on whether wild-type genes or mutants with multiple nucleotide changes will be more problematic during subsequent analysis.

Conclusion

The methods outlined in this chapter are by no means meant to be a comprehensive examination of the many variations available for introducing point mutations into cloned DNA sequences. The methods outlined, however, give a sample of the available approaches that can be used to mutationally analyze gene function whether by changing a single amino acid of interest or by generating a library of point mutants distributed throughout the gene. While *in vitro* mutagenesis techniques are so widely used as to be commonplace, it is worth carefully considering, particularly when constructing libraries of mutants, which of the general or oligonucleotide-directed methods outlined here is best suited to a specific purpose.

Acknowledgments

We thank D. Bartel, J. Boeke, E. Bolton and S.L. Ooi for useful discussions and sharing unpublished data. B.C. also thanks K. Struhl, in whose laboratory most of these methods were learned.

[12] Insertional Mutagenesis: Transposon–Insertion Libraries as Mutagens in Yeast

By ANUJ KUMAR, SUSANA VIDAN, and MICHAEL SNYDER

Transposons are powerful mutagenic agents by which large numbers of insertion alleles may be quickly and economically generated within a target population of DNA. In yeast, plasmid-based libraries of insertion alleles may be generated using bacterial transposons. In this case, yeast DNA is mutagenized in *Escherichia coli;* transposon-mutagenized DNA is then shuttled back into yeast, either individually or *en masse,* where the insertion alleles substitute for their chromosomal copies. Libraries of these alleles constructed with bacterial transposons are advantageous: prokaryotic transposons exhibit less bias in target site selection than do most of their eukaryotic counterparts, including the yeast transposon Ty1.[1,2] Additionally, many bacterial transposons are subject to transposition immunity, rendering DNA molecules already containing a single transposon insertion (mTn terminus) immune from further insertion.[3,4] Finally, libraries of plasmid-based insertion alleles, generated in *E. coli,* provide a replenishable supply of DNA which may be subsequently introduced into any number of genetic backgrounds. We, therefore, have employed bacterial transposons to mutagenize plasmid-based libraries of yeast genomic DNA, generating libraries of "single-hit" insertion alleles suitable for further functional analysis.

These transposon-based insertional libraries are exceptionally useful as a means of screening for desired phenotypes and expression patterns. As compared to mutations generated by classical methods of chemical treatment and ultraviolet irradiation, transposon insertions are easy to identify: the transposon-encoded sequence serves as a marker or tag through which sites of insertion may be rapidly identified by PCR amplification or plasmid rescue. Additionally, transposon insertion alleles can be uniquely informative: a single insertion may be sufficient to generate a gene disruption, reporter gene fusion, epitope-tagged allele, and conditional mutation. Transposons, if engineered appropriately, may also serve as gene traps identifying novel coding sequences.

In this chapter, we provide comprehensive protocols for the use of insertional libraries as mutagens in yeast. From the protocols presented here, a plasmid-based library of transposon-mutagenized yeast DNA may be used to generate and identify yeast mutants of interest; transposon insertion sites within these mutants may be

[1] N. L. Craig, *Annu. Rev. Biochem.* **66,** 437 (1997).

[2] S. E. Devine and J. D. Boeke, *Genes Dev.* **10,** 620 (1996).

[3] M. K. Robinson, P. M. Bennett, and M. H. Richmond, *J. Bacteriol.* **129,** 407 (1977).

[4] L. K. Arciszewska, D. Drake, and N. L. Craig, *J. Mol. Biol.* **207,** 35 (1989).

0076-6879/02 $35.00

characterized by vectorette PCR as described. Also, by employing insertional libraries carrying a specially designed multipurpose transposon, insertions may be modified in yeast to generate corresponding epitope-tagged alleles for a variety of functional studies.

Transposon–Insertion Libraries

By transposon mutagenesis, we have generated a series of plasmid-based insertional libraries,[5–7] each derived from a genomic library of yeast DNA mutagenized by a modified prokaryotic transposon. All insertional libraries (along with additional reagents) may be requested from the TRIPLES database[8] at ygac.med.yale.edu. As these insertional libraries facilitate similar studies, we will focus on a single representative library generated with the multipurpose transposon mTn-3xHA/*lacZ* (Fig. 1).

This Tn*3*-derived minitransposon (mTn) is multifunctional in that it can be used to generate a variety of mutant alleles from a single insertion.[6] The mTn-3xHA/*lacZ* transposon carries the reporter gene *lacZ;* therefore, transposon insertions may be used to generate *lacZ* gene fusions. As the mTn-encoded *lacZ* reporter lacks both its start codon and upstream promoter, production of β-galactosidase (β-Gal) is dependent on transposon insertion within a transcribed and translated region of the yeast genome—typically corresponding to an in-frame fusion with yeast protein coding sequence. The *lacZ* reporter in mTn-3xHA/*lacZ* is terminated by a series of stop codons, so that mTn insertion also creates a gene truncation. In addition, mTn-3xHA/*lacZ* contains two *lox* elements, one located near each mTn end. One *lox* site is internal to a sequence encoding three copies of an epitope from the influenza virus hemagglutinin protein (the HA epitope). These *lox* sites are target sequences for the site-specific recombinase Cre; Cre-*lox* recombination may be used in yeast to reduce the full-length 6-kb transposon to a small 93-codon read-through insertion encoding three copies of the HA epitope (the HAT tag). In this manner, mTn-mediated disruption alleles may be converted to epitope-tagged alleles in yeast.[6,9,10]

[5] N. Burns, B. Grimwade, P. B. Ross-Macdonald, E.-Y. Choi, K. Finberg, G. S. Roeder, and M. Snyder, *Genes Dev.* **8,** 1087 (1994).

[6] P. Ross-Macdonald, A. Sheehan, G. S. Roeder, and M. Snyder, *Proc. Natl. Acad. Sci. U.S.A.* **94,** 190 (1997).

[7] P. Ross-Macdonald, P. S. R. Coelho, T. Roemer, S. Agarwal, A. Kumar, R. Jansen, K.-H. Cheung, A. Sheehan, D. Symoniatis, L. Umansky, M. Heidtman, F. K. Nelson, H. Iwasaki, K. Hager, M. Gerstein, P. Miller, G. S. Roeder, and M. Snyder, *Nature* **402,** 413 (1999).

[8] A. Kumar, K.-H. Cheung, P. Ross-Macdonald, P. S. R. Coelho, P. Miller, and M. Snyder, *Nucleic Acids Res.* **28,** 81 (2000).

[9] P. Ross-Macdonald, A. Sheehan, C. Friddle, G. S. Roeder, and M. Snyder, *Methods Enzymol.* **303,** 512 (1999).

[10] A. Kumar, S. A. des Etages, P. S. R. Coelho, G. S. Roeder, and M. Snyder, *Methods Enzymol.* **328,** 550 (2000).

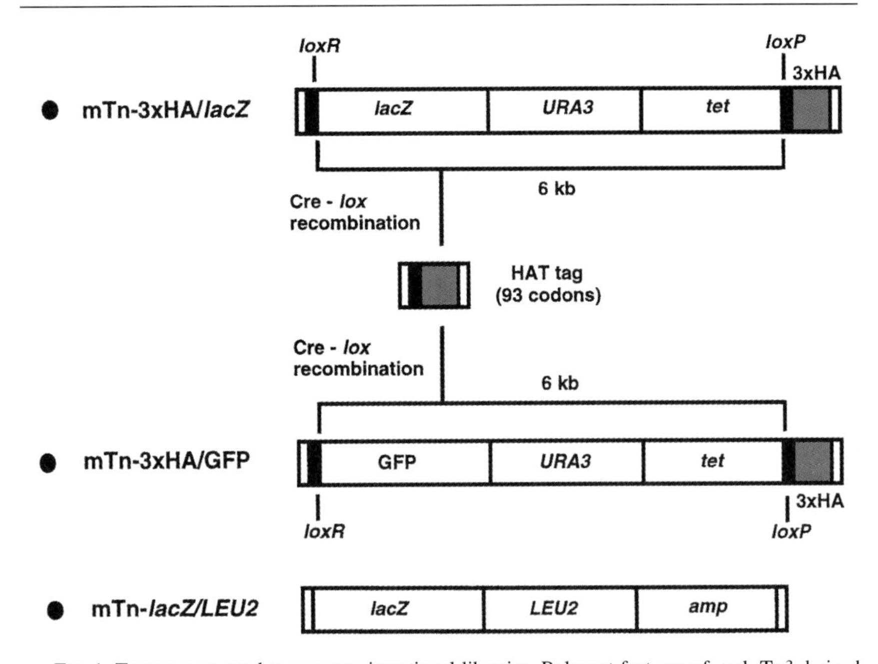

FIG. 1. Transposons used to generate insertional libraries. Relevant features of each Tn*3*-derived transposon are illustrated (not drawn to scale). Transposons mTn-3xHA/lacZ and mTn-3xHA/GFP may be reduced by Cre-*lox* recombination in yeast to a HAT tag insertion element. Both mTn-3xHA/*lacZ* and mTn-*lacZ*/*LEU2* carry *lacZ* reporter genes lacking a start codon and promoter; mTn-3xHA/GFP carries a full-length GFP reporter not suitable for use as a gene trap. All libraries and reagents can be requested from our Web page at ygac.med.yale.edu.

Insertional libraries have been generated with several transposons functionally similar to mTn-3xHA/*lacZ* (Fig. 1). Carrying sequence encoding full-length green fluorescent protein (GFP) in place of β-Gal, mTn-3xHA/GFP is otherwise identical to mTn-3xHA/*lacZ*. It is not, however, suitable as a gene trap and may provide a less sensitive reporter. The Tn*3*-derived mTn-*lacZ*/*LEU2* transposon may be used as a gene trap producing reporter gene fusions and gene disruption alleles[5]; however, lacking *lox* sites and epitope sequence, mTn-*lacZ*/*LEU2* cannot be used to generate epitope-tagged alleles. Most recently, a multipurpose transposon similar to mTn-3xHA/*lacZ* has been constructed from the bacterial transposon Tn*7*.[11] The Tn*7* transposition system possesses several advantages as a mutagenic agent. Tn*7* mutagenesis may be applied *in vitro*[12]; additionally, Tn*7* possesses a different

[11] A. Kumar, M. C. Biery, R. Sarnovsky, N. L. Craig, and M. Snyder, unpublished, 2001.

[12] M. C. Biery, F. J. Stewart, A. E. Stellwagen, E. A. Raleigh, and N. L. Craig, *Nucleic Acids Res.* **28,** 1067 (2000).

(and less pronounced) bias in target site selection than does Tn*3*.[1,12,13] Used in conjunction, Tn*3*- and Tn*7*-based insertional libraries may mutually complement each other, providing a means by which a comprehensive set of insertions may be generated throughout the yeast genome.

Irrespective of the library used, yeast insertional mutants may be generated and screened as follows.

Method and Applications

Insertional mutants are generated in yeast by means of shuttle mutagenesis[14] (Fig. 2). Transposon insertions are generated in a plasmid-based library of yeast genomic DNA; insertion alleles are subsequently introduced ("shuttled") into yeast by standard DNA transformation methods. Each plasmid-borne fragment of genomic DNA, carrying a single mTn insertion, will integrate at its corresponding genomic locus by homologous recombination. Yeast strains carrying a chromosomal mTn insertion may be assayed for desired phenotypes; a diploid yeast strain may be employed to analyze haploin-sufficient phenotypes resulting from insertions in essential genes. Transposon insertions within translated sequences can be identified by β-Gal assay, providing a means by which novel genes may be identified. Precise sites of transposon insertion may be detected by vectorette PCR or other methods. Strains containing in-frame multipurpose transposon insertions can be analyzed further by inducing Cre-*lox* recombination in yeast. Resulting epitope-tagged proteins may be immunoprecipitated or localized by immunofluorescence microscopy. Transposon-encoded epitope insertions are of further utility as a potential source of conditional mutations; mutants exhibiting partial gene function are particularly helpful in analyzing essential genes.

Protocols

Amplifying Library DNA in E. coli

The following protocols assume use of the mTn-3xHA/*lacZ* insertional library, although similar approaches are applicable to any transposon-mutagenized library. The mTn-3xHA/*lacZ* library is derived from a plasmid library (containing 50 genome equivalents) generated by *Sau*3A 1 partial digestion of genomic DNA isolated from a cir^0rho^0 yeast strain (lacking 2-μm plasmids and mitochondrial DNA). Aliquots of library DNA are distributed in 10 pools as dried-down solutions, which may be suspended in an appropriate volume of TE buffer, pH 8.0 (e.g., at a concentration of approximately 100 ng/μl).

[13] C. J. Davies and C. A. Hutchison, *Nucleic Acids Res.* **23,** 507 (1995).

[14] H. S. Seifert, E. Y. Chen, M. So, and F. Heffron, *Proc. Natl. Acad. Sci. U.S.A.* **83,** 735 (1986).

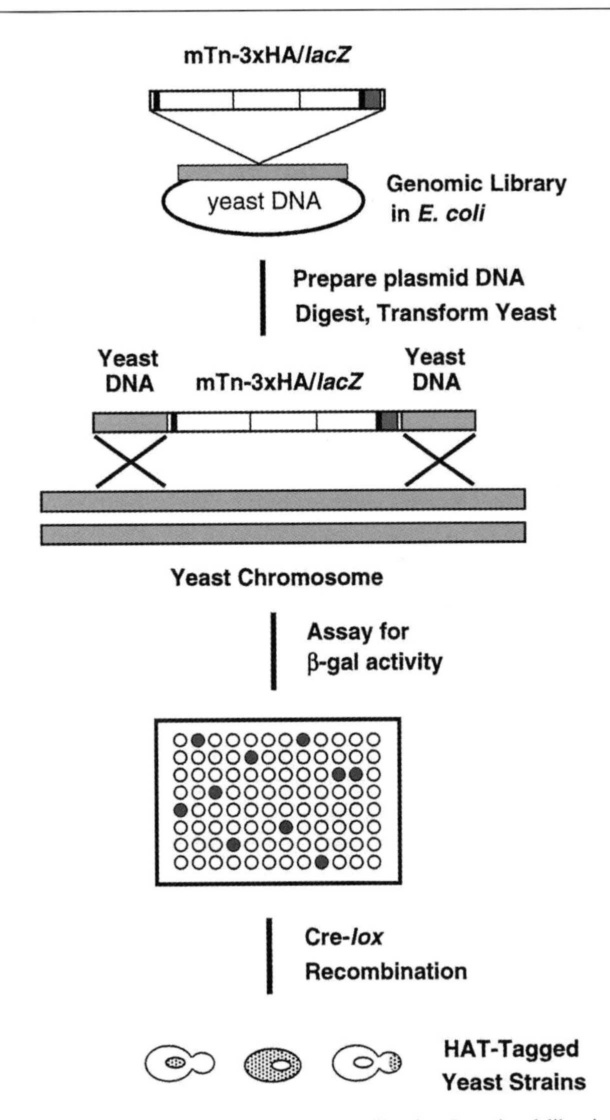

FIG. 2. Insertional mutagenesis with transposon–insertion libraries. Insertional libraries may be used to generate yeast mutants and screen for desired phenotypes and expression patterns by the steps indicated. Transposon insertions can be converted to HAT tags by means of Cre-*lox* recombination in yeast. HAT-tagged proteins may be immunolocalized or immunoprecipitated; HAT tags can often generate conditional mutations as well.

1. Introduce a suitable amount of insertion library DNA into any tetracycline- and kanamycin-sensitive *E. coli* strain by standard transformation procedures. Select transformants on LB medium supplemented with tetracycline (3 μg/ml) and kanamycin (40 μg/ml) using plates 14 cm in diameter. Approximately 10,000 transformants should be obtained per pool (approximately 100,000 in total) following overnight growth at 37°.

2. Elute transformant colonies as follows: place 6 ml of LB medium onto each plate and scrape cells into a homogenous suspension. Dilute an aliquot of this eluate into LB medium supplemented with tetracycline (3 μg/ml) and kanamycin (40 μg/ml) to yield a culture of nearly saturated cell density. Incubate at 37° with aeration for 2–3 hr.

3. Isolate plasmid DNA by any standard minipreparation or large-scale protocol.

Transforming Yeast

Yeast transformations are performed using a modification of the lithium acetate/single-stranded DNA/polyethylene glycol (PEG) method described by Chen *et al.*[15]

1. Digest a small aliquot of plasmid DNA (e.g., 1 μg) with *Not*I. Subsequently, analyze a portion of the reaction mixture by agarose gel electrophoresis to ensure release of mTn-mutagenized yeast DNA from the plasmid vector. Upon electrophoresis, a distinct 2.1-kb band (corresponding to the vector) and broad 8-kb band should be visible: the broad 8-kb band consists of 2- to 3-kb inserts of yeast genomic DNA carrying the 6-kb mTn construct. Store the remaining reaction mixture for later use in step 4.

2. Grow a 10-ml culture of any desired *ura3* yeast strain to mid-log phase (a density of 10^7 cells/ml or OD_{600} of approximately 1) maintaining appropriate selection if applicable. Ideally, choose a diploid yeast strain to screen for desired patterns of gene expression. To screen for disruption phenotypes, a haploid strain is often used; from previous studies (S. A. des Etages and M. Snyder, unpublished, 2001), we estimate that 10% of transposon insertions in essential genes are viable. For the eventual analysis of HAT-tagged proteins, choose a *ura3 leu2* strain (as the pGAL-*cre* plasmid is marked with *LEU2*).

3. Pellet cells in a clinical tabletop centrifuge at 1100*g* for 5 min. Wash once with 5 volumes of One-Step Buffer [0.2 *M* lithium acetate, 40% (w/v) PEG 4000, 100 m*M* 2-mercaptoethanol].

4. Resuspend cells in 1 ml One Step Buffer supplemented with 1 mg denatured salmon sperm DNA. Add 100-μl aliquots from this suspension to 0.1–1 μg

[15] D. C. Chen, B. C. Yang, and T. T. Kuo, *Curr. Genet.* **21,** 83 (1992).

*Not*I-digested plasmid DNA from step 1. Use a small quantity of transforming DNA in order to minimize generation of transformants containing more than one insertion. Vortex, and incubate at 45° for 30 min.

5. Pellet cells and subsequently suspend in 400 μl SC −Ura medium. Spread 200-μl aliquots onto SC −Ura plates, and incubate at 30° for 3–4 days. Up to 1000 transformants may be recovered per microgram of transforming DNA. To ensure 95% coverage of the genome (without regard to in-frame reporter activity), screen 30,000–50,000 colonies. To identify in-frame insertions within at least 95% of all yeast genes, screen approximately 180,000–200,000 transformants for β-Gal activity.

Screening for β-Galactosidase Activity

1. To maximize detection of *lacZ* fusions expressed at low levels, patch transformant colonies onto YPD plates (supplemented with 80 μg/ml adenine if using an *ade2* host strain) at a density of up to 100 colonies per plate.

2. Place a sterile disk of Whatman 3MM filter paper (Clifton, NJ) onto a plate of SC –Ura medium; repeat for as many plates as needed. Replicate transformant cells onto filter-covered plates and incubate overnight at 30°. Alternative growth conditions (e.g., growth on sporulation medium) may be substituted as desired.

3. Following overnight growth, lift filters from plates and place in the lid of a 9-cm glass petri dish. Place this lid inside a closed 15-cm petri dish containing chloroform. Incubate for 10–30 min.

4. Place filters colony-side up onto fresh X-Gal plates [5-bromo-4-chloro-3-indolyl-β-D-galactopyranoside (X-Gal, 120 μg/ml), 0.1 *M* $NaPO_4$ (pH 7.0), 1 m*M* $MgSO_4$ in 1.6% (w/v) agar]. Incubate inverted at 30° for up to 3 days. After several days of growth, β-Gal levels can be reliably estimated from the observed intensity of blue staining. We typically observe β-Gal activity in 12–16% of transformants.

Identifying Insertion Sites by Vectorette PCR

Genomic sites of transposon insertion may be identified within strains of interest by several approaches. Genomic DNA immediately adjacent to a transposon insertion can be recovered in *E. coli* by plasmid rescue.[5,16,17] Alternatively, insertion sites may be identified through direct genomic sequencing of mTn-mutagenized strains using a transposon-specific primer.[18] PCR-based techniques are also applicable as a means of detecting transposon insertions: most notably,

[16] P. Ross-Macdonald, N. Burns, M. Malcynski, A. Sheehan, S. Roeder, and M. Snyder, *Meth. Mol. Cell. Biol.* **5,** 298 (1995).

[17] H. U. Mosch and G. R. Fink, *Genetics* **145,** 671 (1997).

[18] J. Horecka and Y. Jigami, *Yeast* **16,** 967 (2000).

the vectorette PCR method of Riley *et al.*[19] has been successfully utilized for this purpose.[9]

In vectorette PCR (Fig. 3A), genomic DNA is digested with a blunt-end restriction endonuclease possessing a 4- to 6-base-pair recognition sequence. Digested DNA fragments are ligated to a pair of annealed primers containing a nonhomologous central region; these primer pairs form "anchor bubbles" flanking each genomic fragment. PCR is then performed using a primer complementary to the transposon and a primer identical to sequence within the anchor bubble. During the initial round of amplification, only the mTn primer can bind its template; however, during subsequent cycles, the anchor bubble primer can anneal to the extended mTn primer, resulting in selective amplification of DNA sequence adjacent to the point of transposon insertion.

The vectorette PCR protocol provided below should yield approximately 200–400 ng of product, constituting sufficient template for two to three sequencing reactions.

1. Prepare genomic DNA by any standard protocol (e.g., the Zymolyase-based method of Philippsen *et al.*[20]); care should be taken to obtain high-quality DNA, as this is critical to successful PCR amplification. Digest 5 μg of yeast genomic DNA with a blunt-end restriction endonuclease (such as *Alu*I) in a total volume of 20 μl. After overnight digestion, the enzyme is heat inactivated by incubating 20 min at 65°.

2. Primers ABP1 and ABP2 (Fig. 3B) are annealed to each other to form the adaptor anchors by mixing 1 pmol of each primer in 200 μl of annealing buffer containing 10 m*M* Tris, 10 m*M* $MgCl_2$, and 50 m*M* NaCl. The primer mixture is heated for 5 min at 95° and allowed to slowly cool to 37°.

3. The adaptors are ligated to the DNA fragments by adding 1 μl of the annealed primers, 0.25 μl of 10 m*M* ATP, 3 μl of 10× restriction buffer used in the digestion, 24.75 μl H_2O, and 1 μl (400 U) of T4 DNA Ligase to the 20 μl restriction digest mixture from step 1. The ligation reaction is incubated overnight at 16°.

4. A standard 100 μl PCR reaction is set up using 5 μl from the ligation mixture, 2.5 μl each of primers UV and M13(-47) (Fig. 3B) at 20 μM, 5 U of *Taq* polymerase, and 1 μl of dNTPs (at 20 m*M* each dNTP) in a final volume of 100 μl. The PCR program consists of one cycle of 2 min at 92°, followed by 35 cycles of 20 sec at 92°, 30 sec at 67°, and 45 to 180 sec at 72° with a final extension of 90 sec at 92°.

[19] J. Riley, R. Butler, D. Ogilvie, R. Finniear, D. Jenner, S. Powell, R. Anand, J. C. Smith, and A. F. Markham, *Nucleic Acids Res.* **18,** 2887 (1990).

[20] P. Philippsen, A. Stotz, and C. Scherf, *Methods Enzymol.* **194,** 169 (1991).

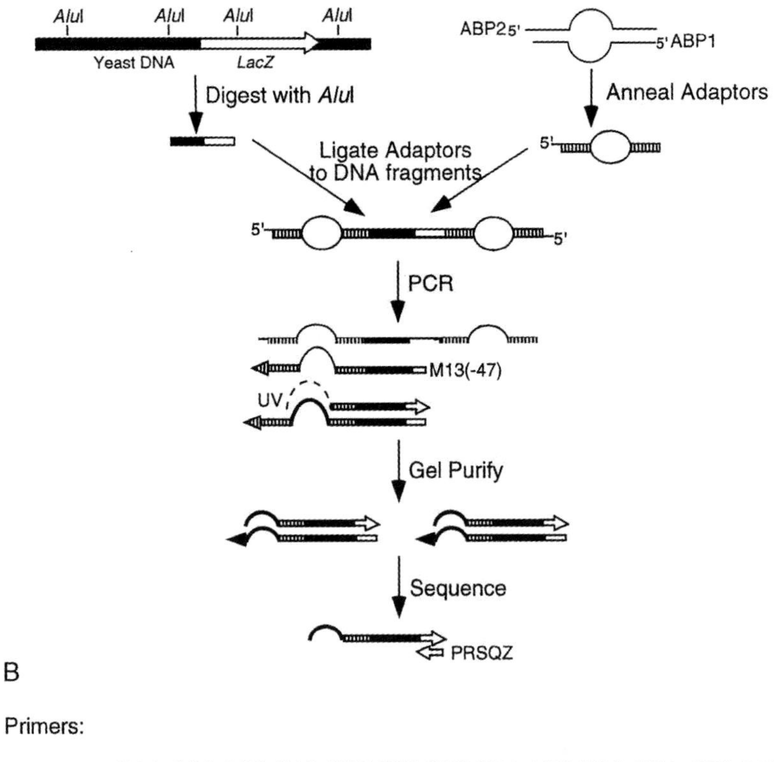

ABP1	GAA GGA GAG GAC GCT GTC TGT CGA AGG TAA GGA ACG GAC GAG AGA AGG GAG AG
ABP2	GAC TCT CCC TTC TCG AAT CGT AAC CGT TCG TAC GAG AAT CGC TGT CCT CTC CTT C
UV	CGA ATC GTA ACC GTT CGT ACG AGA ATC GCT
M13-47	CGC CAG GGT TTT CCC AGT CAC GAC
PRSQZ	CGA CGG GAT CCC CCT TAA CG

FIG. 3. Vectorette PCR. (A) Yeast genomic DNA is cut with a blunt-end restriction endonuclease that cuts both in the yeast sequence and transposon. Adaptors are formed by annealing primers ABP1 and ABP2 to each other, forming an anchor sequence for PCR. The ABP2/ABP1 annealed primers are ligated to the blunt ends of the cut DNA. Proper orientation of the adaptors is ensured by a two-nucleotide overhang in ABP2. PCR is carried out with primers M13(-47) and UV which allow for the amplification of sequences 5′ of the transposon insertion point. M13(-47) is complementary to sequences in the mTn-encoded *LacZ* reporter and, upon extension, corresponds to sequence immediately upstream of the transposon. UV is identical to sequences found in ABP2 and is complementary to sequences found in the M13(-47) extended fragment. The PCR products are gel purified and sequenced using primer PRSQZ, which is complementary to *LacZ*. (B) Suggested primer sequences are shown 5′ to 3′.

5. Each PCR product is gel purified using standard protocols into a final volume of 30 μl TE. 10 μl of the purified product is sufficient for one sequencing reaction with primer PRSQZ (Fig. 3B).

Producing Epitope-Tagged Strains by Cre-lox Recombination

Strains bearing an in-frame transposon insertion may be used to derive corresponding HAT-tagged strains by Cre-*lox* recombination in yeast. The phage P1 Cre recombinase can be expressed exogenously from plasmid pGAL-*cre* (available at ygac.med.yale.edu); on this plasmid, *cre* is under transcriptional control of the *GAL* promoter, so that galactose induction in yeast may be used to drive *cre* expression.[6] Following induction on galactose, cells having undergone Cre-mediated recombination (and loss of the mTn-encoded *URA3* marker) may be selected on medium containing 5-fluoroorotic acid (5-FOA).[21] Although both mTn-encoded *lox* sites are targets of the Cre recombinase, their sequences have been engineered to be slightly divergent, thereby reducing the frequency with which these sites spontaneously recombine. From previous experience, galactose induction results in Cre-mediated excision of the *URA3* marker in over 90% of cells analyzed.

1. Transform the mTn-mutagenized *ura3 leu2* host strain with pGAL-*cre* (*amp*, *ori*, *CEN*, *LEU2*); subsequently, select transformants on SC −Leu −Ura dropout medium.

2. To derepress the *GAL* promoter, inoculate transformants into 2 ml SC −Leu −Ura medium with 2% raffinose as its carbon source. Incubate at 30° with aeration until the culture has grown to saturation.

3. Dilute cultures 100-fold into SC −Leu medium with 2% galactose as its carbon source. As a control, dilute an aliquot of the same culture 100-fold into 2 ml SC −Leu medium with 2% glucose as its carbon source. Grow cultures for 2 days at 30° with aeration.

4. If visible growth is apparent, dilute cultures 100-fold in sterile water and withdraw a 10-μl aliquot. If no growth is apparent, withdraw a 10-μl aliquot from the undiluted culture. Spot onto a 5-FOA plate, and isolate single colonies by streaking the droplet. Dilute cultures grown in 2% glucose 100-fold in sterile water, withdraw a 10-μl aliquot, spot, and streak onto a 5-FOA plate. Incubate 5-FOA plates at 30° until growth is visible on those plates inoculated with strains grown in galactose. As an alternative, cultures may be plated onto SC medium and replicated onto SC −Ura dropout medium. Incubate at 30° approximately 2 days. Cultures grown in galactose should yield approximately 100-fold more cells on SC −Ura medium than identical cultures grown in glucose.

[21] R. S. Sikorski and J. D. Boeke, *Methods Enzymol.* **194,** 302 (1991).

5. Single colonies from strains having lost the mTn-encoded *URA3* marker (exclusively following galactose induction) may be saved as stock in 15% glycerol at −70°. PCR analysis can be used to confirm the position of the HAT tag; the complete sequence encoding this tag may be viewed at ygac.med.yale.edu.

Screening with Insertional Libraries

By the protocols presented here, transposon-based insertional libraries may be easily used to perform genetic screens: insertion alleles are transformed *en masse* into yeast for the subsequent identification of mutants exhibiting a phenotype of interest. By this approach, Erdman *et al.*[22] screened 91,200 transformants carrying mTn-encoded *lacZ* insertions for β-Gal activity in the presence and absence of α factor mating pheromone; this study uncovered 20 genes previously uncharacterized for their role in yeast mating differentiation. Mosch and Fink[17] have employed transposon mutagenesis in a genetic screen to identify genes involved in filamentous growth, thereby identifying 16 novel targets of the filamentation signaling pathway in yeast.

Alternatively, transposon–insertion libraries may be used to generate genome-wide collections of defined insertion alleles. Plasmid-borne insertion alleles may be introduced individually into yeast by high-throughput DNA transformations. Using the transposon as a gene trap, yeast mutants carrying insertions within gene-coding sequences can be selected. Assuming insertion alleles have integrated by homologous recombination in these strains, precise genomic sites of transposon insertion may be identified from the DNA sequence of each corresponding plasmid. Of course, linkage of the phenotype to the insertion should be checked by genetic analysis. Plasmid-borne alleles of interest may be permanently stored, thereby generating a streamlined collection of insertions within experimentally identified yeast genes. This collection will provide a means by which gene function in yeast may be investigated through a survey of 6200 defined mutants, rather than through a screen of several hundred thousand uncharacterized mutations. These plasmid-based collections promise to significantly expedite the functional analysis of yeast genes by fundamentally altering the manner in which genetic screens are performed.

Acknowledgments

Many members of the Snyder laboratory, past and present, have contributed to the development and evolution of these protocols; credit is due all these individuals. A.K. is supported by a postdoctoral fellowship from the American Cancer Society. S.V. is supported by an NSF Minority Postdoctoral Fellowship.

[22] S. Erdman, M. Malczynski, and M. Snyder, *J. Cell Biol.* **140,** 461 (1998).

[13] Tn7-Mediated Mutagenesis of *Saccharomyces cerevisiae* Genomic DNA *in Vitro*

By NURJANA BACHMAN, MATTHEW C. BIERY, JEF D. BOEKE, and NANCY L. CRAIG

Introduction

Transposons have been used as insertional mutagens in many different organisms. *In vivo* transposition-based mutagenesis techniques have been developed for various bacterial organisms, yeast, *Drosophila melanogaster,* and zebrafish. The advent of reconstituted transposition systems has allowed the development of *in vitro* techniques for mutagenesis of specific DNA targets by the bacterial transposons Tn*7*, Tn*5*, Tn*552*, Mu, and *Saccharomyces cerevisiae* Ty1. The products of the reactions can be used for a variety of applications, including large-scale genome sequencing efforts, as well as the generation of large libraries of insertional mutations *in vitro,* which can then be introduced into the genome of the organism by transformation and homologous recombination. To date, primarily prokaryotic organisms have been mutagenized by *in vitro* mutagenesis: for example, the Tn*7* mutagenesis of the *Haemophilus influenzae* genome,[1] the Tn*5* mutagenesis of *Proteus mirabilis, Escherichia coli, Salmonella typhimurium,* and others,[2,3] and the Ty1 *in vitro* mutagenesis of *Haemophilus influenzae*[4] and *Leishmania tarentolae*.[5,6]

This chapter describes *in vitro* insertional mutagenesis of the complete genome of *Saccharomyces cerevisiae* by the bacterial transposon Tn*7*. The *in vitro* transposition reaction of Tn*7* is accomplished with purified components, allowing the mutagenesis of any purified DNA fragment. We review the application of this *in vitro* transposition technique using total yeast genomic DNA to create a library of insertional mutants, which are incorporated into the genome by transformation and homologous recombination. After phenotypic analysis of the resulting mutant collection, the mutagenic insertion site can be determined rapidly and easily. The randomness, coverage, and efficiency of this method are also discussed.

[1] M. L. Gwinn, A. E. Stellwagen, N. L. Craig, J. F. Tomb, and H. O. Smith, *J. Bacteriol.* **179,** 7315 (1997).

[2] R. Belas, D. Erskine, and D. Flaherty, *J. Bacteriol.* **173,** 6289 (1991).

[3] I. Y. Goryshin, J. Jendrisak, L. M. Hoffman, R. Meis, and W. S. Reznikoff, *Nat. Biotechnol.* **18,** 97 (2000).

[4] K. A. Reich, L. Chovan, and P. Hessler, *J. Bacteriol.* **181,** 4961 (1999).

[5] L. A. Garraway, L. R. Tosi, Y. Wang, J. B. Moore, D. E. Dobson, and S. M. Beverley, *Gene* **198,** 27 (1997).

[6] S. Devine, S. Beverley and J. D. Boeke, unpublished results, 1995.

0076-6879/02 $35.00

Tn*7* is a bacterial DNA transposon that transposes via a cut-and-paste mechanism.[7,8] The Tn7 reaction is mediated by the transposon-encoded proteins, TnsA, TnsB, TnsC, TnsD, and TnsE. TnsABC form the core recombination machinery, and TnsD and E are alternative target site selecting proteins. If insertion is mediated by TnsABC+D *in vivo,* transposition occurs into the single *attTn7* site in the *E. coli* genome.[9] If transposition is mediated by TnsABC+E, transposition occurs into many non-*attTn7* sites, particularly into plasmids.[10] TnsD or TnsE binds the target DNA and TnsC functions as a linker protein between the target selector and the transposase complex, TnsAB.[11] TnsC is an ATPase that is likely to be in the ATP-bound conformation when bound to TnsD or E and the target DNA. In its ATP-bound state, TnsC facilitates the recruitment of the Tn7 donor and the transposase complex that ultimately catalyzes the transposition reaction. TnsAB generates a double-strand break flanking both ends of the transposon in the donor and catalyzes the attack of the transposon's 3′-hydroxyl ends on the target DNA. The sites of attack are staggered by 5 base pairs on the target DNA, and the transposon becomes covalently integrated into the target DNA. A single-stranded region of 5 bases flanks each end of the newly inserted transposon and must be filled in by host-encoded proteins, resulting in a 5-bp direct repeat flanking the insertion.

The components of the transposition reaction have been purified and the TnsABC+D transposition reaction has been reconstituted *in vitro*.[11] TnsC gain-of-function mutants (TnsC*) have been identified that are able to promote transposition in the presence of TnsABC*, i.e., the absence of TnsD or TnsE.[12] These mutants abolish *attTn7* target site selectivity of the reaction without compromising the reaction's efficiency, allowing many Tn*7* insertions at numerous sites. To mutagenize the yeast genome *in vitro,* we used the TnsABC^{A225V} system, based on its efficiency and lack of sequence selectivity.[13]

Figure 1 depicts the yeast genome *in vitro* mutagenesis strategy schematically. The *in vitro* transposition reaction is carried out using purified yeast genomic DNA as the target for transposition, effectively mutagenizing it by random insertions. The mini-Tn7 element contains both yeast and *E. coli* selectable markers, a nutritional marker for yeast (*LEU2* or *TRP1*) and a kanamycin resistance marker (*kan*) for selection in *E. coli,* as well as an *E. coli* plasmid origin of replication (*ori*). The resulting mutagenized DNA is then transformed back into yeast and colonies are

[7] N. L. Craig, *Curr. Top. Microbiol. Immunol.* **204,** 27 (1996).

[8] N. L. Craig, *in* "Mobile DNA II" (N. L. Craig, R. Craigie, M. Gellert, and A. Lambowitz, eds.). ASM Press, 2001.

[9] R. L. McKown, K. A. Orle, T. Chen, and N. L. Craig, *J. Bacteriol.* **170,** 352 (1988).

[10] C. A. Wolkow, R. T. DeBoy, and N. L. Craig, *Genes Dev.* **10,** 2145 (1996).

[11] R. J. Bainton, K. M. Kubo, J. N. Feng, and N. L. Craig, *Cell* **72,** 931 (1993).

[12] A. E. Stellwagen and N. L. Craig, *Genetics* **145,** 573 (1997).

[13] M. C. Biery, F. J. Stewart, A. E. Stellwagen, E. A. Raleigh, and N. L. Craig, *Nucleic Acids Res.* **28,** 1067 (2000).

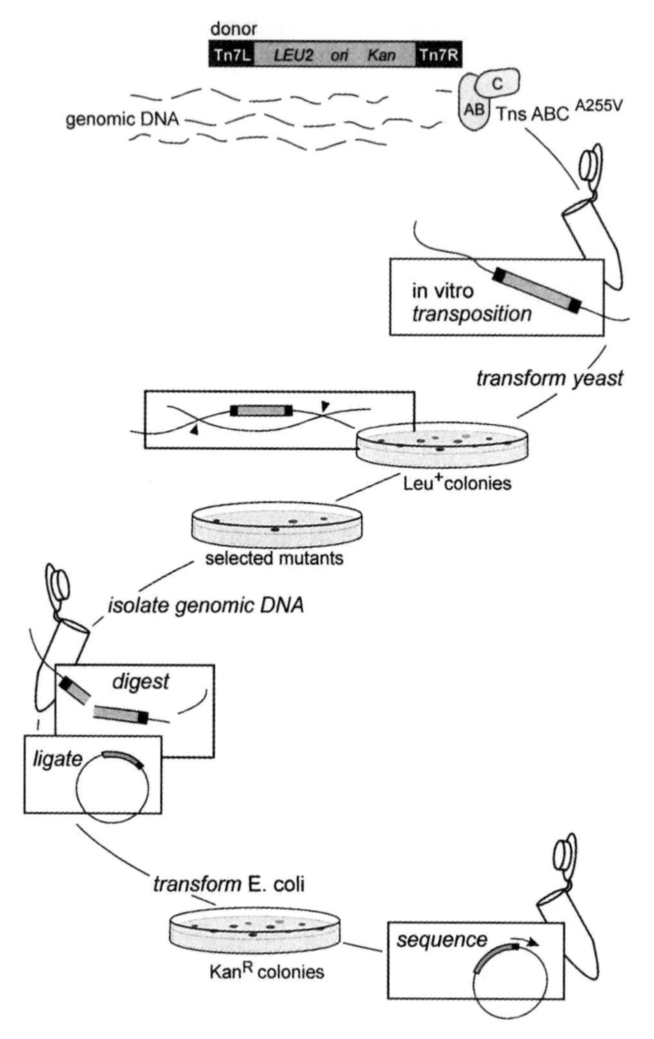

FIG. 1. Schematic of the *in vitro* mutagenesis technique and gene identification strategy. The Tn7 donor plasmid is incubated with target DNA and catalytic proteins to perform an *in vitro* transposition reaction. The resulting mutagenized DNA is transformed into yeast. The transformants are plated onto a selective medium, and the transformants can then be assayed for the phenotype of choice. Genomic DNA is isolated from the mutants of interest and cut with a restriction enzyme that cuts both within the transposon and in the flanking genomic DNA, leaving at least the *kan* resistance marker and the *E. coli* origin of replication intact. These fragments are ligated intramolecularly to form a circular plasmid, which can be transformed into *E. coli* plated on kanamycin-containing medium. The plasmids can be prepared and the sequence flanking the Tn7 insertion can be determined by sequencing from an oligonucleotide primer complementary to the right end of the transposon. The identity of the gene can be determined by a BLASTN search of the *Saccharomyces* Genome Database (SGD).

selected for the ability to grow on selective media (e.g., SC-Leu). The mutagenized yeast colonies can then be assayed for any phenotype of choice.

To determine the identity of any mutagenized gene, genomic DNA from the selected mutant colony is isolated and cut with a restriction enzyme that liberates one end of the minitransposon and some adjacent yeast genomic DNA. This DNA is then circularized by ligation and transformed into *E. coli,* where it can replicate, by virtue of the *ori* segment and be selected for by the presence of *kan*. The sequence of the yeast DNA into which the Tn7 insertion occurred is then determined merely by sequencing the resulting plasmids using primers derived from the right or left end of Tn7. The sequence can be compared to the yeast genome database and the identity of the mutagenized gene determined.

This technique can be used to screen the *S. cerevisiae* genome. For 10-fold coverage of the 6000 genes in yeast, 60,000 colonies must be generated, each harboring one insertion. An optimized transformation protocol for most yeast strains can generate at least 6×10^3 Leu^+ colonies per μg DNA; therefore approximately 10 μg of yeast DNA mutagenized by Tn7 is required. Using donor/target DNA ratios optimized for maximal *in vitro* transposition efficiency, the product of 30 standard Tn7 transposition reactions generates 10 μg of mutagenized yeast genomic DNA. Thus, simply scaling up the described *in vitro* Tn7 transposition reaction yields a mutagenized yeast genome that can be easily analyzed.

Materials

Plasmids

Plasmids pNB1 and 3 (Fig. 2) are both derived from pMCB43. pMCB43 is made by adding the *E. coli* origin of replication πAN7 (GenBank L08875) to the *Spe*I site of pMCB31.[13]

The pNB1 plasmid is made by inserting a *TRP1*-containing *Bam*HI–*Bam*HI fragment derived from pJEF981[14] into the *Bam*HI site of pMCB43.

pNB3 contains the *Bam*HI–*Pst*I *LEU2* fragment from pJJ282[15] cloned into the *Bam*HI site of pMCB43.

Yeast and Bacterial Strains

YNB1: *MATα leu2Δ1 his3Δ200 ura3-167 met15Δ0 ade2Δ::hisG*
YNB19: *MAT***a** *leu2Δ1 his3Δ200 ura3-167 met15Δ0 ade2Δ::hisG lys2Δ0 RDN1::Ty1-MET15 TELV::ADE2*
BY4732[16]: *MAT***a** *his3Δ200 ura3Δ0 met15Δ0 trp1Δ63*
DH5α cells are used for all steps in bacteria.

[14] J. D. Boeke, H. Xu, and G. R. Fink, *Science* **239,** 280 (1988).
[15] J. S. Jones and L. Prakash, *Yeast* **6,** 363 (1990).
[16] C. B. Brachmann, A. Davies, G. J. Cost, E. Caputo, J. Li, P. Hieter, and J. D. Boeke, *Yeast* **14,** 115 (1998).

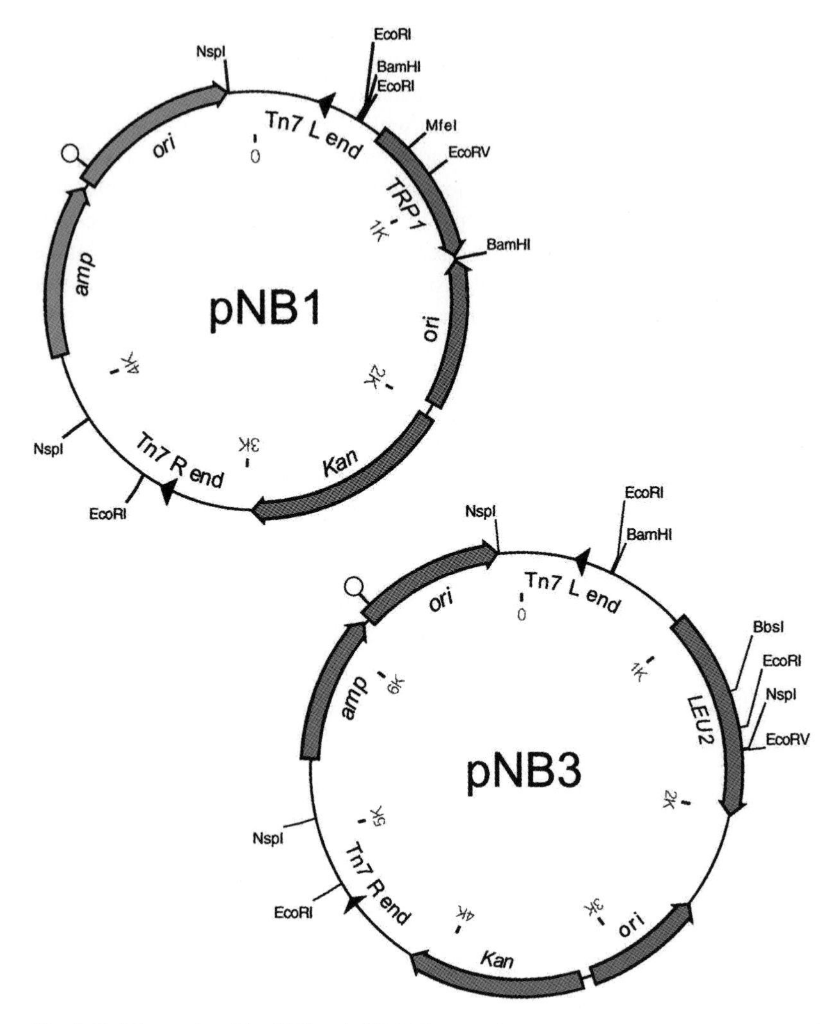

FIG. 2. Tn*7* donor plasmids pNB1 and pNB3. The complete sequences and plasmids are available from New England Biolabs (www.neb.com).

In principle, any *S. cerevisiae* strain with appropriate selectable marker mutations can be used with this technique. Transformed strains are haploid to allow for phenotypic expression of the mutagenized gene. Ideally, the strain for genomic DNA isolation should lack the yeast 2μ plasmid. We suggest that, if possible, it should also be a ρ^0 strain (lacking mitochondrial DNA), as the presence of these DNA species can generate false positives (see section titled "Marker Instability").

Method and Results

Preparation of Yeast Genomic DNA for Mutagenesis

Genomic DNA is prepared from strains by a cesium chloride centrifugation gradient technique derived from a method published in this series.[17]

1. Grow a 10 ml overnight culture in YPD (2% w/v yeast extract, 1% w/v peptone, 2% w/v dextrose) until saturated.
2. The following day, dilute to 1 liter and grow at 30° until there are a total of 2.5×10^8 cells.
3. Spin down cells at 4° for 20 min at 2500 rpm in a GS3 and wash with 100 ml of cold 50 m*M* Na_2EDTA (pH 8). Resuspend in 100 ml 50 m*M* Tris-HCl (pH 9.5) + 2% 2-mercaptoethanol and incubate at room temperature for 15 min.
4. Repeat spin. Resuspend in 10 ml of 1 *M* sorbitol, 1 m*M* Na_2EDTA, and 50 μg/ml Zymolyase 100T and incubate at 37° until >95% of the cells are spheroplasted (30 min to 1 hr). The percent digestion can be determined by mixing cells with 1% sarkosyl and examining under a phase-contrast microscope. Spheroplasted cells lyse when exposed to detergent and leave behind an empty membrane ("ghost") that appears darker than intact cells.
5. Spin down spheroplasts in a Sorvall RT 6000B or other tabletop centrifuge for 20 min at 2500 rpm and resuspend in 2.5 ml of lysis buffer (0.1 *M* Tris-HCl, 0.1 *M* Na_2EDTA, 0.15 *M* NaCl, 2% 2-mercaptoethanol). To aid the spheroplasting process, cells can be frozen in liquid nitrogen and thawed on ice.
6. Dilute 3-fold into lysis buffer containing 4% sarkosyl, pH 9.5, and incubate at 45° for 20 min. Add 25 ml lysis buffer with 4% sarkosyl, pH 8.0, and incubate at 70° for 15 min.
7. Add RNase A [prepared by boiling for 10 min at 10 mg/ml in 50 m*M* potassium acetate (pH 5.5)] to a final concentration of 0.1 mg/ml and incubate at 45° for 1 hr.
8. To degrade the proteins, add Pronase (nuclease-free; CalBiochem, San Diego, CA) to a final concentration of 1.33 mg/ml in two aliquots. Begin by adding 39 mg of Pronase and incubating lysates at 45° for 1 hr. Then add another aliquot of 39 mg Pronase and incubate another hour at 45°. After these incubations are complete, shift the lysates to 70° for 15 min.
9. Remove proteins by adding an equal volume of chloroform/isoamyl alcohol (24 : 1). Rock until white emulsion forms and spin for 30 min at 10,000 rpm in glass tubes in a Sorvall SS-34 rotor at 4°.
10. Transfer supernatant to a beaker and precipitate DNA by adding 10 ml of 3 *M* sodium acetate and 250 ml of 100% ethanol. Place at −20° overnight.

[17] M. D. Rose and J. R. Broach, *Methods Enzymol.* **194,** 195 (1991).

11. If DNA is sufficiently concentrated, spool the DNA on a glass rod and place in an Eppendorf tube. If it cannot be spooled, pellet by spinning at 3000 rpm in a Sorvall SS-34 rotor for 10 min at 4°.

12. Dry the pellet.

13. Resuspend in 13 ml of TE [10 m*M* Tris-HCl (pH 8), 1 m*M* Na_2EDTA]. (An additional phenol/chloroform extraction and DNA precipitation step can be added here.)

14. Allow the DNA solution to rehydrate overnight at 4°.

15. Add 10 g cesium chloride per 8 ml, balance, seal tubes, and spin in a Ti70. 1 rotor at 46,200 rpm for 40 hr.

16. Carefully remove tubes from rotor. Puncture the top of the tube with a syringe needle to provide an air vent.

17. Puncture the bottom of the tube and collect 5 drop fractions.

18. Run 5 μl of each fraction on a 0.6% agarose gel with 2.5 μg/ml ethidium bromide, and visualize the DNA under ultraviolet light.

19. Pool fractions with the highest DNA concentration, determine total volume, and add 2 volumes of water and 3 volumes of 100% ethanol. Incubate at −20° for 20 min, then spin at 4° at 9000 rpm in a Sorvall SS-34 for 15 min. Wash pellet with 3 ml 70% ethanol. Dry pellet and resuspend in the original volume of distilled H_2O. Determine concentration of DNA spectrophotometrically.

20. Dilute DNA to 1 μg/μl.

21. Sonicate DNA until fragments are approximately 3–5 kb in length. Using a Branson sonifier model 450, sonicate 50 μl of 1 μg/μl solution with a microtip using 100% duty cycle, output 2. Turn on power in 3-sec pulses for 9 sec. Keep all tubes in ice–water bath during sonication. Monitor 1 μl by electrophoresis on a 0.8% agarose gel with an appropriate molecular weight standard.

The sonication step is very important. Very long fragments (longer than 10 kb) are inefficient targets for *in vitro* transposition, and tiny fragments (smaller than 3 kb) do not allow adequate coverage of the genome. Our optimized methods for a Branson sonifier model 450 are described here. However, modifications may be necessary depending on the concentration and volume of DNA, as well as the sonifier used. Moreover, highly purified DNA is required for this technique, but cesium chloride centrifugation may not be necessary. Less labor-intensive methods may be used, provided that the sonication step produces fragments in the suggested size range. We have found that excessive target fragment length is more inhibitory to the *in vitro* transposition reaction than the use of genomic DNA purified by methods other than cesium banding.

In Vitro Transposition Reactions

To achieve the desired 10-fold coverage of the genome, thirty 100-μl reactions are recommended. Proteins and DNA should be diluted to the appropriate concentration before beginning. Proteins can be purified by methods previously described.[13] Alternatively, proteins can be purchased commercially (New England Biolabs, Beverly, MA).

1. For each 100 μl reaction, mix the following components:

1× (μl)	Reaction mix
10	250 m*M* HEPES
1	250 m*M* Tris-HCl
0.5	1% Bovine serum albumin (BSA)
2	100 m*M* ATP
0.2	1 M Dithiothreitol (DTT)
3.3	TnsD buffer*
72.5	Distilled H_2O
89.5	per reaction

* 50 m*M* Tris-HCl (pH 7.5), 2 m*M* DTT, 500 m*M* KCl, 1 m*M* Na_2EDTA, 25% glycerol.

2. Dilute purified, sonicated *S. cerevisiae* genomic DNA to 400 ng/μl with sterile, deionized water. Add 1 μl to the reaction mix (400 ng).
3. Add 1 μl (50 ng) $TnsC^{A225V}$.
4. Incubate 20 min at 30°. The volume should be 91.5 μl.
5. Add:
   1.5 μl TnsA (40 ng)
   1.0 μl TnsB (25 ng)
   4.0 μl 375 m*M* magnesium acetate
   2.0 μl donor Tn7 plasmid (pNB1 or pNB3) DNA (100 ng)

The final volume of the reaction should be 100 μl.

6. Incubate 45 min at 30°.
7. Stop the reaction by extracting the sample with phenol/chloroform to remove proteins. For ease of handling, bring volume up to 200 μl with water, then add 100 μl phenol, 100 μl chloroform/isoamyl alcohol (24 : 1). Vortex and spin at maximum speed in an Eppendorf 5417 C centrifuge for 5 min. Transfer the supernatant carefully to a clean tube. Add 200 μl chloroform/isoamyl alcohol (24 : 1), vortex, spin, and transfer supernatant again. Add 20 μl 3 *M* sodium acetate, 5 μg tRNA, and 300 μl 100% ethanol and place at −20° for at least 30 min.
8. Spin DNA for 15 min at 4° to precipitate DNA. Wash pellet with 200 μl 70% ethanol and then dry pellet.
9. Resuspend DNA in 50 μl TE or sterile water (10 ng/μl).

The Tn*7 in vitro* transposition reaction is very efficient. A Southern blot of the reaction digested with *Nde*I and probed for the right end of Tn*7* shows efficient

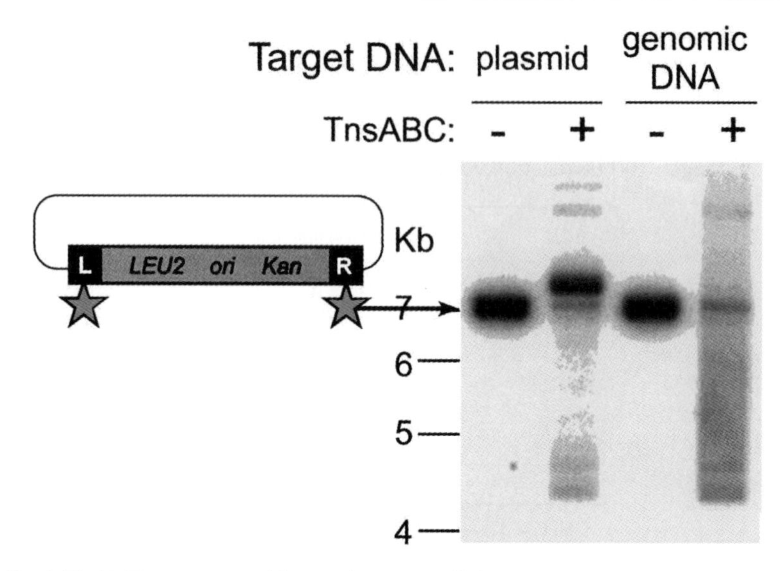

FIG. 3. The Tn*7 in vitro* transposition reaction occurs efficiently into both a plasmid target and purified, sheared, yeast genomic DNA. This is a Southern blot of the reaction products with pNB3 cut with *Nde*I and probed with a mixture of two ^{32}P-labeled oligonucleotides, one complementary to each end of the Tn7 transposon. We used the left end primer 5′-ATAATCCTTAAAAACTCCATTTCCACCCCT-3′ and the right end primer 5′-ACTTTATTGTCATAGTTTAGATCTATTTTG-3′. The reactions were performed with and without the catalytic proteins. The stars indicate the location of the oligonucleotide probes and the arrow the size of the unreacted donor plasmid. The smears represent the population of transposition products.

conversion of the donor band into other species with different molecular weight, using either a plasmid target or sonicated total yeast genomic DNA as the target (Fig. 3). Moreover, the transposition efficiency of the mini-Tn7 with the yeast phenotypic marker is no different from that of unmarked elements (data not shown).

Yeast Transformation

Transformation efficiency must be optimized for each strain used. The maximum transformation efficiency obtained for strain YNB19 (GRF167 background) was 6×10^3 Leu^+ colonies per microgram of DNA mutagenized with pNB3. A modified lithium acetate transformation protocol was used to achieve maximal transformation efficiency. See TRAFO (http://www.umanitoba.ca/faculties/medicine/biochem/gietz/Trafo.html)[18] for procedures to optimize the transformation efficiency of yeast strains.

1. Grow strain in 10 ml YPD until A_{600} is >5.
2. Start a 500 ml YPD culture at an A_{600} of 0.1.

[18] R. D. Gietz and R. A. Woods, *Biotechniques* **30,** 816, 822, 828, passim (2001).

3. Grow to an A_{600} of 0.5.

4. Spin down cells, wash once with 1/2 volume of distilled H_2O, and once with 0.1 *M* lithium acetate in TE.

5. Resuspend in 1/200 volume of 0.1 *M* lithium acetate in TE.

6. Aliquot 100 μl of cells per reaction. Add 10 μl mutagenized DNA (100 ng), and 10 μl boiled, sheared herring sperm (or other carrier) DNA (10 mg/ml). Each of the 30 transposition reactions requires 5 separate transformation reactions, resulting in 150 independent transformations.

7. Incubate at 30° for 15 min.

8. Add 600 μl of polyethylene glycol (PEG) solution to each reaction [40% polyethylene glycol, molecular weight 3350 (Sigma, St. Louis, MO), 0.1 *M* lithium acetate in TE].

9. Incubate at 30° for 30 min.

10. Add 70 μl dimethyl sulfoxide (DMSO).

11. Incubate at 42° for 15 min.

12. Spin cells down at room temperature for 1 min at 2000 rpm, wash with distilled H_2O, resuspend in sterile water, and plate 1/2 of the transformation reaction on each selective plate (e.g., SC-Leu). Yield should be approximately 200 colonies per plate, resulting in 60,000 total colonies.

13. Screen colonies for desired mutant phenotype and collect mutants.

Plasmid Recovery and Target Gene Identification

This procedure is very similar to that described previously in this series.[19]

1. Streak out mutants to single colonies on a selective plate (e.g., SC-Leu).

2. Pick a single colony and grow a 10-ml culture overnight in YPD.

3. Purify genomic DNA: spin down cells in 10-ml cultures, wash in 5 ml water, and resuspend in 1 ml 1 *M* sorbitol/0.1 *M* Na_2EDTA (pH 8.0).

4. Pellet and resuspend in 0.3 ml 1 *M* sorbitol/0.1 *M* Na_2EDTA (pH 8.0) containing 14 m*M* 2-mercaptoethanol and 0.5 mg/ml lyticase. Incubate at 37° for 20 min to 1 hr, or until spheroplasting is at least 50% (mix with 10% SDS and check under microscope).

5. Pellet 5–10 sec and resuspend in 400 μl TE.

6. Add 100 μl fresh lysis buffer (280 m*M* Na_2EDTA, 400 m*M* Tris base, 2% SDS), mix by inversion, and incubate at 65° for 30 min.

7. Add 100 μl of 5 *M* potassium acetate (not buffered), mix by inversion, and incubate on ice for at least 45 min.

8. Pellet for 15 min at 4° in a microcentrifuge.

9. Transfer supernatant to a new 1.5-ml Eppendorf tube and fill it with 95% ethanol. Invert 10 times.

[19] F. Winston, F. Chumley, and G. R. Fink, *Methods Enzymol.* **101,** 211 (1983).

10. Pellet 5–10 seconds in a microcentrifuge. Wash pellet with 500 μl 70% ethanol. Dry pellet and resuspend in 100 μl of water or TE.

Note: Steps 11–14 can be scaled down 20-fold if desired.

11. Digest 15 μl of genomic DNA with a restriction enzyme that leaves at least the *kan* gene, the *E. coli* origin of replication, and one end of the transposon intact. If recovery of both ends is desired, use an enzyme that does not cut within the transposon at all. With the *LEU2*-marked plasmid, we use *Nsp*I to recover just the right end. Table I is a list of suggested enzymes for use with either pNB1 or pNB3. If the transposon is excised with an enzyme that does not cut within the Tn7 element, sequencing from both ends can be done to observe the 5 bp duplication generated by transposition. However, sequencing from one end is sufficient for determining the point of insertion and the mutant gene's identity. The total reaction volume should be large, at least 10 times greater than the volume of the genomic DNA added. Digest at least 5 hr.

12. Heat inactivate the restriction enzyme at 65° for 20 min and ethanol precipitate DNA. If using an enzyme that cannot be heat-inactivated, phenol/chloroform treat the sample before precipitation.

13. Resuspend DNA in 200 μl ligation mixture [179 μl distilled H_2O, 20 μl 10× T4 DNA ligase buffer (NEB), and 1 μl T4 DNA ligase (NEB)]. Incubate at 16° overnight.

14. Transform 10 μl of the ligation reaction into 150 μl competent DH5α or other *Escherichia coli* strain and select on LB plates containing 25 μg/ml kanamycin.

15. Prepare plasmids from several colonies. A *Bgl*II digest should give an internal Tn*7* diagnostic band of 2070 bp with both pNB1 and pNB3.

16. Perform sequencing reactions using a primer of the sequence 5′-ACTTTATTGTCATAGTTTAGATCTATTTTG-3′ for sequencing from Tn*7* right end. If the flanking sequences from both ends of the transposon are desired, the following primer can be used to sequence from the Tn*7* left end: 5′-ATTTTCGTATTAGCTTACGACGCTACACC-3′.

17. Perform a BLASTN search on the *Saccharomyces* Genome Database (SGD) (http://genome-www.stanford.edu/Saccharomyces/) to determine the identity of the gene that has been disrupted.

Initial Assay of Insertions into 13 kb Fragment of Yeast Genomic DNA

As a pilot study to determine whether the Tn*7 in vitro* reaction would efficiently generate recoverable insertions into yeast genomic DNA, we used a cloned 13 kb fragment of yeast genomic DNA containing *URA3* and genes flanking it as a target for the Tn7 mutagenesis using the *TRP1*-marked Tn*7* donor plasmid. This mutagenized DNA was transformed into haploid yeast and 24 transformants were

TABLE I
RESTRICTION ENZYMES TO DETERMINE SITE OF INSERTION

Enzyme	Number of sites[a,b]	Average fragment size	pNB1	pNB3
*Acl*I	3469	3479	[c]	[c]
Afl II	1867	6465		R[d]
*Age*I	1487	8117		
*Bam*HI	1600	7543	R	R
*Bbs*I	7463	1617		R
*Bcl*I	3807	3170		
*Bpm*I	2196	5496		R
*Bsa*I	2671	4519		
*Bse*RI	2555	4724		R
*Bsg*I	1513	7977	R	
*Bsr*BI	1526	7909		
*Bsr*GI	2899	4163		R
*Bst*BI	4264	2830		
*Bst*Z17I	2634	4582	R	
*Btr*I	2106	5731		
*Dra*I	8653	1395		
*Eco*RI	4219	2861	R	R
*Eco*RV	4065	2969	R	R
*Fau*I	4789	2520		
*Hpa*I	2540	4752		
*Kpn*I	1832	6588		R
*Mfe*I	5504	2193	R	R
*Msc*I	2893	4172		
*Nco*I	2311	5223		
*Nde*I	3225	3742		
*Nsp*I	7756	1556		R
*Pci*I	2562	4711		R
*Psi*I	5727	2107		R
*Pvu*II	2058	5865		
*Sac*I	1366	8835		
*Sap*I	1841	6556		
*Sca*I	2967	4068		
*Sna*BI	2046	5899		
*Sph*I	1440	8381		

[a] Number of these restriction enzyme sites in the entire *S. cerevisiae* genome.
[b] Michael Cherry, personal communication, 1999.
[c] Noncutters for the entire minitransposon donor (for recovery of both ends) in pNB1 or pNB3.
[d] Enzymes for recovery of the right end (R) only, leaving *kan* and *ori* intact.

analyzed. The site and orientation of insertion was determined by plasmid recovery and sequencing and appears random (Fig. 4). All 25 of the insertions occurred into the 13 kb fragment, as expected. Of the 25, 24 were sequenced from the right end; 6 were sequenced from both the left and right end, and in all of these cases the 5 bp duplication characteristic of Tn7 insertion was observed. One of 25 could not

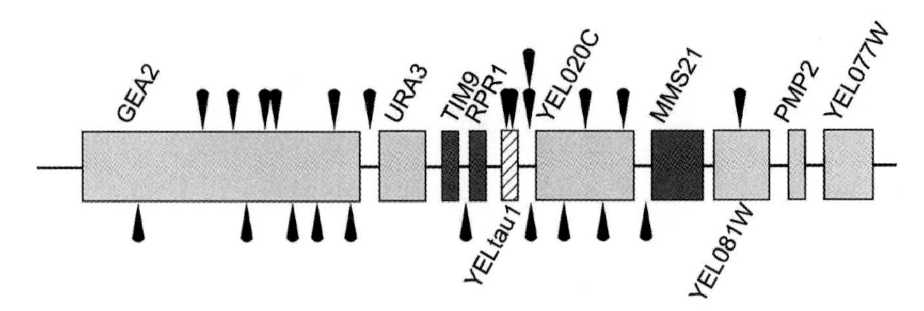

FIG. 4. The Tn*7 in vitro* transposition reaction occurs randomly into a 13 kb cloned yeast genomic fragment from *S. cerevisiae* chromosome V. Arrowheads denote positions of insertions and the direction indicates the orientation. The solid gray rectangles are open reading frames. The dark gray boxes represent essential genes. The hatched box is the LTR of a retrotransposon.

be recovered even after multiple attempts, because transformations into *E. coli* did not yield transformants.

This analysis indicated that the transposon does not have a bias for specific target sites, as the insertions occurred with a wide distribution into both protein-coding regions and intergenic regions. As expected, insertions into the coding region of essential genes were not observed because a haploid strain was transformed. Transposition events near the extreme ends of the target fragment were also not recovered, probably because insertions near the ends would be lost due to the requirement for homologous recombination with genomic DNA.

Insertions into Total Genomic DNA

We used total genomic DNA as a target for a pilot experiment before beginning a genetic screen. Using total yeast genomic DNA purified and sonicated as described above, we performed 5 *in vitro* reactions and transformed the resulting products into BY4732.[16] We analyzed genomic DNA from 23 Trp^+ colonies that had Tn7 insertions. We estimated the number of double insertions by performing a Southern blot on genomic DNA cut with *Eco*RI and probing for the *TRP1* coding sequence. The number of double insertions, whether by cotransformation or insertion of two transposons into the same piece of DNA, was low: only 2 of 23 isolates showed two bands that hybridized with the *TRP1* probe, and the rest showed exactly one band. We did not determine whether the double insertions were on the same DNA molecule in this experiment. Such linked insertions would be unlikely, however, based on Tn7's innate immunity (see below, "Conclusions"). Because high-efficiency transformation protocols were used for these experiments, it is more likely that the double insertions occurred by contransformation of two independent recombinant molecules.

Genetic Screen

To assess the randomness, the coverage, and the practicality of the *in vitro* mutagenesis technique at the genomic level, we performed a genetic screen using the *LEU2*-marked mini-Tn7. Thirty *in vitro* Tn7 reactions were completed using sonicated total yeast genomic DNA as the target, as described; 70,000 Leu^+ transformants were generated and screened for the expression of a normally silent *GAL1-HIS3* reporter gene. The basis for this analysis was the observation that an intact tRNA gene, in this case, *SUP53,* represses expression of a neighboring gene transcribed from a weak promotor by RNA polymerase II, in this case, the *GAL1–HIS3* gene, in a position-dependent manner.[20] When placed adjacent to one another on a plasmid, *SUP53* suppresses expression of the *GAL1–HIS3* reporter, thereby preventing growth of strains containing this plasmid on medium lacking histidine. Our screen consisted of simply replica-plating the 70,000 Leu^+ transformants to SC-His containing galactose, and selecting for growth. The His^+ colonies must therefore have a defect in *SUP53*-mediated repression. One hundred and twenty colonies were isolated and retested. Twenty-two of these colonies gave a convincing His^+ phenotype on retest. These were mated and heterozygous mutant diploids were sporulated to assess the linkage between the transposon marker (*LEU2*) and the mutant phenotype (His^+). In only 1 mutant did the *LEU2* marker not segregate 2 : 2. It showed a combination of 0 : 4 (70%) and 4 : 0 (30%) tetrads, suggesting the presence of a Tn7:: *LEU2* that had not integrated into the genome. In 5 mutants *LEU2* was not linked to the His^+ phenotype. In 14 of the remaining 16 mutants, the identity of the mutagenized gene was determined by sequencing from the Tn7 right end, as described above. For 2 mutants, plasmid rescue was unsuccessful using *Nsp*I to liberate the Tn7 right end and genomic flanking sequence.

Our screen addressed the efficiency and randomness of the Tn7 technique when applied to the complete genome. The results indicate the plausibility and success of the application, as a large number of colonies were generated, the sites of insertion were easily determined, and a total of 11 different sites identified. The only gene previously known to be involved in *SUP53*-mediated repression, *CBF5,*[21] is an essential gene and was not recovered because our screen was performed in a haploid strain. Our results implicate the nonsense-mediated RNA decay (NMRD) pathway as important to the phenotype that we examined.[22] We recovered single insertions in 2 of the genes in the NMRD pathway, Upf1 and Upf2, and multiple insertions at different nucleotide positions in a third gene in the pathway, Upf3 (details of this screen will be published elsewhere). The fact that the screen identified multiple

[20] M. W. Hull, J. Erickson, M. Johnston, and D. R. Engelke, *Mol. Cell. Biol.* **14,** 1266 (1994).

[21] A. Kendall, M. W. Hull, E. Bertrand, P. D. Good, R. H. Singer, and D. R. Engelke, *Proc. Natl. Acad. Sci. U.S.A.* **97,** 13108 (2000).

[22] N. Bachman and J. D. Boeke, unpublished results, 2001.

genes in the same pathway suggests that the genomic coverage of the mutagenesis under the conditions described is within the optimal range.

Marker Instability

The effectiveness of the Tn*7 in vitro* mutagenesis technique depends on homologous recombination of the mutagenized DNA fragment with genomic DNA, leading to replacement of the wild-type allele. The auxotrophic marker can be followed phenotypically to determine whether the mutagen has been effectively integrated into the genome. An integration event will result in 2 : 2 segregation of the marker in a tetrad assay, and a stable phenotype (all colonies contain the marker) when plated after outgrowth on the nonselective medium YPD. The observation of a mutant with an unstable Leu^+ phenotype (0 : 4 in the tetrad segregation assay) lead us to examine whether the observed instability represented a general phenomenon. To assess the frequency of this event, a simple plasmid-loss assay was used. Twenty-four random Leu^+ colonies (not selected for a mutant phenotype) were picked and assayed for their ability to lose the Leu^+ phenotype by growing them on YPD and then replica plating them back to SC-Leu. Surprisingly, 25% of the colonies picked lost the Leu^+ phenotype at a high frequency, indicating that in these transformants, the *LEU2* marker had not been integrated into the genome. Analysis of insertions recovered from these unstable Leu^+ strains showed two insertions into mitochondrial DNA, two into the yeast 2μ plasmid, one into an rDNA repeat, and one into a subtelomeric region. Each of these cases represents an insertion into circular DNA or repeated DNAs containing an ARS sequence, which could replicate episomally. Because of the high frequency of marker instability, we recommend that the genomic DNA donor yeast strain contain neither 2μ plasmid DNA nor mitochondrial DNA. Alternatively, a period of outgrowth in nonselective medium prior to plating would significantly reduce the frequency of unstable Leu^+ transformants, though it could introduce propagation bias.

Conclusion

Tn7-mediated *in vitro* mutagenesis of yeast genomic DNA is a new addition to the expanding repertoire of methods available for genetic manipulation of yeast. As all methods, it is optimal for certain applications. This is a powerful technique for the generation of a library of null mutants that is near-random and complete. In addition, it allows rapid determination of the mutant's identity. The collection of mutants created by Tn7 mutagenesis is likely to complement those generated by other methods, because of the unique properties of Tn7 and its mechanism. The coverage of the Tn7-mediated mutagenesis is comprehensive by virtue of the efficiency of the transposition reaction *in vitro,* the low target-site selectivity conferred by the $TnsC^{A225V}$ protein, and the avoidance of linked double insertions offered by Tn7's innate transposition immunity.

A major virtue of the Tn*7 in vitro* transposition reaction is its efficiency. Other methods are limited by the transposition efficiency of the element used. For example, when Ty*1* or Tn*5* are used, only approximately 0.1% of the donor material is converted into the final insertion product,[23] whereas using the *in vitro* Tn*7* reaction, up to 90% can be converted (Fig. 3).

The insertions into the genome by this method are likely to be more random than those generated by other methods for several reasons. First, every transposon system has some target-site preferences. Tn*3*, Tn*5*, and Mu, for example, have notable target site preferences, even *in vitro*.[24–26] Although wild-type Tn*7* is one of the most target-selective transposons, the discovery of TnsC mutants that abolish target site selectivity allow the use of Tn*7* as an insertional mutagen.[12] Target site-selectivity has been analyzed for the $\text{TnsC}^{\text{A225V}}$ mutant and it was determined that the only target bias was a slight bias for the presence of an AT-rich region upstream of the insertion site.[13]

Importantly, there is no library propagation step with this genomic DNA mutagenesis technique. Bias can be introduced when a library is propagated, as some mutagenized strains will grow faster than others. Most *in vivo* techniques, such as the popular method developed by Snyder and colleagues,[27] require a propagation step in *E. coli* to generate enough of the mutagenized library to transform into yeast, and to allow it to be a regeneratable resource. Using Tn7 *in vitro* mutagenesis, the "library" is the mutagenized DNA, which can be stored. The library is never transformed into *E. coli,* ensuring against propagation bias and allowing the expression and recovery of genes toxic to *E. coli*. Once transformed, the yeast are plated directly on selective medium. One potential disadvantage of the *in vitro* technique is that the mutagenized DNA is a finite resource. The yeast transformants are the final collection of mutants, and this collection can be regenerated by growth, but doing so will introduce propagation bias. Ideally, the collection of mutagenized yeast transformants should be used only once, and not pooled, reproduced, and reused.

Because of its innate transposition immunity, the use of Tn*7* increases the likelihood of thorough mutagenic coverage of the genome. Once Tn*7* inserts, it marks the site of insertion and the surrounding area, signaling that no further integration events can occur in the immediate vicinity of the primary insertion.

[23] J. D. Boeke, *in* "Mobile DNA II" (N. L. Craig, R. Craigie, M. Gellert, and A. Lambowitz, eds.). ASM Press, 2001.

[24] C. J. Davies and C. A. Hutchison III, *Nucleic Acids Res.* **23,** 507 (1995).

[25] I. Y. Goryshin, J. A. Miller, Y. V. Kil, V. A. Lanzov, and W. S. Reznikoff, *Proc. Natl. Acad. Sci. U.S.A.* **95,** 10716 (1998).

[26] M. Mizuuchi and K. Mizuuchi, *Cold Spring Harb. Symp. Quant. Biol.* **58,** 515 (1993).

[27] P. Ross-Macdonald, P. S. Coelho, T. Roemer, S. Agarwal, A. Kumar, R. Jansen, K. H. Cheung, A. Sheehan, D. Symoniatis, L. Umansky, M. Heidtman, F. K. Nelson, H. Iwasaki, K. Hager, M. Gerstein, P. Miller, G. S. Roeder, and M. Snyder, *Nature* **402,** 413 (1999).

Immunity is mediated by TnsC, and the reactions involving the $TnsC^{A255V}$ mutant retain this characteristic, even *in vitro*.[28] Target site immunity is a feature of the *in vitro* mutagenesis technique that decreases the likelihood of two insertions into the same target piece of DNA, decreasing the chances of double insertions that would be linked and difficult to analyze. Each target DNA molecule, therefore, should contain a single transposition event, which emphasizes the importance of generating small target molecules by sonication.

The Tn*7 in vitro* system uses a bacterial transposon in yeast, which provides several practical advantages. Because the Tn*7* transposon is bacterial and the Tns proteins are not encoded by the yeast genome, there is no possibility of *trans* complementation of the transposition reaction in yeast. In addition, the origin of replication on the donor plasmid only works in bacteria. Consequently, there is no risk of amplification of the donor plasmid and expression of the auxotrophic marker in yeast without an insertion event. This technique does not rely on conditional origins of replication as some other *in vitro* techniques do.[13] Moreover, the lack of homology in the yeast genome to the transposon donor prevents entry into the genome by homologous recombination. The only false positives we observed were insertions into yeast sequences that can replicate autonomously in yeast (see above, "Marker Instability").

A major advantage of this technique is its flexibility, conferred by the fact that the mutagenesis is done *in vitro*. Any DNA can be mutagenized: the entire yeast genome, a single chromosome, or a single gene. Moreover, saturating mutagenesis can be completed in any strain background. This offers a distinct advantage over the collection of yeast deletions,[29] which are usually analyzed in the strain background in which they were made. Added flexibility comes from the fact that the Tn7 donor transposon can be readily modified. Additional markers, frame-shift signals, in-frame fusions, and linker scanning insertions[13] can all be added to the donor transposon, which is ultimately inserted into the genome.

Last, but perhaps most important, is the practical ease of this method, in particular the identification of the mutagenized gene. The presence of both the *E. coli* origin of replication and the *kan* drug resistance marker greatly simplifies the gene identification process. No subcloning or complementation analysis is necessary. The Tn*7* ends offer unique sequences that provide convenient sites for sequencing primers, and having the complete *S. cerevisiae* genome sequence as a resource makes gene identification almost immediate. In addition, the commercial availability of the Tns catalytic proteins (New England Biolabs, Beverly, MA) allows investigators to perform the transposition reactions without having to purify the necessary components.

[28] A. E. Stellwagen and N. L. Craig, *EMBO J.* **16,** 6823 (1997).

[29] E. A. Winzeler, D. D. Shoemaker, A. Astromoff, H. Liang, K. Anderson, B. Andre, R. Bangham, R. Benito, J. D. Boeke, H. Bussey *et al., Science* **285,** 901 (1999).

The Tn*7* system is similar in some ways to the miniTn*3* system that has been developed by Mike Snyder and colleagues.[27] The Tn*3* system uses *Not*I fragments from the yeast genome to make a library on an *E. coli* plasmid backbone. A *LEU2*-marked miniTn*3* is transposed into the library, which is then digested with *Not*I to liberate the fragments. The mutagenized fragments are transformed into yeast and integrated into the genome by homologous recombination. The strength and major utility of the newer version of the Snyder system is the generation of a yeast library of in-frame fusions with a 3× HA tag. These can be used to determine the localization of many proteins without making each one individually. The information generated by these studies is available at http://ygac.med.yale.edu/triples/triples.htm.[30] It is easy to determine the point of insertion using the miniTn*3* system, as the Tn*3* end provides a unique sequence from which the insertion can be sequenced. The Tn*3* donor plasmid does not contain a bacterial origin of replication, however, which must be added if plasmid recovery is desired. In addition, Tn*3* has notable target site selectivity,[24] and the yeast library must be propagated in *E. coli*. See Kumar and Snyder (this series)[31] for a complete discussion of the technique.

As with any insertional mutagenesis, the Tn*7* mutagenesis is not an effective way to generate conditional alleles. In addition, because of its dependence on phenotypic expression, the technique must be completed in haploid yeast strains and will therefore be ineffective as a tool to study essential genes, although some insertions may disrupt control regions rather than coding regions of essential genes and can still generate viable yeast strains.[32]

In conclusion, the Tn*7* system is a very simple and powerful tool for the mutagenesis of any target DNA.

Acknowledgments

We thank Michael Cherry for kindly providing the number of restriction enzyme sites in the yeast genome. We thank Bang Wong for assistance with Fig. 1 and Eric Bolton for critical reading of the manuscript. Supported in part by NIH Grant CA77812 to J.D.B., M.C.B., and N.L.C. are employees of the Howard Hughes Medical Institute. Under a licensing agreement between the Johns Hopkins University and New England Biolabs, N.L.C. is entitled to a share of the royalty received by the University from the sales of the licensed technology. The terms of this agreement are being managed by the University in accordance with its conflict-of-interest policies.

[30] A. Kumar, K. H. Cheung, P. Ross-Macdonald, P. S. Coelho, P. Miller, and M. Snyder, *Nucleic Acids Res.* **28,** 81 (2000).

[31] A. Kumar, S. A. des Etages, P. S. Coelho, G. S. Roeder, and M. Snyder, *Methods Enzymol.* **328,** 550 (2000).

[32] J. S. Smith, C. B. Brachmann, L. Pillus, and J. D. Boeke, *Genetics* **149,** 1205 (1998).

[14] Vector Systems for Heterologous Expression of Proteins in *Saccharomyces cerevisiae*

By MARTIN FUNK, RAINER NIEDENTHAL, DOMINIK MUMBERG, KAY BRINKMANN, VOLKER RÖNICKE, and THOMAS HENKEL

Introduction

Over the past decade the genetic and biochemical analysis of the yeast *Saccharomyces cerevisiae* has provided detailed insights into a great variety of regulatory processes that ensure the biological homeostasis of eukaryotic cells. In particular, *S. cerevisiae* has become a model organism for the study of several basic eukaryotic cellular processes such as gene expression, protein trafficking, and cell cycle control. Many genes from mammalian cells have been shown to complement conditional lethal mutations of yeast genes involved in critical steps of these cellular processes. In fact, several of these yeast mutants were used to isolate functional homologs from mammalian cells by heterologous complementation. In general yeast has been very successfully used over the years for the heterologous expression and purification of proteins from different sources including therapeutic proteins. Yeast can be used as a powerful tool in high-throughput assays for functional genomics such as two-hybrid assays or for drug screening. The further development of these approaches including isolation, expression, and structure–function analysis of cDNAs requires continuous efforts to improve the cloning and ectopic as well as regulatable expression or coexpression of relevant proteins in yeast. During recent years we have developed a system of more than 80 compact expression vectors for the convenient cloning of genes and their heterologous expression in most strains of *S. cerevisiae* (Fig. 1).[1–3]

These vectors are based on the pRS series of centromeric or 2μ plasmids carrying the *HIS3*, *TRP1*, *LEU2*, *URA3*, and *G418* resistance marker.[4–6] We have extended the available set of selection markers by inclusion of the *MET15* and *ADE2* markers described by Brachmann *et al.*[7]

The expression cassettes of our plasmids comprise a distinct promoter, a cloning array with six to nine unique restriction sites, and the *CYC1* terminator.

[1] D. Mumberg, R. Müller, and M. Funk, *Nucleic Acids Res.* **22,** 5767 (1994).

[2] D. Mumberg, R. Müller, and M. Funk, *Gene* **156,** 119 (1995).

[3] V. Rönicke, W. Graulich, D. Mumberg, R. Müller, and M. Funk, *Methods Enzymol.* **283,** 313 (1997).

[4] R. S. Sikorski and P. Hieter, *Genetics* **122,** 19 (1989).

[5] T. W. Christianson, R. S. Sikorski, M. Dante, J. H. Shero, and P. Hieter, *Gene* **110,** 119 (1992).

[6] A. Wach, A. Brachat, R. Pöhlmann, and P. Philipppsen, *Yeast* **10,** 1793 (1994).

[7] C. B. Brachmann, A. Davies, G. J. Cost, E. Caputo, J. Li, P. Hieter, and J. D. Boeke, *Yeast* **14,** 115 (1998).

0076-6879/02 $35.00

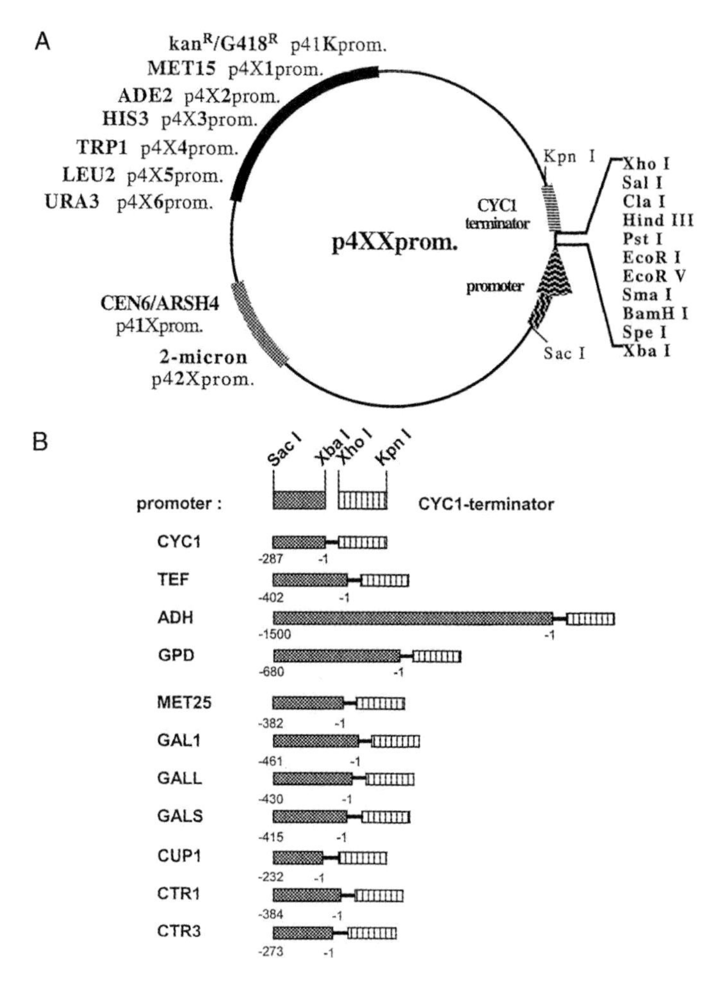

FIG. 1. Structure of plasmids from the expression vector collection. (A) Schematic map of the vectors constructed. The nomenclature is based on the plasmids described by T. W. Christianson, R. S. Sikorski, M. Dante, J. H. Shero, and P. Hieter, *Gene* **110,** 119 (1992) and C. B. Brachmann, A. Davies, G. J. Cost, E. Caputo, J. Li, P. Hieter, and J. D. Boeke, *Yeast* **14,** 115 (1998). For example plasmid p424GALL is based on the vector pRS424 carrying the *TRP1* gene and the 2 micron origin of replication and carries the *GALL* promoter. The kan^R/G418^R cassette codes for a dominant marker gene that can be used in strains lacking conventional markers (see plasmid construction). Shown are the restriction sites of the polylinker (bold face type) located between the terminator and the promoter (arrow). (B) Maps of the different promoters (shaded box) and the *CYC1* terminator (striped box). The expression properties of the constitutive *CYC1, TEF, ADH,* and *GPD* promoters are described by D. Mumberg, R. Müller, and M. Funk, *Gene* **156,** 119 (1995). Expression levels of the inducible *GAL1, GALL,* or *GALS* promoters and the repressible *MET25* promoter are described by D. Mumberg, R. Müller, and M. Funk, *Nucleic Acids Res.* **22,** 5767 (1994) and those of the *CUP1, CTR1,* and *CTR3* promoters in S. L. Labbe and D. J. Thiele, *Methods Enzymol.* **306,** 145 (1999). Numbers at the boxes represent the regions of the promoters cloned by PCR relative to the start codon (+1 would be A of the ATG codon).

One class of promoters used is considered to be constitutive, including a weakened *CYC1* promoter, the ADH1 promoter, and the stronger *TEF2* or *GPD* promoter. The second class includes the regulatable *MET25* promoter, which is controlled by methionine repression, as well as the *GAL1* promoter, which is subject to a strong glucose repression and can be induced by galactose. Labbe and Thiele[8] extended the set of available regulatable promoters by inclusion of the copper-inducible *CUP1* promoter or the copper-repressible *CTR1* and *CTR3* promoter. Together, these vectors allow the regulated expression or coexpression of up to seven different proteins at various levels and provide a powerful tool for the analysis of dosage-dependent effects in *S. cerevisiae*.

In this report, we describe the conversion of several of these expression vectors to the GATEWAY format (Invitrogen Corp., Carlsbad, CA) and based on this a vector collection for expression of epitope-tagged proteins. We describe the expression properties of some promoters present in the destination vectors. We analyzed the properties of the vector collection for expression of epitope-tagged proteins with special emphasis on the effect of the epitope tags on the protein levels and activity obtained for the heterologous expression of the green fluorescent protein (GFP).

Conversion of Yeast Expression Vectors into GATEWAY Format

The advance in genomic technologies and information from systematic studies of dozens of genomes including the human genome results in a high demand for technologies facilitating the cloning and expression of up to thousands of genes. With respect to this the GATEWAY system (Invitrogen) was shown to be a powerful methodology that facilitates cloning of genes, protein expression, and analysis of gene function by replacing the use of restriction endonucleases and ligase for cloning with site-specific *in vitro* recombination.[9–12] The underlying reactions, which are based on the bacteriophage lambda (λ) recombination sites and enzymes, are very rapid, robust, accurate, and able to be automated. The cloning and transfer of genes to different vectors is performed in two steps starting with the transfer of the gene of interest into the so-called entry vector followed by shuffling of the gene to different so-called destination vectors (for example, yeast expression vectors). In most cases this is achieved by first cloning a PCR

[8] S. L. Labbe and D. J. Thiele, *Methods Enzymol.* **306,** 145 (1999).

[9] J. L. Hartley, G. F. Temple, and M. A. Brasch, *Genome Res.* **10,** 1788 (2000).

[10] A. J. Walhout, R. Sordella, X. Lu, J. L. Hartley, G. F. Temple, M. A. Brasch, N. Thierry-Mieg, and M. Vidal, *Science* **287,** 52 (2000).

[11] A. J. Walhout, G. F. Temple, M. A. Brasch, J. L. Hartley, M. A. Lorson, S. van den Heuvel, and M. Vidal, *Methods Enzymol.* **328,** 575 (2000).

[12] J. Reboul, P. Vaglio, N. Tzellas, N. Thierry-Mieg, T. Moore, C. Jackson, T. Shin-i, Y. Kohara, D. Thierry-Mieg, J. Thierry-Mieg, H. Lee, J. Hitti, L. Doucette-Stamm, J. L. Hartley, G. F. Temple, M. A. Brasch, J. Vandenhaute, P. E. Lamesch, D. E. Hill, and M. Vidal., *Nat. Genet.* **27,** 227 (2001).

(polymerase chain reaction) fragment flanked by *attB* recombination sites into a donor vector with *attP* recombination sites via a so-called BP reaction in order to generate an entry clone flanked by *attL* recombination sites. Second, the gene is then transferred via a so-called LR reaction to one or multiple destination vectors with *attR* recombination sites in order to generate different expression clones.

We converted 32 vectors carrying the constitutive TEF or GPD promoter or the regulatable *MET25* or *GAL1* promoter into the GATEWAY format (Fig. 2). Conversion of the vectors was achieved by homologous recombination of the Rfc cassette into the respective expression vector as described in "Procedures and Vector Constructions." The Rfc cassette comprises the chloramphenicol resistance marker and the *ccdB* gene coding for the toxic gene product used as a negative selection marker and is flanked by the *attR*1/2 recombination sites. The expression cassettes we constructed were validated by analyzing the levels of GFP when expressed from the different promoters of the cassettes. The GFP gene was introduced into the different p416 destination vectors by an LR reaction from an entry clone carrying the codon optimized GFP variant described by Cormack *et al.*[13] The resulting expression clones were transformed into yeast and the relative GFP activity was determined by FACS (fluorescence activated cell sorter) analysis (Fig. 3A).[14] Yeast cells carrying the TEF or GPD promoter constructs were grown in minimal medium with 2% glucose. Cells carrying the *GAL1* promoter constructs were pregrown in minimal medium with 2% raffinose as a carbon source and then shifted for 5 hr to medium containing 2% raffinose and 2% galactose. Cells carrying the *MET25* promoter constructs were pregrown in minimal glucose medium in the presence of 1 m*M* methionine and then shifted for 5 hr to glucose medium in the absence of methionine. Results obtained for the overall observed GFP activities are in the same range as the *lacZ* activities described for the parental promoter constructs by Mumberg *et al.*,[1,2] showing that the *attB* sites present in the expression cassettes do not per se influence the promoter activity. The yeast destination vectors described above can be used for the cloning and expression of single genes, for the parallel automated cloning of up to thousands of different genes, and for shuffling of gene pools or entire cDNA libraries into different yeast expression vectors by single reactions.

Destination Vector Collection for Expression of Epitope-Tagged Proteins

Next we constructed a collection of destination vectors for expression of epitope-tagged proteins based on the vectors p426MET25 and p426GAL1 (Fig. 4).

[13] B. P. Cormack, G. Bertram, M. Egerton, N. A. Gow, S. Falkow, and A. J. Brown, *Microbiology* **143,** 303 (1997).

[14] R. K. Niedenthal, L. Riles, M. Johnston, and J. H. Hegemann, *Yeast* **30,** 773 (1996).

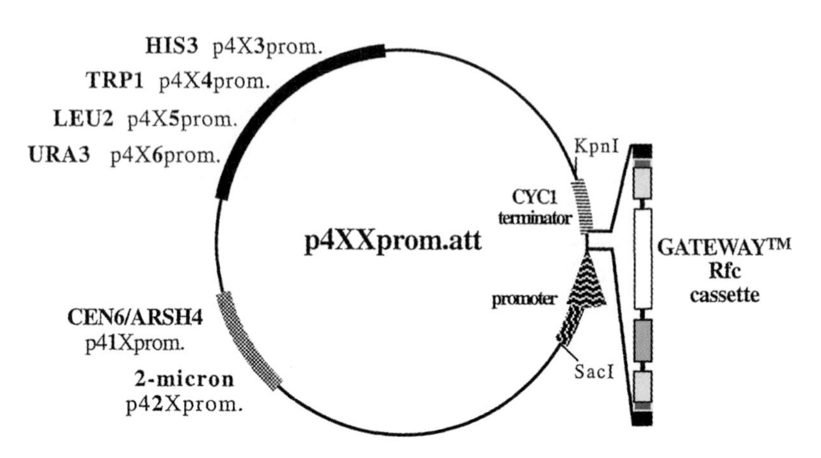

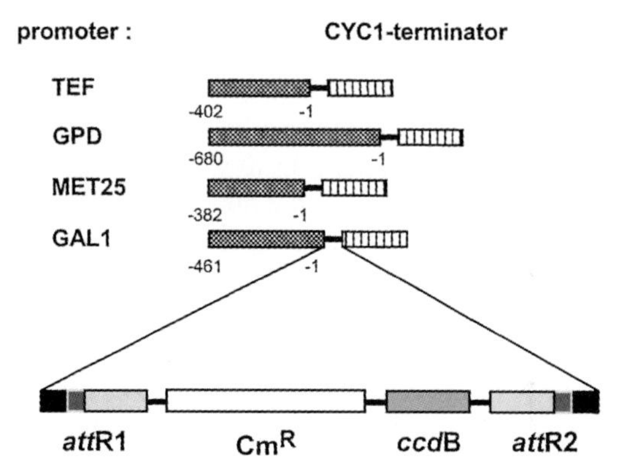

FIG. 2. Structure of GATEWAY compatible destination vectors. (A) Schematic map of the destination vectors constructed by introducing the GATEWAY Rfc cassette between the respective promoter and the *CYC1* terminator. The nomenclature is according to the one described in Fig. 1. (B) Structure of the constitutive *TEF* and *GPD* promoter, of the inducible *GAL1* promoter, or of the repressible *MET25* promoter followed by the Rfc cassette and the *CYC1* terminator. The Rfc cassette comprises a chloramphenicol resistance marker (Cm^R) and the toxic *ccdB* gene flanked by the *attR1* and *attR2* recombination sites. The destination vectors can be used in the *in vitro* GATEWAY LR recombination reaction, replacing the Rfc cassette by the insert from a entry clone flanked by *attL*1/2 sites. Construction of the expression cassettes was carried out as described in the section "Procedures and Vector Constructions."

A

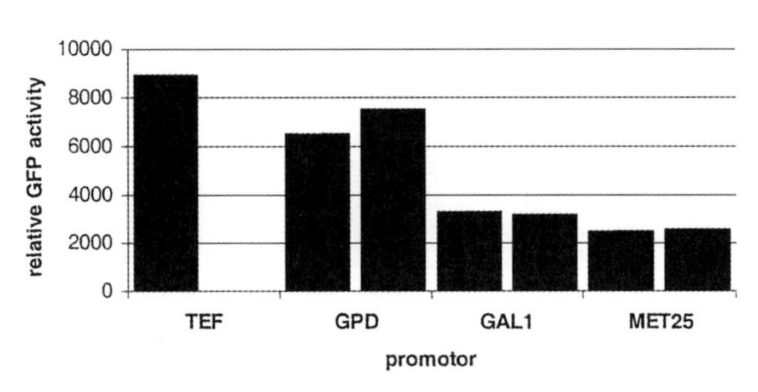

B

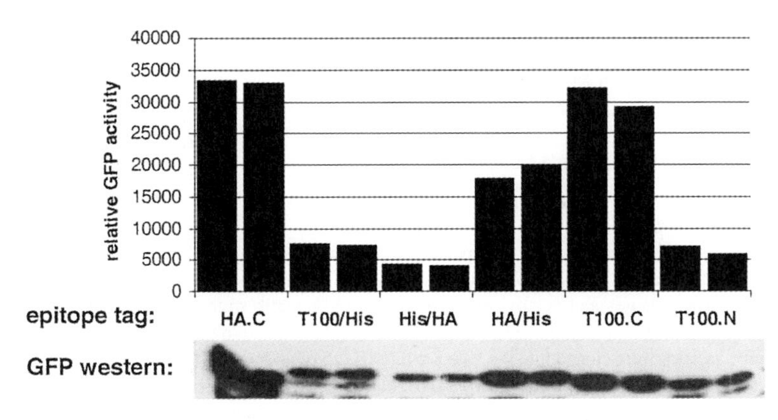

FIG. 3. GFP expression levels from different promoters or with different epitope tags. (A) GFP expression levels from constructs based on the vector pRS416 carrying the different promoters described in Fig. 2B. The GFP activity of one or two clones from the respective plasmid transformants was determined by FACS analysis. (B) GFP expression levels from constructs based on the vector p426GAL1 carrying the different promoters described in Fig. 2C. The relative GFP activity of one or two clones from the respective plasmid transformants was determined by FACS analysis and the corresponding protein levels were analyzed by a GFP Western blot. Western blot analysis of cells expressing GFP and quantification of GFP activity was performed as described in the section "Procedures and Vector Constructions."

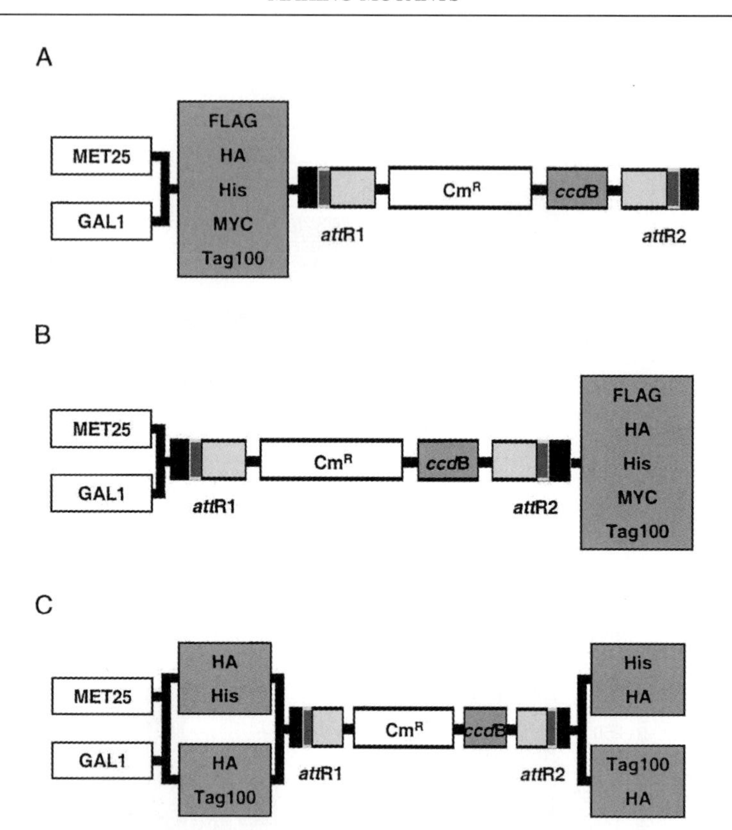

FIG. 4. Expression cassettes for epitope-tagged proteins. Expression cassettes for proteins fused to different epitope tags at the N terminus (A), C terminus (B), or both termini (C) were constructed as described in procedures and vector constructions. (A) Schematic structure of the cassettes for the N-terminal fusion to the FLAG, HA, His MYC, or Tag100 tag followed by the Rfc cassette. The different cassettes were cloned into the vector p426MET25 or p426GAL1. The nomenclature of these vectors has the following structure: 426GAL1HAN.att designates a vector based on p426GAL1 with an N-terminal HA epitope tag pictured. (B) Analogous cassettes for the C-terminal fusion to the epitope tags. (C) Cassettes that allow the expression of proteins tagged at both termini. The HA tag can be combined with the His tag or the Tag100 tag in both possible orientations placing the HA tag at the N terminus or at the C terminus. The nomenclature of these vectors has the following structure: 426MET25HA/His.att designates a vector based on p426GAL1 with an N-terminal HA and C-terminal His epitope tag.

These vectors allow the expression of proteins fused to the broadly used FLAG, HA, His MYC, and Tag100 epitope tags.[15–17] Tags can be fused either to the

[15] A. Cravchik and A. Matus, *Gene* **137,** 139 (1993).

[16] B. L. Brizzard, R. G. Chubet, and D. L. Vizard, *Biotechniques* **16,** 730 (1994).

[17] M. R. Martzen, S. M. McCraith, S. L. Spinelli, F. M. Torres, S. Fields, E. J. Grayhack, and E. M. Phizicky, *Science* **286,** 1153 (1999).

N terminus (Fig. 4A) or the C terminus of the expressed protein. Double-tagged constructs (Fig. 4C) can be used for expression, isolation, and oriented binding of His-tagged proteins to Ni-NTA HisSorb plates or Ni-NTA Magnetic Beads (Qiagen), and bound protein can be quantified by using the Tag100 or HA antibody. Details of the epitope tag sequences, of the flanking sequences, and of the reading frame relative to the *attR* sequences are shown in Table I.

For validation of different selected epitope tags we analyzed the expression levels of GFP from the corresponding expression cassettes in the vector backbone p426GAL1. Cells carrying the *GAL1* promoter expression constructs were pregrown in minimal medium with 2% raffinose as carbon source and then shifted for 5 hr to medium containing 2% raffinose and 2% galactose. Expression levels were analyzed by GFP Western blot and the relative GFP activity was determined by FACS analysis (Fig. 3B). In general, we observed that the tagging of GFP with different epitopes can have a dramatic effect on the GFP activity, reducing it as much as sevenfold. The parallel GFP Western blot clearly shows that this decrease in activity correlates with a decrease in the amount of GFP protein. Intriguingly, placing a HA or Tag100 epitope at the N terminus of GFP led to the highest decrease in activity. These effects could be explained by an influence of the tag on the stability of the GFP protein when placed at the N terminus or by a negative effect on the mRNA stability or translation efficiencies of the mRNA when the tag sequences are located in the very 5′ region of the GFP mRNA.

Despite these issues all vectors express considerable amount of protein. All single epitope tags and combinations of tags tested were recognized by their specific anti-tag antibody in Western blot analysis when fused to a standard test protein (data not shown). Therefore the collection of vectors we have constructed is very useful for heterologous expression of proteins and for monitoring of the expression by the corresponding epitope tag. The design of the system permits the coexpression of up to five differently tagged proteins and the analysis of the expression properties of every single protein used. The double-tagged vectors are suitable for parallel high-throughput expression and purification of immobilized His-tagged proteins from yeast that could be used for studying protein/protein interactions or for functional genomics applications such as genomic biochemistry.[18]

Procedures and Vector Constructions

Expression plasmids are transformed into the yeast strain YPH499 (*a*; *lys2-801*^{amber}; *ade2-101*^{ochre}; *leu2-Δ1*; *trpΔ63*; *his3Δ200*; *ura3Δ52*)[3] as described by Gietz and Schiestl[19] and cells are grown in minimal medium[20] supplemented with

[18] P. A. Kolodziej and R. A. Young, *Methods Enzymol.* **194,** 508 (1991).

[19] R. D. Gietz and R. H. Schiestl, *Yeast* **7,** 253 (1991).

[20] F. Sherman, G. R. Fink, and J. B. Hicks, "Methods in Yeast Genetics." Cold Spring Harbor Laboratory Press, Cold Spring Harbor, NY, 1986.

TABLE I
EPITOPE TAG SEQUENCES OF EXPRESSION CASSETTES[a]

Tag	
N-terminal tags	
FLAG	GAATTGGAAGGAGATAGAACC**ATGGATTACAAGGATGACGACGATAAGATC**ACAAGTTTGTACAAAAAA
	M D Y K D D D D K I *T S L Y K K*
HA	GAATTGGAAGGAGATAGAACC**ATGTACCCTTATGATGTGCCAGATTATGCC**ACAAGTTTGTACAAAAAA
	M Y P Y D V P D Y A *T S L Y K K*
His	GAATTGGAAGGAGATAGAACC**ATGAGAGGATCGCATCACCATCACCATCAC**ACAAGTTTGTACAAAAAA
	M R G S H H H H H H *T S L Y K K*
MYC	GAATTGGAAGGAGATAGAACC**ATGGAGCAGAAACTCATCTCTGAAGAGGATCTG**ACAAGTTTGTACAAAAAA
	M E Q K L I S E E D L *T S L Y K K*
Tag100	GAATTGGAAGGAGATAGAACC**ATGGAAGAGACTGCACGTTTCCAGCCGGGTTATCGTTCT**ACAAGTTTGTACAAAAAA
	M E E T A R F Q P G Y R S *T S L Y K K*
C-terminal Tags	
FLAG	TTCTTGTACAAAGTGGTT**GATTACAAGGATGACGACGATAAGATC**CTCGAGTCATGTAATTAG
	F L Y K V V **D Y K D D D D K I** L E S C N *
HA	TTCTTGTACAAAGTGGTT**TACCCTTATGATGTGCCAGATTATGCC**CTCGAGTCATGTAATTAG
	F L Y K V V **Y P Y D V P D Y A** L E S C N *
His	TTCTTGTACAAAGTGGTT**AGAGGATCGCATCACCATCACCATCAC**CTCGAGTCATGTAATTAG
	F L Y K V V **R G S H H H H H H** L E S C N *
MYC	TTCTTGTACAAAGTGGTT**GAGCAGAAACTCATCTCTGAAGAGGATCTG**CTCGAGTCATGTAATTAG
	F L Y K V V **E Q K L I S E E D L** L E S C N *
Tag100	TTCTTGTACAAAGTGGTT**GAAGAGACTGCACGTTTCCAGCCGGGTTATCGTTCT**CTCGAGTCATGTAATTAG
	F L Y K V V **E E T A R F Q P G Y R S** L E S C N *

[a] The DNA and protein sequences for the N- and C-terminal epitope tags (bold face type) are shown. In the case of the N-terminal tags the first six of nine amino acids coded by the *attR*1 sequence are shown in italics. In the case of the C-terminal tags the last six of nine amino acids coded by the *attR*2 sequence (italic) and the five amino acids following the different tags are shown. Construction of the corresponding expression cassettes is described in the section "Procedures and Vector Constructions."

the appropriate carbon source or the corresponding amount of methionine. For shift experiments precultures are grown to a density between 1 and 3 $OD_{600\,nm}$, washed once with water, and diluted 1:10 into the new medium. For Western blot analysis 5×10^7 cells are boiled in SDS-loading buffer[21] and protein samples are separated on 10% SDS–PAGE gels. Immunodetection with the anti-GFP monoclonal antibody (BD Biosciences Clontech, Palo Alto, CA) is done according to Ausubel and Frederick.[22] FACS analysis of the GFP expression constructs is done as described by Niedenthal *et al.*[14]

All plasmids described by Mumberg *et al.*[1,2]; Roenicke *et al.*[3]; and Labbe and Thiele[8] can be requested from the ATCC (Manassas, VA). Details for the plasmid construction of the expression vectors shown in Fig. 1 are given by Mumberg *et al.*[1,2]; Roenicke *et al.*[3]; and Labbe and Thiele.[8] The vectors carrying the *MET15* or *ADE2* marker are constructed by *in vivo* homologous recombination in yeast[23] followed by the reisolation of the resulting vectors. For this purpose the expression cassettes of the different vectors described by Mumberg *et al.*[1,2] are PCR-amplified including a sequences flanking the multiple cloning array of the pBluescript vector. These PCR fragments are then used for homologous recombination with the *Bam*HI linearized vector DNAs from the plasmids pRS411, pRS421, pRS412, and pRS422 described by Brachmann *et al.*[7]

Yeast expression vectors carrying no epitope tag are converted into GATEWAY destination vectors by *in vivo* homologous recombination for the expression cassettes. This construction is achieved by generating PCR fragments of the Rfc cassette flanked 5′ by a 40-bp sequence homologous to the respective promoter and 3′ by a 40 bp sequence homolgous to the *CYC1* terminator.

Yeast expression vectors carrying epitope tags are converted into GATEWAY destination vectors in two steps. First the sequences coding for the HA,[17] FLAG,[16] MYC,[15] Tag100 (Qiagen), or His (RGS epitope, Qiagen) epitopes described in Table I are added to the Rfc cassette by PCR, thereby introducing the restriction sites used for *Mfe*I/*Xho*I subcloning into the vector pSL1180.[24] The *Mfe*I/*Xho*I fragments from p*xyz* derivatives are cloned into the *Eco*RI/*Xho*I cut vector p426GAL1 or p426MET25 resulting in the different yeast destination vectors.

The GFP gene is introduced into the different destination vectors by an LR reaction from an entry clone carrying the yeast codon optimized GFP variant described by Cormack *et al.*[13]

Acknowledgments

We are grateful to S. Kunz, B. Bathke, and S. Schaberg for excellent technical assistance. We also wish to thank B. Nave for reading the manuscript.

[21] A. Horvath and H. Riezman, *Yeast* **10,** 1305 (1994).

[22] I. Ausubel and M. Frederick, "Current Protocols in Molecular Biology." Wiley, New York, 1991.

[23] J. Brosius, *DNA* **8,** 759 (1989).

[24] C. K. Raymond, T. A. Pownder, and S. L. Sexson, *Biotechniques* **26,** 134 (1999).

[15] Cloning-Free Genome Alterations in *Saccharomyces cerevisiae* Using Adaptamer-Mediated PCR

By ROBERT J. D. REID, MICHAEL LISBY, and RODNEY ROTHSTEIN

Introduction

The budding yeast *Saccharomyces cerevisiae* is an appealing eukaryotic model system in part due to its amenability to genome manipulation. Because linear DNA readily undergoes homologous recombination in yeast, precise alterations of genes in the chromosome can be done with relative ease.[1] For example, targeted gene disruptions to produce null alleles are performed routinely in yeast and provide a vital tool for understanding gene function.[2,3] Allele replacement methods allow the introduction of novel mutations into any gene of interest.[4] Similarly, gene fusions that append functional epitopes such as green fluorescent protein (GFP) or an immunogenic tag are another important tool for studying gene function.[5] These methods all rely on the efficient homologous recombination of yeast to make specific changes to genes at their endogenous chromosomal loci.

In recent years, many genome manipulation techniques have been revised to take advantage of PCR. Below we describe PCR-based techniques for gene disruption, allele replacement, and epitope fusions using special primers we call adaptamers. Adaptamers are chimeric primers containing sequences at their 5′ end that facilitate the fusion of any two pieces of DNA by PCR. This principle is illustrated in Fig. 1. The two DNA fragments to be fused are each amplified with an adaptamer and a standard PCR primer. The 5′ sequence tags on the adaptamers are reverse and complementary to each other. In a second round of PCR, the two amplified DNA fragments are mixed; the complementary ends anneal and are then extended by the polymerase. Primers are also included that bind at the termini of the fused DNA fragments leading to exponential amplification of the fused product.

In each of the techniques described below, adaptamer technology is used to mediate the fusion of a dominant selectable marker to DNA fragments that provide homology for chromosomal targeting. On transformation of these PCR products into yeast, homologous recombination yields the targeted genomic integration

[1] T. L. Orr-Weaver, J. W. Szostak, and R. J. Rothstein, *Proc. Natl. Acad. Sci. U.S.A.* **78,** 6354 (1981).
[2] D. Shortle, J. E. Haber, and D. Botstein, *Science* **217,** 371 (1982).
[3] R. J. Rothstein, *Methods Enzymol.* **101,** 202 (1983).
[4] R. Rothstein, *Methods Enzymol.* **194,** 281 (1991).
[5] M. E. Petracek and M. S. Longtine, *Methods Enzymol.* **350,** [25], 2002 (this volume).

0076-6879/02 $35.00

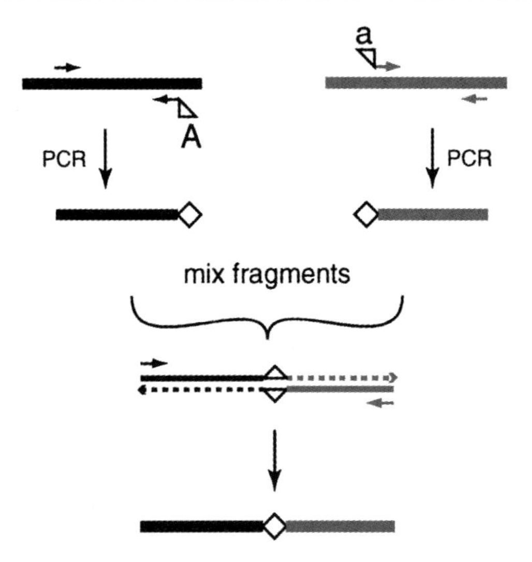

FIG. 1. Adaptamer-directed PCR fusions. Two different double-stranded DNA sequences are represented by thick gray and black lines. PCR primers are illustrated by arrows. Adaptamers are shown as arrows with triangles at their 5′ ends representing the complementary sequence tags A and a. PCR amplification with these adaptamers incorporates the 5′ sequence tags into double-stranded DNA (diamonds). In a second PCR reaction the first PCR products are mixed; the incorporated complementary sequence tags anneal and are extended by the polymerase (dashed lines). In successive PCR cycles, distal primers (arrows) amplify the fused product.

with high efficiency. Thus adaptamers obviate the need for traditional cloning in *Escherichia coli* to build integrating vectors and offer a considerable savings of time. In addition, these methods each incorporate directly repeated sequences flanking the selectable marker. This permits reuse of the genetic marker after its deletion from the genome by direct repeat recombination.

Gene Disruption Strategies

One-step gene disruption in yeast is a technique used to replace a functional gene with a dominant selectable marker in a process requiring homologous recombination.[3] Traditionally this involves interrupting a cloned gene on a plasmid by inserting a dominant selectable marker using standard cloning techniques (Fig. 2A). The plasmid is cut with restriction enzymes to produce a linear DNA containing the selectable marker and several hundred base pairs of DNA on each side of the selectable marker. On transformation into yeast, the homologous DNA flanking the marker promotes two recombination events that replace the copy of the gene on the chromosome with the inactive copy in a single step (Fig. 2B). The dominant marker allows selection of recombinants after transformation. Although the

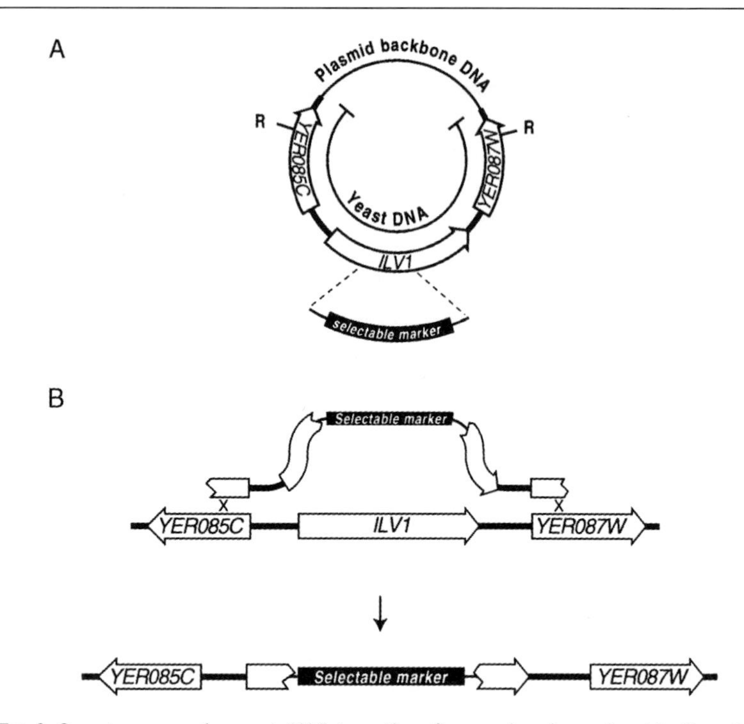

FIG. 2. One-step gene replacement. (A) Interruption of a gene cloned on a plasmid with a selectable marker using standard cloning techniques is represented by dashed lines. ORFs are indicated by open arrows, yeast intergenic sequences are illustrated by thick black lines, and plasmid DNA is indicated by a thin line. The restriction sites to linearize the gene disruption fragment are marked R. (B) Two homology-directed recombination reactions (X) occur to produce a gene disruption in a single step.

recombination events are efficient, this method requires time-consuming cloning steps in *E. coli* to produce the desired integrating DNA. In addition, interruption of the gene depends on available restriction sites. In many cases this leaves a partial fragment of an open reading frame (ORF) and may not provide a true null allele. PCR methods significantly expedite construction of integrating DNA.

PCR-based methods of gene disruption streamline the traditional approach by replacing cloning in *E. coli* with PCR amplification. The simplest of these approaches is sometimes referred to as a "microhomology" or "long primer" method. The homologous DNA is synthesized as 40 to 60 nucleotide 5′ extensions on primers designed to amplify a selectable marker (Fig. 3A). After amplification, the PCR products are transformed into yeast and the 40 to 60 base pairs of homology on each end of the DNA promotes recombination with genomic DNA to disrupt the gene (Figure 3B).[6–8]

This method greatly simplifies one-step gene disruptions since no cloning is necessary. In addition, it is possible to create precise deletions of ORFs to

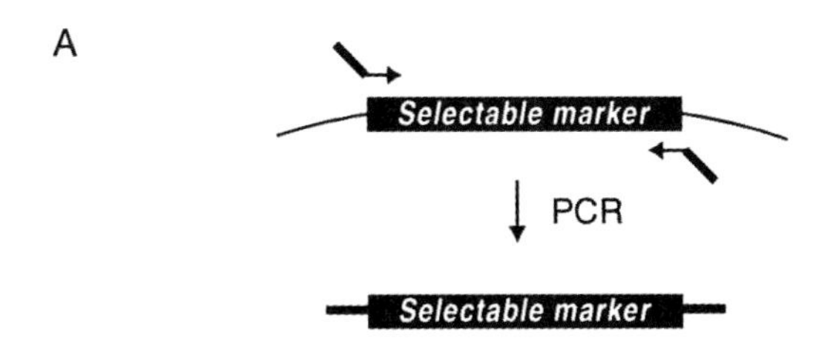

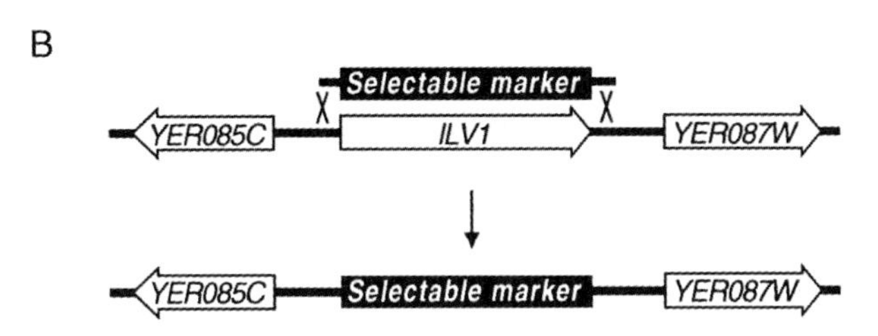

FIG. 3. "Microhomology" one-step gene disruptions. (A) A selectable marker is PCR amplified using chimeric primers containing 5′ sequences with homology to genomic DNA immediately upstream and downstream of the ORF to be disrupted. (B) Integration mediated by short homology regions flanking the selectable marker.

ensure production of the null alleles. However, the length of homology to the genome is limited by the amount of DNA that can be reasonably synthesized on the 5′ ends of the primers. Recombination is relatively inefficient with such short homology regions requiring several micrograms of DNA to produce only a few transformants.[7] Increasing the length of homologous DNA to $\geq$600 bp on each side of a selectable marker increases recombination efficiency producing 100- to 1000-fold more transformants per microgram of DNA.[8]

The adaptamer-directed gene disruption method described below combines the efficient genome integration made possible by longer homologous DNA fragments and the rapid construction methods facilitated by PCR.

Adaptamer-Directed Gene Disruptions

The adaptamer-directed gene disruption method relies on the same principles of homologous recombination as the gene disruption methods described above. However, in this strategy yeast intergenic regions are amplified to provide 200 to 500 bp of homology on each side of the gene to be disrupted. The homologous

[6] A. Baudin, O. Ozier-Kalogeropoulos, A. Denouel, F. Lacroute, and C. Cullin, *Nucleic Acids Res.* **21,** 3329 (1993).

[7] A. Wach, A. Brachat, R. Pohlmann, and P. Philippsen, *Yeast* **10,** 1793 (1994).

[8] P. Manivasakam, S. C. Weber, J. McElver, and R. H. Schiestl, *Nucleic Acids Res.* **23,** 2799 (1995).

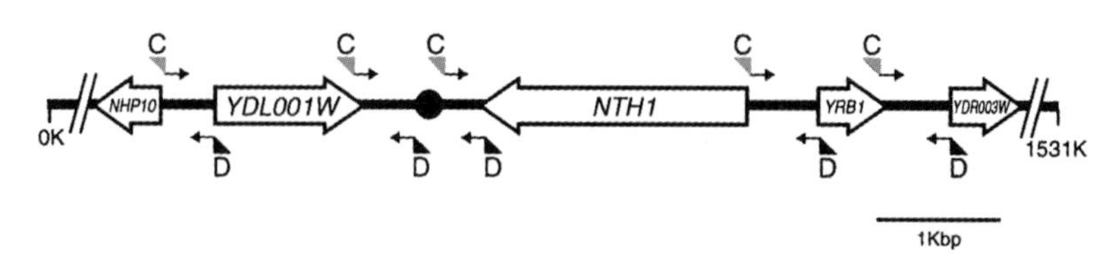

FIG. 4. Orientation of adaptamers on an 8 kilobase pair region of yeast chromosome IV. A map of 15-kb pairs near the centromere of chromosome IV. The black circle represents the centromere. White arrows indicate known or predicted open reading frames. Adaptamers (not to scale) are shown as black arrows with gray (C) or black (D) triangles representing the standard 5′ sequence tags. [Adapted from R. J. D. Reid, I. Sunjevaric, M. Keddache, and R. Rothstein, *Yeast,* in press (2002).]

DNA fragments are then fused to a selectable marker via complementary adaptamers. Upon transformation into yeast, these DNA fragments recombine with the homologous genomic locus to disrupt the gene. A set of 6361 primer pairs was designed to amplify yeast intergenic regions. These contain common 20 bp 5′ sequence tags that are not homologous to *S. cerevisiae* genomic DNA.[9] These intergenic adaptamers can be specifically suited for the task of gene disruptions and are available from Research Genetics (Huntsville, AL). Figure 4 shows a map of intergenic adaptamers on an 8 kb segment of yeast chromosome IV. The intergenic adaptamers are uniformly oriented, independent of gene orientation, so that all forward primers contain the same 5′ sequence tag called C and all reverse primers contain a different 5′ sequence tag called D (see Table I for sequences).

As outlined in Figure 5, two rounds of PCR are necessary to create a gene disruption cassette by this method. In the first round, intergenic DNA fragments and the selectable marker are amplified in four PCR amplifications. Amplification of intergenic regions using the C and D adaptamer incorporates the sequence tags into the ends of those PCR products (Fig. 5A). Approximately 2 ng genomic DNA is added to 20 μl reactions containing 0.5 μM of the specific intergenic adaptamers, 200 μM of each dNTP, and 1.5 units *Taq* polymerase. Genomic DNA is prepared by standard methods for use as template DNA.[10] Intergenic adaptamers were designed to have a minimal annealing temperature of 52°.[11] Amplification is performed using the following cycle conditions: 94° for 3 min followed by 30 cycles of 94° for 30 sec, 52° for 30 sec, 72° for 1 min, and finally 72° for 5 min. In rare cases where amplification using intergenic adaptamers has given poor or inconsistent yields using the above conditions, the annealing temperature was reduced to 45° for the first eight cycles followed by 22 cycles using the 52° annealing temperature.

[9] V. Iyer, P. Brown, M. Keddache, and R. Rothstein, personal communication, 1998.

[10] C. S. Hoffman and F. Winston, *Gene* **57,** 267 (1987).

[11] V. Iyer, personal communication, 1999.

TABLE I
PCR PRIMERS AND ADAPTAMERS[a]

Primer sequence	Name	Description
ccgctgctaggcgcgccgtg...	C	5′ Nonhomologous sequence tag used for all forward intergenic adaptamers
gcagggatgcggccgctgac...	D	5′ Nonhomologous sequence tag used for all reverse intergenic adaptamers
CTTGACGTTCGTTCGACTGATGAGC	kli5′	*K. lactis URA3* internal 5′ primer
GAGCAATGAACCCAATAACGAAATC	kli3′	*K. lactis URA3* internal 3′ primer
gtcagcggccgcatccctgcCCTCACTAAAGG GAACAAAAGCTG	d-Kl	3′ *K. lactis URA3* "d" adaptamer. Nonhomologous region is reverse and complement of D
cacggcgcgcctagcagcggTAACGCCA GGGTTTTCCCAGTCAC	c-Kl	5′ *K. lactis URA3* "c" adaptamer. Nonhomologous region is reverse and complement of C
GTCGACCTGCAGCGTACG	U2	5′ Primer for amplification of intergenic DNA from the yeast deletion strains
CGAGCTCGAATTCATCGAT	D2	3′ Primer for amplification of intergenic DNA from the yeast deletion strains
cgtacgctgcaggtcgacgggccc GTGTCACCATGAACGACAATTC	u2-Kl[b]	5′ *K. lactis URA3* "u2" adaptamer. Nonhomologous region is reverse and complement of U2
atcgatgaattcgagctcgatcgat GTGATTCTGGGTAGAAGATC	d2-Kl[c]	3′ *K. lactis URA3* "d2" adaptamer. Nonhomologous region is reverse and complement of D2
ggaattccagctgaccaccatg...	A[d]	5′ Nonhomologous sequence tag included on all forward ORF adaptamers
gatccccgggaattgccatg...	B	5′ Nonhomologous sequence tag included on all reverse ORF adaptamers
catggcaattcccggggatcGTGATTCTGGGT AGAAGATCG	b-Kl	5′ *K. lactis URA3* "b" adaptamer. Nonhomologous region is reverse and complement of B
catggtggtcagctggaattccCGATGATGTAG TTTCTGGTT	a-Kl	3′ *K. lactis URA3* "a" adaptamer. Nonhomologous region is reverse and complement of A
gttcttctcctttactcat...	g1	5′ Nonhomologous sequence tag of the GFP 5′ end adaptamer
ggatgaactatacaaaTAA...	g2[e]	5′ Nonhomologous sequence tag of the GFP 3′ end adaptamer
ATGAGTAAAGGAGAAGAAC	GFPstart-F	5′ GFP primer. Reverse and complement of the g1 sequence tag
TTTGTATAGTTCATCCATGC	GFPend-R	3′ GFP primer. Underlined sequence is reverse and complement of the nonhomologous 5′ section of the g2 sequence tag

[a] All sequences are listed in the 5′ → 3′ direction. Lowercase sequences denote nonhomologous 5′ segments or sequence tags on adaptamers.
[b] The underlined sequence is a *Apa*I restriction site.
[c] The underlined sequence is a *Cla*I restriction site.
[d] The underlined sequence is the ORF start codon.
[e] The underlined sequence is a stop codon necessary for C-terminal GFP fusions.
[Adapted from R. J. D. Reid, I. Sunjevaric, M. Keddache, and R. Rothstein, *Yeast,* in press (2002).]

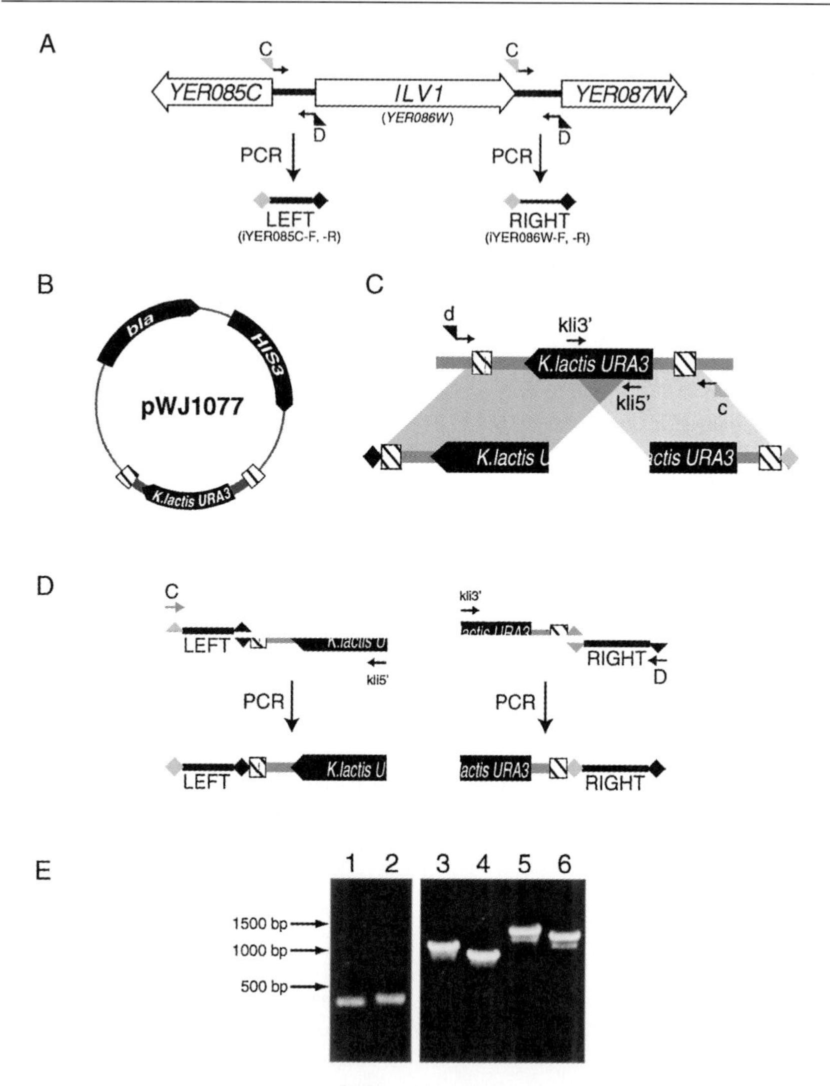

FIG. 5. Adaptamer-directed gene disruptions. (A) Amplification of intergenic regions flanking the *ILV1* gene on chromosome V. Two PCR reactions amplify intergenic DNA containing the adaptamer tags (diamonds). (B) Plasmid pWJ1077 containing the *K. lactis URA3* gene and the 143 bp direct repeats. (C) Direct repeats flanking the *K. lactis URA3* ORF are represented as hatched boxes and were made by PCR amplifying a 143 bp sequence 5′ to the *URA3* ORF and cloning it into *Cla*I and *Apa*I restriction sites on the 3′ side of the ORF in plasmid pWJ1077. Amplification of overlapping segments in two PCR reactions is indicated by shading. (D) Fusion PCR reactions using the left intergenic DNA and the 3′ section of *URA3* in one reaction and the right intergenic region and the 5′ portion of *URA3* in the second reaction. The upper diagrams illustrate annealing of single strands mediated by complementary sequence tags while the bottom cartoons illustrate fused PCR products. (E) Example

The selectable marker is amplified in two overlapping segments from plasmid pWJ1077 (Figs. 5B and 5C). We use the orthologous *URA3* gene from *Kluyveromyces lactis* as a selectable marker because it complements a *S. cerevisiae ura3* mutant and can also be counterselected using 5-fluoroorotic acid (5-FOA).[12] Additionally, the sequence homology between the *K. lactis URA3* and the *S. cerevisiae URA3* is approximately 70% at the nucleotide level.[13] Therefore, recombination of *K. lactis URA3* sequence with the endogenous *URA3* allele is greatly decreased.[14,15] We incorporated 143 bp direct repeats into the *URA3* plasmid so the marker can be recycled after integration (see below). The 3′ section of *URA3* is amplified with adaptamer d and internal primer kli5′, while the 5′ section of *URA3* is amplified with adaptamer c and internal primer kli3′ (see Table I for sequences). Adaptamers c and d have 5′ sequence tags that are the reverse and complement of the 5′ tags on the C and D intergenic adaptamers. PCR of selectable marker segments is performed using 2 ng pWJ1077 plasmid DNA as a template with the same conditions described above, and substituting 55° for the annealing temperature.

A second round of PCRs is required to fuse the selectable marker fragments to the amplified intergenic regions via the complementary sequence tags on the adaptamers. The left intergenic region is fused to the 3′ *URA3* segment through annealing of the d and D sequence tags, while the right intergenic region is fused to the 5′ *URA3* segment by the annealing of the c and C sequence tags (Fig. 5D). Conditions for these PCR fusions are as follows: 10 to 25 ng of the intergenic PCR amplified DNA is mixed in a 50 μl reaction with approximately equimolar amounts of the appropriate *K. lactis URA3* DNA fragment. Typically, the template DNA fragments for fusion are simply diluted 50- to 100-fold from the first round of PCRs into the fusion reaction mixture and do not require purification. Fusion PCRs also contain 0.5 μM primers, 200 μM of each dNTP, and 3.8 units of *Taq* polymerase.

[12] J. D. Boeke, J. Trueheart, G. Natsoulis, and G. R. Fink, *Methods Enzymol.* **154,** 164 (1987).

[13] J. R. Shuster, D. Moyer, and B. Irvine, *Nucleic Acids Res.* **15,** 8573 (1987).

[14] A. M. Bailis and R. Rothstein, *Genetics* **126,** 535 (1990).

[15] S. D. Priebe, J. Westmoreland, T. Nilsson-Tillgren, and M. A. Resnick, *Mol. Cell. Biol.* **14,** 4802 (1994).

of PCR fragments used to produce an *ILV1* gene disruption. The 352 bp left and 380 bp right intergenic regions for the *ILV1* gene (lanes 1 and 2) were amplified using wild-type genomic DNA from the W303 strain background as a template. The 1095 bp *K. lactis URA3* 3′ and 946 bp *URA3* 5′ DNAs (lanes 3 and 4) were amplified using plasmid pWJ1077 as a template. Two μl of each reaction was loaded and run on a 0.8% electrophoresis gel. DNA was visualized by ethidium bromide staining. Fusion reactions were performed by diluting the PCR products from the first reactions 100-fold into a 50 μl PCR. Two μl of each of the amplified 1447 bp *ILV1*-left fused to *URA3*-3′ and the 1326 bp *ILV1*-right fused to *URA3*-5′ were loaded and run on the same gel (lanes 5 and 6). [Adapted from R. J. D. Reid, I. Sunjevaric, M. Keddache, and R. Rothstein, *Yeast,* in press (2002).]

Cycle conditions are 94° followed by 30 cycles of 94° for 30 sec, 55° for 30 sec, 72° for 1.5 or 2 min (depending on total length), and finally 72° for 10 min.

DNA fragments for a gene disruption are produced in two overlapping parts for several reasons. First, every amplified intergenic region is bounded by the same C and D sequence tags; thus the split marker approach isolates the complementary c and d ends of the *URA3* gene into two separate PCR fusion reactions, ensuring proper orientation of the fused product. Second, PCR fusions can be accomplished using the 20-mer C or D sequences as primers rather than the specific adaptamers for each intergenic region (Fig. 5D). This simplifies the setup of PCRs when DNAs for many gene disruptions are produced at once. Finally, while the length an integrating DNA sequence can be long, the split marker approach generates shorter individual fusion PCR fragments, thereby increasing the reliability of the amplifications.

Typical products produced from first round of PCR and subsequent PCR fusions are shown in Fig. 5E. The fused DNA fragments can be taken directly from the PCR amplification mixtures and used for yeast transformations. The three recombination events required to reconstitute the selectable *URA3* marker and replace the targeted ORF are illustrated in Fig. 6. The homologous intergenic regions

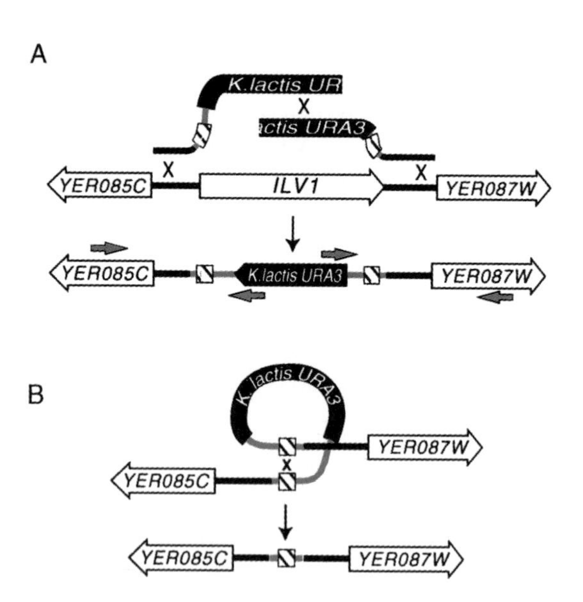

FIG. 6. Integration of two PCR fragments to disrupt a gene. (A) Three recombination events are required to replace a gene in a single step using two DNA fragments. Thick gray arrows represent primer binding sites for PCR analysis of integrations. (B) Direct repeat recombination results in "pop-out" of the *URA3* marker leaving a small fragment of non-*S. cerevisiae* DNA in place of the ORF. [Adapted from R. J. D. Reid, I. Sunjevaric, M. Keddache, and R. Rothstein, *Yeast,* in press (2002).]

target recombination to the intended locus while the *URA3* segments reconstitute the selectable marker,[16] resulting in a precise disruption of the ORF. Yeast transformations are performed using the standard lithium acetate procedure as described.[17] Approximately 300 ng of each fusion DNA fragment is added to competent cells using conditions that give 10^6 transformants/μg of circular plasmid DNA. The number of transformants in various gene disruption experiments ranges from 10 to over 200 using these amounts of DNA.

Often the integrated selectable marker is used to follow segregation of the null mutant during crosses to other strains. In this case, it is best to verify gene disruptions using two PCRs. Internal *K. lactis URA3* primers are paired with primers that bind in the left and right genomic DNA outside of the intergenic regions to amplify across each recombination junction (gray arrows in Fig. 6A). However, it may also be desirable to recycle the selectable marker. This is achieved by recombination between the 143 bp direct repeats flanking URA3 which leads to a pop-out (Fig. 6B). This is typically done by growing overnight cultures of a yeast gene disruption strain without selection for the *URA3* marker and then plating 100–200 μl of the saturated culture onto 5-FOA plates to select for the pop-out recombinants. Verification of markerless gene disruptions can be done using a single PCR reaction that amplifies across the disruption since this is now a short DNA sequence. For the gene disruptions performed thus far using this method, 11 of 14 yielded a gene disruption efficiency of 70–90%. Lower gene disruption efficiencies (about 20%) occur when a particular gene disruption produces a slow growth phenotype. In such cases, it is more efficient to perform gene disruption in a diploid strain.[18]

Intergenic PCRs and PCR fusions are simple to perform. Disruption of any yeast gene requires four specific intergenic adaptamers to amplify the flanking DNA, but all other components are common. The *K. lactis* DNA fragments can be amplified in quantity and used as a resource for many different PCR fusions. Purification of intergenic PCR products is not necessary between the first and second rounds of PCR except in rare cases where PCR products contain contaminating nonspecific DNA fragments. In addition, PCR products from the fusion reactions can be added directly to yeast transformation reactions. Thus all PCRs and the yeast transformations for a gene disruption experiment can usually be accomplished in a day. Although this method incurs an added cost for the extra PCRs compared to the "long primer" gene disruption method, an overall increase in efficiency can be realized because fewer transformants need to be screened to identify successful gene disruptions. This becomes important when multiple gene disruption experiments are carried out at once.

[16] C. Fairhead, B. Llorente, F. Denis, M. Soler, and B. Dujon, *Yeast* **12,** 1439 (1996).
[17] R. H. Schiestl and R. D. Gietz, *Curr. Genet.* **16,** 339 (1989).
[18] R. J. D. Reid, I. Sunjevaric, M. Keddache, and R. Rothstein, *Yeast,* in press (2002).

Intergenic adaptamers have standard names based on the name of the adjacent chromosome feature to the "left" of the intergenic region. This is illustrated in Fig. 5 where the adaptamers to amplify the "left" intergenic region for *ILV1* (iYER085C-F and -R) are named after the YER085C ORF and the adaptamers to amplify the "right" intergenic region (iYER086W-F and -R) are named after the systematic name of the *ILV1* ORF. Although the nomenclature is standard, it is complicated by the fact that adaptamer pairs do not exist for every intergenic region because of overlapping ORFs. In addition, intergenic regions $\geq$1.5 kb are split into sections of $\leq$1.2 kb amplicons.

Lists of all the intergenic adaptamers and sequences can be obtained by contacting Research Genetics. In addition, maps of all 16 yeast chromosomes showing intergenic adaptamer binding sites, the systematic name for each set of intergenic primers, and the length of the amplified DNA can be viewed on our Web site (http://rothsteinlab.hs.columbia.edu/projects/primersearch.html).

Moving "Bar-Coded" Gene Disruptions into a New Strain

A worldwide consortium of yeast laboratories recently completed construction of an arrayed library of yeast gene disruption strains for more than 5000 genes.[19] This library represents a unique resource for an experimental organism and is a valuable addition to yeast researchers' toolbox. The deletion consortium strains are constructed by the "microhomology" PCR technique using kanamycin resistance as the selectable marker. The kanMX4 cassette was PCR amplified and attached to 45 bp of homology from each side of the targeted coding region. The 5′ homology is adjacent to and includes the ORF start codon while the 3′ homology is adjacent to and includes the ORF stop codon. Thus a precise disruption of each ORF was made. Additionally, the long primers contain two unique gene-specific sequence tags on the 5′ (UPTAG) and 3′ (DOWNTAG) ends of each disruption cassette providing distinct "bar-code" identifiers for each gene disruption. As an example, the structure of the *ILV1* locus is shown in Fig. 7A along with the structure of the *ilv1* deletion strain produced by the consortium.

In Fig. 7 the UPTAG (checkered box) and DOWNTAG (black box) contain a unique 20 bp sequence specific to each strain flanked by standard 18 bp primer binding sites for amplification of these tags so that all the bar codes in a population can be PCR-amplified using the same pairs of primers. Hybridization of the amplified DNA to a microarray gene chip that contains the bar codes for all the gene disruption strains can be used to monitor the presence or absence of a particular disruption strain in a population of cells. This has been used to identify gene disruptions in a population that are enriched for or selected against in response to various growth conditions.[19]

[19] E. A. Winzeler, D. D. Shoemaker, A. Astromoff, H. Liang, K. Anderson, B. Andre, R. Bangham, R. Benito, J. D. Boeke, H. Bussey *et al., Science* **285,** 901 (1999).

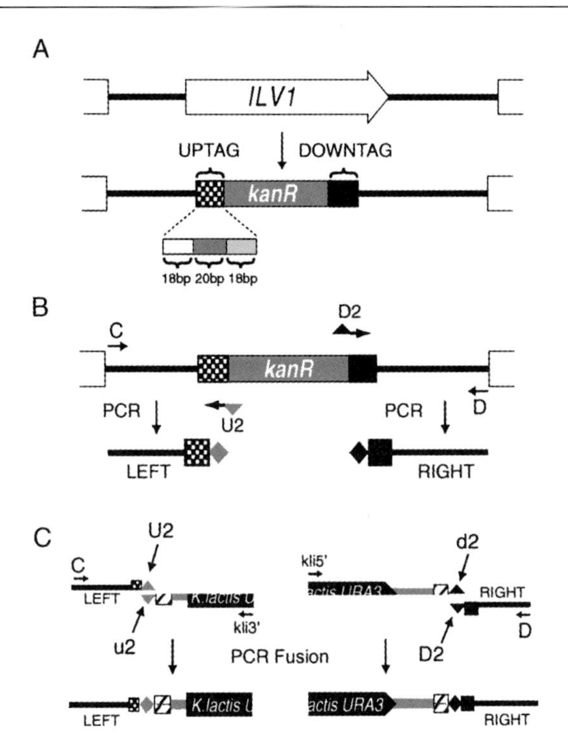

FIG. 7. Transfer of gene disruptions from consortium strains to a new strain. (A) Construction of a gene disruption by the yeast deletion consortium was performed using 45 bp homology to a gene of interest. *ILV1* is used as an example. Unique identifiers on the 5′ side of the ORF (UPTAG) and 3′ side of the ORF (DOWNTAG) are indicated by the checkered and black boxes. Each tag contains a unique 20 bp identifying tag flanked by common 18 bp primer binding sites for amplification of the unique sequence from any deletion strain. This is shown in detail for the UPTAG sequence (also see Table I for primer sequences). (B) The 18 bp primer binding sites flanking the unique identifying tags are used in combination with appropriate intergenic primers to amplify the identifying tag along with the intergenic DNA (see text). (C) Fusion of selectable markers to intergenic regions is mediated by the 18 bp U2 or D2 adaptamers. [Adapted from R. J. D. Reid, I. Sunjevaric, M. Keddache, and R. Rothstein, *Yeast,* in press (2002).]

These types of population biology experiments are likely to be useful to a number of researchers who work with other yeast strain backgrounds. Therefore we have developed an adaptamer gene disruption method with the specific goal of transferring gene disruptions containing the unique tags into a new strain. We have moved a number of these tagged gene disruptions into the W303 strain background[20] and find that the method is as efficient as the general method described above.

[20] B. J. Thomas and R. Rothstein, *Genetics* **123,** 725 (1989).

A simple way to transfer the gene disruptions from the consortium strains to a new strain is to choose primers that PCR amplify across the gene disruption to produce a DNA molecule containing the selectable marker, the unique tags, and sufficient homologous DNA to effect recombination. The DNA can then be transformed to select for G418-resistant recombinants. However, the KanMX selectable marker is not recyclable and thus cannot be reused if additional disruptions need to be performed in that strain.

A simple adaptation of the adaptamer-directed gene disruption method described above allows transfer of the bar-coded gene disruptions into a new strain while replacing the kanamycin resistance cassette with the recyclable *K. lactis URA3* marker. As shown in Fig. 7B, the intergenic regions flanking the gene disruption are amplified using intergenic adaptamers and the common primers flanking the UPTAG and DOWNTAG, U2 and D2, respectively. *K. lactis URA3* is amplified in overlapping sections as described above (see Fig. 4B). In this case, the 5′ sequence tags on the adaptamers are the reverse and complement of U2 and D2 and are referred to as u2 and d2, respectively. Figure 7C shows PCR fusion using these adaptamer sequences to generate DNA fragments for gene disruption.

UPTAG and DOWNTAG sequences in the deletion strains have the same orientation as the start and stop codons of the disrupted gene and the C and D intergenic adaptamers all have the same orientation along the chromosome. Therefore amplification of a Watson strand gene disruption (e.g., *MCD1* in Fig. 4) is accomplished using the C intergenic adaptamer with the U2 adaptamer and the D intergenic adaptamer with the D2 adaptamer. For a Crick strand gene disruption (e.g., *NTH1* in Figure 4), the C intergenic adaptamer is used with the D2 adaptamer and the D intergenic adaptamer is used with the U2 adaptamer.

Amplification of intergenic regions from the consortium strains is performed using the same PCR conditions described above. Most of the amplification of yeast intergenic regions in our lab have been performed using relatively clean DNA preparations.[10] However, it is also possible to perform these amplifications by colony PCR.[21] This is useful where gene disruptions are moved from many different strains.

Allele Replacements Using Adaptamers

The ability to introduce a specific mutant allele into the yeast genome is an invaluable genetic tool. A mutant allele in one strain can be moved into a different strain background by a standard genetic cross. However, this requires successive backcrosses to ensure that the new mutant strain becomes congenic to the desired background. Allele replacement by homologous recombination obviates the need for multiple backcrosses because the new allele is introduced directly into the

[21] A. C. Ward, *BioTechniques* **13,** 350 (1992).

desired strain. For novel mutations constructed *in vitro,* allele replacement methods facilitate the stable integration of the mutant allele into the genome.

Traditional allele replacement methods require that the mutant allele be subcloned into a yeast integrating vector that contains a selectable/counterselectable marker such as *URA3*. The integrating plasmid is linearized with a restriction enzyme that cuts once within the mutant allele. Transformation of this linear DNA into yeast effects homologous recombination to yield an integrated direct repeat in which a mutant and a wild-type copy of the targeted gene are separated by *URA3*. Direct repeat recombination pops out the marker, leaving a single allele. The efficiency of recovering the mutant allele is 50% or less and depends on the location of the mutation with respect to the ends of the DNA repeats. We have developed an adaptamer-based allele replacement technique in which mutant alleles are PCR amplified and fused to a selectable marker for genome integration.[22] A major advantage of the adaptamer method is the high probability that both integrated repeats contain the mutant lesion. This increases the efficiency with which the mutant allele is recovered after the selectable/counterselectable marker is recycled.

The adaptamer-based allele replacement method described below uses the complete ORF for DNA homology to target integration. This was designed to take advantage of the complete set of commercially available adaptamers that can amplify every yeast ORF (Research Genetics, Huntsville, AL).[23] All forward ORF adaptamers contain the 19 bp A sequence tag followed by 20 to 25 nucleotides of homology to a specific ORF starting at the ATG (see Table I). All reverse adaptamers contain the 20 bp B sequence tag followed by 20 to 25 nucleotides of homology at the 3′ end of the ORF and includes the stop codon. Amplification using these adaptamers produces a precise copy of the ORF with the A and B tags appended to the 5′ and 3′ ends, respectively (Fig. 8A).

The *K. lactis URA3* selectable marker is amplified in two overlapping sections as described earlier (Fig. 5C). Fusion of the selectable marker to the amplified ORF is performed in two parallel reactions. The 5′ *K. lactis URA3* fragment is fused to the ORF via the complementary b and B sequence tags while the 3′ *URA3* fragment is fused via the complementary a and A sequence tags (Fig. 8B). Integration into the genome is mediated by the homologous ORF DNA, while the *URA3* segments recombine to reconstitute the selectable marker (Fig. 8C).

There are three possible recombination products depending on the position of the crossovers between the incoming DNA and the chromosome. Either both alleles are incorporated (as shown in Fig. 8D) or one mutation is incorporated and the other is not (there are two types, one of which is shown in Fig. 8E). In fact it is most common to recover the class in which both integrated direct repeats contain the mutation. We have shown that when the transferred mutation lies $\geq$80 bp from

[22] N. Erdeniz, U. H. Mortensen, and R. Rothstein, *Genome Res.* **7,** 1174 (1997).

[23] J. R. Hudson, Jr., E. P. Dawson, K. L. Rushing, C. H. Jackson, D. Lockshon, D. Conover, C. Lanciault, J. R. Harris, S. J. Simmons, R. Rothstein, and S. Fields, *Genome Res.* **7,** 1169 (1997).

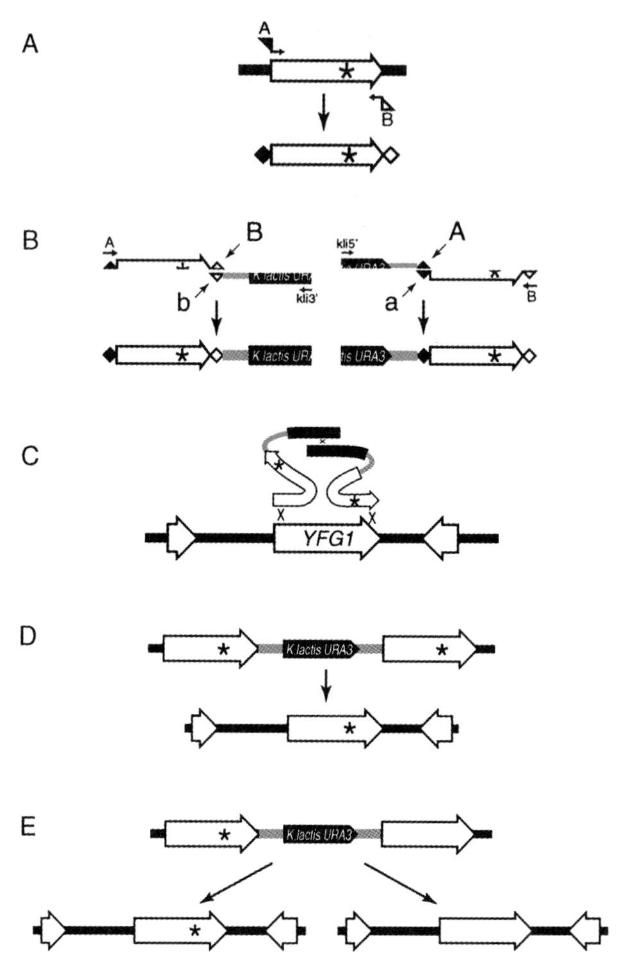

FIG. 8. Allele replacement using adaptamers. (A) Amplification of ORFs is accomplished using adaptamers designed to precisely amplify every yeast ORF from start to stop codons. The forward and reverse adaptamers contain 5′ sequence tags referred to as A and B, respectively. A mutation in the amplified ORF is indicated by an asterisk. (B) Fusion to a selectable marker is performed in two reactions generating DNAs with overlapping segments of the selectable marker. (C) Genome integration is mediated by homologous ORF sequences. (D) Integration producing two mutant copies of the ORF as direct repeats. Pop-out of the selectable marker results in a single mutated copy of the ORF in the genome. (E) Integration resulting in one mutant and one wild-type copy of the ORF. Pop-out of the selectable marker can result in the mutant or the wild-type copy of the allele integrated into the genome.

the end of the fragment, the crossover occurs between the mutation and the end of the DNA greater than 95% of the time.[22] This is most likely due to the fact that crossing over is stimulated at or near the ends of the incoming DNA.[24]

This adaptamer-based allele replacement method is an efficient way to introduce an allele into a new strain background without cloning in *E. coli*. The method as presented above uses complete ORFs as homologous DNA to promote genome integration. This is advantageous since the premade adaptamers for allele replacement of any ORF can be obtained at a modest cost. Furthermore, the *URA3*-marked intermediate maintains gene function as the ORF is not disrupted. This is important when alleles of essential genes must be transferred. The drawback to this allele replacement method is that random mutations can be introduced by PCR. Thus, it is important to limit the number of cycles of amplification and to use a high-fidelity polymerase. In cases where ORFs are long, it is prudent to design new adaptamers that limit the total length of amplified DNA. This is not necessarily problematic for allele replacement of essential genes. As long as the homologous DNA includes the N terminus and the promoter or the C terminus with the stop codon, a functional copy of the gene is maintained upon integration. Here as well, there is a high probability of incorporating the mutation into both repeats.

In-Frame Gene Fusions Using Adaptamers

Attaching functional epitopes to a gene can be an important route to understanding protein function. Fusion of immunogenic tags can facilitate protein purification and biochemical analysis of protein–protein interactions. Indirect immunofluorescence methods using such epitopes make it possible to study subcellular localization of a protein and even colocalization with other proteins provided that distinct immunogenic tags are available. Microhomology-based integration methods have been used to introduce such changes directly into the yeast genome, obviating the need for a cloned gene.[7] In fact, a set of plasmids has been constructed containing several kinds of useful epitopes adjacent to selectable markers.[5,25] Amplification of the epitopes and markers in these plasmids with primers containing homology to a gene can be used to integrate and construct epitope fusions in the yeast genome.

One drawback to introducing epitopes by the microhomology methods is that, along with the desired epitope, most of these protocols also integrate a selectable marker and affect gene structure. This imposes limits on how these epitopes can be used. For instance, insertions on the 5′ end of an ORF to produce an N-terminal protein fusion must also include a heterologous promoter to drive gene expression. If expression from the endogenous promoter is important, then it is necessary to use a C-terminal epitope fusion—which may or may not compromise protein

[24] R. Rothstein, *Cold Spring Harb. Symp. Quant. Biol.* **49,** 629 (1984).

[25] M. S. Longtine, A. McKenzie III, D. J. DeMarini, N. G. Shah, A. Wach, A. Brachat, P. Philippsen, and J. R. Pringle, *Yeast* **14,** 953 (1998).

function. This problem is solved by integrating the epitope as a direct repeat flanking a selectable marker.[26] Recombination between the repeats deletes the marker and other heterologous sequences. However, integrating epitopes using microhomology is still inefficient.

We have developed an adaptamer-directed method in which fusions can be made to any functional sequence without leaving a selectable marker in the genome. This is accomplished by using the epitope sequence as a direct repeat flanking the selectable marker. The epitope is fused to genomic DNA by first amplifying this DNA with appropriate adaptamers. We have thus far applied these techniques to produce fusions of cyan fluorescent protein (CFP) and yellow fluorescent protein (YFP) to Rad52 to investigate its relocalization in response to DNA damage.[27] Below we describe the method for adding C-terminal or N-terminal CFP and YFP tags to any gene.

The cloning of GFP from *Aequorea victoria* has provided a useful tool for cell biological studies in *S. cerevisiae* and other organisms.[28] Subsequent mutagenesis of GFP has resulted in a number of enhanced variants of the protein having altered fluorescent properties. In particular, blue- and red-shifted versions, CFP and YFP, respectively, that have nonoverlapping excitation and emission spectra have been produced. These fluorophores can be visualized independently in a living cell, thus facilitating colocalization experiments. In addition, with appropriate filters, CFP and YFP can potentially be used to study protein–protein interactions *in vivo* by means of fluorescence resonance energy transfer (FRET).[29] Here we utilize the W7 (CFP) and 10C (YFP) clones obtained from R. Tsien (University of California, San Diego, CA).[29,29a]

We have constructed two plasmids each for CFP and YFP that can be used to create fusions to any gene of interest (Fig. 9).[21,28,29,29b,30] One plasmid contains YFP next to the first two-thirds segment of *URA3*. The second plasmid contains the last two-thirds of *URA3* followed by the YFP ORF such that when the overlapping segments of *URA3* recombine, the YFP sequences are oriented as direct repeats (Fig. 10A). The method as illustrated in Fig. 10[16,21,22,26] shows an in-frame fusion of YFP to the 3′ end of the ORF. Two PCRs are performed to amplify the YFP-*URA3* DNA fragments. The first plasmid is amplified with the forward YFP primer GFPstart-F and the internal kli3′ *URA3* primer. The second plasmid is amplified with the internal kli5′ *URA3* primer and the reverse YFP primer GFPend-R. Parallel PCRs are performed to amplify a 200 to 300 bp section of the 3′ end of a coding

[26] B. L. Schneider, W. Seufert, B. Steiner, Q. H. Yang, and A. B. Futcher, *Yeast* **11,** 1265 (1995).

[27] M. Lisby, R. Rothstein, and U. H. Mortensen, *Proc. Natl. Acad. Sci. U.S.A.* **98,** 8276 (2001).

[28] J. Abelson, "Green Fluorescent Protein." Academic Press, San Diego, 1999.

[29] R. Heim and R. Y. Tsien, *Curr. Biol.* **6,** 178 (1996).

[29a] M. Ormo, A. B. Cubitt, K. Kallio, L. A. Gross, R. Y. Tsien, and S. J. Remington, *Science* **273,** 1392 (1996).

[29b] M. Keddache and R. Rothstein, personal communication, 1998.

[30] T. W. Christianson, R. S. Sikorski, M. Dante, J. H. Shero, and P. Hieter, *Gene* **110,** 119 (1992).

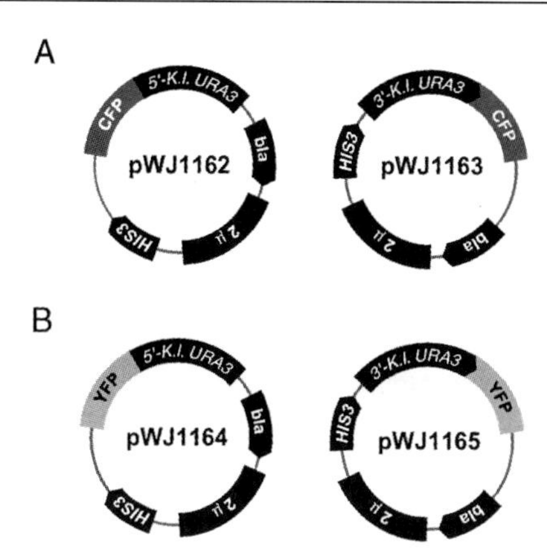

FIG. 9. Construction of CFP/YFP-tagging vectors. DNA sequences encoding either the blue- or red-shifted version (W7, 10C) of GFP were amplified by PCR from the corresponding pRSETB vectors [J. Abelson, "Green Fluorescent Protein." Academic Press, San Diego, 1999; R. Heim and R. Y. Tsien, *Curr. Biol.* **6,** 178 (1996)]. These DNA fragments were fused by PCR to either the 5′ or 3′ two-thirds of *K. lactis URA3*, which was amplified from pWJ716 [A. C. Ward, *BioTechniques* **13,** 350 (1992)]. The resulting PCR products were cloned into the SacII site of pRS423 [T. W. Christianson, R. S. Sikorski, M. Dante, J. H. Shero, and P. Hieter, *Gene* **110,** 119 (1992)] to make vectors for CFP/YFP-tagging. (A) Vector maps of pWJ1162 and pWJ1163 for CFP tagging. (B) Vector maps of pWJ1164 and pWJ1165 for YFP tagging.

region and 200 to 300 bp directly adjacent to that ORF. Adaptamers are used for each of these PCRs in order to fuse these segments to the YFP-*URA3* fragments in the second round of PCRs (Fig. 10B). When C-terminal fusions are made, the stop codon from the gene is included in the DNA amplified from the intergenic region and omitted in the DNA amplified from the ORF (see DFxTag2 adaptamer in Table I). PCR conditions are identical to those used for the gene disruption methods described above. Cotransformation and integration of these sequences produces a precise in-frame fusion of YFP to the targeted ORF. Subsequent recombination between YFP direct repeats pops out the marker, leaving only the inserted YFP coding sequences (Fig. 10C).

The above method describes C-terminal YFP fusions, but it can, in principle, be used to fuse any C-terminal epitope. For N-terminal or internal epitope fusions, the adaptamers must be designed so that fusion to the ORF maintains the desired reading frame. The PCRs and transformations described can easily be performed in 1 to 2 working days. In general, the success rate of the procedure is high with approximately 50% of the candidate clones being correct after pop-out of the *URA3* marker. Among the incorrect clones, about half lack an integration at the target site and the other half have acquired mutations during the PCRs, which underscores the importance of using a high-fidelity *Taq* polymerase. We also applied this method to short tags such as 6xHIS and FLAG. Because of the small size of these tags, a region of the flanking noncoding genomic DNA was incorporated as part of the

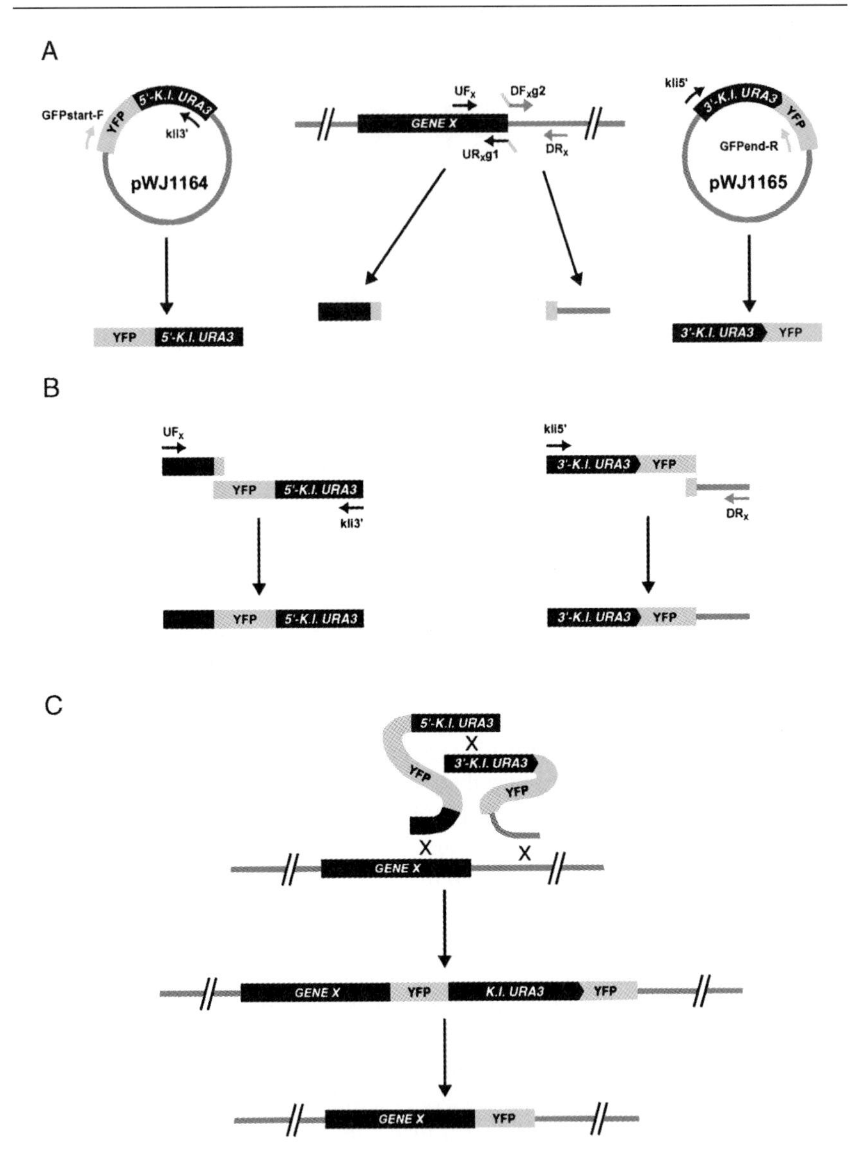

FIG. 10. General strategy for CFP/YFP-tagging of yeast proteins. Using appropriately designed primers, CFP and YFP can be targeted to any site in the yeast genome. This figure describes the fusion of YFP to the 3′ end of gene X. The procedure involves two rounds of PCR and a yeast transformation. (A) PCR amplification of target sequences. The first round of PCR amplifies 300 to 500 bp DNA sequences upstream and downstream of the target site using primer pairs UF$_X$/UR$_X$g1 and DF$_X$g2/DR$_X$, respectively. The UR$_X$g1 and DF$_X$g2 primers contain 19 bp 5′ sequence tags complementary to the 5′ and 3′ ends of YFP, respectively. Since YFP is fused to the 3′ end of gene X in this example, a *TAA* stop codon has been added to g2. The stop codon is omitted when making N-terminal and internal fusions of CFP/YFP. YFP-*URA3* sequence cassettes were PCR amplified from pWJ1164 and pWJ1165

direct repeat since efficient recombinational pop-out of the *URA3* marker requires a minimum of approximately 100 bp of DNA homology.[29b,31]

Summary

Each of the adaptamer-directed genome manipulation methods is predicated on the fact that recombination between two DNAs is enhanced by increasing the length of homology. Many of the current PCR-based genome manipulation techniques rely on very short homologies to promote recombination. In these cases homology length is dictated by the technical limits of oligonucleotide synthesis. Adaptamers circumvent this problem since long homology regions are produced in a first round of PCR, and then fused to the selectable marker in a second round of PCR via complementary sequence tags on the adaptamers. Furthermore, many of the techniques described here rely on preexisting and commercially available adaptamer sets that can be obtained inexpensively rather than designing new primers for every experiment. Although a cost is incurred when performing multiple PCR amplifications, the increase in recombination efficiency is dramatic. Finally, the adaptamer-mediated PCR fusion methodology is versatile and can be applied to varied genome manipulations.

Acknowledgments

We thank Marisa Wagner, Bilyana Georgieva, and Uffe Mortensen for critical reading of the manuscript. This work was supported by the National Institutes of Health Grant HG00193 to R. Reid, a grant from the Alfred Benzon Foundation to M. Lisby, the National Institutes of Health Grants HG01620 and GM50237 to R. Rothstein, and Merck Genome Research Institute Grant 5 to R. Rothstein.

[31] N. Sugawara and J. E. Haber, *Mol. Cell. Biol.* **12,** 563 (1992).

---

using primer pairs GFPstart-F/kli3′ and kli5′/GFPend-R. (B) Fusion of the YFP coding sequence to target sequences. The sequence tags (g1 and g2) facilitate the fusion of the target sequences to YFP-*URA3* sequences using adaptamer technology [A. C. Ward, *BioTechniques* **13,** 350 (1992); N. Erdeniz, U. H. Mortensen, and R. Rothstein, *Genome Res.* **7,** 1174 (1997)] and the primer pairs UF_X/kli3′ and kli5′/DR_X. Approximately 100 ng of each PCR fragment was used in the fusion reactions. (C) Integration by homologous recombination. The two PCR fragments (500 ng of each) were cotransformed into yeast for integration by homologous recombination using the lithium acetate method [C. Fairhead, B. Llorente, F. Denis, M. Soler, and B. Dujon, *Yeast* **12,** 1439 (1996)]. The recombination event results in a YFP direct repeat flanking an intact *K. lactis URA3* sequence which allows transformants to be selected on SC-Ura. Transformants were restreaked on SC-Ura and single colonies picked into 2 ml YPD and grown overnight before plating on 5-FOA to select for deletion of the *URA3* marker by pop-out recombination between the two flanking YFP sequences. A clean fusion of YFP to gene X is left in the genome. [Adapted from B. L. Schneider, W. Seufert, B. Steiner, Q. H. Yang, and A. B. Futcher, *Yeast* **11,** 1265 (1995).]

[16] *kar*-Mediated Plasmid Transfer between Yeast Strains: Alternative to Traditional Transformation Methods

By BILYANA GEORGIEVA and RODNEY ROTHSTEIN

Introduction

The budding yeast *Saccharomyces cerevisiae* is an excellent system for the study of many aspects of eukaryotic biology. Despite the diverse nature of the questions asked, almost all of these studies use molecular tools and often require the transfer of exogenous DNA into the yeast cell. Several techniques for high-efficiency transformation of plasmid DNA into yeast have been developed and used successfully over the years. These include spheroplast transformation, transformation by electroporation, and lithium acetate transformation.[1–3] One of the major drawbacks of these methods is the many and often time-consuming steps that are necessary to prepare the host strain, introduce the exogenous plasmid, and identify successful transformants. This becomes an important issue when the transformation of a plasmid into a large number of host strains is desired. Modifications that scale down the lithium acetate method have been successfully used for large-scale transformation of the same plasmid into a set of strains.[4] However, transformation efficiency is compromised and this method may not be useful for all purposes.

The problem becomes even more acute when multiple distinct plasmids need to be transferred from one strain to another since this requires the extra steps of DNA isolation from yeast and plasmid rescue in bacteria. In the age of genome-scale manipulations of yeast, an easier and less time-consuming method of plasmid introduction or transfer is desirable. Here we describe a method of *kar*-mediated transfer that allows the quick and efficient transfer of plasmids from one yeast strain to another or to a number of different strains. The method is based on yeast mating and uses a particular mutation in the karyogamy pathway, *kar1*.[5] *kar1* mutant strains initiate mating and proceed through conjugation without nuclear fusion, and therefore haploid progeny can be recovered.[5] During a *kar* mating heterokaryons with mixed cytoplasm are formed and nuclear DNA such as

[1] A. Hinnen, J. B. Hicks, and G. R. Fink, *Proc. Natl. Acad. Sci. U.S.A.* **75,** 1929 (1978).

[2] D. M. Becker and L. Guarente, *Methods Enzymol.* **194,** 182 (1991).

[3] R. H. Schiestl and R. D. Gietz, *Curr. Genet.* **16,** 339 (1989).

[4] P. Uetz, L. Giot, G. Cagney, T. A. Mansfield, R. S. Judson, J. R. Knight, D. Lockshon, V. Narayan, M. Srinivasan, P. Pochart, A. Qureshi-Emili, Y. Li, B. Godwin, D. Conover, T. Kalbfleisch, G. Vijayadamodar, M. Yang, M. Johnston, S. Fields, and J. M. Rothberg, *Nature* (*London*) **403,** 623 (2000).

[5] J. Conde and G. R. Fink, *Proc. Natl. Acad. Sci. U.S.A.* **73,** 3651 (1976).

0076-6879/02 $35.00

plasmids and whole chromosomes can be transferred between the two parents.[6,7] With the proper selection of nuclear genetic markers, plasmid transfer can be made directional.

Glossary

karyogamy	Process of nuclear congression and fusion during conjugation
kar mutants	Mutants defective in the karyogamy pathway
kar mating	Mating in which one of the parents carries a particular *kar* mutation, in this context, *kar1Δ15*
recipient strain	In a *kar* mating, the haploid parent (nucleus) that is selected as a host for the transferred material (plasmid)
donor strain	Haploid strain that carries the material (plasmid) that is transferred to the recipient via a *kar* mating
cytoductants	Haploid progeny from a *kar* mating with the nucleus of the recipient and mixed cytoplasm from both parents
chromoductants	Haploid progeny from a *kar* mating with the nucleus of the recipient and a chromosome(s) transferred from the donor
plasmoductants	Haploid progeny from a *kar* mating with the nucleus of the recipient and a plasmid from the donor

Background and Principles of Method

Most laboratory yeast strains exist as *MAT***a** or *MATα* haploids. A mating between *MAT***a** and *MATα* yeast produces diploids. This process begins when haploid cells of the opposite mating type adhere to one another after the mutual exchange of pheromones. This is followed by the orderly removal of cell walls and plasma membrane to complete cell fusion.[8] One characteristic feature of all fungi including yeast is that during the process of conjugation, the nuclear envelope remains intact and the resulting diploid nucleus is the product of the direct fusion of the two parental nuclei.[9]

A number of mutations that block the karyogamy pathway have been identified and studied. One such mutation is *kar1-1,* which prevents nuclear fusion in approximately 90–95% of matings.[5,10] *KAR1* is essential for mitotic growth because of its

[6] D. C. Sigurdson, M. E. Gaarder, and D. M. Livingston, *Mol. Gen. Genet.* **183,** 59 (1981).

[7] S. K. Dutcher, *Mol. Cell. Biol.* **1,** 245 (1981).

[8] L. Marsh and M. D. Rose, *in* "The Molecular and Cellular Biology of the Yeast *Saccharomyces*" (J. R. Broach, J. R. Pringle, and E. W. Jones, eds.), Vol. III, p. 827. Cold Spring Harbor Laboratory Press, Cold Spring Harbor, NY, 1997.

[9] B. Byers, *in* "The Molecular Biology of the Yeast *Saccharomyces:* Life Style and Inheritance" (J. N. Strathern, E. W. Jones, and J. R. Broach, eds.), p. 59. Cold Spring Harbor Laboratory Press, Cold Spring Harbor, NY, 1981.

[10] M. D. Rose and G. R. Fink, *Cell* **48,** 1047 (1987).

role in the initial stage of spindle pole body duplication.[10] Kar1 has three functional domains that separate its mitotic function from its role during nuclear fusion.[11] An amino-terminal domain is important for the nuclear fusion function and a deletion of this protein domain gives an allele, *kar1Δ15,* that is viable and defective only in nuclear fusion.[11] *kar1Δ15* mutation is unilateral in that its defect is observed even in a mating to a wild-type strain.[11]

Matings in which either parent is *kar1Δ15* mutant are unproductive since they generate diploid nuclei at a very low frequency.[11] The majority of the products from such matings are cytoductants and, in practice, they can be selected when the nuclei and the cytoplasm of each parent are marked genetically.[12,13] Another characteristic of *kar* mating, of particular interest to the technique described here, is the occasional transfer of genetic material from one nucleus to the other.[6,7] This property of *kar* mating has been used for the directional transfer of yeast artificial chromosomes from one strain to another host of interest.[14,15] We have developed this technique further for the directional transfer of plasmids from one yeast strain to another.

The most important prerequisite for a directional transfer of plasmids is the proper design of the strains. The goal is to select haploid progeny that contain the desired parental nucleus and that have acquired the plasmid of interest. One consideration is that the recipient be auxotrophic for the marker selecting the plasmid. The other consideration is that the parental genotypes allow selection for the recipient and against the donor nucleus. The *CAN1* and *CYH2* genes are suitable for this purpose. *CAN1* encodes the arginine permease and mutations that disrupt its function confer recessive resistance to canavanine, a toxic arginine analog.[16] *CYH2* encodes the essential L29 ribosomal protein and is the target of the peptidyltransferase inhibitor, cycloheximide.[17–19] Recessive mutations in *CYH2* confer resistance to high doses of this drug.[17] The recipient strain must carry recessive drug-resistance markers, e.g., $can1^R$ and $cyh2^R$, whereas the donor strain must have the corresponding dominant drug-sensitive alleles, $CAN1^S$ and $CYH2^S$. Selection on cycloheximide and canavanine-containing medium ensures that only the $can1^R$ $cyh2^R$ recipient nucleus survives in a *kar* mating. The configuration of markers just described not only allows selection of plasmoductants but also allows counter-selection of the plasmid donor nucleus and the rare diploids that arise during the *kar* mating. Use of two drug-sensitive markers is most efficient in selecting against

[11] E. A. Vallen, T. Y. Scherson, T. Roberts, K. van Zee, and M. D. Rose, *Cell* **69,** 505 (1992).
[12] V. Berlin, J. A. Brill, J. Truehart, J. D. Boeke, and G. R. Fink, *Methods Enzymol.* **194,** 774 (1991).
[13] M. D. Rose, *Annu. Rev. Microbiol.* **45,** 539 (1991).
[14] Y. Hugerat, F. Spencer, D. Zenvirth, and G. Simchen, *Genomics* **22,** 108 (1994).
[15] F. Spencer, Y. Hugerat, G. Simchen, O. Hurko, C. Connelly, and P. Hieter, *Genomics* **22,** 118 (1994).
[16] E. Gocke and T. R. Manney, *Genetics* **91,** 53 (1979).
[17] W. Stocklein and W. Piepersberg, *Antimicrob. Agents Chemother.* **18,** 863 (1980).
[18] H. M. Fried and J. R. Warner, *Nucleic Acids Res.* **10,** 3133 (1982).
[19] D. Cooper, D. V. Banthorpe, and D. Wilkie, *J. Mol. Biol.* **26,** 347 (1967).

the donor nucleus since the frequency of spontaneous mutation to double drug resistance is extremely low. Either parent can provide the *kar1Δ15* allele because of the unilateral nature of this mutation.

Strains, Media, and Drugs

Most of the experiments described here use strains with the W1588 background that is isogenic to W303 (*MAT***a** or α *ade2-1 can1-100 his3-11,15 leu2-3,112 trp1-1 ura3-1*),[20] but is *RAD5*. However, any haploid yeast strain that mates well and contains the appropriate genetic markers can be used. For example, the two-hybrid strain PJ69-4A (*MAT***a** *trp1-901 leu2-3,112 ura3-52 his3-200 gal4Δ gal80Δ LYS2::GAL1-HIS3 ade2-101::GAL2-ADE2 met2::GAL7-lacZ KAR1 CYH2 CAN1 RNR1*)[21] was successfully used as the donor strain in the experiment described in Fig. 1. Introduction of the genetic markers can be achieved through standard genetic manipulations or through allele replacement methods.[22–25] In addition, spontaneous canavanine or cycloheximide resistant mutants can be selected after plating $CAN1^S$ or $CYH2^S$ strains on drug-containing medium. This approach is quite easy for selecting can^R colonies as the frequency of mutation from $CAN1^S$ to $can1^R$ is approximately 10^{-6}. Selecting cyh^R colonies is not as easy because the mutation frequency from $CYH2^S$ to cyh^R is approximately 10^{-8}. This is due to the fact that spontaneous cyh^R mutants are almost always altered at a single amino acid in Cyh2 protein, Q38.[26–28] Selection for spontaneous cyh^R colonies occasionally results in slow-growing colonies and such candidates should not be used.

Because of the mating defect of the *kar1Δ15* mutant, genetic manipulations with this allele will be explained in some detail. Since *kar* matings are inefficient in generating diploids, one can "enrich" for diploids by extending the mating time during strain construction. Instead of the normal "overnight" incubation for matings of wild-type haploids, the incubation period for *kar1Δ15* strains should be extended to at least 2 days at 30° on rich medium. It is important that diploids be selected by complementation of recessive auxotrophic markers in the haploid parents. When a strain construction scheme requires several crosses for the

[20] B. J. Thomas and R. Rothstein, *Genetics* **123,** 725 (1989).

[21] P. James, J. Halladay, and E. A. Craig, *Genetics* **144,** 1425 (1996).

[22] F. Sherman, G. R. Fink, and J. B. Hicks, "Methods in Yeast Genetics." Cold Spring Harbor Laboratory Press, Cold Spring Harbor, NY, 1986.

[23] R. J. Rothstein, *Methods Enzymol.* **101,** 202 (1983).

[24] N. Erdeniz, U. H. Mortensen, and R. Rothstein, *Genome Res.* **7,** 1174 (1997).

[25] R. J. D. Reid, M. Lisby, and R. Rothstein, *Methods Enzymol.* **350,** [15], 2002 (this volume).

[26] W. Stocklein, W. Piepersberg, and A. Bock, *FEBS Lett.* **136,** 265 (1981).

[27] N. F. Kaufer, H. M. Fried, W. F. Schwindinger, M. Jasin, and J. R. Warner, *Nucleic Acids Res.* **11,** 3123 (1983).

[28] I. Sunjevaric, unpublished results, 2000.

introduction of genetic markers, it is helpful to add the *kar1Δ15* allele last. This avoids successive crosses requiring extended periods of mating due to the *kar1* mutation.

Compared to wild type, the mating deficiency of the *kar1Δ15* mutant provides an easy phenotype for scoring this mutation in tetrad analysis. In a time-limited mating assay,[10] plates with dissected tetrads are mated to appropriately marked *MAT***a** and *MATα* mating testers for 2–3 hr at 30° on YPD and subsequently replica plated on synthetic dextrose (SD) plates to select diploids. Because of the short incubation, only the *KAR1* spores undergo a robust mating and give rise to patches of complementing diploid clones on the SD plates. The *kar1Δ15* spores rarely produce diploids in this short mating and these are seen as occasional papillae on the SD plate. The mating type of the *kar1* mutant spores can be determined separately by mating the spore clones to the mating testers for 1–2 days before replicating them to the SD plates. After extended incubation, the *kar1Δ15* spores produce enough diploids to allow scoring of mating type by growth on SD plates.

In the experiments described in this chapter, all plasmids used in the transfers are 2μ-based vectors. In other experiments, we have shown that *CEN-ARS*-based vectors also transfer efficiently between strains via *kar* mating.

Standard yeast media and techniques for strain manipulations were used as described previously.[22] Because the method is based solely on mating and replica plating, the only equipment needed is a replicating block and sterile velvets. All matings are done at 30° on YPD medium. Freshly grown strains, 2–3 days old, give best results.

Cycloheximide and canavanine at final concentrations of 1 and 60 μg/ml, respectively, are added to autoclaved medium after cooling to 55°. Canavanine is a toxic arginine analog and it should be added to synthetic complete medium that lacks arginine (SC-Arg).

General Protocol

Depending on the experiment, *kar* transfer can be done from many donors into one recipient, or from one donor into many recipients. In either case, the multiple strains are patched on a master plate of the appropriate medium and incubated for 1–2 days until patches, approximately 6–8 mm in diameter, are formed. In the case of uneven growth, strains can be grown for a longer period of time to produce enough cells in each patch.[28] The single strain, regardless of whether it is the recipient or donor, is spread on the appropriate medium to create a uniform lawn that covers the entire surface of the plate. Subsequently this lawn can be propagated via replica plating. One lawn is used for mating to each master plate containing multiple strains. Mating is performed by replica-plating the two plates together on a fresh YPD plate and incubating for 6–12 hr at 30°. The time of mating can be varied such that the number of papillae seen on the drug selection medium is optimized. Because *kar1* mutants are impaired, but not completely deficient for mating, the shortest mating time that efficiently gives plasmoductants should be used. The matings are next replica-plated to SC-Arg+Can+Cyh medium that also lacks the nutrient for the prototrophic marker of the transferred plasmid and are grown for 4–5 days at 30°, or until plasmoductants are seen at the mating control patch. A second replica to selective medium is important to ensure that the nonplasmoductant cells on the plates are dead. This step eliminates background and it is very important if the plasmoductants are tested for another phenotype by an additional replica plating.

Transfer of Library or Set of Plasmids to Strain of Interest: Test for Complementation

In this section, we describe the *kar*-mediated transfer of multiple distinct plasmids to a single strain in a complementation experiment. A library of PCR-mutagenized *RNR1* in a *TRP1*-marked 2μ plasmid was initially tested for altered protein interactions in the two-hybrid host PJ69-4A.[29] In the next step, each individual plasmid from the mutagenized *rnr1* library was analyzed for functional complementation in a new host that carried a temperature-sensitive allele of the *RNR1* gene (Fig. 1). The two-hybrid plasmid donor strain was *MAT***a** *trp1* $CYH2^S$ $CAN1^S$, and the recipient strain was *MAT*α $kar1\Delta15$ *trp1* $cyh2^R$ $can1^R$ $rnr1^{ts}$. To perform the *kar* transfer of the mutagenized *rnr1* plasmids, the plasmid donor strains were patched on SC-Trp master plates and grown for 1–2 days. The recipient strain was spread as a uniform lawn on YPD plates. By replica-plating, each donor plate was mated for 6 hr at 30° to the recipient on a fresh YPD plate. The mating plates were next replica-plated to SC-Trp-Arg+Can+Cyh medium and grown for 4–5 days at 30° until papillae of plasmoductants were seen at the donor patches. These plasmoductants were replica-plated once more to SC-Trp-Arg+Can+Cyh medium and subsequently tested for functional complementation by analysis of their ability to grow after a temperature shift to 37°, the nonpermissive temperature for the $rnr1^{ts}$ allele.

The technique described above is a generic method to transfer any set of distinct plasmids into a single assay strain. For example, this method can be utilized for cloning by complementation. In this case, the donor is a set of strains carrying an ordered library of yeast sequences. The single recipient strain is a newly isolated unknown mutant. The arrayed library could then be used for complementation cloning of the mutant if its phenotype can be scored by replica plating. Since the initial step of arraying such a library is time consuming and laborious, it is worthwhile only if it will be used multiple times. Otherwise traditional cloning by transformation is more efficient.

[29] B. Georgieva, X. Zhao, and R. Rothstein, *Cold Spring Harb. Symp. Quant. Biol.* **65,** 343 (2000).

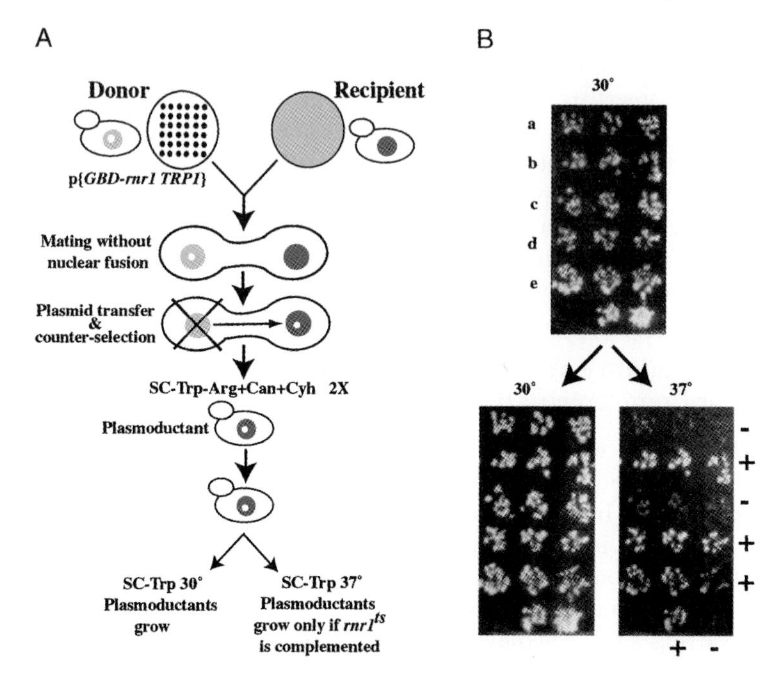

FIG. 1. *kar*-mediated plasmid library transfer and functional complementation test. (A) Schematic representation of the experimental design. The donor strain, PJ69-4A, is *MAT***a** *CAN1*^{S} *CYH2*^{S} and carries a library of PCR-mutagenized *rnr1* plasmids. The recipient strain is *MAT*α *kar1*Δ*15 can1*^{R} *cyh2*^{R} *rnr1*^{ts}. The plasmid is represented as a white dot. For complete experimental details see the text. (B) Complementation of a *rnr1*^{ts} allele. Each row, a–e, contains papillae of plasmoductants of three isolates of the same *rnr1* mutant candidate. The two patches on the bottom are controls: the left patch, indicated by "+" symbol, is a donor strain that carries a plasmid with the wild-type *RNR1* gene, a positive control for complementation; the right patch, indicated by a "−" symbol, is a donor strain with the empty vector, a negative control for complementation. The rows marked with a "+" show complementation (b, d, and e), while the rows marked with a "−" indicate lack of complementation (a and c).

Transfer of Same Plasmid to Set of Strains: Test for Synthetic Phenotype

This section describes the use of *kar*-mediated plasmid transfer when moving the same plasmid into a number of different strains. This is very useful in mutant screens that are based on a plasmid-borne assay or phenotype, e.g., mutant searches for synthetic lethal[30] (SL) or "synthetic viable" (SV) interactions.[31] *S*ynthetic *l*ethal *m*utations (*slm*) cause a synergistic growth defect under certain

[30] J. E. Kranz and C. Holm, *Proc. Natl. Acad. Sci. U.S.A.* **87,** 6629 (1990).

[31] X. Zhao, personal communication, 1998.

conditions when combined with a preexisting mutation, for example, *y*our *f*avorite *g*ene (*yfg*Δ). Typically such screens rely on a strain that lacks the chromosomal copy of *YFG* (*yfg*Δ) but contains a plasmid that harbors the wild-type copy of *YFG* and a marker that is both selectable and counterselectable, e.g., *URA3* (p{*YFG URA3*}). After mutagenesis and the initial round of selection, colonies that are incapable of losing p{*YFG URA3*} are isolated. However, it is important to show that the *slm*-dependent plasmid retention is due to *YFG* and not to any other elements on the plasmid. This requires demonstration that the *slm* phenotype can be reproduced with a new plasmid carrying *YFG* and marked with something other than *URA3,* e.g., p{*YFG LEU2*}. The new p{*YFG LEU2*} as well as the empty vector p{*LEU2*} must be introduced in parallel in each *slm* candidate. Upon counterselection against p{*YFG URA3*}, a true *slm* strain can lose p{*YFG URA3*} in the presence of p{*YFG LEU2*}, but not in the presence of p{*LEU2*}. Traditionally, p{*YFG LEU2*} and p{*LEU2*} are introduced through transformation methods. Depending on the scale of the screen, the number of transformation reactions can be several hundred, thus making the analysis quite laborious.

The method of *kar*-mediated plasmid transfer simplifies the problem of multiple transformations of the same plasmid by providing an alternative means for plasmid introduction. Through two parallel matings and a few steps of replica-plating, both the p{*YFG LEU2*} plasmid and the empty vector p{*LEU2*} can be transferred from two different donor strains to a large number of *slm* candidates. A scheme that depicts the *kar* plasmid transfer as a step in a SL screen is shown in Fig. 2. The *URA3* gene in the starting plasmid p{*YFG URA3*} allows selection on SC-Ura and counterselection on 5-fluoroorotic acid (5-FOA).[32] Each master plate with *slm* candidates is mated independently to two donor strains that carry either the p{*YFG LEU2*} plasmid or the p{*LEU2*} empty vector. Two successive rounds of replica-plating on SC-Leu-Arg+Cyh+Can medium selects plasmoductants. To test the synthetic lethal interaction, the two plates with plasmoductants of either the p{*YFG LEU2*} or the empty vector are replica-plated to SC-Leu+5-FOA to shuffle out the starting p{*YFG URA3*} plasmid.[33] There are several classes of mutant candidates: Because true *slm* candidates require the function of *YFG*, only plasmoductants that transferred the p{*YFG LEU2*} plasmid, but not the empty vector, grow on 5-FOA-containing medium. On the other hand, plasmoductants of false *slm* candidates grow regardless of whether they transferred the p{*YFG LEU2*} plasmid or the empty vector. Any candidates that successfully transferred the p{*YFG LEU2*} plasmid and the empty vector, but fail to grow on SC-Leu+5-FOA plates after counterselection of the p{*YFG URA3*} plasmid, represent mutants that have SL interaction with other elements on the starting p{*YFG URA3*} plasmid and are of no interest. Finally, it is possible that a *slm*

[32] J. D. Boeke, F. Lacroute, and G. R. Fink, *Mol. Gen. Genet.* **197,** 342 (1984).

[33] R. S. Sikorski and J. D. Boeke, *Methods Enzymol.* **194,** 302 (1991).

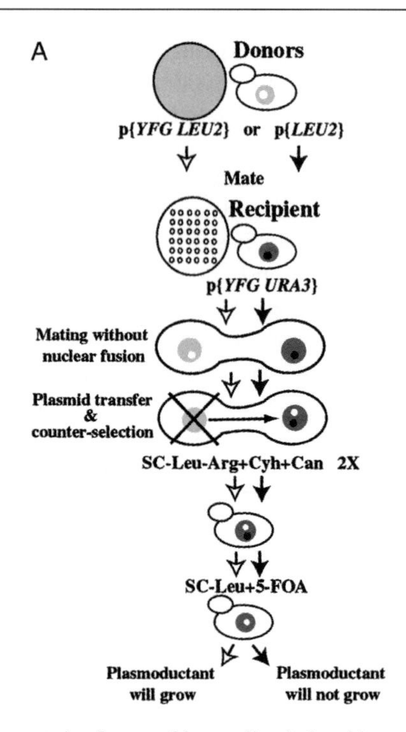

FIG. 2. Schematic representation for use of *kar*-mediated plasmid transfer as a step in a synthetic lethal (SL) screen. The recipient strains are SL candidates of the genotype *yfg*Δ *slm can1*^{R} *cyh2*^{R}. After mutagenesis, these candidates require the plasmid p{*YFG URA3*} for viability (represented as a black dot). The genotype of the donor strain is *kar1*Δ*15 CAN1*^{S} *CYH2*^{S}. In this experiment there are two different donor strains, one that carries the new plasmid p{*YFG LEU2*} and another one that carries the empty vector p{*LEU2*} (both represented as a white dot). Two separate matings between the recipients and each donor strain are made on YPD. One mating is to the donor strain carrying the p{*YFG LEU2*} plasmid and the other is to the donor strain with the vector p{*LEU2*} (designated by the open and filled arrowheads, respectively). Successful plasmoductants are selected after two rounds of replica plating to SC-Leu-Arg+Cyh+Can. Plasmoductants from each mating are counterselected on SC-Leu+5-FOA medium to assay for their ability to grow in the absence of the original plasmid p{*YFG URA3*}. If the SL interaction is real, only plasmoductants that contain the new plasmid p{*YFG LEU2*} but not the empty vector p{*LEU2*} can grow after shuffling out p{*YFG URA3*}.

or a nonspecific mutation renders the candidate mating-incompetent and plasmoductants cannot be recovered. In this case traditional transformation methods are used.

A similar scheme can be employed to facilitate screening for mutations that show an SV interaction. In many cases a loss-of-function mutation in an essential gene is inviable unless another mutation in a different gene bypasses its defect. This second mutation is called a suppressor when it has been identified *after* the original mutation in the essential gene. However, if a potential suppressor mutation is *first* introduced into a strain and mutations of essential genes are sought that are now viable, this interaction is said to be "synthetic viable." Mutations in essential genes that show such interaction are called *s*ynthetic *v*iable *m*utation (*svm*) mutants. Another way of thinking about SV interactions is that the removal of one gene's function permits growth of an otherwise inviable mutation in an essential gene.

An experiment to identify *svm* mutants can be based on a plasmid assay. In this scheme, mutants are sought that show synthetic viable interaction with deletion of a potential suppressor, e.g., *y*our *f*avorite *s*uppressor (*yfs*Δ). The starting strain is *yfs*Δ $can1^R$ $cyh2^R$ and carries a plasmid with *YFS* and a color marker such as *ADE3,* which allows plasmid visualization in an *ade2 ade3* background. For example, in one common design, yeast cells that maintain the plasmid grow to form red colonies while those that lose the plasmid are white.[30] After mutagenesis of the starting strain, *yfs*Δ *svm* candidates must lose the plasmid with *YFS* to be alive. Since the plasmid is not essential for viability, it could be lost for many other reasons following mutagenesis. Random loss vs obligatory loss of the plasmid is determined based on colony color. Red–white sectored colonies denote random loss, whereas solid white colonies indicate obligatory loss. However, it is important to show that the observed obligatory plasmid loss is due to a SV interaction with *YFS.* If mutant candidates are transformed with a new plasmid with *YFS,* e.g., p{*YFS URA3*}, or the corresponding empty vector, p{*URA3*}, true *yfs*Δ *svm* mutants give transformants only with the empty vector and not with the p{*YFS URA3*}plasmid. This analysis requires two transformation reactions per *yfs*Δ *svm* candidate.

A *kar*-mediated plasmid transfer simplifies the introduction of the new p{*YFS URA3*} plasmid and the empty vector plasmid described above. In contrast to the SL mutants, true *svm* candidates will not grow when provided with a wild-type copy of *YFS* but will grow when provided with the empty vector (Fig. 3). In this experiment, recipient *yfs*Δ *svm* candidates are patched in duplicate on the master plate. The two donor strains of the p{*YFS URA3*} plasmid and of the vector control are spread as separate lawns on SC-Ura plates. Matings between the recipient and the two donors are made on YPD for 12–14 hr at 30°. Plasmoductants are analyzed for growth after two successive rounds of selection on SC-Ura-Arg+Cyh+Can. Acting as a control, patches that give rise to p{*URA3*} plasmoductants demonstrate the candidate's ability to mate and to stably maintain a plasmid. Furthermore, patches that fail to give plasmoductants with the p{*YFS URA3*} plasmid identify true *yfs*Δ *svm* strain. On the other hand, strains that give plasmoductants with both plasmids represent false *yfs*Δ *svm* candidates.

In Fig. 3, only the control patches show a true *yfs*Δ *svm* mutant phenotype since these strains give plasmoductants only with the empty vector but not with p{*YFS URA3*}. All other patches are false *yfs*Δ *svm* candidates since they grow after transferring either plasmid. The duplicate patches that did not give plasmoductants with either plasmid may represent a nonmating *yfs*Δ *svm* candidate or a

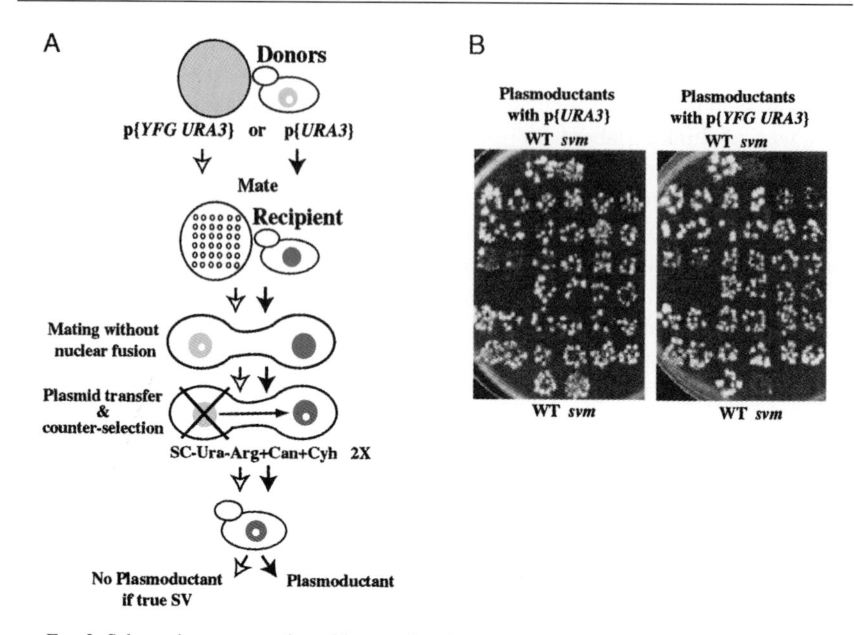

FIG. 3. Schematic representation of *kar*-mediated transfer in a screen for SV mutants. (A) Experimental design. The recipient strains are SV candidates of the genotype *MAT***a** *yfs*Δ *svm can1*^{R} *cyh2*^{R} that showed obligatory loss of a plasmid containing *YFG*. These candidates were patched in duplicate and mated independently to donor strains carrying either the p{*YFG URA3*} plasmid (open arrowhead) or the empty vector p{*URA3*} (filled arrowhead; both plasmids are represented as a white dot). The genotype of the donor strains was *MAT*α *kar1*Δ*15 CAN1*^{S} *CYH2*^{S}. Matings were done for 12–14 hr at 30° on YPD. After two rounds of selection on SC-Ura-Arg+Can+Cyh, plasmoductants were analyzed for their ability to grow after transfer of the p{*YFG URA3*} plasmid or the empty vector. True *yfs*Δ *svm* candidates cannot grow after transferring p{*YFG URA3*}, whereas they can grow after transferring the empty vector. (B) Results from an experiment of *kar*-mediated plasmid transfer in a screen for SV mutants. In this experiment *yfs*Δ *svm* candidates were tested for their ability to grow when given the plasmid p{*YFG URA3*} (right photo) or the empty vector p{*URA3*} (left photo). There are two patches for each *yfs*Δ *svm* candidate. In addition, the two patches on the top and the bottom of the plate are controls for the experiment. The left control patch is a wild-type strain that tests for mating and plasmoduction and that grows when either plasmid is transferred. The control patch on the right shows a strain with a known SV interaction with *yfs*Δ. This is a true *svm* candidate because it can grow after transfer of the empty vector (photo on the left), but not after transfer of the p{*YFG URA3*} plasmid (photo on the right). The two patches unable to form plasmoductants with either plasmid (left middle of the photo) probably represent a strain with a mating defect or plasmid stability mutation.

strain with a plasmid stability mutation. Such candidates can be examined using traditional transformation methods (for the nonmaters) or by standard genetic analysis after a cross to a *YFS* strain. In this experiment, *kar*-mediated transfer was used successfully to rigorously test the validity of a presumptive synthetic viable interaction with *yfs*Δ. This analysis helped eliminate further examination of false *yfs*Δ *svm* candidates.

Considerations

The process of *kar*-mediated plasmid transfer has one inherent problem. Transfer of yeast chromosomes from the donor nucleus to the recipient nucleus is observed at a frequency of 2–5% for individual chromosomes.[7,15] Therefore plasmoductants may also contain extra chromosomes. This is only a problem if single plasmoductant colonies are picked, propagated, and used in further experiments. For the applications described in this chapter, any potential chromosome transfer is irrelevant for several reasons. First, in all cases, the experimental conditions provide an opportunity to look at multiple independent plasmoductants in a patch. Second, in all cases, plasmoductants in at least two individual patches are observed to score the desired phenotype. By meeting these two criteria, the possibility that the observed phenotype is affected by potential chromosome transfer events is virtually eliminated. Third and most importantly, the resulting plasmoductants are not used for further experimentation since the plasmids of interest are rescued from the original donor strain.

Summary and Concluding Remarks

We have described a method, based entirely on yeast mating and replica plating, that allows the transfer of plasmids from one strain to another and eliminates the need for DNA isolation and transformation. Because this method uses a *kar1* mutation defective only in nuclear fusion, as well as markers that select against rare diploids, only haploid progeny are recovered. The method of *kar*-mediated plasmid transfer is simple, fast, efficient, and does not require any special equipment or elaborate technical manipulations. In addition, it allows the transfer of plasmids into strains that normally do not transform well. Furthermore, depending on the strain background, only minor genetic changes of the strain may be necessary to meet the required marker criteria. Since the donor and recipient strains do not have to be in a common genetic background, this provides the flexibility to transfer plasmids between different laboratory strains. Because of these features, *kar*-mediated transfer is an excellent solution for plasmid library transfers or for the repeated transfer of the same plasmid into a number of host strains.

Acknowledgments

We thank Marisa Wagner, Robert Reid, Uffe Mortensen, Michael Lisby, Justin Weinstein, and Erika Shore for critical reading of this manuscript. We thank Ivana Sunjevaric and Xiaolan Zhao for permission to cite unpublished material. This work was supported by Grants HG01620 and GM50237 from the National Institutes of Health.

[17] Gene Disruption

By MARK JOHNSTON, LINDA RILES, and JOHANNES H. HEGEMANN

Introduction

Gene inactivation is usually the first step one takes on the road to dissection of gene and protein function. In addition to revealing the phenotype caused by loss of gene function, gene deletion mutants provide reagents for testing and manipulating altered versions of a gene, and for further genetic analysis. Methods for disrupting genes were one of the first additions to the yeast geneticist's toolbox following development of DNA-mediated transformation of the organism.[1]

The ability to engineer specific DNA sequence changes into the yeast genome is enabled by the fact that nearly all DNA recombination in yeast is homologous (i.e., occurring between identical or nearly identical sequences).[2,3] Thus, an incoming DNA molecule is targeted with high fidelity to its corresponding location in the genome. This phenomenon was first exploited to introduce mutations into the genome in two steps: a circular plasmid carrying the altered gene was first integrated into the genome by homologous recombination, generating transformants carrying two copies of the target gene separated by plasmid sequences (Fig. 1A). This was followed by removal of the plasmid sequences, along with the wild-type copy of the gene, by a second homologous recombination event between the duplicated sequences, leaving the altered copy of the gene behind in the genome.[1]

A major advance soon followed with the discovery that homologous recombination is greatly stimulated by a double-strand break in the incoming yeast DNA sequence.[4,5] This led to development of the simple one-step gene disruption technique (Fig. 1B), in which a selectable gene is inserted into a linear DNA fragment carrying the gene to be disrupted. The DNA flanking each side of the selectable gene recombines with the genome, inserting the selectable gene into the target gene, thereby disrupting it.[6] This technique, which stood us in good stead for about a decade, was significantly improved when it was realized that only a very small amount of yeast DNA sequence is necessary for homologous recombination (fewer than 50 base pairs!). (In retrospect, it seems this possibility should have been recognized earlier, since we knew for some time that very short sequences

[1] S. Scherer and R. W. Davis, *Proc. Natl. Acad. Sci. U.S.A.* **76,** 4951 (1979).
[2] R. H. Schiestl, M. Dominska, and T. D. Petes, *Mol. Cell. Biol.* **13,** 2697 (1993).
[3] R. H. Schiestl and T. D. Petes, *Proc. Natl. Acad. Sci. U.S.A.* **88,** 7585 (1991).
[4] T. L. Orr-Weaver, J. W. Szostak, and R. J. Rothstein, *Proc. Natl. Acad. Sci. U.S.A.* **78,** 6354 (1981).
[5] T. L. Orr-Weaver, J. W. Szostak, and R. J. Rothstein, *Methods Enzymol.* **101,** 228 (1983).
[6] R. J. Rothstein, *Methods Enzymol.* **101,** 202 (1983).

0076-6879/02 $35.00

A. Two-step gene disruption

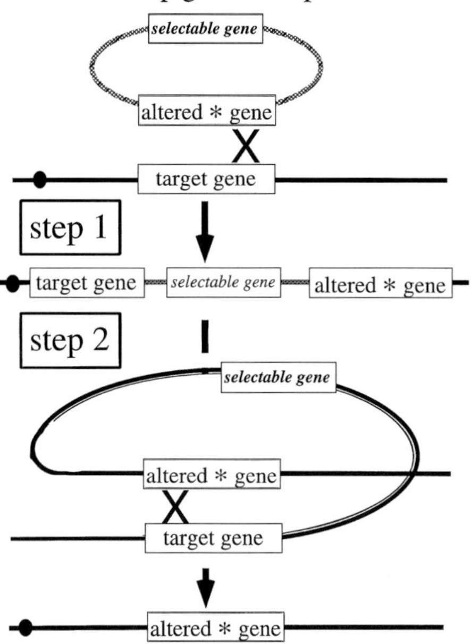

B. One-step gene disruption

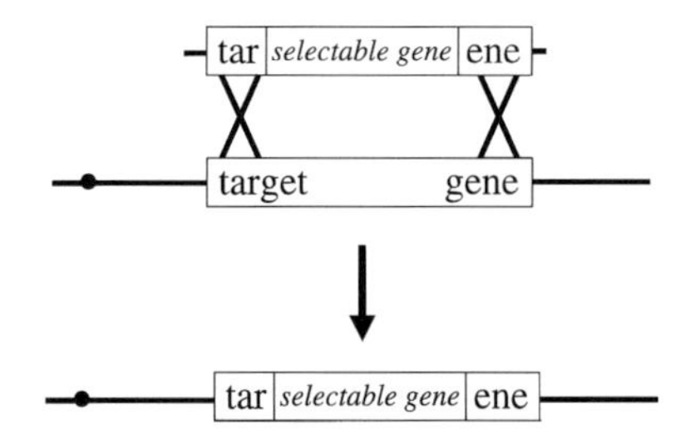

FIG. 1. (A) Two-step gene disruption. An altered form of the target gene (e.g., carrying a single base mutation) present on a circular plasmid with a selectable gene (e.g., *URA3*) is integrated into the yeast genome by homologous recombination, selecting for inheritance of the selectable gene (e.g., Ura^+). This generates a duplication of the target gene, with one copy carrying the mutation, the two copies separated by the plasmid sequences, and the selectable gene. In the second step, loss of the gene on the plasmid is screened or selected for (e.g., by selecting loss of *URA3* by resistance to 5-FOA). This occurs by a second homologous recombination, which, if it occurs on the other side of the mutation from the first recombination, leaves the mutation behind in the genome (a recombination event on the same side of the mutation as the first recombination simply reverses the plasmid integration event, leaving the target gene in the genome unaltered). (B) One-step gene disruption. The target gene has been interrupted by a selectable gene and released from a plasmid by cleaving at restriction endonuclease sites flanking the gene. Recombination between the ends of this DNA fragment {which are recombinagenic [T. L. Orr-Weaver, J. W. Szostak, and R. J. Rothstein, *Proc. Natl. Acad. Sci. U.S.A.* **78,** 6354 (1981); T. L. Orr-Weaver, J. W. Szostak, and R. J. Rothstein, *Methods Enzymol.* **101,** 228 (1983)]} replaces the target gene with the disrupted gene.

can engage in homologous recombination with the yeast genome.[7]) This made it possible to generate the DNA fragment required for disruption of a gene by the PCR (polymerase chain reaction), amplifying the selectable gene with oligonucleotides that include 30–50 base pairs of sequence flanking the sequences to be deleted.[8,9] This quick and simple method, called PCR-mediated gene disruption, quickly took root in the yeast community[10,11] and has since been used on a large scale.[12–14] It has been employed to delete nearly all of the approximately 6000 genes in the yeast genome (see below). The same method has also been used to disrupt genes in several non-*Saccharomyces* yeast species.[15–20]

Scope of This Article

Because of its ease and precision, PCR-mediated gene disruption is the preferred method for disrupting and deleting genes in yeast. We will therefore limit this article to a description of this technique. Essentially identical methods can be used to modify proteins and genes in various ways (e.g., tag proteins with epitopes, place a gene under the control of a regulated promoter), and these methods and the available reagents are described in another article in this volume.[20a] Readers interested in earlier incarnations of the one-step gene disruption method should refer to its excellent description in a previous volume in this series.[21]

We will first present the principles of gene disruption techniques before describing in detail the methods and protocols.

General Approach

To disrupt a gene, one simply transforms yeast cells with a DNA fragment consisting of a "gene disruption cassette" that provides a selectable phenotype

[7] R. P. Moerschell, S. Tsunasawa, and F. Sherman, *Proc. Natl. Acad. Sci. U.S.A.* **85,** 524 (1988).
[8] J. McElver and S. Weber, *Yeast* **8,** S627 (1992).
[9] A. Baudin, O. Ozier-Kalogeropoulos, A. Denouel, F. Lacroute, and C. Cullin, *Nucleic Acids Res.* **21,** 3329 (1993).
[10] F. Langle-Rouault and E. Jacobs, *Nucleic Acids Res.* **23,** 3079 (1995).
[11] M. C. Lorenz, R. S. Muir, E. Lim, J. McElver, S. C. Weber and J. Heitman, *Gene* **158,** 113 (1995).
[12] E. A. Winzeler, D. D. Shoemaker, A. Astromoff, H. Liang, K. Anderson, B. Andre, R. Bangham, R. Benito, J. D. Boeke, H. Bussey *et al., Science* **285,** 901 (1999).
[13] R. Niedenthal, L. Riles, U. Guldener, S. Klein, M. Johnston, and J. H. Hegemann, *Yeast* **15,** 1775 (1999).
[14] K.-D. Entian, T. Schuster, J. H. Hegemann, D. Becher, H. Feldmann, U. Güldener, R. Götz, M. Hansen, C. P. Hollenberg, G. Jansen *et al., Mol. Gen. Genet.* **262,** 683 (1999).
[15] R. Kaur, S. S. Ingavale, and A. K. Bachhawat, *Nucleic Acids Res.* **25,** 1080 (1997).
[16] R. B. Wilson, D. Davis, and A. P. Mitchell, *J. Bacteriol.* **181,** 1868 (1999).
[17] R. B. Wilson, D. Davis, B. M. Enloe, and A. P. Mitchell, *Yeast* **16,** 65 (2000).
[18] R. de Hoogt, W. H. Luyten, R. Contreras, and M. D. De Backer, *Biotechniques* **28,** 1112 (2000).
[19] C. Gonzalez, G. Perdomo, P. Tejera, N. Brito, and J. M. Siverio, *Yeast* **15,** 1323 (1999).
[20] J. Wendland, Y. Ayad-Durieux, P. Knechtle, C. Rebischung, and P. Philippsen, *Gene* **242,** 381 (2000).
[20a] M. E. Petracek and M. S. Longtine, *Methods Enzymol.* **350,** [25], 2002 (this volume).
[21] R. Rothstein, *Methods Enzymol.* **194,** 281 (1991).

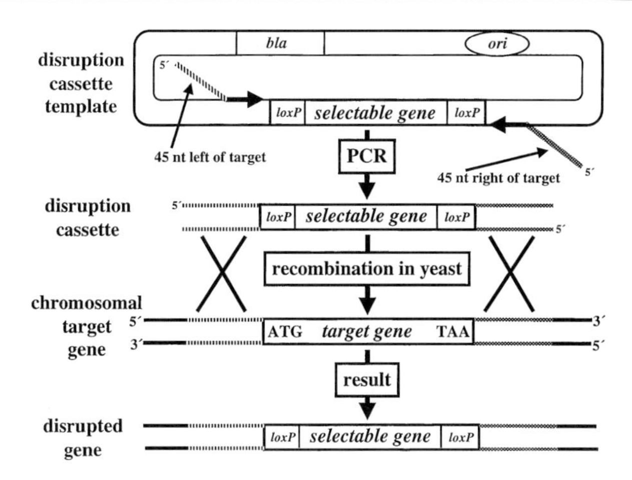

FIG. 2. Gene disruption approach. The gene disruption cassette (a gene whose inheritance can be selected for in yeast, usually flanked by sequences such as *loxP*, which allow its subsequent removal from the yeast genome) is amplified in the PCR using primers whose 5′ sequence are the 30–50 nucleotides that flank the target gene (dashed and stippled lines). The linear disruption cassette is introduced into yeast cells, where it recombines with the chromosome, replacing the target gene.

(usually drug resistance or prototrophy), surrounded by 30–50 base pairs of sequence flanking the sequence to be deleted. This DNA fragment is produced in the PCR using oligonucelotides with their 3′ 18–22 nucleotides complementary to sequences in the template flanking the disruption cassette, and their 5′ 30–50 nucleotides being from one or the other side of the sequence to be deleted (Fig. 2). The PCR product can be used directly (i.e., without purification) to transform yeast cells, and recombinants that have inherited the disruption cassette are selected. Cells that have correctly integrated the disruption cassette are identified by detecting PCR products generated using primers complementary to sequence within the cassette ("cassette B" and "cassette C" in Fig. 3) and primers flanking the site of integration of the selectable marker ("A" and "D" in Fig. 3). A PCR product of the expected size will be obtained only if the disruption cassette inserted into the genome by homologous recombination.

Requirements and Properties of the Method

Because only a small amount of sequence flanking the disruption cassette is available for homologous recombination, it is desirable to prevent recombination between the chromosome and sequences within the selectable gene on the disruption cassette. If this is not done, in the majority of transformants the disruption cassette will have recombined with the chromosomal copy of the selectable gene and repaired its mutation without disrupting the target gene.[22] The simplest and

[22] P. Manivasakam, S. C. Weber, J. McElver, and R. H. Schiestl, *Nucleic Acids Res.* **23,** 2799 (1995).

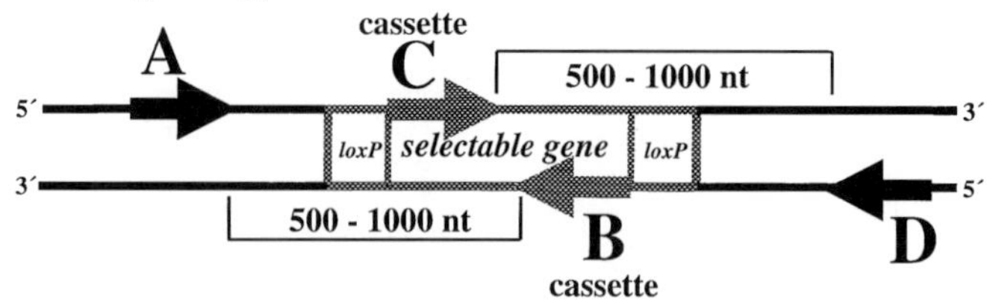

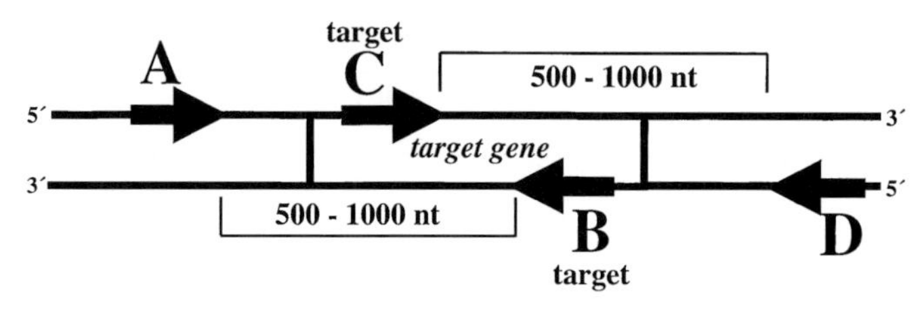

FIG. 3. Verification of gene disruption. A successful gene disruption is verified by testing for PCR products using the oligonucleotide primers shown with DNA (or cells) of the disruption mutant as template. Primer pairs A + cassette B and cassette C + D will yield a PCR product only if the gene is correctly disrupted. The deletion of the target should also be verified by confirming that primer pairs A + target B and target C + D do not yield a PCR product. The oligonucleotides should be chosen such that the PCR products are between 500 and 1000 bp in length. *S. cerevisiae* chromosomal sequences are indicated by a solid black line; the gene disruption cassette is indicated with stippled lines.

best way to avoid this is to use one of the disruption cassettes listed in Table I (see p. 296) that contain no *S. cerevisiae* DNA sequences (indicated in boldface type in Table I). Several of these with different selectable genes encoding resistance to various drugs or prototrophy for various amino acids and nucleotide bases are available, and they should suffice for nearly all purposes. The plasmids indicated in bold in Table I on page 296, plus the plasmids indicated in bold in Table IV are available from EUROSCARF as a set with the name "DEL-MARKER-SET."

If one insists on using a disruption cassette containing *S. cerevisiae* sequences, recombination between the disruption cassette and its chromosomal counterpart can be prevented by disrupting the gene in a yeast strain missing all sequences present in the disruption cassette. Yeast strains are available that carry deletion mutations that remove more sequence than is present in several popular gene disruption cassettes (Table II, see p. 298).

The relatively low frequency of recombination provided by the short sequences flanking the disruption cassette also means that if the template used in the PCR to produce the disruption cassette is a plasmid that is able to replicate autonomously in yeast cells (because it contains an ARS sequence), enough of it is present in the PCR that is used to transform yeast cells that a significant number of the transformants (sometimes a majority) will inherit the plasmid rather than the disruption cassette. Although this can be minimized by cleaving the plasmid with a restriction enzyme that does not cut within the disruption cassette, a better solution is to use as template a plasmid that is not able to replicate autonomously in yeast cells (nor recombine with the chromosome, because it carries no *S. cerevisiae* sequences). All of the recommended disruption cassettes listed in Table III (see p. 300) (in boldface type) cannot be inherited by yeast cells.

How much sequence flanking the disruption cassette is necessary for efficient homologous recombination? As few as 15 base pairs on each side of the disruption cassette is sufficient, but in that case less than 4% of the transformants have the disruption cassette correctly integrated into the genome.[22] Disruption cassettes with 60 or more base pairs flanking them correctly integrate into the genome approximately 90% of the time in most cases (for the rare exceptions, which are probably due to particular chromatin structures, using 80 to 100 base pairs of homology should solve the problem). To reduce the length of the oligonucleotides, most investigators choose to use about 45 base pairs of flanking sequence, which yields approximately 80% correct integration of the disruption cassette. A reduction to 30 base pairs of flanking sequence still yields about 50% correct integration, but further reduction to 25 base pairs reduces the frequency of correct integration of the disruption cassette by an order of magnitude.[22] All of these frequencies of correct integration of the disruption cassette are reduced substantially if the disruption cassette is able to recombine with its genomic counterpart (because of a point mutation, rather than a large deletion, in the chromosomal copy of the disruption cassette).[22]

The short sequence flanking the gene disruption cassette provides for only a small number of yeast transformants [generally a 10^4-fold reduction relative to the number of transformants obtained with an autonomous CEN plasmid (e.g., pRS316); typical yields are 5–50 colonies, but this of course depends on the transformation efficiency of the recipient strain]. This can be problematical in cases where the recipient strain is not efficiently transformed with DNA, or when a large number of genes are being disrupted. In these cases it is desirable to use sequences flanking the gene disruption cassette that are longer than 50 base pairs to increase the frequency of homologous recombination and greatly improve the efficiency of gene disruption. In cases where the recipient yeast strain may contain a significant number of DNA sequence polymorphisms (e.g., with clinical or industrial *S. cerevisiae,* one of which was found to differ from the laboratory strain by as much as 1 in 160 base pairs[23]), it is essential to use longer sequences (several hundred base pairs) flanking the disruption cassette.

[23] E. A. Winzeler, D. R. Richards, A. R. Conway, A. L. Goldstein, S. Kalman, M. J. McCullough, J. H. McCusker, D. A. Stevens, L. Wodicka, D. J. Lockhart, and R. W. Davis, *Science* **281,** 1194 (1998).

TABLE I
REMOVABLE DISRUPTION CASSETTES[a]

Plasmid name	Sequence database accession number	Selectable gene	Selectable phenotype	Notes	Refs.[b]
		Cre-mediated marker removal			
pUG6/EUROSCARF P30114	AF298793	*kan Tn 903*	Geneticin (G418) resistance	Each gene flanked by *loxP* sites, allowing induction of marker loss by expression of Cre recombinase	(*1*)
pUG66/EUROSCARF P30116	AF298794	*ble Tn 5*	Phleomycin resistance		(*2*)
pUG27/EUROSCARF P30115	AF298790	*his5*$^+$ *S. pombe*	His$^+$ (complements *S. cerevisiae his3*)		(*2, 3*)
pUG72/EUROSCARF P30117	AF298788	*URA3 K. lactis*	Ura$^+$		(*2, 3*)
pUG73/EUROSCARF P30118	AF298792	*LEU2 K. lactis*	Leu$^+$		(*2*)
pUG-LYS2		*LYS2*	Lys$^+$		(*3*)
pRKO		*kan Tn 903*	Geneticin (G418) resistance	For gene disruption reversible by Cre	(*4*)
		Mitotic recombination-mediated marker removal			
pAG61/EUROSCARF P30112		*URA3 C. albicans*	Ura$^+$	Each gene flanked by ~400 nt direct repeats from *Ashbya gossypii*	(*5*)
pAG35/EUROSCARF P30107		*nat1 S.noursei*	Nourseothricin resistance	*URA3* loss can be *selected* as 5FOA^R; loss of drug^R cassettes must be *screened* after growth in the absence of drug	(*6*)
pAG31/EUROSCARF P30108		*pat S. viridochromogenes*	Bialaphos resistance		(*6*)
pAG34/EUROSCARF P30109		*hph E. coli*	Hygromycin B resistance		(*6*)

pFA6a-kanMX3	AJ002684	*kan Tn 903*	Geneticin (G418) resistance		(*7, 8*)
pMPY-ZAP or pDR941 or pNKY51		*URA3*	Ura+	Genes flanked by "short" *hisG* repeats	(*9*)
				URA3 loss can be *selected* as 5FOA^R	(*10, 11*)
pHUKH3,4		kan *Tn 903, URA3*	Geneticin (G418) resistance and Ura+		(*12*)
pJH184		*GAP1*	Growth on L-citrulline	Requires *gap1* mutant strain; *GAP1* loss *selected* as D-histidine^R	(*13*)
			Flp-mediated marker removal		
pWKW *et al*		*kan Tn 903*	Geneticin (G418) resistance	Genes flanked by Flp recombinase recognition sites	(*14*)
pXUX *et al*		*URA3 K. lactis*	Ura+		(*14*)
pAT526		*LEU2*	Leu+		(*15*)

[a] Selectable genes are from *S. cerevisiae,* unless otherwise noted.

[b] *Key to References:* (*1*) U. Güldener, S. Heck, T. Fielder, J. Beinhauer, and J. H. Hegemann, *Nucleic Acids Res.* **24,** 2519 (1996); (*2*) U. Güldener, J. Heinisch, D. Voss, G. Koehler, and J. H. Hegemann, *Nucleic Acids Res.,* in preparation (2001); (*3*) D. Delneri, G. C. Tomlin, J. L. Wixon, A. Hutter, M. Sefton, E. J. Louis, and S. G. Oliver, *Gene* **252,** 127 (2000); (*4*) T. H. Cheng, C. R. Chang, P. Joy, S. Yablok, and M. R. Gartenberg, *Nucleic Acids Res.* **28,** E108 (2000); (*5*) A. L. Goldstein, X. Pan, and J. H. McCusker, *Yeast* **15,** 507 (1999); (*6*) A. L. Goldstein and J. H. McCusker, *Yeast* **15,** 1541 (1999); (*7*) A. Wach, A. Brachat, R. Pohlmann, and P. Philippsen, *Yeast* **10,** 1793 (1994); (*8*) A. Wach, A. Brachat, C. Alberti-Segui, C. Rebischung, and P. Philippsen, *Yeast* **13,** 1065 (1997); (*9*) B. L. Schneider, B. Steiner, W. Seufert, and A. B. Futcher, *Yeast* **12,** 129 (1996); (*10*) E. Alani, L. Cao, and N. Kleckner, *Genetics* **116,** 541 (1987); (*11*) K. Replogle, L. Hovland, and D. H. Rivier, *Yeast* **15,** 1141 (1999); (*12*) M. C. Earley and G. F. Crouse, *Gene* **169,** 111 (1996); (*13*) B. Regenberg and J. Hansen, *Yeast* **16,** 1111 (2000); (*14*) F. Storici, M. Coglievina, and C. V. Bruschi, *Yeast* **15,** 271 (1999); (*15*) A. Toh-e, *Curr. Genet.* **27,** 293 (1995).

TABLE II
YEAST STRAINS

Strain name	Genotype	Parent strain	Notes	Ref.[a]
BY4743/ATCC	*MATa ura3Δ0 his3Δ1 leu2 Δ0 met15Δ0* LYS2 / *MAα ura3Δ0 his3Δ1 leu2Δ0 MET15 lys2Δ0*		Diploid strain used to make YKO collection	(*1*)
BY4705/ATCC 200869	*MATα ade2Δ::hisG his3Δ200 leu2Δ0 lys2Δ0 met15Δ0 trp1 Δ63 ura3Δ0*	S288C/ FY1679	Also many other strains with nearly all combinations of these mutations	(*1*)
FY1679 EUROSCARF #10000D	*MATa ura3-52 trp1Δ63 leu2Δ1 his3Δ200 GAL2* / *MATα ura3-52 TRP1 LEU2 HIS3 GAL2*	S288C	Diploid strain	
YPH499/ATCC 204679	*MATa ura3-52 lys2-801(amber) ade2-101(ochre) his3Δ200 trp1Δ63 leu2Δ1*	S288C		(*2*)
DRY1725	*MATa arg4Δ0::hisG his3-11,15 lys2Δ::hisG trp1Δ0::hisG ura3Δ0*	W303	Also many other strains with nearly all combinations of these mutations	(*3*)
BMA64 EUROSCARF #20000D	*MATa ura3-52 2 trp1Δ2 leu2-3,112 his3-11 ade2-1 can1-100* / *MATα ura3-52 trp1Δ2 leu2-3,112 his3-11 ade2-1 can1-100*	W303	Diploid strain	
CEN.PK/EUROSCARF #30000D	*MATa leu2-3,112 ura3-52 trp1-289 his3-Δ1 MAL2-8^c SUC2* / *MATα leu2-3,112 ura3-52 trp1-289 his3-Δ1 MAL2-8^c SUC2*	Unknown	Diploid strain	(*4*)

[a] *Key to References:* (*1*) C. B. Brachmann, A. Davies, G. J. Cost, E. Caputo, J. Li, P. Hieter, and J. D. Boeke, *Yeast* **14,** 115 (1998); (*2*) R. S. Sikorski and P. Hieter, *Genetics* **122,** 19 (1989); (*3*) K. Replogle, L. Hovland, and D. H. Rivier, *Yeast* **15,** 1141 (1999); (*4*) K.-D. Entian, *in* "Yeast Mutants and Plasmid Collections" (M. F. Tuite, ed.). Academic Press, London, 1998.

Perhaps the simplest way to generate gene disruption cassettes carrying longer sequences flanking the gene to be disrupted is simply to amplify the cassette from genomic DNA of a strain in which the gene is already disrupted (using primers "A" and "D" in Fig. 3). The PCR can be primed with oligonucleotides complementary to sequences well outside of the gene, generating a gene disruption cassette flanked by hundreds of base pairs on each side of the gene to be disrupted. Since disruptants of nearly all yeast genes are readily available (in the YKO collection; more on this below), this is probably the first approach one should consider when seeking to disrupt a gene in yeast. In addition, about 1300 yeast genes disrupted with *kanMX* (encoding resistance to the drug geneticin; see below) have been cloned in an *E. coli* plasmid (available from EUROSCARF; see below), and these serve as an excellent resource for gene disruption cassettes carrying several hundred base pairs of sequence flanking the gene to be disrupted. Gene disruption cassettes with long flanking sequences can also be produced in several sequential PCRs.[24–26]

Repeated Gene Disruption

It is often necessary to disrupt several genes in a yeast strain. This can be done in two ways. First, genes can be deleted sequentially using different gene disruption cassettes providing different selectable phenotypes. Cassettes of several genes commonly missing in auxotrophic laboratory strains (*URA3, HIS3, TRP1, LEU2, ADE2, LYS2, MET15*), as well as five different genes encoding drug resistance [geneticin (G418), phleomycin, nourseothricin, bialaphos, hygromycinB] are available (Tables I and III).

Second, the gene disruption cassette can be removed from the genome so it can be used again to disrupt another gene. There are two ways to do this. Probably the most efficient is to use a gene disruption cassette flanked by *loxP* sites. Expression of Cre recombinase (see Table IV (p. 302) and Fig. 4 (p. 303)) in a strain disrupted with one of these cassettes removes it in 80–90% of the cells.[27,28] In fact, several disruption cassettes can be removed by Cre at the same time[29] (though the unnecessarily high level of the Cre recombinase that was used in this case resulted in undesirable intra- and interchromosomal recombination events). The Flp recombinase achieves the same result, but the gene disruption cassettes for this system are not

[24] D. C. Amberg, D. Botstein, and E. M. Beasley, *Yeast* **11,** 1275 (1995).

[25] J. Nikawa and M. Kawabata, *Nucleic Acids Res.* **26,** 860 (1998).

[26] I. Eberhardt and S. Hohmann, *Curr. Genet.* **27,** 306 (1995).

[27] U. Güldener, S. Heck, T. Fielder, J. Beinhauer, and J. H. Hegemann, *Nucleic Acids Res.* **24,** 2519 (1996).

[28] U. Güldener, J. Heinisch, G. J. Koehler, D. Voss, and J. H. Hegemann, *Nucleic Acids Res.* **30,** in press (2002).

[29] D. Delneri, G. C. Tomlin, J. L. Wixon, A. Hutter, M. Sefton, E. J. Louis, and S. G. Oliver, *Gene* **252,** 127 (2000).

TABLE III
NONREMOVABLE DISRUPTION CASSETTES

Plasmid name	Sequence database accession number	Selectable gene	Selectable phenotype	Notes	Ref.[a]
pRS406/Stratagene 217406	U03446	*URA3*	Ura^+	Each gene is flanked by the same pBluescript polylinker sequences, so the same oligonucleotides can be used to amplify each disruption cassette	*(1)*
pRS403/Stratagene 217403	U03443	*HIS3*	His^+		
pRS402/ATCC 87477	U93717	*ADE2*	Ade^+		
pRS405/Stratagene 217405	U03445	*LEU2*	Leu^+		
pRS404/Stratagene 217404	U03444	*TRP1*	Trp^+		
pRS401/ATCC 87473	U93714	*MET15*	Met^+		
pRS400	U93713	*kan Tn903*	Geneticin (G418) resistance		
pDR827		*ARG4*	Arg^+		(2)
YDp-H		*HIS3*	His^+	Each gene is flanked by the same pUC9 polylinker sequences, so the same oligonucleotides can be used to amplify each disruption cassette	*(3)*
YDp-K		*LYS2*	Lys^+		
YDp-L		*LEU2*	Leu^+		
YDp-U		*URA3*	Ura^+		
YDp-W		*TRP1*	Trp^+		

pFA6-kanMX2	S78175	*kan Tn903*	Geneticin (G418) resistance		(*4, 5*)
pFA6-kanMX4	AJ002680	*kan Tn903*	Geneticin (G418) resistance	Disruption cassette used to make the YKO collection	(*4, 5*)
pFA6a-kanMX6	AJ002682	*kan Tn903*	Geneticin (G418) resistance		(*5, 6*)
pAG25		*nat1 S. noursei*	Nourseothricin resistance		(*7*)
pAG29		*pat S. viridochromogenes*	Bialaphos resistance		(*7*)
pAG32		*hph E. coli*	Hygromycin B resistance		(*7*)
pAG60		*URA3 C. albicans*	Ura^+		(*8*)
pFA6-AgLEU2MX1		*LEU2 A. gossypii*	Leu^+	Poor *LEU2* expression; colonies require 1 week to form	(*4*)
pFA6a-His3MX6	AJ002681	*his5*$^+$ *S. pombe*	His^+(complements *S. cerevisiae his3*)		(*5, 6*)
pFA6a-TRP1		*TRP1*	Trp^+		(*6*)

[a] *Key to References:* (*1*) C. B. Brachmann, A. Davies, G. J. Cost, E. Caputo, J. Li, P. Hieter, and J. D. Boeke, *Yeast* **14,** 115 (1998); (*2*) K. Replogle, L. Hovland, and D. H. Rivier, *Yeast* **15,** 1141 (1999); (*3*) G. Berben, J. Dumont, V. Gilliquet, P. A. Bolle, and F. Hilger, *Yeast* **7,** 475 (1991); (*4*) A. Wach, A. Brachat, R. Pohlmann, and P. Philippsen, *Yeast* **10,** 1793 (1994); (*5*) A. Wach, A. Brachat, C. Alberti-Segui, C. Rebischung, and P. Philippsen, *Yeast* **13,** 1065 (1997); (*6*) M. S. Longtine, A. McKenzie III, D. J. Demarini, N. G. Shah, A. Wach, A. Brachat, P. Philippsen, and J. R. Pringle, *Yeast* **14,** 953 (1998); (*7*) A. L. Goldstein and J. H. McCusker, *Yeast* **15,** 1541 (1999); (*8*) A. L. Goldstein, X. Pan, and J. H. McCusker, *Yeast* **15,** 507 (1999).

TABLE IV
PLASMIDS EXPRESSING Cre OR Flp FOR CASSETTE REMOVAL

Plasmid name	Sequence database accession number	Recombinase	Selectable marker for yeast	Notes	Ref.[a]
pSH47/EUROSCARF P30119	AF298782	Cre	*URA3*	Cre expressed from *GAL1* promoter	(*1*)
pSH62/EUROSCARF P30120	AF298785		*HIS3*		
pSH63/EUROSCARF P30121	AF298789		*TRP1*		
pSH65/EUROSCARF P30122	AF298780		*ble*		
Yep351-cre-cyh			*CYH*^{R}	High-copy plasmid produces very high Cre levels	(*2*)
pAT399		Flp	*URA3*	Flp expressed from *GAL7* promoter	(*3*)

[a] *Key to References:* (*1*) U. Güldener, J. Heinisch, D. Voss, G. Koehler, and J. H. Hegemann, *Nucleic Acids Res.,* in preparation (2001). (*2*) D. Delneri, G. C. Tomlin, J. L. Wixon, A. Hutter, M. Sefton, E. J. Louis, and S. G. Oliver, *Gene* **252,** 127 (2000). (*3*) A. Toh-e, *Curr. Genet.* **27,** 293 (1995).

as well developed.[30,31] Alternatively, several gene disruption cassettes flanked by a directly repeated sequence are available.[32–37] Mitotic recombination between these repeated sequences excises the cassette and one copy of the repeat, leaving one copy of the repeat behind in the genome. The main problem with this approach is that mitotic recombination is relatively rare, so a large number of colonies (about 1000) need to be analyzed before one missing the disruption cassette is found. (This is not a problem, however, for gene disruption cassettes carrying *URA3*, because cells that have lost this gene can be selected.) Another problem with this approach is that the single copy of the repeated sequences flanking the disruption cassette that remains in the genome can serve as a site of homologous recombination for the same cassette when it is used again to disrupt another gene.

Variations of the Method

Several variations of the standard gene disruption approach have been developed that enable additional feats of gene engineering. All of these methods take

[30] F. Storici, M. Coglievina, and C. V. Bruschi, *Yeast* **15,** 271 (1999).
[31] A. Toh-e, *Curr. Genet.* **27,** 293 (1995).
[32] E. Alani, L. Cao, and N. Kleckner, *Genetics* **116,** 541 (1987).
[33] K. Replogle, L. Hovland, and D. H. Rivier, *Yeast* **15,** 1141 (1999).
[34] B. L. Schneider, B. Steiner, W. Seufert, and A. B. Futcher, *Yeast* **12,** 129 (1996).
[35] A. Wach, A. Brachat, C. Alberti-Segui, C. Rebischung, and P. Philippsen, *Yeast* **13,** 1065 (1997).
[36] A. L. Goldstein and J. H. McCusker, *Yeast* **15,** 1541 (1999).
[37] A. L. Goldstein, X. Pan, and J. H. McCusker, *Yeast* **15,** 507 (1999).

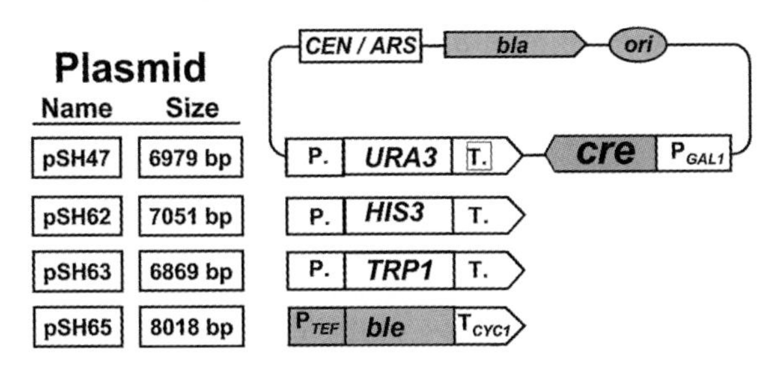

FIG. 4. Plasmids for expression of Cre for removal of gene disruption cassettes flanked by *loxP*. Autonomously replicating, stable plasmids expressing Cre are available with four different selectable genes [U. Güldener, J. Heinisch, D. Voss, G. Koehler, and J. H. Hegemann, *Nucleic Acids Res.* **30,** in press (2002).]

advantage of the flexibility of the PCR. In addition to the method mentioned above for adding in several PCRs several hundred base pairs of sequence homology flanking the gene disruption cassette,[24,25] the PCR can be used to construct plasmids that can be used to disrupt a gene and at the same time fuse it to a reporter gene,[38] and to construct plasmids for disrupting as well as cloning a gene.[39] Multiple PCRs have been employed in a scheme to move mutations between strains, essentially by a modification of the two-step gene disruption method.[40] Finally, a modified gene disruption cassette has been used to construct genes whose expression can be regulated at the level of mRNA splicing.[41] These variations of the gene disruption method are of limited utility because they are seldom necessary, and in any case the same ends can usually be achieved by simpler means. Therefore, we will not provide details of these methods.

Problems and Pitfalls

Genes That Are Difficult to Disrupt

Duplicated Genes. We have found that genes located in duplicated regions of the genome are often difficult to disrupt. When a gene of interest lies in a duplicated region of the genome, it is necessary to check the sequences immediately flanking the gene used for recombination of the gene disruption cassette to ensure that they are not repeated elsewhere in the genome. If the ORF has unique flanking sequence on at least one side, the gene should be able to be deleted using the methods described. It may be necessary to use for recombination sequences farther upstream or downstream of the gene. The sequence of the deletion confirmation primers (Fig. 3) should also be checked to ensure they are not duplicated in the genome.

[38] M. Maftahi, C. Gaillardin, and J. M. Nicaud, *Yeast* **12,** 859 (1996).

[39] L. Mallet and M. Jacquet, *Yeast* **12,** 1351 (1996).

[40] N. Erdeniz, U. H. Mortensen, and R. Rothstein, *Genome Res.* **7,** 1174 (1997).

[41] T. H. Cheng, C. R. Chang, P. Joy, S. Yablok, and M. R. Gartenberg, *Nucleic Acids Res.* **28,** E108 (2000).

Genes Flanked by Simple Sequence. We have encountered difficulty disrupting genes flanked by long stretches of simple sequence. For example, the 45 base pairs immediately downstream of YHR115c includes 27 base pairs of poly(AT), and a gene disruption cassette flanked by these sequences yielded no transformants. A new deletion oligonucleotide that included the immediate 18 nucleotides of unique flanking sequence, all of the poly(AT) tract (32 nucleotides), and a further 18 nucleotides of unique sequence enabled disruption of this gene. Another example is YHR071w, which has a poly(A) tract upstream of the coding sequence and was only able to be deleted if the unique sequence further upstream of this was used to direct recombination of the disruption cassette.

Using Longer Homology. A few ORFs we have attempted to disrupt produced few transformants using the usual 45 bp of flanking homology, for no obvious reason. A gene disruption cassette flanked by 90 base pairs of homology gave a 3- to 5-fold increase in the number of transformants in these cases. However, in some cases longer regions of sequence homology give rise to pseudo-stable transformants (unpublished observations, 2000), probably because the longer sequences include an element that allows the disruption cassette to be maintained for several generations, presumably after its circularization.

Collateral Mutations

Between 5% and 10% of gene disruptants carry an unlinked mutation that causes an obvious growth defect. We have even found a few disruptants that carry two collateral mutations that cause a visible growth defect. These are identified by observing a growth phenotype (no growth, or slow growth) that does not segregate with the gene disruption cassette in tetrads of a transformant heterozygous for the gene disruption. Since only one-quarter of all gene disruptions cause an easily identified growth phenotype on YPD medium, the frequency of collateral mutations must be far higher. To avoid this problem, it is best to work with a diploid homozygous for the disruption, constructed by mating two independently generated haploid disruptants, so that any collateral mutations (which are likely to be recessive) are complemented. If one must work with haploid disruptants, it would be prudent to cross the disruptant back to wild-type several times to segregate the gene disruption mutation away from the collateral mutation(s).

Gene Disruption Accompanied by Gene Duplication

Gene deletion is accompanied by a duplication of the ORF about 8% of the time. In many cases the duplicated gene appeared to be linked to the gene disruption[42] (unpublished observations, 2000). Some of these events are due to duplication of

[42] T. R. Hughes, C. J. Roberts, H. Dai, A. R. Jones, M. R. Meyer, D. Slade, J. Burchard, S. Dow, T. R. Ward, M. J. Kidd, S. H. Friend, and M. J. Marton, *Nat. Genet.* **25,** 333 (2000).

the entire chromosome; some are duplications of only a portion of the chromosome including the disrupted gene. Thus, the absence of the deleted ORF must be confirmed by a PCR assay.

Materials

YKO Collection of Yeast Gene Disruptions

Before attempting to disrupt a gene, one should first consider if a disruption mutant that has already been made can be used. Nearly all of the approximately 6000 yeast genes have been disrupted, and collections of these mutants are available (the yeast gene knockout, or YKO Collection). Available are (1) a collection of ~4800 homozygous diploids of nonessential gene mutants, (2) a collection of ~4800 haploids (both mating types) of nonessential gene mutants, and (3) a collection of ~1000 heterozygous diploids of essential gene mutants. In each mutant a gene is deleted from the ATG translational START codon through the translational STOP codon and replaced with *kanMX*. In addition, each mutant is tagged with two unique 20-nucleotide sequences (one on the left-hand side of the disrupted gene, one on the right-hand side) which can be used to uniquely identify each strain.[12] The sequence tags ("molecular bar codes") that identify each mutant can be searched at http://www-deletion.stanford.edu/cgi-bin/tag_sequences/tagsequence.cgi.

The YKO collections can be obtained from:

Research Genetics, Inc.
2130 Memorial Parkway
Huntsville, AL 35801 USA
Tel. U.S. or Canada: 1-800-533-4363
Tel. Europe: 00800 5456 5456
Tel. Worldwide: 001-256-533-4363
Fax U.S. or Canada: 256-536-9016
Fax Europe: 00800 7890 7890
Fax Worldwide: 001-256-536-9016
E-mail: info@resgen.com.
http://www.resgen.com/products/YEASTD.php3

EUROSCARF
Institute for Microbiology
Johann Wolfgang Goethe-University Frankfurt
Marie-Curie-Strasse 9; Building N250
D-60439 Frankfurt, Germany
Fax: +49-69-79829527
E-mail: Euroscarf@em.uni-frankfurt.de
http://www.rz.uni-frankfurt.de/FB/fb16/mikro/euroscarf/col_index.html

ATCC (American Type Culture Collection)
10801 University Boulevard
Manassas, VA 20110-2209 USA
Tel. (800) 638-6597
E-mail: sales@atcc.org
http://www.atcc.org/cydac/cydac.cfm

If a particular gene disruption needs to be placed into a different strain, it is probably best simply to amplify the *kanMX* disruption cassette from the mutant in the YKO collection, using as primers for the PCR oligonucleotides complementary to sequences a few hundred base pairs upstream and downstream of the disrupted gene. The long flanking sequence homology should ensure efficient gene disruption in the host strain. In some cases it may prove difficult to amplify the gene disruption cassette from the genome (for unknown reasons). We have successfully circumvented this problem by separately amplifying each half of the disruption cassette, with the products overlapping in the *kanMX* sequences [using primers "A" and "cassette B" to generate one product; "cassette C" and "D" to generate the other product (Fig. 3)]. When these two PCR products are introduced into yeast cells, three homologous recombination events replace the target gene with the disruption cassette. Alternatively, the cloned gene disruption cassette can be obtained from EUROSCARF (if available; the ~1300 that have been cloned can be searched at http://www.rz.uni-frankfurt.de/FB/fb16/mikro/euroscarf/col_index.html).

Gene Disruption Cassettes

Several gene disruption cassettes carrying a variety of selectable genes are available. The most useful of them are highlighted in boldface type in Tables I and III and diagrammed in Figs. 5 and 6. The most useful of these carry genes encoding resistance to several drugs that inhibit yeast growth; several others carry genes required for amino acid or nucleotide base prototrophy that are commonly mutated in most standard laboratory strains (Table II). Several of these genes are from non-*Saccharomyces* yeast species, which prevents them from recombining with the *Saccharomyces* genome, ensuring that the recombination is targeted to the gene to be disrupted. The *S. cerevisiae* gene disruption cassettes (Fig. 6) should be used with strains carrying a complete deletion of that gene (Table III) to ensure proper targeting of the gene disruption event. Increasingly popular are the several gene disruption cassettes that carry genes for resistance to drugs that inhibit yeast growth, because these can be used to disrupt genes in any yeast strain (particularly useful for strains which do not carry auxotrophic markers, such as industrial or clinical strains). The most commonly used of these drug resistance cassettes are the ones containing *kanMX,* encoding resistance to geneticin (G418), which has

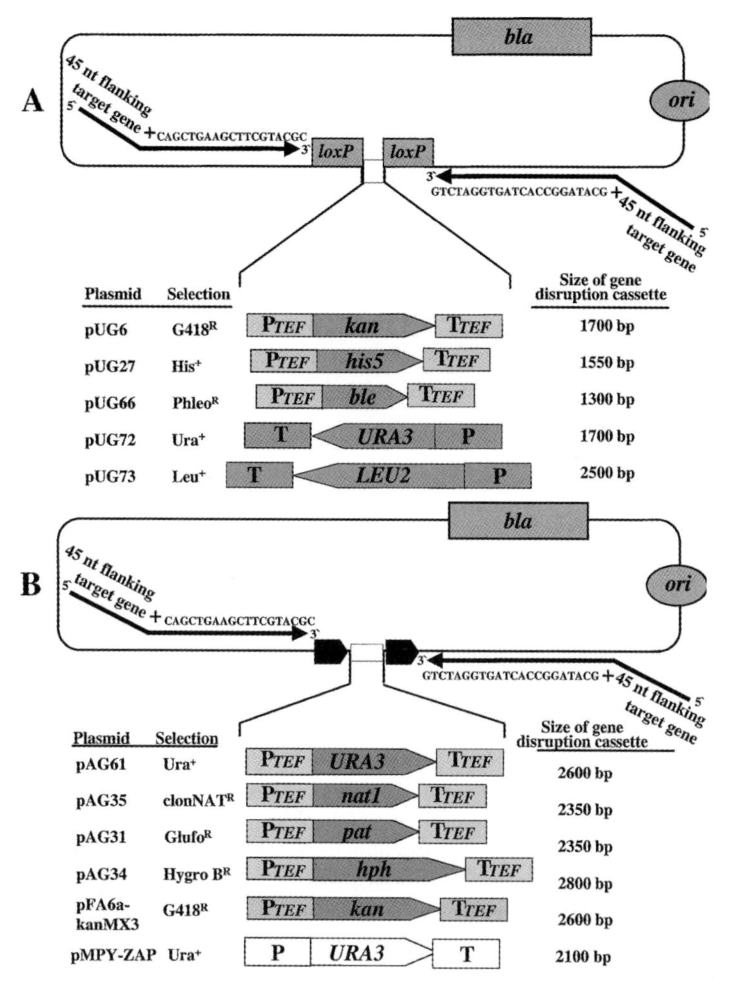

FIG. 5. (A) Gene disruption cassettes removable by Cre recombinase. (B) Gene disruption cassettes removable by mitotic recombinaton. All the selectable genes are from organisms other than *S. cerevisiae* and thus will not recombine with the *S. cerevisiae* genome (except for pMPY-ZAP, whose selectable gene is from *S. cerevisiae*). Each selectable gene is expressed from the *Ashbya gossypii TEF2* promoter (P_{TEF}) and terminator (T_{TEF}), except for the two *K. lactis* genes shown in (A) and pMPY-ZAP shown in (B), which come with their own promoter (P) and terminator (T). The direct repeat (DR) flanking each gene is a 470 bp fragment of the 3′-region of the *Ashbya gossypii LEU2* gene [A. Wach, A. Brachat, R. Pohlmann, and P. Philippsen, *Yeast* **10,** 1793 (1994)], except for pMPY-ZAP, in which it is 400 nucleotides of the *hisG* gene of *S. typhimurium* [B. L. Schneider, B. Steiner, W. Seufert, and A. B. Futcher, *Yeast* **12,** 129 (1996); E. Alani, L. Cao, and N. Kleckner, *Genetics* **116,** 541 (1987)]. Each gene shown was inserted between two loxP sites in the pUG6 vector backbone (Fig. 3A) or the two DR sites in the pFA6 vector backbone (B), and therefore each disruption cassette can be amplified with the same PCR primers (except for pMPY-ZAP; see below). The size of the PCR product is indicated. For pMPY-ZAP, the selectable gene is *URA3* from *S. cerevisiae,* and the sequences of the 3′ nucleotides of the primers to amplify the disruption cassette are 5′ AGGGAACAAAAGCTGG 3′ (upstream of *URA3*) and 5′ CTATAGGGCGAATTGG 3′ (downstream of *URA3*) [see B. L. Schneider, B. Steiner, W. Seufert, and A. B. Futcher, *Yeast* **12,** 129 (1996)].

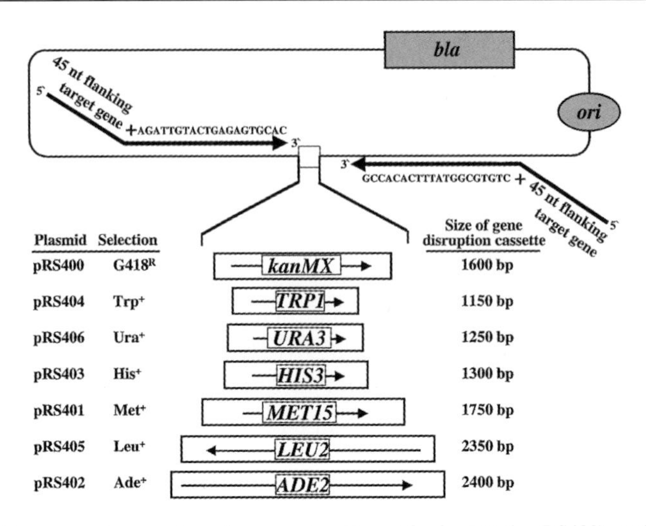

FIG. 6. Nonremovable gene disruption cassettes. Except for *kanMX* (in pRS400), each selectable gene is from *S. cerevisiae,* which requires that these gene disruption cassettes be used with yeast strains completely lacking these sequences (listed in Table II).

been shown to be selectively neutral for yeast cells.[43] The other drug resistance cassettes are also likely to be selectively neutral.[36]

In most cases a removable gene disruption cassette will be the one of choice, because it provides the most flexibility for later manipulation of the resultant strain. The *loxP*-containing cassettes are probably the easiest to use,[28] although the *URA3* cassettes removable by mitotic recombination[32–34] are also convenient, because cells that have lost URA3 can be easily selected.[44,45] Loss of all other cassettes by mitotic recombination, which is a relatively rare event (~1/1000 cells), must be screened for.

Several of the gene disruption cassettes are in families that can be amplified in the PCR by the same set of oligonucleotide primers because they are inserted at an identical location in the same plasmid. The pRS and the YDp series of nonremovable cassettes comprise such a family; the pUG series of Cre-removable cassettes and the pAG series of mitotic recombination-removable drug resistant cassettes make up another such family.

Gene disruption cassettes are also available for deleting a gene and fusing its promoter to a reporter of gene expression (Table V). The cassette carrying *lacZ* as a reporter of gene expression is probably the easiest to use because of the ease

[43] F. Baganz, A. Hayes, D. Marren, D. C. Gardner, and S. G. Oliver, *Yeast* **13,** 1563 (1997).
[44] J. D. Boeke, J. Trueheart, G. Natsoulis, and G. R. Fink, *Methods Enzymol.* **154,** 164 (1987).
[45] J. D. Boeke, F. LaCroute, and G. R. Fink, *Mol. Gen. Genet.* **197,** 345 (1984).

TABLE V
CASSETTES FOR GENE DISRUPTION WITH FUSION OF PROMOTER TO REPORTER GENE

Plasmid name	Selectable gene	Selectable phenotype	Reporter	Ref.[a]
pBM2983	*HIS3* *S. cerevisiae*	His^+	GFP	(*1*)
pFA6a–GFP(S65T)-kanMX	*kan Tn903*	Geneticin (G418) resistance	GFP	(*2, 3*)
pFA6a–GFP(S65T)-HIS3MX6	$his5^+$ *S. pombe*	His^+(complements) *S. cerevisiae his3*)	GFP	(*2*)
pFA6a–GFP(S65T)-TRP1	*TRP1*	Trp^+	GFP	(*2*)
pFA6a,b,c-lacZMT-kanMX3	*kan Tn903*	Geneticin (G418) resistance	*lacZ*	(*2, 3*)
PYGFPgN	*kan Tn903*	Geneticin (G418) resistance	*GFP*	(*4*)

[a] *Key to References:* (*1*) R. K. Niedenthal, L. Riles, M. Johnston, and J. H. Hegemann, *Yeast* **12,** 773 (1996); (*2*) A. Wach, A. Brachat, C. Alberti-Segui, C. Rebischung, and P. Philippsen, *Yeast* **13,** 1065 (1997); (*3*) M. S. Longtine, A. McKenzie III, D. J. Demarini, N. G. Shah, A. Wach, A. Brachat, P. Philippsen, and J. R. Pringle, *Yeast* **14,** 953 (1998); (*4*) B. Prein, K. Natter, and S. D. Kohlwein, *FEBS Lett.* **485,** 29 (2000).

and simplicity of the assay for β-galactosidase. Several cassettes with GFP as the reporter of gene expression are available, but are probably less desirable because the assay for GFP is more cumbersome.[13,46] The GFP cassettes are excellent, however, for fusing GFP to any protein, by techniques essentially identical to those used to disrupt genes[46,47] (see article elsewhere in this volume[20a]).

Chemicals and Enzymes

Geneticin (G418): Life Technologies, Rockville, MD
Hygromycin B: Roche; Calbiochem-Novabiochem, La Jolla, CA
ClonNAT (nourseothricin): Werner BioAgents, Jena-Cospeda, Germany; E-mail: WeBioAge@aol.com
Bialaphos (phosphinothricin): Shinyo Sangyo Co., Lte., Tokyo, Japan; Fax: 81-3-5449-3598); or glufosinate: Sigma-Aldrich, St. Louis, MO; or Finale, a commercial herbicide formulation of glufosinate, which is much cheaper and can be bought at any garden supply store. The concentrated formula contains ~6% glufosinate.
Phleomycin (Cayla, France)

[46] R. K. Niedenthal, L. Riles, M. Johnston, and J. H. Hegemann, *Yeast* **12,** 773 (1996).
[47] B. Prein, K. Natter, and S. D. Kohlwein, *FEBS Lett.* **485,** 29 (2000).

(Detailed information about Geneticin, hygromycin B, ClonNAT, and glufosinate can also be found at http://www.duke.edu/web/microlabs/mccusker/Resources/drugsupply.html.)

Taq DNA Polymerase (5 units/μl with 10× buffer): Roche, but enzyme from other suppliers should be satisfactory
KlenTaqLA (with 10× buffer): (Clontech), but enzyme from other suppliers should be satisfactory
Lithium acetate dihydrate: Sigma
PEG-3350: Polyethylene glycol (molecular weight 3350) Sigma
Fish sperm carrier DNA (10 mg/ml, sheared): Roche
DMSO: Dimethyl sulfoxide, Sigma

Media

YPD, SD: See elsewhere in this volume.[48]
SD/YPD/G418 plates: The active concentration of G418 (Geneticin) varies with each lot (500 to 800 μg/mg, w/w). Add Geneticin powder to 65° autoclaved media/2% agar to a final *active* concentration of 200 μg/ml. Dry plates for several days at room temperature. *Note:* For SD media, replace the ammonium sulfate with 1 g/liter glutamic acid (monosodium salt, Sigma).[41]
YPD/Hygromycin B plates: Add hygromycin B powder to final concentration of 300 μg/ml to 65° autoclavedYPD/2% agar.
YPD/Nourseothricin plates: Add ClonNAT powder to final concentration of 100 μg/ml to 65° autoclavedYPD/2% agar.
SDP/Phosphinothricin plates: SDP (1.7 g/liter Difco Yeast Nitrogen Base without $(NH_4)_2SO_4$ and without amino acids, 1 g/liter L-proline, and 20 g/liter dextrose)
Add *either* 200 μg/ml bialaphos powder *or* 600–800 μg/ml glufosinate powder *or* ~2 ml of Finale (not necessary to filter sterilize; concentration may need to be adjusted) to 65° autoclaved SDP/2% agar.
YPD/Phleomycin plates: Add Phleomycin powder to a final concentration of 7.5 μg/ml to 65° autoclavedYPD/2% agar.

Buffers and Solutions

Lithium acetate: 1 *M* solution; filter sterilize.
PEG (prepare fresh): Mix 15 g PEG-3350 with 16.5 ml water. Dissolve with gentle heat and stirring, filter sterilize, Parafilm, and use within 1 week.

[48] F. Sherman, *Methods Enzymol.* **350,** [1], 2002 (this volume).

Lithium acetate/PEG solution (prepare fresh): In a sterile, graduated 15 ml tube add 3 ml of 500 m*M* lithium acetate to 12 ml PEG solution; mix well. Use the same day.

Methods and Protocols

PCR Primer Design

Disruption Cassette Primers. The sequences of the 19 and 22 nucleotides 3′ of the PCR primers for amplifying the most useful disruption cassettes are shown in Figs. 5A, B and 6 (in the case of pMPY-ZAP, they are listed in the legend to Fig. 3). The same primers will amplify all the gene disruption cassettes shown in Fig. 5; a different pair of primers will amplify all the gene disruption cassettes shown in Fig. 6. The sequences flanking the target gene are added to the 5′ end of these sequences: (a) 30–50 nucleotides of the sequence of the top strand immediately upstream of the sequences to be deleted (the nucleotides immediately before the ATG in the example shown in Fig. 2) are added to the 5′ end of one PCR primer; (b) 30–50 nucleotides of the sequence of the bottom strand immediately downstream of the sequences to be deleted (the nucleotides immediately after the translational stop codon in the example shown in Fig. 2) are added to the 5′ end of the other PCR primer.

Disruption Confirmation Primers. The PCR primers flanking the disrupted gene should be chosen so that the product produced in conjunction with the PCR primers within the disruption cassette (cassette B and cassette C primers in Fig. 3) or in conjunction with the PCR primers within the target gene (target B and target C in Fig. 3) is 300–1000 nucleotides in length. The primers should be 17 to 28 nucleotides and have melting temperatures of $65° \pm 2°$. The sequences of the primers used to produce the YKO collection can be found at http://www.deletion.stanford.edu/cgi-bin/deletion/search3.pl. Using primers with these characteristics and the protocols below, we have found that about 90% of the confirmation tests are successful.

The sequences of the cassette B and cassette C primers for disruption cassettes containing *kanMX* are:

kanB = 5′ CTGCAGCGAGGAGCCGTAAT 3′
kanC = 5′ TGATTTTGATGACGAGCGTAAT 3′

Preparative PCR of Gene Disruption Cassette

A. From Plasmid DNA

1. Make a cocktail minus the oligonucleotide primers.
   1 Reaction
   5 μl 10× KlenTaqLA buffer

4 μl 10 mM dNTP mix (2.5 mM each dNTP)
2 μl 1 ng/μl template DNA
34 μl H_2O
1 μl KlenTaqLA
46 μl Total

2. Dispense 46 μl aliquots into PCR tubes.
3. Add 2 μl of 25 μM upstream disruption primer.
4. Add 2 μl of 25 μM downstream disruption primer.
5. Add 4 μl H_2O (no primers) to the negative control tube.
6. Allow the PCR block to reach 94° before adding PCR tubes. *Keep everything on ice until placing tubes in thermocycler.*

Hot start	94°	3.0 min—hot start is very important!
Denature	98°	5 sec
Anneal/Extend	62°	6 min/cycle, 20 cycles

7. Assay 2 μl on a 1% agarose gel (concentration should be approximately 50 ng/μl). Reactions should be stored at −20°.
8. Transform yeast with remaining product (no purification or precipitation required).

B. From Gene Disruption Cassette Present in Genome (i.e., Disruption Mutant). The gene disruption cassette can be amplified as one product, using oligonucleotide primers flanking the disrupted gene (primers A and D in Fig. 3). However, when using yeast colonies as template, we have found that PCR product is more reliably obtained by amplifying the disruption cassette as two products, one extending from upstream of the disrupted gene (using primer A in Fig. 3) to the 3′ end of *KanMX* (primer = TTAGAAAACTCATCGAGCATC), and the other extending from the 5′ end of *KanMX* (primer = ATGGGTAAGGAAAAGACTCAC) to downstream of the disrupted gene (primer D in Fig. 3). These two PCR products efficiently transform yeast cells to Kan^R by 3 recombination events: two flanking the gene, and one within *KanMX* coding sequences.

1. Use as template yeast cells freshly grown on YPD (no more than 2 days old, and *never* refrigerated). Gently touch the surface of the colony (avoid touching the agar) with a plastic pipette tip attached to a 20-μl pipettor (set at 20 μl), place the tip into a 50-μl PCR reaction mix (make a cocktail with extra volume for pipetting error), and pipette up and down once. (Cells should be just barely visible on the end of the pipette tip. Too many cells, or the presence of agar will inhibit the reaction.) *Keep everything on ice!*

Reaction 1: Set up a 50-μl PCR as described in step 1 (except substituting Taq DNA Polymerase for KlenTaq) under "Preparative PCR of Gene Disruption Cassette," using as primers:

5.0 μl	10 μ*M* upstream primer (primer A; see description above)
5.0 μl	10 μ*M* *kanB* (cassette B) primer

Reaction 2: Like reaction 1, except different primers:

5.0 μl	10 μ*M* downstream primer (primer D; see description above)
5.0 μl	10 μ*M* *kanC* (cassette C) primer

2. Immediately transfer from ice to preheated (94°) PCR block.

Hot start*	94°	3.0 min
Denature	94°	1.5 min
Anneal	57°	2.0 min
Extend*	72°	3.0 min
Final extension	72°	3.0 min

*We have observed that the hot start and the long extension time is very important for success of the PCR.

35 cycles—a large number of cycles is very important!

3. Assay 10 μl from each of the two PCR reactions on an agarose gel. (*Note:* Because less PCR product is made when using cells as template, more of the PCR product is used for the gel assay than in the case of preparative PCR where a plasmid is used as template.)

4. Transform yeast with the remaining product (no purification or precipitation necessary).

Yeast Transformation

A. For Gene Disruption Cassette Generated from Plasmid Template

1. Dilute an overnight (fully grown) culture of the recipient yeast strain 1 : 50 (40 μl in 2 ml) into YPD.
2. OD_{600} of culture ~4 hr later should be 1–5.
3. Inoculate a flask containing 125 ml of YPD prewarmed to 30° with the appropriate volume of cells. This volume depends on the doubling time of the strain and on the number of hours of growth. For BY4743, the diploid strain used in the Yeast Genome Deletion Project, a volume of 4-hr culture that contains 7.5×10^4 cells gives the desired OD_{600} after 18 hr of growth.
4. Prepare 1 liter of sterile 100 m*M* lithium acetate solution.
5. Prepare fresh PEG.
6. Prepare fresh lithium acetate/PEG solution.
7. Boil carrier DNA for 10 min; place on ice.

8. Prepare cells:

(a) Harvest cells at $OD_{600} \sim 2.0$.
(b) Centrifuge 3000*g* for 5 min at 4°.
(c) Wash cells twice in 0.5 volume 100 m*M* lithium acetate.

9. Resuspend cells in 1/100 volume 100 m*M* lithium acetate, multiplied by the OD_{600}, e.g. (for $OD_{600} = 2.0$), 1.25 ml × 2.0 = 2.5 ml.

10. Add 1/9 volume boiled carrier DNA. Mix well.

11. Add 48 μl PCR product to each tube.

12. Add 50 μl of carrier DNA/cell mixture to each tube. Mix well. Incubate 15 min at 30°.

13. Add 300 μl lithium acetate/PEG solution. Mix well. Incubate 30 min at 30°.

14. Add 34 μl DMSO. Mix well.

15. Place tubes in a 42° water bath for 15 min.

16. Pellet the cells by spinning at 10,000–12,000 rpm at room temperature for 15 sec in a microfuge.

17. Remove PEG carefully with a P-1000 tip.

18. Add 200 μl YPD to each tube and mix well.

19. Incubate at 30° for 3 hr.

20. Spread on selectable media plates.

21. Incubate 3–5 days at 30°. Store at 4°. Expect to obtain 10–100 transformants on each plate. Control: 1 ng of an autonomous plasmid (e.g., pRS316) should yield 500–1000 transformants.

B. For Gene Disruption Cassette Generated from Yeast Genomic Template. Follow steps 1–10 above. Then:

11. Add 40 μl of *each* PCR product (80 μl total) to a 1.5-ml Eppendorf tube.

12. Add 100 μl of carrier DNA/cell mixture. Mix well. Incubate 15 min at 30°.

13. Add 600 μl lithium acetate/PEG solution. Mix well. Incubate 30 min at 30°.

14. Add 68 μl DMSO to each tube. Mix well.

15. Place tube in a 42° water bath for 15 min.

16. Pellet the cells by spinning at high speed for 15 sec in a microfuge.

17. Remove PEG carefully with a P-1000 tip.

18. Add 400 μl YPD and mix well.

19. Incubate at 30° for 3 hr.

20. Divide in half and spread 200 μl on each of two selectable media plates.

21. Incubate 3–5 days at 30°. Expect to obtain 10–100 transformants on each plate.

Verification of Correct Gene Disruption by PCR

1. Colony purify the transformants on selective agar plates and grid 4 from each transformant to a YPD master plate. Use the wild-type strain as a negative control.

2. Use freshly grown yeast cells (no more than 2 days old, and *never* refrigerated) as template for the PCR. Gently touch the surface of the colony (avoid touching the agar) with a plastic pipette tip attached to a 20-μl pipettor (set at 20 μl). Place the tip into a 50-μl PCR reaction mix (make a cocktail with extra volume to allow for pipetting error) and pipette up and down once. (Too many cells or the presence of agar will inhibit the reaction.) Keep everything on ice!

3. The following reactions test a single gene knockout in a haploid strain (see Fig. 3):

Absence of ORF	Presence of knockout
Pair 1. Primer "A"	Pair 2. Primer "A"
Primer "target B"	Primer "cassette B"
Pair 3. Primer "target C"	Pair 4. Primer "cassette C"
Primer "D"	Primer "D"

While it is probably sufficient to test just one of each pair of primers (pair 1 and pair 2 or pair 3 and pair 4), it is prudent to test all four.

1. Set up a 50 μl PCR as described in step 1 (except substituting Taq DNA Polymerase for KlenTaq) under "Preparative PCR of Gene Disruption Cassette," using the primers listed above.
2. Immediately transfer from ice to preheated (94°) PCR block.

35 Cycles—a large number of cycles is very important!

Hot start*	94°	3.0 min
Denature	94°	1.5 min
Anneal	57°	2.0 min
Extend*	72°	3.0 min
Final extension	72°	3.0 min

*We have observed that the hot start and the long extension time is very important for success of the PCR.

3. Add 5.0 μl of loading dye to each reaction and run 20 μl on an agarose gel.
4. Expect 80–90% of the transformants to pass these tests.

[18] Synthetic Dosage Lethality

By VIVIEN MEASDAY and PHILIP HIETER

Introduction

Identification of proteins that interact to perform a common function is crucial to understanding the mechanisms of cellular processes. Both genetic and biochemical methods are used to uncover an interaction between two proteins. In yeast, many genetic screens have been developed which begin with the mutation of a gene of interest and look for a phenotype (such as cell lethality) in combination with mutation or overexpression of another gene (reviewed in Ref. 1). For example, a dosage suppression screen employs a conditionally lethal mutant to isolate a gene which, when overexpressed, will rescue lethality of the mutant at the nonpermissive condition. Synthetic lethal screens select for two nonallelelic and nonessential mutations that are lethal for the cell only when both are present. Both dosage suppression and synthetic lethal screens have proven very useful in identifying genes encoding interacting proteins. This chapter discusses a screen called synthetic dosage lethality (SDL) which is based on features of both dosage suppression and synthetic lethal screening to identify interacting proteins. An SDL interaction occurs when overexpression of a cloned wild-type (reference) gene is lethal in a target mutant strain but is viable in a wild-type strain. Here we describe the SDL method and an example of its use to study budding yeast kinetochore function. SDL is a generally applicable technique for studying multiple cellular processes in both budding yeast and other organisms.

Principles of Synthetic Dosage Lethality

SDL is based on the premise that increasing levels of a "reference" protein has no detrimental effect on the growth of a wild-type strain but may cause lethality in a "target" mutant strain with reduced activity of an interacting protein. Overexpression of most yeast genes is not toxic to wild-type strains.[2] An SDL assay is a useful method to test for an interaction between two proteins. A cloned gene, or reference gene, is overexpressed in a mutant strain and in an isogenic wild-type strain. The mutant may potentially carry a target mutation in a protein that functions in a common cellular process to the reference protein. On induced overexpression of

[1] E. M. Phizicky and S. Fields, *Microbiol. Rev.* **59,** 94 (1995).
[2] H. Liu, J. Krizek, and A. Bretscher, *Genetics* **132,** 665 (1992).

0076-6879/02 $35.00

the reference gene, growth is scored at a variety of temperatures. An SDL event is defined as mutant cell lethality on reference gene overexpression at conditions where cells harboring vector alone are viable. In addition, wild-type cells overexpressing the reference gene must be viable.

The method of SDL can be applied to at least four basic screening procedures to uncover interacting proteins. First, SDL can be used as a direct test to assay for an interaction between a reference gene and a target mutation. For example, overexpression of the Cdk inhibitor protein Far1p is lethal to a strain defective in G_1 (*cdc34* mutant) but not G_2 (*cdc16* mutant) ubiquitination-dependent proteolysis.[3] This is consistent with the ubiquitination and targeting of Far1 for degradation in G_1 phase.[3] Second, SDL can be used as a secondary screen to uncover mutants of interest from a primary mutant collection by overexpression of a protein with a particular biological role. This screen may specifically identify mutants from the collection that are defective in a common function with the reference gene. The SDL assay presented here successfully isolated a kinetochore mutant from a collection of chromosome transmission fidelity mutants by overexpression of known kinetochore proteins.[4] Third, SDL can be adapted to a primary genetic screen to isolate mutants which interact with the reference protein. In this experiment, a wild-type strain is first mutagenized, an inducible reference gene introduced, and colonies isolated which are lethal only when the reference gene is overexpressed. Finally, a variation of an SDL screen can be performed on a single yeast target mutant of interest. A mutant strain is transformed with an overexpression library and screened for genes that when overexpressed cause cell lethality. For example, previous studies have shown that yeast strains with reduced APC (anaphase-promoting complex) activity cannot tolerate high levels of an APC substrate.[5] Putative APC substrates could be identified by isolating genes that when overexpressed cause cell death in an *apc* mutant. It may be important to ascertain which mutant alleles of a particular gene are sensitive to gene dosage. Mutants that are sensitive to overexpression of a protein can first be selected, or generated and then used in a screen to identify novel proteins which cause cell lethality when overexpressed.

There are a few theories that hypothesize why SDL screens identify genetically interacting proteins. The mutated gene, or target mutation, may encode a defective protein with a number of flaws. For example, the mutant may be a full-length protein with reduced activity, a truncated protein with partial activity, or a completely nonfunctional protein. If a target mutation is a partial loss-of-function mutation,

[3] S. Henchoz, Y. Chi, B. Catarin, I. Herskowitz, R. J. Deshaies, and M. Peter, *Genes Dev.* **11,** 3056 (1997).

[4] E. S. Kroll, K. M. Hyland, P. Hieter, and J. J. Li, *Genetics* **143,** 95 (1996).

[5] S. Irniger, S. Piatti, C. Michaelis, and K. Nasmyth, *Cell* **81,** 269 (1995).

the overexpressed reference protein may titrate out the remainder of the activity, resulting in cell lethality. In the case of a genomic deletion, or missense mutation near the start codon, no protein is expressed. Increasing the dosage of the reference protein may titrate out another protein that is required for cell viability in the knockout or mutant strain (essentially a synthetic lethal effect). If the function of the reference protein is to regulate the mutant protein, then overexpression of the regulatory factor could be detrimental in a strain already compromised because of mutation. Conversely, if the normal function of the mutant protein is to regulate the overexpressed protein, then increasing the dosage of the reference protein in a strain defective for its regulatory factor could be lethal. The exact molecular mechanism resulting in SDL in each scenario will depend on the pathway being investigated, the nature of the target mutation, and the function of the overexpressed reference protein.

Methods

Synthetic Dosage Lethality (SDL) Assay

A general scheme for a synthetic dosage lethality assay is diagrammed in Fig. 1. Vector alone and vector containing a reference gene are transformed separately into wild-type and mutant strains under conditions that do not allow reference gene expression (Fig. 1A). In the example shown, gene expression from the *GAL1* promoter is induced on galactose plates and repressed on glucose plates. Once transformants have been isolated and colony purified under noninducing conditions, overexpression of the reference gene is induced by streaking yeast onto plates containing appropriate media (galactose) (Fig. 1B). Growth of the yeast is assessed at various temperatures after a specified number of days in the incubator. A synthetic dosage lethal event occurs when a mutant exhibits lethality (no visual colonies on the streak) on overexpression of a reference gene at a particular temperature, but is viable with vector alone (Fig. 1C). An SDL event is valid only if the wild-type strain remains viable upon reference gene induction (Fig. 1C). In addition to scoring lethality, mutant strains that display a slow growth phenotype upon reference gene induction but not vector alone should be noted. If overexpression of vector alone and reference gene display the same phenotype in a strain (both either viable or lethal) we consider no SDL genetic interaction to have occurred (Fig. 1C).

The temperatures used to assay for an SDL phenotype depend on the mutant strain being studied. For example, a temperature-sensitive (ts) mutant should not be assessed at its nonpermissive temperature for synthetic dosage lethality. However, a ts mutant can be tested for dosage suppression at its nonpermissive temperature; in this case vector only is lethal whereas overexpression of the reference gene

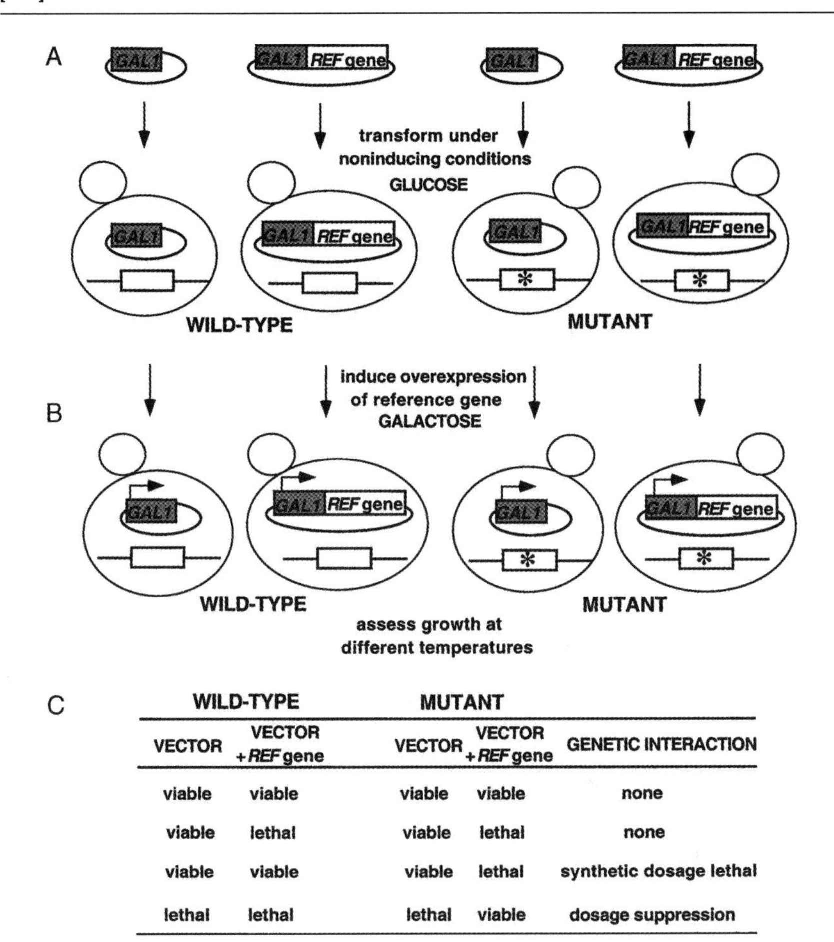

WILD-TYPE		MUTANT		
VECTOR	VECTOR + *REF* gene	VECTOR	VECTOR + *REF* gene	GENETIC INTERACTION
viable	viable	viable	viable	none
viable	lethal	viable	lethal	none
viable	viable	viable	lethal	synthetic dosage lethal
lethal	lethal	lethal	viable	dosage suppression

FIG. 1. Synthetic dosage lethality assay (A) Wild-type (gene represented by a rectangle flanked by two lines) and mutant cells (a target mutation in a gene is marked with an asterisk) are transformed with vector alone or vector plus reference (REF) gene under noninducing conditions (glucose medium). (B) Transformants are streaked onto galactose medium to induce expression (marked with an arrow) of the reference gene. (C) Growth is assessed at a variety of temperatures and the genetic interactions are scored.

is viable (Fig. 1C). If conditionally lethal mutants are used for an SDL assay, both a dosage suppression and SDL screen can be carried out at the same time. In this scenario, an SDL phenotype (induced loss of viability) is assayed at the permissive temperature for growth and dosage suppression (induced rescue of viability) is assessed at the nonpermissive temperature for growth.

Technical Tips

When transforming a large number of yeast mutant strains, we found growing each strain in liquid media to be cumbersome. We decided to use a quick and simple method for yeast transformation developed by the Gietz lab called the TRAFO method: http://www.umanitoba.ca/faculties/medicine/biochem/gietz/Quick.html.[6] Yeast strains are struck out on a plate, then used directly from the plate for transformation without growing in liquid medium. This method enables many transformations to be performed per day.

Once the strains have been transformed, it is important to ensure that petite mutants have not arisen during the colony purification. Petite mutants, which have defective mitochondria, grow poorly on galactose plates; thus it is possible to mistake a petite phenotype for an SDL phenotype. Petites form smaller colonies than wild type on glucose media—however, the mutant strains used for an SDL screen may also have a smaller colony size. A conclusive method to eliminate petite strains is to streak the yeast onto 3% glycerol plates: petite mutants will not grow whereas nonpetite strains will.

Mutant Collection

The primary phenotype used to isolate the mutant collection depends on the interest of the researcher—we had previously isolated a collection of point mutants that are unable to faithfully transmit a nonessential chromosome fragment (see below).[7] A collection of yeast genomic deletions could be used as the starting point for an SDL screen. Ongoing studies by the yeast deletion consortium to analyze phenotypes of gene knockout strains may be very useful to define a group of mutants to subject to an SDL screen. For example, the TRIPLES database (http://ygac.med.yale.edu/triples/triples.htm) offers phenotypes of mutants containing a transposon insertion at identified genomic loci—all insertion strains known to be sensitive to benomyl (a drug which destabilizes microtubules) would serve as a good starting point to screen by SDL for proteins potentially involved in microtubule function.

Inducible Gene Overexpression

The plasmid vector we chose to conditionally overexpress reference proteins is p415GEU2, which is a centromere based plasmid that contains the *GAL1* promoter.[4] Although we have not tried other inducible promoters—such as other

[6] R. D. Gietz and R. A. Woods, *in* "Molecular Genetics of Yeast: Practical Approaches" (J. A. Johnston, ed.), p. 121. Oxford University Press, Oxford, 1994.
[7] F. Spencer, S. L. Gerring, C. Connelly, and P. Hieter, *Genetics* **124,** 237 (1990).

GAL promoters, the *MET25* promoter, or the *CUP1* promoter[8,9]—they could conceivably be used for SDL assays. In addition to choosing a promoter, the copy number of the plasmid should be considered. The SDL assays performed in our lab have used both a single copy (centromere based) plasmid and a multicopy (2-μm based) plasmid.[4,10] An SDL phenotype may only be detected if the reference gene is expressed at a certain dosage. Therefore, it is worthwhile to test reference gene expression in both a single and a multicopy plasmid and under different promoters, depending on the mutant being examined.

Prior to starting an SDL screen, expression of the reference protein must be tested in a wild-type strain. Overexpression of the reference protein should not produce a noticeable phenotype in a wild-type strain. The choice of reference protein is entirely dependent on the program of study. A reference protein that is a member of a macromolecular complex may serve as a good starting point for an SDL screen; macromolecular complexes are sensitive to changes in dosage of their components.[11] In the SDL screen described below, two components of the inner yeast kinetochore and one component of the outer yeast kinetochore were chosen for overexpression. SDL tests have also been performed with Orc6p, which is a subunit of a macromolecular complex that initiates DNA replication. Increasing the dosage of *ORC6* was shown to be specifically detrimental to strains carrying mutations in DNA replication proteins, but not to strains carrying mutations in kinetochore components.[4]

Examples of Synthetic Dosage Lethality

Direct SDL Tests to Characterize Overexpression of Kinetochore Proteins in Kinetochore Mutant Strains

We performed direct SDL assays using proteins and mutants involved in kinetochore function to test the validity of the SDL method. The budding yeast kinetochore is composed of conserved centromere DNA elements and associated proteins. An essential protein complex called CBF3, which binds to the CDEIII DNA element, consists of four essential subunits and comprises the inner kinetochore proteins (reviewed in Refs. 12 and 13). We chose two CBF3 proteins, Ctf13p and Ndc10p/Ctf14p, to inducibly overexpress in an SDL assay because of their essential function at the kinetochore. We also inducibly overexpressed a member of the outer kinetochore called Ctf19p (reviewed in Refs. 12 and 13). The vector plus

[8] J. O. Mascorro-Gallardo, A. A. Covarrubias, and R. Gaxiola, *Gene* **172,** 169 (1996).
[9] D. Mumberg, R. Muller, and M. Funk, *Nucleic Acids Res.* **22,** 5767 (1994).
[10] K. M. Hyland, J. Kingsbury, D. Koshland, and P. Hieter, *J. Cell Biol.* **145,** 15 (1999).
[11] J. Rine, *Methods Enzymol.* **194,** 239 (1991).
[12] A. L. Pidoux and R. C. Allshire, *Curr. Opin. Cell Biol.* **12,** 308 (2000).
[13] J. Ortiz and J. Lechner, *Protoplasma* **211,** 12 (2000).

TABLE I
SYNTHETIC DOSAGE LETHALITY ASSAYS WITH CHARACTERIZED KINETOCHORE PROTEINS AND MUTANTS[a]

Mutant	Temperature (°C)	*p415GEU2* GLU[e]	GAL[f]	*pGAL1-CTF13*[b] GLU	GAL	*pGAL1-CTF14*[c] GLU	GAL	*pGAL1-CTF19*[d] GLU	GAL
ctf13-30	25	+	+	+	+	+	+	+	+
	30	+	+	+	+	+	+	+	+
	35	−	−	−	+	−	−	−	−
ctf14-42	25	+	+	+	−	+	+	+	−
	30	+	+	+	−	+	+	+	−
	35	−	−	−	−	−	+	−	−
ctf19-58	25	+	+	+	−	+	−	+	+
	30	+	+	+	−	+	−	+	+
	35	+	+	+	−	+	−	+	+

[a] +, Visible colonies on streak; −, no visible colonies on streak. Mutants per Ref. 7.
[b] *pGAL1-CTF13* (pKF88), SDL interactions published [E. S. Kroll, K. M. Hyland, P. Hieter, and J. J. Li, *Genetics* **143,** 95 (1996)].
[c] *pGAL1-CTF14* (pKH2).[14]
[d] *pGAL1-CTF19* (pKH21), data reproduced with permission from Table III from *J. Cell Biol.* **145,** 21 (1999).
[e] Four days of growth on glucose (GLU) plates.
[f] Six days of growth on galactose (GAL) plates.

reference gene constructs used were *pGAL1-CTF13* (pKF88),[4] *pGAL1-CTF19* (pKH21),[10] and *pGAL1-NDC10/CTF14* (pKH2).[14]

Overexpression of the kinetochore proteins was tested in three strains containing mutations in the same proteins (*ctf13-30^{ts}*, *ctf14-42^{ts}*, and *ctf19-58*).[7] This control assay served two purposes. One was to confirm expression of *CTF13* and *NDC10/CTF14* from the *GAL1* promoter by testing for rescue of *ctf13-30^{ts}* and *ctf14-42^{ts}* strains, respectively, at their nonpermissive temperatures (Table I).[4] Previously, expression of *CTF19* from pKH21 was shown to rescue the sectoring phenotype of a *ctf19* deletion strain.[10] The second purpose of the assay was to test for an SDL phenotype among well-characterized mutant strains carrying mutations in known kinetochore proteins. Overexpression of *CTF13* on galactose plates, as previously published, was found to be lethal to *ctf14-42^{ts}* and *ctf19-58* mutants at all temperatures tested (25°, 30°, 35°) whereas vector alone had no effect on growth (Table I).[4] Induction of *NDC10/CTF14* expression was SDL in a *ctf19-58* mutant strain at all temperatures tested but was not detrimental in the *ctf13-30^{ts}* mutant. Overexpression of *CTF19* was also SDL in a *ctf14-42^{ts}* mutant at all temperatures tested but did not affect a *ctf13-30^{ts}* mutant.[10] This series of direct

[14] pKH2 was created by cloning a 3.3 kb *Cla*I–*Pvu*II fragment containing *NDC10/CTF14* into *Cla*I–*Sma*I digested p415GEU2.

tests demonstrates that the SDL assay can identify a genetic interaction between a kinetochore protein and a kinetochore protein mutation and that not all genetic interactions are reciprocal. For example, overexpression of *CTF13* is lethal to a *ctf14-42* mutant, but overexpression of *NDC10/CTF14* is not lethal to a *ctf13-30* mutant (Table I).

SDL Screen to Identify Novel Kinetochore Proteins in Chromosome Transmission Fidelity Mutant Collection

The chromosome transmission fidelity (*ctf*) mutants were originally isolated as mutants that lose a nonessential chromosome fragment at a higher frequency than wild-type strains.[7] Some of the *ctf* mutants have been cloned and found to encode kinetochore proteins (Ctf13p, Ndc10p/Ctf14p, Ctf19p),[10,15] cohesion proteins (Ctf7p/Eco1p and Ctf8p),[16–18] and a DNA replication protein (Ctf10p/Cdc6p).[4] We decided to use the SDL assay as a method to screen through the remainder of the unknown *ctf* mutants to identify those more likely to be defective in kinetochore function. We chose to overexpress the three kinetochore proteins characterized above (*CTF13, NDC10/CTF14, CTF19*) in six *ctf* complementation groups (containing multiple alleles) and 31 single allele mutants that have not been assigned to complementation groups. The results of the SDL screen will be published elsewhere[19]; however, data concerning one of the *ctf* mutants isolated in this screen will be discussed.

Our SDL screen demonstrated that four out of five total *ctf5* alleles showed sensitivity to overexpression of the kinetochore proteins (Table II). *ctf5-31* was SDL on *NDC10/CTF14* overexpression at all temperatures tested while *ctf5-70* was SDL on increased dosage of both *CTF13* and *NDC10/CTF14* at all temperatures tested. *ctf5-110* and *ctf5-131* grew more slowly on *CTF13* and *NDC10/CTF14* overexpression at 30° and displayed an SDL phenotype at 35°. One *ctf5* allele, *ctf5-163*, displayed no SDL phenotypes. The *ctf5* allele specificity of the SDL phenotype is illustrated in Fig. 2. At 30°, *ctf5-70* is clearly nonviable on galactose plates with increased dosage of *CTF13* and *NDC10/CTF14*, whereas *ctf5-131* is still able to grow, but not as well as the wild-type strain. The sensitivity that *ctf5* mutants display to overexpression of kinetochore proteins suggests that these mutants may be defective in kinetochore function.

[15] K. F. Doheny, P. K. Sorger, A. A. Hyman, S. Tugendreich, F. Spencer, and P. Hieter, *Cell* **73,** 761 (1993).

[16] M. L. Mayer, S. P. Gygi, R. Aebersold, and P. Hieter, *Mol. Cell* **7,** 959 (2001).

[17] R. V. Skibbens, L. B. Corson, D. Koshland, and P. Hieter, *Genes Dev.* **13,** 307 (1999).

[18] A. Toth, R. Ciosk, F. Uhlmann, M. Galova, A. Schlieffer, and K. Nasmyth, *Genes Dev.* **13,** 320 (1999).

[19] V. Measday, D. W. Hailey, I. Pot, S. A. Givan, K. M. Hyland, G. Cagney, S. Fields, T. N. Davis, and P. Hieter, *Genes Dev.* **16,** 101 (2002).

TABLE II
ctf5 MUTANT DISPLAYING SYNTHETIC DOSAGE LETHAL INTERACTIONS WITH KINETOCHORE PROTEINS[a]

Mutant	Temperature (°C)	*p415GEU2*		*pGAL1-CTF13*[b]		*pGAL1-CTF14*[c]		*pGAL1-CTF19*[d]	
		GLU[e]	GAL[f]	GLU	GAL	GLU	GAL	GLU	GAL
ctf5-31	25	+	+	+	+	+	−	+	+
	30	+	+	+	+	+	−	+	+
	35	+	+	+	+	+	−	+	+
ctf5-70	25	+	+	+	−	+	−	+	+
	30	+	+	+	−	+	−	+	+
	35	+	+	+	−	+	−	+	+
ctf5-110	25	+	+	+	+	+	+	+	+
	30	+	+	+	(+)	+	(+)	+	+
	35	+	+	+	−	+	−	+	+
ctf5-131	25	+	+	+	+	+	+	+	+
	30	+	+	+	(+)	+	(+)	+	+
	35	+	+	+	−	+	−	+	+
ctf5-163	25	+	+	+	+	+	+	+	+
	30	+	+	+	+	+	+	+	+
	35	+	+	+	+	+	+	+	+

[a] +, Visible colonies on streak; −, no visible colonies on streak; (+) small colony size on streak. Mutants per Ref. 7.
[b] *pGAL1-CTF13* (pKF88) [E. S. Kroll, K. M. Hyland, P. Hieter, and J. J. Li, *Genetics* **143,** 95 (1996)].
[c] *pGAL1-CTF14* (pKH2); see Ref. 14.
[d] *pGAL1-CTF19* (pKH21) [K. M. Hyland, J. Kingsbury, D. Koshland, and P. Hieter, *J. Cell Biol.* **145,** 15 (1999)].
[e] Four days of growth on glucose (GLU) plates.
[f] Six days of growth on galactose (GAL) plates.

In order to identify the gene locus mutated in *ctf5* strains, the *CTF5* gene was cloned by complementation of the chromosome fragment (CF) loss phenotype originally used to isolate the *ctf* mutant collection.[7] The rescuing genomic clones that were isolated contained three overlapping open reading frames (ORFs): *MCM21*, *YDR319C*, *SWA2* (rescue coordinates: Chromosome IV—1103280bp to 1108078bp). Interestingly, Mcm21p has been identified as a kinetochore protein that interacts with Ctf19p and Okp1p (outer kinetochore protein 1) and that coimmunoprecipitates with centromere DNA.[19] Although Mcm21p has not been shown to immunoprecipitate with Ctf13p or Ndc10p/Ctf14p, the Mcm21p interacting protein, Okp1p, immunoprecipitates with Ndc10p/Ctf14p and Mcm21p centromere binding requires functional Ndc10p/Ctf14p.[20] Therefore, Mcm21p was a likely candidate to be mutated in *ctf5* strains. The other two overlapping ORFs in the

[20] J. Ortiz, O. Stemmann, S. Rank, and J. Lechner, *Genes Dev.* **13,** 1140 (1999).

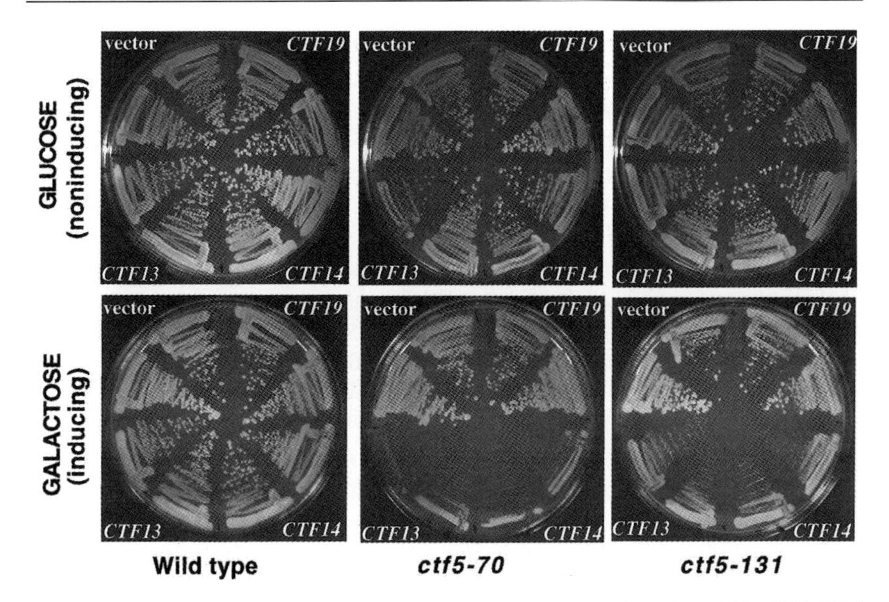

FIG. 2. *ctf5* Mutants are synthetic dosage lethal on overexpression of *CTF13* and *NDC10/CTF14*. Yeast strains—wild-type (YPH277), *ctf5-70* and *ctf5-131*—were transformed with vector alone (*p*415GEU2), *pGAL1-CTF13* (pKF88), *pGAL1-NDC10/CTF14* (pKH2), and *pGAL1-CTF19* (pKH21) under noninducing conditions (glucose). Single colonies were isolated and struck onto glucose plates where plasmid gene expression is repressed and grown at 30° for 4 days. The same isolates were struck onto galactose plates to induce plasmid gene expression and grown at 30° for 6 days. Plates are divided into quadrants containing two isolates of each transformant: upper left quadrant is vector alone, p415GEU2; upper right, pKH21; lower left, pKF88; lower right, pKH2.

rescue clones, *YDR319C* and *SWA2*, do not encode proteins with any known role in kinetochore function. We were able to demonstrate genetic linkage between a deletion of *MCM21* and the *ctf5* mutation.

Discussion

We have described a method termed synthetic dosage lethality, or SDL, which is useful for identifying genetic interactions between a mutant protein and an overexpressed protein involved in a common pathway. We have discussed the use of SDL both for direct assays between two known proteins and as a method to screen for novel proteins in a collection of mutants that are defective in a specific function. The specific SDL screen described here was designed to uncover mutants from a large collection that were defective in kinetochore function.

SDL is applicable to a wide variety of biological processes. In addition, SDL assays are able to distinguish between mutations in proteins that share a general common phenotypic defect but do not have the same function in the cell.

For example, the chromosome transmission fidelity mutant collection from our lab contains mutants defective in DNA replication, cohesion, kinetochore activity, and potentially other functions. Overexpression of the *CTF13* kinetochore protein was specifically detrimental to *ctf* kinetochore mutants but not to *ctf* DNA replication or cohesion mutants.[4] It was also found that the SDL method identified a broader spectrum of genetic interactions than dosage suppression upon overexpression of *CTF13*.[4]

The SDL assay can be sensitive to allele specific interactions. In our screen, not all of the five *ctf5/mcm21* alleles displayed the same phenotype with respect to SDL interactions. For example, *ctf5-70* was SDL at 25° on *CTF13* and *NDC10/CTF14* overexpression, whereas *ctf5-163* was unaffected (Table II). Previously, it was found that increasing the dosage of *ORC6* was toxic in a *cdc6-1* mutant but suppressed the growth defect of a *cdc6-103* mutant.[4] This suggests that all alleles of a particular mutant should be tested when performing an SDL screen. It is also noteworthy that overexpression of the outer kinetochore protein, *CTF19*, did not perturb any of the *ctf5/mcm21* mutants despite the fact that Mcm21p coimmunoprecipitates with Ctf19p.[19] Therefore, more than one choice of reference gene may be prudent prior to starting an SDL screen.

The method of synthetic dosage lethal screening may be broadly applicable to other systems/organisms. Mammalian cell lines that have been mutagenized to select for a particular phenotype (such as insensitivity to certain growth factors) may be a good starting point for an SDL screen. Retroviral cDNA libraries could be useful instead of cDNA transfections for testing expression of a random population of cDNAs. The transfer efficiencies of retroviral vectors are higher than those of cDNA transfections and the integrated viruses can be recovered, thus facilitating identification of the cDNA inserts (reviewed in Ref. 21). Many *Drosophila melanogaster* mutant collections have been isolated using X-irradiation, P-element insertion, and other methods, and thus an SDL assay could conceivably be performed in this organism. As more tools are developed to create mutants and inducibly express genes in other organisms, SDL may serve as a good method to identify interacting proteins by genetic screening in a variety of systems.

Acknowledgments

We thank Kristin Baetz, Kathy Hyland, Katsumi Kitagawa, Melanie Mayer, and Isabelle Pot for critical reading of this manuscript and Scott Givan for help with graphics. V.M. was supported by an NCIC Postdoctoral Fellowship; P.H. was supported by a CIHR Senior Scientist Award. The kinetochore work was supported by NIH Grant CA16519 to P.H.

[21] G. R. Stark and A. V. Gudkov, *Hum. Mol. Genet.* **8,** 1925 (1999).

Section III

Genomics

[19] *Saccharomyces* Genome Database

By LAURIE ISSEL-TARVER, KAREN R. CHRISTIE, KARA DOLINSKI, REY ANDRADA, RAMA BALAKRISHNAN, CATHERINE A. BALL, GAIL BINKLEY, STAN DONG, SELINA S. DWIGHT, DIANNA G. FISK, MIDORI HARRIS, MARK SCHROEDER, ANAND SETHURAMAN, KANE TSE, SHUAI WENG, DAVID BOTSTEIN, and J. MICHAEL CHERRY

Introduction

The goal of the *Saccharomyces* Genome Database (SGD) is to provide information about the genome of this yeast, the genes it encodes, and their biological functions. The genome sequence of *S. cerevisiae* provides the structure around which information in SGD is organized; value is added to the sequence by careful biological annotation drawn from a number of sources. SGD curates and stores information about budding yeast DNA and protein sequences, genetics, cell biology, and the associated community of researchers. SGD also provides search and analysis tools designed to help researchers mine the data for pieces or patterns of biological information relevant to their interests. A continuing challenge for the staff of SGD is to present up-to-date information about yeast genes in a format that is intuitive and useful to biomedical researchers, while responding to the needs of this community by providing resources and tools for exploring the data in new ways.

This chapter describes the organization of SGD, the sources of the data stored in SGD, some methods for retrieving information from the database, connections SGD has with outside databases and non-yeast research communities, and SGD's repository of yeast community information. This is not a complete overview of the database, as SGD contains hundreds of tools, including specialized sequence analysis programs, and new tools are always being added. As of this writing, several new tools and Web interfaces for DNA and protein sequences are being developed, along with enhanced database navigation methods. New tools for comparison of the *S. cerevisiae* genome with other fungal genomes will also soon be available. To explore the resources currently available at SGD, visit the Web site at http://genome-www.stanford.edu/Saccharomyces/.

Locus-Centered Organization of SGD

The systematic sequencing project[1] defined the set of genes and non-open reading frame (ORF) features (centromeres, tRNAs, etc.) around which information in SGD is organized. SGD's Locus pages display basic information about a locus and provide links to further information and resources (Fig. 1). The basic information

0076-6879/02 $35.00

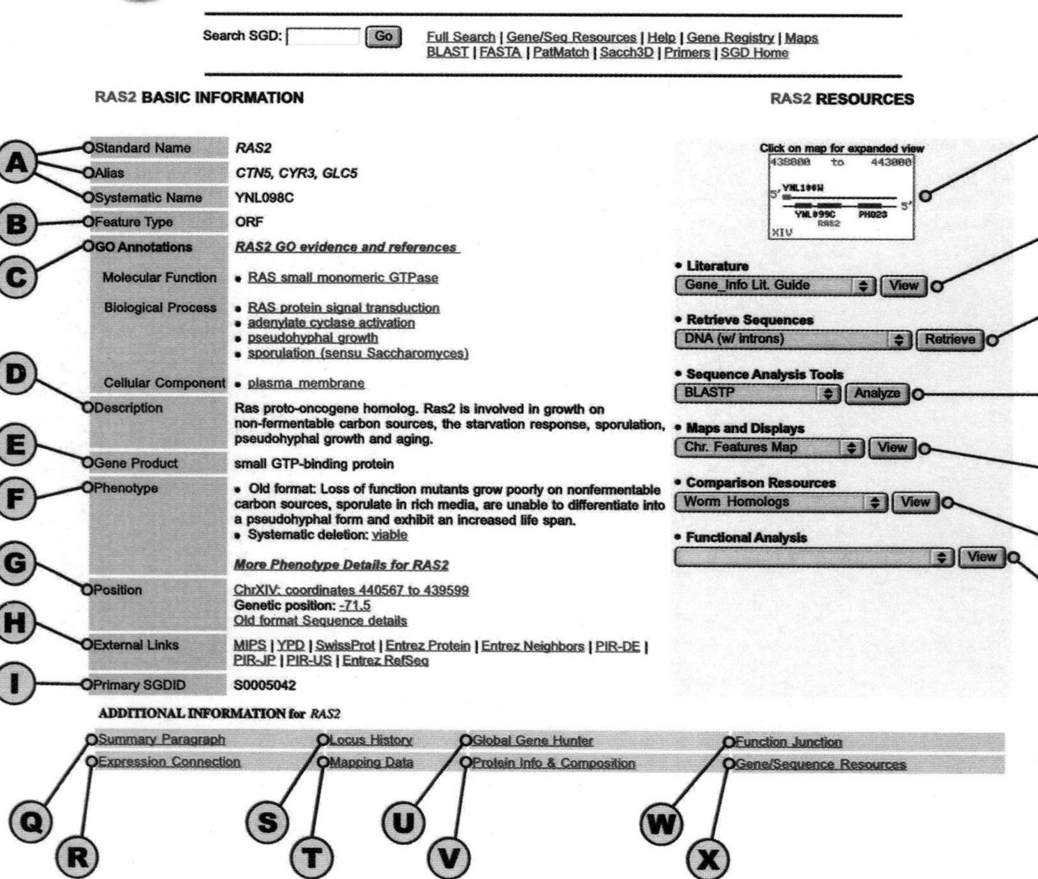
SGD
RAS2/YNL098C
Help
Search SGD:
Go
Full Search | Gene/Seq Resources | Help | Gene Registry | Maps
BLAST | FASTA | PatMatch | Sacch3D | Primers | SGD Home
RAS2 BASIC INFORMATION
Standard Name
RAS2
Alias
CTN5, CYR3, GLC5
Systematic Name
YNL098C
Feature Type
ORF
GO Annotations
RAS2 GO evidence and references
Molecular Function
RAS small monomeric GTPase
Biological Process
RAS protein signal transduction
adenylate cyclase activation
pseudohyphal growth
sporulation (sensu Saccharomyces)
Cellular Component
plasma membrane
Description
Ras proto-oncogene homolog. Ras2 is involved in growth on non-fermentable carbon sources, the starvation response, sporulation, pseudohyphal growth and aging.
Gene Product
small GTP-binding protein
Phenotype
Old format: Loss of function mutants grow poorly on nonfermentable carbon sources, sporulate in rich media, are unable to differentiate into a pseudohyphal form and exhibit an increased life span.
Systematic deletion: viable
More Phenotype Details for RAS2
Position
ChrXIV: coordinates 440567 to 439599
Genetic position: -71.5
Old format Sequence details
External Links
MIPS | YPD | SwissProt | Entrez Protein | Entrez Neighbors | PIR-DE | PIR-JP | PIR-US | Entrez RefSeq
Primary SGDID
S0005042
RAS2 RESOURCES
Click on map for expanded view
Literature
Gene_Info Lit. Guide
View
Retrieve Sequences
DNA (w/ introns)
Retrieve
Sequence Analysis Tools
BLASTP
Analyze
Maps and Displays
Chr. Features Map
View
Comparison Resources
Worm Homologs
View
Functional Analysis
View
ADDITIONAL INFORMATION for RAS2
Summary Paragraph
Locus History
Global Gene Hunter
Function Junction
Expression Connection
Mapping Data
Protein Info & Composition
Gene/Sequence Resources
A
B
C
D
E
F
G
H
I
J
K
L
M
N
O
P
Q
R
S
T
U
V
W
X

FIG. 1. The *RAS2* Locus Page. (A) The names and aliases for a locus are listed, with an indication of whether a name is standard or reserved. (B) Feature Type indicates whether a locus is an ORF, TyORF, LTR, tRNA, RNA gene, or centromere. (C) Gene Ontology terms assigned to a gene are listed, and each term links to a page showing all yeast genes annotated to that term (see Fig. 5). There is also a link from the locus page to a table that lists the references that were used for assigning each GO term, and the type of evidence from that reference (direct assay, genetic interaction, etc.) that supported the assignment. (D) The Description field lists general information about the gene. (E) The Gene Product field lists the protein or gene product that the gene encodes. (F) Mutant phenotypes of a gene are listed. Clicking on a phenotype leads to a table of other yeast genes that share the phenotype. (G) The Position field lists the chromosomal coordinates of the gene as well as its genetic position. The chromosomal coordinates link to the chromosomal features map, and the genetic position links to the combined physical and genetic map. (H) External links provide additional sources of information about a gene. These links take users outside of SGD. (I) SGDIDs are unique database identifiers for *S. cerevisiae* loci. (J) The mini ORF map shows the chromosomal features located near the locus. Loci encoded on the Watson and Crick strands are shown separately and color coded according to strand and feature type. (K) The pull-down Literature menu gives users access to SGD-curated items (the Literature Guide and Gene Summary) as well as an external link to a PubMed search for that gene name. (L) Users may specify a sequence to retrieve, choosing from options that include DNA (with introns), coding sequence, ORF translation, and 6-frame translation. Other options include custom retrieval specified by the user, retrieving all associated sequences, and retrieving lists of restriction fragment sizes. (M) Sequence Analysis Tools include BLASTN, BLASTP, FASTA nt, and FASTA aa analyses. This menu also allows retrieval of a restriction map of the sequence or the Design Primers tool. (N) Maps and Displays retrieves several locus-centered maps or tables, including a chromosomal features map or table, the Physical & Genetic Map, Physical Map, and the Physical/Genetic Map Ratios. (O) Comparison Resources allow users to retrieve information about genes in other organisms that show significant similarity to the yeast gene described on this locus page. (P) Users can retrieve gene-specific results from a fast-growing list of published functional analysis studies. (Q) The Gene Summary Paragraph is a summary of published biological information for a gene and its product that is designed to familiarize both yeast and non-yeast researchers with the general facts and important subtleties regarding a locus. (R) Expression Connection allows users to search the results of several microarray studies for gene expression data. (S) Locus History provides notes about the locus, which may alert the user to contradictory information in the literature or to potentially confusing gene names. For reserved gene names, the Locus History includes the reservation date and expiration date. (T) Two-point genetic mapping data tables are available for many loci. (U) Global Gene Hunter searches several online databases for locus-specific information. (V) The Protein Info and Composition page provides a great deal of information about protein sequence, chemistry, and more. (W) Function Junction searches functional analysis project sites for locus-specific results. (X) Gene/Sequence Resources allows users to access biological information, table/map displays, and sequence analysis and retrieval options for a locus.

about a locus includes the standard gene name, the systematic ORF name, and any aliases; Gene Ontology[2,3] annotations describing the gene product's molecular functions, biological processes, and cellular components; additional brief information about the locus and gene product; phenotype information; the position of the locus on the genetic and physical maps; and links to information about the locus in other databases. The assembled resources available for a locus include links to the scientific literature (SGD's Literature Guide, a PubMed search for that locus name, and a curator-composed Gene Summary if available); several options for locus-specific DNA or protein sequence retrieval; sequence analysis tools; genetic and physical map displays; genome comparison resources for finding homologs of the locus in other organisms; and a large set of functional analysis links which connect the user to data for that locus from several published functional analysis studies. Also present on the locus page is a clickable mini-ORF map, showing the locus and any adjacent chromosomal features. Arrayed along the bottom of the locus page is a third series of links: the locus' Gene Summary provides a brief, in-depth description of the gene and its gene product; Locus History contains gene nomenclature information, a list of researchers associated with the locus, a history of updates to the locus sequence or coordinates, and other pertinent information; Global Gene Hunter allows users to locate information about the locus from several different online databases; Function Junction searches for results specific to that locus from several genome-wide functional analysis projects; Expression Connection simultaneously searches the results of several published microarray studies for gene expression results for the locus; Mapping Data provides genetic and physical map information for the locus; Protein Information and Composition contains data generously provided by the YPD[4]; Researchers provides a list of Colleagues associated with the locus; and Gene/Sequence Resources gives users many options for accessing information about a locus' sequence, map position, biology, and more. As research into yeast genes provides new and different types of data, the locus information will evolve to reflect those changes.

Sources of Information in SGD

SGD obtains biological data from many different sources. Curators, who are Ph.D. biologists, are trained to maintain and validate information within the

[1] A. Goffeau, B. G. Barrell, H. Bussey, R. W. Davis, B. Dujon, H. Feldmann, F. Galibert, J. D. Hoheisel, C. Jacq, M. Johnston, E. J. Louis, H. W. Mewes, Y. Murakami, P. Philippsen, H. Tettelin, and S. G. Oliver, *Science* **274,** 546, 563 (1996).

[2] Gene Ontology Consortium, *Genome Res.* **11,** 1425 (2001).

[3] Gene Ontology Consortium, *Nat. Genet.* **25,** 25 (2000).

[4] M. C. Costanzo, J. D. Hogan, M. E. Cusick, B. P. Davis, A. M. Fancher, P. E. Hodges, P. Kondu, C. Lengieza, J. E. Lew-Smith, C. Lingner, K. J. Roberg-Perez, M. Tillberg, J. E. Brooks, and J. I. Garrels, *Nucleic Acids Res.* **28,** 73 (2000).

database, process the information to check for accuracy and consistency, and then post the data for public view. The majority of the data in SGD are freely available for download via ftp (ftp://genome-ftp.stanford.edu/pub/yeast/). SGD information sources include the yeast genome systematic sequencing project, published literature, individual users, genome-wide functional studies, and scientific databases.

Systematic Sequencing Project

The genome sequence of *S. cerevisiae,* completed in 1996,[1] is at the heart of SGD. Curators keep the genome sequence up-to-date by incorporating individual corrections made by researchers as reported in the literature or communicated directly to SGD curators, and by incorporating the more extensive changes that result from larger-scale resequencing efforts. These sequence updates are performed in collaboration with MIPS,[5] and in consultation with the original systematic sequencing groups. A compendium of the changes made to the original genome sequence is available in sequence update tables at the SGD Web site (http://genome-www.stanford.edu/Saccharomyces/sequenceupdates.html). The systematic sequencing effort also defined a set of known and predicted open reading frames (ORFs); curators at SGD revise that set of ORFs by incorporating previously unpredicted (often small) ORFs that researchers have experimentally identified and designating as "questionable" those ORFs that research indicates are unlikely to be transcribed genes. SGD strives to provide laboratory researchers and computational biologists with an accurate, up-to-date representation of the yeast genome. Generally, this represents a synthesis of the systematic sequencing entries in GenBank. However, at times the SGD-provided sequence is ahead of the international DNA databanks.

Published Literature

The biological information about yeast genes found in SGD is largely derived from the published literature. Weekly searches of PubMed incorporate newly published papers describing named *S. cerevisiae* genes into SGD and are critical to keeping the database current. SGD adds value to this resource by assigning papers associated with a gene to categories according to the topics that each paper covers, thus creating a comprehensive Literature Guide (previously called Gene Info) for each gene. This Literature Guide helps direct researchers to particularly relevant publications. Curators also use published literature to choose Gene Ontology[2,3] terms to describe gene products and to compose Gene Summary paragraphs (see below). Besides acting as a valuable resource for yeast biologists, SGD's literature services help researchers from outside the community take advantage of the wealth

[5] H. W. Mewes, D. Frishman, C. Gruber, B. Geier, D. Haase, A. Kaps, K. Lemcke, G. Mannhaupt, F. Pfeiffer, C. Schuller, S. Stocker, and B. Weil, *Nucleic Acids Res.* **28,** 37 (2000).

of experimental evidence available for *S. cerevisiae,* by allowing them to quickly focus on their particular area of interest.

Individual Users

Much of the data contained in SGD has come directly from individual users. The SGD curators daily incorporate information sent by researchers to update and improve all aspects of the database. There are two main routes by which users communicate information to SGD: Web-based forms and email to the curators.

Web-Based Forms. There are forms on SGD's Web site that allow users to submit information directly to curators. One of the most commonly used is the Gene Registry form (http://genome-www4.stanford.edu/cgi-bin/SGD/registry/geneRegistry). In 1994 Robert Mortimer transferred the task of maintaining the nomenclature of *S. cerevisiae* genes, the Gene Name Registry, to SGD. Yeast researchers can reserve a gene name or register a published gene name by submitting a completed Gene Registry form. This form accepts an explanation of the acronym, the identity of the corresponding open reading frame (ORF), any aliases for that gene, a description of the encoded gene product, any phenotypes associated with mutations in the gene, references associated with the gene, and any other comments the researcher would like listed with the gene entry in SGD. Curators process the form first by searching several databases to ensure that the proposed gene name has not been previously used for another *S. cerevisiae* gene, and then by reviewing the submitted data to ensure that the resulting database entries will be clear and useful to all users.

The Colleague Submission/Update form (http://genome-www4.stanford.edu/cgibin/SGD/colleague/colleagueSearch) is another frequently used form on SGD's Web site. This form allows users to create or update their contact information in SGD, thus providing other members of the research community with telephone and mail information, Web page addresses, a description of research interests, lists of co-workers, and other relevant data.

E-mail to Curators. SGD encourages the submission of information via e-mail to the yeast curator address (*yeast-curator@genome.stanford.edu*). Users enrich SGD by providing updated information about genes and sequences, and by making suggestions about the content and features provided. User scrutiny of database contents benefits the community by ensuring that SGD remains accurate and up-to-date. Questions submitted by researchers about retrieving particular kinds of data from SGD and other resources have been an important source of inspiration for the development of new tools. Pattern matching and sequence retrieval tools have been created, and existing tools have been improved, in response to user requests.

Genome-Wide Studies

Genome-wide analyses of genes and gene products have created enormous datasets rich with information about individual genes. As the number of large-scale experiments has grown, so has the need to create intuitive and useful methods for accessing and analyzing the results of those experiments. Consequently, one of SGD's high priorities is to find ways to assimilate increasing numbers of datasets from gene expression, systematic deletion, and functional analysis projects, to make them readily available and easy to navigate. To present the data from genome-wide analyses SGD either incorporates the data into its own database or provides users with gene-specific links to external databases.

Other Databases

Other scientific databases are an important source of information for SGD. These databases include those that provide specifics about yeast genes (such as descriptions of gene products, sequences, intron/exon boundaries, promoters, and tRNAs), those that provide a framework for classifying gene products, and those that provide information about homologs of yeast genes in other organisms. Some of the external databases that SGD relies on most heavily include PubMed, GenBank, YPD, MIPS, and SwissProt.[4–8]

A database collaboration that has contributed greatly to the biological information available in SGD is the Gene Ontology (GO).[2,3] SGD is one of the founding members of GO, a collaboration among several model organism databases whose objective is to produce shared, structured vocabularies for the biological description of gene products in any organism. Consortium members are developing three independent networks of terms, collectively called Gene Ontology, in which biological concepts and the relationships between them are specified. One ontology describes the molecular functions a gene product carries out, another describes the biological processes in which a gene product is involved, and the third describes the cellular components where a gene product is found. The short phrases that describe a gene product's function, process, and cellular component are called GO terms. In addition to developing the ontologies, the member databases of the GO Consortium are using GO terms to annotate gene products in their respective model organisms and contributing the annotations to a shared central resource.

Accessing Biological Information at SGD

There are several possible starting points and paths a researcher might use to get biological information about yeast genes. Options for identifying genes of interest and discovering connections between genes include retrieving information such as sequences, expression patterns, phenotypes, associated key words, or Gene

[6] A. Bairoch and R. Apweiler, *Nucleic Acids Res.* **28,** 45 (2000).

[7] D. L. Wheeler, C. Chappey, A. E. Lash, D. D. Leipe, T. L. Madden, G. D. Schuler, T. A. Tatusova, and B. A. Rapp, *Nucleic Acids Res.* **28,** 10 (2000).

[8] K. Dolinski, C. A. Ball, S. A. Chervitz, S. S. Dwight, M. A. Harris, S. Roberts, T. Roe, J. M. Cherry, and D. Botstein, *Yeast* **14,** 1453 (1998).

Ontology annotations relevant to the user's interests. Once the appropriate genes have been identified, the user's options for exploring the genes' biology begin at the locus pages. In particular the user should read Gene Summary paragraphs, explore the literature through the Literature Guide, and browse the hyperlinks to other databases provided as external links. For navigating easily through the database, the SGD Search box at the top of almost every page allows users to do a quick search using query terms such as gene names, Colleague names, GO terms, gene product names, and more.

Starting Points

To start learning about the tools provided by SGD we recommend the Resource Guide, available at the URL http://genome-www.stanford.edu/Saccharomyces/resource_guide.html (Fig. 2). This provides a listing of resources for investigating gene information, the scientific literature, sequence analysis options, bench-top tools, genetic and physical maps, and genome-wide functional analysis studies. There are innumerable strategies researchers might use to explore yeast genes, and SGD hopes to aid its users in as many of these as possible. A few likely starting points for identifying yeast genes relevant to a researcher's interests are described below.

Sequence. Because genes with similar sequences often perform similar molecular functions (although perhaps in different pathways), sequence comparisons provide a powerful method of identifying genes with common functions. Similarity to a characterized yeast gene product can give significant clues about the function of an uncharacterized gene, whether from yeast or from another organism.

Sequence comparisons against *S. cerevisiae* sequences can be performed at the SGD Web site (http://genome-www2.stanford.edu/cgi-bin/SGD/nph-blast2sgd). Users input any DNA or protein query sequence and choose from among several BLAST options. The *S. cerevisiae* DNA datasets that can be queried include the complete genomic sequence including mitochondrial DNA, ORF coding DNA, intergenic DNA, ORF upstream flanking sequences, and the set of all *S. cerevisiae* DNA sequences found in GenBank.[7] This set includes the individual results of the systematic sequencing efforts as well as those from the many laboratories which represent the yeast community. Protein datasets that can be queried include translations of all *S. cerevisiae* ORFs, and all *S. cerevisiae* protein sequences from GenPept, PIR, and Swissprot.[6,7,9] Several parameters can be modified to customize the BLAST search, and results can be returned to the user in a variety of formats. The results can be viewed immediately or optionally sent via e-mail to an address the user defines.

[9] W. C. Barker, J. S. Garavelli, H. Huang, P. B. McGarvey, B. C. Orcutt, G. Y. Srinivasarao, C. Xiao, L. S. Yeh, R. S. Ledley, J. F. Janda, F. Pfeiffer, H. W. Mewes, A. Tsugita, and C. Wu, *Nucleic Acids Res.* **28,** 41 (2000).

Gene Information

SGD Resource	Primary Application	Description
Gene Summary Paragraph	Overview an unfamiliar gene	A synopsis of published information on a gene, including selected references, written in natural language by SGD Scientific Curators.
Locus Page	Find basic gene information and use as a "launch pad" for many SGD tools	Concise, basic information on a gene product, phenotypes, mapping, sequence, functional analysis, expression data, and more. Access to many popular SGD tools and services with the gene name/sequence already entered as a default.
Global Gene Hunter	Retrieve gene information from several databases	Simultaneous retrieval of information for a given locus from the following eight databases: SGD, Genbank, PubMed, Sacch3D, Swiss-Prot, MIPS, Yeast Protein Database (YPD), Protein Information Resource (PIR)

Literature

SGD Resource	Primary Application	Description
Gene Summary Paragraph	Key references for general facts about a gene	A synopsis of published information on a gene, including selected references, written in natural language by SGD Scientific Curators.
Literature Guide	An annotated list of publications on a gene	Published literature on a gene, grouped according to biological topics. References are generated by PubMed searches and then reviewed and categorized by SGD Scientific Curators

Sequence Analysis & Comparison

SGD Resource	Primary Application	Dataset	Data Returned	Description
BLAST	Find sequence similarity	Yeast	Alignment	A very fast search algorithm that identifies similar protein or DNA sequences
FASTA	Find sequence similarity	Yeast	Alignment	A slower search algorithm that identifies similar protein or DNA sequences and can produce different results than BLAST
Genome-wide Similarity View	Overview of similar yeast genes	Yeast	Graphic	Displays all ORFs in the S. cerevisiae genome that show similarity to a query ORF's DNA, based on a Smith-Waterman protein sequence comparison.
PatMatch	Find short DNA/protein sequence matches	Yeast	Graphic	A pattem matching program that allows ambiguous characters, but not gaps. Works well for short sequences (e.g. motifs).
Worm Homologs	Identify Yeast/worm homologs	Yeast & Worm	graphic & alignment	Reports similarity between yeast and worm genes based on the comparison of the entire complement of predicted proteins from *C. elegans* and *S. cerevisiae* (analysis described in detail in Chervitz et al., (1998). *Science* 282:2022-2028).
Mammalian Homologs	Identify Mammalian homologs to yeast genes	Yeast & Mammals	alignment	Serves up pre-existing BLAST reports comparing each yeast peptide sequence against all unique human, mouse, rat, cow, and sheep protein sequences in GenBank.

Bench-top Tools

SGD Resource	Primary Application	Description
Design Primers	Pick PCR or sequencing primers	Recommends primers appropriate for either PCR or sequencing of a given gene or DNA sequence, within parameters set by the user (end points, Tm, GC/AT ratios, etc.).
Yeast Genome Restriction Analysis	Find restriction sites	Generates a restriction map of a specified DNA sequence. The restriction map may include all enzymes, or a subset of enzyme types (3' overhangs, 5' overhangs, blunt ends, or enzymes that cut once or twice).

Maps and Displays

SGD Resource	Primary Application	Display Features	Description
Genomic View	Overview yeast chromosomes, Access other maps	• Relative size of chromosomes • Location of centromeres and of select marker genes	Provides a broad overview of chromosomal features and a gateway to other map displays.
Features Map	Locate any chromosomal feature and identify neighboring features	• Readable format • Comprehensive display for specified chromosomal regions • SAGE tags	Graphic representation of a region of chromosomal DNA. Includes the locations of ORFs, centromeres, tRNAs, RNA genes, Ty transposons, LTR elements, rRNAs and snRNAs.
Physical Map	Locate ATCC clones	• ATCC clones • Comprehensive display for specified chromosomal regions	Graphic representation of a region of chromosomal DNA including all the features found on the Features Map (see above), with the addition of ATCC clones
Combined Physical and Genetic Map	Overview entire chromosome, compare mapping and sequencing data	• Simultaneous display of mapping data (cM) and sequencing data (Kbp) • View either entire chromosome or specified region	Graphic representation of a yeast chromosome, displaying all genetically and/or physically mapped ORFs.

Functional Analysis

SGD Resource	Primary Application	Description
Protein Info & Composition	Synopsis of protein information	Retrieves information about proteins from YPD.
SAGE Query (Simple)	Find data on the transcription levels of a gene. Query by: • gene name • chromosome map	Reports data on the expression profiles of genes analyzed using the SAGE technique (Serial Analysis of Gene Expression). Data are available for thousands of genes in log phase growth, S phase arrest, and G2/M phase arrest. (analysis described in detail in Velculescu, et al., (1997) *Cell* 88:243-251).
SAGE Query (Advanced)	Find data on the transcription levels of a gene. Query by: • gene name • tag sequence • expression levels during specific phases • other specified parameters	Reports data on the expression profiles of genes analyzed using the SAGE technique (Serial Analysis of Gene Expression). Data are available for thousands of genes in log phase growth, S phase arrest, and G2/M phase arrest. (analysis described in detail in Velculescu, et al., (1997) *Cell* 88:243-251).
Worm Homologs	Predict function by identifying Yeast/worm homologs	Reports similarity between yeast and worm genes based on the comparison of the entire complement of predicted proteins from *C. elegans* and *S. cerevisiae* (analysis described in detail in Chervitz et al., (1998). *Science* 282:2022-2028).
Function Junction	Retrieve data from several functional analysis projects	Simultaneous retrieval of functional analyses for a given locus from the following six project sites: SGD SAGE Query, Yeast Cell Cycle Analysis Project, Yeast PathCalling, YGAC Triples Database, Worm-Yeast Protein Comparison, Yeast Protein Function Assignment
Expression Connection	Retrieve data from several microarray experiments	Simultaneous retrieval of yeast gene expression data for a given locus from several publically available microarray experiments

FIG. 2. The SGD Resource Guide provides a listing of resources for investigating gene information, the scientific literature, sequence analysis options, benchtop tools, genetic and physical maps, and genome-wide functional analysis studies.

Comparisons using a *S. cerevisiae* sequence as the initial query sequence are easy to implement using SGD's flexible retrieval tools. Users can choose to retrieve the sequence of any locus via a pop-up menu from the appropriate locus page, or they may begin with a tool (Gene/Sequence Resources) that allows customization of sequence retrieval for a desired locus or for a chosen region of yeast DNA by specifying the chromosomal coordinates. Gene or ORF sequences can be customized by choosing whether to include introns or flanking sequences of user-specified lengths. Of course options include the ability to retrieve the reverse complement of a specified DNA sequence. Researchers can also retrieve all sequences associated with a particular locus, including the systematic ORF sequence, sequences from mapped cosmids, and individual GenBank entries. Protein sequences encoded by the systematic ORFs are available; in addition, restriction maps are available with 6-frame translations of a specified sequence.

Expression Pattern. Because genes involved in the same or related processes may have coordinated regulation of expression, searching for genes that share similar (or diametrically opposed) expression patterns may provide clues about the roles of those genes in the cell. SGD provides access via a tool called Expression Connection[10] to many published genome-wide expression studies that can be queried to identify genes whose expression is coordinated (positively or negatively) with a query gene or ORF. Users may query a single dataset or several at once. Users may also browse the clustered expression data in a given dataset to scan for genes with an expression pattern, for example, one resulting from a stimulus or correlated with a cell-cycle phase or developmental program, such as sporulation[11] (Fig. 3). In addition to being labeled with the appropriate gene names, expression profiles also show Gene Ontology annotations that can give researchers clues to the biology of genes that have similar expression patterns. These annotations are particularly useful for uncharacterized genes that fall within a group of genes with correlated expression, as the annotations of characterized genes with similar expression patterns may hint at the cellular roles for the uncharacterized genes.

Phenotype. One of the first ways yeast genes were named and grouped was according to common mutant phenotype, partly because related mutant phenotypes could indicate that genes participate in a common process, and partly because they were isolated during comprehensive screening experiments. SGD allows users to retrieve lists of genes that share the same mutant phenotype. Included in SGD's display of phenotype data are the results of the systematic deletion project, in which each of the *S. cerevisiae* ORFs was deleted and the resulting strains analyzed.[12]

Text Search. There are occasions when a user may be interested in finding out biological information about a specific topic rather than beginning with a specific sequence or experimental result. For instance, the user may want to obtain information on the general subject of "chitin." In this case, a text search of the database for "chitin" will retrieve all information associated with this word. Examples of

[10] C. A. Ball, H. Jin, G. Sherlock, S. Weng, J. C. Matese, R. Andrada, G. Binkley, K. Dolinski, S. S. Dwight, M. A. Harris, L. Issel-Tarver, M. Schroeder, D. Botstein, and J. M. Cherry, *Nucleic Acids Res.* **29,** 80 (2001).

[11] S. Chu, J. DeRisi, M. Eisen, J. Mulholland, D. Botstein, P. O. Brown, and I. Herskowitz, *Science* **282,** 699 (1998).

[12] E. A. Winzeler, D. D. Shoemaker, A. Astromoff, H. Liang, S. Whelen Dow, S. H. Friend, C. J. Roberts, T. Ward, R. W. Davis *et al., Science* **285,** 901 (1999).

Expression during sporulation for SPO1/YNL012W

Help

Search SGD: [ ] Go

Full Search | Gene/Seq Resources | Help | Gene Registry | Maps
BLAST | FASTA | PatMatch | Sacch3D | Primers | SGD Home

Scale : (fold repression/induction)
>2.8 1:1 >2.8
repression induction

Click on a color strip to see data for that gene.

Up to 20 similar genes are shown, with a Pearson correlation of > 0.8 to the query gene

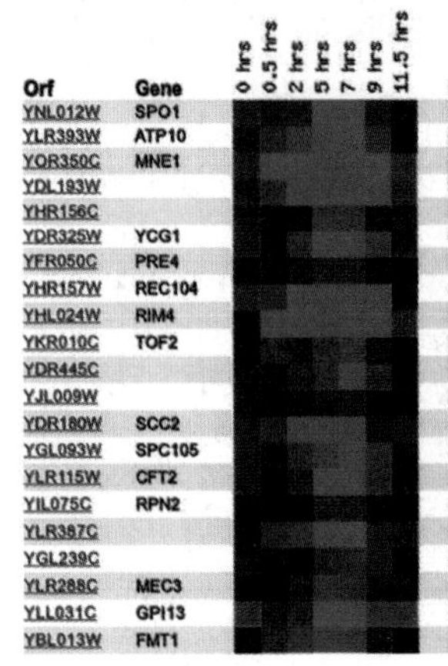

Orf	Gene	Process	Function	Component
YNL012W	SPO1	meiosis	phospholipase	nucleus
YLR393W	ATP10	protein complex assembly	molecular_function unknown	mitochondrial membrane
YOR350C	MNE1	biological_process unknown	molecular_function unknown	not yet annotated
YDL193W		biological_process unknown	molecular_function unknown	endoplasmic reticulum*
YHR156C		biological_process unknown	molecular_function unknown	not yet annotated
YDR325W	YCG1	mitotic chromosome condensation	molecular_function unknown	condensin
YFR050C	PRE4	ubiquitin-dependent protein degradation	multicatalytic endopeptidase	20S core proteasome
YHR157W	REC104	meiotic recombination	molecular_function unknown	cellular_component unknown
YHL024W	RIM4	not yet annotated	molecular_function unknown	not yet annotated
YKR010C	TOF2	DNA topological change	molecular_function unknown	cell
YDR445C		biological_process unknown	molecular_function unknown	not yet annotated
YJL009W		biological_process unknown	molecular_function unknown	not yet annotated
YDR180W	SCC2	mitotic sister chromatid cohesion	molecular_function unknown	cohesin
YGL093W	SPC105	microtubule nucleation	structural protein of cytoskeleton	spindle pole body
YLR115W	CFT2	not yet annotated	not yet annotated	not yet annotated
YIL075C	RPN2	ubiquitin-dependent protein degradation	molecular_function unknown	nucleus*
YLR387C		biological_process unknown	molecular_function unknown	not yet annotated
YGL239C		biological_process unknown	molecular_function unknown	not yet annotated
YLR288C	MEC3	not yet annotated	not yet annotated	not yet annotated
YLL031C	GPI13	GPI anchor synthesis	phosphoethanolamine N-methyltransferase	endoplasmic reticulum
YBL013W	FMT1	protein synthesis initiation*	methionyl-tRNA formyltransferase	mitochondrion

** : indicates that more than one annotation exists for the gene.*

See the Summary of the Gene Ontology annotations for this group

Expression during sporulation for SPO1/YNL012W

Visit the Website
Browse clustered data

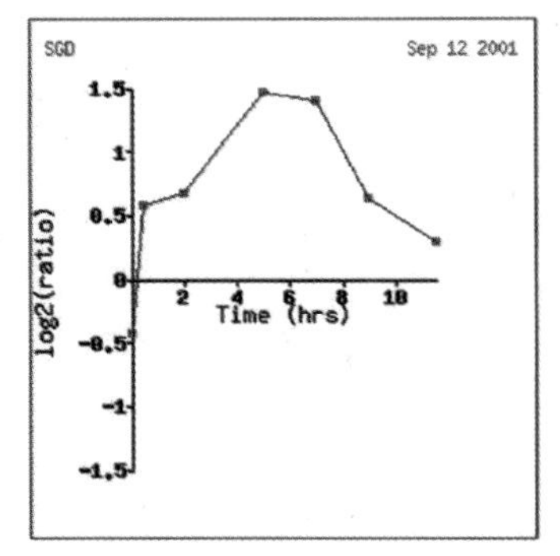

FIG. 3. Data showing the 20 genes with expression most similar to *SPO1* during sporulation [S. Chu, J. DeRisi, M. Eisen, J. Mulholland, D. Botstein, P. O. Brown, and I. Herskowitz, *Science* **282,** 699 (1998)]. Expression information like this is most easily accessed using SGD's Expression Connection tool, or from the appropriate locus page.

the types of information that can be retrieved using a text search include locus, sequence, descriptions, phenotype, gene product, GO terms, paper abstracts, and colleague information. The Text Search is available from the "Full Search" page. Because text searches provide the ability to scan many different types of biological data at once and can therefore be very powerful, SGD has made an effort to store its data in a way that allows it to be efficiently queried. At a simple level, the association of such items as gene product and GO terms with a locus entry means that a text query may associate a locus with a given biological topic. Similarly, the association of keywords with colleagues allows one to find which colleagues might be doing research on a specific subject (for instance, querying for "chitin" brings back several colleague entries in which this word is listed as a keyword on the colleague page). On a more complex level, Gene Summary paragraphs are written in a markup language, hidden from the users, which serves to break the paragraph down into different biological topics. One purpose for designing the paragraphs this way is so that a text query can bring back a specific section of text that contains the search word and is already marked as being relevant to a specific biological topic. In early 2002 this feature will become more powerful, after the installation of a new text processing system.

After Search: Information about Genes

As mentioned above, SGD's locus pages provide concise information about genes and gene products, including Gene Ontology annotations, phenotype descriptions, and links to a variety of resources. A few of the most useful resources SGD provides for exploring the biology of yeast genes are discussed below.

Gene Summaries. A Gene Summary is a short synopsis of the published biological information about a gene and its product and is designed to familiarize yeast and non-yeast researchers with the general facts and important subtleties regarding a locus (Fig. 4). The SGD curators compose Gene Summary paragraphs using natural language and a controlled vocabulary based on the Gene Ontology described above. A few recent publications are selected so the resulting paragraph is a snapshot of the current understanding of the gene, rather than an exhaustive review.

The first instance of each Gene Ontology term used in a Gene Summary is marked by curators so that it can serve as a link to a list of other genes that have been annotated to the same term (Fig. 5). Each sentence of a Gene Summary is also marked by curators according to the topics covered in the sentence, so that the summaries can be easily searched and parsed according to content.

Literature. SGD contains a set of those research papers (from PubMed and other sources) that are relevant to yeast biology. As an ongoing process, curators create and update a Literature Guide for each yeast gene that has been described

RAS2 Gene Summary

Help

Search SGD: Go

Full Search | Gene/Seq Resources | Help | Gene Registry | Maps
BLAST | FASTA | PatMatch | Sacch3D | Primers | SGD Home

RAS2 Gene Summary

RAS2 Literature Guide | RAS2 Locus Info

RAS2 encodes a homolog of the mammalian oncogene RAS and is highly related to the yeast *RAS1* gene (1). Ras2p is a small GTP-binding protein localized to the plasma membrane due to modification of its C-terminus with palmitoyl and farnesyl groups (2). Ras2p regulates processes such as sporulation, pseudohyphal growth and the nitrogen starvation response through its effects on yeast adenylate cyclase (encoded by the *CYR1* gene). In the activated, GTP-bound form Ras2p directly stimulates the production of cAMP by adenylate cyclase (3). Cdc25p binds to and activates Ras2p by directly stimulating the exchange of GDP for GTP (4). Conversely, the redundant proteins Ira1p and Ira2p inactivate Ras2p by stimulating hydrolysis of GTP to GDP (5).

Date: 1999-03 04 JW

Reference	Genes Addressed
1) Kataoka, T., *et al.* (1984) Genetic analysis of yeast RAS1 and RAS2 genes. *Cell* 37(2):437-45 SGD Curated Paper · PubMed · Cell	HIS3 | MET4 | RAS2 |
2) Bhattacharya S., *et al.* (1995) Ras membrane targeting is essential for glucose signaling but not for viability in yeast. *Proc Natl Acad Sci U S A* 92(7):2984-8 SGD Curated Paper · PubMed · PNAS	
3) Broek D., *et al.* (1985) Differential activation of yeast adenylate cyclase by wild-type and mutant RAS proteins. *Cell* 41(3):763-9 SGD Curated Paper · PubMed · Cell	RAS2
4) Lai CC., *et al.* (1993) Influence of guanine nucleotides on complex formation between Ras and CDC25 proteins. *Mol Cell Biol* 13(3):1345-52 SGD Curated Paper · PubMed · MBC	BUD5 | CDC25 | RSR1 | YPT1
5) Parrini MC., *et al.* (1996) Determinants of Ras proteins specifying the sensitivity to yeast Ira2p and human p120-GAP. *EMBO J* 15(5):1107-11 SGD Curated Paper · PubMed · EMBO	IRA2 | RAS2

FIG. 4. The Gene Summary for *RAS2* includes links to lists of genes that share GO terms, and to the locus pages of genes mentioned in the paragraph. There is also a list of references used in composing the summary. The reference display includes the list of yeast genes addressed in each publication.

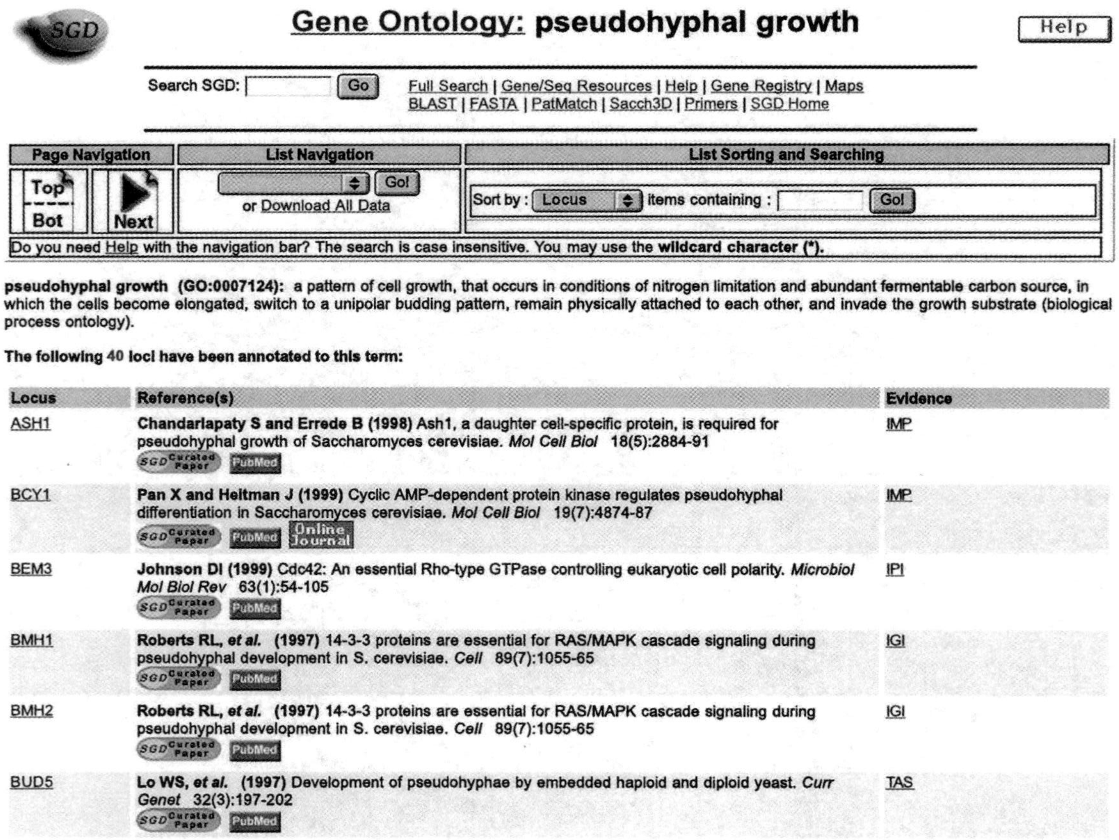

SGD

Help

Gene Ontology: pseudohyphal growth

Search SGD: Go Full Search | Gene/Seq Resources | Help | Gene Registry | Maps
BLAST | FASTA | PatMatch | Sacch3D | Primers | SGD Home

Page Navigation	List Navigation	List Sorting and Searching
Top / Bot / Next	Go! or Download All Data	Sort by : Locus items containing : Go!

Do you need Help with the navigation bar? The search is case insensitive. You may use the **wildcard character (*)**.

pseudohyphal growth (GO:0007124): a pattern of cell growth, that occurs in conditions of nitrogen limitation and abundant fermentable carbon source, in which the cells become elongated, switch to a unipolar budding pattern, remain physically attached to each other, and invade the growth substrate (biological process ontology).

The following 40 loci have been annotated to this term:

Locus	Reference(s)	Evidence
ASH1	**Chandarlapaty S and Errede B (1998)** Ash1, a daughter cell-specific protein, is required for pseudohyphal growth of Saccharomyces cerevisiae. *Mol Cell Biol* 18(5):2884-91 SGD Curated Paper PubMed	IMP
BCY1	**Pan X and Heitman J (1999)** Cyclic AMP-dependent protein kinase regulates pseudohyphal differentiation in Saccharomyces cerevisiae. *Mol Cell Biol* 19(7):4874-87 SGD Curated Paper PubMed Online Journal	IMP
BEM3	**Johnson DI (1999)** Cdc42: An essential Rho-type GTPase controlling eukaryotic cell polarity. *Microbiol Mol Biol Rev* 63(1):54-105 SGD Curated Paper PubMed	IPI
BMH1	**Roberts RL, *et al.* (1997)** 14-3-3 proteins are essential for RAS/MAPK cascade signaling during pseudohyphal development in S. cerevisiae. *Cell* 89(7):1055-65 SGD Curated Paper PubMed	IGI
BMH2	**Roberts RL, *et al.* (1997)** 14-3-3 proteins are essential for RAS/MAPK cascade signaling during pseudohyphal development in S. cerevisiae. *Cell* 89(7):1055-65 SGD Curated Paper PubMed	IGI
BUD5	**Lo WS, *et al.* (1997)** Development of pseudohyphae by embedded haploid and diploid yeast. *Curr Genet* 32(3):197-202 SGD Curated Paper PubMed	TAS

FIG. 5. Clicking on the GO term "pseudohyphal growth" in the *RAS2* Gene Summary brings the user to this page (only the top of the page is shown here). The GO term is defined on this page, and a list of other yeast genes that have been annotated to this term is shown. As always, each GO term assignment is documented by its association with a reference, and the appropriate evidence code.

in a publication (Fig. 6). By categorizing papers according to topics addressed (e.g., cellular location, protein sequence features), the guides are intended to help researchers search through the literature about a given gene quickly and efficiently. Each paper in SGD is searched for the mention of all gene names in its title or abstract. Accessing a paper allows the user to identify a group of related genes.

Other Databases. SGD provides gene-specific connections to many databases that contain important information about yeast biology. These databases provide further information about the yeast genome (MIPS), gene products (YPD and SwissProt), gene sequences (GenBank), and more.[4–8]

Connecting to Larger Biological Community

As researchers studying yeast and other organisms deepen their understanding of biological processes and the roles specific genes play in those processes, the ability to make sophisticated comparisons among different research organisms becomes increasingly important. SGD is taking several steps to facilitate the flow of biological information to and from the yeast community: annotating yeast genes to a universal framework, composing Gene Summaries to describe yeast genes, and making the database easily accessible for data mining by other scientific databases.

Annotating Yeast Genes

By annotating yeast genes to the universal framework provided by the Gene Ontology Consortium described above, comparisons of the molecular functions, biological processes, and cellular components of gene products can be made within and across species bounds[2,3] (Fig. 7). In combination with sequence comparisons, these annotations provide a powerful tool for studying similarities and differences in the biology of different organisms. Because of the wealth of information available about yeast genes, other model organism databases and members of their research communities can draw great benefit from comparisons with yeast. Yeast researchers can derive similar benefit from comparisons with other organisms.

Gene Summaries

As previously described, curator-composed Gene Summaries are brief descriptions of the current state of knowledge about individual yeast genes. The summaries are written with a target readership of those educated in biology but not necessarily with a yeast background. It is particularly hoped that these summaries will provide researchers studying other organisms a convenient entree to the body of knowledge compiled by yeast scientists. In composing the summaries, SGD curators emphasize any known relationships between the yeast genes and genes from other organisms. In some cases, such as yeast genes with a human disease gene homolog,

SGD

ACT1 Literature Guide

Help

Search SGD: [ ] Go

Full Search | Gene/Seq Resources | Help | Gene Registry | Maps
BLAST | FASTA | PatMatch | Sacch3D | Primers | SGD Home

ACT1 LITERATURE TOPICS
(formerly Gene Info)

Genetics/Cell Biology
- Cellular Location
- Function/Process
- Genetic Interactions
- Mutants/Phenotypes
- Regulation of

Nucleic Acid Information
- DNA/RNA Sequence Features
- Mapping
- RNA Levels and Processing
- Transcription
- Translational Regulation

Protein Information
- Protein Physical Properties
- Protein-protein Interactions
- Protein/Nucleic Acid Structure
- Substrates/Ligands/Cofactors
- Protein Sequence Features

Related Genes/Proteins
- Non-Yeast Related Genes/Proteins
- Yeast Related Genes/Proteins

Research Aids
- Alias
- Other Features
- Strains/Constructs
- Techniques and Reagents
- Genome-wide Analysis

Curated Literature
- Selected Review
- Reviews
- List of all Curated References

Additional Information
- References Not Yet Curated
- Archived Literature
- ▶ Literature Curation Summary

ACT1 Locus Info

ACT1 Literature Curation Summary

Curated References for ACT1: 150
References Not Yet Curated: 14

Selected Review:

Ayscough KR and Drubin DG (1996) ACTIN: general principles from studies in yeast. *Annu Rev Cell Dev Biol* 12():129-60

Note: The literature for this gene has been reviewed in the reference(s) listed under Selected Review. Due to the extensive literature available for this gene, only references published since 1999-01-28 have been curated. Earlier references can be found under Archive of older references and older reviews can be found under the Reviews topic.

Number of Other Genes referred to in ACT1 Literature: 242

Date of last curation: 2001-09-11
Date of last PubMed Search: 2001-09-11

Other ACT1 Literature Resources:

PubMed Search
Expanded PubMed Search

the homologous gene named in the summary is hotlinked to a database outside SGD where readers can learn more about that related gene.

Use of SGDIDs

Another way in which SGD facilitates the free exchange of biological information among databases is by making the database easy to retrieve information from, and reliable to connect with. One component of this strategy is the use of SGDIDs, unique identifiers for elements of the genome. Using SGDIDs as accession numbers for genes and other features of the *S. cerevisiae* genome prevents problems when loci change names; the SGDIDs are a stable means of connecting entries in our database with entries in other databases, remaining unaffected by nomenclature changes.

Yeast Community Information at SGD

SGD was designed to be a centralized resource for the yeast community, and an important part of that role is to provide a forum for the collection and display of community-related information.

Colleague Information

One feature designed to facilitate communication among yeast researchers is a searchable database of colleague information. SGD users may choose to enter their contact information, a description of research interests, a list of collaborators, and relevant Web site addresses. SGD also provides a separate list of links to several yeast laboratories located around the world (http://genome-www4.stanford.edu/cgi-bin/SGD/colleague/yeastLabs/yeastLabs.pl). This list provides links to each laboratory's Web site, and to colleague information for the laboratory's Principal Investigator. In addition to these links, the above list also displays key words and gene names associated with the Principal Investigator's research. The list's search option allows users to search the list using the Principal Investigator's name, institution, gene name, or key words.

FIG. 6. SGD's Guide to the Literature for *ACT1*. The left-hand column of all of Literature Topics pages lists the various categories of biological information that were found for that locus in the PubMed abstracts. There are 32 different topics that are currently in use. A topic will be missing from the list of Literature Topics if no abstract associated with the locus has made reference to that kind of information. This column functions as a navigation bar between the individual topics and additional information including the Literature Curation Summary. The Literature Curation Summary is the starting page to access the Literature Topics. It gives the curation status, with the numbers of curated and uncurated references, the date of last curation, and the date of the last systematic search of PubMed. Any notes or information specific to the curation of the locus are found on this page, as are a link to SGD's Gene Summary Paragraph, if available, and links to PubMed to search for references that mention the locus.

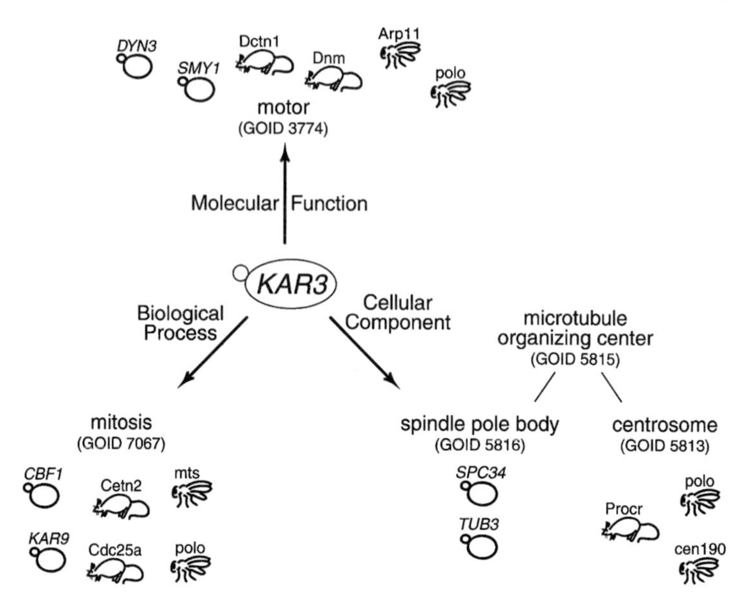

FIG. 7. GO connects across species boundaries. The controlled vocabularies of the Gene Ontology project provide a way to identify genes in multiple species with similar annotations. Here, the yeast gene, *KAR3,* is used as an example. *KAR3* is annotated to GO terms, each with a unique identifier (GOID), in each of the three ontologies: Molecular Function, Biological Process, and Cellular Component. As genes in other organisms (here we show only mouse and fly) are annotated using the same controlled vocabularies, use of these GO terms allows identification of other genes from yeast and other organisms which are involved in the same functions, processes, or structures. One benefit of the organization of the GO vocabularies and annotations is highlighted by the Cellular Component ontology annotations in this example. *KAR3* is annotated to the term "spindle pole body," an instance of the parent term "microtubule organizing center." In flies and mice, the microtubule organizing center is a different structure, the centrosome, represented by the GO term "centrosome." The ontologies show that the yeast genes *KAR3, SPC34,* and *TUB3,* the mouse gene *Procr,* and the fly genes *polo* and *cen190* are all involved in organizing microtubules.

Meetings and Community Resources

Upcoming yeast conferences and courses are listed at SGD, with links for further information and registration. For some past meetings, abstracts can be searched and lists of participants are available. SGD maintains a list of other Web sites that could be of use to yeast researchers, including the yeast "Virtual Library" of Web sites, and several databases and functional analysis Web sites. SGD also provides a searchable archive of the BioSci Yeast Newsgroup.

[20] Three Yeast Proteome Databases: YPD, PombePD, and CalPD (MycoPathPD)

By CSILLA CSANK, MARIA C. COSTANZO, JODI HIRSCHMAN, PETER HODGES, JANICE E. KRANZ, MARY MANGAN, KATHY E. O'NEILL, LAURA S. ROBERTSON, MAREK S. SKRZYPEK, JOAN BROOKS, and JAMES I. GARRELS

Introduction

Proteome's fungal databases, YPD (for *Saccharomyces cerevisiae*), PombePD (for *Schizosaccharomyces pombe*), and CalPD (for *Candida albicans*), work together to present a comprehensive view of the molecular and cell biology of the genes and proteins of the fungal world.[1] YPD, CalPD, and PombePD are part of a group of databases collectively called the BioKnowledge Library which also includes databases for complex organisms such as the worm *Caenorhabditis elegans* (WormPD), a model for development of a multicellular organism, and a protein survey database for humans (Public Human PSD). YPD, PombePD, WormPD, and Public HumanPSD are available to nonprofit organizations through our Web site, www.incyte.com.

YPD, CalPD, PombePD, and WormPD are model organism volumes of the BioKnowledge Library and are united by a single Protein Report structure (Fig. 1). Each Protein Report presents comprehensive and up-to-date information for the genes and proteins of each species. Although most sequence databases rely on automated methods of data retrieval and annotation, the content of Proteome databases is derived from sequence data evaluated and chosen by trained sequence editors and from the scientific literature by teams of curators and editors who read the full text of scientific manuscripts. Curation at Proteome is the process of extracting information from research articles in the fields of genetics, biochemistry, cell biology, structural biology, evolutionary biology, developmental biology, and related fields.

The information on Protein Report Web pages is organized into four main sections. First is the Title Line, which introduces key features of the protein. Next, a Protein Properties Table presents those features of a protein that may be described using a controlled vocabulary, information on the gene and protein sequence, and precomputed BLAST search results against several species. A third section of the Protein Report consists of bulleted free-text annotations. Information that cannot

[1] M. C. Costanzo, M. E. Crawford, J. E. Hirschman, J. E. Kranz, P. Olsen, L. S. Robertson, M. S. Skrzypek, B. R. Braun, K. L. Hopkins, P. Kondu, C. Lengieza, J. E. Lew-Smith, M. Tillberg, and J. I. Garrels, *Nucleic Acids Res.* **29,** 75 (2001).

0076-6879/02 $35.00

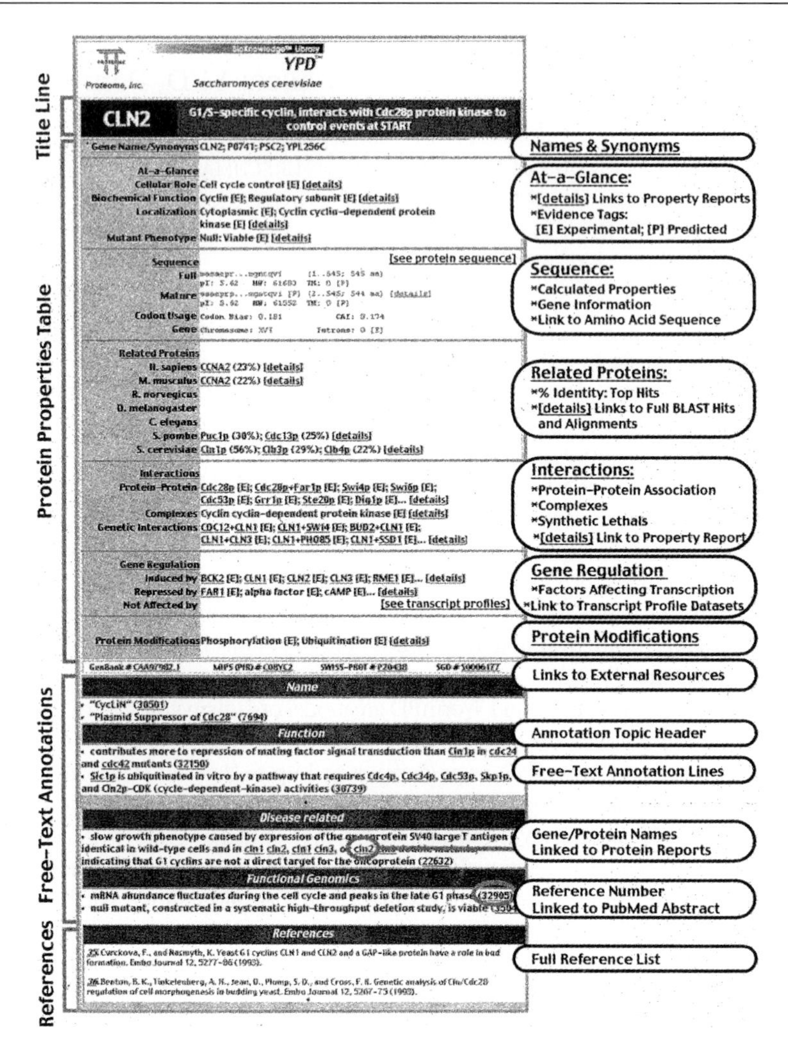

FIG. 1. A Protein Report Web page. An abbreviated version of the YPD Protein Report for the *CLN2* gene is shown. Information extracted by expert curators from the full literature is organized into four main sections. The Title Line presents a concise description of the protein. The Protein Properties Table presents information on a protein that may be described using a controlled vocabulary and that is searchable from the Full Search page. The free-text annotations describe additional features and experimental details of a gene/protein. Free-text annotation lines may be sorted into more than 30 separate topics (see Table II). The *CLN2* Protein Report has been shortened in this figure to display key features. The actual *CLN2* report contains 450 annotation lines separated into 23 topics covering more than 200 references. The Reference section lists all references used on the Protein Report page, and the Proteome Reference number links to the PubMed abstract for each paper. Links throughout the Protein Report lead to accessory Web pages. Protein Reports can be quickly accessed using the Quick Search Form.

be described using a controlled vocabulary is described in this section. Research findings and experimental data are presented in sorted free-text annotation lines, with links to the appropriate abstract(s) in the PubMed database at the National Library of Medicine, National Center for Biotechnology Information (NCBI).[2] A comprehensive reference list is the last section of the Protein Report. This Protein Report structure common to the different volumes of the BioKnowledge Library allows the user to navigate through the proteins of different species with ease and efficiency.

The Protein Report for each protein/gene of a species may be reached through the database homepage (Fig. 2). Throughout the database homepage are hyperlinks, which lead to search forms, precompiled lists of proteins separated into indices, and documentation for each database. To search the database for Protein Reports, three types of search form are provided. Use the Quick Search form (Fig. 3) to search by *Gene Name* or *Keyword;* use the Full Search form (Fig. 4) to search by multiple criteria, including controlled vocabulary properties and calculated sequence properties. Use the Sequence Search form (Fig. 5) to search by amino acid sequence. Search Results pages (Fig. 6) with hit lists from queries are linked by gene/protein name to their respective Protein Reports. The Protein Report Web page is also a gateway to additional detailed information on the protein, via hyperlinks to accessory web pages (for examples, see Figs. 7–9).

In this chapter we provide detailed coverage of the YPD database, which has served as the prototype for the creation of all other volumes of the BioKnowledge Library. We then present the unique features of the newer fungal databases PombePD and CalPD. CalPD is expected to be incorporated into MycoPathPD, a new volume that will include genes and proteins of *C. albicans* and other human pathogenic fungi. YPD is recommended as a good initiation site for newcomers. Table I provides answers to a list of frequently asked questions (FAQs). The BioKnowledge Library is a dynamic database that is updated regularly with the most recent information on the proteins and genes of different species; the information presented in this chapter reflects the database as of February 2001.

Overview of YPD Database: The First Volume of the Bioknowledge Library

YPD, the Yeast Proteome Database, grew out of a need to better identify and understand yeast proteins on two-dimensional gels. Dr. James I. Garrels, then director of the QUEST Protein Database Center at Cold Spring Harbor Laboratory, began the collection of yeast protein names and properties as a spreadsheet for use by the yeast community. After founding Proteome, Inc., in 1995, Dr. Garrels

[2] D. L. Wheeler, D. M. Church, A. E. Lash, D. D. Leipe, T. L. Madden, J. U. Pontius, G. D. Schuler, L. M. Schriml, T. A. Tatusova, L. Wagner, and B. A. Rapp, *Nucleic Acids Res.* **29,** 11 (2001).

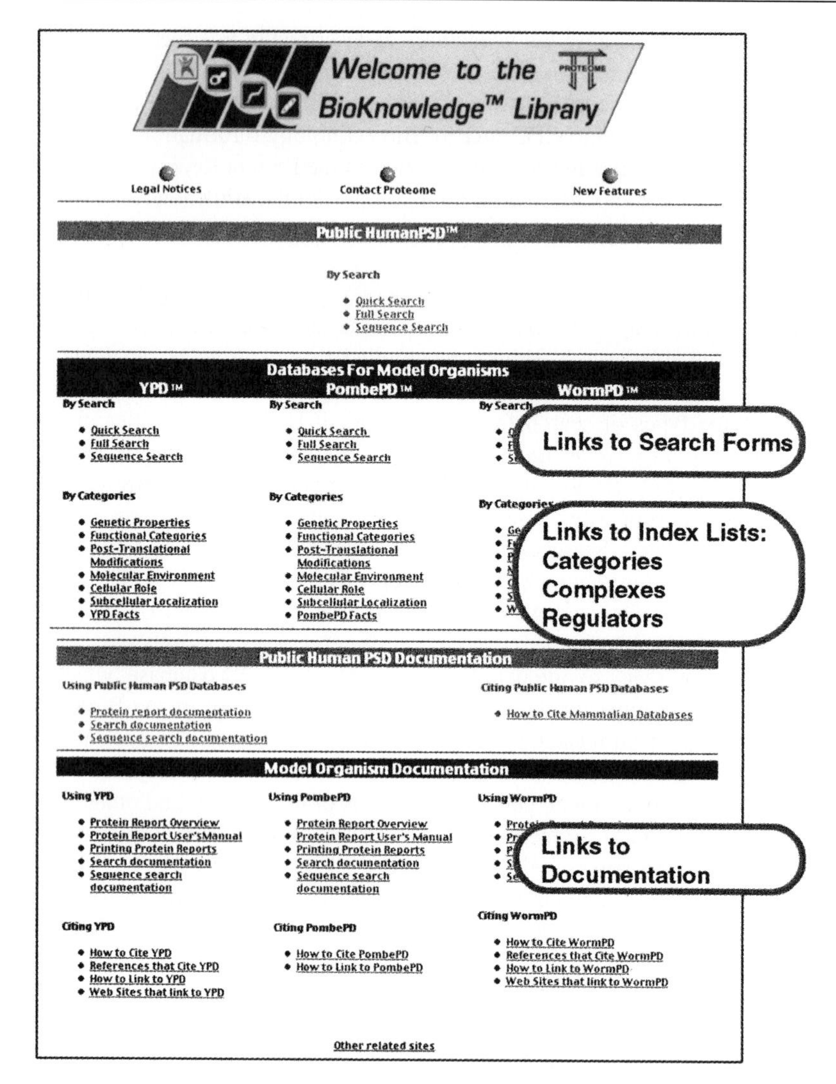

FIG. 2. The Proteome Public Database Homepage: An entry point to the BioKnowledge Library. The public homepage includes links into the YPD, PombePD, WormPD, and Public HumanPSD volumes of the BioKnowledge Library. CalPD is available to users by request. *By Search* provides links to three types of search forms for searching by Gene Name, Keyword, Category, and amino acid sequence. *By Indices* provides links to Index pages with unique lists of proteins grouped by various criteria such as complex membership. Documentation links are also provided.

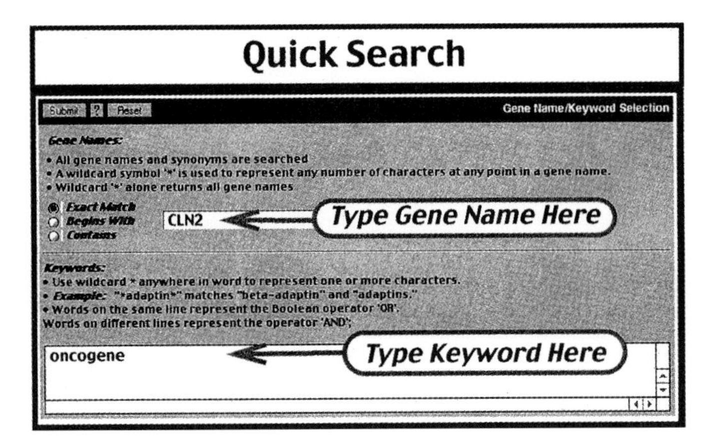

FIG. 3. The Quick Search Form. *Gene Name* searches the Gene Name/Synonym block of the Protein Reports. *Keyword* searches for words found in Title Lines and free-text annotation lines.

continued to support YPD as a freely available database on the proteome.com Web site. In late 1996 commercial subscriptions began, but Web access to YPD and other model organism databases remained available to academic users. In late 2000, Proteome, Inc., was acquired by Incyte Genomics; however, Proteome continues to operate as an independent product line with several model organism databases and a public human database accessible to nonprofit organizations.

As the first volume of Proteome's BioKnowledge Library, YPD is the prototype database and is the most information-rich of all the volumes of the BioKnowledge Library, reflecting the intensive study of *S. cerevisiae* over the past decades. As of February 2001, YPD consisted of 6237 Protein Reports, covering all known *S. cerevisiae* open reading frames (ORFs) recognized by the *Saccharomyces* Genome Database (SGD[3]) with the exception of those that encode RNAs or transposon proteins.

The most prominent feature that sets YPD and the model organism volumes of the BioKnowledge Library apart from other databases is full curation of the literature. The information in YPD is derived from careful reading and extraction of detailed scientific data by Ph.D. scientists trained in yeast biochemistry and molecular and cellular biology. Impartiality is emphasized, as the role of the curators is not to evaluate the quality of the data, but to understand the results and details and to organize that information for the pages of YPD. Curators look for information in all sections of a research paper, including the figures, the tables, and the results

[3] C. A. Ball, H. Jin, G. Sherlock, S. Weng, J. C. Matese, R. Andrada, G. Binkley, K. Dolinski, S. S. Dwight, M. A. Harris, L. Issel-Tarver, M. Schroeder, D. Botstein, and J. M. Cherry, *Nucleic Acids Res.* **29,** 80 (2001).

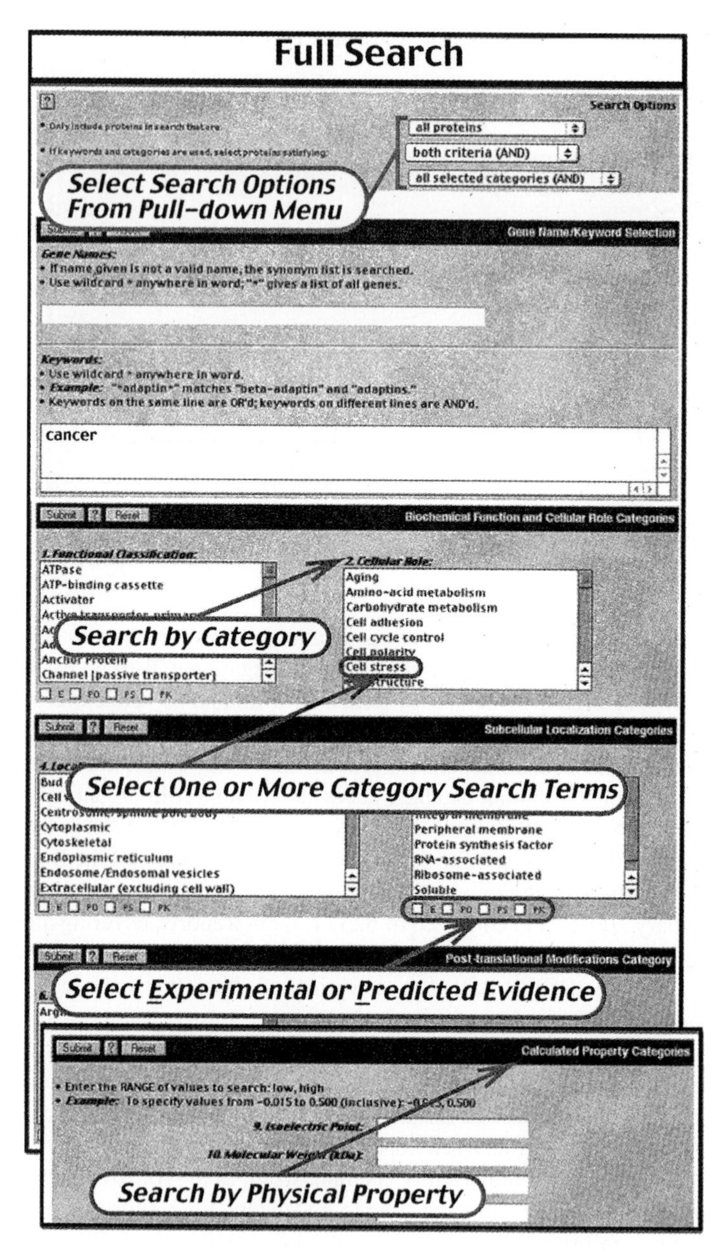

FIG. 4. The Full Search form. *Gene Name* searches the Gene Name/Synonym block of the Protein Reports. *Keyword* searches for words found in Title Lines and free-text annotation lines. *Categories* searches for information found in the controlled vocabulary Protein Properties Table of the Protein Report. *Search Options* allows searching by multiple criteria. The type of evidence for the information may also be selected. Evidence tags: [E], experimentally derived; [PS], predicted by sequence similarity; [PO], predicted by a method other than sequence similarity; [PK], predicted by BioKnowledge Transfer, a proprietary Proteome process (see text for details).

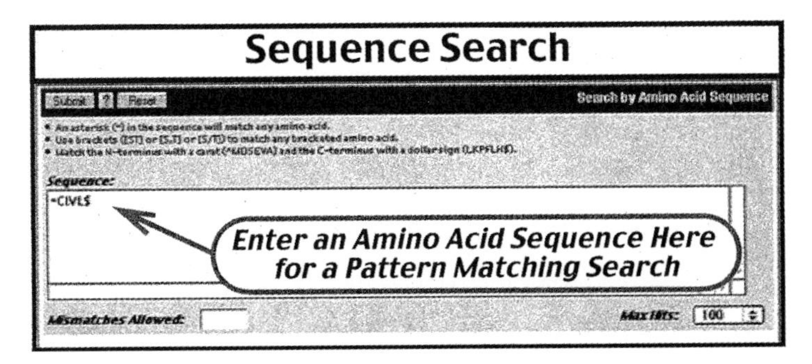

FIG. 5. The Sequence Search form. A Pattern Matching Sequence search.

section, in order to capture how each protein in a cell functions. The information is then written so that each free-text annotation line stands alone. The flexibility of using free-text enables the curator to capture even the subtlest of research results, ensuring comprehensive coverage for each protein. If new data are in conflict with prior results and there is no generally accepted explanation for the conflict, both

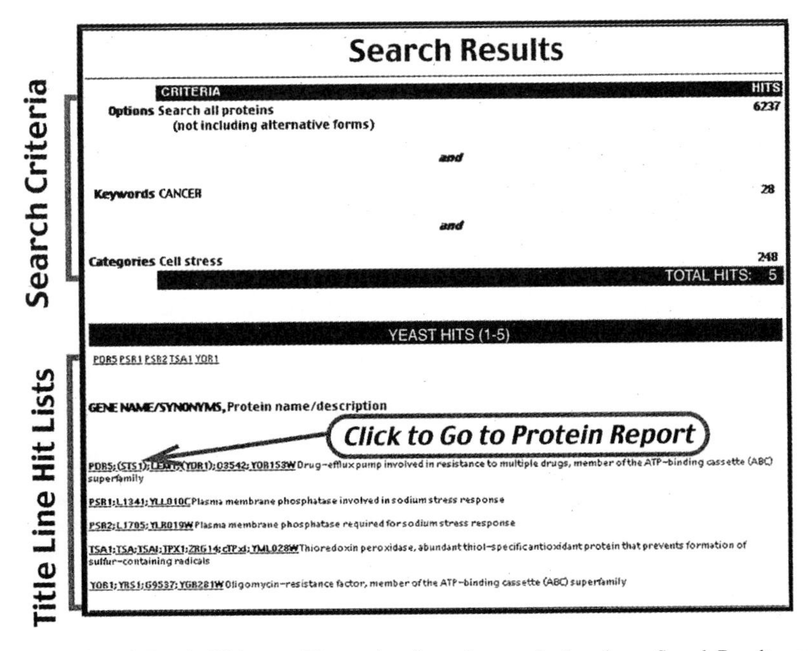

FIG. 6. Search Results Web page. The results of searches are displayed on a Search Results page. Both the total number of proteins searched and the number of hits are shown. The Gene Name/Synonym and descriptive Title Lines are shown for each hit. The Gene Name links to the Protein Report for that hit.

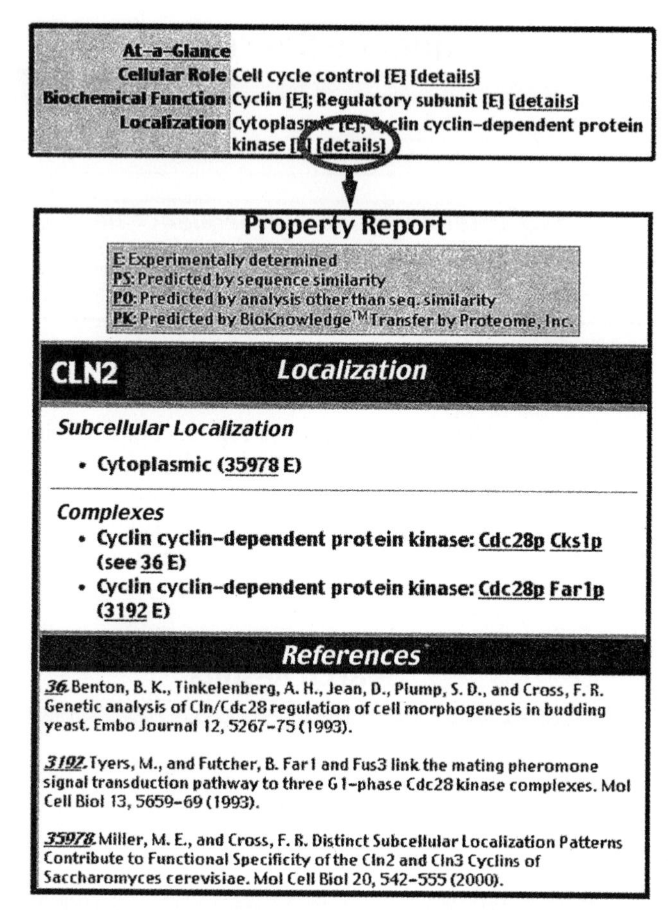

FIG. 7. Property Report pop-up window. Clicking on [details] links found embedded in the At-a-Glance, Sequence, Gene Regulation, and Interactions blocks of Protein Reports leads to Property Reports. Property Reports show additional information and the source of information and also define evidence tags [E] and [P].

results are reported; occasionally authors are contacted to try to resolve major differences. Editing teams review the work submitted by curators and a quality control team monitors the database regularly; however, nothing supersedes the quality assurance that results from the highly valued feedback received from our users who write the manuscripts from which we obtain information. Curation is an ongoing process, so Title Lines, Protein Properties, and free-text annotations in the pages of YPD are updated continually.

There is a wealth of knowledge about *S. cerevisiae* proteins in the scientific literature. More than 18,500 papers, covering nearly all of the molecular and cellular

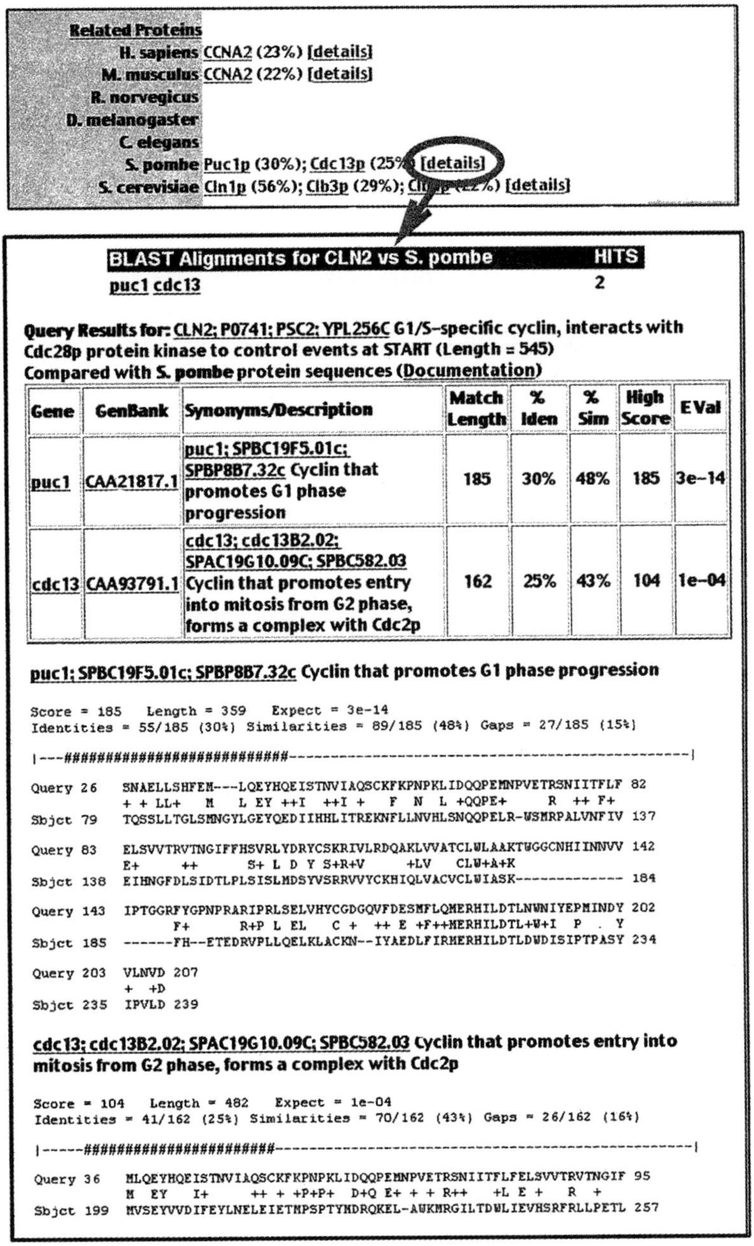

Related Proteins
H. sapiens CCNA2 (23%) [details]
M. musculus CCNA2 (22%) [details]
R. norvegicus
D. melanogaster
C. elegans
S. pombe Puc1p (30%); Cdc13p (25%) [details]
S. cerevisiae Cln1p (56%); Clb3p (29%); Cl[illegible] (22%) [details]

BLAST Alignments for CLN2 vs S. pombe — **HITS**
puc1 cdc13 — 2

Query Results for: CLN2; P0741; PSC2; YPL256C **G1/S-specific cyclin, interacts with Cdc28p protein kinase to control events at START (Length = 545)**
Compared with S. pombe protein sequences (Documentation)

Gene	GenBank	Synonyms/Description	Match Length	% Iden	% Sim	High Score	EVal
puc1	CAA21817.1	puc1; SPBC19F5.01c; SPBP8B7.32c Cyclin that promotes G1 phase progression	185	30%	48%	185	3e-14
cdc13	CAA93791.1	cdc13; cdc13B2.02; SPAC19G10.09C; SPBC582.03 Cyclin that promotes entry into mitosis from G2 phase, forms a complex with Cdc2p	162	25%	43%	104	1e-04

puc1; SPBC19F5.01c; SPBP8B7.32c Cyclin that promotes G1 phase progression

```
Score = 185   Length = 359   Expect = 3e-14
Identities = 55/185 (30%) Similarities = 89/185 (48%) Gaps = 27/185 (15%)

|---###########################-----------------------------------------------|

Query 26    SNAELLSHFEM---LQEYHQEISTNVIAQSCKFKPNPKLIDQQPEMNPVETRSNIITFLF 82
            + + LL+   M   L EY ++I   ++I +    F  N  L +QQPE+     R  ++ F+
Sbjct 79    TQSSLLTGLSMNGYLGEYQEDIIHHLITREKNFLLNVHLSNQQPELR-WSMRPALVNFIV 137

Query 83    ELSVVTRVTNGIFFHSVRLYDRYCSKRIVLRDQAKLVVATCLWLAAKTWGGCNHIINNVV 142
            E+      ++        S+ L D Y S+R+V      +LV   CLW+A+K
Sbjct 138   EIHNGFDLSIDTLPLSISLMDSYVSRRVVYCKHIQLVACVCLWIASK------------- 184

Query 143   IPTGGRFYGPNPRARIPRLSELVHYCGDGQVFDESMFLQMERHILDTLNWNIYEPMINDY 202
                  F+       R+P L EL   C +  ++ E +F++MERHILDTL+W+I  P  . Y
Sbjct 185   ------FH--ETEDRVPLLQELKLACKN--IYAEDLFIRMERHILDTLDWDISIPTPASY 234

Query 203   VLNVD 207
            +  +D
Sbjct 235   IPVLD 239
```

cdc13; cdc13B2.02; SPAC19G10.09C; SPBC582.03 Cyclin that promotes entry into mitosis from G2 phase, forms a complex with Cdc2p

```
Score = 104   Length = 482   Expect = 1e-04
Identities = 41/162 (25%) Similarities = 70/162 (43%) Gaps = 26/162 (16%)

|-----#########################-----------------------------------------------|

Query 36    MLQEYHQEISTNVIAQSCKFKPNPKLIDQQPEMNPVETRSNIITFLFELSVVTRVTNGIF 95
            M  EY   I+      ++ + +P+P+  D+Q E+ + + R++   +L E +   R  +
Sbjct 199   MVSEYVVDIFEYLNELEIETMPSPTYMDRQKEL-AWKMRGILTDWLIEVHSRFRLLPETL 257
```

FIG. 8. BLAST Alignments pop-up window. Complete precomputed BLAST alignments for Cln2p from YPD versus polypeptides from the PombePD database for *S. pombe* are shown. This page is accessed through the [details] link in the Related Proteins block next to the species name. The Related Proteins block on the Protein Report shows the top hits of the protein of the page against the species indicated. The BLAST alignments pop-up window shows all hits (with a maximum number of 250).

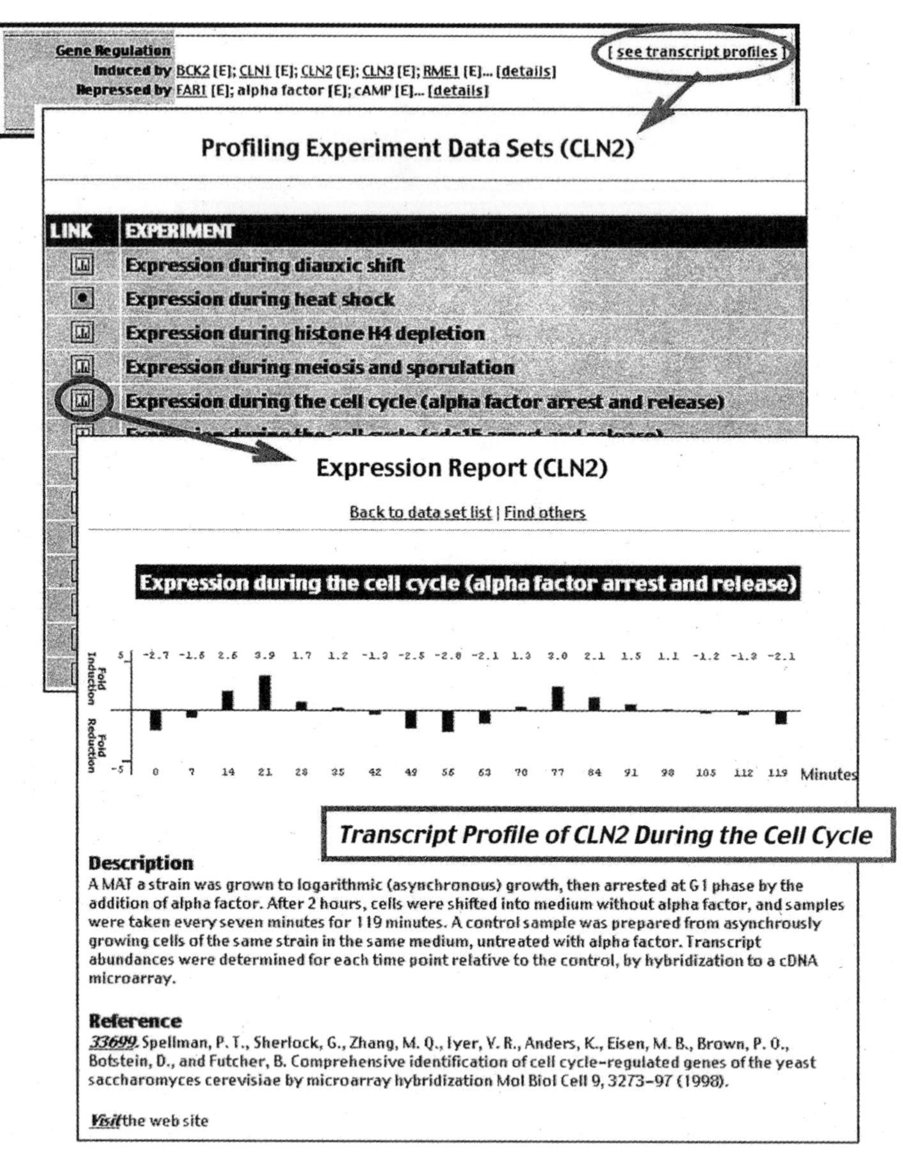

FIG. 9. Transcript Profiles: Expression Report pop-up window. A YPD Expression Report showing a graphical representation of the pattern of *CLN2* transcript abundance during the cell cycle. The [see transcript profiles] link in the Gene Regulation block of the Protein Report leads to a Profiling Experiment Data Sets window with links to Expression Reports. The change in expression under defined conditions is displayed as a fold induction or reduction relative to the control strain, control treatment, or zero time point. A description of the experiment and the reference from which the data was obtained are given. Links are also provided to Web sites associated with the research article.

TABLE I
FREQUENTLY ASKED QUESTIONS

Question	Answer
Why isn't my paper covered?	We strive for complete coverage of all publications relevant to our databases. Typically current literature is curated and appears in the Proteome databases within 1 to 2 months of publication. If you do not find your paper, it may be because it is currently being curated or it was inadvertently overlooked. In either case, we appreciate your writing us at ypd@proteome.com to notify us
What is the difference between a property and an annotation?	Properties in Proteome databases are features that can either be calculated (e.g., molecular weight) or described by a controlled vocabulary (e.g., biochemical function). Annotations are succinct statements of published observations or results that are not amenable to such controlled vocabulary. In Proteome databases, "properties" are located in the top section of the Protein Report, while "annotations," divided into separate topics, are located below the properties. See Fig. 1
How often is information updated?	Every week, except for the precomputed BLAST reports which are updated every 2 weeks
How do you handle conflicting results?	We consider ourselves reporters, not referees. When conflicting results are found, they often are due to experimental conditions, strain background, or other unanticipated variables. Often such conflicts can be enlightening (e.g., a certain gene is essential in one strain background but not in another because the second strain contains an active redundant gene). Thus we try to indicate that this result conflicts with another reported and annotated result and position both of these annotation lines next to each other. The user can investigate further by checking the references
What exactly is the difference between "subcellular localization" and "molecular environment"?	Subcellular localization refers to an actual organelle or physical part of a cell (such as Golgi, nuclear pore, or plasma membrane), whereas "molecular environment" refers to a more general biochemical location of the protein (such as "RNA-associated," "peripheral membrane," or "soluble")
Why can't I find any Ty- or 2-micron-encoded genes?	Proteome databases do not include transposon- or plasmid-encoded genes, because their presence is highly variant between strains. Gene products that function only as RNA, such as tRNAs, rRNA, snRNAs, and snoRNAs, are not included in Proteome databases. Similarly, chromosomal DNA sequence elements, including promoters, centromeres, telomeres, and origins of replication, are also not included

(continued)

TABLE I (*continued*)

Question	Answer
I found an old paper from 1962 that contains a lot of information on a genetic locus named "abc1," but I cannot find this gene in YPD. Why not?	We actually do collect such information, but we do not make it visible to users because these loci have not been associated with a particular gene sequence. In fact, we have more than 500 such records. When the DNA sequence of such a locus is identified and published, then we are able to associate it with the information we have captured about the "genetic locus" and we make this information available.
I know my favorite protein is strongly related to a protein from *E. coli,* but I don't see it in the Related Proteins section. Why not?	To keep the Related Proteins section of manageable size, we could not include all known sequences. To enhance the utility and integration of all volumes of the BioKnowledge Library, we include all organisms for which we have a Proteome database; also, since the original focus was for eukaryotes, we include *D. melanogaster* complete set of predicted proteins. Currently we do not have information on *E. coli* proteins in our database, so they are not included in this section. However, published similarities to proteins of many additional organisms may be noted in an annotation in the "Related to" topic.
Why do some genes not have entries for certain transcript profile experiments?	Such high-throughput studies that examine the entire genome at once rarely get informative data for 100% of the genes. Any given experiment is typically lacking data for some small number of genes. We convert all raw data available from each study into our standard format of "X-fold reduced or induced."
I am beginning my graduate studies and need some information on how to grow yeast. Can you help?	No, but other chapters in this *Methods in Enzymology* volume can.

biological knowledge on yeast proteins, are summarized in YPD at this time, making it a rich source of information relevant to the prediction of function for similar, uncharacterized proteins of other organisms. Each YPD Protein Report also presents the results of a BLAST search against the proteins of *Homo sapiens, Rattus norvegicus, Mus musculus, Drosophila melanogaster, C. elegans, S. pombe,* and *S. cerevisiae*. Information about the *S. cerevisiae* protein may indicate similar functions or characteristics in these putative homologs, or conversely, an uncharacterized *S. cerevisiae* protein may have similarity to a characterized protein of another organism. If a protein is conserved between species, it is of great value to easily access information about potential homologs. Using this rationale, we have developed other model organism volumes for the BioKnowledge Library, including WormPD for *C. elegans,* PombePD for *S. pombe,* and CalPD for *C. albicans*.

Getting to Know the Protein Report Web Page

Structure and Organization of Protein Report

Information on each protein is presented in Web-page format as a Protein Report. The Protein Report consists of a Title Line description of the protein, a Protein Property Table, an annotation section, and a comprehensive list of references (Fig. 1). Links embedded in each Protein Report take the user to Property Reports (Fig. 7), PubMed abstracts, external resources, transcript profiles, the amino acid sequence of the protein, documentation pages, or to Protein Reports for other genes in the BioKnowledge Library. The links are presented as underlined text throughout the Protein Report page.

The following sections describe the different components of the YPD Protein Report.

Title Line. The Title Line is shown at the top of the page and presents a concise description of the protein. The user will find information on the principal roles or functions of the protein in the cell, relevant domains and sequence motifs, family membership, and cross-species similarities.

Gene Name/Synonyms. This block presents a list of all known gene names. The Proteome Gene Name is shown first. All other gene names used in the literature are listed next as synonyms, and nonunique synonyms are in parentheses. The last name is usually the systematic ORF (Open Reading Frame) name from the genome-sequencing project. If applicable, the penultimate synonym derives from the cosmid number used in the genome-sequencing project. The Proteome Gene Name is generally the same name used in SGD.

Controlled Vocabulary, Evidence Tags [E] and [P], [details] Links, and Property Reports. Protein Properties in the Protein Properties Table of the Protein Report are described using controlled vocabulary terms. The complete lists and definitions of controlled vocabulary terms are available under *Using YPD> Protein Report User's Manual* on the database homepage (Fig. 2). These terms are utilized throughout the BioKnowledge Library and permit direct comparisons of proteins across the databases. The Protein Property categories are searchable from the Full Search page (Fig. 4), which the user can access via the database homepage.

Evidence tags [E] and [P], [details] links, and Property Reports (Fig. 7) are key components of many of the distinct blocks of the Protein Property Table. Within the Protein Property Table (below the Gene Names/Synonyms block), the properties are followed by one of the evidence tags [**E**] or [**P**] and also by the [details] link to Property Reports. The evidence tag [**E**] indicates to the user that the information presented was experimentally derived, while [**P**] indicates that the information presented was predicted. The [details] link leads to a Property Report containing further relevant information. Within the Property Report window, additional information is added for the Predicted evidence tag [**P**]. [**PS**] indicates that properties are predicted from sequence similarity; [**PO**] indicates prediction through criteria other than sequence similarity (for example, a protein shown experimentally to act as a transcription factor is predicted to be localized to the nucleus); [**PK**] indicates prediction by BioKnowledge Transfer, a proprietary Proteome process (based on Pfam and BLAST analyses and information captured within the BioKnowledge Library), which is performed by specially trained curators. Such predicted information is often used for comparative genomics. For example, several proteins that show a high level of similarity at the sequence level may be predicted to have the same **Biochemical Function** (e.g., "Cyclin") even though this function may have been determined experimentally for only one of the proteins. The Property Report also lists references (assigned a Proteome reference number) from which the information was obtained, and provides a link to the PubMed abstract through the number in parentheses.

At-a-Glance. The At-a-Glance block presents a quick summary of key properties of the protein that are described using a controlled vocabulary. The controlled vocabulary terms used in At-a-Glance are divided into the following categories: **Cellular Role, Biochemical Function, Localization,** and **Mutant Phenotype.** Multiple terms may be used for a single protein if multiple roles or functions have been experimentally demonstrated or predicted.

Cellular Role lists the major biological processes involving the protein, for example, "Cell cycle control." **Biochemical Function** lists the principal structural, regulatory, or enzymatic functions of the protein, for example, "Cyclin" and "Regulatory subunit." **Localization** lists the major ***Subcellular Localization,*** which refers to an organelle or physical part of the cell (e.g., "Cytoplasmic" or "Golgi"). It also lists ***Molecular Environment,*** which refers to a more general biochemical location in the cell (e.g., "RNA-associated," "Actin-cytoskeleton associated," "Soluble") and ***Complex*** membership of the protein (e.g., "Cyclin cyclin-dependent protein kinase"). For proteins that have multiple or variable localizations, each is listed. **Mutant Phenotype** in YPD indicates the phenotype of the null mutant: "Viable" or "Lethal" under normal growth conditions (rich medium at 30°). The controlled vocabulary terms in At-a-Glance are also listed under the search categories of the Full Search form (Fig. 4).

Sequence. The Sequence block presents information calculated from the peptide sequence and includes additional information on the gene encoding the protein. For identification the **Full** category shows the seven N-terminal amino acids, the seven C-terminal amino acids, and the number of amino acids of the initially translated protein product. The link [see protein sequence] leads to the full-length amino acid sequence. The **Mature** category shows the seven N-terminal and seven C-terminal amino acids after known or predicted processing events, and the final amino acid length after processing is specified. For all peptides in *S. cerevisiae,* when the N-terminal penultimate residue is glycine, alanine, serine, cysteine, threonine, proline, or valine, then the prediction is made that the N-terminal methionine is removed.[4,5] If the mature form is N-acetylated, this is indicated by "Ac-" on the N terminus of the mature sequence, and the calculated molecular weight and isoelectric point are changed accordingly. The [details] link at the end of the **Mature** line leads to a pop-up Property Report with more details and the references for this information.

Isoelectric point **pI** (determined according to Bjellqvist *et al.*[6]), molecular weight **MW** (calculated by the method of Gribskov and Devereux[7]), and the number of predicted transmembrane segments **TM** (determined according to Kyte and Doolittle[8]) are calculated for both the full-length peptide and the mature sequence if processing is known to occur.

For YPD, the **Codon Usage** category presents both **Codon Bias**[9] and Codon Adaptation Index, **CAI.**[10] Both are calculated from the DNA sequence of the open reading frame for the full-length form and offer measurements of the predicted relative expression of a protein. For **Codon Bias,** higher values predict higher levels of expression. For **CAI** highly expressed proteins have predicted values greater than 0.5, moderately expressed proteins have **CAI** values between 0.2 and 0.5, and poorly expressed proteins have **CAI** values between 0.1 and 0.2.

The **Gene** category shows the **chromosome** number on which the gene resides, and the **Introns** category indicates the number of introns (either predicted or experimentally determined) in the gene. Each of these features can be searched from the Full Search query page.

The polypeptide sequence may be viewed as a Property Report pop-up window by clicking on [see protein sequence] on the right of the page. This Property Report has the Protein ID (PID) number linked to the GenBank, GenPept (NCBI), record from which it is obtained. Most YPD protein sequences are obtained from GenBank.[2] When a single gene is represented by multiple sequences in GenBank, the sequence displayed from the Protein Report is generally from the genome-sequencing project, unless a subsequent report has corrected an error in the genomic sequence. Some sequences that are not in GenBank, primarily those of small open reading frames, are added to YPD directly from the SGD chromosomal sequences. We have also incorporated the updated sequence for chromosome III,

[4] S. Huang, R. C. Elliott, P. S. Liu, R. K. Koduri, J. L. Weickmann, J. H. Lee, L. C. Blair, P. Ghosh-Dastidar, R. A. Bradshaw, K. M. Bryan *et al., Biochemistry* **26,** 8242 (1987).

[5] R. P. Moerschell, Y. Hosokawa, S. Tsunasawa, and F. Sherman, *J. Biol. Chem.* **265,** 19638 (1990).

[6] B. Bjellqvist, G. J. Hughes, C. Pasquali, N. Paquet, F. Ravier, J. C. Sanchez, S. Frutiger, and D. Hochstrasser, *Electrophoresis* **14,** 1023 (1993).

[7] M. Gribskov and J. Devereux, "Sequence Analysis Primer." Oxford University Press, Oxford, 1994.

[8] J. Kyte and R. F. Doolittle, *J. Mol. Biol.* **157,** 105 (1982).

[9] J. L. Bennetzen and B. D. Hall, *J. Biol. Chem.* **257,** 3026 (1982).

[10] P. M. Sharp and W. H. Li, *Nucleic Acids Res.* **15,** 1281 (1987).

which was determined by laboratories at the Munich Information Center for Protein Sequences (MIPS).[11]

Related Proteins. The Related Proteins block presents the top results from precomputed BLAST searches using the protein of the page as a query against the protein sequences found in the Proteome databases with the addition of *Drosophila melanogaster* sequences from the LocusLink database at NCBI.[2] The [details] link at the end of any species list leads to a BLAST Alignments pop-up window (Fig. 8) with the complete set of results for that species. A maximum of 250 alignments is shown, sorted by E-value, with BLAST results presented in tabular form at the top of the page. The pairwise amino acid sequence alignments are shown below the BLAST results table. Directly above each pairwise comparison is a Position Line, which displays the position of the pairwise alignment (with ### symbols) relative to the full length of the query proteins (indicated by dashed lines). Here, the Gene/Protein Names are linked to Protein Reports for the species searched (if available). At the time of this writing, BLAST 2.0.10.[12,13] is used to calculate the alignments. SEG and COIL filtering options[14,15] are used to mask regions of low complexity and coiled-coil domains, respectively, which can generate false positives. Sequence hits are refined using the Smith–Waterman algorithm[16] to retrieve the hits with an E-value of less than or equal to le-3. (The most current information on how BLAST results are compiled may be found in the BLAST documentation link from the Related Proteins Property Report page.) Only alignments that demonstrate a minimum of 20% identity or 40% similarity are shown.

Interactions. The Interactions property block is divided into three categories. **Protein–Protein** presents a list of proteins that have been demonstrated or are strongly predicted to interact physically with the protein of the page. The [details] link presents the user with a pop-up Property Report listing the interacting proteins and the method(s) by which the interactions was detected (for example, coimmunoprecipitation, two-hybrid assay, high-throughput two-hybrid analysis). Similarly, proteins that have been demonstrated or predicted to be in multiprotein complexes are shown in the **Complexes** category. **Genetic Interaction** shows ***synthetic lethality*** arising from the combination of two otherwise nonlethal mutations within a single haploid cell. Lethality is defined here as no growth on rich medium at 30°. Other types of genetic interactions are recorded in the **Genetic Interactions** topic of the annotations section (see below).

[11] W. C. Barker, J. S. Garavelli, Z. Hou, H. Huang, R. S. Ledley, P. B. McGarvey, H. W. Mewes, B. C. Orcutt, F. Pfeiffer, A. Tsugita, C. R. Vinayaka, C. Xiao, L.-S. Yeh, and C. Wu, *Nucleic Acids Res.* **29,** 29 (2001).

[12] S. F. Altschul, W. Gish, W. Miller, E. W. Myers, and D. J. Lipman, *J. Mol. Biol.* **215,** 403 (1990).

[13] S. F. Altschul, T. L. Madden, A. A. Schaffer, J. Zhang, Z. Zhang, W. Miller, and D. J. Lipman, *Nucleic Acids Res.* **25,** 3389 (1997).

[14] J. C. Wootton and S. Federhen, *Methods Enzymol.* **266,** 554 (1996).

[15] A. Lupas, M. Van Dyke, and J. Stock, *Science* **252,** 1162 (1991).

[16] M. S. Waterman, "Introduction to Computational Biology: Maps, Sequences and Genomes." Chapman & Hall, London, 1995.

Gene Regulation. The Gene Regulation property block provides information on factors that affect mRNA abundance, directly or indirectly, for the gene of the page. Effectors of gene expression may be proteins, small molecules, environmental conditions, developmental stage, cell cycle stage, or cell type, and they are classified as inducers, as repressors, or as having no effect. A [see transcript profiles] link leads to a pop-up window with a list of Profiling Experimental Data Sets for the gene (Fig. 9). Data sets are of two types: single point and multipoint. Single-point data sets are linked to an Expression Report pop-up window showing the fold induction or reduction upon the single conditions, while multipoint data sets are linked to an Expression Report pop-up window with a graphical representation of the expression pattern (Fig. 9). Full references to the experimental details and links to the authors' Web sites are provided. For further information on transcript profiling data sets, see the "Functional Genomics" part of this chapter.

Protein Modification. The Protein Modification block lists posttranslational modifications such as methylation, glycosylation, or ubiquitination that can affect the function or stability of proteins. Controlled vocabulary permits these terms to be selected as search criteria on the Full Search page, enabling a search for all proteins known or predicted to be modified in a particular way. The [details] link leads to additional information.

Free-Text Annotations. The annotations section, below the Property Table, presents a comprehensive review of published information about a protein and provides in-depth experimental results displayed in short free-text annotation lines. Annotation lines are sorted into more than 30 distinct topics (Table II). A Proteome

TABLE II

FREE-TEXT ANNOTATION TOPICS USED IN YPD, CALPD, AND POMBEPD

Name	Turnover
Function	Abundance
Transcript Information	Purified
Catalytic Activity	Antibodies
Pathway	Sequence Reference
Related To	Gel Mobility
Mutations	Structure
Mutant Phenotype	Motif(s)
Functional Complementation	Disease Related
Overproduction	Amino Acid Configuration
Genetic Interactions	Drug Effects
Physical Interactions	Functional Genomics
Domains	Relation to Adhesion (CalPD)
Localization	Relation to Morphology (CalPD)
Maturation	Relation to Virulence (CalPD)
Modifications	Other
Regulation	

Reference Number is included after each line in parentheses, and one click will bring the user to the PubMed abstract at NCBI. In addition, proteins or gene names in an annotation line are hyperlinked to the Protein Report for that gene if it is covered in any volume of the BioKnowledge Library accessible to the user. For example, the well-studied protein Cln2p has more than 450 annotation lines separated into 23 annotation topics covering more than 200 references (an abbreviated version is shown in Fig. 1).

References. This section lists all references cited in the Protein Property Table and annotation sections of the Protein Report. A Proteome Reference Number (in parentheses) is assigned to each reference, and this number remains the same throughout the BioKnowledge Library. The full text of these references is curated by Proteome scientists, as described above in the Introduction.

Exercise: Navigating the CLN2 Protein Report

This section provides an interactive tour of the Protein Report. As an exercise while you go through this page, think about how you would purify Cln2p. If you do not already have access to YPD, see the Access Information section of this chapter.

(1) Select *By Search*> *Quick Search* from the database homepage (Fig. 2) to get to Quick Search (Fig. 3). Type "CLN2" (case insensitive) in the *Gene Name* text box. Click *Submit* to enter your query and retrieve the Search Results page. Click on the underlined gene name CLN2 to view the *CLN2* Protein Report (Fig. 1). *What does the Title Line tell you about CLN2?*

(2) At the top of the Protein Property Table, select At-a-Glance to view documentation.

(3) Click on a [details] link in the At-a-Glance section to view a Property Report (Fig. 7). *What are evidence tags and how do they contribute to your interpretation of the information?*

(4) In the Sequence block, click on [see protein sequence] to view the amino acid sequence for Cln2p. The GenBank number at the top of the page links to the GenPept report (NCBI) from which the nucleotide sequence may be retrieved (see Dbsource on the GenPept Report). *What is the length of the full length Cln2p polypeptide? What is its molecular weight? Would you expect Cln2p to be a highly abundant protein (look at the Codon Usage)?*

(5) In the next block, Related Proteins, click on a protein name of the *S. pombe* hit II pucl. The PombePD Protein Report for this protein appears. *How many controlled vocabulary terms in the At-a-Glance section are shared between pucl and CLN2? Are any different? Why are controlled vocabulary terms useful?* Go back to the CLN2 page by hitting the *Back* button on your browser.

(6) On the CLN2 Protein Report, click on the [details] link for *S. pombe.* A page "BLAST alignments for CLN2 vs. *S. pombe*" appears (Fig. 8). *Find the*

protein that has the highest Match Length vs the length of the query CLN2. Which S. pombe protein is the better candidate for a functional homolog of Cln2p?

(7) In the Interactions block, explore the [details] links. *Can the information you find here help you in purifying full-length Cln2p? How many proteins are known to interact with Cln2p? What methods have been used to determine these interactions? Under what conditions would cells die if they lack CLN2?*

(8) To the right of the Gene Regulation block, click on [see transcript profiles] to open the Profiling Experiment Data Sets window with links to multi-point and single point transcript profile data sets. Click on the experiment "Expression during the cell cycle (alpha factor arrest and release)." This opens the Expression Report for this experiment with a graphical representation of the pattern of *CLN2* transcript abundance during the cell cycle. *From the Profile, can you tell how long it takes to complete one round of the cell cycle?*

(9) Below the Protein Properties section there are links to the different public databases. Click on these links to explore the different resources available to you.

(10) The annotation section, which follows, is composed of free-text annotations. Click on the number in parentheses at the end of an annotation line. This retrieves the PubMed abstract of the paper from which this information was derived. Click on an underlined gene name to go to the Protein Report for that protein. Select your Browser's *Back* button or replace CLN1 or CLN3 in the URL with CLN2 to return to the CLN2 page. *How many factors regulate Cln2p activity in cells? Would these factors affect your purification strategy? What information can you find in the free-text annotations that you would not find in the Protein Properties Table?*

(11) Scroll down the page until the Reference section appears. Click on a Proteome Reference Number. *In what year was CLN2 isolated and named? (If you have trouble scrolling through the reference list for this information, look under the annotation topic "Name."*)

(12) Print out the Protein Report page. In order to visualize Title Lines and headers, when printing in black and white for Apple Macintosh computers, choose the "black and white" setting, not "color/grayscale" from the print options; for IBM-compatible computers print grayscale or black and white, not "pure black and white."

Searching YPD

Search Features

Proteome's Search Engine Web pages may be accessed from the database homepage (Fig. 2) or from the ***Navigate:*** scroll box at the top of each Protein Report. The Quick Search (Fig. 3) and Full Search (Fig. 4) forms both permit searching by *Gene Name* and *Keyword,* and the Full Search form also allows the user to limit searches by controlled vocabulary category (Fig. 4). The *Gene Name* field searches the Gene Name/Synonyms information as shown on the Protein Property Table. Submitting an entry in *Keyword* searches Title Lines and the free-text annotations of the Protein Report for your entry; Category searches query for controlled vocabulary terms within the Protein Properties Table of the Protein Report. The Sequence Search form (Fig. 5) allows searches of the Proteome database by pattern matching. Users may press *Submit* or *Reset* buttons anywhere on the search forms. The question mark (?) button leads to Search Documentation. Searches generate a Search Results Web page (Fig. 6) that lists the Gene Name, Synonyms, and Title Line description of each hit. Each hit is linked to its Protein Report.

Search by Gene Name. To search by Gene Name select *By Search>Quick Search* or *Full Search* on the database homepage (Fig. 2). The selected Search Web page will appear (Fig. 3; Fig. 4). Type a gene name into the *Gene Name* text box of the search form. Wild-card symbols, hyphens, and either lowercase or uppercase letters may be used. The default radio button is set to *Exact Match.* You may select the radio button *Begins With* to find proteins that begin with the term entered. *Contains* allows the user to find matches that include the search term anywhere in the word.

Search by Keyword. To search by Keyword, select *By Search>Quick Search* or *Full Search* on the database homepage. The selected Search Form web page will appear (Fig. 3; Fig. 4). Type search words into the *Keyword* text box. Terms on the same line retrieve proteins that fit either criterion (OR). Terms on separate lines retrieve all proteins that fit both criteria (AND). The wild card symbol "*" can be placed anywhere in the word. The *Keyword* query searches Title Lines and the free-text annotations of the Protein Report for your entry.

Search by Gene Name, Keyword, and Category. To access the Full Search page (Fig. 4) select *By Search>Full Search* on the database homepage. The Full Search page permits the user to search by several criteria at once. The top of the Full Search form has a *Search Options* field that allows the user to select options from pull-down menus. You may search "all proteins" or restrict your search to "characterized," "uncharacterized," or "predicted by similarity" proteins only. Define your search by selecting all selected criteria (AND) or any selected criteria (OR). The Full Search page allows searching by controlled vocabulary protein properties that are organized into categories. These properties are the same as those that appear in the Protein Property Table of the Protein Report. Most search categories have scroll boxes with lists of protein property terms. Select terms from the following scroll boxes: *Functional Classification, Cellular Role, Mutant Phenotype, Localization, Molecular Environment, Post-Translational Modifications, Presence of Intron,* and *Chromosome Number*. In the *Calculated Property Categories* enter a range of values in the text boxes for *Isoelectric Point, Molecular Weight, Codon Adaptation Index,* and *Codon Bias* to include these in a search. Enter an integer or a range of integers in the

Predicted Motif Category>Potential Transmembrane Segments to look for potential membrane spanning proteins.

Search by Amino Acid Sequence. To access the Sequence Search page (Fig. 5) select *By Search>Sequence Search* on the database homepage. To conduct a Pattern Matching search against all proteins in the volume of the BioKnowledge Library selected, a sequence is entered or pasted into the *Search by Amino Acid Sequence>Sequence* text box. This tool is especially useful for motif searches. An asterisk (*) in the sequence will match any amino acid. Ambiguous sites are entered in brackets ([ST] or [S,T] or [S/T]) to match any of the bracketed amino acids. The N terminus can be specified with a caret (^MDSEVA) and the C-terminus with a dollar sign (LKPFLH$). The user can also stipulate the number of mismatches permitted, and can select up to 250 hits to be returned.

Exercise: BioKnowledge Library Search Engines

1. From the Quick Search page (Fig. 3), type CLN* into the *Gene Name* text box. Click *Submit.* A Search Results page appears, showing that 6237 proteins were searched in YPD and four hits were found whose accepted SGD Gene Names are CLN1, CLN2, CLN3, and PCL2. *Why does PCL2 appear in the hit list*? Choose the *Back* button on your browser to return to the Quick Search page. Press *Reset.*

2. In the *Keywords* text box, type "invasive" and "filament*" on one line. Click *Submit.* The wild card at the end of "filament" allows "filamentation" and "filamentous" to be included in the search. A Search Results page appears with a hit list of more than 200 proteins. Now return to the Quick Search form, click *Reset,* enter the same two keywords on separate lines, and click *Submit.* A much smaller hit list appears (30 genes in February 2001).

3. Go to the Full Search page (Fig. 4). In *Search Options>If keywords and categories are used, select proteins satisfying:* select "both criteria (AND)" from the pull-down menu. For *Search Options>Between categories, select proteins that match:* the default is set at "all selected categories (AND)." Type "cancer" into the *Keyword* text box. Then from *Biochemical Function and Cellular Role Categories>Cellular Role* select "cell stress" from the scroll-down menu. Press *Submit* anywhere on the page. A Search Results Page appears (Fig. 6). Return to the Full Search page by pressing the *Back* button on your browser.

4. Go to the Sequence Search page. Type "ELVIS" into the *Search by Amino Acid Sequence>Sequence* text box; click *Submit.* A Sequence Search Results page appears with a total of three hits, which contain the string "ELVIS." The results page presents the Gene Name/Synonym list, Title Lines, and the amino acid sequences of the hits with the string ELVIS highlighted in green. Press the *Back* button on your browser to return to the Sequence Search Form, and now allow for one mismatch by entering "1" in the *mismatches allowed* text box. A hit list of 100 proteins appears. This time matches are highlighted green and mismatches appear in red.

Functional Genomics and YPD

Functional genomics may be defined as an approach directed toward uncovering known or predicted properties for large numbers of genes/proteins within a single experiment. Unlike traditional, hypothesis-driven research, functional genomics studies do not usually focus on individual proteins; instead their goal is to identify unanticipated features of large groups of proteins or genes, possibly even all, in a single experiment. For this purpose, experimental design must allow large-scale, high-throughput data gathering, which frequently includes automation while preserving the reliability and reproducibility of results. Statistical and computational analyses of the data are frequently used to assess the significance of findings.

YPD collects data from functional genomics studies published in peer-reviewed journals, including data from any supplementary on-line material provided by the authors. Currently, YPD contains information derived from several types of functional genomics experiments. These include: (a) mutant phenotypes revealed in a genome-wide gene deletion project[17]; (b) protein–protein interactions revealed by systematic, genome-wide two-hybrid analysis[18]; (c) transcriptional regulation, as determined by microarray experiments (for instance[19,20]), or by serial analysis of gene expression (SAGE[21]); (d) protein quantification, as determined by two-dimensional protein electrophoresis.[22]

Functional genomics experimental results are summarized in the annotation topic, **Functional Genomics,** and also in certain sections of the Protein Property Table of each Protein Report; the microarray data are presented in a special feature called Transcript Profiles accessible from the Gene Regulation block of the Protein Property Table. The **Functional Genomics** annotation topic comprises free-text annotation lines that describe both experimental observations and methods as well as broader implications suggested by the authors, such as transcriptional co-regulation of a group of genes. The Protein Properties Table includes data from large-scale gene deletion and protein–protein interaction studies. Results of gene deletions, either viability of a null mutant or lethality of a null mutation, are added to **Mutant Phenotype** in the At-a-Glance block of the Protein Report. Proteins

[17] E. A. Winzeler, D. D. Shoemaker, A. Astromoff, H. Liang, K. Anderson, B. Andre, R. Bangham, R. Benito, J. D. Boeke, H. Bussey *et al., Science* **285,** 901 (1999).

[18] P. Uetz, L. Giot, G. Cagney, T. A. Mansfield, R. S. Judson, J. R. Knight, D. Lockshon, V. Narayan, M. Srinivasan, P. Pochart, A. Qureshi-Emili, Y. Li, B. Godwin, D. Conover, T. Kalbfleisch, G. Vijayadamodar, M. Yang, M. Johnston, S. Fields, and J. M. Rothberg, *Nature* **403,** 623 (2000).

[19] S. A. Jelinsky and L. D. Samson, *Proc. Natl. Acad. Sci. U.S.A.* **96,** 1486 (1999).

[20] P. T. Spellman, G. Sherlock, M. Q. Zhang, V. R. Iyer, K. Anders, M. B. Eisen, P. O. Brown, D. Botstein, and B. Futcher, *Mol. Biol. Cell.* **9,** 3273 (1998).

[21] V. E. Velculescu, L. Zhang, W. Zhou, J. Vogelstein, M. A. Basrai, D. E. Bassett, Jr., P. Hieter, B. Vogelstein, and K. W. Kinzler, *Cell* **88,** 243 (1997).

[22] J. Norbeck and A. Blomberg, *Yeast* **16,** 121 (2000).

interacting in large-scale two-hybrid analyses are indicated in the **Protein–Protein** category of the Interactions block. References and the method, "High-throughput two-hybrid analysis" appear in the [*details*]>*Property Report* pop-up window.

YPD presents the results of transcript profiles as Expression Reports (Fig. 9) for each gene from large scale microarray experimental data sets. The [see transcript profiles] link in the Gene Regulation block on Protein Reports leads to a Profiling Experiment Data Sets pop-up window. Lists of available profile experiments link to single data point or multipoint experiments (such as a time course or concentration curve). Data sets include comparisons of mutants to wild-type strains, effects of environmental stresses or drug treatments, expression responses in signal transduction pathways, and expression during processes such as mitosis, meiosis, or the pseudohyphal transition (see YPD for individual references). The Expression Reports display the expression value or a multipoint graph. To allow comparison between experiments and between genes of different expression levels, the change in expression under defined conditions in displayed as a fold induction or reduction relative to the control strain, control treatment, or zero time point, as provided by the authors. A short text description summarizes the experimental protocol, a reference is provided, and a link to the authors' Web site provides access to the primary data and the authors' presentation or interpretation. We recognize that it would be valuable to quickly survey all published transcript profiles for any given gene to find which conditions affect gene expression, and are currently modifying the Profiling Experiment Data Sets page to display the fold induction/reduction values, providing just such a summary.

Not only does YPD serve as a resource for access to published functional genomic data, but in addition YPD offers informative Title Lines and the classification of proteins into categories which can be useful to researchers for the interpretation of their own functional genomic experiments. For example, resources such as the Spotfire.net decision analytic platform (Spotfire, Inc., Göteborg, Sweden, and Cambridge, MA) may be used to segregate yeast microarray results into different YPD protein property categories and also permit the identification of genes by Proteome Gene Name and Title Line. YPD Title Lines give succinct, up-to-date protein descriptions, bringing meaningful interpretation to hit lists. YPD protein properties allow categorization of expression data for rapid discovery of data trends. Researchers can then drill down to Proteome's full review of the published literature by connecting to YPD Protein Report Web pages.

Extending YPD Format to Other Fungi: PombePD, CalPD, and MycoPathPD

Schizosaccharomyces pombe Proteome Database, PombePD

PombePD for the proteins of *S. pombe* was added to the BioKnowledge Library in the summer of 2000. The *S. pombe* genome contains several hundred genes that are conserved among plants or animals but missing in *S. cerevisiae,* including components of the signalosome and spliceosome.[23] Thus PombePD nicely complements YPD as a model organism database (see also Forsburg[24]). In addition, *S. pombe* also serves as an excellent model system for other fungi, including the human pathogen *Pneumocystis carinii*. For example, preliminary analysis of *P. carinii* protein sequences shows that for those proteins that have significant similarity to other known proteins, 66% are most closely related to *S. pombe* proteins, while only 20% are most closely related to *S. cerevisiae* proteins (see the *Pneumocystis carinii* Genome Project and Resource Site, http://biology.uky.edu/Pc/home2.html).

Similar to YPD, PombePD is organized in the Protein Report Web page format with information about each protein summarized in the controlled-vocabulary Protein Property Table and the free-text annotation section that are both easily searchable and fully interlinked to the other volumes of the BioKnowledge Library. PombePD comprehensively covers the *S. pombe* literature and is updated weekly. The layout of the PombePD Protein Report is similar to that of YPD, with an expanded mutant phenotype property in the top At-A-Glance section. In addition to listing viable and lethal null mutant phenotypes, this property captures information about a wide array of morphological and growth phenotypes. These mutant phenotypes are listed and defined in *Using PombePD>Protein Report User's Manual,* accessible from the database homepage. Included are general mutant phenotypes such as "cold sensitive," as well as phenotypes specific to *S. pombe* such as the "cut" phenotype. For each mutant phenotype, the type of mutation (null, gain-of-function, reduction-of-function) is stated. Experimental evidence tags and references are given for each mutant phenotype listed.

The *S. pombe* genome sequence was almost entirely complete in February 2001 (see the *S. pombe* Sequencing Group at the Sanger Centre, UK; www.sanger.ac.uk). The genome is about 14 Mb, roughly the same size as that of *S. cerevisiae,* and contains approximately 5000 genes. PombePD is based on the Pompep genome sequence from the Sanger Centre and is regularly compared with Pompep to ensure that PombePD contains a complete, nonredundant set of proteins. In addition, sequence analysts incorporate any new *S. pombe* sequence entries from GenBank daily. All known synonyms for a gene are listed at the top of each page, with the systematic sequencing name in the last position. PombePD can be searched using any synonym or the systematic sequencing name. The systematic sequencing names are in the format SPBC4C3.05c, where "SP" stands for *S. pombe,* "B" represents chromosome II, c4C3 is the cosmid name, "05" marks the fifth open reading frame of the cosmid, and "c" stands for complementary strand.

[23] L. Aravind, H. Watanabe, D. J. Lipman, and E. V. Koonin, *Proc. Natl. Acad. Sci. U.S.A.* **97,** 11319 (2000).

[24] S. L. Forsburg, *Trends Genet.* **15,** 340 (1999).

Candida albicans Proteome Database, CalPD, and Proteome Database for Fungal Pathogens of Humans, MycoPathPD

CalPD, containing information on the major opportunistic human pathogenic fungus *C. albicans,* is a volume of the BioKnowledge Library developed to aid antifungal research. *C. albicans* serves as a model organism for studies of fungal infection. This is largely because *C. albicans* and *S. cerevisiae* are closely related organisms, which has allowed molecular genetic tools developed for *S. cerevisiae,* such as directed gene replacement, to be adapted for use in *C. albicans*. The user may thus find it useful to navigate between YPD and CalPD when exploring *C. albicans* proteins that have relatives in *S. cerevisiae*. CalPD will be incorporated into MycoPathPD, a new volume that will include the genes and proteins of a collection of human pathogenic fungi. One reason for this expansion stems from a need to develop broad-spectrum antifungals.

As for several other pathogenic fungi, *C. albicans* adhesins, proteases, lipases, and different morphological forms contribute to pathogenicity (see Madhani and Fink[25]; Molero *et al.*[26]; Scherer and Magee[27] for reviews). Yet ironically this protein-specific information can be particularly difficult to find for such clinically relevant organisms. For example, we have found information in the literature in about 1200 papers relevant to specific genes or proteins of *C. albicans,* among the nearly 14,000 papers on "Candida albicans" in the PubMed database. Thus to advance molecular work, there is a critical need to organize the sequence information in concert with the protein-specific information extracted from the much larger clinical literature, in the context of knowledge about proteins of other fungi and model organisms. The need for direct comparisons with other pathogenic fungi has spurred an effort to add information on additional pathogenic fungi to CalPD under the umbrella of a new all-inclusive database, MycoPathPD, to be released during the year 2001. MycoPathPD will include genes from the following additional species: *Aspergillus* spp. (*fumigatus, flavus,* and *niger*); *Blastomyces dermatitidis*; other *Candida* spp. (*dubliniensis, glabrata, guilliermondii, krusei, lusitaniae, parapsilosis, pseudotropicalis,* and *tropicalis*); *Coccidioides immitis; Cryptococcus neoformans; Histoplasma capsulatum;* and *Pneumocystis carinii.* As with all volumes of the BioKnowledge Library, the format of CalPD is consistent with that of YPD. Some differences are discussed below, and the format of CalPD will be easily extended to MycoPathPD.

C. albicans is the first major fungal pathogen for which a genomic sequence has become publicly available. The genomic sequence of clinical isolate SC5314 has been determined at the Stanford DNA Sequencing and Technology Center

[25] H. D. Madhani and G. R. Fink, *Trends Cell Biol.* **8,** 348 (1998).

[26] G. Molero, R. Diez-Orejas, F. Navarro-Garcia, L. Monteoliva, J. Pla, C. Gil, M. Sanchez-Perez, and C. Nombela, *Int. Microbiol.* **1,** 95 (1998).

[27] S. Scherer and P. T. Magee, *Microbiol. Rev.* **54,** 226 (1990).

(http://www-sequence.stanford.edu/group/candida/) to 10.4 X coverage. Since *C. albicans* is an obligate diploid, however, assembly of the sequence has been particularly challenging. As of February 2001 assembly was still in progress, but a preliminary set of 9168 open-reading-frame sequences was made available to the public. Since this set of open reading frames contains overlapping genes as well as divergent alleles of the same gene, the final complement of genes defined after completion and analysis of the assembly is expected to be significantly smaller.

CalPD Protein Reports use *C. albicans* sequences available from GenBank for individually sequenced *C. albicans* genes (about 525 entries in February 2001), but in addition, open reading frame sequences released by the Stanford DNA Sequencing and Technology Center are being added to CalPD. Unlike the situation in *S. cerevisiae* and *S. pombe,* protein information not linked to a gene has represented a significant proportion of the literature for *C. albicans*. Therefore, in contrast to YPD and PombePD, in which all Protein Reports are sequence-based, CalPD includes Protein Reports not linked to a sequence. These Protein Reports concern proteins that have been characterized biochemically but whose gene sequence has not been determined. Other unique aspects of CalPD include the following: As in PombePD, the **Mutant Phenotype** property of CalPD contains detailed information about phenotypes of null and other mutations and includes the effects of the mutant on virulence. Special additions to CalPD include the **Cellular Role** "Virulence," which is used to describe proteins with direct or indirect roles in virulence, and the annotation topics **Relation to Adhesion, Relation to Morphology,** and **Relation to Virulence,** which organize information about the involvement of the protein of the page in adherence to host cells, in morphology and morphological transitions, and in virulence, respectively.

Access Information

Access to YPD, PombePD, WormPD, and Public HumanPSD are available for complimentary online use to nonprofit organizations through the Proteome Home Page (www.incyte.com). This access is available through registration to verifiable local area networks that belong to nonprofit institutes.

Future of Proteome's BioKnowledge Library

Proteome, now part of Incyte Genomics, is committed to continued improvement and maintenance of the BioKnowledge Library. In the future, the BioKnowledge Library will encompass new databases and new features for existing databases. The model for curation and presentation of functional genomic data developed for YPD will be extended to Proteome's other databases. We hope we may continue to provide the users with databases that are both informative and pleasurable to use. Mining the information landscape and converting scattered information into collected knowledge, organized and presented within the unified structure of the BioKnowledge Library, continues to be our mission.

Acknowledgments

We thank the *Saccharomyces* Genome Database (SGD), the Munich Information Center for Protein Sequences (MIPS), the Sanger Centre, the Stanford DNA Sequencing and Technology Center, our academic advisors, and all our contributors. We thank Scott MacDonald for assistance with the figures. Thanks also to Ann Fancher. Very special thanks goes to the yeast geneticists, cellular and molecular biologists, and all scientists throughout the world who have provided feedback on the databases. Construction and maintenance of the databases described here were supported in part by the following grants to J. Garrels from the National Institutes of Health, SBIR Program: #R43 GM54110-01, -02, -03, #R43 AI43728-01, -02, -03, and #R43 GM59559-01, -02, and -03.

[21] Database Resources Relevant to Yeast Biology

By FRAN LEWITTER

There is a wealth of information about yeast biology in many publicly available databases. Some resources offer general information about genomic, biological, and functional aspects of yeast biology. Other sites provide access to more specialized information such as introns or protein–protein interactions. A third class of resources consists of information gathered from gene expression experiments. There are also sites with information about species of yeast other than *Saccharomyces cerevisiae*. The final class of resources discussed here, although not limited to yeast, contains a substantial amount of useful information about yeast.

Many sites provide access to the data only through the World Wide Web (see Table I for URLs); others allow the data to be downloaded to a local computer for further analysis. Here I provide an overview of many of these resources highlighting unique features of each database and methods of access to the information. New resources are introduced on an ongoing basis. I will briefly address strategies for keeping current.

General Genome Databases

Saccharomyces Genome Database (SGD) and Yeast Proteome Database (YPD)

Two of the resources used most heavily by yeast biologists are SGD and YPD. The SGD project collects information and maintains a database of the molecular biology of the yeast *Saccharomyces cerevisiae*. The database includes a variety of

0076-6879/02 $35.00

TABLE I
URLs for Yeast Database Resources

Database	URL address
General Yeast Genome Databases	
SGD	http://genome-www.stanford.edu/Saccharomyces/
YPD	http://www.proteome.com/databases/index.html
Specialized Databases	
Database of yeast protein functional assignment	http://www.doe-mbi.ucla.edu/people/marcotte/yeast.html
MIPS	http://mips.gsf.de/proj/yeast/
Protein–protein interaction	http://portal.curagen.com/extpc/com.curagen.portal.servlet.Yeast
Saccharomyces cerevisiae Gene Index (ScGI)	http://www.tigr.org/tdb/scgi/
Triples Database	http://ygac.med.yale.edu/default.stm
YIDB: the Yeast Intron DataBase	http://www.embl-heidelberg.de/ExternalInfo/seraphin/yidb.html
YIPD: Yeast Intron Database	http://www.cse.ucsc.edu/research/compbio/yeast_introns.html
Expression Data Resources for Specific Experiments	
Effectors of developmental MAP kinase cascade revealed by expression signatures of signaling mutants	http://staffa.wi.mit.edu/fink_public/mapk/
Exploring Metabolic and Genetic Control of Gene Expression on Genomic Scale	http://cmgm.stanford.edu/pbrown/explore/
Genome-Wide Expression	
CTD-Phosphatase	http://web.wi.mit.edu/young/CTD_phosphatase/
Holstege	http://web.wi.mit.edu/young/expression/
Genomic Expression Programs in Response of Yeast Cells to Environmental Changes	http://genome-www.stanford.edu/yeast_stress/
Global Response of *Saccharomyces cerevisiae* to Alkylating Agent	http://www.hsph.harvard.edu/geneexpression/index_1.html
Location and Function of DNA-binding Proteins	http://web.wi.mit.edu/young/location/
Nucleosome-Dependent Gene Expression and Silencing in Yeast	http://web.wi.mit.edu/young/chromatin/
Phosphate Metabolism in Yeast	http://cmgm.stanford.edu/pbrown/phosphate/
Ploidy Regulation of Gene Expression	http://staffa.wi.mit.edu/fink_public/ploidy/
Redundant Roles for TFIID and SAGA Complexes in Global Transcription	http://web.wi.mit.edu/young/TFIID_SAGA/
Regulatory Networks Revealed by Transcriptional Profiling of Damaged *Saccharomyces cerevisiae* Cells	http://www.hsph.harvard.edu/geneexpression/index1.htm
Remodeling of yeast genome expression in response to environmental change	http://web.wi.mit.edu/young/environment
Transcriptional Program of Sporulation in Budding Yeast	http://cmgm.stanford.edu/pbrown/sporulation/index.html
Yeast A Kinases Differentially Regulate Iron Uptake and Respiration	http://web.wi.mit.edu/young/PKA/
Yeast Cell Cycle Analysis Project	http://genome-www.stanford.edu/cellcycle/
Yeast Evolution Project	http://genome-www.stanford.edu/evolution/
General Gene Expression Databases for Yeast	
ChipDB	http://chipdb.wi.mit.edu
ExpressDB	http://arep.med.harvard.edu/cgi-bin/ExpressDByeast/EXDStart
Stanford Microarray Database	http://genome-www.stanford.edu/microarray
Non-*Saccharomyces cerevisiae* Databases	
Candida albicans	http://sequence-www.stanford.edu/group/candida/index.html
Comparative genomics of 13 yeast species	http://cbi.labri.u-bordeaux.fr/Genolevures/Genolevures.php3
Schizosaccharomyces pombe Gene Index (SpGI)	http://www.tigr.org/tdb/spgi/
General databases with specific yeast information	
Entrez	http://www.ncbi.nlm.nih.gov/Entrez/
Keggs (pathways DB)	http://www.genome.ad.jp/kegg/kegg2.html
How to keep current	
World-Wide Web Virtual Library for Yeast	http://genome-www.stanford.edu/Saccharomyces/VL-yeast.html
NAR Database Issue	http://nar.oupjournals.org/

genomic and biological information and is maintained and updated frequently by SGD curators. There are many search tools available at the SGD Web site and many subsets of data are available to download to local computers. YPD concentrates more on protein information about the yeast genome. It is also professionally curated and updated weekly. For more information about either SGD or YPD, refer to the individual chapters in this volume.[1]

Specialized Databases

The resources discussed above provide access to general genomic information, a wealth of annotation, and many analytical tools for access to the information. Other yeast resources are less comprehensive but nevertheless provide useful specialized information. A brief description of these resources follows.

Database of Yeast Protein Functional Assignment

This resource provides access to information published in several articles from David Eisenberg's laboratory at UCLA (Los Angeles, CA). They have assigned putative functions to yeast proteins based on comparative genome analysis known as protein phylogenetic profiles and on genome expression data. Links to the original publications are available at this site. The site provides searching by open reading frame (ORF) name and links to relevant information.

MIPS

The Munich Information Center for Protein Sequences (MIPS-GSF) maintains a resource for yeast as well as other organisms. Starting with the genomic structure of yeast, the database provides information about open reading frames (ORFs), RNA genes, and other genetic elements. Annotations such as functional properties, homologies, and structures are accompanied by displays of genetic, biochemical, and cell biological knowledge extracted from the literature. Relevant citations and corresponding abstracts are integrated into the MIPS reference database.

The MIPS database can be accessed on the Web through an online retrieval system. The resource provides functional classification information on different mutant phenotypes. It supplies a synopsis of functional descriptions of genetic elements and proteins. Yeast genes are categorized by function, protein complexes, protein classes, mutant phenotypes, interaction patterns, and their subcellular localization. In addition to viewing genes by functional classification, a user can

[1] L. Issel-Taver, K. R. Christie, K. Dolinski, R. Andrada, R. Balakrishnan, C. A. Ball, G. Binkley, S. Dong, S. S. Dwight, D. G. Fisk *et al., Methods Enzymol.* **350,** [19], 2002 (this volume); C. Csank, M. C. Costanzo, J. Hirschman, P. Hodges, J. E. Kranz, M. Mangan, K. E. O'Neill, L. S. Robertson, M. S. Skrzypek, J. Brooks, and J. Garrels, *Methods Enzymol.* **350,** [20], 2002 (this volume).

search the database using a gene name as input. All relevant information about that gene is returned to the individual.

Protein–Protein Interaction Data

This site presents data generated in a collaboration between Stanley Fields' Laboratory at the University of Washington (Seattle, WA) and CuraGen Corporation (New Haven, CT) and published in early 2000. They have completed a two-hybrid analysis of the protein-coding genes of the yeast genome and identified pairs of proteins which are likely to form stable complexes *in vivo*. The site can be browsed by gene or keyword, as well as by an entire list. Interactions are shown as a figure with secondary interactions also shown.

Saccharomyces cerevisiae Gene Index (ScGI)

The gene index for each organism is available at the TIGR (The Institute for Genomics Research, Rockville, MD) Web site. Each gene index integrates research data from international EST sequencing and gene research projects. The goal of these species-specific gene indexes is to represent a nonredundant view of the genes and data on their expression patterns, cellular roles, functions, and evolutionary relationships. The databases are searchable online, or the data can be downloaded after signing a license agreement.

TRIPLES Database

This database was developed at Yale University (New Haven, CT) by the laboratories of Michael Snyder and Shirleen Roeder. TRIPLES is an acronym for transposon-insertion phenotypes, localization, and expression in *Saccharomyces,* a methodology used for ongoing functional analysis of the yeast genome. Using a novel transposon-tagging approach, they have analyzed disruption phenotypes, gene expression, and protein localization on a genome-wide scale in *Saccharomyces*. The Web site provides searching by CloneID or Gene Name.

YIDB: Yeast Intron DataBase

The Yeast Intron DataBase (YIDB) developed by Pascal J. Lopez and Bertrand Séraphin at EMBL in Heidelberg, Germany contains currently available information about all introns encoded in the nuclear and mitochondrial genomes of the yeast *Saccharomyces cerevisiae*. Introns are divided according to their mechanisms of excision as pre-mRNA introns, tRNA introns, the HAC1 intron, group I introns, and group II introns. The data are arranged in tables for each type of intron. This is not a searchable database but is a comprehensive compilation of intron sequences and associated information.

YIPD (Ares lab Yeast Intron Database)

This site contains information about the spliceosomal introns of yeast. It is developed and maintained by the laboratory of Manuel Ares at University of California at Santa Cruz. Data are presented in large tables or various search forms can be used to select a subset of data. The data returned consists of a list of genes containing introns, with short descriptions of gene function or structure. In addition each entry has a link to a page describing the intron or introns in the gene, as well as gene-specific links to SGD and to YPD.

Expression Data Resources for Specific Experiments in Yeast

Many laboratories are using microarray and related technologies to study the expression of genes in yeast. Although summaries of the results of such experiments are published in the literature, most of the voluminous supporting data are not published. However, usually a Web site is built to provide access to the underlying data. The site might provide a number of search and display options and usually provides a mechanism for downloading the data. A list of some of these sites is found in Table I.

General Gene Expression Databases for Yeast

ChipDB

ChipDB is a publicly accessible database developed and maintained by Richard Young's laboratory at Whitehead Institute (Cambridge, MA). It includes data generated from experiments in yeast on oligonucleotide arrays. Gene reports summarize changes in the expression of the gene across multiple experiments, and also contain links to external databases such as SGD and YPD. A set-based search tool allows users to compare a list of genes affected in an experiment against all other sets in the database, including chromosomal location, MIPS functional categories, or results from other experiments. There are plans to expand ChipDB to include data from the Young laboratory genome-wide location analysis experiments on DNA microarrays.

ExpressDB

ExpressDB[2] is a relational database for maintaining yeast RNA expression data and is developed and maintained by George Church's laboratory at Harvard Medical School (Boston, MA). It contains publicly available data from many laboratories using diverse technologies such as oligonucleotide arrays, DNA microarrays, and serial analysis of gene expression (SAGE). The data are available

[2] J. Aach, W. Rindone, and G. M. Church, *Genome Res.* **10,** 431 (2000).

for online querying via Web forms or for download to a local computer. Online querying requires a multistep process. First a Web page presents a list of datasets currently included in ExpressDB. Select the datasets you wish to work with and then select the information items and measures of interest. These will be the items and measures that will be presented in a query form. From the query form you can select search conditions (e.g., if you select Gene Name as a field you can limit your search to Gene Names beginning with FLO, for example). When the query has finished processing, there will be links to the data subsets you have selected through your search.

Stanford Microarray Database (SMD)

SMD is a repository for microarray data generated from experiments at Stanford University.[3] It stores raw and normalized data from microarray experiments and provides Web interfaces for researchers to retrieve, analyze, and visualize the data. In addition, it connects the expression data to relevant biological information through links to SGD and other publicly available databases. Although SMD contains data from several different organisms, it is possible to limit a search to a specific organism (e.g., yeast). In addition to searching the database, the data are available for download. Furthermore, all of the source code that is used by SMD will be made freely available to academic researchers who wish to set up their own database using SMD's model. Table specifications for implementing the database are available at the SMD site.

Non-*Saccharomyces cerevisiae* Databases

In addition to the wealth of information about Saccharomyces cerevisiae, there is a growing amount of information about other species of yeast. These sites may be helpful for those interested in comparative genomics. A list of some of these sites can be found in Table I.

General Databases with Specific Yeast Information

Entrez

The Entrez system is a text-based system that provides integrated access to multiple molecular biology databases and provides links to relevant information for entries that match the search criteria. This system was developed by and is available through a Web site at the National Center for Biotechnology Information

[3] G. Sherlock, T. Hernandez-Boussard, A. Kasarskis, G. Binkley, J. C. Matese, S. S. Dwight, M. Kaloper, S. Weng, H. Jin, C. A. Ball, M. B. Eisen, P. T. Spellman, P. O. Brown, D. Botstein, and J. M. Cherry, *Nucleic Acids Res.* **29,** 152 (2001).

(NCBI, Bethesda, MD). Through Entrez you can search the scientific literature, nucleotide and protein sequence databases, protein structure databases, and other related resources. For each search you can limit your query to a specific species by using the Preview/Index feature.

The KEGG Database

KEGG (Kyoto Encyclopedia of Genes and Genomes) is a knowledge base for systematic analysis of gene functions and links to genomic information with higher order functional information. This is developed and maintained by Minoru Kanehisa and colleagues at Kyoto University, Japan. There is information in KEGG on many organisms; however, it is easy to limit the search to just yeast. In addition to searching the database, you can browse a table of all metabolic pathways for a particular organism and then view the genes in the pathway. The database is updated regularly and provides valuable functional information.

Keeping Up to Date

One of the biggest challenges is keeping current in a field. New databases frequently appear on the Web while some databases on the Web are no longer maintained and become obsolete. It is important to know what resources to use. When visiting a site, try to identify when the information was last updated. Also, visit some of the prominent sites in the field—the ones that are public repositories for yeast information (e.g., SGD). They will have links to other databases of interest. You could also use a general web search engine such as Google (http://www.google.com) and search for "yeast database." There is also the Yeast page of the WWW Virtual Library that lists many yeast resources and is maintained at Stanford. One further place to learn about new databases is from the Special Database issue of Nucleic Acids Research (NAR) published as the first issue of each calendar year. The contents are freely available on the NAR Web page.

Acknowledgments

I thank Alex Andalis, John Barnett, Ezra Jennings, Steve Johnston, and Bing Ren for helpful discussions and critical reading of the manuscript.

[22] Searching Yeast Intron Data at Ares Lab Web Site

By LESLIE GRATE and MANUEL ARES, JR.

Introduction

It must be obvious to every geneticist by now that the future will be consumed by the need to understand how the elemental properties of genes so elegantly described in the past half-century come together with the environment to produce the subtle differences that are key to the fitness of the organism. This will require a partial abandonment of the reductionism so favored since Mendel, to be replaced by the adoption of a more synthetic view that addresses the molecular underpinnings of complex phenotypes, penetrance, expressivity, and the small contributions of many genes. Although many of us were trained to design experiments about single genes, or at the most two interacting genes, our students and researchers need more. More in this case is a healthy computational philosophy and experience.

We have tried to embrace this in our own small way by setting up a searchable database containing information concerning the introns found in the genome of *Saccharomyces cerevisiae*. Since one of us (MA) has training in genes but not computers, and the other (LG) has training in computers but not genes, this effort has been a cultural compromise. Despite its lack of sophistication and dotcom sheen, the database has found many uses in our laboratory and has been accessed by yeast geneticists, splicers, and bioinformaticists the world around. In the following pages we explain the browsing and search capabilities of the site, and how to read and interpret the findings.

Getting Into the Site

Probably the best way to use this chapter is to sit at the computer with the book open, as you go through the descriptions of the different searches. Although some readers may find computers intimidating, there is really no way to damage equipment or files using a Web browser, so no big mistakes can be made. Explore! Experiment! The site can be found by following the "Ares Lab Yeast Intron Database" link from the Ares lab home page at http://ribonode.ucsc.edu. The site can also be accessed from the *Saccharomyces* Genome Database (SGD[1]) by clicking on their "Yeast WWW Sites" link and scrolling down to their "Yeast Introns" link. Alternatively, access the site directly by typing "http://www.cse.ucsc.edu/research/compbio/yeast_introns.html" into the location window of your favorite browser, and hit return.

[1] C. A. Ball, K. Dolinski, S. S. Dwight, M. A. Harris, L. Issel-Tarver, A. Kasarskis, C. R. Scafe, G. Sherlock, G. Binkley, H. Jin *et al., Nucleic Acids Res.* **28,** 77 (2000).

METHODS IN ENZYMOLOGY, VOL. 350
0076-6879/02 $35.00

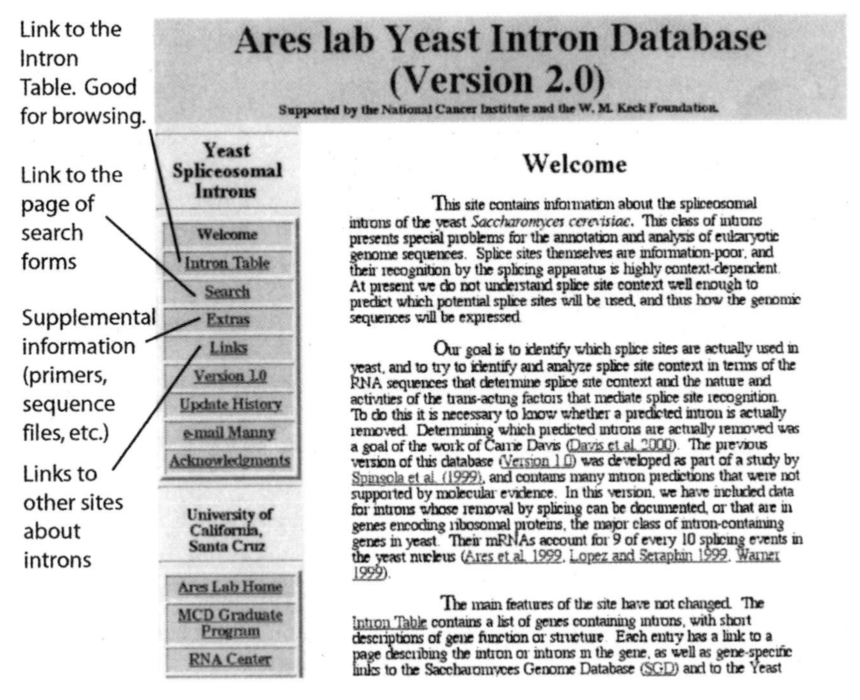

FIG. 1. Front page of the Ares Lab Yeast Intron Database.

The front page is shown in Fig. 1. In addition to some text about introns and where the information comes from, this page has four important links, two of which we will discuss at length. The "Intron Table" link (Fig. 1) will load a large document that includes a table that has the yeast genes with introns (Fig. 2). The next link on the navigation bar, "Search" links to a page (Fig. 3) that contains a set of links to different types of searches that can be performed (discussed below). The "Extras" link goes to a page that contains additional information of interest, such as available PCR primers for detecting splicing of individual introns, text files of intron sequences and alignments, and various graphs and other data related to introns in yeast. The "Links" page includes links to other Web sites concerning introns, of note, Seraphin's yeast site[2] and Kent's *Caenorhabditis elegans* Introneratoror.[3] Click on the "Intron Table" link and look at the Intron Table page.

Browsing the Intron Table

The Intron Table page is shown in Fig. 2. Each entry has a link to an "Ares Lab Intron Report" which is represented by the entry number. The entries in the table are

[2] P. J. Lopez and B. Seraphin, *Nucleic Acids Res.* **28,** 85 (2000).

[3] W. J. Kent and A. M. Zahler, *Nucleic Acids Res.* **28,** 91 (2000).

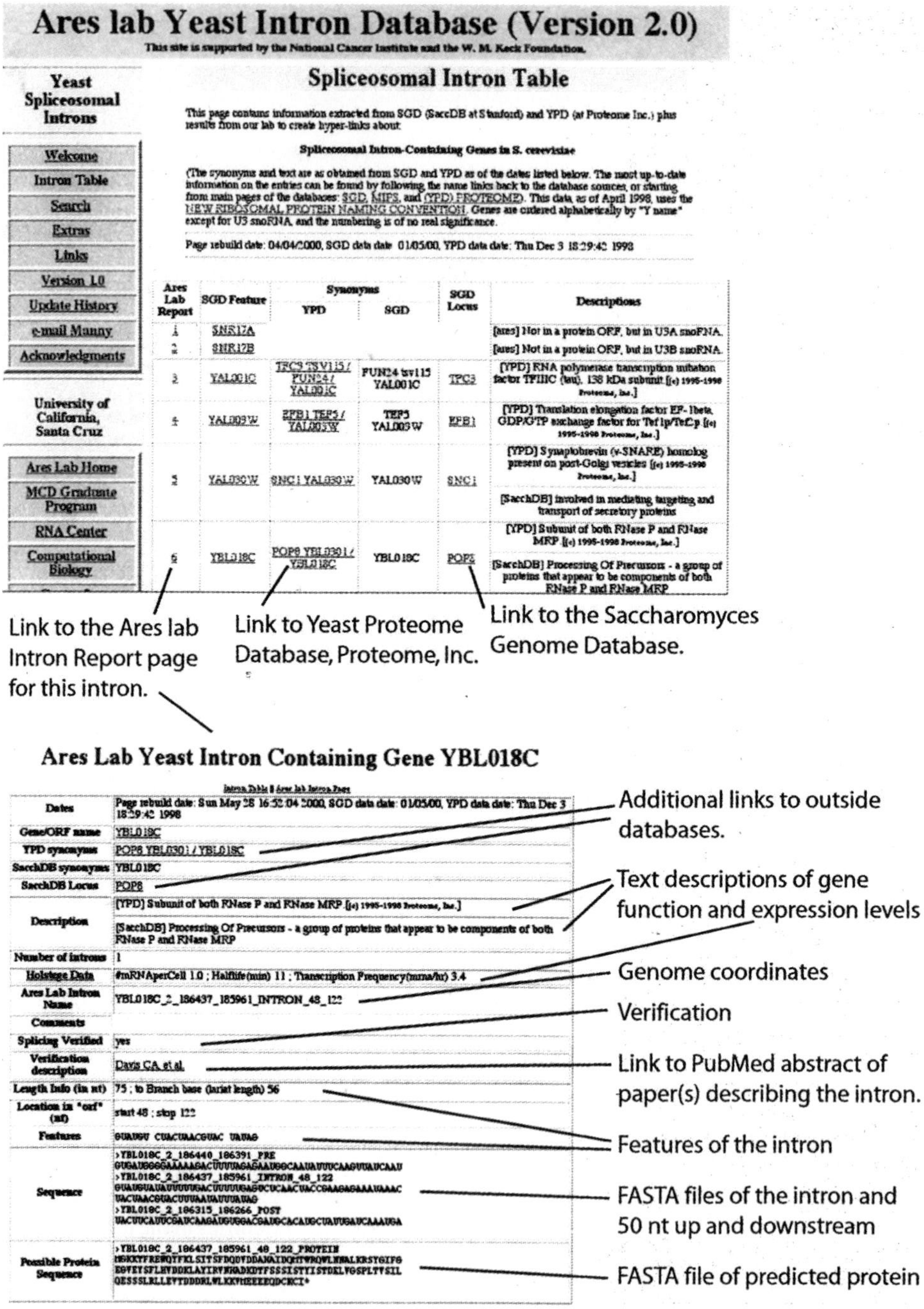

Ares lab Yeast Intron Database (Version 2.0)

This site is supported by the National Cancer Institute and the W. M. Keck Foundation.

Yeast Spliceosomal Introns

Welcome
Intron Table
Search
Extras
Links
Version 1.0
Update History
e-mail Manny
Acknowledgments

University of California, Santa Cruz

Ares Lab Home
MCD Graduate Program
RNA Center
Computational Biology

Spliceosomal Intron Table

This page contains information extracted from SGD (SaccDB at Stanford) and YPD (at Proteome Inc.) plus results from our lab to create hyper-links about:

Spliceosomal Intron-Containing Genes in S. cerevisiae

(The synonyms and text are as obtained from SGD and YPD as of the dates listed below. The most up-to-date information on the entries can be found by following the name links back to the database sources, or starting from main pages of the databases: SGD, MIPS, and (YPD) PROTEOME). This data, as of April 1998, uses the NEW RIBOSOMAL PROTEIN NAMING CONVENTION. Genes are ordered alphabetically by "Y name" except for U3 snoRNA, and the numbering is of no real significance.

Page rebuild date: 04/04/2000, SGD data date 01/05/00, YPD data date: Thu Dec 3 18:29:42 1998

Ares Lab Report	SGD Feature	Synonyms		SGD Locus	Descriptions
		YPD	SGD		
1	SNR17A				[ares] Not in a protein ORF, but in U3A snoRNA.
2	SNR17B				[ares] Not in a protein ORF, but in U3B snoRNA.
3	YAL001C	TFC3 TSV115 / FUN24 / YAL001C	FUN24 tsv115 YAL001C	TFC3	[YPD] RNA polymerase transcription initiation factor TFIIIC (tau), 138 kDa subunit [(c) 1995-1998 Proteome, Inc.]
4	YAL003W	EFB1 TEF5 / YAL003W	TEF5 YAL003W	EFB1	[YPD] Translation elongation factor EF-1beta, GDP/GTP exchange factor for Tef1p/Tef2p [(c) 1995-1998 Proteome, Inc.]
5	YAL030W	SNC1 YAL030W	YAL030W	SNC1	[YPD] Synaptobrevin (v-SNARE) homolog present on post-Golgi vesicles [(c) 1995-1998 Proteome, Inc.] [SacchDB] involved in mediating targeting and transport of secretory proteins
6	YBL018C	POP8 YBL0301 / YBL018C	YBL018C	POP8	[YPD] Subunit of both RNase P and RNase MRP [(c) 1995-1998 Proteome, Inc.] [SacchDB] Processing Of Precursors - a group of proteins that appear to be components of both RNase P and RNase MRP

Link to the Ares lab Intron Report page for this intron.

Link to Yeast Proteome Database, Proteome, Inc.

Link to the Saccharomyces Genome Database.

Ares Lab Yeast Intron Containing Gene YBL018C

Intron Table | Ares lab Intron Page

Dates	Page rebuild date: Sun May 28 16:52:04 2000, SGD data date: 01/05/00, YPD data date: Thu Dec 3 18:29:42 1998
Gene/ORF name	YBL018C
YPD synonyms	POP8 YBL0301 / YBL018C
SacchDB synonyms	YBL018C
SacchDB Locus	POP8
Description	[YPD] Subunit of both RNase P and RNase MRP [(c) 1995-1998 Proteome, Inc.] [SacchDB] Processing Of Precursors - a group of proteins that appear to be components of both RNase P and RNase MRP
Number of introns	1
Holstege Data	#mRNAperCell 1.0 ; Halflife(min) 11 ; Transcription Frequency(mrna/hr) 3.4
Ares Lab Intron Name	YBL018C_2_186437_185961_INTRON_48_122
Comments	
Splicing Verified	yes
Verification description	Davis CA et al.
Length Info (in nt)	75 ; to Branch base (lariat length) 56
Location in "orf" (nt)	start 48 ; stop 122
Features	GUAUGU CUACUAACGUAC UAUAG
Sequence	>YBL018C_2_186440_186391_PRE GUGAUGGGGAAAAAGACUUUUAGAGAAUGGCAAUAUUUCAAGUUAUCAAU >YBL018C_2_186437_185961_INTRON_48_122 GUAUGUAUAUUUUUGACUUUUUGAGUCUCAACUACCGAAGAGAAAUAAAC UACUAACGUACUUUAAUAUUUAUAG >YBL018C_2_186315_186266_POST UACUUCAUUCGAUCAAGAUGUGGACGAUGCACAUGCUAUUGAUCAAAUGA
Possible Protein Sequence	>YBL018C_2_186437_185961_48_122_PROTEIN MSKKTFREWQTFKLSITSFDQDVDDANAIDQHTWRQWLNNALKRSYGIFG EGVEYSFLHVDDKLAYIRVNHADKDTFSSSISTYISTDKLVGSPLTVSIL QESSSLRLLEVTDDDRLWLKKVMEEEEQDCKCI*

Additional links to outside databases.

Text descriptions of gene function and expression levels

Genome coordinates

Verification

Link to PubMed abstract of paper(s) describing the intron.

Features of the intron

FASTA files of the intron and 50 nt up and downstream

FASTA file of predicted protein

FIG. 2. The Intron Table and the Intron Report pages.

listed in alphabetical order by ORF name or "Y name" (e.g., YAL001C, meaning: Y, yeast; A, first chromosome; L, left arm; three digit number, assigned to ORF; C, the Crick strand), with the exception of the snoRNA genes U3 genes *SNR17A* and *SNR17B*. These are listed first, RNA coming before protein. Each entry contains a link to the SGD locus page for that gene, as well as information concerning synonyms, gene names (as opposed to ORF names), and a short description of the function of the gene product.

Clicking on the entry number link in the column "Ares lab report" produces a page for the gene in that entry, shown in the lower part of Fig. 2. This report shows much of the data in the underlying database for the intron-containing gene specified at the top of the page. The first four rows of the report reiterate the information and links on the Intron Table, including the short text descriptions. Additional information such as number of introns, expression level of the gene in the experiment by Holstege *et al.*,[4] the genome coordinates of the intron, any comments we had, and whether or not splicing has been verified experimentally or is predicted are included in the next rows. A link to the Genbank file (if one is available), or the PubMed abstract of the paper describing molecular evidence for splicing, or the method of prediction is included next. Next, physical features of the intron are presented. Length, position relative to the AUG of the ORF (except for 5′ UTR introns), and the starting, ending, and branchpoint region sequences of the intron are listed.

In the row labeled Sequence, there are three FASTA files (a "FASTA file" is a standard file format for presenting sequence data) associated with each intron. These include sequences 50 nucleotides (nt) upstream of the 5′ splice site (ending with the label "PRE"), the sequence of the intron itself, and the first 50 nt following the intron (labeled "POST"). These sequences can be copied for use with other kinds of sequence searches and alignment programs. Be aware that precise transcription initiation sites for most yeast genes are unknown, and our arbitrary use of 50 nt upstream of the intron does not imply that initiation occurs more than 50 nt upstream of the intron in every case. An example would be the U3 genes in which the first exon is only 16 nucleotides.[5] The last line, "Possible Protein Sequence," shows a protein prediction from sequence that has had the intron removed. It is accurate for genes with a single intron in the coding sequence, but the program we use for this may generate faulty predictions for genes with two introns or with introns in the 5′ UTR. Therefore, the protein predictions should be used with care.

The Intron Table page is a large (~200 K) document, and clicking back and forth from it to the report pages can be slow, even with a fast connection. The best way to avoid reloading this large page is to open it once, and open a new

[4] F. C. Holstege, E. G. Jennings, J. J. Wyrick, T. I. Lee, C. J. Hengartner, M. R. Green, T. R. Golub, E. S. Lander, and R. A. Young, *Cell* **95,** 717 (1998).

[5] E. Myslinski, V. Segault, and C. Branlant, *Science* **247,** 1213 (1990).

browser window in which to view the report. Usually the right button on a three-button mouse will open a link in a new window. With a single-button mouse, hold down the button with the cursor over the link until the pop-up menu appears on the screen near the cursor. Select "New window with this link" or "Open link in new window" and release the mouse button. The linked report page should open in a new window that is displayed over the top of the Intron Table. When done viewing the report, close the new window and return to the Intron Table window. This avoids having to reload the Intron Table page each time. A little bit of practice will make these operations go more smoothly.

Other features of most browsers are also useful, for example, the "Find..." function available under the "Edit" menu. Selecting "Find..." opens a small dialog box with a space in which to type what you want to find in the open page. We use this to find things in the Intron Table without scrolling up and down forever. Try it by searching for the actin intron: type "actin" in the dialog box and click "Find." The browser searches the text for "actin" and highlights the first occurrence of this string of letters. If we start at the top of the Intron Table, the first occurrence of "actin" is in the *ARP2* gene entry (actin-related protein). Using "Find again" to go to the next occurrence, we find SAC6 (actin filament bundling protein), and finally we come to *ACT1* itself. Other intron-containing genes that work with actin can be found by continuing the process to the end of the Table. This approach is good for browsing for key words as well as for finding a particular gene in the Table by name.

Searching the Intron Database

The simple searches offered by the browser are limited to short text strings (i.e., sequences of letters and numbers) and only in the page displayed. More complex queries can be generated using four types of searches available by clicking "Search" on the navigation bar on the left of most pages from the site. The Search page is shown in Fig. 3 and basically contains links to each of the searches. The first is the Intron Table Text Query, which allows selected properties of introns to be defined and returns a smaller table identical in format to the large Intron Table, but which only contains the entries that match the query. The Intron Splice Signals page can be used to identify introns that have particular branchpoint or splice site sequences of interest. The YAG Query page allows specifics concerning the 3′ splice site and the region between the branchpoint and the 3′ splice site to be captured. Finally, the Intron Sequence Search allows identification of introns containing a particular nucleotide sequence of interest. Below we will describe examples of how to use these search functions.

Go to the Intron Database Text search page (Fig. 4) from the "Search" page by clicking on the "Intron Table Text Query" link. This page provides a more sophisticated means to find and organize a subset of introns of interest. One can search

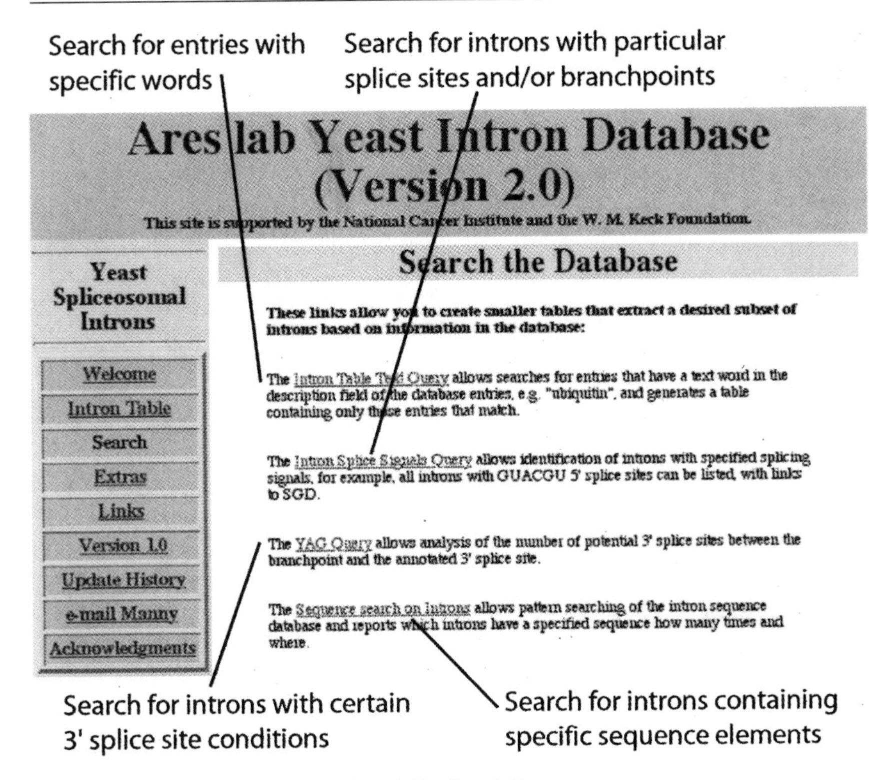

FIG. 3. The Search Page.

for introns by different properties, such as description text words, gene names, comments, intron number, and the length of different parts of the intron itself. The example described in Fig. 4 shows the output from a search for “alternative splicing” in the comments box. Submitting this request (by typing “alternative splicing” and clicking on the “Submit intron table query” button) returns a page that has a small table identical in form to the large Intron Table, but which only contains the entries that fit the terms requested. In this case there are two yeast genes known to have alternative splicing, *MTR2* and *SRC1*.[6] Each of these appears in a row as it would in the large Table, but without the other introns. As with the large Intron Table, clicking on the entry number generates the intron report page for that entry. This type of search allows rapid winnowing of lists of introns to obtain those that fit certain criteria of interest.

Multiple constraints can be imposed on the search, because the form treats each piece of information submitted in the different boxes as “AND” rather than

[6] C. A. Davis, L. Grate, M. Spingola, and M. Ares, *Nucleic Acids Res.* **28,** 1700 (2000).

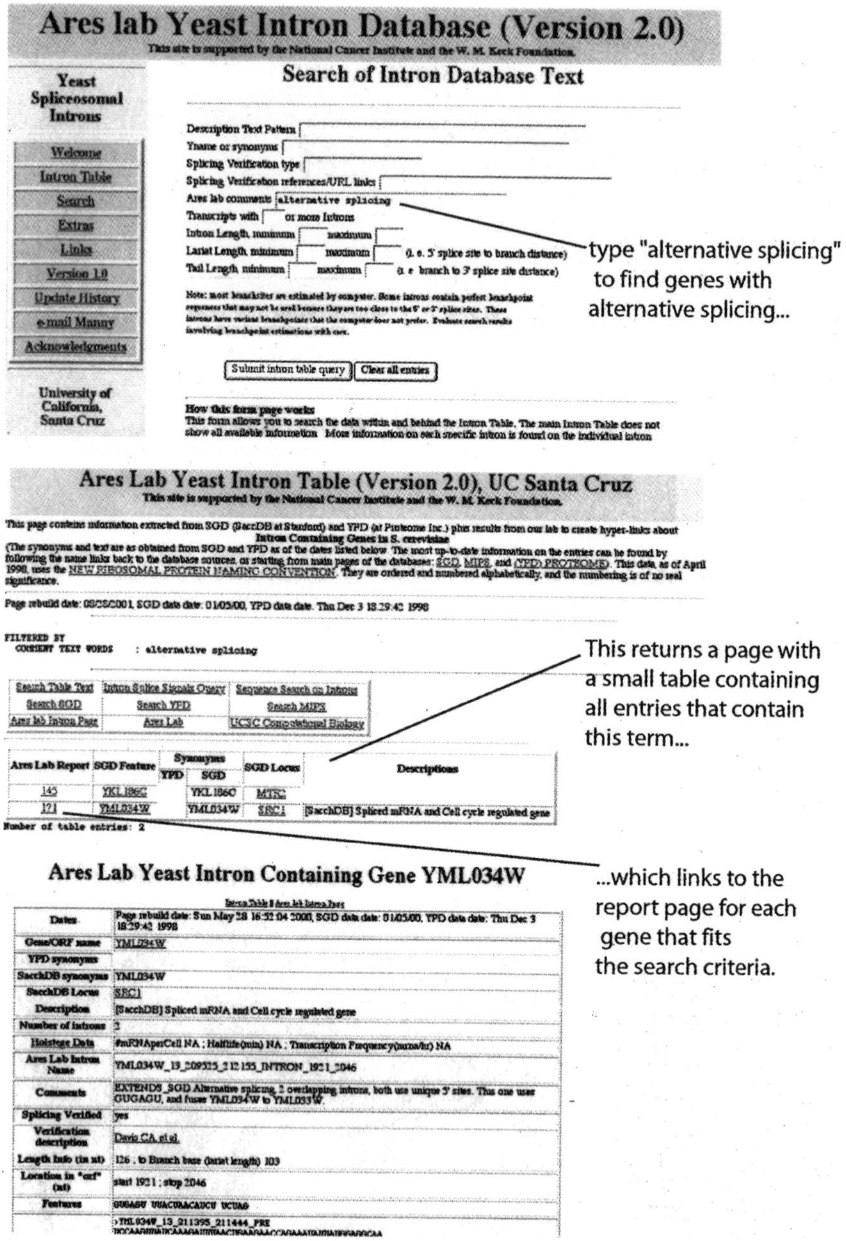

Ares lab Yeast Intron Database (Version 2.0)

This site is supported by the National Cancer Institute and the W. M. Keck Foundation.

Yeast Spliceosomal Introns

Welcome
Intron Table
Search
Extras
Links
Version 1.0
Update History
e-mail Manny
Acknowledgments

University of California, Santa Cruz

Search of Intron Database Text

Description Text Pattern
Yname or synonyms
Splicing Verification type
Splicing Verification references/URL links
Ares lab comments `alternative splicing`
Transcripts with [ ] or more Introns
Intron Length, minimum [ ] maximum [ ]
Lariat Length, minimum [ ] maximum [ ] (i. e. 5' splice site to branch distance)
Tail Length, minimum [ ] maximum [ ] (i. e. branch to 3' splice site distance)

Note: most branchsites are estimated by computer. Some introns contain perfect branchpoint sequences that may not be used because they are too close to the 5' or 3' splice sites. These introns have variant branchpoints that the computer does not prefer. Evaluate search results involving branchpoint estimations with care.

[Submit intron table query] [Clear all entries]

How this form page works
This form allows you to search the data within and behind the Intron Table. The main Intron Table does not show all available information. More information on each specific intron is found on the individual intron

type "alternative splicing" to find genes with alternative splicing...

Ares Lab Yeast Intron Table (Version 2.0), UC Santa Cruz

This site is supported by the National Cancer Institute and the W. M. Keck Foundation.

This page contains information extracted from SGD (SaccDB at Stanford) and YPD (at Proteome Inc.) plus results from our lab to create hyper-links about **Intron Containing Genes in S. cerevisiae**
(The synonyms and text are as obtained from SGD and YPD as of the dates listed below. The most up-to-date information on the entries can be found by following the name links back to the database sources, or starting from main pages of the databases: SGD, MIPS, and (YPD) PROTEOME). This data, as of April 1998, uses the NEW RIBOSOMAL PROTEIN NAMING CONVENTION. They are ordered and numbered alphabetically, and the numbering is of no real significance.

Page rebuild date: 05/28/2001, SGD data date: 01/05/00, YPD data date. Thu Dec 3 18:29:42 1998

```
FILTERED BY
  COMMENT TEXT WORDS    : alternative splicing
```

Search Table Text	Intron Splice Signals Query	Sequence Search on Introns
Search SGD	Search YPD	Search MIPS
Ares lab Intron Page	Ares Lab	UCSC Computational Biology

Ares Lab Report	SGD Feature	Synonyms		SGD Locus	Descriptions
		YPD	SGD		
145	YKL186C		YKL186C	MTR2	
171	YML034W		YML034W	SRC1	[SacchDB] Spliced mRNA and Cell cycle regulated gene

```
Number of table entries: 2
```

This returns a page with a small table containing all entries that contain this term...

Ares Lab Yeast Intron Containing Gene YML034W

Intron Table || Ares lab Intron Page

Dates	Page rebuild date: Sun May 28 16:52:04 2000, SGD data date: 01/05/00, YPD data date: Thu Dec 3 18:29:42 1998
Gene/ORF name	YML034W
YPD synonyms	
SacchDB synonyms	YML034W
SacchDB Locus	SRC1
Description	[SacchDB] Spliced mRNA and Cell cycle regulated gene
Number of introns	2
Holstege Data	#mRNAperCell NA ; Halflife(min) NA ; Transcription Frequency(mrna/hr) NA
Ares Lab Intron Name	YML034W_13_209525_212155_INTRON_1921_2046
Comments	EXTENDS_SGD Alternative splicing, 2 overlapping introns, both use unique 5' sites. This one uses GUGAGU, and fuses YML034W to YML033W.
Splicing Verified	yes
Verification description	Davis CA et al.
Length Info (in nt)	126 ; to Branch base (lariat length) 103
Location in "orf" (nt)	start 1921 ; stop 2046
Features	GUGAGU UUACUAACAUCU UCUAG
	>YML034W_13_211395_211444_PRE [illegible]

...which links to the report page for each gene that fits the search criteria.

FIG. 4. Text searches to obtain subsets of genes of interest.

"OR." For example, try typing "UTR" in the comments box and submitting the form. This will return all genes with introns in the 5′ UTR. Note the number of introns. Now try typing "ribosomal protein" in the "Description Text Pattern" box and "UTR" in the comments box. This will only return ribosomal protein genes that have introns in their 5′ UTRs. Thus, a search that retrieves too many results can be made more restrictive. To make a search less restrictive, one can use a partial word as a text pattern (since a "text pattern" is not exactly the same as a word). For example, the text pattern "ubi" recalls genes encoding ubiquitin, ubiquitin-conjugating enzymes, and ubiquinol cytochrome *c* reductase subunits. This can sometimes have unintended effects. Try typing "actin" in the "Description Text Pattern" box and clicking on the "Submit intron table query" button. Most of the results returned will be of interest; however, note the presence of *YIP2* on the list. Why is this gene here? It is not similar to actin or known to be involved in actin function. The computer found the sequence "actin" in the text description "Ypt interacting protein." This shows that humans will always remain important in evaluating results from computational processes, and that each result should be considered innocent until proven guilty.

An intron feature commonly of interest is the sequence of the splice sites and branchpoint. The set of yeast introns has highly conserved splicing signals[7]; thus deviations are of interest. Go back to the "Search" page (Fig. 4) and then click on the "Intron Splice Signals Query" link. The form page for the Splice Site Query is shown in Fig. 5. To find all introns with the nonstandard branchpoint GACUAAC (rather than the most common UACUAAC sequence) type "gacua" into the "Bpre" box on the form. The form will appear with "a" in the Branch box, since all yeast introns are thought to use A as the branched nucleotide. Leaving the other boxes open is the same as asking for anything at those positions. Click on the "Submit intron query" button. As shown in Fig. 5, the search identifies 10 introns that contain GACUAAC as the best match to the consensus. Each intron is listed with its intron signals and a link to its report page. Having the other splicing signals displayed shows, for example, that *YGL251C* and *YLR211C* also have unusual 5′ splice sites, and that *YDL115C* has an unusual 3′ splice site. The search can be extended to reveal splice site and branchpoint context as well. For example, to search for all introns containing four U residues immediately upstream of a GACUAAC branchpoint, one would type "uuuugacua" into the "Bpre" box and submit the form. The five introns that fit this description would then be returned. The other boxes work in the same way, allowing the intronic context of splice signals to be explored.

We became interested in what the intron collection might tell us about 3′ splice site selection in yeast, and we developed a search that allows identification of the number of potential 3′ splice site–like sequences that might exist between the

[7] M. Spingola, L. Grate, D. Haussler, and M. Ares, *RNA* **5,** 221 (1999).

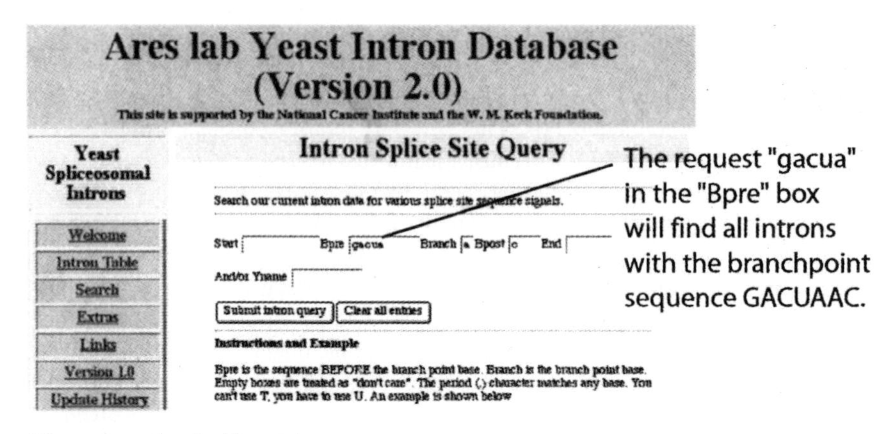

The report looks like this:

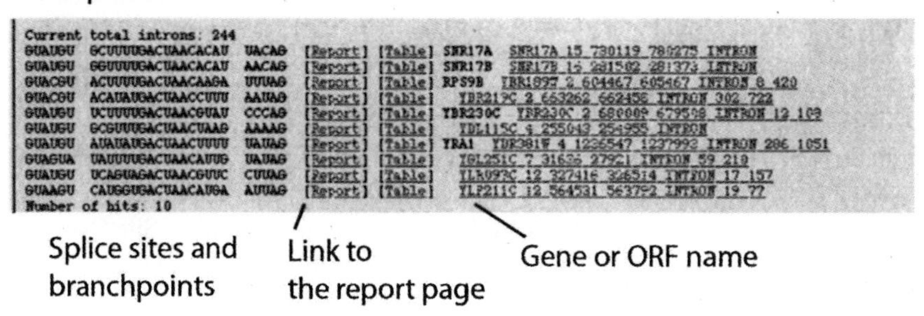

FIG. 5. Splice site query.

branchpoint and the true 3′ splice of each intron. Go back again to the “Search” page and click on the “YAG Query” link. This is the YAG Query form page and it is shown in Fig. 6. In the first box, type the nucleotide sequence pattern to search for between the branchpoint and the 3′ splice site. In the example shown, “[uc]ag” (use square brackets, not braces or parentheses!), the computer will search for all pyrimidine-A-G sequences found between the branchpoint and the 3′ splice site. If nothing is put into the “Report full sequences . . .” box, then only the counts of the number of introns in each class will be returned. To see the full report (bottom of Fig. 6), put “0” (the number zero) in that box. If you are only interested in introns that have one or more occurrence of the sequence, put “1.” In our search here, we have also restricted the introns we want to look at to be the ones that end in the unusual AAG 3′ splice site. This is specified by including the last three bases in the desired introns in the “Optional specific end of intron pattern” box. One can specify more or fewer bases using this box as well (e.g., “UUAAG”), but it does not accept wild-card characters or bracket requests.

The output of the search is shown in the bottom half of Fig. 6. The first line contains the number of occurrences of the pattern, the next line has links to the report for

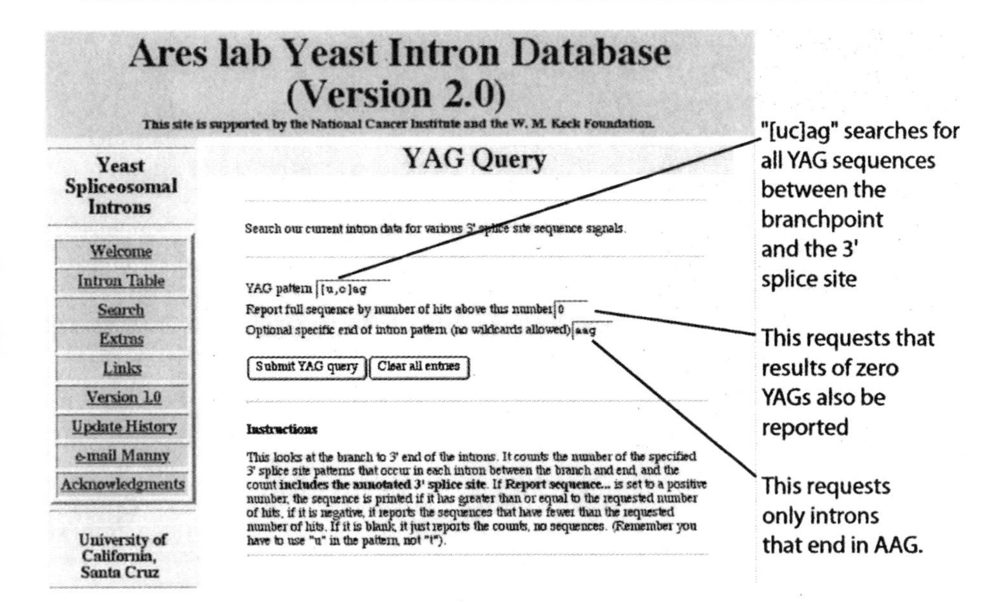

The output looks like this:

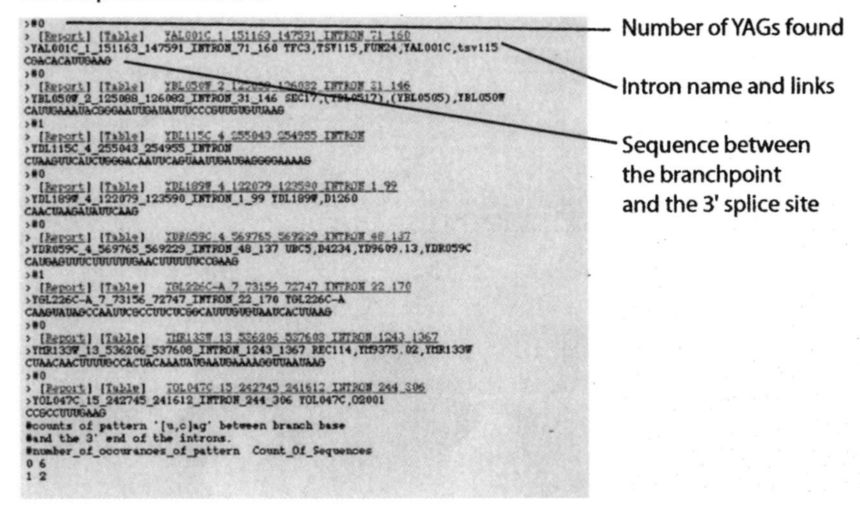

FIG. 6. Using the YAG query.

that intron, including the gene name(s), and the last line has the sequence between the branchpoint and the 3′ splice site. The search shows that two genes, *YDL115C* and *YGL226C-A*, have introns that skip a YAG sequence in favor of an AAG sequence. In the case of *YGL226C-A*, the UAG is only 9 nt from the branchpoint, a distance less than that found in the intron with the shortest such distance (*MATa1* second intron, 10 nt). Care must be used in evaluating the sequence by eye, since in about 16 introns there is a CAG immediately after the branched residue, and these are unlikely to represent 3′ splice sites. (Can you find these? *Hint:* Go back to the Intron Splice Signals Query page and type "cag" into the "Bpost" box and submit the form.) The search for sequences on the YAG Query page is not limited to 3′ splice site-like sequences. Actually any sequence can be requested using the "YAG pattern" box on the form. We have also used this search page to evaluate the existence of pyrimidine tracts and other sequences in the branchpoint-3′ splice site interval of yeast introns.

The final search is the Intron Sequence Search. Go back once more to the Search page and click on the "Sequence search on introns" link. This page (Fig. 7)

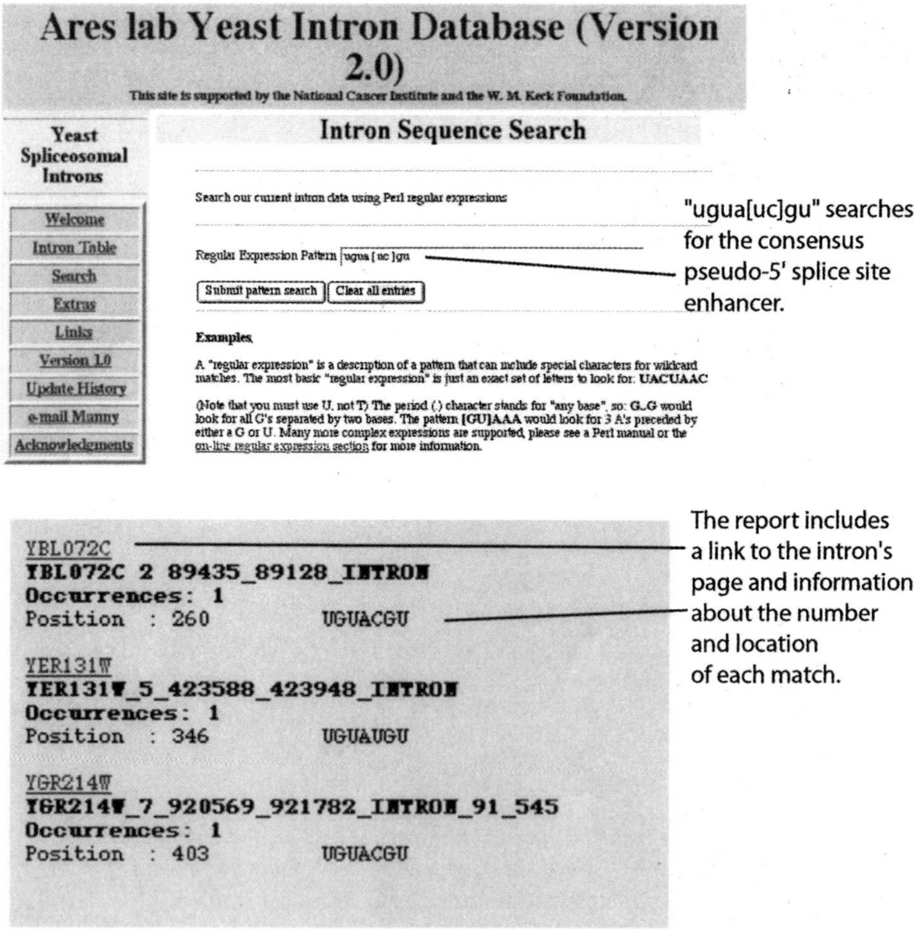

FIG. 7. Intron sequence search.

allows searches for any sequence anywhere in the intron. Type the sequence you are interested in finding into the box labeled "Regular Expression Pattern." You must use U instead of T! (Is our RNA bias showing?) This form accepts wild-card characters and bracket requests (square only!) within a string of nucleotides, such as "ugua[uc]gu." This particular request searches for either of two sequences: UGUAUGU or UGUACGU. Inserting a period "." means "any base" and is formally the same as "[agcu]." To get any pyrimidine at a particular position, as above, type "[uc]"; to get any purine, type "[ag]"; or for only G or C, use "[gc]." The "caret" character "^" can be used to specify "NOT" inside the brackets, so that if a nucleotide is excluded from a position this would be indicated by [^c], meaning "any nucleotide except C," or "A or U or G." To exclude two nucleotides, type "[^ag]." Note that "[^ag]" means the same to the computer as "[cu]" (you have just learned a little of the computer language Perl).

In the example in Fig. 7, we searched for the consensus sequence found in an *in vitro* evolution experiment that identified a pseudo-5′ splice site as an enhancer of splicing efficiency in yeast.[8] We wanted to see if such sequences might be common in yeast introns. To begin the search, we used the strictest definition of the consensus, which is essentially the same as the 5′ splice site consensus sequence except that it has a U upstream of the first G: UGUAYGU, which we express for the search as "ugua[uc]gu." Clicking on the "Submit pattern search" button, the computer returns a page containing information on three introns. Note that the natural 5′ splice site is excluded because the intron sequence data in the database includes no exon nucleotides, and thus none has a U upstream, at least in the data being searched. Many more introns are returned if the first U is deleted from the request. The results provide a link to the report page, the intron coordinates, the number of occurrences of the sequence, and the position of the sequence found. These three introns might be good candidates in which to test the idea that pseudo-5′ splice site sequences contribute to splicing in natural yeast introns. Additional searches with more relaxed consensus sequences may reveal additional candidates.

Database Errors, Programming Bugs, and Interpretational Caveats

The Ares Lab Intron Database is a work in progress. We originally devised it for our own use, but found it so useful that we thought others might want to use it as well. We try to maintain the accuracy of the data, but there is a lot of it underneath. (One can think of the distinction between a searchable database and a publication as similar to the distinction between performance art and painting. Each time you access the database, we are putting on a show, which is a different responsibility than one has after painting a picture.) Although the data are useful for gaining broad-brush impressions of the intron family in yeast, and the choice nugget will occasionally be found, all specific results of importance should be

[8] D. Libri, A. Lescure, and M. Rosbash, *RNA* **6,** 352 (2000).

confirmed by comparing the findings with information in other databases. SGD, PubMed, and GenBank employ individuals who are responsible for constantly updating information of relevance to the data in our collection. Database errors exist in all databases, so beware! Also note that future versions of the database may return results slightly different from those presented in the figures, because of updated information.

There are a few programming bugs yet in the system as well. The Intron Sequence Search report gives spurious position information if there is more than one occurrence of a particular sequence (although the first listed position is correct, and the other occurrences are present) and we are working to fix that. One limitation we cannot really surmount involves the identification of the branchpoint. We specify this position based on looking at the sequence and in many cases it is a guess. Molecular analysis of branchpoints is challenging, and there are few hard data for most introns. Finally one must be careful in interpretation of the findings. An example is the abundance of U tracts in introns. These are easy to find, and some introns have many, many short runs of U. The results of such a search may be impressive but the software assesses no significance to these results. The investigator must ask, How big is this intron? Given the G+C content of the yeast genome, how often might I expect to observe U runs of this length in a sequence of this size? What is the probability that my observation could be due to chance? In the future when all possible experiments have been done and all data is archived in searchable structures, all we will need do to test a hypothesis is to submit a form. Until then we can at least use our current databases to sharpen a few of our experimental rationales and hone some of our conclusions.

Acknowledgments

We thank Chuck Sugnet, Tyson Clark, Carrie Davis, and Marc Spingola for help keeping the database tidy. We also thank Haller Igel for comments on the manuscript. This work was supported by the W. M. Keck Foundation grant to the RNA Center at Santa Cruz, and by grants to M.A. from the National Institutes of Health (CA 77813 and GM 40478).

[23] Yeast Genomic Expression Studies Using DNA Microarrays

By AUDREY P. GASCH

Introduction

The technological developments in the area of genome science have revolutionized the methods of biological exploration, allowing organisms to be studied on a genomic scale. Characterizing whole-genome expression using DNA microarrays provides a snapshot of an organism's genome in action by revealing the relative transcript levels of thousands of genes at a time. Exploration of the dynamic nature of genomic expression programs throughout the natural life cycle of cells presents a variety of biological insights, hinting at the underlying cellular processes that mediate or respond to changes in gene expression.

In a seminal study by DeRisi *et al.*,[1] DNA microarrays were used to follow the global changes in gene expression during the natural transition of yeast cells from fermentative metabolism to respiration, known as the diauxic shift.[1] Despite decades of study on the metabolic pathways involved, the work demonstrated the potential of genomic expression profiling by revealing many previously unrecognized features of the metabolic transition. Nearly 30% of the yeast genome (approximately 1700 genes) was significantly altered in expression as the available glucose in the medium was depleted, and subsets of genes displayed distinct temporal patterns of transcript variation. DeRisi *et al.*[1] demonstrated that genes displaying similar expression patterns were functionally related and regulated by upstream sequence motifs common to the promoters of the genes. This result established the enormous potential of genomic expression characterization in implicating hypothetical functions for uncharacterized genes, because of their similarity in expression with genes of known functions, while suggesting modes of transcriptional regulation by pointing to signaling factors known to act through the conserved promoter elements common to coregulated genes.

Since then, genomic expression studies have revealed a wealth of information about yeast biology. Characterization of transcript fluctuation during the yeast cell cycle not only identified approximately 700 cell-cycle regulated genes, but also painted a more complete picture of the biological processes that occur during each cell-cycle phase.[2,3] In addition, the identification of groups of coregulated

[1] J. L. DeRisi, V. R. Iyer, and P. O. Brown, *Science* **278,** 680 (1997).

[2] R. J. Cho, M. J. Campbell, E. A. Winzeler, L. Steinmetz, A. Conway, L. Wodicka, T. G. Wolfsberg, A. E. Gabrielian, D. Landsman, D. J. Lockhart, and R. W. Davis, *Mol. Cell* **2,** 65 (1998).

[3] P. T. Spellman, G. Sherlock, M. Q. Zhang, V. R. Iyer, K. Anders, M. B. Eisen, P. O. Brown, D. Botstein, and B. Futcher, *Mol. Biol. Cell.* **9,** 3273 (1998).

0076-6879/02 $35.00

cell-cycle genes led to the subsequent identification of the signaling factors that regulate their expression and the promoter elements through which they execute their control.[3–5] Investigation of the temporal changes in gene expression during sporulation of diploid cells helped to more accurately define the temporal phases of the developmental process, while identifying thousands of genes, many of them completely uncharacterized, that were involved in meiosis.[6] A separate study characterizing genomic expression patterns dependent on ploidy further clarified cellular processes and gene expression patterns that were affected by genomic copy number.[7]

Microarray studies have presented a complex picture of developmental processes in yeast, and they have also revealed the intricacies of cellular responses to environmental changes. Comparison of the genomic expression programs triggered by diverse environmental stresses revealed many condition-specific features in the gene expression responses,[8] but at the same time indicated that the majority of the genomic expression responses to each unique environment was in fact part of a general environmental stress response, common to all conditions.[8,9] Subtleties in the expression profiles, however, revealed that this general stress response was regulated by different upstream signaling factors in response to different experimental conditions, demonstrating the complexities in the regulation of yeast genomic expression.[8,10]

Exploration of global gene expression in wild-type cells has revealed the cells' natural expression programs during development and environmental responses; however, aberrant expression in mutant strains has been equally informative. For example, Hughes *et al.* generated a large compendium of single-gene deletion mutants and demonstrated that the genomic expression programs resulting from many of the individual gene deletions were indicative of the deleted gene's cellular role, as cells lacking functionally related genes displayed similar anomalies in genomic expression relative to wild-type cells.[11] Many other studies have identified

[4] G. Zhu, P. T. Spellman, T. Volpe, P. O. Brown, D. Botstein, T. N. Davis, and B. Futcher, *Nature* **406,** 90 (2000).

[5] V. R. Iyer, C. E. Horak, C. S. Scafe, D. Botstein, M. Snyder, and P. O. Brown, *Nature* **409,** 533 (2001).

[6] S. Chu, J. DeRisi, M. Eisen, J. Mulholland, D. Botstein, P. O. Brown, and I. Herskowitz, *Science* **282,** 699 (1988). [Published erratum appears in *Science* **282,** 1421 (1998)].

[7] T. Galitski, A. J. Saldanha, C. A. Styles, E. S. Lander, and G. R. Fink, *Science* **285,** 251 (1999).

[8] A. P. Gasch, P. T. Spellman, C. M. Kao, O. Carmel-Harel, M. B. Eisen, G. Storz, D. Botstein, and P. O. Brown, *Mol. Biol. Cell* **11,** 4241 (2000).

[9] H. C. Causton, B. Ren, S. S. Koh, C. T. Harbison, E. Kanin, E. G. Jennings, T. I. Lee, H. L. True, E. S. Lander, and R. A. Young, *Mol. Biol. Cell* **12,** 323 (2001).

[10] A. P. Gasch, M. X. Huang, S. Metzner, D. Botstein, S. J. Elledge, and P. O. Brown, *Mol. Biol. Cell* **12,** 2987 (2001).

[11] T. R. Hughes, M. J. Marton, A. R. Jones, C. J. Roberts, R. Stoughton, C. D. Armour, H. A. Bennett, E. Coffey, H. Dai, Y. D. He, M. J. Kidd, A. M. King, M. R. Meyer, D. Slade, P. Y. Lum, S. B. Stepaniants, D. D. Shoemaker, D. Gachotte, K. Chakraburtty, J. Simon, M. Bard, and S. H. Friend, *Cell* **102,** 109 (2000).

abnormalities in gene expression in strains harboring mutations in signaling factors such as protein kinases and transcription factors, revealing targets of those signaling molecules while implicating specific conditions under which those signaling pathways are active.[1,8,10–21]

The use of microarrays to characterize genomic expression programs is expanding to study the evolution of yeasts by exploring variation of genomic expression responses in natural *Saccharomyces cerevisiae* isolates[22] and strains evolved under experimental conditions.[23] As more complete genome sequences emerge, including that of *Candida albicans,*[24] *Cryptococcus neoformans,*[25] the fission yeast *Schizosaccharomyces pombe,*[26] and other yeast species,[27] comparison of genomic expression programs within and across different yeasts will illuminate the physiological processes within each organism and reveal evolutionary differences in the regulation and orchestration of genomic expression programs.

The advent of DNA microarrays has allowed the exploration of gene expression on a whole-genome scale. DNA microarrays consist of DNA fragments that have been immobilized onto a glass support.[28] Unlike some other commercially available

[12] H. D. Madhani, T. Galitski, E. S. Lander, and G. R. Fink, *Proc. Natl. Acad. Sci. U.S.A.* **96,** 12530 (1999).

[13] U. S. Jung and D. E. Levin, *Mol. Microbiol.* **34,** 1049 (1999).

[14] C. J. Roberts, B. Nelson, M. J. Marton, R. Stoughton, M. R. Meyer, H. A. Bennett, Y. D. He, H. Dai, W. L. Walker, T. R. Hughes, M. Tyers, C. Boone, and S. H. Friend, *Science* **287,** 873 (2000)

[15] R. Casagrande, P. Stern, M. Diehn, C. Shamu, M. Osario, M. Zuniga, P. O. Brown, and H. Ploegh, *Mol. Cell* **5,** 729 (2000).

[16] K. J. Travers, C. K. Patil, L. Wodicka, D. J. Lockhart, J. S. Weissman, and P. Walter, *Cell* **101,** 249 (2000).

[17] T. J. Lyons, A. P. Gasch, L. A. Gaither, D. Botstein, P. O. Brown, and D. J. Eide, *Proc. Natl. Acad. Sci. U.S.A.* **97,** 7957 (2000).

[18] N. Ogawa, J. DeRisi, and P. O. Brown, *Mol. Biol. Cell* **11,** 4309 (2000).

[19] C. Gross, M. Kelleher, V. R. Iyer, P. O. Brown, and D. R. Winge, *J. Biol. Chem.* **275,** 32310 (2000).

[20] F. Posas, J. R. Chambers, J. A. Heyman, J. P. Hoeffler, E. de Nadal, and J. Arino, *J. Biol. Chem.* **275,** 17249 (2000).

[21] J. C. Kapteyn, B. ter Riet, E. Vink, S. Blad, H. De Nobel, H. Van Den Ende, and F. M. Klis, *Mol. Microbiol.* **39,** 469 (2001).

[22] D. Cavalieri, J. P. Townsend, and D. L. Hartl, *Proc. Natl. Acad. Sci. U.S.A.* **97,** 12369 (2000).

[23] T. L. Ferea, D. Botstein, P. O. Brown, and R. F. Rosenzweig, *Proc. Natl. Acad. Sci. U.S.A.* **96,** 9721 (1999).

[24] http://www-sequence.stanford.edu/group/candida

[25] http://www-sequence.stanford.edu/group/C.neoformans

[26] http://www.sanger.ac.uk/Projects/S_pombe

[27] J. Souciet, M. Aigle, F. Artiguenave, G. Blandin, M. Bolotin-Fukuhara, E. Bon, P. Brottier, S. Casaregola, J. de Montigny, B. Dujon, P. Durrens, C. Gaillardin, A. Lepingle, B. Llorente, A. Malpertuy, C. Neuveglise, O. Ozier-Kalogeropoulos, S. Potier, W. Saurin, F. Tekaia, C. Toffano-Nioche, M. Wesolowski-Louvel, P. Wincker, and J. Weissenbach, *FEBS Lett.* **487,** 3 (2000).

[28] D. Shalon, S. J. Smith, and P. O. Brown, *Genome Res.* **6,** 639 (1996).

arrays, in which oligonucleotides are synthesized on the solid surface,[29,30] microarrays are constructed by spotting DNA fragments onto a pretreated microscope slide. Hybridization of a fluorescently labeled nucleic acid solution to the microarray allows the rapid identification and quantitation of all molecules in the solution, by exciting and detecting the fluorescent molecules bound to their immobilized DNA complement. In actuality, accurate measurement of absolute nucleic acid levels is confounded by a number of experimental variables. To avoid these limitations, microarrays utilize competitive hybridization of two, differentially labeled probe samples. By measuring the relative abundance of molecules in the two samples, the effects of subtle variation in the hybridization conditions are minimized, resulting in highly reproducible measurements. In addition to providing reliable measurements, microarrays are relatively affordable and straightforward to use. They can be applied to any technique in which identification by nucleic acid hybridization can be utilized, including measuring relative transcript abundance to characterize genomic expression patterns.

This chapter will focus on the use of microarrays to characterize genomic expression programs in yeast, although many of the details can be applied to microarray uses beyond gene expression studies. Because previous sources have extensively described the construction and use of microarrays,[31–33] this chapter will focus on details that are specific to yeast microarray experiments, as well as critical features of microarray use. Readers are encouraged to consult the listed resources that complement this chapter.

Microarray Construction

The construction of microarrays involves four general steps: generation of the target DNA to be spotted, preparation of glass microscope slides, printing of the arrays, and processing the final product before use. Microarray spotting machines can be purchased commercially; however, construction of the microarrayer developed in Brown's laboratory at Stanford University (Stanford, CA)[28] is relatively straightforward, thanks to detailed instructions provided by DeRisi and members of the Brown laboratory.[32,33] The information contained on these Web sites provides specific instructions and tips for the printing process, which will not be covered here.

[29] S. P. Fodor, R. P. Rava, X. C. Huang, A. C. Pease, C. P. Holmes, and C. L. Adams, *Nature* **364,** 555 (1993).

[30] A. C. Pease, D. Solas, E. J. Sullivan, M. T. Cronin, C. P. Holmes, and S. P. Fodor, *Proc. Natl. Acad. Sci. U.S.A.* **91,** 5022 (1994).

[31] M. Eisen and P. O. Brown, *Methods Enzymol.* **303,** 179 (1999).

[32] http://cmgm.stanford.edu/pbrown

[33] http://www.microarrays.org

Generation of Target DNA for Spotting

Any DNA solution can be spotted onto microarrays; however, the most common yeast microarrays to date consist of PCR (polymerase chain reaction) products representing every predicted open reading frame in the *S. cerevisiae* genome. Genomic sequences can be amplified using standard PCR conditions and sequence-specific primers designed to amplify *S. cerevisiae* open reading frames (ORF) as well as intergenic sequences (Research Genetics, Huntsville, AL, and at a current price of $38,000 per primer set). For other genomes for which the complete sequence is not yet known, genomic or cDNA libraries can be amplified using common, plasmid-specific primers, as has been done for other organisms.[34] Standard 100 μl PCR reactions are performed in 96-well plates using 1 μM primer, 40 ng yeast genomic DNA from the S288C sequenced strain (Research Genetics), 2 mM magnesium chloride, and *Taq* polymerase conditions. The amplification begins with a 10 min denaturation at 92°, followed by 35 cycles of 92° for 30 sec, 56° for 45 sec, and 72° for 3.5 min. After verification of the reaction products by gel electrophoresis and reamplification of failed reactions, the DNA is precipitated as previously described.[31–33] The PCR products are resuspended overnight in 50 μl of 3× SSC (or alternatively a larger volume of double-distilled, deionized water to facilitate robotic transfer) and then distributed into nonsterile, v-bottom 384-well plates (Genetix, Hampshire, UK) in 4 μl per well aliquots to yield 12 replicate plate sets. (Be aware that other brands of 384-well plates or plates that have been sterilized by irradiation are prone to warping and may display variable well spacing, causing problems during the printing process.) After lyophilization in a speed vacuum, the dried DNA pellets can be stored at room temperature for months. If the original PCR products were resuspended in water, the plates should be resuspended one time in 4 μl 3× SSC before use. Subsequently, the dried DNA pellets should be resuspended overnight in 4 μl doubly distilled H_2O before each use, and after use the plates can be dehydrated, stored, and reused for at least 5 printings. (After ~4 printings from the same plate set, the DNA should be resuspended one time in 4 μl 1.5× SSC to replace the salt lost during previous printings.) Each print run yields 140–250 microarrays, depending on the microarrayer used, and reuse of the print plates allows the production of at least 750 microarrays per plate set. Thus, each genome amplification yields approximately 10,000 microarrays.

Recently, 70-mer oligonucleotides representing each of the predicted open reading frames in *S. cerevisiae* have been designed (Operon Technologies, Alameda, CA, current price $35,000 for enough sample to generate ~5000 microarrays), providing an alternative source of target DNA to be spotted onto the microarrays. The sequence chosen for each oligonucleotide is designed to minimize sequence similarity between the set of 70-mers, thereby reducing cross-hybridization of

[34] R. E. Hayward, J. L. DeRisi, S. Alfadhli, D. C. Kaslow, P. O. Brown, and P. K. Rathod, *Mol. Microbiol.* **35,** 6 (2000).

the spots on the microarrays. In addition, because the concentration of all the 70-mers in the printing set is roughly equal, the DNA mass in the microarray spots is normalized, unlike spotted PCR products. Although the cost per microarray is slightly higher when printing oligonucleotides (~$7 per microarray) versus PCR products (<$1 per microarray), the initial monetary investment is substantially less (since much of the cost of PCR amplification is accounted for by PCR reagents), and the labor required to prepare the genomic sequences is drastically reduced. These features make the spotting of 70-mers an attractive alternative to PCR amplification of the yeast genome.

Preparing Slides before Printing

A critical step in making good microarrays is generating high-quality microscope slides for the arrays. Although various types of slides are commercially available, generation of polylysine-coated slides is simple and reliable. Glass microscope slides (Gold Seal, Portsmouth, NH) can be loaded into metal slide racks and processed in 400-ml glass slide chambers (Shandon Lipshaw, Pittsburgh, PA). The first step is to thoroughly clean the slide surface before coating. For each rack of 30 slides, dissolve 35 g sodium hydroxide in 140 ml of doubly distilled H_2O, then add 210 ml 95% (v/v) ethanol and mix thoroughly. (Never use 100% ethanol in the array-making process, as benzene contaminants can carry through the process and fluoresce on the microarrays.) Place the slide rack containing the slides into the solution and incubate for 2 hr with slight agitation on a platform shaker. The slides should then be rinsed 5–7 times in ~400 ml doubly distilled H_2O per slide rack, a step that appears to be critical for proper slide generation. Exposure of the slides to air should be minimized during this process. Next, prepare the polylysine solution by mixing 35 ml poly (L-lysine) (Sigma, St. Louis, MO), 35 ml filtered PBS buffer, and 280 ml doubly distilled H_2O per slide rack in a plastic cylinder. Place the slide rack into the solution and incubate for 30 min at room temperature with slight agitation on a platform shaker. Rinse the slides one time by plunging ~5 times in 350 ml doubly distilled H_2O per rack, then dry the slides by centrifugation at 500 rpm for 5 min in a tabletop centrifuge. Before use, the slides must be aged for ~14 days, to achieve optimal hydrophobicity of the slide surface, in a plastic slide box (slide boxes with rubber surfaces should not be used as emissions from the rubber can deteriorate the polylysine surface). The aging process can be accelerated by baking the slides for 5–10 min at 50° in an oven after centrifuge drying. Before use, the slides should be tested by printing a small number of DNA spots onto representative slides from each batch, hybridizing fluorescently labeled probe to the microarrays, then scanning the finished arrays to verify adequate hybridization signal in the spots and low background signal elsewhere on the arrays. It is feasible to generate 4–6 racks of slides at a time, and it is advisable to prepare multiple batches of slides before microarray printing to ensure a sufficient number of high-quality slides.

Processing Microarrays after Printing

Another critical step in the construction of microarrays is postprocessing of the microarrays before hybridization. After the DNA is spotted onto the microarrays, the unbound lysine groups on the slide surface must be chemically blocked to prevent nonspecific binding of probe DNA to the array.

Because postprocessing removes the salt deposited during printing, and thus prevents visualization of the DNA spots, the first step of the process is to mark the boundary of the spots on each microarray. Place the microarray DNA face-down over two spaced supports, such that the sides of the microarray are supported and the DNA spots are free from any surface. Etch the boundary of the microarray spots onto the back of the slide with a glass etcher. The etched boundary allows accurate placement of the fluorescent probe and the coverslip for microarray hybridization.

The microarrays must then be rehydrated to evenly distribute the mass of DNA within each spot. Place the array DNA face-down in a slide hydrating chamber (Sigma) over a room-temperature solution of 1× SSC, and incubate until the spots on the array are visibly hydrated and glistening (5–15 min). Overhydration of the microarrays can cause the spots to bleed together, and therefore the process should be monitored and terminated if the spots are in danger of contaminating others. The arrays are then rapidly dried by placing the array, DNA face-up, onto a preheated metal block at 80° for 3–5 sec, or until the vapor has evaporated from the slide. The DNA can be cross-linked to the slide in a UV cross-linker at 60 mJ by laying the individual microarrays DNA face-up in the cross-linker.

Before performing the chemical step, gather the reagents, including 1 liter of preheated doubly distilled H_2O in a beaker large enough to hold a slide rack, one empty glass slide chamber, one glass slide chamber filled with 95% ethanol, and chemical reagents. Place the slides into a metal rack and set aside. The critical chemical blocking step must be initiated rapidly and carried out exactly as described here: in a separate flask, add 6 g succinic anhydride (Aldrich, Milwaukee, WI) to 325 ml 1-methyl-2-pyrrolidinone (Aldrich) and stir gently with a stir bar (succinic anhydride is stable long-term if stored under vacuum; 1-methyl-2-pyrrolidinone should be clear before mixing, and yellow solvent should not be used). Immediately after the last grains of succinic anhydride dissolve, add 25 ml 1 *M* sodium borate pH 8.0 (made by dissolving boric acid in doubly distilled H_2O and adjusting the pH with sodium hydroxide). Immediately pour the mixed solution into the *empty* glass chamber, then swiftly plunge the rack of microarrays *into* the solution. After a 15 min incubation with gentle agitation on a platform shaker, quickly remove the rack of arrays from the blocking solution and plunge into preheated doubly distilled H_2O for cleansing and denaturation: for microarrays containing PCR products, plunge the slide rack into just-below-boiling water for 2 min, and for oligonucleotide-containing microarrays, plunge into 55° water for only 5 sec to minimize loss of DNA from the microarray slide. Finally, plunge the slide rack into 95% ethanol about 5 times and dry the slides by centrifugation at 500 rpm for 3 min. The arrays are ready for hybridization and can be stored at room temperature for at least 3–6 months.

Planning Experiment

The key to a successful genomic expression experiment is proper planning and preparation. This section describes important considerations in the setup of genomic expression studies and includes examples of common oversights in experimental design and execution.

Strains and Growth Conditions

Although many laboratories often use their favorite strains, careful consideration should be given to the strain background used for microarray experiments. Different laboratory strains can have significantly different genotypes, leading to strain-dependent differences in genomic expression responses. When comparing the results of genomic expression experiments, particularly when comparing wild-type and mutant responses, an isogenic strain background must be used. Strains that have unnecessary auxotrophic markers should be avoided, as these additional phenotypes can complicate cellular responses and confound the analysis of genomic expression patterns.

Genomic expression can also be affected by growth conditions, including culture system (shaker flasks vs chemostat), culture aeration, growth temperature, and medium. Synthetic medium offers an advantage over rich yeast-extract medium, in that nutrient concentrations can be precisely controlled. The choice of synthetic vs rich medium can significantly affect genomic expression responses to environmental stimuli.[35] The experimenter should be aware that these features may affect genomic expression programs and that variations in the growth conditions may result in somewhat different gene expression patterns.

Experimental Design

Perhaps the most important aspect of setting up genomic expression studies is choosing experimental conditions for which there are as few variables as possible. Often, experimental variables are overlooked, such as extensive cell handling during the experiment, the effects of a drug carrier solution on the cells, pleiotropic effects of the chosen stimulus, and progression of the cells into the diauxic shift during the experiment (see below). These features are to be avoided as they confound data analysis and can lead to poor experimental reproducibility. If these features cannot be avoided (for example, extensive culture handling), a

[35] A. P. Gasch, O. Carmel-Harel, G. Storz, D. Botstein, and P. O. Brown, unpublished data (2000).

separate control experiment must be performed to identify the genomic expression responses triggered by these unavoidable variables.

Choosing the appropriate dose of a stimulus and assessing the timing of the cellular response often requires prior investigation. The desired dose can be assessed by performing dosage curves and monitoring the response of cellular indicators, such as changes in the expression of representative genes, cell morphology, growth rate, cell-cycle arrest, and cell viability. Conditions that result in high cell death should be avoided, as the gene expression results are often difficult to interpret.

The timing of the cellular response must also be determined to ensure that the full genomic expression response is observed. Timing of the peak gene expression response can vary widely depending on the experimental conditions: under some conditions the peak response occurs and subsides within minutes of stimulus exposure, and in other situations the majority of the response occurs hours later. The temporal response can be determined by following representative cellular indicators over time, as well as by analyzing genomic expression in a trial time course.

Conducting time-course experiments with multiple time points, rather than single-time point microarray studies, offers a number of advantages. Time-course experiments reveal additional levels of detail, including the choreography of primary and secondary gene expression responses. Time courses are particularly useful when comparing the responses of different strains to a given stimulus, since the timing of the response may be slightly different in each strain. In addition to these advantages, characterizing a gene's expression over multiple time points provides multiple data points per gene, and the nonrandom pattern of gene expression seen for well-measured spots increases the confidence in the accuracy of the individual measurements.

Choosing Microarray Reference Sample

One must also give consideration to the type of reference that will be used on the microarrays, to ensure that enough of this sample is collected during the experiment. In many yeast microarray experiments, mRNA collected from cells growing in logarithmic phase, just before the culture is exposed to the desired stimulus (the so-called "time zero" cells), is used as the reference sample. In other cases a pool of the collected mRNA samples is used as the reference, and even genomic DNA can be used. The required attribute of the reference is that it give significant hybridization signal in all of the microarray spots, so that the relative two-color fluorescence on the microrarrays can be accurately measured. Regardless of the reference used, it is important that the same reference be used on all microarrays in a given time course experiment in order to accurately compare the results within the time course.

Conducting Experiment

Beginning Experiment

The experiment must be conducted on cells that remain in the same growth phase throughout the experiment, to prevent confounding genomic expression changes. For example, one of the most common problems in yeast genomic expression studies is progression of the cells into the diauxic shift. When the concentration of fermentable sugars in the medium becomes limiting, the metabolic framework of the cell shifts dramatically, leading to alterations in the expression of thousands of genes.[1] The timing of the diauxic shift varies depending on the strain and the exact growth conditions, as cells may consume glucose at rates dependent on their genotype as well as environmental features. Therefore, it is important to investigate the timing of the diauxic shift under each experimental system. This can be accomplished by following the glucose concentration in the medium in a trial experiment to be sure that the glucose concentration does not change appreciably over the course of the experiment.

To start the yeast culture, inoculate fresh medium of appropriate volume from a saturated, overnight culture, and allow the subculture to recover from stationary phase at least two doublings. Begin the experiment at a low cell density, and ensure that the projected cell density at the end of the experiment is well within log phase, to avoid the diauxic shift. As a precautionary measure, the glucose concentration in the culture medium can be measured by saving aliquots of the medium for subsequent analysis (Boehringer Mannheim, Basel, Switzerland). For long experiments in which the cells may approach the diauxic shift, a chemostat system should be used to maintain the cells in a steady state of growth.

Recording Experimental Details

Interpreting genomic expression data is often clarified by additional experimental details; however, it is sometimes difficult to predict *a priori* which details will be useful. Therefore, it is advisable to record as many experimental details as possible. For example, recording the growth rate and cell viability over the course of the experiment, collecting cell images that may indicate changes in cell morphology and cell-cycle progression, recording relevant nutrient levels in the medium, quantifying drug concentrations when possible, and documenting the timing and method of sample collection can aid in the subsequent analysis of the genomic expression patterns. Any anomalies in the experiment should also be recorded.

Sample Collection

Cell culture can be collected by a variety of methods. Cells can be recovered by centrifugation of 50 ml culture samples at 500 rpm for ~3 min at room temperature.

The cell pellet is immediately resuspended in 3–5 ml Lysis buffer [10 m*M* Tris-Cl pH 7.4, 10 m*M* EDTA, 0.5% sodium dodecyl sulfate (SDS)] in a 50-ml centrifuge tube, flash frozen in liquid nitrogen, and stored at −80° until RNA preparation. An alternative to the centrifugation method is filtration of the culture through a 1 μm sterile filter, allowing rapid collection of the cells. Immediately after the filtration, submerge the entire filter with recovered cells in 3–5 ml Lysis buffer, flash freeze, and store at −80° until RNA isolation.

Collection of the cells should be done in a quick and controlled manner, introducing as few variables into the experiment as possible. Cells should not be washed or unnecessarily handled before collection. Ideally, the temperature of sample collection should be close to the growth temperature; because this is not always possible, it is recommended that cells be collected at room temperature for cultures grown at standard laboratory temperatures (25–30°). Collection at extreme temperatures, such as on ice, can induce temperature shock in the cells and should be avoided.

RNA Isolation and Probe Generation

The fluorescent probe that will be hybridized to the array is generated by reverse transcription of the RNA isolated from the cell samples and incorporation of fluorescent or modified nucleotides into the nascent cDNA. Either total RNA or poly(A)-selected RNA can be labeled, through a variety of methods described below. Standard RNA handling techniques should be used throughout the RNA isolation procedures, including wearing gloves at all times and using RNase-free plasticware. Use of RNA inhibitor-treated plasticware and solutions is optional. Repeated freezing and thawing of the RNA should be avoided, and thawed samples should be maintained on ice to prevent RNA degradation.

RNA Isolation

The frozen cell samples are lysed by standard acid–phenol lysis at 65°: to each frozen cell lysate add 1 vol of water-saturated phenol (do not allow the lysate to thaw before adding phenol) and incubate at 65° for 30–45 min with periodic vortexing. Centrifuge the emulsion at 10,000*g* for 15 min at 4° to remove the cellular debris; extract the aqueous phase again with water-saturated phenol, then once with chloroform. The total RNA is precipitated from the aqueous phase by adding 0.1 vol 3 *M* sodium acetate and 2.5 vol ethanol, and incubated >1 hr at −20°. After centrifugation at 20,000*g* for 30 min at 4°, the washed and dried RNA pellet should be resuspended in ~0.5 to 1 ml TE buffer. The final concentration, estimated by measuring the absorbance at 260 nm, should be 1–2 mg/ml, and the ratio of absorbance at 260 nm vs 280 nm should be 2.0–2.3 for protein-pure RNA. The RNA can be aliquoted and stored at −80°.

For isolation of polyadenylated RNA, various kits are available. The mRNA can also be recovered using oligo(dT)-cellulose (Ambion, Austin, TX).[36] Prepare the cellulose by washing 1 g of cellulose 2–3 times in 2× NETS buffer (200 m*M* NaCl, 20 m*M* Tris-Cl pH 7.4, 20 m*M* EDTA, 0.4% SDS), and resuspending in 10 ml 2× NETS buffer. To bind the sample to the resin, dilute 1 mg of total RNA into 1 ml of TE and combine with 1 ml of the suspended oligo(dT)-cellulose in a 2-ml disposable column (Bio-Rad, Hercules, CA). Allow the column to rotate on a small rotor for 1 hr at room temperature to allow the polyadenylated RNA to bind the resin. To begin column washes, break the bottom of each column to allow liquid to flow through, and mount the columns onto a 15-ml Falcon tube. Wash each column 7 times with 1 ml 1× NETS buffer. Transfer the column to a fresh Falcon tube for elution, then elute 2 times with 700 μl 1× ETS buffer (10 m*M* Tris-Cl, pH 7.4, 10 m*M* EDTA, 0.2% SDS) heated to 70°. To remove any granules of cellulose that escape into the eluate, extract the eluate once with 1 vol chloroform, then precipitate the RNA by adding 0.1 vol 3 *M* sodium acetate and 1 vol isopropanol, and incubate overnight at −20°. Collect the RNA by centrifugation and resuspend the washed pellet in 20–30 μl TE to yield a concentration of ~1 μg/μl, quantifiable by measuring the absorbance at 260 nm. The purity of the mRNA can be assessed by electrophoresis of an equal mass of total RNA and the purified polyadenylated RNA; although the two prominent bands of ribosomal RNA will likely remain in the polyadenylated sample, an enrichment of polyadenylated RNA is evident by a smear ranging from ~500 bp up the gel. Significant intensity at the bottom of the gel (<300 bp) indicates RNA degradation; the sample should not be used for microarray probe generation.

In some cases it may be necessary to amplify the mass of mRNA to generate enough template for probe preparation. Amplification from total RNA or poly(A)-isolated RNA involves generating a double-stranded cDNA that is subsequently transcribed from a linker sequence that has been incorporated into the cDNA, using T7 polymerase. Complete details on the amplification protocol can be found elsewhere.[32,33,37]

Microarray Probe Generation

Fluorescent cDNA probe can be generated through a variety of methods. The probe can be generated from mRNA, total RNA, or DNA by incorporating fluorescent or modified nucleotides into the cDNA product. The protocols for generating probe using fluorescent nucleotides (Cy-dUTP labeling) or modified nucleotides (aminoallyl labeling) are listed below. Regardless of the protocol, consistency in

[36] Protocol developed by Joe DeRisi (1997).

[37] E. Wang, L. D. Miller, G. A. Ohnmacht, E. T. Liu, and F. M. Marincola, *Nat. Biotechnol.* **18,** 457 (2000).

the labeling should be maintained. When labeling many samples to be compared, it is advisable to create a cocktail of reaction mix and perform the labeling reactions at the same time.

Reverse Transcription Using Cy-dUTP

Two fluorescently labeled probes are generated for each microarray experiment, incorporating one of two modified dUTP nucleotides (Cy3-dUTP or Cy5-dUTP) through reverse transcription using StrataScript reverse transcriptase (Stratagene, La Jolla, CA). Combine either 15 μg of total RNA or 2 μg mRNA with 5 μg of an "anchored" oligo(dT) primer of the sequence 5′ $(T)_{15}$-V-N 3′, where V is any nucleotide but T and N is any nucleotide. The two additional nucleotides on the 3′ end of the oligo(dT) primer promote priming at the junction between the transcript and the poly(A) tail. Adjust the volume with TE buffer to 15.5 μl. Denature the RNA mixture by incubating at 70° for 10 min, then immediately place on ice. In the meantime, a reaction cocktail is prepared and stored on ice: for each reaction, add to the cocktail 3 μl 10× StrataScript buffer, 3 μl 100 m*M* dithiothreitol (DTT), 0.5 μl 50× dNTP mixture (10 m*M* dTTP, 25 m*M* dATP, dCTP, dGTP), 2 μl doubly distilled H_2O, and 3 μl StrataScript reverse transcriptase. To each denatured RNA mixture, add 3 μl 1 m*M* Cy3-dUTP or Cy5-dUTP (Amersham, Buckinghamshire, England) and 11.5 μl of the reaction cocktail for a final reaction volume of 30 μl. Incubate the reaction at 42° for 2 hr. After the reaction has completed, the unincorporated fluorescent nucleotides must be removed from the reaction. Dilute the reaction mixture with 500 μl TE buffer, and add to a Microcon 30 spin filter (Amicon, Danvers, MA). Spin in a microfuge approximately 9 min until the volume is reduced to <5 μl. Collect the concentrated probe by inverting the filter into a fresh tube, then adjust the final concentration of the probe to 10 μl with TE buffer. (Unincorporated Cy-dUTP can be collected from the flow through for reuse, as described elsewhere.[32]) The probe can be stored at this stage for a few days at 4° but should not be frozen.

The labeling reaction can also be primed with random primers, as opposed to or in combination with the oligo(dT) primer that specifically binds to the poly(A) RNA tail. The protocol is nearly identical to the protocol described for using oligo(dT) primer, except that 5 μg of random octamers is used [either alone or in addition to 5 μg of the anchored oligo(dT) primer] and that the mixed reaction must be incubated for 10 min at room temperature before transferring to 42°.

In addition to labeling RNA samples, DNA can also be used to generate fluorescent probes using Klenow polymerase.[38] Adjust the volume of a 2 μg DNA sample to 21 μl with TE buffer. Add 20 μl of 2.5× reaction cocktail (125 m*M* Tris-HCl pH 6.8, 12.5 m*M* magnesium chloride, 25 m*M* 2-mercaptoethanol, 0.75 μg/μl

[38] Protocol developed by Jonathan Pollack (2000).

random octamer primers). Incubate the solution at 94° for 5 min to denature the probe, then chill on ice. Add 5 μl of a 10× dNTP mixture (0.6 m*M* dTTP, 1.2 m*M* dATP, dCTP, dGTP), 3 μl of 1 m*M* Cy3-dUTP or Cy5-dUTP (Amersham), and 1 μl (20–50 units/μl) Klenow DNA polymerase for a final reaction volume of 50 μl. Incubate the reaction for 2 hr at 37°, then quench the reaction by adding 5 μl 0.5 *M* EDTA. As described above, purify the probe by dilution and filtration in a Microcon 30 filter.

Reverse Transcription Using Aminoallyl-Modified Nucleotides

An alternative and more economical method of generating fluorescent probes is to chemically couple the fluorescent dye to the cDNA after reverse transcription. By incorporating aminoallyl-modified nucleotides into the nascent cDNA, activated dyes can be subsequently coupled to the amino moieties on the cDNA. A variety of fluorescent dyes can be attached to the amino moieties, including NHS ester Cy3 and Cy5 (Amersham). The NHS ester Cy dyes should be suspended in 16 μl fresh DMSO, divided into 1-μl aliquots (providing dye for one labeling reaction), and dried completely in a speed vacuum. Because the dyes are labile in solution, it is important to be sure the samples are completely dry. The dried dyes can be stored under desiccation at 4°. Alternatively, each tube of dye can be suspended in 8–12 μl dimethyl sulfoxide (DMSO) and used immediately for coupling, using 1-μl dye per reaction.

The aminoallyl-labeling method consists of three steps: cDNA synthesis, coupling of fluorescent dyes to the cDNA, and quenching of the reaction to prevent cross coupling of the dyes once the samples are combined.[39] For the reverse transcription, combine 2 μg mRNA or 15 μg total RNA with 5 μg anchored oligo(dT) primer in a volume of 15.5 μl, denature the RNA at 70° for 10 min, and place the sample on ice. To the denatured RNA solution add 14.5 μl of the reaction cocktail: for each reaction, add to the cocktail 3 μl 10× StrataScript buffer, 3 μl 100 m*M* DTT, 0.5 μl 50× dNTP mixture [10 m*M* 5-(3-aminoallyl)-2′-deoxyuridine 5′-triphosphate (Sigma), 15 m*M* dTTP, 25 m*M* dATP, dCTP, dGTP], 5 μl doubly distilled H_2O, and 3 μl StrataScript reverse transcriptase (Stratagene). The final reaction volume should be 30 μl. Allow the reaction to proceed at 42° for 2 hr. After completion of the cDNA synthesis, the remaining RNA can be hydrolyzed by adding 10 μl 1 *N* sodium hydroxide and 10 μl 0.5 *M* EDTA. After a 15 min incubation at 65°, the solution is neutralized by adding 25 μl 1 *M* HEPES pH 7.5 or 10 μl 1 *N* hydrochloric acid. At this point the samples can be stored at −20° for months.

Before the chemical coupling step, all Tris must be removed from the reaction. Dilute the neutralized reaction mixture into 500 μl doubly distilled H_2O in a

[39] Protocol developed by Rosetta Inpharmatics and modified by Joe DeRisi (2000).

Microcon 30 spin filter (Amicon) and centrifuge for 8–10 min until the volume is <10 μl. The filtration is repeated 2 more times, each time by adding 500 μl doubly distilled H_2O to the filter and repeating the centrifugation. After recovering the concentrated cDNA into a fresh tube, dry the sample in a speed vacuum or continue the Microcon filtration until the volume is ~4 μl. The dried or liquid samples can be stored at −20°.

To couple the fluorescent dyes to the modified cDNA, resuspend the dried cDNA in 9 μl 0.05 *M* sodium bicarbonate buffer pH 9.0, or add 4.5 μl of 0.1 *M* sodium bicarbonate buffer to 4.5 μl of the concentrated sample for a final concentration of 0.05 *M* sodium bicarbonate (verify the pH of the buffer before each use), and incubate at room temperature for 10 min to resuspend. After mixing well, transfer this solution to the dried aliquot of activated Cy3 or Cy5 dye or to 1 μl of the freshly suspended dye and mix well. Allow the reaction to proceed for 1 hr at room temperature in the dark.

Finally, the reaction must be quenched to prevent cross-coupling of the Cy dyes when the labeled samples are combined. To each reaction, add 4.5 μl 4-*M* hydroxylamine and incubate for 15 min at room temperature in the dark. After quenching the reaction, combine the Cy3- and Cy5-labeled samples and purify the mixture using a QIAquick PCR purification kit (Qiagen, Westburg, The Netherlands), as described in the kit protocol. The eluate should be concentrated by microcon filtration or drying in a speed vacuum, and the sample should be resuspended in doubly distilled H_2O for a final volume of 10 μl and prepared for hybridization as described below. Faint color in the concentrated sample indicates that the labeling reaction was successful. When color is not evident, the fluorescence can be checked by spotting <0.5 μl of the probe onto a glass microscope slide and scanning the spot to verify fluorescence.

Microarray Hybridization and Data Acquisition

Microarray Hybridization

To prepare the probe for hybridization to a ~22 × 22 mm microarray, combine 5 μl of each of the Cy-dUTP labeled samples or 10 μl of the combined aminoallyl-labeled sample, 1 μl of 10 mg/ml poly(A) carrier (Sigma), 2.3 μl 20× SSC, and 0.36 μl 10% SDS for a final volume of 14 μl (if the microarray is larger than 22 × 22 mm add to the 10 μl of probe solution 2.6 μl poly(A), 5.0 20× SSC, and 0.78 μl 10% SSD and adjust the final volume to 30 μl with water). Denature the probe by heating at 94° for 3 min, then cool to room temperature by rapidly transferring to a microfuge and spinning for 10 sec (avoid adding bubbles during handling). Do not chill the probe on ice as salt will precipitate from the solution.

Prepare the hybridization chamber (Corning): place the microarray into the chamber and add ~20 μl 3× SSC to each of the hybridization wells in the chamber and an additional ~30 μl of 3× SSC divided into isolated drops on the edges of the microarray. Insufficient volume will result in inadequate chamber hydration during hybridization, causing the probe to dry out under the coverslip. Select multiple microscope slide coverslips of the appropriate size (Corning) that appear free of dust, chips, and scratches, and prop them for easy access with a tweezers.

Adding the probe to the microarray and sealing with the coverslip requires some experience and personal technique. It is advisable to practice the following procedure before hand. To the center of the microarray, deposit 12–15 μl of the denatured and cooled probe for 22 × 22 mm arrays, or 28–30 μl probe for 22 × 44 mm microarrays. Bubbles can prevent proper hybridization, and therefore they should be avoided and removed when possible. Large bubbles can be popped with the corner of a coverslip, but great care should be taken not to touch the microarray surface. Quickly place the coverslip directly over the deposited probe and gently drop or place the coverslip over the array. The coverslip should not be moved so as to prevent scratching of the microarray surface; however, in unavoidable cases it can be gently positioned by sliding. Close the hybridization chamber and gently and quickly submerge the chamber in a 65° water bath. For optimal signal, hybridize the arrays for 7–16 hr. For consistent results, the hybridization conditions, including hybridization time, should be maintained for all microarrays.

Microarray Washing

After the hybridization is complete, the unbound fluorescent probe and detergent are removed from the microarrays in a series of three washes of increasing stringency. Prepare three wash solutions in glass slide chambers: for the first wash, dilute 18 ml 20× SSC and 1 ml 10% SDS with doubly distilled H_2O to a final volume of 350 ml; the second wash is identical to the first wash only without any SDS; for the final wash add 1 ml 20× SSC to 350 ml doubly distilled H_2O. After removing the arrays from the water bath, disassemble the hybridization chamber and quickly set the array into a metal slide rack submerged in the first wash. Gently and slowly move the slide rack under the wash solution until the coverslip slides off, taking care not to scratch the microarray surface with the coverslip. When the coverslip has separated, remove each microarray individually from the first wash, allow the solution to drain from the slide, and place the slide into a separate slide rack submerged in the second wash solution. Residual SDS can increase the background fluorescence on the arrays, so minimizing carryover of the first wash is recommended. When disassembling multiple hybridizations, the microarrays can remain in the second wash until all chambers have been disassembled. Plunge the rack in the second wash for a few seconds, then remove the entire slide rack from the second wash and transfer to the final, low-stringency wash. Incubate the slides in the final wash for 5 min with slight agitation on a platform shaker, then dry the slides by centrifugation at 500 rpm for 3 min at room temperature in a tabletop centrifuge before scanning. The arrays can be stored overnight in the dark at room temperature before scanning; however, longer storage may result in diminished Cy5 signal on the arrays.

Data Acquisition

Retrieving the microarray data involves a "scanning" process, in which the fluorescent dyes hybridized to the array are excited with a laser and the emitted fluorescence for each of the dyes is recorded. Among the most popular of the scanning laser devices available for purchase is the GenePix scanner from Axon Instruments (Union City, CA). Dual lasers simultaneously excite the Cy3 and Cy5 dyes at 532 nm and 635 nm, respectively, and the emitted fluorescence for each of the dyes is measured independently (different dye-specific laser and filter sets are also available). The parameters of the scanner can be adjusted during scanning, as described in the GenePix manual. The microarrays should be scanned at settings that maximize the intensity of the spot fluorescence signal within the range of signal detection, rather than settings that result in normalized signal between the two channels. Normalization of the signal intensities should be done mathematically after scanning and data extraction, as described below.

Once the microarray has been scanned, the data are extracted through a procedure known as "gridding." The GenePix software that accompanies the scanner can be used to place a grid of circles over the microarray fluorescence image, such that each spot is encircled. The fluorescence signal corresponding to each dye is then quantified: the local background signal measured just outside of each grid circle is subtracted from the signal measured within the circle, providing background-subtracted values for each DNA spot. Flawed spots, including those surrounded by abnormally high background signal, spots obscured by dust, or areas of bled and contaminated spots, can be removed from the dataset by manually "flagging" bad spots. Because spots with weak fluorescent signal are difficult to quantitate, they are automatically flagged by the GenePix gridding software.

A critical aspect of extracting microarray data is correct identification of each microarray spot. Identifying the spots with a unique ORF or sequence name requires deconvolution of the list of DNAs in the 384-well plates, generated from the original 96-well plates of PCR products or 70-mer oligonucleotides, that were printed onto the microarrays. This deconvolution process involves two steps: converting the list of DNAs in 96-well format into a 384-well format, and then converting the list of DNAs in the 384-well plates into the list of ordered spots on the microarrays. Conversion of the DNA sample order in the 96-well format to the 384-well configuration is dependent on the configuration used to transfer four 96-well plates into each 384-well plate, whereas deconvoluting the order of DNAs in 384-well format to the microarray spot order is dependent on the number of tips used in the printing and the order that the plates were spotted onto the arrays. Both of these deconvolutions can be performed using the GenePix software. In both cases, separate lists of DNA identifications are created for each 96-well or 384-well plate to identify the DNA in the consecutive plate wells. The required parameters (including plate transfer configuration or microarray printing details) are set and the separate plate lists are selected in the appropriate order. The resulting spot list can be associated with the microarray grid before data extraction as described in the GenePix manual. A note of caution is that it is *imperative* that the microarray spot identification be correct. When creating a new spot list for each set of microarrays, careful verification of the list should be done, ideally by hybridizing an identical sample to a newly printed array and a previously verified microarray to be sure that the data agree. Correct spot identification can also be checked by verifying the expected expression ratios for internal control spots in the experiment, including auxotrophic markers, gene deletions, or gene overexpression present in one of the labeled samples hybridized to the microarrays.

Before recording the relative background-corrected fluorescence within each spot, the fluorescence intensity corresponding to the two dyes must be normalized. Because the two fluorescent dyes are scanned and detected independently, the overall signal in the two channels must be normalized to each other. One method of normalization is based on the assumption that most genes will not change in expression in that experiment. The fluorescence signal can be normalized, based on this assumption, by adjusting the overall fluorescence intensity in one of the channels by a normalization factor, such that the relative signal intensity in most of the microarray spots is equal. Another method of signal normalization is to dope the Cy3 and Cy5 labeling reactions with known quantities of an empirically determined set of control DNA fragments that are also present on the microarray. The fluorescence signal in the channels can then be adjusted to achieve the known ratios for those control microarray spots.

Databases

Microarray experiments quickly generate massive datasets that require efficient means to store, explore, and retrieve the data. Efficient databases allow the storage of each set of microarray files (including the scan images, grid file, and extracted data file), details regarding the microarray used (printing details including print batch and identification of the microarray spots), and specific details of the experiment (strain genotype, medium, culture conditions, and specific experimental details). A publicly available database can be downloaded at http://www.microarrays.org/software.html.[33,40]

[40] J. DeRisi, M. Diehn, M. B. Eisen, and P. T. Spellman (2000).

Data Analysis and Interpretation

The existing challenge is extracting biological insights from the massive genomic expression datasets generated by microarray studies. A variety of computational methods can be used in the analysis of genomic expression data. Effective interpretation of the data requires not only computational techniques, but also a constant awareness of the experimental conditions. Before formalizing hypotheses about the cellular signals that provoke gene expression changes, the experimenter should consider the primary vs secondary cellular responses to the experimental conditions, possible pleiotropic effects of the cells' environment, and the specificity of the observed gene expression responses to the experimental conditions.

In many cases, interpreting the genomic expression program triggered by one set of experimental conditions is clarified by comparing the results to other conditions. For example, to identify cell cycle-regulated transcripts, Spellman *et al.* analyzed genomic expression data recovered from four different experiments, in which cells were synchronized by different means, to identify the cell cycle-regulated genes common to all four experiments while filtering out genes whose expression was specifically affected by the mechanism of synchronization.[3] In another study by Gasch *et al.* that followed the genomic expression responses of cells to different environmental changes, it became apparent only when the data from the different experiments were compared that much of each of the genomic expression responses was not specific to the individual conditions, but rather was part of a general stress response that was triggered by all of the experiments.[8]

Troubleshooting Microarrays

Measuring the relative spot fluorescence on microarrays with bright spots of high signal-to-noise is quantitative and highly reproducible; however, a variety of common flaws reduce the reliability of quantitation. The most frequently encountered microarray blemishes are summarized below, with suggested techniques for improving microarray quality.

Weak Spot Signal

Weak global fluorescence signal most commonly arises from inefficient labeling of the sample. RNA of insufficient quality or quantity prevents proper probe generation, and a first step toward identifying weak signal problems is to check the quality of the RNA sample. Probe fluorescence can be increased simply by increasing the amount of sample labeled in the reaction; when the sample is limiting, methods such as amplification may be required to generate enough template for the reaction. Alternate labeling protocols may also increase the probe intensity. For example, mRNA labeled with random primers or a combination of random primers and the specific oligo(dT) primer results in brighter probe than using the oligo(dT) primer alone. Alternate labeling systems also increase the signal intensity by increasing the number of fluorescent dye molecules incorporated into the probe (Genisphere, Montvale, NJ).

Other microarray flaws result in weak local signal. Circular areas of weak signal may reflect the presence of bubbles during the hybridization. Areas of weak signal near the perimeter of the microarray, outlined by extremely bright fluorescence intensity, arise when the probe dries out during hybridization because of insufficient chamber hydration. This issue can be eliminated by increasing the volume of 3× SSC added to the microarray chamber during hybridization. Streaks of low signal across the microarray result from scratches, most often introduced during removal of the microarray coverslip.

High Background Signal

Just as spots with weak signal are more difficult to quantitate, spots surrounded by high background signal are equally susceptible to measurement error. High global background can result from bad postprocessing of the microarrays before hybridization, since fluorescent probe DNA may bind to the microarray surface if the unbound polylysine groups are not sufficiently blocked. Another suspect when global background is high is the probe mixture: inappropriate concentrations of the probe components, particularly SDS, can increase the global background signal on the microarrays.

High local background can result in a variety of impressive but undesirable patterns on the arrays. Residual SDS, introduced during washings of the hybridized microarrays, is detected as swirly patterns of green fluorescence. Residual SDS can sometimes be removed after scanning by further washing the arrays in a warm solution of the final wash. More problematic are the bright, multicolored background patterns that arise due to bad polylysine-coated glass. These patterns, often accompanied by areas of very high background fluorescence and no spot fluorescence, may reflect polylysine-coated glass that is over-aged. It is strongly advised that test hybridizations be performed on representative polylysine slides from each batch before investing the effort to print complete microarrays.

Comet Tails, Speckles, and Other Blemishes

Another problem caused by bad polylysine slides or ineffective postprocessing is comet tails trailing off of the DNA spots. These fluorescent tails likely result when DNA from the microarray spots migrates across the slide. Although the exact cause of this blemish is unclear, its source may be related to the chemical postprocessing step. The chemical blocking step should be initiated rapidly, and the rack of slides should be plunged into the blocking solution, rather than the solution being poured over the dry slides. The comet tail phenomenon may be exacerbated when the polylysine-coated glass is insufficiently aged, resulting in poor hydrophobicity of the surface. Poor hydrophobicity also promotes the bleeding of spots on the microarray during printing. Bleeding between spots can be minimized by reducing the rehydration time in the postprocessing procedure, and by increasing the salt concentration in the hybridization chamber from 1× to 3× SSC.

Speckles of high fluorescence on the microarrays can reduce the accuracy of measurements and also interfere with automatic gridding software. Dust on the microarrays is most often the culprit. Although dust particles may be fluorescent themselves, fluorescent probe can also bind to the particles. Care should be taken when handling the microarrays, and they should be stored in a dust-free environment. Be sure all equipment is free of dust, including glassware, centrifuges, and the microarrayer itself.

Another source of speckled arrays is precipitation of the probe components before or during hybridization. Precipitation of salt and SDS can occur, especially if the probe is chilled (or is allowed to freeze during speed vacuuming). Therefore, the probe should not be stored once the salt and detergent have been added, and the solution should not be chilled below room temperature after denaturation.

Summary

The exploration and characterization of yeast genomic expression programs is providing a wealth of information about yeast biology, as well as other organisms. The intriguing biology of yeast species invites characterization of genomic expression patterns to illuminate the details of cellular physiology. In addition to its value as an interesting organism, yeast maintains its role as an excellent model in which to characterize genomic expression programs. Microarray studies are quickly spreading to plant, animal, and microbial organisms that remain in the early stages of characterization. The extensive knowledge of yeast biology, as well as the relative ease with which yeast studies can be performed and controlled, facilitates interpretation of the genomic expression data. Importantly, existing information about yeast biology, including functional annotations for each gene, is captured and efficiently presented in databases such as the *Saccharomyces* Genome Database (SGD),[41] the Munich Information Center Yeast Genome Database (MIPS),[42] the Yeast and Pombe Protein Databases (YPD and PPD, respectively),[43,44] and others.

[41] C. A. Ball, K. Dolinski, S. S. Dwight, M. A. Harris, L. Issel-Tarver, A. Kasarskis, C. R. Scafe, G. Sherlock, G. Binkley, H. Jin, M. Kaloper, S. D. Orr, M. Schroeder, S. Weng, Y. Zhu, D. Botstein, and J. M. Cherry, *Nucleic Acids Res.* **28,** 77 (2000).

[42] H. W. Mewes, D. Frishman, C. Gruber, B. Geier, D. Haase, A. Kaps, K. Lemcke, G. Mannhaupt, F. Pfeiffer, C. Schuller, S. Stocker, and B. Weil, *Nucleic Acids Res.* **28,** 37 (2000).

[43] M. C. Costanzo, J. D. Hogan, M. E. Cusick, B. P. Davis, A. M. Fancher, P. E. Hodges, P. Kondu, C. Lengieza, J. E. Lew-Smith, C. Lingner, K. J. Roberg-Perez, M. Tillberg, J. E. Brooks, and J. I. Garrels, *Nucleic Acids Res.* **28,** 73 (2000).

[44] M. C. Costanzo, M. E. Crawford, J. E. Hirschman, J. E. Kranz, P. Olsen, L. S. Robertson, M. S. Skrzypek, B. R. Braun, K. L. Hopkins, P. Kondu, C. Lengieza, J. E. Lew-Smith, M. Tillberg, and J. I. Garrels, *Nucleic Acids Res.* **29,** 75 (2001).

A number of databases also allow the exploration of published genomic expression studies, including the "Expression Connection" at SGD and the Microarray Global Viewer (yMGV) organized by Marc *et al.*[45] Consulting these databases to retrieve known details about gene function and regulation vastly facilitates interpretation of the genomic expression data, allowing biological hypotheses to be formulated and tested. These hypotheses can be applied to other organisms that may execute genomic expression programs similar to those seen in yeast. Furthermore, as more genomic expression studies in multiple organisms emerge, large-scale data comparisons can be conducted, within and across organisms. Incorporating the results of yeast studies into such comparisons is certain to increase our understanding about the function, regulation, and evolution of genomic expression programs.

Acknowledgment

A.P.G. is a postdoctoral researcher at the E. O. Lawrence Berkeley National Laboratory and was supported by a grant from the National Institutes of Health (HL07279) under Department of Energy contract DE-AC0376SF00098.

[45] P. Marc, F. Devaux, and C. Jacq, *Nucleic Acids Res.* **29,** E63 (2001).

[24] Transcriptome Analysis of *Saccharomyces cerevisiae* Using Serial Analysis of Gene Expression

By MUNIRA A. BASRAI and PHILIP HIETER

Introduction

A "transcriptome" provides information about the identity and expression level of genes for a defined population of cells. While the genome (the genetic content of a cell) remains stagnant, the transcriptome (mRNA population) is dynamic and is affected by changes in environmental (growth conditions) and internal (development, differentiation, and cell cycle stage) factors. Therefore, a description of the transcriptome of any cell would provide information for functional characterization of a process or a pathway. In the past, RNA–DNA hybridization measurements have provided some general features of gene expression. These results suggest there may be at least three classes of transcripts, with either high, medium, or low levels of expression allowing the estimation of the number of transcripts per cell.[1]

[1] J. O. Bishop, J. G. Morton, M. Rosbash, and M. Richardson, *Nature* **250,** 199 (1974).

0076-6879/02 $35.00

These expression data provided little information about the specific genes that were members of each class. Some of the techniques that permit a limited analysis of global gene expression patterns include differential display,[2] subtractive hybridization,[3] and microarray technologies.[4,5] Microarrays are being used extensively for the comprehensive gene expression analysis of a defined set of genes.

In this chapter we describe the technique and application of serial analysis of gene expression (SAGE), an approach that allows the rapid, detailed, systematic, and quantitative analysis of thousands of transcripts.[6] SAGE can be used to determine global or differential gene expression under a set of defined conditions. Unlike other methods, SAGE provides an absolute level of gene expression for all transcribed genes in a defined population of cells and does not depend on prior availability of sequence information. SAGE also does not require specialized equipment beyond access to a sequencing apparatus and a PCR (polymerase chain reaction) machine. We used SAGE to define the transcriptome of *Saccharomyces cerevisiae* during normal growth and cell cycle arrest in S or G_2/M phases of the cell cycle.[7] This organism was chosen because it is used widely to define the biochemical and physiologic parameters underlying eukaryotic cellular functions and its genome is completely sequenced.[8] SAGE studies done with *S. cerevisiae* cells grown in oleate-containing media have led to the identification of about 100 differentially regulated genes.[9] In mammalian cells, SAGE analysis has provided valuable information about the global gene expression level and also to determine if the transcription of a given gene is specific to a particular tissue, developmental or cell cycle stage, or pathological condition.[10,11]

In addition to describing the SAGE method we also discuss an important application of the SAGE technique, namely, the identification of previously unidentified transcripts from NORFs (nonannotated open reading frames). SAGE analysis allowed us to identify at least 302 NORFs from *S. cerevisiae*.[7,12] These NORFs

[2] P. Liang and A. B. Pardee, *Science* **257,** 967 (1992).

[3] S. M. Hedrick, D. I. Cohen, E. A. Nielsen, and M. M. Davis, *Nature* **308,** 149 (1984).

[4] D. D. Bowtell, *Nat. Genet.* **21,** 25 (1999).

[5] M. B. Eisen and P. O. Brown, *Methods Enzymol.* **303,** 179 (1999).

[6] V. E. Velculescu, L. Zhang, B. Vogelstein, and K. Kinzler, *Science* **270,** 484 (1995).

[7] V. Velculescu, L. Zhang, W. Zhou, J. Vogelstein, M. A. Basrai, D. E. Bassett, Jr., P. Hieter, B. Vogelstein, and K. Kinzler, *Cell* **88,** 243 (1997).

[8] A. Goffeau, B. G. Barrell, H. Bussey, R. W. Davis, B. Dujon, H. Feldman, F. Galibert, J. D. Hoheisel, C. Jacq, M. Johnston, E. J. Louis, H. W. Mewes, Y. Murakami, P. Phillipsen, H. Tetteli, and S. G. Oliver, *Science* **274,** 546 (1996).

[9] A. J. Kal, A. J. van Zonneveld, V. Benes, M. van den Berg, M. G. Koerkamp, K. Albermann, N. Strack, J. M. Ruijter, A. Richter, B. Dujon, W. Ansorge, and H. F. Tabak, *Mol. Biol. Cell.* **10,** 1859 (1999).

[10] L. Zhang, W. Zhou, V. E. Velculescu, S. E. Kern, R. H. Hruban, S. R. Hamilton, B. Vogelstein, and K. W. Kinzler, *Science* **276,** 1268 (1997).

[11] B. St. Croix, C. Rago, V. Velculescu, G. Traverso, K. E. Romans, E. Montgomery, A. Lal, G. J. Riggins, C. Lengauer, B. Vogelstein, and K. W. Kinzler, *Science* **289,** 1197 (2000).

[12] M. A. Basrai, V. E. Velculescu, K. W. Kinzler, and P. Hieter, *Mol. Cell. Biol.* **19,** 7041 (1999).

may correspond to small previously unrecognized ORFs (<99 amino acids) or large ORFs (>99 amino acids) that may have been overlooked because of possible sequencing errors. The genome sequencing efforts of *S. cerevisiae* have annotated about 6275 ORFs representing all ORFs larger than 100 contiguous amino acids.[8,13] However, identification of genes encoded by small ORFs (<100 amino acids) based on sequence analysis alone has been severely limited by high false positive rates and traditional functional screens have been hampered similarly by the small target size for mutagenesis.[14] Evidence from several microorganisms suggests that a significant fraction of genomes are encoded by small genes. For example, the *Escherichia coli* genome encodes 381 proteins of fewer than 100 amino acids in length from a total of 4288 annotated ORFs (8.9%),[15] and random protein sequencing in the fully sequenced cyanobacterium *Synechocystis* revealed that 11.8% of the total proteins were encoded by ORFs of <100 codons.[16] Extrapolating such studies to yeast would suggest that there may be as many as 800 small open reading frames in the entire yeast genome, of which only 177 have been identified.[17] A subset of small open reading frames will likely encode conserved proteins in all organisms including humans. In *S. cerevisiae,* these small proteins include mating pheromones, proteins involved in energy metabolism, proteolipids, chaperonins, stress proteins, transporters, transcriptional regulators, nucleases, ribosomal proteins, thioredoxins, and metal ion chelators. In multicellular organisms, there is a rich diversity of short peptides including many hormones, antibacterial defensins, cecroporins, and magainins. There are also small ORFs encoding transporter proteins, homeobox proteins, transcription factors, and kinase regulatory subunits reported in the nematode *Caenorhabditis elegans*.[18] We performed the first systematic analysis of NORFs in the yeast genome and characterized one of the NORFs, *NORF5/HUG1*.[12] Our findings highlight the importance of the development and application of new technologies in the total-genome-sequence era to fully understand the genetic complexity of an organism.

Principles of SAGE

The SAGE technology is based on two basic principles.[6] First, a short sequence tag (9–11 bp) contains sufficient information to uniquely identify a transcript, provided that it is derived from a defined location within that transcript. For example, a 10-bp sequence can distinguish 1,048,576 transcripts given a random nucleotide distribution at the tag site. Second, many 10-bp transcript tags can be concatenated

[13] B. Dujon, D. Alexandraki, B. Andre, W. Ansorge, V. Baladron, J. P. Ballesta, A. Banrevi, P. A. Bolle, M. Bolotin-Fukuhara, and P. Bossier, *Nature* **369,** 371 (1994).

[14] M. A. Basrai, P. Hieter, and J. D. Boeke, *Genome Res.* **7,** 768 (1997).

[15] http://genetics/wisc.edu/

[16] http://www.kazusa.or.jp/tech/sazuka/cyano/proteome.html

[17] http://www.mips.biochem.mpg.de/

[18] http://www.sanger.ac.uk/Projects/C_elegans

into a single molecule and then sequenced, revealing the identity of multiple tags simultaneously. This allows efficient analysis of tags in a serial manner by the sequencing of multiple tags within a single clone.

The expression pattern of any population of transcripts can be quantitatively evaluated by determining the abundance of individual tags and identifying the gene corresponding to each tag. The method takes advantage of the ability of a type IIS restriction enzyme to cleave DNA at a defined site up to 20 bp away from its recognition site. The bases adjacent to the recognition site provide the specificity for the site of interest among a large number of different transcripts. Ligation of the tags permits rapid identification of up to 40 individual tags from each lane on an automated sequencing gel. A detailed protocol for the technique is described in the Methods section. An overview that summarizes the steps is provided for quick reference. Briefly, double-stranded cDNA is synthesized from mRNA by means of a biotinylated oligo(dT) primer (Fig. 1[6], steps 1 and 2). The cDNA is then cleaved with an anchoring enzyme (AE) such as *Nla*III with a 4-bp recognition site expected to cut DNA about every 256 base pairs (Fig. 1, step 3). The most 3′ end of the cleaved cDNA is then isolated by binding to streptavidin beads (Fig. 1, step 4). The cDNA is then divided into half and ligated to one of two linkers containing a type IIS restriction site, tagging enzyme (TE) such as *Bsm*FI (Fig. 1, step 5). The TE cleaves DNA at a defined distance up to 20 bp away from their asymmetric recognition site. After digestion with the TE, the linker with a short piece of cDNA is released from the beads (Fig. 1, step 6). The ends of the resulting cDNA are converted to blunt ends and the two pools of released tags are ligated to each other (Fig. 1, steps 7 and 8). Ligated tags serve as templates for polymerase chain reaction (PCR) amplification with primers specific to each linker (Fig. 1, step 9). This step provides orientation to the tag sequence. The amplification products contain a ditag (two tags) linked tail to tail, flanked by sites for the AE. In the final sequencing template, this results in a 4-bp punctuation (the AE site) per ditag. The PCR product is cleaved with the AE thereby releasing the ditags that are isolated and purified (Fig. 1, steps 10 and 11). The resulting ditags are concatenated by ligation, cloned, and sequenced (Fig. 1, steps 12 and 13). The number of tags to be analyzed will depend on the individual application. Use of a different AE will ensure that virtually all transcripts can be identified. It is important to note that the ditags are formed before the amplification hence eliminating potential artifacts introduced by PCR. Repeated ditags due to PCR are excluded from the final analysis. The probability of two tags being coupled in the same ditag is small even for highly transcribed genes.

Materials

1. Yeast Strains and Media

We determined the expression profiles of *S. cerevisiae* genes required for normal growth and during progression through the cell cycle.[7] SAGE libraries

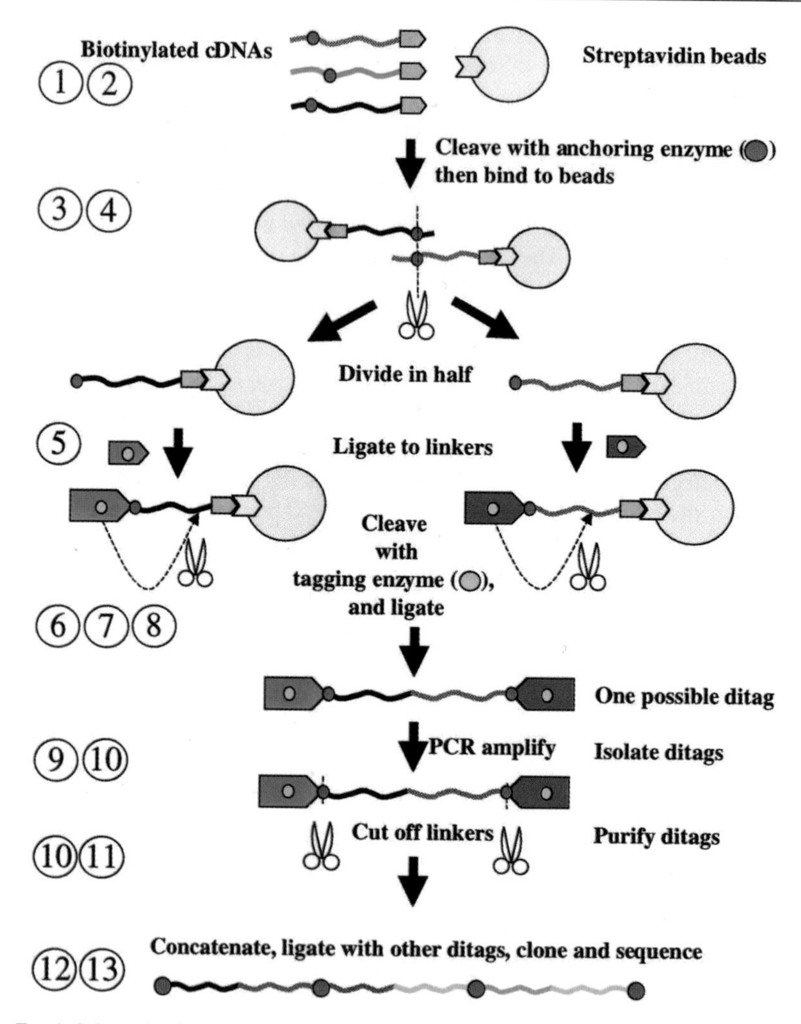

FIG. 1. Schematic of SAGE. The anchoring enzyme is *Nla*III (pink circle) and the tagging enzyme is *Bsm*FI (green circle in the linkers). The numbers on the right-hand side correspond to the steps in the SAGE protocol as described in detail in the text. From V. E. Velculescu, L. Zhang, B. Vogelstein, and K. Kinzler, *Science* **270,** 484 (1995).

were generated from yeast cells in three states: logarithmic phase, S phase arrested, and G_2/M phase arrested. The source of transcripts for all experiments was *S. cerevisiae* strain YPH499 (*MATa ura3-52 lys2-801 ade2-101 leu2Δ1 his3Δ200 trp1Δ63*).[19] Logarithmically growing cells were obtained by growing yeast cells

[19] R. S. Sikorski and P. Hieter, *Genetics* **122,** 19 (1989).

to early log phase (3×10^6 cells/ml) in YPD[20] medium supplemented with 6 m*M* uracil, 4.8 m*M* adenine, and 24 m*M* tryptophan at 30°. For arrest in the G_1/S phase of the cell cycle, hydroxyurea (HU) (0.1 *M*) was added to early log phase cells, and the culture was incubated an additional 3.5 hr at 30°. For arrest in the G_2/M phase of the cell cycle, nocodazole (15 μg/ml) was added to early log phase cells and the culture was incubated for an additional 100 min at 30°. Harvested cells were washed once with water prior to freezing at −70°. The growth states of the harvested cells were confirmed by microscopic and flow cytometric analyses.[21]

2. *Special Reagents and Solutions*

2.1. *RNA Reagents*

RNAgents—Total RNA Isolation Kits (Promega, Madison, WI)
MessageMaker mRNA kit (Gibco-BRL, Gaithersburg, MD)
Superscript Choice System cDNA Synthesis Kit (Gibco-BRL)

2.2. *Magnetic Beads*

Dynabeads M-280 Streptavidin Slurry (Dynal, Oslo, Norway)
Magnet (Dynal)

2.3. *Enzymes*

*Bsm*FI (NEB, Beverly, MA).
*Nla*III (NEB)—ship on dry ice and store at −80°.
*Sph*I (NEB).
Klenow (Pharmacia/USB, Swampscott, MA).
T4 ligase high concentration (5 U/μl) (Gibco-BRL).
T4 ligase regular concentration (1 U/μl) (Gibco-BRL).
Platinum *Taq* (BRL, Gaithersburg, MD).

2.4. *Miscellaneous*

Glycogen (Boerhinger Mannheim, Indianapolis, IN).
pZERO-1 plasmid (Invitrogen).
10 m*M* dNTP mix (Gibco-BRL).
Dimethyl sulfoxide (DMSO; Sigma, St. Louis, MO).
7.5 *M* Ammonium acetate (Sigma).

[20] M. D. Rose, F. Winston, and P. Hieter, "Methods in Yeast Genetics." Cold Spring Harbor Laboratory Press, Cold Spring Harbor, NY, 1990.

[21] M. A. Basrai, J. Kingsbury, D. Koshland, F. Spencer, and P. Hieter, *Mol. Cell. Biol.* **16,** 2838 (1996).

2.5. Solutions

2.5.1. 2× B + W Buffer: 10 m*M* Tris-HCl (pH 7.5) 1 m*M* EDTA, 2.0 *M* NaCl. Store at room temperature.

2.5.2. LoTE: 3 m*M* Tris-HCl (pH 7.5), 0.2 m*M* EDTA (pH 7.5) in distilled water. Store at 4°.

2.5.3. PC8: 480 ml phenol (warm to 65°), 320 ml of 0.5 *M* Tris-HCl (pH 8.0), 640 ml of chloroform. Add in sequence, shake, and place at 4°. After 2–3 hr, shake again. After another 2–3 hr, aspirate aqueous layer. Aliquot and store at −20°.

2.5.4. 10× PCR Buffer: 166 m*M* $(NH_4)_2SO_4$, 670 m*M* Tris-HCl (pH 8.8), 67 m*M* $MgCl_2$, 100 m*M* 2-mercaptoethanol. Distribute into 0.5 ml aliquots and store at −20°.

3. Linker and Primer Sequences

High-quality linkers are crucial to several steps in the SAGE method. All linkers and primers 1 and 2 should be obtained gel purified from the oligonucleotide synthesis company (Integrated DNA Technologies, Coralville, IA).

Linker 1A : 5′ TTT GGA TTT GCT GGT GCA GTA CAA CTA GGC TTA ATA GGG ACA TG 3′
Linker 1B: 5′ TCC CTA TTA AGC CTA GTT GTA CTG CAC CAG CAA ATC C[amino mod. C7] 3′
Linker 2A: 5′ TTT CTG CTC GAA TTC AAG CTT CTA ACG ATG TAC GGG GAC ATG 3′
Linker 2B: 5′ TCC CCG TAC ATC GTT AGA AGC TTG AAT TCG AGC AG[amino mod. C7] 3′

Primers 1 and 2 should be ordered with the addition of 2 sequential biotins on the 5′ end.

Primer 1: 5′ GGA TTT GCT GGT GCA GTA CA 3′
Primer 2: 5′ CTG CTC GAA TTC AAG CTT CT 3′
Biotinylated oligo(dT)
5′ [biotin]T_{18}
M13 Forward: 5′ GTA AAA CGA CGG CCA GT 3′
M13 Reverse: 5′ GGA AAC AGC TAT GAC CAT G 3′

4. Gel Electrophoresis and Miscellaneous Methods

4.1. 12% PAGE (for Isolating PCR Products and Ditags). Solution: 10.5 ml of 40% (w/w) polyacrylamide (19 : 1 acrylamide : bisacrylamide) (Bio-Rad, Hercules,

CA), 23.5 ml of distilled water, 700 μl of 50× Tris–acetate buffer (Quality Biologicals, Gaithersburg, MD), 350 μl of 10% (w/v) APS, 30 μl of TEMED. Mix the ingredients and add to a vertical gel apparatus (Owl Scientific Products, Portsmouth, NH, Model No. P9DS with 1.5-mm spacers). Add comb and let the gel polymerize for at least 30 min. Run gel at 160 V for 2–2.5 hr.

4.2. 8% PAGE (for Separating Concatemers). Solution: 7 ml of 40% polyacrylamide (37.5 : 1 acrylamide : bisacrylamide) (Bio-Rad), 27 ml of distilled water, 700 μl of 50× Tris–acetate buffer (Quality Biologicals), 350 μl of 10% APS, 30 μl of TEMED. Mix the ingredients and add to a vertical gel apparatus (Owl Scientific Products, Model No. P9DS with 1.5-mm spacers). Add comb and let the gel polymerize for at least 30 min. Run gel at 130 V for 2.5 hr.

4.3. PC8 Extraction. Add equal volume PC8 (see Section 2.5.3) to sample and vortex for several seconds. Spin for 2 min at 14,000 rpm at 4° in microcentrifuge and then transfer aqueous (top) layer to a new microcentrifuge tube.

4.4. Kinasing Reaction for Linkers. The two linkers, Linker 1B and 2B, are kinased in separate tubes. First, dilute Linker 1B and 2B to 350 ng/μl. Add 9 μl of diluted Linker 1B to tube 1 and 9 μl of Linker 2B to tube 2. To both tubes add the following reagents: 6 μl of LoTE, 2 μl of kinase buffer (10×, NEB), 2 μl of ATP (10 m*M* stock), and 1 μl of T4 polynucleotide kinase (10 U/μl, NEB). Incubate the reaction at 37° for 30 min and heat inactivate the reaction at 65° for 10 min. Then, mix 9 μl Linker 1A with 20 μl kinased Linker 1B (final concentration 200 ng/μl) and 9 μl Linker 2A with 20 μl kinased Linker 2B (final concentration 200 ng/μl). To anneal linkers, heat to 95° for 2 min, then place at 65° for 10 min, 37° for 10 min and room temperature for 20 min, store at −20°. The efficiency of the kinase reaction should be tested by self-ligating about 200 ng of each linker pair and running on 20% Novex (Invitrogen, Carlsbad, CA) gel: Kinased linkers should allow linker–linker dimers (80–100 bp) to form after ligation, while unkinased linkers will prevent self-ligation. Only linker pairs that self-ligate >70% should be used in further steps.

4.5. Ethidium Bromide Dot Quantitation. Use any solution of pure DNA to prepare the following standards: 0 ng/μl, 1 ng/μl, 2.5 ng/μl, 5 ng/μl, 7.5 ng/μl, 10 ng/μl, and 20 ng/μl in LoTE. Use 1 μl of sample DNA to make 1/5, 1/25, and 1/125 dilutions in LoTE. Add 4 μl of each standard or 4 μl of each diluted sample to 4 μl ethidium bromide (stock of 1 μg/ml). Spot 8 μl of each sample on a sheet of plastic wrap placed on a UV transilluminator and photograph using UV light. Estimate DNA concentration by comparing the intensity of the sample to the standards.

4.6. Testing Biotinylation of Biotin–Oligo(dT). Obtain biotin–oligo(dT) gel-purified from oligo synthesis company. Test biotinylation by adding several hundred nanograms of biotin–oligo(dT) to 1 μg streptavidin (Sigma). Incubate several minutes at room temperature. Both the unbound oligonucleotide and that bound to streptavidin are run on a 20% Novex gel. If the oligonucleotide is well biotinylated, the entire amount of oligonucleotide should be shifted to a higher molecular weight in the lane containing the streptavidin. Alternatively, increasing amounts of oligonucleotide (from several hundred nanograms to several micrograms) can be incubated with and without separate aliquots of 100 μl of Dynabeads (Dynal). After 15 min, the beads are separated from the supernatant using a magnet, the supernatant is removed, and DNA quantitation is performed at OD_{260}. At low amounts of oligo, when bead-binding capacity is not saturated, the ratio of unbound oligo to the total oligo will indicate the percent of oligo that is not biotinylated.

Protocol

Described below is a detailed SAGE protocol. Please refer to the Web site[22] for updates and recommendations as well as trouble shooting guides. Alternatively, one can purchase an I-SAGE kit commercially available from Invitrogen which includes all the reagents and instructions for SAGE libraries (http://www.invitrogen.com/Content/expressions/E082002.pdf).

1. Preparation of mRNA

Total RNA from yeast cells grown under the experimental conditions is prepared using the hot phenol method as described.[23] mRNA is obtained using the MessageMaker Kit (Gibco-BRL) following the manufacturers protocol. Northern blot analysis is performed as described.[24] Five hundred μg of total RNA usually provides 5–20 μg of poly(A) RNA. The integrity of the RNA should be checked by gel electrophoresis and Northern hybridization.

2. cDNA Synthesis

Before cDNA synthesis, it is important to check the biotinylated oligo(dT) for biotinylation by streptavidin gel shift assays (see Section 4.6 above). cDNA synthesis is done using the Superscript Choice System for cDNA synthesis (Gibco-BRL) with a gel-purified 5′ biotinylated oligo(dT) following the protocol of Gibco-BRL except that we use 2.5 μg biotinylated oligo(dT) instead of the oligo(dT) that is provided with the kit.

2.1. First Strand Synthesis. Mix 2.5 μl of biotinylated oligo(dT) (stock, 1 μg/μl) with 4.5 μl of poly(A) RNA (5 μg) and heat to 70° for 10 min and place on ice. Spin tube in a microcentrifuge at 10,000 rpm for 1 min. For first strand synthesis, add the following reagents: 4 μl of 5× first strand buffer, 2 μl of

[22] http://www.sagenet.org

[23] P. Leeds, S. W. Peltz, A. Jacobson, and M. R. Culbertson, *Genes Dev.* **5,** 2303 (1991).

[24] W. S. El-Deiry, T. Tokino, V. E. Velculescu, D. B. Levy, R. Parsons, J. M. Trent, D. Lin, W. E. Mercer, K. W. Kinzler, and B. Vogelstein, *Cell* **75,** 817 (1993).

0.1 *M* dithiothreitol (DTT), 1 μl of 10 m*M* dNTP, 1 μl of diethyl pyrocarbonate (DEPC)-treated water, and 5 μl of Superscript (Gibco-BRL, Gaithersburg, MD) reverse transcriptase. Incubate the reaction for 1 hr at 37°. Remove 1 μl for gel analysis.

2.2. Second Strand Synthesis. Add the following reagents: 19 μl of first strand reaction product, 94 μl of DEPC-treated water, 30 μl of 5× second strand synthesis buffer, 3 μl of 10 m*M* dNTPs, 3 μl of *E. coli* DNA ligase, 4 μl of *E. coli* DNA polymerase I, and 1 μl of *E. coli* RNase H. Mix the ingredients, incubate at 16° for 2 hr, and then place on ice. Add T4 DNA polymerase (2 μl containing 10 units) and incubate the reaction at 16° for 5 min. Then add 10 μl 0.5 *M* EDTA (pH 7.5) and heat the reaction to 70° for 10 min.

After second strand synthesis, add 50 μl of distilled water and extract twice with phenol : chloroform (1 : 1, pH 8.0) being careful not to disturb the interface. Ethanol precipitate the resulting cDNA as follows. To 200 μl of the extracted sample, add 133 μl of 7.5 *M* ammonium acetate, 3 μl of glycogen, and 777 μl of 100% ethanol, centrifuge at 14,000 rpm for 2 min at 6° and wash the pellet twice with 70% (v/v) ethanol, and then resuspend in 20 μl of LoTE. It is very important to check the integrity of the cDNA by gel electrophoresis. Run 1 μl of the first strand and 2 μl of the second strand synthesis on a gel. The second strand synthesis product should yield a uniform intense smear ranging from several hundred bp to over 10 kb.

3. Cleavage of Biotinylated cDNA with Anchoring Enzyme (NlaIII)

It is important to store *Nla*III at −80° as *Nla*III has a short half-life. Mix the following components: 10 μl of sample cDNA (half of the total cDNA), 163 μl of LoTE, 2 μl of BSA (100× stock from NEB), 20 μl of Buffer 4 (NEB), and 5 μl of *Nla*III (NEB, 10 U/μl). The reaction is incubated at 37° for 1 hr. Extract with an equal volume of PC8, ethanol precipitate the DNA by adding to 200 μl sample, 133 μl 7.5 *M* ammonium acetate, 3 μl glycogen, and 777 μl 100% ethanol, centrifuge and wash the pellet twice with 70% ethanol, centrifuge at 14,000 rpm for 2 min at 4°, remove ethanol, and resuspend in 20 μl LoTE.

4. Binding Cleaved Biotinylated cDNA to Magnetic Beads

Note there are 2 tubes for each sample. Add 100 μl Dynabead M-280 streptavidin slurry (Dynal, 10 mg/ml) to each of two 1.5 ml microcentrifuge tubes. Use the magnet to immobilize beads and remove supernatant. Wash beads once as follows: add 200 μl of 1× B + W buffer, mix, immobilize the beads, and remove wash. To the beads add 100 μl of 2× B + W, 90 μl distilled water, and 10 μl of the cleaved cDNA to each tube. Incubate 15 min at room temperature. Mix intermittently and wash the beads as above three times with 200 μl of 1× B + W and once with 200 μl LoTE, removing the wash each time. Proceed immediately to step 5 below.

5. Ligating Linkers to Bound cDNA

Linkers 1B and 2B must be kinased and annealed to their complementary linker 1A and 2A, respectively, before ligation. Kinasing can be performed chemically at the time of oligonucleotide synthesis or enzymatically as described in Methods Section 4.4. Linker kinasing should be tested by self-ligation (see Section 4.4).

The two sets of linkers, 1A and 1B and 2A and 2B, are ligated in separate tubes. After removal of LoTE from step 4, ligate the linkers as follows. To each of tubes 1 and 2, add Dynabeads bound to cDNA fragments and 25 μl of LoTE. Then add 5 μl of annealed Linker 1A and 1B (200 ng/μl) to tube 1 and 5 μl of annealed Linker 2A and 2B (200 ng/μl) to tube 2, and 8 μl of 5$\times$ ligase buffer (Gibco-BRL) to each tube. The Dynabead slurry is resuspended by mixing the tubes gently. The tubes are heated at 50° for 2 min, followed by 15 min at room temperature. T4 DNA ligase (2 μl of 5 U/μl stock, Gibco-BRL) is added and incubated for 2 hr at 16° with gentle mixing at intermittent intervals. After ligation, each reaction is washed 3 times with 200 μl of 1$\times$ B + W buffer. Contents are transferred to a new tube and washed once with 200 μl 1$\times$ B + W buffer and twice with 200 μl of 1$\times$ Buffer 4 (NEB).

6. Release of cDNA Tags Using Tagging Enzyme

The cDNA tags are released from the magnetic beads as follows. The following are added to the washed beads contained in each of tube 1 and 2 : 86 μl of LoTE, 10 μl of Buffer 4 (10$\times$ stock, NEB), 2 μl of bovine serum albumin (BSA, 100$\times$ stock, NEB), and 2 μl of *Bsm*FI (2 U/μl, NEB). Reaction is incubated at 65° for 1 hr after gentle mixing. Use the magnet to immobilize beads and remove supernatant (see Section 4.3). Extract the supernatant with an equal volume of PC8 and ethanol precipitate the cDNA tags using 200 μl of sample, 133 μl of 7.5 *M* ammonium acetate, 3 μl of glycogen, and 1 ml of 100% ethanol. Spin for 30 min at 4°. Wash the pellet twice with 70% ethanol, centrifuge, remove ethanol, and resuspend the cDNA tags in 10 μl of LoTE.

7. Blunt Ending Released cDNA Tags

Although we have used T4 DNA polymerase successfully for blunt ending cDNA tags in the past, we have found that klenow polymerase provides more consistent results. Add the following to each microcentrifuge tube after tagging enzyme-derived cDNA fragments have been obtained in step 6 above. To 10 μl of sample in tubes 1 and 2, and 10 μl of second strand synthesis buffer (5$\times$ stock, Gibco-BRL, cDNA synthesis kit), 2.5 μl of dNTP (10 m*M*, Gibco-BRL), 23.5 μl of distilled water, and 3 μl of Klenow (1 U/μl, USB). The reaction is incubated at 37° for 30 min and then the reaction volume is brought to 200 μl with LoTE and extracted with an equal volume of PC8. The reaction mixture is ethanol precipitated as follows. To 200 μl sample add 133 μl 7.5 *M* ammonium acetate, 3 μl glycogen, and 1 ml of 100% ethanol. Spin for 30 min at 4°, wash the pellet twice with 70% ethanol, centrifuge, remove ethanol, and resuspend in 6 μl LoTE.

8. *Ligating Blunt-Ended Tags to Form Ditags*

Blunt-ended samples 1 and 2 are ligated to each other in the following way. The first microcentrifuge tube is the ligation reaction; the second serves as a control for contamination in later PCR steps (no ligase control). To 2 μl of blunt-ended samples 1 and 2 add 1.2 μl of ligase buffer (5× Gibco-BRL), 0.8 μl of T4 DNA ligase (5 U/μl, Gibco-BRL) in the ditag reaction tube and 0.8 μl of distilled water in the negative control tube. The reaction is incubated overnight at 16°. Next day add 14 μl of LoTE to ligation mixture and PCR amplify as in step 9 below or store at −20°.

9. *PCR Amplification of Ditags*

Amplify ditags using Primers 1 and 2 (see Section 3 in Materials). Optimize amplification by using different dilutions of template (1 μl of 1/50, 1/100, and 1/200 dilutions of ligation product per PCR reaction). Perform10 PCR reactions using a cocktail containing the following for each reaction: 5 μl of PCR Buffer (10×, see Section 2.5.4), 3 μl of DMSO (Sigma), 7.5 μl of dNTPs (10 m*M* stock, Gibco-BRL), 1 μl each of Primers 1 and 2 (stock 350 ng/μl), 30.5 μl of distilled water, 1 μl of Platinum *Taq* (5 Units/μl, Gibco BRL), and 1 μl of ligation product (various dilutions). Add 30 μl of mineral oil to the PCR reaction and perform PCR using the following cycling parameters optimized for the HYBAID PCR machine: one cycle at 94° for 1 min, 26–30 cycles of 94° for 30 min, 55° for 1 min and 70° for 1 min and one cycle at 70° for 5 min. The control (no ligase sample) should be amplified for 35 cycles. The appropriate cycle number is critical for isolating an adequate amount of DNA for SAGE. Too few cycles will result in a low yield and may cause problems with subsequent steps. Too many cycles will give erratic results and can also result in low yields. Therefore, we recommend trying various cycle numbers (for example, 26, 28, 30) to determine the optimal number. We have found batch-to-batch variability in dNTPs. Occasionally, at the above dNTP concentrations, PCR reactions are inhibited. We recommend optimizing dNTP concentrations before performing large-scale amplifications. Typically, for the optimization step we use final dNTP concentrations from 0.8 to 2.0 m*M* in 0.2 m*M* increments (4 μl to 10 μl, in 1 μl increments of the 10 m*M* dNTP stock solution in the above reaction mixture). Use the highest dNTP concentration that leads to robust amplification of the 102-bp band (the ditag PCR product).

The efficiency of amplification is assessed by gel electrophoresis as follows. Run 10 μl from each reaction on a 12% polyacrylamide gel or prepoured 20% polyacrylamide gel (Novex), using a 20 bp ladder as a marker (10 μl of a 1 : 5

dilution of the stock solution from GenSura, San Diego, CA). Amplified ditags should be 102 bp in size; a background band of equal or lower intensity occurs around 80 bp. All other background bands should be of substantially lower intensity. The no-ligase samples should not contain any amplified product of the size of the ditags even at 35 cycles. After PCR conditions have been optimized, large-scale PCR (two to three 96-well plates containing 50 μl reactions/well) can be performed. We usually use a 300-reaction PCR premix which we aliquot into three 96-well plates (Omniplate 96; Marsh Biomedical Products, Rochester, NY) of 50 μl PCR reactions/well.

10. Isolation of Ditags and Digestion with Anchoring Enzyme

Pool the PCR reactions into eight microfuge tubes (approximately 450 μl each). Extract with an equal volume of PC8, transfer aliquots of 300 μl aqueous solution to twelve 1.6 μl microcentrifuge tubes, and ethanol precipitate as follows: 300 μl sample, 133 μl 7.5 *M* ammonium acetate, 5 μl glycogen, and 1 ml of 100% ethanol. Wash the pellet twice with 70% ethanol and resuspend each pellet in 18 μl LoTE (216 μl LoTE total). At this step, one may wish to quantitate the DNA sample, which should be in the range of 60 to 100 μg total.

In an alternative approach pool PCR products into one 50-ml conical tube. Add an equal volume of PC8 (approximately 13 ml) and centrifuge in swinging-bucket rotor (Sigma 4K15, rotor 11156/13115) at 5100 rpm for 10 min at 4°. Transfer aqueous phase to a new 50 ml tube and ethanol precipitate as follows: 11.5 ml of product, 5.1 ml of 7.5 *M* ammonium acetate, 191 μl of glycogen, and 38.3 ml of 100% ethanol. Vortex vigorously and spin in fixed-angle rotor (Sigma 4K15, rotor 12169) at 10,000 rpm for 30 min. Wash the pellet with 5 ml of 70% ethanol, vortex, centrifuge for 5 min in swinging-bucket rotor at 5100 rpm, and resuspend in 216 μl LoTE. Add 54 μl of 5× sample buffer to this sample (270 μl total volume). Load 10 μl of the sample into each well of 9 wells of three 12% polyacrylamide 10-well gels (10-μl samples are loaded in 27 total lanes with extra long micropipette tips) (see Section 4.1). The marker on each gel contains 10 μl of a 20-bp ladder. Run the gel for 3 hr and 20 min at 160 V. Soak gel using SYBR Green I stain (Molecular Probes, Eugene, OR) at 1 : 10,000 dilution (add 20 μl SYBR Green 1 to 200 ml of 1× TAE Buffer in foil-wrapped container for 15 min). Visualize on UV box using SYBR green filter.

Amplified ditags should run at 102 bp while a background band runs at about 80 bp. Cut out only amplified ditags from gel and place 3 cut-out bands in a 0.5-ml microcentrifuge tube (nine 0.5-ml tubes total) whose bottom has been pierced with a 21-gauge needle to form a small hole of about 0.5 mm diameter. Place the 0.5-ml microcentrifuge tubes in 2.0 ml siliconized microcentrifuge tubes (Ambion) and centrifuge in microfuge at full speed for 2 min (this serves to break up the cut-out bands into small fragments at the bottom of the 2.0-ml microcentrifuge tubes).

Discard 0.5-ml tubes and add 250 μl of LoTE and 50 μl of 7.5 *M* ammonium acetate to each 2.0-ml tube. Tubes can remain at this point at 4° overnight. Then, vortex each tube and place at 65° for 15 min.

Place 5 μl of LoTE on the membrane of each of 18 SpinX tubes. Transfer contents of each 2.0-ml tube to 2 SpinX microcentrifuge tubes (9 tubes transferred to 18 SpinX microcentrifuge tubes). Spin each SpinX in microcentrifuge for 5 min at full speed. Combine sets of two eluates (300 μl total) and transfer to a 1.6-ml tube. Ethanol precipitate the eluates as follows: 300 μl sample, 5 μl glycogen, 133 μl of 7.5 *M* ammonium acetate, and 1 ml of 100% ethanol. Spin in a microcentrifuge at full speed for 15 min. Wash the pellet twice with 75% ethanol and resuspend each DNA sample in 10 μl LoTE. Pool samples into one tube (90 μl total) and dot quantitate (see Section 4.5). Total amount of DNA at this stage should be 10 to 20 μg.

Digest purified PCR products with *Nla*III by adding the following to the sample tube. To 90 μl of PCR products in LoTE add 226 μl of LoTE, 40 μl of NEB Buffer 4 (10×), 4 μl of BSA (100×, NEB), and 40 μl of *Nla*III (10 U/μl, NEB). Incubate at 37° for 1 hr.

11. Purification of Ditags by Linker Purification

Biotinylation of Primer 1 and Primer 2 allows the removal of linker products following *Nla*III digestion by use of streptavidin-linked magnetic beads. Historically, we removed the linkers by gel purifying the ditags and have generated many libraries without the use of biotinylated primers. However, we have found that addition of this extra purification step ensures complete removal of linkers that can poison the subsequent ditag concatamerization reaction by capping the concatamer ends and rendering the concatamers unclonable. In our experience the combination of these two purification techniques is better than the use of either alone. Removal of linkers by streptavidin beads can be done either before or after gel purification. We prefer the former because it allows us to visualize the purity of the ditags following removal of linkers by streptavidin-linked magnetic beads.

Linker purification is performed as follows. During the *Nla*III digestion, 1600 μl of Dynal streptavadin magnetic beads are placed into two tubes (800 μl each) and the beads are prewashed 3 times with 800 μl 1× B&W of wash buffer (see Section 2.5.1) containing 1× BSA (NEB). After adding the last wash, aliquot 200 μl into each of eight tubes (two for each of four purification cycles). Label four tubes A through D, do this twice for each set (e.g., A1, B1, C1, D1 and A2, B2, C2, D2). To the completed *Nla*III digest, add 400 μl of 2× B&W wash buffer and 4 μl of 100× BSA. Using a magnet, remove wash buffer from the first 2 tubes containing beads, i.e., A1 and A2. Aliquot the 800 μl digest mix to both tubes (i.e., 400 μl into A1 and A2) (purification cycle 1). Mix end over end at room temperature for 15 min (Labquake rotator, Barnstead/Thermolyne, Dubuque, IA). Place on magnet for 2 min. At same time place the second set of 200 μl aliquot of beads onto magnet. Remove wash buffer from second 200 μl aliquot of beads, i.e., B1 and B2. Transfer supernatant from first set of tubes, i.e., A1 and A2 to B1 and B2, respectively. Immediately add 200 μl of rinse (1× B + W, 1× BSA, Methods 2.5.1) to first aliquot of beads (tubes A1 and A2). Pipette back and forth several times to remove residual ditags. Mix end over end A1, A2, B1, and B2 at room temperature for 15 min. Remove supernatant from C1 and C2. Transfer supernatant from B1 and B2 to C1 and C2, respectively. Transfer supernatant from A1 and A2 to B1 and B2, respectively. Mix for 15 min. Remove supernatant from D1 and D2. Transfer C1 and C2 to D1 and D2, respectively. Transfer B1 and B2 to C1 and C2. Mix end over end for 15 min. Collect supernatant from D1 and D2 and transfer to new tubes. Transfer supernatant from C1 and C2 to D1 and D2, respectively. Mix end over end for 15 min. Consolidate supernatants (about 550 μl in each of 2 tubes). We have noted that ditags, although very stable in the presence of high salt, occasionally melt in the presence of low salt, especially if the temperature rises above room temperature or the DNA pellets are allowed to dry. Therefore, the following steps are performed on ice until samples are placed in the high-salt TAE gel buffer. Also, pellets are resuspended in TE instead of LoTE.

The linkers are then extracted with an equal volume of PC8. This is done by pooling the aqueous phases and transferring into 5 tubes followed by ethanol precipitation in dry ice as follows: 200 μl of sample, 66 μl of 7.5 *M* ammonium acetate, 5 μl of glycogen, and 825 μl of 100% ethanol. Vortex and place in dry ice/methanol bath for 15 min. Then warm at room temperature for 2 min until solution has melted, spin at 4° for 15 min at 14,000 rpm. Wash the pellet once with cold 75% ethanol, remove ethanol traces with a needle-nose pipette tip, and resuspend pellet in each tube in 6 μl of cold TE (not LoTE). Pool resuspended DNA samples into one tube (30 μl total). On ice add 7.5 μl 5× sample buffer (37.5 μl total volume). Load this sample into 4 lanes of a 12% polyacrylamide gel (10-well) and run at 160 V for 2.5 hr. Stain gel using SYBR Green I stain, at 1 : 10,000 dilution.

The 24–26 bp bands containing the purified linkers from 4 lanes are cut out from the gel. Two cut-out bands are placed in each of two 0.5-ml microcentrifuge tubes whose bottoms have been pierced with a 21-gauge needle. Each tube is placed in a 2.0 ml siliconized microcentrifuge tube and spun in a microfuge at 14,000 rpm at 4° for 2 min. The 0.5 ml tubes are discarded and to 2.0 ml tubes the following are added: 250 μl TE and 50 μl 7.5 *M* ammonium acetate. Vortex the tubes and place at 37° (not 65°) for 15 min. Longer incubations (even overnight) can be performed but do not appear to result in significantly higher yields. Use four SpinX tubes as above to isolate eluate. Ethanol precipitate in three tubes (200 μl each) as follows: 200 μl sample, 66 μl of 7.5 *M* ammonium acetate, 5 μl of glycogen, and 825 μl of 100% ethanol. Place in dry-ice/methanol bath for 10 min and spin at 4° for 15 min at 14,000 rpm. Wash the pellet twice with cold 75% ethanol. Resuspend each DNA sample in 2.5 μl cold TE (7.5 μl total).

12. Ligation of Ditags to Form Concatemers

The length of the ligation time depends on quantity and purity of ditags. Typically, several hundred nanograms of ditags are isolated and produce large concatemers when the ligation reaction is carried for 1–3 hr at 16° (lower quantities or less pure ditags will require longer ligations). Mix the following: 7 μl of pooled, purified ditags, 2 μl of ligation buffer (5×), and 1 μl of T4 DNA ligase (5 U/μl, Gibco-BRL). The reaction is incubated at 16° for 1–3 hr. Add 2.5 μl of sample buffer (5×) to the ligation reaction, heat the sample at 65° for 5 min, and then place it on ice. The concatemers are separated by polyacrylamide gel electrophoresis. In the first lane of an 8% polyacrylamide gel (described earlier), load 10 μl of a 1-kb ladder (25 ng/μl) as a marker. Skipping one lane, load the entire concatenated sample into the third well (one lane). Samples are run at 130 V for 3 hr. Stain gel with SYBR Green I 1 : 10,000 dilution as described previously. Visualize on UV box using SYBR Green filter. The concatemers will form a smear on the gel ranging from about 100 bp to several kilobases. We usually isolate the region 600–1200 bp and 1200 bp to 2500 bp.

The concatemers are purified as follows. Place each of these gel pieces into a 0.5-ml microcentrifuge tube (2 tubes total) whose bottom has been pierced with a 21-gauge nedle. Place the tubes in a 2.0-ml microcentrifuge tube and spin in microfuge at full speed for 2 min. Discard 0.5-ml tubes and add 300 μl LoTE to 2.0-ml tubes. Vortex the tubes, and place at 65° for 15 min. If necessary, this incubation can be extended to overnight, but yields are not significantly increased. Transfer contents of each tube to two SpinX microcentrifuge tubes (four SpinX tubes total). Centrifuge the SpinX tubes in microcentrifuge for 5 min at 14,000 rpm at 4°. Pool eluates from two SpinX tubes into one 1.5-ml tube and ethanol precipitate as follows: to 300 μl eluate, add 3 μl of glycogen, 133 μl of 7.5 *M* ammonium acetate, and 1 ml of 100% ethanol. Spin in microcentrifuge at full speed for 15 min. Wash the pellet twice with 70% ethanol, centrifuge at 14,000 rpm for 2 min at 4°, and remove ethanol. Resuspend purified concatemer-containing DNA in 6 μl of LoTE.

One can also use an alternative protocol using agarose gels to separate the concatemers. Prepare a 13-cm long 1.5% agarose gel (we use an apparatus from Owl Scientific). In the first lane of the gel load 10 μl of 1 KB ladder (25 ng/μl) as a marker. After the ligation reaction and subsequent heating and ice step detailed above, skip one lane and load entire concatemer mix onto one lane of a 1 m*M* 20-well gel. Run at 90 V until bromphenol blue dye reaches the end of gel (about 5 hr). Stain for 15–30 min and excise desired fractions. Extract DNA with QIAquick gel extraction kit (Qiagen, Valencia, CA). (*Note:* Follow manufacturer's protocol, except that step 5 on page 25 of the protocol should be skipped.) Increase eluate volume from 50 to 200 μl with LoTE. Extract with PC8 and ethanol precipitate as follows. To 200 μl of sample add 133 μl of 7.5 *M* ammonium acetate, 5 μl of glycogen, and 777 μl of 100% ethanol. Wash pellet twice with 70% ethanol, centrifuge at 14,000 rpm for 2 min at 4°, remove ethanol, and resuspend pellet in 6 μl LoTE.

13. Cloning Concatemers and Sequencing

Concatemers can be cloned and sequenced in a vector of your choice. We currently clone concatemers into a *Sph*I cleaved pZero vector (Invitrogen). This is done using the appropriate controls, and given below is the order of addition to three tubes. To tube 1, add 6 μl of purified concatamer; to tube 2, add 6 μl of water; and to tube 3, add 7 μl of water. To each of the tubes add 1 μl of pZero vector cut with *Sph*I (25 ng/μl), 2 μl of ligase buffer (5×), and only to tubes 1 and 2 add 1 μl of T4 DNA ligase (Gibco-BRL). Incubate the reaction overnight at 16°. Bring sample volume to 200 μl with LoTE. Extract with an equal volume of PC8 and ethanol precipitate as follows. To 200-μl sample add 133 μl of 7.5 *M* ammonium acetate, 5 μl of glycogen, and 777 μl of 100% ethanol. Wash the pellet four times with 70% ethanol, centrifuge at 14,000 rpm for 2 min at 4°, remove ethanol, and resuspend in 10 μl LoTE.

ElectroMAX DH10B cells (Gibco-BRL) are transformed with the DNA (1 μl) by electroporation. Plate 1/10 of the bacterial transformants onto each 10-cm zeocin-containing plate and analyze 12–16 hr later. Plates containing transformants with insert DNA should have hundreds to thousands of colonies while control plates should have zero to tens of colonies. Save all 10 plates for each concatemer ligation reaction. If insert size appears appropriate, these may be used for sequencing as described below. Check insert sizes by PCR. Set up 25-μl PCR reactions using the following per reaction: 2.5 μl of PCR Buffer (10×, see Section 2.5.3), 1.25 μl of DMSO (Sigma), 1.25 μl of dNTP (10 m*M*, Gibco-BRL), 0.5 μl of M13 forward primer (see Section 3) and 0.5 μl of M13 reverse primer (see Section 3), 19 μl of distilled water, and 0.2 μl of Platinum *Taq* (5 U/μl, Gibco-BRL). The PCR cocktail is added to each well of a 96-well PCR plate (Omniplate 96; Marsh Biomedical Products). Use a sterile tip to gently touch the *E. coli* transformant colony and then dip into the PCR mix. Add one drop of oil over the PCR reaction. Perform PCR at following temperatures (optimized for HYBAID PCR machine): 1 cycle at 95° for 2 min; 25 cycles of 95° for 30 min, 56° for 1 min, and 72° for 30 min; and 1 cycle at 70° for 5 min. Run 4 μl of the PCR reaction on a 1.5% agarose gel. For large-scale screenings we use Transferpette multichannel pipettes. The tips of these multichannel pipettes fit into every second well of the 50-slot comb used on our Owl Centipede horizontal apparatus. Consequently, to maintain a sequential loading order for each 96-well plate, we prepare a separate 96-well loading plate as follows. First add sample-loading dye to all the wells of the loading plate. Next, eight tips of the multichannel are used to transfer 4 μl from the first column of odd wells (i.e., A1-H1) of the original PCR plate to the corresponding wells of the loading plate. Next, transfer 4 μl from the second column of odd wells (i.e., A3-H3) to wells A2–H2 of the loading plate. This process is continued until the first six columns of the loading plate (i.e., 1, 2, 3, 4, 5, and 6) are filled with all of the odd columns from the original plate (i.e., 1, 3, 5, 7, 9, 11). Next, the even wells (i.e., 2, 4, 6, 8, and 10) are transferred to the last six columns of the loading plate (i.e., 7, 8, 9, 10, 11, and 12). Subsequently, the gel is loaded using six tips by transferring the first odd row of the loading plate followed by the second odd row, and so on. Finally, the empty wells of the gel (in between the odd wells) are filled in with all of the even rows.

The remaining PCR product that contains the concatemers composed of at least 15 ditags (>616 bp containing 226 bp vector plus 26 bp per ditag $\times$ 15 ditags) is purified by 2-propanol precipitation as follows. Mix 17 μl of PCR reaction with 75 μl of the following premixed 2-propanol solution containing 28 μl of distilled water, 15 μl of $NaClO_4$ (2 *M* stock), and 33 μl of 2-propanol. Spin at maximum speed in a centrifuge containing 96-well buckets. Decant and soak up residual precipitation mix by placing upside-down on paper towel. Rinse once by adding 100 μl 70% ethanol to wells. Spin 5 min at 14,000 rpm at 4°. Decant, blot on paper towel, dry samples (e.g., under laminar flow or in an immobile speed vacuum), resuspend in 25 μl distilled water, and store at −20°.

Sequencing can be performed manually or on automated sequencers. We currently sequence using the ABI-21M13 Dye Primer FS sequencing kit (PerkinElmer, Norwalk, CT) following the manufacturer's protocol. We use about 1/5 of the purified PCR product per sequencing reaction. Load and run on ABI 373 or 377 automated sequencer. Sequence files are analyzed by means of the SAGE program group,[24a] which identifies the anchoring enzyme site with the proper spacing and extracts the two intervening tags and records them in a database.

Results

SAGE Data Analysis

We determined the expression profiles of genes required for normal growth and during progression through the cell cycle in *S. cerevisiae*.[7] SAGE libraries were generated from yeast cells grown logarithmically or arrested in S phase or G_2/M phase of the cell cycle. In total, SAGE tags corresponding to 60,633 transcripts were identified (20,184 from logarithmic phase, 20,034 from S phase arrested, and 20,415 from G_2/M phase arrested cells). We used the 14 bp of each SAGE tag (i.e., the *Nla*III site plus the adjacent 10 bp) to search the yeast genome database on August 7, 1996.[25] Because only coding regions are annotated in the yeast genome and SAGE tags can be derived from 3′ untranslated regions of genes, a SAGE tag was considered to correspond to a particular gene if it matched the ORF or the

[24a] V. E. Velculescu, L. Zhang, B. Vogelstein, and K. W. Kinzler, *Science* **270,** 484 (1995).

[25] http://genome-www.stanford.edu/Saccharomyces/

region 500 bp 3′ of the ORF (locus names, gene names, and ORF chromosomal coordinates were obtained from the Stanford Genome Database and ORF descriptions were obtained from MIPS[17] on August 14, 1996). ORFs were considered genes with known functions if they were associated with a three-letter gene name, whereas ORFs without such designations were considered uncharacterized.

As expected, SAGE tags matched transcribed portions of the genome in a highly nonrandom fashion, with 88% matching ORFs or their adjacent 3′ regions in the correct orientation (χ^2 P value $<10^{-30}$). When more than one tag matched a particular ORF in the correct orientation, the abundance was calculated to be the sum of the matched tags. Tags that matched ORFs in the incorrect orientation were not used in abundance calculations. In instances when a tag matched more than one region of the genome (for example, an ORF and non-ORF region), only the matched ORF was considered. In some cases the 15th base of the tag could also be used to resolve ambiguities.

In addition to identifying transcription from ORFs, SAGE also permitted us to identify transcription from previously unidentified ORFs, present in the intergenic regions.[7,12] Yeast genome intergenic regions were defined as regions outside annotated ORFs or the 500 bp region downstream of annotated ORFs (yeast genome sequence and tables of annotated ORFs were obtained from *Saccharomyces* Genome Database).[25] Based on sequence analysis a total of 9524 putative ORFs of 25–99 amino acids were present in the intergenic regions. Of the 60,633 SAGE tags analyzed, there were 302 unique SAGE tags that matched the genome uniquely, were in the correct orientation, and were expressed at levels greater than 0.3 transcript copies per cell (tag observed at least two times in the SAGE libraries). The 302 unique SAGE tags were either within or adjacent to intergenic ORFs (100 bp upstream or 500 bp downstream of the ORF).

Global Gene Expression

The number of SAGE tags required to define a yeast transcriptome depends on the confidence level desired for detecting low-abundance mRNA molecules. The number of mRNA molecules per cell has been estimated to be 15,000.[26] Hence, 20,000 tags would represent a 1.3-fold coverage, providing a 72% probability of detecting single copy transcripts (as determined by Monte Carlo simulations). The estimate of 15,000 mRNA molecules per cell is reasonably accurate based on expression data for *SUP44/RPS4,* one of the few genes whose absolute mRNA levels have been reliably determined.[27] We determined that the expression level for *SUP44/RPS4* was 63 copies per cell by SAGE, in good accordance with the previously reported value of 75 ± 10 copies/cell by hybridization.[27]

[26] L. M. Hereford and M. Rosbash, *Cell* **10,** 453 (1977).

[27] V. Iyer and K. Struhl, *Proc. Natl. Acad. Sci. U.S.A.* **93,** 5208 (1996).

Analysis of 20,184 tags from logarithmic phase cells identified 3298 unique genes. Analysis of SAGE tags from S phase arrested and G_2/M phase arrested cells revealed similar expression levels for *SUP44/RPS4* (range 52 to 55 copies/cell), as well as for the vast majority of expressed genes. As less than 1% of the genes were expressed at dramatically different levels among these three states, SAGE tags obtained from all libraries were combined and used to analyze global patterns of gene expression. Analysis of ascertained tags at increasing increments revealed that the number of unique transcripts plateau at ~60,000 tags. This suggested that generation of further SAGE tags would yield few additional genes, consistent with the fact that 60,000 transcripts represented a fourfold redundancy for genes expressed as low as one transcript per cell. Likewise, Monte Carlo simulations indicated that analysis of 60,000 tags would identify at least one tag for a given transcript 97% of the time if its expression level were one copy per cell.

Analysis of the 60,000 tags showed that of these 56,291 tags precisely matched the yeast genome and represented transcription from 4665 different genes. This number is in good agreement with the estimate of 3000 to 4000 expressed genes obtained by RNA–DNA reassociation kinetics.[26] These expressed genes included 85% of the genes with characterized functions (1981 of 2340), and 76% of the total genes predicted from analysis of the yeast genome (4665 of 6121). The data represent a relatively complete sampling of the transcriptome in spite of the limited number of physiological states examined and the large number of genes predicted solely on the basis of genome sequence analysis. The expression level of genes varied from 0.3 to more than 200 transcripts per cell. Analysis of the distribution of gene expression levels revealed several abundant classes that were similar to those observed in previous studies using reassociation kinetics.[26] Among the most highly expressed genes (>60 mRNA copies per cell) were the ones that encode for enzymes involved in energy metabolism and protein synthesis. These genes were expressed at similar levels in all three growth states. Key transcripts, such as those encoding enzymes required for DNA replication (e.g., *POL1* and *POL3*), kinetochore proteins (e.g., *NDC10*, *CTF13*, and *SKP1*), and many other proteins, were present at one or fewer copies per cell on average. These levels are consistent with previous qualitative data from reassociation kinetics, which suggested that the largest number of expressed genes was present at 1 or 2 copies per cell. These observations indicate that low transcript copy numbers are sufficient for gene expression in yeast and suggest that yeast possess a mechanism for rigid control of RNA abundance. In a study aimed at defining the yeast proteome it was shown that the SAGE data correlates well with protein abundance and codon bias.[28]

SAGE libraries were made from yeast grown in glucose-rich medium, and hence as expected we observed a high level of expression of genes required for

[28] B. Futcher, G. I. Latte, P. Monardo, C. S. McLaughlin, and J. I. Garrels, *Mol. Cell. Biol.* **19,** 7357 (1999).

metabolism of glucose. These included genes such as *ENO2*, *PDC1*, *PGK1*, *PYK1*, and *ADH1*, which are known to be dramatically induced in the glucose-rich growth conditions. In contrast, glucose repressible genes, such as the *GAL1/GAL7/GAL10* cluster and *GAL3*, were observed to be expressed at very low levels (0.3 or fewer copies per cell). Also, for the yeast strain used in this study, mating type a-specific genes, such as the a factor genes (*MFA1*, *MFA2*)[29] and alpha (α) factor receptor (*STE2*),[30] were all observed to be expressed at significant levels (range 2 to 10 copies per cell), while mating type α-specific genes (*MFa1*, *MFa2*, *STE3*)[31–33] were observed to be expressed at very low levels (<0.3 copies/cell).

Integration of Expression Information with Genomic Map

We integrated the SAGE expression data with existing positional information to generate chromosomal expression maps. These maps were generated using the sequence of the yeast genome and the position coordinates of ORFs obtained from the Stanford Genome Database. Except for a few genes, there did not appear to be any clusters of genes with particularly high or low expression on any chromosome. The genes encoding the histones, H3 and H4, which are known to have coregulated divergent promoters and are immediately adjacent,[34] had very similar expression levels (5 and 6 copies per cell, respectively). The distribution of transcripts among the chromosomes suggested that overall transcription was evenly dispersed, with total transcript levels being roughly linear to genetic content and independent of the coding strand. However, regions within 10 kb of telomeres were not highly transcribed, containing on average 3.2 tags per gene as compared with 12.4 tags per gene for nontelomeric regions. This is consistent with the previously described observations of telomeric silencing in yeast.[35,36] Additionally, internally silenced regions, such as the 10-kb region comprising the cryptic mating-type loci HML and HMR, were observed to be transcriptionally repressed (average of 2.4 tags per gene).

Differential Gene Expression

We analyzed the SAGE data to identify genes whose transcription was induced in HU arrested cells. We sorted the SAGE data to identify genes whose transcription was induced at least twofold in HU arrested cells compared to cycling cells and

[29] S. Michaelis and I. Herskowitz, *Mol. Cell. Biol.* **8,** 1309 (1988).

[30] A. C. Burkholder and L. H. Hartwell, *Nucleic Acids Res.* **13,** 8463 (1985).

[31] D. C. Hagen, G. McCaffrey, and G. F. Sprague, *Proc. Natl. Acad. Sci. U.S.A.* **83,** 1418 (1986).

[32] J. Kurjan and I. Herskowitz, *Cell* **30,** 933 (1982).

[33] A. Singh, E. Y. Chen, J. M. Lugovoy, C. N. Chang, R. A. Hitzeman, and P. H. Seeburg, *Nucleic Acids Res.* **11,** 4049 (1983).

[34] M. M. Smith and K. Murray, *J. Mol. Biol.* **169,** 641 (1983).

[35] D. E. Gottschling, O. M. Aparicio, B. L. Billington, and V. A. Zakian, *Cell* **63,** 751 (1990).

[36] H. Renauld, O. M. Aparicio, P. D. Zierath, B. L. Billington, S. K. Chhablani, and D. E. Gottschling, *Genes Dev.* **7,** 1133 (1993).

TABLE I
SAGE ANALYSIS OF CELLS ARRESTED WITH HYDROXYUREA[a]

Ratio of[b]		Number of SAGE Tags in:[c]				
HU/CYC[c]	TAG	CYC	HU	NOC	TOTAL	Gene/locus[d]
49	CTTCTCTTTT	0	49	0	49	*HUG1*/NORF5/YML058W-A
15.9	TCTAAGTCCG	7	111	9	127	*RNR4*/YGR180C
15	GCAGTAAAGG	1	15	10	26	*ALD5*/YER073W
14	CCATACAGGT	0	14	2	16	NORF?
10	AGAAAGGATA	1	10	8	19	*ECM15*/YBL001C
10	GAGGTGCTGT	1	10	6	17	YBR137W
9.4	GAAAACATCT	9	85	9	103	*RNR2*/YJL026W
9	GGCAACACCT	0	9	1	10	*PHO3*/YBR092C
9	GTTCGAGACA	0	9	3	12	*DOG1*/YHR044C
9	TCCTTCAGTA	0	9	4	13	*BRX1*/YOL077C
8	TTAATTCTGT	2	16	10	28	YGL102C
8	TACGTAAGTT	0	8	3	11	NORF?

[a] The SAGE data was analyzed to identify genes whose transcription was induced at least two-fold in HU arrested cells compared to cycling cells and there was at least a 3-tag difference between the two states.

[b] Ratio of number of SAGE tags in HU-arrested cells divided by number of tags in cycling cells. Zero SAGE tags in the denominator is treated as 1 for calculation of the ratio.

[c] SAGE tags from cycling (CYC), HU-arrested (HU), or nocodazole-arrested (NOC) cells; total represents total of all three states.

[d] Locus name derived from the *Saccharomyces* Genome Database. A SAGE tag that does not match an annotated ORF may represent transcription from a NORF.

there was at least a 3-tag difference between the two states. The results presented in Table I showed that the ratios of HU-induced transcripts varied from 49-fold to 8-fold (number of tags in HU divided by number of tags in cycling cells). This analysis led to the identification of genes such as *RNR2* and *RNR4* known to be upregulated in response to HU.[37] In addition to this, we identified transcription from other known and unknown genes that were previously not reported to be induced in response to HU. Interestingly, one of the *NORF* genes (*NORF5*) was only expressed in S-phase arrested cells and corresponded to the transcript whose abundance varied the most in the three states analyzed (>49-fold). Our studies led to the characterization of the first NORF, namely, *NORF5/HUG1*, as described later.[12] While there were many relatively small differences between the states, overall comparison of the three states revealed surprisingly few dramatic differences. There were only 29 transcripts whose abundance was varied more than 10-fold among the three different states analyzed.

[37] S. J. Elledge, Z. Zhou, J. B. Allen, and T. A. Navas, *Bioessays* **15,** 333 (1993).

TABLE II
ANALYSIS OF SAGE TAGS CORRESPONDING TO NORFs[a]

NORF[b]	Size[c]	Number of SAGE tags in:[d] CYC	HU	NOC	Total	dBEST Hits[e]	*P* Value
NORF1	198	182	84	114	380	—	—
NORF2	243	9	111	114	294	Human cDNA	5.7 e-21
NORF3	189	16	17	33	66	Human cDNA	1.2 e-06
NORF4	177	15	39	5	9	Human cDNA	3.0 e-21
NORF5	204	0	49	0	49	—	—
NORF6	252	25	17	1	43	—	—
NORF7	192	7	18	19	42	Human cDNA	5.7 e-02
NORF8	257	9	10	13	32	Human cDNA	1.6 e18

[a] SAGE tags that corresponded to yeast genome intergenic regions (defined as regions outside annotated ORFs or the 500 bp region downstream of annotated ORFs) were analyzed for NORFs. These SAGE tags matched the genome uniquely, were in the correct orientation either within or adjacent to an intergenic ORF, and were expressed at levels greater than 0.3 transcript copies per cell.
[b] All putative ORFs corresponding to the SAGE tag were analyzed and the NORFs are assigned arbitrary numbers in the order of the abundance of transcription.
[c] All NORFs correspond to ORFs less than 300 bp in size.
[d] SAGE tags from cycling (CYC), HU arrested (HU), or nocodazole-arrested (NOC) cells. Total represents a total of all tags from all three states.
[e] Homology searches for NORFs was done using the BLAST server against the dBEST database; the statistical significance is denoted by the *P* value.

SAGE Analysis Revealing Transcription from NORFs That Are Evolutionarily Conserved

As previously reported SAGE has identified transcripts that correspond to NORFs in the intergenic regions of *S. cerevisiae*.[12,7] Whether any of these NORFs are important for the growth and biology of yeast is unclear. We performed a systematic analysis of the SAGE tags that correspond to the NORFs. Of the 60,633 SAGE tags analyzed, there were 302 unique SAGE tags that were either within or adjacent to intergenic ORFs of <100 amino acids. The 302 SAGE tags were expressed at levels ranging from 0.6 to 94 transcript copies per cell. The 30 most abundant of the transcripts detected by SAGE were observed at least nine times.[38] Northern blot analysis for 4 of the NORFs (NORF1, NORF5, NORF14, and NORF17) has confirmed their transcription in *S. cerevisiae*. Database searches showed that 5 of the 8 highly expressed NORF genes are evolutionarily conserved with mammalian homologs (Table II). In addition, the SAGE data facilitated the addition of 27 new ORFs (<100 amino acids) to the *S. cerevisiae* genome database.[39]

[38] http://www.sagenet.org/NORF/NORF.html
[39] http://www.stanford.edu/Saccharomyces/newORF.html

NORF5/HUG1 was chosen for further analysis because its dramatic expression in HU-treated cells suggested a potential role in transcriptional response after replication arrest and DNA damage.[12]

Transcription of NORF5/HUG1 Induced by Replication Arrest and DNA Damage

NORF5, a putative 68 amino acid protein, corresponds to a previously unidentified ORF transcribed in HU arrested cells. HU, a potent inhibitor of ribonucleotide reductase (*RNR*) which is required for dNTP synthesis, leads to replication arrest in S phase.[37,40] The transcript abundance of NORF5 in logarithmically grown yeast cells was <1 copy/cell, whereas in HU arrested cells it was 37 copies/cell, exhibiting a higher level of differential gene expression in HU arrested cells than any other *S. cerevisiae* gene.[7] Northern blot analysis supported SAGE data as a transcript of approximately 400 bp, corresponding to NORF5, is present in RNA prepared from HU arrested cells (Fig. 2A[7,12]). Consistent with these results, Western blot analysis of the candidate epitope tagged 68 amino acid ORF (Chromosome 13, coordinates 158760–158966) confirmed a protein of about 10 kDa in HU arrested cells. Transcription of NORF5 is also induced in cells exposed to ultraviolet (UV) or gamma radiation (Fig. 2B). The transcriptional induction of NORF5 appears to be specific to replication arrest and DNA damage since there was no induction of NORF5 in cells subjected to heat shock (data not shown) or nocodazole-induced G_2/M arrest (Fig. 2A). On the basis of its transcription pattern, we named the NORF5 gene *HUG1* (hydroxyurea, ultraviolet, gamma induced).[12] We found that following addition of HU, low levels of *HUG1* transcription are detected at earlier time periods of 30 min and 60 min, followed by an almost linear increase until 3.5 hr post-HU addition. We also determined that the DNA damage-dependent transcription of *HUG1* is not restricted to any particular stage of the cell cycle. Cells arrested in G_1 with alpha factor or G_2/M with nocodazole show similar patterns of transcription of *HUG1* as compared to asynchronous populations on exposure to gamma radiation. Therefore, DNA damage-induced transcription of *HUG1* can occur in G_1 and G_2/M phases. We also determined that the DNA damage-specific induction of *HUG1* is due to the alleviation of repression by the Crt1p, Ssn6p, and Tup1p complex.[41]

NORF5/HUG1 as Critical Downstream Target of MEC1-Mediated Pathway for DNA Damage and Replication Arrest

Several checkpoint genes in *S. cerevisiae* are required for transcriptional induction of a large regulon of genes that facilitate DNA repair, cause cell cycle arrest,

[40] M. Huang and S. J. Elledge, *Mol. Cell. Biol.* **17,** 6105 (1997).
[41] M. Huang, Z. Zhou, and S. J. Elledge, *Cell* **94,** 595 (1998).

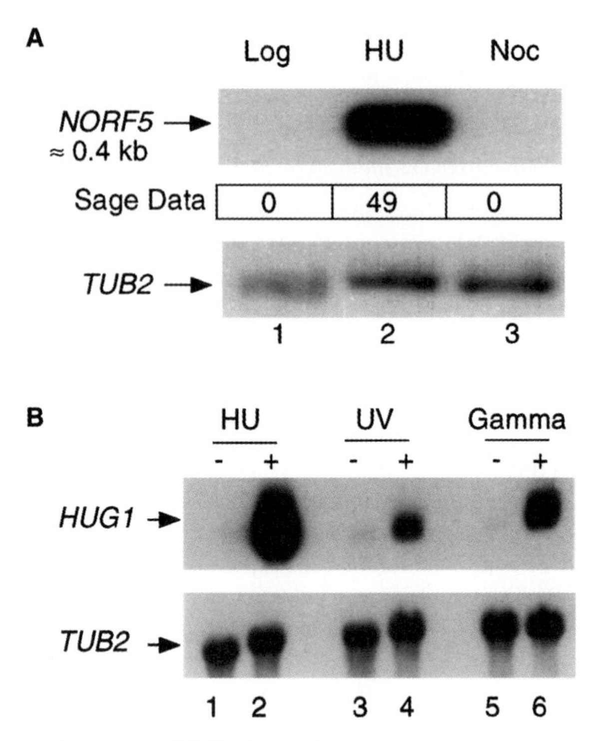

FIG. 2. Transcription of *NORF5/HUG1* is induced by replication arrest and DNA damage [see M. A. Basrai, V. E. Velculescu, K. W. Kinzler, and P. Hieter, *Mol. Cell. Biol.* **19,** 7041 (1999)]. (A) NORF5 transcription is up-regulated in cells arrested with HU. Northern blot analysis using wild-type cells (YPH499) grown logarithmically (lane 1), arrested with hydroxyurea (HU) (lane 2), or Nocodozole (Noc) (lane 3). The expression pattern observed by SAGE is indicated at the bottom (0 : 49 : 0) [see V. Velculescu, L. Zhang, W. Zhou, J. Vogelstein, M. A. Basrai, D. E. Bassett, Jr., P. Hieter, B. Vogelstein, and K. Kinzler, *Cell* **88,** 243 (1997)]. (B) Transcription of *NORF5/HUG1* is hydroxyurea-, ultraviolet-, and gamma-induced. Northern blot analysis using wild-type cells (YPH499) grown logarithmically (lanes 1, 3, and 5) arrested with HU (lane 2), exposed to ultraviolet radiation (UV) (lane 4), or exposed to gamma radiation (lane 6).

and mediate recovery from DNA damage.[42,43] A central component of these checkpoints is *MEC1,* the budding yeast homolog of the hereditary ataxia–telangiectasia ATM gene and a member of the PI3 kinase family.[44,45] Signals of DNA damage normally pass from sensor genes such as *RAD9, RAD17, RAD24, MEC3* and *DDC1*

[42] S. J. Elledge, *Science* **274,** 1664 (1996).
[43] T. Weinert, *Cell* **94,** 555 (1998).
[44] Y. Shiloh, *Annu. Rev. Genet.* **31,** 635 (1997).
[45] V. A. Zakian, *Cell* **82,** 685 (1995).

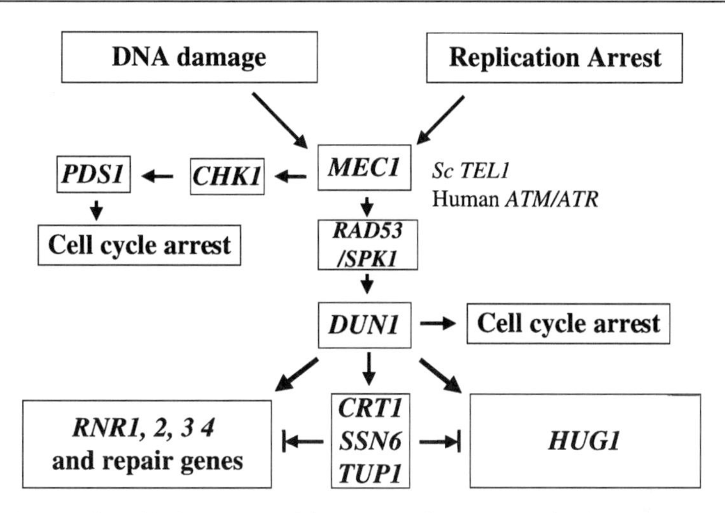

FIG. 3. *HUG1* is a critical component of the *MEC1*-mediated pathway for checkpoint response [see M. A. Basrai, V. E. Velculescu, K. W. Kinzler, and P. Hieter, *Mol. Cell. Biol.* **19,** 7041 (1999)]. Signals received from the sensors for DNA damage and replication arrest are transduced through the kinases *MEC1* and *TEL1* leading to cell cycle arrest in a *RAD53*-independent or -dependent manner. The latter pathway also mediates transcriptional induction, which can be *DUN1* independent or dependent. Transcription of *HUG1* is induced in response to replication arrest and DNA damage in a checkpoint-dependent manner. The DNA damage-specific induction of *HUG1,* which is independent of the cell cycle stage, is due to the alleviation of repression by the Crt1p, Ssn6p, Tup1p complex.

to *MEC1,* leading to phosphorylation of Rad53p, Replication protein A, and potentially other targets, causing cell cycle arrest and transcriptional response. We found that genes in the *MEC1* checkpoint pathway are required for the transcriptional induction of *NORF5/HUG1* in response to replication arrest and DNA damage (Fig. 3). Additional experiments have shown that *NORF5/HUG1* has distinct genetic interactions with *MEC1*. Overexpression of *HUG1* is lethal in the presence of DNA damage and replication arrest (Todd and Basrai, unpublished, 2002), whereas a deletion of *HUG1* rescues the lethality due to a *MEC1* null allele. The data suggest that the small protein Hug1p, the product of a NORF, is a critical mediator of the *MEC1* pathway (Fig. 3).[12] These results validate the idea that the NORFs are biologically relevant and highlight the importance of global approaches such as SAGE to identify a significant number of genes in yeast and other organisms that may be missed by sequence analysis alone.

Several key features of the DNA damage-induced pathways are conserved in human, yeast, and other systems. The *S. cerevisiae MEC1* gene for example, is homologous to the *S. pombe rad3*$^+$, the *D. melanogaster mei-41,* and the human

ATM gene.[46] By analogy, a *HUG1* homolog regulated by ATM or p53 may be present in humans. Identification and characterization of homologs of *HUG1* from other organisms, including humans, may further our understanding of the role of *MEC1* in budding yeast and may allow greater insight into the ATM- and p53-mediated checkpoint pathway in humans.

SAGE Data for Your Favorite Gene

The SAGE transcription data for any *S. cerevisiae* gene, ORF, or chromosomal region can be accessed via the *Saccharomyces* Genome Database as outlined in Fig. 4.[47] There are two forms of SAGE query, a basic query outlined in Fig. 4 and an advanced SAGE query that can be accessed via the same Web site. For example, to query the SAGE data for *NUP1,* type in the gene name in the query form, which gives a chromosomal view of the chromosomal region containing *NUP1* (*YOR097C*) and the adjacent flanking region. SAGE tags are represented by triangles, and the color of these triangles represents the location of the SAGE tag relative to known ORFs as follows: Class 1 tags are within an ORF (orange), Class 2 tags are within 500 bp 3′ of the ORF (violet), Class 4 tags are on the strand opposite to an ORF (yellow), and Class 3 tags represent none of the others (bright pink). Tags that match the genome once have an outer rectangle. Criteria used to categorize a SAGE tag as authentic include uniqueness in the genome and being adjacent to the 3′-most *Nla*III site of the ORF sequence or regulatory sequence corresponding to the ORF. The SAGE data should be verified by Northern blot analysis and/or other approaches. Other possible sources for tags include SAGE tags from transcripts with alternative poly A sites, alternative splicing, or internal priming during cDNA synthesis. Further analysis can be done by double-clicking on the orange triangle located at the 3′ end of *NUP1*. This gives information for the SAGE tag, which includes number of times tag matches the genome and number of times tag was present in logarithmic, S, or G_2/M phases of the SAGE libraries. The SAGE tag abundance data gives information on the class of the SAGE tag (described above), the chromosomal location of the tag, coding strand, ORF name, gene name, and the coordinates of the ORF.

In the example illustrated in Fig. 4 for the top coding strand, we observe unique SAGE tags for the following genes: *RPS7A*, *KTR1*, *RAS1*, and *VAM3*. Note that there are two SAGE tags corresponding to *RAS1*; however, the 3′-most one is probably the authentic one. Also, in case of SAGE tags for *YOR104W* and *YOR105W*, because of the overlapping nature of the ORFs, it is difficult to ascertain which tag may represent an accurate estimate of gene expression. For SAGE tags on the bottom strand, the SAGE tags corresponding to *OST2*, *NUP1*, *RKI1*, and

[46] K. Savitsky, S. Sfez, D. A. Tagle, Y. Ziv, A. Sartiel, F. S. Collins, Y. Shiloh, and G. Rotman, *Hum. Mol. Genet.* **4,** 2025 (1995).

[47] http://genome-www.stanford.edu/cgi-bin/SGD/SAGE/querySAGE

http://genome-www4.stanford.edu/cgi-bin/SGD/SAGE/querySAGE

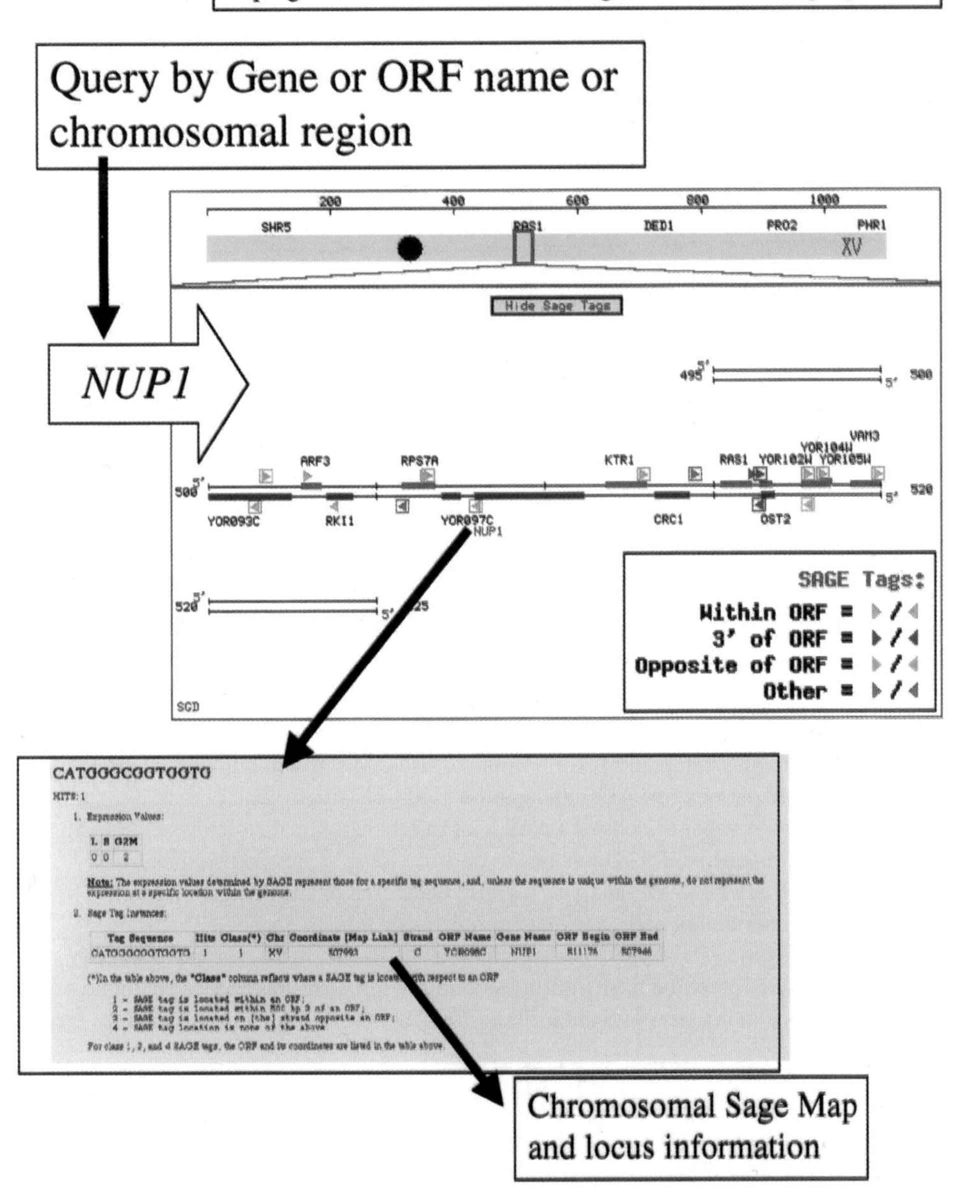

FIG. 4. SAGE data for a favorite gene. The SAGE data is available at the Stanford Genome Database at the indicated Web site. Query for *NUP1* generates a chromosomal view with genes flanking *NUP1* on both strands of DNA. The color of the SAGE tag reflects the location of the tag with respect to a gene. A detailed report on the frequency of the SAGE tags is available as a hyperlink.

YOR093C are unique. The SAGE tags in bright pink located between *KTR1* and *RAS1* on the top coding strand and between *NUP1* and *RKI1* on the bottom strand may correspond to transcription from NORFs. However, these tags are located too close to an annotated ORF. For SAGE tags that may correspond to true NORFs, *Saccharomyces* Genome Database has analyzed them further with information on all potential ORFs in the region with homology search data. Our analysis of SAGE tags for identification of NORFs disregarded tags that were within 500 bp of annotated ORFs.

Limitations of SAGE

Like other genome-wide analysis, SAGE analysis has several potential limitations. For example, our analysis will only detect transcripts that have an *Nla*III site and whose abundance are at least 0.3 copies per cell. Also, we will not detect transcripts that are regulated in response to other growth conditions. Another potential limitation is the unequivocal assignment of some SAGE tags to a gene. This is true for SAGE tags corresponding to regions of the genome with overlapping genes or those with unusually long 3′ untranslated regions. This problem stems from the fact that there are very few data for mRNA sequences in budding yeast. Despite these potential limitations, our results have shown that SAGE provides data for both global and differential gene expression for *S. cerevisiae,* precisely defined at the nucleotide level.

Modifications of SAGE

SAGE analysis has been applied to several other systems, most notably to study gene expression patterns for normal vs diseased states in humans. There are several modifications of SAGE that can be used to evaluate gene expression patterns in small number of cells or a particular microanatomical structure. One such technique, "microSAGE," allows one to perform gene expression studies from as few as 5000 cells.[48] SAGE analysis has also been used to analyze gene expression patterns in endothelial cells purified from primary human cancers[11] or from microdissected kidney tubules.[49] The technique of PCR–SAGE allows one to amplify cDNA mixtures prepared from limiting amounts of RNA and has been used to study gene expression in cerebrovascular tissue[50] and as few as nine human oocytes.[51]

[48] N. A. Datson, J. van der Perk-de Jong, M. P. van der Berg, E. R. de Kloet, and E. Vreugdenhil, *Nucleic Acids Res.* **27,** 1300 (1999).

[49] B. Virlon, L. Cheval, J. M. Buhler, E. Billon, A. Doucet, and J. M. Elalouf, *Proc. Natl. Acad. Sci. U.S.A.* **96,** 15286 (1999).

[50] D. G. Peters, B. Virlon, L. Cheval, J. M. Buhler, E. Billon, A. Doucet, and J. M. Elalouf, *Nucleic Acids Res.* **27,** e39 (1999).

[51] L. Neilson, A. Andalibi, D. Kang, C. Coutifaris, J. F. Strauss III, J. A. Stanton, and D. P. Green, *Genomics* **63,** 13 (2000).

TABLE III
WEB-BASED RESOURCES FOR SAGE

Database site	Web address	Resources
Johns Hopkins Oncology	http://www.sagenet.org	Protocols for SAGE, software, references
Saccharomyces Genome Database	http://genome-www.stanford.edu/Saccharomyces/	Chromosomal location of tags
Molecular Biology of the Cell	http://www.molbiolcell.org/content/vol10/issue6/images/data/1859/DC1/sagedata.tsv	SAGE data for *S. cerevisiae* grown on the fatty acid oleate
NCBI/CGAP	http://www.ncbi.nlm.nih.gov/SAGE	Mapping of tags to unigene clusters to SAGE tags and vice versa; download SAGE data from various sources
CEA Saclay	htttp://www-dsv.cea.fr/thema/get/sade.html	SAGE adaptation for down-sized extracts (SADE)
University of Tokyo	http://www.prevent.m.u.tokyo.ac.jp/.SAGE.html	Data for gene expression from human monocytes, macrophages, and dendritic cells
University of Rochester	http://www.urmc.rochester.edu/smd/crc/swindex.html	Database for SAGE analysis for human skeletal muscle
MD Anderson Cancer Center	http://sciencepark.mdanderson.org/ggeg/	SAGE data for effect of estrogen, cancer and normal breast cells, and mouse mammary gland
Compugen	htttp://www.labonweb.com	Gene to tag mapping
Academic Medical Center, Amsterdam	http://bioinfo.amc.sara.nl/HTM/credits.html	Tag to unigene cluster mapping generating a human transcriptome map
Genzyme	http://www.genzyme.com/sage/prodserv.htm	Services for SAGE analysis including library preparation and data analysis
Invitrogen	http://www.invitrogen.com/Content/expressions/E082002.pdf	I-SAGE kit for preparation of SAGE libraries

Conclusions

The completion of the genome sequence from model organisms and humans has led to rapid advancements in genome-wide approaches to define gene function. Some of these studies attempt to answer the question, what defines a gene? This is especially true of coding regions that either are embedded in a sea of non-coding sequences or overlap with other genes. SAGE has led to the identification of transcription from previously unidentified NORFs that may correspond to small ORFs. Some of the NORFs have also been identified by random tagged mutagenesis.[52–54]

Another question is, how many genes must an organism possess and transcribe in order to support its life? Analysis of transcriptomes from a variety of physiologic states should provide a minimum set of genes whose expression is required for normal growth, and another set composed of genes that will be expressed only in response to specific environmental stimuli, or during specialized processes. SAGE provides a snapshot of the genes that are being transcribed under a particular growth condition in a defined population of cells. One such study done in humans includes the analysis of 3.5 million transcripts (SAGE tags) from 84 different libraries representing 19 normal and disease tissue types.[55] The availability of the human genome sequence will make it possible to generate a transcriptome map that enables one to evaluate the expression level of "your favorite gene," like the one available for *S. cerevisiae*. The SAGE data have been used to assign SAGE tags to UniGene clusters and their chromosomal position.[56] Several web sites listed in Table III provide access to SAGE protocols, modifications, applications, and data. The long-term objective is to use the transcriptome data from different cells and from different individuals to gain a better understanding of gene function in normal, developmental, and disease states.

Acknowledgments

The authors thank Victor Velculescu for performing the yeast SAGE analysis and Ken Kinzler and Bert Vogelstein for support and advice. We acknowledge Jacob Vogelstein for NORF analysis, and Mike Cherry and the Stanford Genome Database staff for annotation of new ORFs based on SAGE analysis. We are grateful for the figures to Oliver Kerscher, who along with Marian Nau, Victor Velculescu, and the members of the Basrai laboratory made valuable comments on this article.

[52] A. Kumar, K.-H. Cheung, P. Ross-Macdonald, P. S. R. Coelho, P. Miller, and M. Snyder, *Nucleic Acids Res.* **28,** 81 (2000).

[53] A. Kumar, S. A. des Etages, P. S. Coelho, G. S. Roedder, and M. Snyder, *Methods Enzymol* **328,** 550 (2000).

[54] P. Ross-Macdonald, P. S. Coelho, T. Roemer, S. Agarwal, A. Kumar, R. Jansen, K. H. Cheung, A. Sheehan, D. Symoniatis, L. Umansky, M. Heidtman, F. K. Nelson, H. Iwasaki, K. Hager, M. Gerstein, P. Miller, G. S. Roeder, and M. Snyder, *Nature* **402,** 413 (1999).

[55] V. E. Velculescu, S. L. Madden, L. Zhang, A. E. Lash, J. Yu, C. Rago, A. Lal, C. J. Wang, G. A. Beaudry, K. M. Ciriello *et al., Nat. Genet.* **23,** 387 (1999).

[56] H. Caron, B. van Schaik, M. van der Mee, F. Baas, G. Riggins, P. van Sluis, M. C. Hermus, R. van Asperen, K. Boon, P. A. Voute, S. Heisterkamp, A. van Kampen, and R. Versteeg, *Science* **291,** 1289 (2001).

[25] PCR-Based Engineering of Yeast Genome

By MARIE E. PETRACEK and MARK S. LONGTINE

Introduction

The use of *Saccharomyces cerevisiae* as a model organism continues to increase as more genome sequences from higher eukaryotes become available and investigators who have previously concentrated on studying genes in larger eukaryotes now address the function of potential yeast homologs. High rates of homologous recombination in *Saccharomyces cerevisiae* allow precisely targeted modification of specific yeast chromosomal genes, an attractive feature of this model organism. Single-step PCR-based methods have been developed that allow more facile and economical gene deletion, epitope and/or protein tagging, or promoter substitution than traditional methods.

Homologous Recombination in Yeast

When introduced into the nucleus of *S. cerevisiae,* DNA lacking an origin of DNA replication (also called an autonomously replicating sequence or ARS) and carrying homology to the yeast genome is integrated into the genome by homologous recombination (reviewed in Ref. 1). This homologous recombination is most efficient when the transforming DNA has 100–500 bp of homology with the yeast genome, although as little as ~30–40 bp of homology also directs homologous recombination, while the rates of nonhomologous recombination are low. Double-stranded DNA breaks increase the rate of recombination, and, if a DNA molecule contains more than one region of homology to the yeast genome, the location of a double-stranded break can be used to target the site of recombination.[1]

Uses of Homologous Recombination

Homologous recombination is commonly used to target modification of the yeast genome. Often one of the first steps in studying a yeast gene is to delete the gene using homologous recombination and to investigate the phenotype of the resulting strain. Other targeted modifications include placing genes under the control of heterologous regulatable or constitutive promoters, introducing epitope/protein tags to the target protein (useful for protein localization, protein purification, and biochemical studies), or introducing a mutation into the target gene. The analysis of a modified gene in the genome has at least two significant advantages over the

[1] R. Rothstein, *Methods Enzymol.* **194,** 281 (1991).

0076-6879/02 $35.00

analysis of the same gene carried on a plasmid. First, even when grown under selective conditions, plasmids are absent from at least 5–10% of the cells in a population due to mitotic missegregation.[2] The presence of cells that lack the plasmid can complicate analyses. Second, if a mutated allele has significantly reduced function, selective pressures for increased gene expression can result in increased plasmid copy number of low copy number (CEN) plasmids, resulting in a masked or altered mutant phenotype.

Traditional Methods of Gene Modification by Homologous Recombination

Traditionally, recombinant DNA techniques requiring one or more DNA cloning steps have been necessary to construct a plasmid that carries a selectable marker (and other sequences, if desired) flanked by regions of homology to the chromosomal target locus.[1] Following digestion of the plasmid with a restriction enzyme, the released linear DNA fragment is transformed into yeast, and cells that have incorporated the DNA into the genome by homologous recombination are isolated by growth on medium that selects for expression of the selectable marker gene. Depending on the choice of the flanking DNA used and the other sequences included, this method allows for a variety of gene modifications. For example, if the flanking regions are homologous to the 5′ and 3′ ends of an open reading frame (ORF), then after transformation and homologous recombination the ORF is replaced with the selectable marker gene, generating a null allele. If the flanking regions are homologous to the 3′ end of an ORF, and the DNA includes sequences encoding an in-frame epitope or protein tag, the integration event will result in the expression of a protein with a C-terminal tag. Integration targeted to the 5′ end of a gene can be used to replace the endogenous promoter with a regulatable or constitutive promoter, with or without concomitant N-terminal tagging.

In addition to being labor intensive, traditional methods frequently rely on available restriction enzyme sites for plasmid construction, often yielding DNA fragments that result in suboptimal modifications. For example, a partial deletion of a target gene, generated by using convenient restriction enzyme sites, may result in the expression of a truncated protein that retains some normal function, or gains some abnormal function. Finally, a given plasmid is useful for only one type of modification of the target locus and additional modifications require the construction of additional plasmids.

PCR-Based Methods for Modification of Chromosomal Genes

To overcome problems from traditional recombinant DNA methods described above, facile PCR (polymerase chain reaction)-based methods have been developed

[2] J. C. Schneider and L. Guarante, *Methods Enzymol.* **194,** 373 (1991).

for *S. cerevisiae* genome modification (reviewed in Refs. 3 and 4). These PCR-based methods of gene modification require significantly less labor and less time than traditional, plasmid-based approaches. In addition, unlike traditional methods, it is not necessary to have the gene of interest on a plasmid prior to modification of the chromosomal gene. This situation is common and occurs, for example, when a yeast gene of potential interest is identified by computer database searches, by biochemical methods, or by proteomic approaches. PCR-based methods allow a rapid initial investigation of such genes to determine if they warrant further investigation. Because of the modular nature of many of the PCR plasmid templates, a common primer can often be used to carry out multiple modifications, allowing economical use of primers (Table I).

Short Flanking Homology (SFH) Method

The Short Flanking Homology (SFH) method[5–10] is extremely simple, requiring only a single PCR reaction starting with a known sequence for the gene of interest, a plasmid template, and two gene-specific oligonucleotide primers (Figs. 1A–1C). The ~59 nt forward (F) primer is designed to have ~40 nt of homology at its 5′ end to the target locus and ~19 nt of homology to the plasmid template at its 3′ end, while the ~59 nt reverse (R) primer is designed to have ~40 nt of homology at its 5′ end to the target locus and ~19 nt of homology to the plasmid template at its 3′ end. The 3′ ends of the primers are usually designed to anneal to a polylinker region of the plasmid template that flanks a PCR module including a selectable marker and, if included in the template, other modification sequences such as those encoding an epitope or protein tag and/or a promoter. Amplification of this module by PCR yields linear, double-stranded DNA molecules carrying a selectable marker with ~40 bp of homology at each end to the chromosomal target locus. Following transformation of the resulting PCR product into yeast, cells that have undergone homologous recombination are isolated by growth on medium that selects for the selectable marker. Typically, if selectable markers are used that lack homology to the *S. cerevisiae* genome (see

[3] A. Wach, A. Brachat, C. Rebischung, S. Steiner, K. Pokorni, S. te Heesen, and P. Philippsen, *in* "Methods in Microbiology" (A. Brown and M. Tuite, eds.), Vol. 26. Academic Press, New York, 1998.

[4] S. D. Kohlwein, *Microsc. Res. Tech.* **51,** 511 (2000).

[5] J. McElver and S. Weber, *Yeast* **8** (special issue), S267 (1992).

[6] A. Baudin, O. Ozier-Kalogeropoulos, A. Denouel, F. Lacroute, and C. Cullin, *Nucleic Acids Res.* **21,** 3329 (1993).

[7] A. Wach, A. Brachat, R. Pohlmann, and P. Philippsen, *Yeast* **10,** 1793 (1994).

[8] M. C. Lorenz, R. S. Muir, E. Lim, J. McElver, S. C. Weber, and J. Heitman, *Gene* **158,** 113 (1995).

[9] P. Manivasakam, S. C. Weber, J. McElver, and R. H. Schiestl, *Nucleic Acids Res.* **23,** 2799 (1995).

[10] D. Lafontaine and D. Tollervey, *Nucleic Acids Res.* **24,** 3469 (1996).

TABLE I

PCR PRIMERS USED TO AMPLIFY TRANSFORMATION MODULES[a]

Primer	Purpose	Primer sequence
F1	Deletion	5′-(gene-specific sequence) CCAGCTGAAGCTTCGTACGC-3′[b]
F2	C-terminal tagging	5′-(gene-specific sequence) CGG ATC CCC GGG TTA ATT AA-3′[c]
F3	C-terminal tagging	5′-(gene-specific sequence) GC CAG CTG AAG CTT CGT ACG-3′[c]
R1	Deletion/C-terminal tagging	5′-(gene-specific sequence) GCATAGGCCACTAGTGGATC-3′[b,c]
F4	C-terminal tagging	5′-(gene-specific sequence) TCC ATG GAA AAG AGA AGA-3′[c]
R2	C-terminal tagging	5′-(gene-specific sequence) TACGACTCACTATAGGGCGA-3′[c]
F5	Promoter replacement (with or without tag)	5′-(gene-specific sequence) GAATTCGAGCTCGTTTAAAC-3′[d]
R3	Promoter replacement (no tag)	5′-(gene-specific sequence) **CAT** TTTGAGATCCGGGTTTT-3′[e]
R4	Promoter replacement (no tag)	5′-(gene-specific sequence) **CAT** TGTATATGAGATAGTTG-3′[e]
R5	Promoter replacement (3HA tagging)	5′-(gene-specific sequence) GCA CTG AGC AGC GTA ATC TG-3′[f]
R6	Promoter replacement (GST tagging)	5′-(gene-specific sequence) ACG CGG AAC CAG ATC CGA TT-3′[f]
R7	Promoter replacement (GFP tagging)	5′-(gene-specific sequence) TTT GTA TAG TTC ATC CAT GC-3′[f]
F6	Promoter replacement (no tag)	5′-(gene-specific sequence) CAGCTGAAGCTTCGTACGC-3′[d]
R8	Promoter replacement (no tag)	5′-(gene-specific sequence) **CAT** ATAGGCCACTAGTGGATCTG-3′[g]
F7	N-terminal GFP tagging	5′-(gene-specific sequence) CGGCCGCCAGGG-3′[h]
R9	N-terminal GFP tagging	5′-(gene-specific sequence) **CAT** TTT GTA CAA TTC ATC CAT ACC ATG-3′[h]

[a] The primer combinations used for various manipulations with different PCR module templates are indicated in Fig. 2. The reading frames used for primers to introduce protein tags that must be maintained are indicated by spaces between the letters. Indicated in boldface are the complements of start codons. Primers indicated here are designed to provide the maximum possible number of modifications of the templates in Fig. 2 with the minimum number of primers. Thus, in some cases the primers are different from those used in the original publications (see Fig. 2 for references) and may result in the PCR amplification of several additional nucleotides.

[b] For deletions, the gene-specific sequence of the forward primer typically is chosen to end just upstream of the start codon while those of the reverse primer end just downstream of the stop codon.

[c] For C-terminal tagging of full-length proteins, the gene-specific sequence of the forward primer typically is chosen to end just upstream of the stop codon while that of the reverse primer ends just downstream of the stop codon. For C-terminal truncation with tagging, the gene-specific sequence of the forward primer is chosen from the region where the truncation is desired. For C-terminal truncation without tagging, a stop codon must be included in the 3′ end of the gene-specific sequence in the forward primer.

[d] The gene-specific sequence can be chosen to end upstream of the start codon of the target gene resulting in a deletion of promoter sequences, if it seems unlikely to affect the expression of adjacent genes, or chosen to end just 5′ to the start codon of the target gene, resulting in no deletion.

[e] The gene-specific sequence is chosen to end just downstream of the target gene start codon. For N-terminal truncations without tagging, the gene-specific sequence of the reverse primer is chosen from the region where the truncation is desired.

[f] For N-terminal tagging of full-length proteins, the gene-specific sequence of the reverse primer is chosen to include the N-terminal codons of the target gene, with or without including its start codon.

[g] The gene-specific sequences are chosen to end just downstream of the stop codon of the target gene.

[h] For N-terminal GFP tagging, the gene-specific sequence of the forward primer is chosen to end just upstream of the start codon of the target gene while the gene-specific sequence of the reverse primer is to end at the start codon of the target gene, directing in-frame integration of sequence encoding GFP.

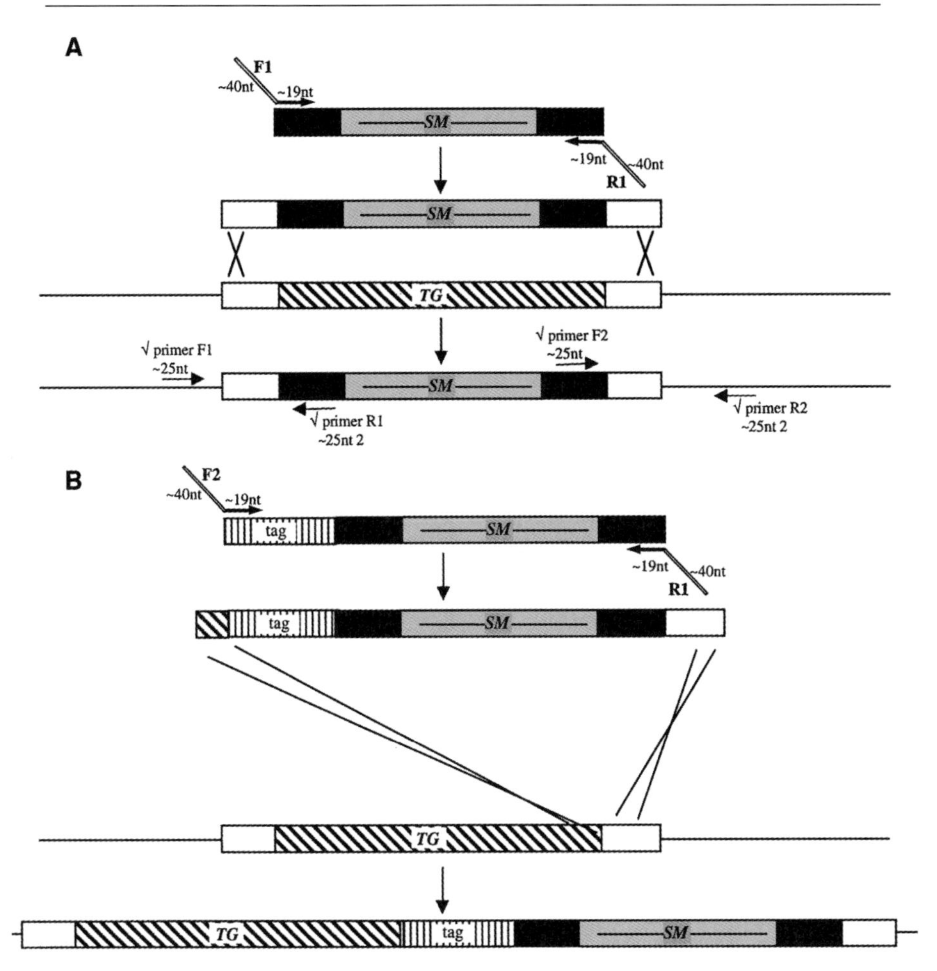

FIG. 1. PCR-mediated gene modification. (A) Gene deletion by SFH. The 5′ end of the forward PCR primer (F1) contains ~40 nucleotides (nt) of homology to the region just upstream of the start codon of the target gene while the 3′ end contains ~19 nt of 3′ homology to the plasmid template. The 5′ end of the reverse primer (R1) contains ~40 nt of homology to the region just downstream of the stop codon of the target gene while the 3′ end contains ~19 nt of homology to the plasmid template. After PCR amplification, the resulting PCR product is transformed into yeast and cells that have integrated the PCR module into the genome are selected on medium that selects for the presence of the selectable marker gene. Recombination into the target locus, directed by the flanking regions of the PCR product, is verified by two PCR reactions that check both junctions formed by the integration event. The 5′ junction is checked using an ~25 nt gene-specific primer (√ primer F1) which anneals upstream of the site of integration and an ~25 nt primer that anneals within the PCR module (√ primer R1). The 3′ junction is checked using a second ~25 nt gene-specific primer (√ primer R2) that anneals downstream of the site of integration and a second ~25 nt primer that anneals within the PCR module (√ primer F2). (B) C-terminal epitope/protein tagging by SFH. The 5′ end of the forward PCR primer (F2) contains ~40 nt homology to the region just upstream of the stop codon of the target gene

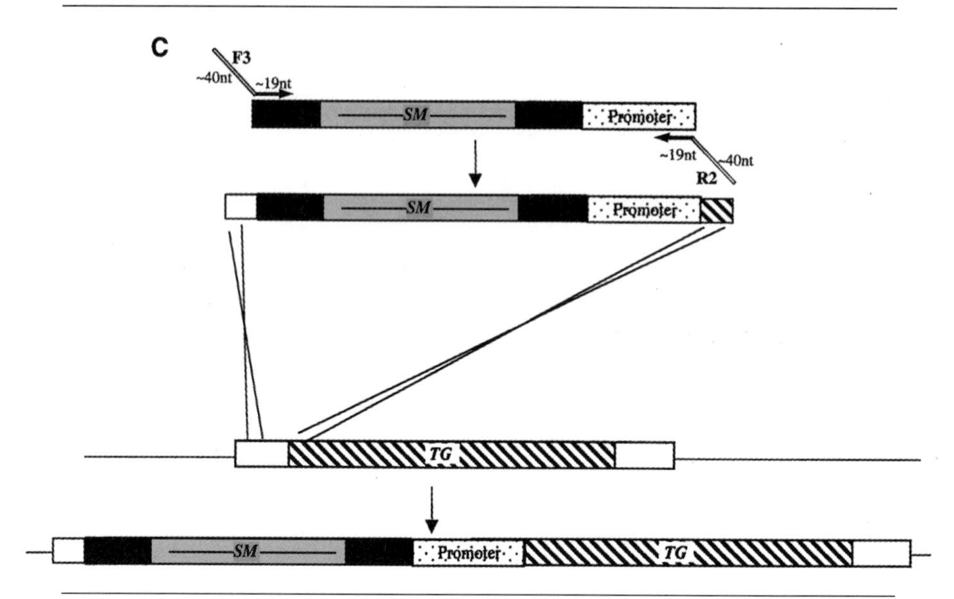

while the 3′ end contains ~19 nt homology to a plasmid template designed to maintain the reading frame between the target gene and the epitope/protein tag. The 5′ end of the reverse PCR primer (R1) contains ~40 nt of homology to the region just downstream of the stop codon of the target gene and ~19 nt of homology to the plasmid template. If a compatible template is used, the reverse primer can be identical to the reverse primer used for gene deletion (R1). Selection of transformants and verification of the site of integration are as described in (A) with appropriate modification of the gene-specific check primers. (C) Promoter replacement by SFH. The 5′ end of the forward PCR primer (F3) contains ~40 nt of homology to the region just upstream of the start codon of the target gene while the 3′ end contains ~19 nt of homology to the plasmid template. Alternatively, the forward primer can be designed to result in the deletion of nucleotides in the promoter region of the target gene. The 5′ end of the reverse PCR primer (R2) contains ~40 nt of homology to the region just downstream of the start codon of the target gene (and may also include the start codon) while the 3′ end contains ~19 nt of homology to the plasmid template. Selection of transformants and verification of the site of integration are as described in (A), with appropriate modification of the gene-specific check primers. (D) Gene deletion by LFH. Two primer pairs and three PCRs are required. For the first primer pair (F5′ and R5′), F5′ contains ~40 nt of homology to a region ~200–500 bp upstream of the start codon of the target gene. The 5′ end of R5′ contains ~30 nt of homology to a plasmid template upstream of the selectable marker gene while the 3′ end contains ~20 nt of homology to a region directly upstream of the start codon of the target gene. For the second primer pair (F3′ and R3′), the 5′ end of F3′ contains ~30 nt of homology to a plasmid template downstream of the selectable marker gene while the 3′ end contains ~20 nt of homology to a region directly downstream of the stop codon of the target gene. R3′ contains ~40 nt of homology to a region ~200–500 bp downstream of the stop codon of the target gene. Two separate PCRs are carried out using these primer pairs and, as template, the target gene on a plasmid or yeast chromosomal DNA. The resulting PCR products are purified away from the initial template and primers and used as primers in a third PCR reaction using the desired plasmid template. Selection of transformants and verification of the site of integration are as described in (A). SFH, short flanking homology; LFH, long flanking homology; SM, selectable marker gene; TG, target gene; F, forward; R, reverse. Intersecting lines represent homologous recombination events. Arrowheads indicate the 3′ ends of oligonucleotide primers.

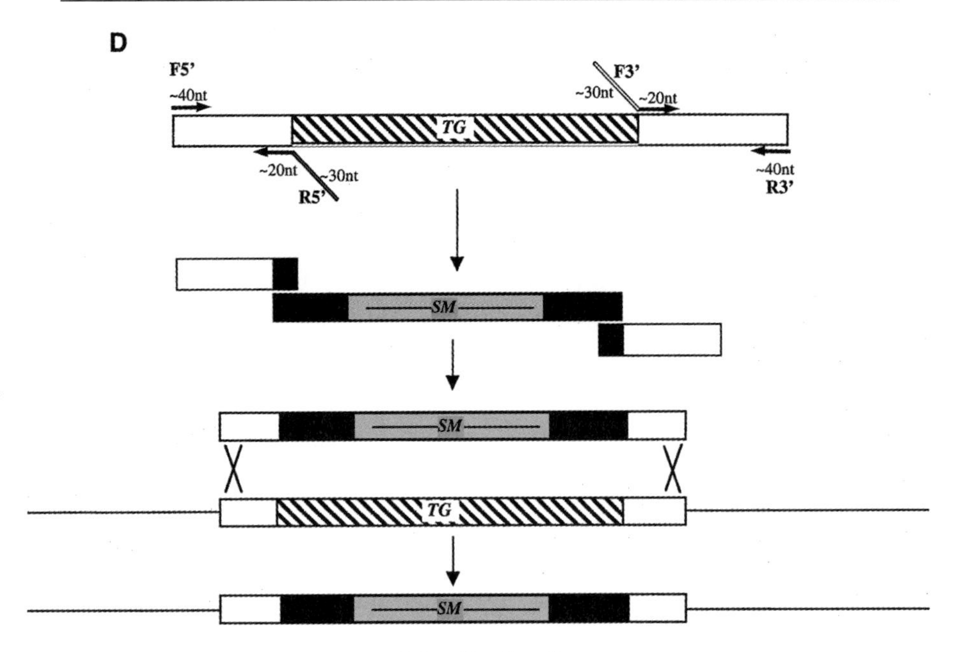

FIG. 1. (*Continued*)

below), SFH results in ~90–95% of the transformants carrying the PCR module integrated at the target locus.

Long Flanking Homology (LFH) Method

In some cases the frequency of integration of SFH PCR products is low, making it difficult to obtain the modified gene. Possible reasons for SFH failure include sequence heterogeneity between *S. cerevisiae* strains, closely related gene family members, or chromatin structure surrounding the target gene that limits access to the recombination machinery (see elsewhere in this volume[10a] for further discussion).

The long flanking homology (LFH) method, which increases the length of flanking homology to the yeast genome, has been extremely useful in modifying genes in *S. cerevisiae* and other organisms that are recalcitrant to modification by SFH.[11–15] The prevalent LFH approach[11] involves three PCR reactions (Fig. 1D).

[10a] M. Johnston, L. Riles, and J. H. Hegemann, *Methods in Enzymol.* **350,** [17], 2002 (this volume).
[11] A. Wach, *Yeast* **12,** 259 (1996).
[12] J. Nikawa and M. Kawabata, *Nucleic Acids Res.* **26,** 860 (1998).
[13] B. M. Pearson, Y. Hernando, and M. Schweizer, *Yeast* **14,** 391 (1998).
[14] C. Gonzalez, G. Perdomo, P. Tejera, N. Brito, and J. M. Siverio, *Yeast* **15,** 1323 (1999).
[15] M. D. Krawchuk and W. P. Wahls, *Yeast* **15,** 1419 (1999).

Generally, each of these PCR reactions is designed to amplify ~250–500 bp of yeast DNA, optimizing homologous integration. The two PCR products are then gel-purified away from the original template and primers and used as primers in a second PCR to amplify a PCR module, resulting in a PCR product with long regions of homology to the target locus flanking the PCR module to be integrated.

Elements of Plasmid Templates for PCR-Based Gene Modification

Selectable Markers

For efficient recombination of an SFH PCR product into the target locus, it is important that the selectable marker lacks significant homology to an endogenous marker gene, thus preventing homologous recombination with the *S. cerevisiae* marker gene. This problem can be circumvented in three ways. First, if the selectable marker used is an *S. cerevisiae* gene, a strain carrying a complete deletion of the corresponding chromosomal gene should be used as a recipient. Second, heterologous auxotrophic markers can be used that complement an *S. cerevisiae* auxotrophic mutation, but that have sufficient nucleotide divergence to avoid homologous recombination with the corresponding *S. cerevisiae* gene. Examples include the *S. kluyveri HIS3* gene and the *Kluveromyces lactis TRP1* and *URA3* genes, which complement mutations in the *S. cerevisiae HIS3*, *TRP1*, and *URA3* genes, respectively. Because these heterologous marker genes lack homology to the *S. cerevisiae* genome, they facilitate efficient integration into the target locus even in strains that do not carry complete deletions of the corresponding endogenous auxotrophic marker gene. Third, heterologous dominant drug resistance (DDR) markers have been developed for use in *S. cerevisiae*. These genes do not contain homology to the *S. cerevisiae* genome, and integration by homologous recombination is efficiently directed to the chromosomal target locus by the regions of flanking homology.

Both auxotrophic and DDR markers are widely used, with selection for auxotrophic markers being, in general, somewhat simpler than selection for DDR markers. However, DDR markers have two potential advantages over auxotrophic marker genes. First, DDR markers can be used for gene modification in strains in which auxotrophic markers are generally not available, such as many strains used in industry, or in which auxotrophic mutations affect the process being analyzed, such as sporulation or pseudohyphal growth. Second, auxotrophic mutations and marker genes can result in a nonneutral phenotype under some conditions,[16,17]

[16] F. Baganz, A. Hayes, D. Marren, D. C. Gardner, and S. G. Oliver, *Yeast* **13,** 1563 (1997).

[17] E. A. Winzeler, D. D. Shoemaker, A. Astromoff, H. Liang, K. Anderson, B. Andre, R. Bangham, R. Benito, J. D. Boeke, H. Bussey *et al., Science* **285,** 901 (1999).

suggesting caution in their use in certain applications. In contrast, the widely used *kan*^{r}-DDR marker gene that confers resistance to geneticin (G418) appears to be selectively neutral.[16] Although not yet demonstrated, this may also be the case for other DDR markers.

In some situations, it may be desirable to reuse a selectable marker gene to carry out multiple modifications. Two systems for recycling markers are available in the plasmids shown in Fig. 2. First, some selectable marker genes are flanked by 470 bp direct repeats derived from the 3′ end of the *Ashbya gossypii LEU2* gene.[7] Following integration into the genome, the selectable marker gene is spontaneously excised by recombination between the repeats at a rate of 10^{-3} to 10^{-4}.[7] Recombination events can be identified by replica plating to rich medium to detect loss of the marker gene, or, if the *URA3* counterselectable marker is used,[18] by replica plating to medium containing 5′-FOA.[19] One potential problem with this system is that the excision of the marker gene is unregulated and will occur spontaneously, even when the presence of the marker is desired as a means to track the modified locus, suggesting the advisability of maintaining continuous selection for the marker gene. Second, the Cre/*loxP* system has been used for excision of selectable marker genes after integration into the genome. In this system, marker genes are flanked by ~34 nt *loxP* sites. Introduction of a plasmid carrying the Cre recombinase gene into the modified strain and induction of Cre expression results in efficient removal of the selectable marker gene.[20–22] This approach is potentially very useful and can be used to remove multiple marker genes in a single excision step.[22] However, chromosomal rearrangements are possible during Cre-mediated excision, perhaps due to activation of cryptic genomic *loxP* sites[23] and/or recombination between *loxP* sites of multiple integrated PCR modules.[22] Thus, it is essential to check for chromosomal rearrangements following Cre-mediated marker excision. Figure 2A shows plasmid template modules for gene deletion (see also Ref. 10a).

C-Terminal and N-Terminal Epitope/Protein Tags

Protein tagging, like gene deletion, is a major application of PCR-based gene modification. Protein tagging has many uses, including subcellular protein localization, detection of a protein during biochemical fractionation, and purification of

[18] A. L. Goldstein, X. Pan, and J. H. McCusker, *Yeast* **15,** 507 (1999). Published Erratum: *Yeast* **15,** 1297 (1999).

[19] R. S. Sikorski and J. D. Boeke, *Methods Enzymol.* **194,** 302 (1991).

[20] U. Güldener, S. Heck, T. Fielder, J. Beinhauer, and J. H. Hegemann, *Nucleic Acids Res.* **24,** 2519 (1996).

[21] A. De Antoni and D. Gallwitz, *Gene* **246,** 179 (2000).

[22] D. Delneri, G. C. Tomlin, J. L. Wixon, A. Hutter, M. Sefton, E. J. Louis, and S. G. Oliver, *Gene* **252,** 127 (2000).

[23] B. Sauer, *J. Mol. Biol.* **223,** 911 (1992).

A

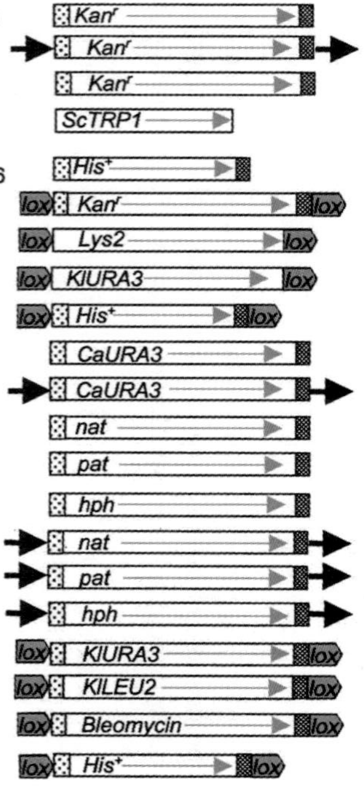

Plasmid	Reference	Primers[a]	~ PCR product size[b]	Sequence[c]	Recycleable?
pFA6a-kanMX6	61	F1 & R1	1622	i	No
pFA6-KanMX3	7	F1 & R1	~2465	n.a.	Yes, DR
pFA6-KanMX4	7	F1 & R1	1625	AJ002680	No
pFA6a-TRP1	61	F1 & R1	1094	i	No
pFA6a-His3MX6	61	F1 & R1	1509	i	No
pUG6	20	F1 & R1	1693	AF298793	Yes, Cre/Lox
pUG-Lys2	22	F1 & R1	~4748	n.a.	Yes, Cre/Lox
pUG-KlURA3	22	F1 & R1	~2048	n.a.	Yes, Cre/Lox
pUG-SpHis5	22	F1 & R1	~1548	n.a.	Yes, Cre/Lox
pAG60	18	F1 & R1	1628	j	No
pAG61	18	F1 & R1	2564	j	Yes, DR
pAG25	66	F1 & R1	1391	j	No
pAG29	66	F1 & R1	1370	j	No
pAG32	66	F1 & R1	1844	j	No
pAG35	66	F1 & R1	2327	j	Yes, DR
pAG31	66	F1 & R1	2306	j	Yes, DR
pAG34	66	F1 & R1	2780	j	Yes, DR
pJJH726	d	F1 & R1	1072	AF298788	Yes, Cre/Lox
pJJH727	d	F1 & R1	2508	AF298792	Yes, Cre/Lox
pUG66	d	F1 & R1	1264	AF298794	Yes, Cre/Lox
pUG27	d	F1 & R1	1534	AF298790	Yes, Cre/Lox

B

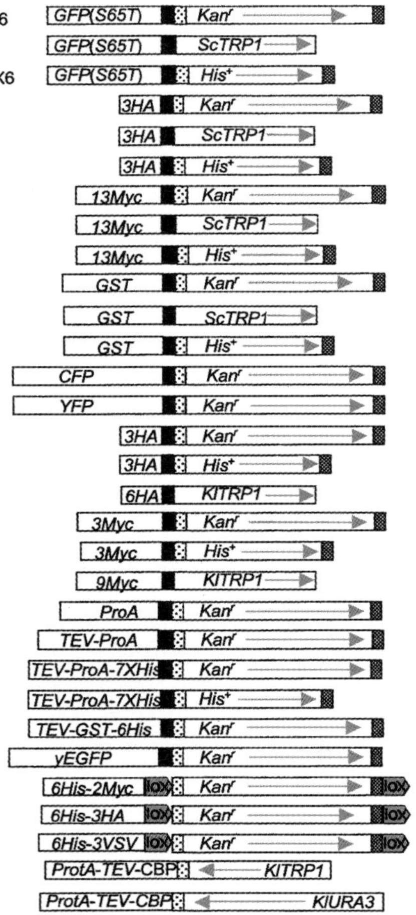

pFA6a-GFP(S65T)-kanMX6	GFP(S65T) Kan^r	61	F2 & R1	2356	AJ002682[i]	No
pFA6a-GFP(S65T)-TRP1	GFP(S65T) ScTRP1	61	F2 & R1	2013	[i]	No
pFA6a-GFP(S65T)-His3MX6	GFP(S65T) His$^+$	61	F2 & R1	2423	AJ002683[i]	No
pFA6a-3HA-kanMX6	3HA Kan^r	61	F2 & R1	1930	[i]	No
pFA6a-3HA-TRP1	3HA ScTRP1	61	F2 & R1	1407	AJ002682[i]	No
pFA6a-3HA-His3MX6	3HA His$^+$	61	F2 & R1	1817	[i]	No
pFA6a-13Myc-kanMX6	13Myc Kan^r	61	F2 & R1	2357	[i]	No
pFA6a-13Myc-TRP1	13Myc ScTRP1	61	F2 & R1	1834	[i]	No
pFA6a-13Myc-His3MX6	13Myc His$^+$	61	F2 & R1	2244	[i]	No
pFA6a-GST-KanMX6	GST Kan^r	61	F2 & R1	2297	[i]	No
pFA6a-GST-TRP1	GST ScTRP1	61	F2 & R1	1974	[i]	No
pFA6a-GST-His3MX6	GST His$^+$	61	F2 & R1	2384	[i]	No
pFA6a-CFP-kanMX6	CFP Kan^r	[e]	F2 & R1	2536	[k]	No
pFA6a-YFP-kanMX6	YFP Kan^r	[e]	F2 & R1	2380	[l]	No
pYM1	3HA Kan^r	[f]	F3 & R1	1950	n.a.	No
pYM2	3HA His$^+$	[f]	F3 & R1	1794	n.a.	No
pYM3	6HA KlTRP1	[f]	F3 & R1	1432	n.a.	No
pYM4	3Myc Kan^r	[f]	F3 & R1	1971	n.a.	No
pYM5	3Myc His$^+$	[f]	F3 & R1	1815	n.a.	No
pYM6	9Myc KlTRP1	[f]	F3 & R1	1600	n.a.	No
pYM7	ProA Kan^r	[f]	F3 & R1	2229	n.a.	No
pYM8	TEV-ProA Kan^r	[f]	F3 & R1	2253	n.a.	No
pYM9	TEV-ProA-7XHis Kan^r	[f]	F3 & R1	2265	n.a.	No
pYM10	TEV-ProA-7XHis His$^+$	[f]	F3 & R1	2109	n.a.	No
pYM11	TEV-GST-6His Kan^r	[f]	F3 & R1	2371	n.a.	No
pYM12	yEGFP Kan^r	[f]	F3 & R1	2390	n.a.	No
pU6H2myc	6His-2Myc lox Kan^r lox	21	F3 & R1	1784	AJ132965	Yes, Cre/Lox
pU6H3HA	6His-3HA lox Kan^r lox	21	F3 & R1	1808	AJ132966	Yes, Cre/Lox
pU6H3VSV	6His-3VSV lox Kan^r lox	21	F3 & R1	1826	AJ132967	Yes, Cre/Lox
pBS1479	ProtA-TEV-CBP KlTRP1	29	F4 & R2	1676	[m]	No
pBS1539	ProtA-TEV-CBP KlURA3	29	F4 & R2	2223	[n]	No

FIG. 2. (Continued)

C

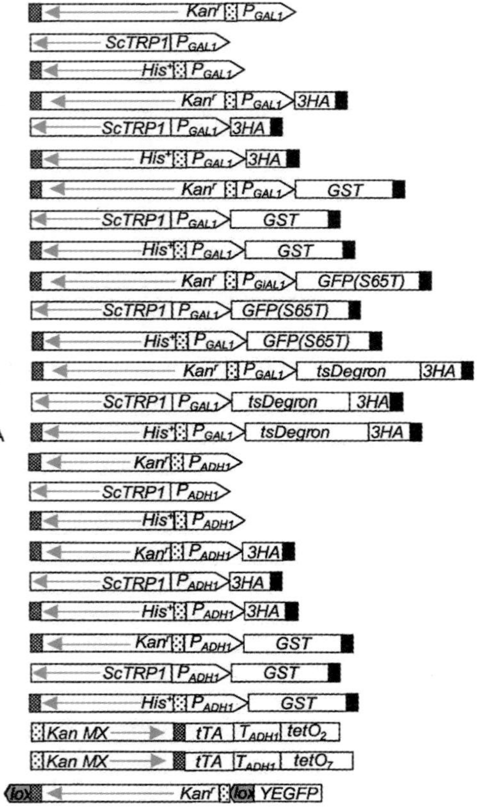

Plasmid	Cassette		Primers	Size		
pFA6a-kanMX6-PGAL1	Kan^r P_{GAL1}	61	F5 & R3	2081	i	No
pFA6a-TRP1-PGAL1	ScTRP1 P_{GAL1}	61	F5 & R3	1558	i	No
pFA6a-His3MX6-PGAL1	His^+ P_{GAL1}	61	F5 & R3	1968	i	No
pFA6a-kanMX6-PGAL1-3HA	Kan^r P_{GAL1} 3HA	61	F5 & R5	2201	i	No
pFA6a-TRP1-PGAL1-3HA	ScTRP1 P_{GAL1} 3HA	61	F5 & R5	1678	i	No
pFA6a-His3MX6-PGAL1-3HA	His^+ P_{GAL1} 3HA	61	F5 & R5	2088	i	No
pFA6a-kanMX6-PGAL1-GST	Kan^r P_{GAL1} GST	61	F5 & R6	2768	i	No
pFA6a-TRP1-PGAL1-GST	ScTRP1 P_{GAL1} GST	61	F5 & R6	2245	i	No
pFA6a-His3MX6-PGAL1-GST	His^+ P_{GAL1} GST	61	F5 & R6	2655	i	No
pFA6a-kanMX6-PGAL1-GFP	Kan^r P_{GAL1} GFP(S65T)	61	F5 & R7	2807	i	No
pFA6a-TRP1-PGAL1-GFP	ScTRP1 P_{GAL1} GFP(S65T)	61	F5 & R7	2284	i	No
pFA6a-His3MX6-PGAL1-GFP	His^+ P_{GAL1} GFP(S65T)	61	F5 & R7	2694	i	No
pFA6a-kanMX6-tsDegron-3HA	Kan^r P_{GAL1} tsDegron 3HA	g	F5 & R5	3034	o	No
pFA6a-TRP1-tsDegron-3HA	ScTRP1 P_{GAL1} tsDegron 3HA	g	F5 & R5	2511	o	No
pFA6a-His3MX6-tsDegron-3HA	His^+ P_{GAL1} tsDegron 3HA	g	F5 & R5	2921	o	No
pFA6a-kanMX6-PADH1	Kan^r P_{ADH1}	40	F5 & R4	1995	p	No
pFA6a-TRP1-PADH1	ScTRP1 P_{ADH1}	40	F5 & R4	1472	p	No
pFA6a-His3MX6-PADH1	His^+ P_{ADH1}	40	F5 & R4	1839	p	No
pFA6a-kanMX6-PADH1-3HA	Kan^r P_{ADH1} 3HA	40	F5 & R5	2115	p	No
pFA6a-TRP1-PADH1-3HA	ScTRP1 P_{ADH1} 3HA	40	F5 & R5	1592	p	No
pFA6a-His3MX6-PADH1-3HA	His^+ P_{ADH1} 3HA	40	F5 & R5	1959	p	No
pFA6a-kanMX6-PADH1-GST	Kan^r P_{ADH1} GST	40	F5 & R6	2682	p	No
pFA6a-TRP1-PADH1-GST	ScTRP1 P_{ADH1} GST	40	F5 & R6	2159	p	No
pFA6a-His3MX6-PADH1-GST	His^+ P_{ADH1} GST	40	F5 & R6	2526	p	No
pCM224	Kan MX tTA T_{ADH1} $tetO_2$	34	F6 & R8	3906	q	No
pCM225	Kan MX tTA T_{ADH1} $tetO_7$	34	F6 & R8	4124	r	No
pYGFPgn	lox Kan^r lox YEGFP	h	F7 & R9	2339	n.a.	Yes, Cre/Lox

FIG. 2. Plasmid template modules for PCR-based gene modification. Diagram indicates elements incorporated into the PCR product using suggested forward (F) and reverse (R) primers indicated in the Primer column and Table I. In all cases, the forward primer is oriented with its 3′ end toward the right, while the reverse primer is oriented with its 3′ end toward the left. (A) PCR modules for gene deletion. (B) PCR modules for C-terminal tagging/truncation. (Stippled boxes, *A. gossypii TEF* promoter; hatched boxes, *A. gossypii TEF* terminator; black boxes, *S. cerevisiae ADH1* terminator; black arrowheads, 470 bp direct repeats from *A. gossypii LEU2* gene; gray pentagons, *loxP* sites. Gray arrows indicate the directions of transcription of the selectable marker genes. Selectable marker genes: *Kan*^{r}*, E. coli* kanamycin resistance gene; *ScTRP1, S. cerevisiae TRP1* gene; *His+*, *Kluverymyces lactis HIS3* gene; *CaURA3, Candida albicans URA3* gene; *nat,* nourseothricin resistance gene; *pat,* bialophos resistance gene; *hph*, hygromycin B resistance gene; *Bleomycin,* bleomycin resistance gene; *KlTRP1, K. lactis TRP1* gene; *KlURA3, K. Lactis URA3* gene. Epitope/protein tags: GFP(S65T), S65T variant of green fluorescent protein; HA, hemagglutinin epitope; Myc, c-*myc*-encoded epitope; GST, *Schistosoma japonicum* glutathione *S*-transferase; CFP, cyan spectral variant of GFP; yEGFP, yeast enhanced GFP; YFP, yellow spectral variant of GFP. ProA, IgG-binding domain of *S. aureus* protein A; TEV, tobacco etch virus protease cleavage site; His_6 and His_7, multiple histidine residues; VSV, vesicular stomatis virus glycoprotein epitope; CBP, calcium binding protein. P_{GAL1}, *S. cerevisiae GAL1* promoter; P_{ADH1}, truncated *S. cerevisiae ADH1* promoter. *References* (a) See Table I for primer sequences. (b) PCR product sizes are indicated assuming use of the primers indicated in Table I with exactly 40 nt of homology to the target gene. (c) Indicated are the GenBank accession numbers, where available. If not available, then an alternative source is indicated. n.a., not available, contact the authors of the original publication. (d) U. Güldener, J. H. Heinishch, D. Voss, G. Koehler, and J. H. Hegemann. A new set of heterologous gene deletion markers for budding yeast (unpublished, 2000). (e) http://depts.washington.edu/%7Eyeastrc/p_p_home.htm (f) M. Knop, K. Siegers, G. Pereira, W. Zachariae, B. Winsor, K. Nasmyth, and A. Scheibel, *Yeast* **15,** 963 (1999). (g) Rick Heil-Chapdelaine, unpublished, 2002. (h) B. Prein, K. Natter, and S. D. Kohlwein, *FEBS Lett.* **485,** 29 (2000). (i) http://opbs.okstate.edu/~longtine/homepage.html (submitted to GENBANK). (j) http://www.duke.edu/web/microlabs/mccusker/Resources/sequence.html (k) http://depts.washington.edu/%7Eyeastrc/fm_home1.htm (l) http://depts.washington.edu/%7Eyeastrc/fm_home2.htm (m) http://www-db.embl-heidelberg.de/jss/servlet/de.embl.bk.wwwTools.GroupLeftEMBL/ExternalInfo/seraphin/pBS1479.html (n) http://www-db.embl-heidelberg.de/jss/servlet/de.embl.bk.wwwTools.GroupLeftEMBL/ExternalInfo/seraphin/pBS1539.html (o) Contact Rick Heil-Chapdelaine, rhchapde@stcloudstate.edu, Department of Biological Sciences, St. Cloud State University, MS 230, 740 Fourth St. N., St. Cloud, MN 56301. (p) Contact Doug Demarini, Douglas_J_DeMarini@sbphrd.com, Department of Comparative Genetics, SmithKline Beecham Pharmaceuticals, 709 Swedeland Road, King of Prussia, PA 19406. (q) http://www.uni-frankfurt.de/fb15/mikro/euroscarf/data/pCM224.txt (r) http://www.uni-frankfurt.de/fb15/mikro/euroscarf/data/pCM225.txt

the tagged protein and associated proteins by affinity methods. Plasmid templates for PCR-based modification are available for introducing into the N terminus and C terminus of a target protein epitope tags, affinity tags, and green fluorescent protein (GFP) tags (Figs. 2B–2C).

Epitopes, short peptides of ~10 amino acids that are recognized specifically by antibodies, are usually used in tandem repeats resulting in high-affinity recognition by anti-epitope antibodies. Templates are available for PCR-mediated tagging with the widely used hemagglutinin (HA) epitope[24,25] and Myc epitope[26,27] as well as the vesicular stomatis virus glycoprotein (VSV) epitope. Commercially available antibodies that recognize these epitopes are useful in Western blotting, immunofluorescence, and immunopurification of the tagged protein and of associated proteins. Commonly used antibodies include the mouse monoclonal anti-HA antibodies, 12CA5 (Roche, Indianapolis, IN) and HA.11 (Covance Research Products, Richmond, CA), and the rat monoclonal antibody 3F10 (Roche); the anti-myc mouse monoclonal antibody 9E10 (Covance); and the mouse monoclonal anti-VSV antibody P5D4 (Roche). Additional monoclonal and polyclonal antibodies directed against these epitopes are also available from a variety of suppliers.

Affinity tags are proteins or protein domains that can be purified using commercially available matrices. Templates are available for PCR-mediated tagging with several affinity tags, including glutathione *S*-transferase (GST), which allows purification over glutathione-agarose beads, the immunoglobulin G (IgG)-binding domains of protein A (ProtA) from *Staphylococcus aureus,* which allows purification over IgG Sepharose, His_6 and His_7 (hexa- and heptahistidine) tags, which allow purification over nickel/NTA agarose, and the calmodulin-binding peptide (CBP), which allows purification over calmodulin beads. Affinity tags have proven useful for the single-step purification of tagged proteins and associated proteins. Adding flexibility to the use of affinity tags is the commercial availability of anti-affinity tag antibodies that are useful for immunological detection of tagged proteins.

Templates are also available that permit simultaneous tagging with two different affinity/epitope tags, allowing two-step affinity purification and (likely) a more pure protein or protein-complex preparation than can be obtained by a single purification step. In some cases, a cleavage site for tobacco etch protease (TEV) is included, allowing enzymatic cleavage to separate the tagged protein from the

[24] J. Field, J.-I. Nikawa, D. Broek, B. MacDonald, L. Rodgers, I. A. Wilson, R. A. Lerner, and M. Wigler, *Mol. Cell. Biol.* **8,** 2159 (1988).

[25] M. Tyers, G. Tokiwa, and B. Futcher, *EMBO J.* **12,** 1955 (1993).

[26] G. I. Evan, G. K. Lewis, G. Ramsay, and J. M. Bishop, *Mol. Cell. Biol.* **5,** 3610 (1985).

[27] S. Munro and H. R. B. Pelham, *Cell* **48,** 899 (1987).

affinity matrix under native conditions. The tandem affinity purification (TAP) protocol, which uses tandem Protein A and CBP affinity tags separated by a TEV site, shows great promise for efficient purification of tagged proteins and associated proteins.[28,29] The use of mass spectrometry to identify copurified proteins, even when multiple proteins are present, makes affinity purification an extremely powerful technique to identify members of protein complexes.[29a]

Green fluorescent protein (GFP) has been exceptionally useful for investigating the dynamics of protein localization in living cells (reviewed in Refs. 4, 30, and 31). Templates are available for N-terminal and/or C-terminal tagging with GFP and the spectral derivatives, yellow fluorescent protein (YFP) and cyan fluorescent protein (CFP). Although often providing lower signals than GFP, YFP and CFP have enough separation between their excitation and emission spectra that they can be used for simultaneous, real-time imaging of two tagged proteins (see http://depts.washington.edu/%7Eyeastrc/fm_home.htm). YFP and CFP are also useful for characterizing the *in vivo* proximity of two tagged proteins using fluorescence energy transfer (FRET) (e.g., Refs. 32 and 33, and see http://depts.washington.edu/%7Eyeastrc/fm_home.htm).

Promoters

Gene promoters can be replaced using PCR-mediated gene modification. Promoter replacement is useful to overexpress or underexpress a target gene by regulated or constitutive transcription. Shown in Fig. 2C are plasmid templates for PCR-mediated promoter replacement with the carbon-source-regulated *GAL1* promoter and with tetracycline/doxycycline-regulated promoters with two ($tetO_2$) or seven ($tetO_7$) operators.[34,35] The *GAL1* promoter is transcribed at high levels in the presence of galactose and its transcription is dramatically repressed in the presence of glucose (reviewed in Ref. 2). It has been widely used for protein overexpression to facilitate biochemical or phenotypic analysis or subcellular localization. It is important to note, however, that protein overexpression may result in mislocalization and other artifacts. Because it is repressible, the *GAL1* promoter is also useful to examine the phenotype of cells lacking a protein of interest, including those

[28] F. Caspary, A. Shevchenko, M. Wilm, and B. Seraphin, *EMBO J.* **18,** 3463 (1999).

[29] G. Rigaut, A. Shevchenko, B. Rutz, M. Wilm, M. Mann, and B. Seraphin, *Nat. Biotechnol.* **17,** 1030 (1999).

[29a] A. Pandey and M. Mann, *Nature* **405,** 837 (2000).

[30] P. M. Conn (ed.), *Methods Enzymol.* **302** (1999).

[31] K. F. Sullivan and S. E. Kay (eds.), *Methods Cell Biol.* **58** (1999).

[32] M. Damelin and P. A. Silver, *Mol. Cell* **5,** 133 (2000).

[33] M. C. Overton and K. J. Blumer, *Curr. Biol.* **10,** 341 (2000).

[34] G. Belli, E. Gari, M. Aldea, and E. Herrero, *Yeast* **14,** 1127 (1998).

[35] G. Belli, E. Gari, L. Piedrafita, M. Aldea, and E. Herrero, *Nucleic Acids Res.* **26,** 942 (1998).

encoded by essential genes, following repression of transcription by the addition of glucose and dilution/degradation of the protein. However, because of the high rate of transcription from the *GAL1* promoter under inducing conditions, the regulated protein often accumulates to very high levels. This may result, in some cases, in the slow and unsynchronous depletion of the protein following repression of the *GAL1* promoter, which can hamper subsequent phenotypic analyses. To alleviate this potential problem, plasmid templates are available for placing genes under *GAL1* promoter control with simultaneous introduction of an N-terminal Arg-DHFR "ts degron" and 3 HA epitopes.[36,37] The Arg-DHFR ts degron results in a tagged protein that is stable at 23° but is rapidly ubiquitinated and degraded at 37°. Thus, a tagged protein is rapidly removed following repression of transcription by the addition of glucose and a shift of the culture to 37° (e.g., Refs. 38 and 39). Although less widely used than the *GAL1* promoter, the tetracycline/tetracycline operator (*TetO*) system also allows regulated gene expression and phenotypic analysis of essential genes.[34]

Plasmids are also available for promoter replacement with a truncated *ADH1* promoter that provides a high level of constitutive transcription with, or without, N-terminal tagging.[40] Unlike the *GAL1* promoter, the *ADH1* promoter is highly expressed in most laboratory media. Thus, using the *ADH1* promoter eliminates the need to shift carbon sources, which alters metabolism and gene expression[41] and may complicate comparisons between cells overexpressing and underexpressing the target gene.

Variations on Method

Internal Tagging and Introduction of Mutations

In some situations, N- or C-terminal tagging is not useful, either because the function of the target protein is disrupted by the N- or C-terminal tag or because promoter replacement is not desired. A method described by Schneider *et al.*[42] allows PCR-mediated tagging with 3HA or 3Myc epitopes at internal sites. The PCR reaction results in a linear DNA fragment containing flanking regions with

[36] R. J. Dohmen, P. Wu, and A. Varshavsky, *Science* **263,** 1273 (1994).

[37] F. Levy, J. A. Johnston, and A. Varshavsky, *Eur. J. Biochem.* **259,** 244 (1999).

[38] J. A. Tercero, K. Labib, and J. F. Diffley, *EMBO J.* **2,** 2082 (2000).

[39] R. D. Gardner, A. Poddar, C. Yellman, P. A. Tavormina, M. C. Monteagudo, and D. J. Burke, *Genetics* **157,** 1493 (2001).

[40] D. J. DeMarini, E. M. Carlin, and G. P. Livi, *Yeast* **18,** 723 (2001).

[41] A. P. Gasch, P. T. Spellman, C. M. Kao, O. Carmel-Harel, M. B. Eisen, G. Storz, D. Botstein, and P. O. Brown, *Mol. Biol. Cell* **11,** 4241 (2000).

[42] B. L. Schneider, W. Seufert, B. Steiner, Q. H. Yang, and A. B. Futcher, *Yeast* **11,** 1265 (1995).

homology to the yeast target locus provided by the primers and direct repeats encoding the epitope tag that flank a *URA3* selectable marker. After transformation into a haploid strain, the *URA3* transformants are plated on medium containing 5′-FOA, selecting for cells that have undergone recombination between the directly repeated epitope tag sequences, resulting in an in-frame insertion of sequences encoding the epitope tag into the target gene.[42–44] This method has the advantage of allowing epitope tagging at any site in the target gene. However, in this method the initial transformant contains a selectable marker insertion in the target gene, which will likely inactivate the target gene. Thus, an essential gene cannot be tagged by this method in a haploid strain. This difficulty cannot be circumvented by simply transforming a diploid strain, because in this case all the cells that are resistant to 5′-FOA have undergone gene conversion of the modified allele by the wild-type allele present on the homologous chromosome.[42] Therefore, the use of this method is currently restricted to nonessential genes.

Similar PCR-based methods that use primers with direct repeats of regions of the target gene that contain a mutation can be used to create site-specific mutations in genomic genes.[42,45] As discussed above for internal epitope tagging, the initial step in these methods requires disruption of the target gene in a haploid cell, limiting their use to nonessential genes. Another, somewhat more complicated method[46] allows the introduction of mutations into chromosomal genes without gene disruption (see [15], this volume.) Thus, this method can be used to mutagenize essential genes, as long as the mutation itself does not cause lethality.

Truncations

By appropriate choice of the region of homology for the forward and reverse primers to the target locus, the PCR-mediated gene modification can be used to create truncation alleles encoding proteins with either N-terminal or C-terminal truncations, with or without tagging. Sequential N-terminal and C-terminal truncations can be used to express internal regions of the target gene. These truncation alleles can be very useful in structure–function studies, including identification of regions involved in subcellular localization, protein–protein interactions, and protein function.

[43] J. F. Charles, S. L. Jaspersen, R. L. Tinker-Kulberg, L. Hwang, A. Szidon, and D. O. Morgan, *Curr. Biol.* **8,** 497 (1998).

[44] S. Prinz, E. S. Hwang, R. Visintin, and A. Amon, *Curr. Biol.* **8,** 750 (1998).

[45] F. Langle-Rouault and E. Jacobs, *Nucleic Acids Res.* **23,** 3079 (1995).

[46] N. Erdeniz, U. H. Mortensen, and R. Rothstein, *Genome Res.* **7,** 1174 (1997).

Incorporating Elements in Primers (e.g., NLS, NES, Tags)

Possible modifications are not limited to those available on the plasmid templates. Primers can be designed to encode a variety of functional motifs, such as nuclear localization or export signals. The only limitations are that the motif must be relatively short, so that it can be included in an oligonucleotide of a reasonable length, and that the motif must be able to function at the N or C terminus of a protein.

Uses of PCR-Mediated Recombination on Episomal Plasmids

Although we have concentrated our discussion on the modification of the genome, PCR and homologous recombination are also useful for cloning DNA into yeast episomal plasmids and for modification of plasmid-borne genes. As described above for genomic loci, plasmid-borne genes can be modified by transforming PCR modules into a yeast strain with an episomal plasmid that carries the target gene followed by selection for integration of the PCR module. We have used this approach to introduce epitope tags into plasmid-borne genes, to replace one selectable marker with another, and to delete an ORF from a high-copy library plasmid to determine which ORF was responsible for the observed complementing activity.

Specific modifications of plasmid-borne genes can also be made without incorporating a selectable marker. When a plasmid that carries a yeast selectable marker and ARS is linearized by digestion with a restriction enzyme and transformed into yeast, it is unstable and rapidly degraded. However, if the linearized plasmid is cotransformed with a linear donor DNA fragment that contains homology to the sequences that flank the break in the plasmid, homologous recombination will result in a circular plasmid that has incorporated the donor DNA fragment.[47–50] PCR can be used to generate the donor fragment, allowing the rapid cloning of almost any gene from yeast or another organism into a yeast vector. This efficient approach was used in cloning almost all of the yeast ORFs into vectors for two-hybrid analysis.[51] PCR-mediated approaches also have been useful in cloning novel alleles generated by random mutagenesis.[52,53]

Domain-exchange chimeras of related proteins can be made if a plasmid-borne target gene contains an unique restriction enzyme site within the domain to be

[47] S. Kunes, H. Ma, K. Overbye, M. S. Fox, and D. Botstein, *Genetics* **115,** 73 (1987).

[48] H. Ma, S. Kunes, P. J. Schatz, and D. Botstein, *Gene* **58,** 201 (1987).

[49] C. Fusco, E. Guidotti, and A. S. Zervos, *Yeast* **15,** 715 (1999).

[50] C. K. Raymond, T. A. Pownder, and S. L. Sexson, *Biotechnology* **26,** 134 (1999).

[51] J. R. J. Hudson, E. P. Dawson, K. L. Rushing, C. H. Jackson, D. Lockshon, D. Conover, C. Lanciault, J. R. Harris, S. J. Simmons, R. Rothstein, and S. Fields, *Genome Res.* **7,** 1169 (1997).

[52] D. Muhlrad, R. Hunter, and R. Parker, *Yeast* **8,** 79 (1992).

[53] C. F. Kostrub, E. P. Lei, and T. Enoch, *Nucleic Acids Res.* **26,** 4783 (1998).

exchanged.[54,55] In this case, PCR primers with homology flanking the target gene domain of interest are designed to amplify the domain of interest of the related donor gene. Cotransformation of the resulting PCR product with the restriction enzyme-digested, plasmid-borne target gene followed by homologous recombination yields a circular plasmid carrying the domain-exchanged, chimeric gene. Small mutations also can be introduced using a similar approach if the plasmid-borne target gene contains an unique restriction enzyme site. In this case, PCR primers, one of which contains the desired mutation, are designed to amplify a region which spans the restriction enzyme site of the target gene. Cotransformation of the resulting PCR product with the restriction enzyme-digested, plasmid-borne target gene followed by homologous recombination yields a circular plasmid carrying the mutated target gene.

For all of these approaches, it is possible that the yeast transformant will carry both modified and unmodified plasmids. Thus, prior to retransformation into yeast and subsequent analysis, it is important to isolate the correctly modified plasmid. Total yeast DNA, prepared as described below, is transformed into *Escherichia coli,* and plasmids isolated from individual *E. coli* colonies are analyzed to identify the correct recombinant plasmid.

Methods

Step 1: Choosing a Yeast Strain

A strain should be selected by considering: (1) suitability for the process being studied (e.g., if studying sporulation, it is advisable to use a strain that sporulates efficiently), (2) efficiency of transformation, and (3) flexibility in selection with the presence of multiple auxotrophic marker genes (preferably deletion alleles) if these mutations are not detrimental to the process being studied. If strain choice is open, then it is reasonable to consider using strains that carry multiple deletion alleles of auxotrophic marker genes to provide a choice of auxotrophic markers for modifications or introduction of episomal plasmids. Deletions of nearly all of the ~6000 *S. cerevisiae* ORFs are available in the BY4743 strain background[56] (see Ref. 10a). The availability of this valuable resource makes the BY4743 strain worthy of consideration.

In many situations it is desirable to use a specific strain background either because of experimental necessity (e.g., industrial strains) or because of the desire to use a strain background which has been used in previous work, thus allowing

[54] K. R. Oldenburg, K. T. Vo, S. Michaelis, and C. Paddon, *Nucleic Acids Res.* **25,** 451 (1997).

[55] F. R. Papa, A. Y. Amerik, and M. Hochstrasser, *Mol. Biol. Cell* **10,** 741 (1999).

[56] C. B. Brachmann, A. Davies, G. J. Cost, E. Caputo, J. Li, P. Hieter, and J. D. Boeke, *Yeast* **14,** 115 (1998).

direct comparison with previous results without the possible complications due to differences in strain genetic backgrounds. If necessary, the utility of such strains for PCR-based modifications can be improved by using plasmids to create deletion alleles of auxotrophic marker genes.[56–58]

In most cases, it is advisable to carry out modifications in a diploid strain, resulting in a strain heterozygous for the modification. This allows the generation of deletions of essential genes and for subsequent tetrad analysis to verify proper integration and the lack of detectable collateral mutations. In addition, some modifications result in a phenotype that becomes suppressed during continued growth, possibly due to the accumulation of suppressor mutations or altered expression of genes that compensate for the defect. Thus, it is advisable to characterize the phenotype of a strain carrying a modified gene shortly after tetrad dissection, and to verify that any phenotype does not become suppressed during continued growth.

Step 2: Choosing a Template

A plasmid template should be chosen that incorporates the desired modification into the genome and that uses an appropriate selectable marker. If introducing an epitope or protein tag, a major concern is the possible effect of the tag on the localization and/or function of the target protein. Thus, the target protein sequence should be examined to identify possible N- or C-terminal domains whose function would likely be disrupted by the introduction of a tag, precluding tagging at that location. Such motifs can be identified by experimental evidence, by homology with characterized homologs, and/or by computational analyses (e.g., PROSITE[59] http://www.expasy.ch/prosite/ and SMART[60] http://smart.embl-heidelberg.de/).

We have found that in the majority of cases both N-terminal and C-terminal tagging result in apparently fully functional proteins. *A priori,* it is reasonable to suspect that the introduction of a smaller tag is less likely to disrupt function of the tagged protein than the introduction of a larger tag. However, in our experience we have not found a clear correlation between the size of the epitope/protein tag and the effects of the tag on protein function. Thus, if one tag at a particular location affects protein function, another tag may not detectably alter protein function.

Another consideration is whether the investigator wants the gene expressed under control of the endogenous promoter or under control of a heterologous regulated or constitutive promoter (e.g., if protein overexpression is desired). See Ref. 4 for further discussion. The modification templates also can be used in combination. For example, it is possible to introduce a C-terminal tag and verify function of the tagged protein expressed at endogenous levels and, subsequently,

[57] J. Horecka and Y. Jigami, *Yeast* **15,** 1769 (1999).

[58] K. Replogle, L. Hovland, and D. H. Rivier, *Yeast* **15,** 1141 (1999).

[59] K. Hofmann, P. Bucher, L. Falquet, and A. Bairoch, *Nucleic Acids Res.* **27,** 215 (1999).

[60] J. Schultz, R. R. Copley, T. Doerks, C. P. Ponting, and P. Bork, *Nucleic Acids Res.* **28,** 231 (2000).

place the gene under control of an inducible promoter for high-level expression of the tagged protein.[61]

Step 3: Primer Design

In most cases, the design of the primers is dictated by the desired modification. We have successfully modified many regions of the genome basing our primer design solely on that which gives the optimal modification without regard for predicted melting temperature (Tm) or secondary structure of the oligonucleotides. However, modifications should be made in regions that minimize the risk of affecting the adjacent gene expression. For SFH, we routinely use 59 nt forward and reverse primers with 40 bp of homology to the yeast genome and 19 bp of homology to the plasmid template produced at the 100 n*M* scale (IDT, Coralville, IA) that have not been gel purified. For gene deletion, we usually design primers that will result in a precise deletion of the gene from start codon to stop codon (for further discussion of primer design for gene deletions, see Ref. 10a). For C-terminal protein tagging, the forward primer is designed to maintain the reading frame of the tag and the reading frame of the target gene (see Table I for the reading frame of the tags) and the reverse primer contains homology to the region immediately 3′ to the stop codon of the target gene. Homologous recombination occurs efficiently with little or no separation of the homologous regions, and the possibility of affecting the expression of flanking genes is minimized if regions of homology to the yeast genome in the forward and reverse primers are close, or adjacent, to each other. For N-terminal tagging and promoter substitution, the forward primer can be designed to anneal immediately 5′ to the ORF, minimizing the potential for disrupting the expression of an upstream gene. If desired, and if it appears unlikely to affect the expression of an upstream gene, the forward primer can be designed to anneal further 5′ to the ORF, resulting in a deletion of endogenous promoter sequences. The reverse primer is designed to be homologous to the 5′ nucleotides of the ORF.

Step 4: PCR Module Amplification

To amplify the desired PCR module, PCR reactions are done essentially as described previously[61] using the Expand High Fidelity system (Roche). Each 100 μl PCR reaction contains 10 μl Expand buffer with 17.5 m*M* $MgCl_2$, 0.2 m*M* each dNTP, 10 μg BSA, and 2 μM of each primer and 0.1 μg plasmid DNA template. The tubes are placed on ice, and 0.6 μl Expand enzyme mixture is added to each tube. Reactions are run for 20 cycles of 1 min at 94°, 1 min at 55°, and 1 min/kb of the desired product at 68°. The 20 cycles are followed by a 10-min extension at

[61] M. S. Longtine, A. I. McKenzie, D. J. Demarini, N. G. Shah, A. Wach, A. Brachat, P. Philippsen, and J. R. Pringle, *Yeast* **14,** 953 (1998).

68°. PCR-induced mutations are minimized by using the proof-reading Expand enzyme and by limiting the PCR to 20 cycles. To prevent the possible predominance of a PCR-induced mutation that may occur early in one of the PCR reactions, the products of 6–8 100 μl PCR reactions are pooled. The pooled PCR products are precipitated, resuspended in 30 μl H_2O, and stored at −20°.

Step 5: Preparation and Storage of Yeast Cells Competent for Transformation

To prepare yeast cells for DNA transformation, a starter culture is grown to saturation in rich YPD medium or, if the strain carries a plasmid, in the appropriate selectable medium.[62] For our strain background (YEF473[63]), we typically start with cells grown on solid medium and scrape a large inoculum (~250 μl cell volume) into a 25 ml YPD liquid culture, which is grown to saturation (~48 hr at 23°). For non-temperature-sensitive strains, the saturated culture is diluted ~1500-fold (e.g., 350 μl into 500 ml YPD medium), and the culture is incubated with shaking at 30°. For temperature-sensitive strains, the entire 25 ml saturated culture is added to 500 ml of YPD medium, and the culture is incubated with shaking at 23°. When the culture reaches an OD_{600} of ~1 (~24 hr after inoculation), the cells are collected by centrifugation at 5000 rpm at 20° for 5 min and resuspended in 75 ml sterile 0.1 *M* lithium acetate pH 7.5, and again collected by centrifugation at 5000 rpm at 20° for 5 min. The supernatant is removed, and the cells are resuspended in an appropriate volume of 0.1 *M* lithium acetate pH 7.5, such that the final concentration of cells is 10 OD_{600} per 100 μl 0.1 *M* lithium acetate. An equal volume of sterile 30% glycerol is mixed with the cells, the mixture is divided into 500 μl aliquots in microfuge tubes, and the aliquots are placed directly in a −70° freezer, where the cells remain competent for many months.

Many procedures are available for preparing transformation competent yeast cells (e.g., Refs. 64 and 64a). The above method is convenient because cells can be prepared ahead of time and stored at −70°. The ready availability of competent cells saves significant time compared to preparing competent cells each time a transformation is desired. Although the transformation efficiency of cells obtained by this procedure is not as high as that of some other procedures, the resulting cells are sufficiently competent to obtain >20 transformants per transformation.

Step 6: Transformation and Selection of Transformants

A 500-μl aliquot of frozen cells is thawed at 23°, and resuspended in 1 ml 0.1 *M* lithium acetate pH 7.5. Cells are collected by a low speed centrifugation in a microcentrifuge for ~30 sec. The supernatant is removed, and the cells are resuspended in 250 μl 0.1 *M* lithium acetate pH 7.5. Twelve μl fish sperm DNA

[62] C. Guthrie and G. R. Fink (eds.), *Methods Enzymol.* **194** (1991).

[63] E. Bi and J. R. Pringle, *Mol. Cell. Biol.* **16,** 5264 (1996).

[64] D. Gietz, A. St. Jean, R. A. Woods, and R. H. Schiestl, *Nucleic Acids Res.* **20,** 1425 (1992).

[64a] R. D. Gietz and R. A. Woods, *Methods in Enzymol.* **350,** [4], 2002 (this volume).

(Roche), 20 μl dimethyl sulfoxide (DMSO), and 10 μl of the PCR DNA (step 4) are added to the resuspended cells. To this mixture is added 600 μl of a filter-sterilized stock containing 40% polyethylene glycol (molecular weight 3350, Sigma, St. Louis, MO) in 0.1 *M* lithium acetate pH 7.5, 10 m*M* Tris pH 7.5, and 1 m*M* EDTA pH 8.0. A negative-control tube containing no PCR-amplified DNA also should be prepared.

The tubes are incubated with gentle rolling for 1 hr at either 30° for wild-type strains or 23° for Ts- strains. Wild-type strains are heat shocked in a 42° water bath for 15 min, and Ts- strains are heat shocked in a 37° water bath for 15 min. Following heat shock, 800 μl sterile H_2O is added, and the cells collected by a brief centrifugation at low speed in a microfuge. The supernatant is removed, and the cells resuspended in 200 μl sterile H_2O. Transformants expressing selectable markers that complement auxotrophic mutations are selected by plating 100 μl on each of two plates of appropriate selection medium.[62] Yeast transformants expressing the Kan^r gene are selected essentially as described previously.[7,11,61] Briefly, the transformed cells are spread on 2 YPD plates (100 μl per plate). These plates are incubated for 2 to 3 days at 23° or 30°, as appropriate, and replica plated to YPD-G418 plates containing 200 μg/ml G418 (Life Technologies, Gaithersburg, MD). The G418 is added after autoclaving the medium. (The amount of G418 added may need to be adjusted, depending on the activity indicated by the supplier.) To identify stable transformants, after 2 to 3 days of incubation these plates are replica plated to a fresh YPD-G418 plate. Stable transformants are then streaked to YPD-G418 plates and DNA is prepared for PCR confirmation of integration at the target locus (step 7). Previously, G418 selection has worked well on rich medium, but only poorly on standard synthetic medium. However, a modification using monosodium glutamate instead of ammonium sulfate[65] allows efficient G418 selection on synthetic medium. Protocols for selecting transformants using the *nat, pat,* and *hph* DDR markers are described elsewhere (Ref. 66 and http://www.duke.edu/web/microlabs/mccusker/resources.html).

Step 7: Confirmation of Transformants

To determine if the PCR module is integrated at the desired location, both junctions formed upon integration of the PCR module should be verified by PCR using genomic DNA as template (Ref. 3, and see Ref. 10a). For these two PCRs, primer pairs include one primer that anneals within the integrated PCR module and a target-gene specific primer that anneals upstream or downstream of the site of integration (Fig. 1A). If the primers that anneals within the PCR module are homologous to a region present in many modules (e.g., the T_{TEF} or P_{TEF} regions), they can then be used as primers to check different modifications.

[65] T. H. Cheng, C. R. Chang, P. Joy, S. Yablok, and M. R. Gartenberg, *Nucleic Acids Res.* **28,** E108 (2000).

[66] A. L. Goldstein and J. H. McCusker, *Yeast* **15,** 1541 (1999).

Total yeast DNA is prepared essentially as described previously.[67] Briefly, transformants are grown as an ~15 mm-diameter patch on a selective plate. Approximately 50–100 μl of cells are removed from the plate with a toothpick, and the cells are resuspended by swirling in 200 μl of "DNA extraction buffer" (DEB; 1% SDS, 100 m*M* NaCl, 10 m*M* Tris pH 8.0, 1 m*M* EDTA pH 8.0, and 0.2% Triton X-100). Washed glass beads, 500 μl (425–600 microns, Sigma), are added, followed by the addition of 200 μl of a 1 : 1 mix of phenol : chloroform. The tubes are then vortexed at high speed for 4 min. A microcentrifuge tube attachment that holds 12 tubes is useful for preparing multiple samples (Daigger, Vernon Hills, IL). Next, 0.2 ml $T_{10}E_1$ (10 m*M* Tris pH 7.5, 1 m*M* EDTA pH 8.0) is added, and the mixture is vortexed at high speed for 30 sec. The tubes are centrifuged at ~12,000 rpm at 20° for 10 min, and 300 μl of the supernatant is removed to a clean tube containing 750 μl ethanol and briefly vortexed. The yeast DNA is collected by centrifugation at ~12,000 rpm at 20° for 10 min, and the supernatant removed by aspiration. A second brief centrifugation is used to collect any remaining liquid attached to the sides of the tube, and removed by aspiration. The cap of the microfuge tube is left open at 23° for ~10 min, allowing the DNA pellet to dry. The DNA is then resuspended in 40 μl H_2O and stored at −20°.

In our experience, a non-proof-reading polymerase is usually more robust for genomic PCR than a proof-reading polymerase. PCR to identify potential properly integrated transformants is carried out as described[61] in a 50 μl reaction containing 2 μM of each primer, 2 m*M* $MgSO_4$, 5 μg of BSA, 0.2 m*M* of each dNTP, 0.5 μl *Taq* DNA polymerase (Promega, Madison, WI), and 1 μl (~0.5 μg) of yeast genomic DNA. PCR reactions are run for 36 cycles of 1 min at 94°, 1 min at 55°, and 1 min/kb of the desired product at 72°. The 36 cycles are followed by a 10-min extension at 72°.

For transformants that appear to contain the PCR module integrated at the desired locus, as determined by PCR verification of the junctions, sporulation and tetrad analysis should be done to verify that the selectable marker gene segregates 2 : 2 and that any associated phenotypes cosegregate with the selectable marker. This will help avoid complications due to collateral mutations that occasionally occur (see Ref. 10a) and verify that the PCR module is integrated at a single locus. Rarely, during integration of the PCR module there can be a duplication of the wild-type allele (see Ref. 10a). Thus, in addition to verifying both integration junctions, it is advisable to use PCR to verify that the wild-type allele is absent in haploid segregants that carry the selectable marker gene.

If working with an allele encoding a tagged protein, or one expressed under control of a heterologous promoter, it is important to use any available phenotypic assays to verify that the modified target gene provides full function. In addition, it is

[67] C. S. Hoffman and F. Winston, *Gene* **57,** 267 (1987).

useful to determine that a phenotype is present in more than one independent transformant, and, finally, if the mutant phenotype is recessive, to demonstrate that any phenotypes ascribed to the mutation are rescued by the wild-type gene on a plasmid.

[26] ChIP-chip: A Genomic Approach for Identifying Transcription Factor Binding Sites

By CHRISTINE E. HORAK and MICHAEL SNYDER

Introduction

Transcription factors are key regulatory proteins that can influence the expression of hundreds of genes in response to a particular environmental condition or internal cue. The collective set of genes regulated by a transcription factor (or gene targets) therefore defines the state of a cell and can determine cell fate. Among the 6200 predicted proteins in the yeast *Saccharomyces cerevisiae,* there are about 300 transcription factors.[1] Approximately 85% of the yeast transcription factors have been characterized to some extent and some of these are known to play a critical role in cell cycle initiation, pheromone response, mating type switching, pseudohyphal growth, and nutrient and stress response.[1] For most yeast transcription factors no targets are known and for those factors that have been studied, usually 10 targets or fewer have been found. Thus, identifying the gene targets of transcription factors is an important problem, not only for defining the function of a particular factor, but for characterizing the molecular response of a cell.

Efforts have been made to comprehensively identify transcription factor targets by examining transcript profiles in the absence of a particular transcription factor or in a gain-of-function mutant using microarray analysis.[2–8] Although this approach

[1] M. C. Costanzo, J. D. Hogan, M. E. Cusick, B. P. Davis, A. M. Fancher, P. E. Hodges, P. Kondu, C. Lengieza, J. E. Lew-Smith, C. Lingner, K. J. Roberg-Perez, M. Tillberg, J. E. Brooks, and J. I. Garrels, *Nucleic Acids Res.* **28,** 73 (2000).

[2] J. L. DeRisi, V. R. Iyer, and P. O. Brown, *Science* **278,** 680 (1997).

[3] P. Sudarsanam, V. R. Iyer, P. O. Brown, and F. Winston, *Proc. Natl. Acad. Sci. U.S.A.* **97,** 3364 (2000).

[4] J. DeRisi, B. van den Hazel, P. Marc, E. Balzi, P. O. Brown, C. Jacq, and A. Goffeau, *FEBS Lett.* **470,** 156 (2000).

[5] C. W. Yun, T. Ferea, J. Rashford, O. Ardon, P. O. Brown, D. Botstein, J. Kaplan, and C. C. Philpott, *J. Biol. Chem.* **275,** 10709 (2000).

[6] T. J. Lyons, A. P. Gasch, L. A. Gaither, D. Botstein, P. O. Brown, and D. J. Eide, *Proc. Natl. Acad. Sci. U.S.A.* **97,** 7957 (2000).

[7] C. Gross, M. Kelleher, V. R. Iyer, P. O. Brown, and D. R. Winge, *J. Biol. Chem.* **275,** 32310 (2000).

[8] S. Chu, J. DeRisi, M. Eisen, J. Mulholland, D. Botstein, P. O. Brown, and I. Herskowitz, *Science* **282,** 699 (1998).

0076-6879/02 $35.00

is comprehensive as it surveys the whole yeast genome, it is not direct. Alterations in transcript levels may result from secondary effects due to the mutation itself or through other downstream regulatory events. Also, some of the gene targets may not be identified because they do not exhibit changes in expression as a result of compensatory mechanisms or redundant transcription factors that exist within the cell. Direct methods for target identification that examine transcription factor binding to promoter sequences upstream of a gene, such as *in vivo* footprinting[9] or PCR (polymerase chain reaction) analysis of DNA coimmunoprecipitated with a transcription factor,[10,11] are not comprehensive. Typically, only a handful of promoters are examined at a time with these methods.

We have developed an approach in yeast that will comprehensively identify genomic sequences directly bound by transcription factors. This approach has been used to successfully identify targets of the yeast transcription factors SBF (Swi4-Swi6 cell cycle box binding factor) and MBF (*MluI* cell cycle box binding factor) in Iyer *et al.*[12] and Gal4 and Ste12 in Ren *et al.*[13] We have termed this approach ChIP-chip because it involves chromatin immunoprecipitation of the factor of interest and its associated DNA and DNA chip (microarray) probing with the immunoselected DNA (Fig. 1). Briefly, protein–DNA complexes are fixed *in vivo* with formaldehyde and lysed, and the lysate is sonicated to shear DNA. The transcription factor of interest is purified by immunoprecipitation; the associated DNA is extracted and then amplified and labeled for hybridization to a yeast intergenic array. Two important reagents for this method are specific immunoprecipitating antibodies and a yeast intergenic microarray.

Antibodies and Tagged Transcription Factors

The quality of the antibody used for ChIP-chip experiments is important because of the sensitivity of microarray analysis. Nonspecific cross-reactions with chromatin will result in a high level of background hybridization to the microarray. Antibodies should be tested for their ability to specifically immunoprecipitate the transcription factor of interest before being used for chromatin immunoprecipitation.

Very few antibodies against yeast transcription factors are available, but the ChIP-chip approach can be generalized for all yeast transcription factors by using proteins that have been tagged with hemagglutinin, myc, or any other epitope for which there are high quality antibodies available. The proteins can be randomly

[9] J. D. Axelrod and J. Majors, *Nucleic Acids Res.* **17,** 171 (1989).

[10] V. Orlando, H. Strutt, and R. Paro, *Methods* **11,** 205 (1997).

[11] M. H. Kuo and C. D. Allis, *Methods* **19,** 425 (1999).

[12] V. R. Iyer, C. E. Horak, C. S. Scafe, D. Botstein, M. Snyder, and P. O. Brown, *Nature* **409,** 533 (2001).

[13] B. Ren, F. Robert, J. J. Wyrick, O. Aparicio, E. G. Jennings, I. Simon, J. Zeitlinger, J. Schreiber, N. Hannett, E. Kanin, T. L. Volkert, C. J. Wilson, S. P. Bell, and R. A. Young, *Science* **290,** 2306 (2000).

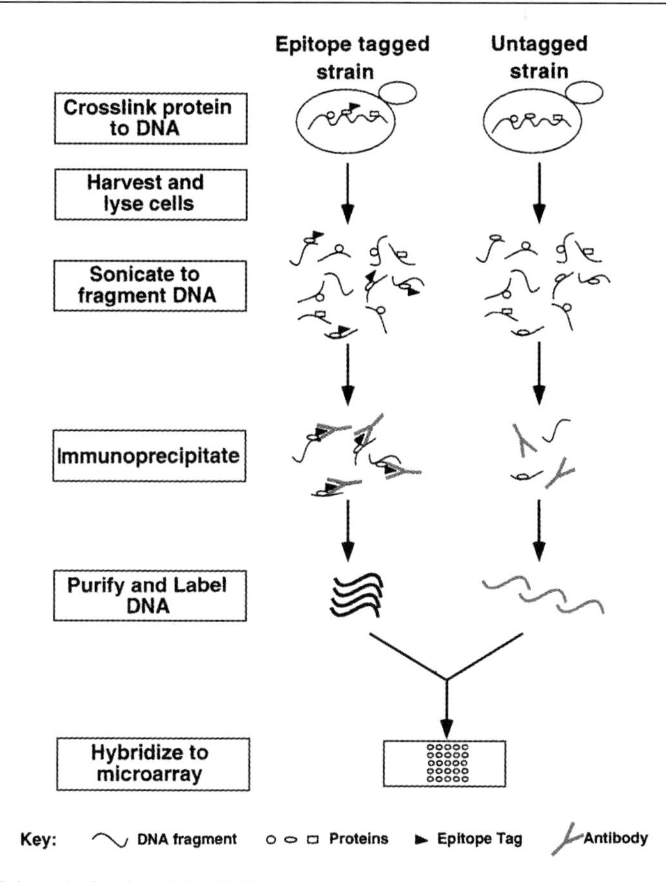

FIG. 1. Schematic drawing of the ChIP-chip technique for epitope-tagged DNA-binding proteins.

tagged using the transposon tagging method described in Ross-MacDonald *et al.*[14] or selectively tagged by the homologous recombination method described by Schneider *et al.*[15] These tagged transcription factors can be immunoprecipitated with commercial antibodies raised against the epitope.

Yeast Intergenic Microarray

Another key component of the ChIP-chip technique is the yeast intergenic microarray. Intergenic regions are the regions between open reading frames (ORFs) or other non-ORF features (rRNA coding sequences, tRNA coding sequences, Ty

[14] P. Ross-MacDonald, A. Sheehan, G. S. Roeder, and M. Snyder, *Proc. Natl. Acad. Sci. U.S.A.* **94,** 190 (1997).

[15] B. L. Schneider, W. Seufert, B. Steiner, Q. H. Yang, and A. B. Futcher, *Yeast* **11,** 1265 (1995).

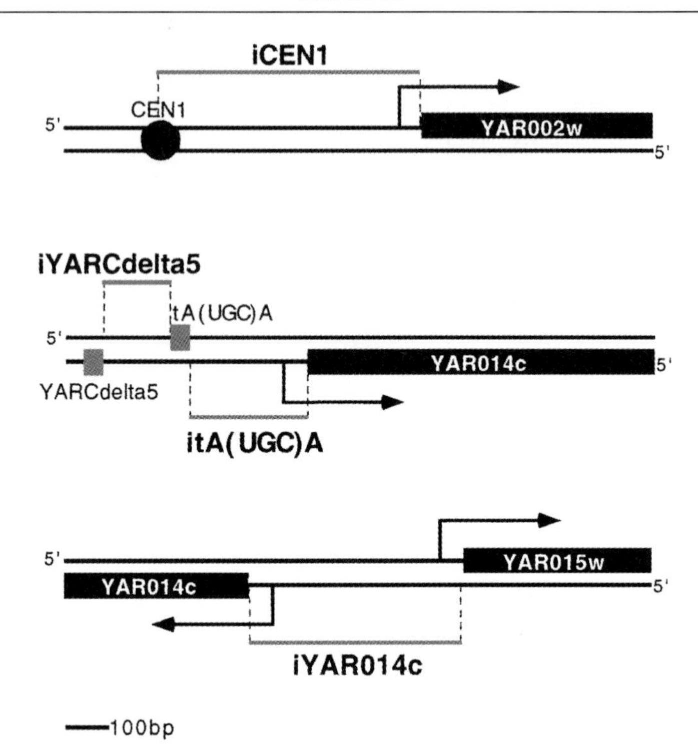

FIG. 2. Examples of intergenic regions on *S. cerevisiae* Chromosome I. Intergenic regions are segments of DNA between ORFs or non-ORF features (such as Ty insertions, centromeric and telomeric sequences, and rRNA and tRNA coding sequences). These sequences are given the name of the SGD designated ORF or non-ORF feature to its left prefixed by the letter "i." The intergenic region iCEN1 is to the left of the centromere CEN1. Intergenic region iYARCdelta5 is to the left of the long tandom repeat of the Ty insertion YARCdelta5. Intergenic region itA(UGC)A is to the left of the tRNA coding sequence for tA(UGC)A, and iYAR014c is to the left of the YAR014c ORF.

insertions and centromeric elements) (Fig. 2). Intergenic regions are designated by prefixing the SGD (*Saccharomyces* Genome Database) designated ORF name or non-ORF feature to its left with an "i." There are approximately 6700 intergenic elements in yeast that range in size from 50 to 5000 bp. Since the intergenic regions generally contain the gene promoter sequences and transcription factor binding sites, use of this type of microarray will provide the most complete survey of transcription factor-associated DNA. However, depending on the extent of DNA shearing, there may be enough sequence overlap between the immunoprecipitated DNA fragments and the affixed ORF probes to allow a traditional ORF microarray to be used for hybridization and target identification.[12] The protocols herein describe a method for yeast transcription factor target identification using the yeast intergenic array.

Protocols

Constructing DNA Microarray of Yeast Intergenic Regions

Constructing microarrays requires an arrayer in a room or chamber that is temperature and humidity controlled. Arrayers are available through several robotics companies, such as Engineering Services Inc. (ESI, Toronto, Canada) or Genemachines (Genemachines, San Carlos, CA), or can be built using the MGuide accessed through the http://cmgm.stanford.edu/pbrown Web site. This Web site also provides links to other general microarray protocols. Further information regarding microarray experiments is provided elsewhere in this volume.[15a]

The yeast intergenic regions can be PCR amplified with primers available through Research Genetics (Huntsville, AL). Regions that are larger than 1500 bp are subdivided into two or more fragments for amplification purposes. The intergenic regions are amplified and arrayed onto slides as described below. These methods are based on previously developed procedures.[2]

1. The forward and reverse primers for PCR amplification from Research Genetics are premixed into 65 96-well plates. Maintain the reactions in the same 96-well arrangement. For a 50-μl reaction, the PCR conditions are as follows: 25 μl 2× Mastermix (Qiagen, Valencia, CA), 4 μl primer mix, 2 μl yeast genomic DNA (100 ng/μl), and 19 μl water. The thermal cycling conditions are as follows: 94° for 5 min, 40 cycles of 92° for 10 sec, 55° for 30 sec, 72° for 2 min, and a final extension step at 72° for 7 min. PCR products are subsequently analyzed by gel electrophoresis on a 1.5% agarose gel (see below).

2. Precipitate the amplification products with 100 μl 100% ethanol and centrifugation at 3000 rpm for 1 hr at 4°. Decant the supernatant, dry the pellets, and resuspend them in 100 μl water. Plates of PCR products can be stored indefinitely at −70°.

3. Transfer DNA from the 65 96-well plates to 19 384-well plates (Whatman Polyfiltronics, Clifton, NJ) for arraying purposes. Using a Tecan robot, aliquot 4 μl of DMSO (dimethyl sulfoxide) into each well of the 384-well plates and then add 4 μl of each amplified intergenic region.

4. Prepare a control plate, which includes repeated 4 μl aliquots of water (no DNA), heterologous DNA (for example, a PCR-amplified region of human β-globin gene) as a control for hybridization specificity, and sonicated total yeast genomic DNA as a positive control. Each control element is mixed with 4 μl DMSO in a 384-well plate.

5. Spot DNA using an arrayer and microspotting pins onto slides. Pins are available from Array-It (TeleChem International, Sunnyvale, CA). Slides coated with γ-aminopropylsilane from Corning Microarray Technology (CMT, Corning, NY), are recommended. While spotting, the temperature and humidity should be maintained at 25° and 50%, respectively.

[15a] A. P. Gasch, *Methods Enzymol.* **350,** [23], 2002 (this volume).

6. Cross-link DNA to the slide surface using a Stratalinker UV cross-linker (Stratagene, La Jolla, CA) at 200 mJ. Store the slides at room temperature in a dry place.

Troubleshooting for PCR of Intergenic Regions. Twenty percent of each PCR product should be analyzed by agarose gel electrophoresis. PCR products should be scored on the basis of their size and extent of amplification. Amplification product sizes are available through the Research Genetics ftp address (ftp://ftp.resgen.com). Incorrectly sized fragments should be noted. Failed and faint PCR reaction products should be repeated using a lower PCR annealing temperature (50°). Correctly sized products from this second round of PCR should be placed in their original position in the 96-well plate after ethanol precipitation.

Alternative Procedures for Microarraying. Polylysine-coated slides can also be used for microarraying, but we have found that the polylysine coat (and adhered spots of DNA) may tear away from the glass surface of the slide rendering the microarray useless. Other methods also involve microspotting with DNA resuspended in 3× SSC, but we have found that DMSO helps prevent evaporation of the solution from the source plate and from the pins during the arraying process and results in more DNA adhering to the slide.

Immunoprecipitation of DNA Associated with Transcription Factor

Protein–DNA interactions can be reversibly cross-linked with formaldehyde,[16,17] and these *in vivo* chromatin complexes can be isolated by immunoprecipitation.[18,19] A schematic of this chromatin immunoprecipitation procedure is shown in Fig. 1. DNA is immunoprecipitated from strains carrying a tagged version of a transcription factor, and because the antibody may cross-react with other components of chromatin, DNA should also be purified from an isogenic untagged strain. If an antibody against the native protein is used, perform a mock immunoprecipitation from a strain lacking the gene for that transcription factor. The following protocol is adapted from methods described by Aparicio,[20] Hecht and Grunstein,[18] Kuo and Allis,[11] and Orlando *et al.*[10] Steps of this procedure may need to be optimized for particular proteins.

1. Grow a 50 ml yeast culture to mid-log phase (OD_{600} of 0.5). The protein–DNA complexes are fixed *in vivo* by adding the yeast culture to 1.4 ml 37% (v/v)

[16] M. J. Solomon and A. Varshavsky, *Proc. Natl. Acad. Sci. U.S.A.* **82,** 6470 (1985).

[17] M. J. Solomon, P. L. Larsen, and A. Varshavsky, *Cell* **53,** 937 (1988).

[18] A. Hecht and M. Grunstein, *Methods Enzymol.* **304,** 399 (1999).

[19] S. Strahl-Bolsinger, A. Hecht, K. Luo, and M. Grunstein, *Genes Dev.* **11,** 83 (1997).

[20] O. M. Aparicio, *in* "Current Protocols in Molecular Biology" (F. M. Ausubel, R. Brent, R. E. Kingston, D. D. Moore, J. G. Seidman, J. A. Smith, and K. Struhl, eds.), p. 21.3.1. John Wiley and Sons, New York, 1999.

formaldehyde solution in a 50-ml conical tube to a final concentration of 1% (v/v) formaldehyde. Fix the cells for 15 min at room temperature, occasionally inverting the tube.

2. Cross-links are quenched by adding glycine to the fixed culture at a final concentration of 125 m*M* and incubating at room temperature for 5 min with occasional inversion.

3. Harvest cells by centrifugation at 3000 rpm for 5 min at room temperature. Discard the supernatant and wash the cells with 10 ml ice-cold 1× TBS (150 m*M* NaCl, 20 m*M* Tris-HCl, pH 7.6). Spin down the cells again at 3000 rpm for 5 min at 4°, resuspend them in 1 ml ice-cold 1× TBS, and transfer them to a 2 ml microcentrifuge tube (the 2 ml microcentrifuge tube allows for maximum bead movement while vortexing). Centrifuge cells at 14,000 rpm for 1 min at 4° and then discard the supernatant. Cells can be kept on ice for several hours at this point. Alternatively, cell pellets can be frozen in liquid nitrogen and stored at −70°.

4. Resuspend the cell pellet in 400 μl ice-cold lysis buffer [0.1% (w/v) deoxycholic acid, 1 m*M* EDTA, 50 m*M* HEPES–KOH, pH 7.5, 140 m*M* NaCl, 1% (v/v) Triton X-100] with protease inhibitors. Add an equal volume of cubic zirconium beads and vortex the mixture at maximum force for 10 min at 4°. Let the samples sit on ice for 15 min before transferring the lysate to a fresh 2 ml microcentrifuge tube. Wash the cubic zirconium beads with 400 μl ice-cold lysis buffer with protease inhibitors and vortex for 1 min. Add the wash to the lysate.

5. Shear chromatin using a Branson 250 Sonifier with a microtip at a power setting of 1.5 and a 100% duty cycle. Sonicate extracts for three 10 sec pulses. In between 10 sec pulses, let samples sit on ice for at least 2 min. This should shear chromatin to a final average size of 500 bp. Sonicators can be calibrated to yield the desired DNA length (see below).

6. Clarify the lysate by centrifugation at 14,000 rpm for 15 min at 4°. Transfer the supernatant to a fresh 1.5 ml microcentrifuge tube.

7. Preclear the extracts to remove fragments of DNA that may adhere to the protein A-Sepharose beads in the immunoprecipitation step. Incubate the extracts with a bed volume of 30 μl protein A-Sepharose beads (Pierce, Rockford, IL) on a rotation wheel for 1 hr at 4°. Centrifuge at 7000 rpm for 2 min and transfer the lysate to a fresh tube.

8. To immunoprecipitate the transcription factor, the primary antibody against the protein of interest or against the epitope tag is added to the lysate (see below for determining antibody amounts). Incubate lysate with the antibody on ice for 3 hr and then add 30 μl bed volume of protein A-Sepharose beads. Incubate on a rotating wheel for 1 hr at 4°. Centrifuge samples for 2 min at 7000 rpm at 4°.

9. Wash the immunoprecipitates four times for 5 min each on a rotation wheel at 4°. Subsequent to each wash, the beads are collected by centrifugation at 7000 rpm for 2 min at 4° and the wash buffer is discarded. The first wash is with 1 ml lysis buffer. The second wash is with 1 ml lysis buffer-500 (0.1% deoxycholic

acid, 1 mM EDTA, 50 mM HEPES–KOH, pH 7.5, 500 mM NaCl, 1% Triton X-100). The beads are then washed with 1 ml LiCl/detergent solution [0.5% (w/v) deoxycolic acid, 1 mM EDTA, 250 mM LiCl, 0.5% (v/v) Nonidet P-40 (NP-40), 10 mM Tris-HCl, pH8] and finally with 1 ml 1× TBS.

10. Elute the immunocomplexes from the sepharose beads by adding 100 μl 1% (w/v) SDS/1× TE and incubating at 65° for 10 min. Spin the beads down briefly and transfer the eluate to a fresh tube. Wash the beads with 150 μl 0.67% (w/v) SDS/1× TE and then add the wash to the eluate.

11. In order to reverse the cross-links, incubate the immunoprecipitates and the total extract material at 65° for at least 6 hr. It is often convenient to extend this step overnight.

12. To further remove protein from the DNA, treat with proteinase K.[16] Add 250 μl proteinase K solution (20 μg glycogen/100 μg proteinase K/1× TE) to each sample and incubate for 2 hr at 37°.

13. DNA is purified by phenol–chloroform extraction and ethanol precipitation as described by Orlando and Paro.[21] Add 55 μl 4 M LiCl and 500 μl phenol/chloroform/isoamyl alcohol [25 : 24 : 1 (v/v)] to each sample and then vortex vigorously for 1 min. Separate phases by centrifugation at 14,000 rpm for 10 min at room temperature. Transfer the aqueous phase to a fresh tube and add 1 ml 100% ethanol to precipitate the DNA. Mix the samples and centrifuge at 14,000 rpm for 20 min at 4°. Discard the supernatant and dry the pellet. Resuspend the DNA in 1× TE and store at −20°.

Troubleshooting. The above procedure should yield approximately 10 ng of immunopurified DNA. The DNA can be quantitated with a fluorimeter or with ethidium bromide spotting and assayed for enrichment of transcription factor bound sequences by a PCR assay[10,11] or microarray hybridization (see below). If too little DNA is purified or the DNA is not enriched for a subset of sequences, several parameters of the chromatin immunoprecipitation procedure can be altered.

a. The extent of cross-linking can be adjusted by changing the concentration of formaldehyde, the time of incubation with the cross-linking agent, or the temperature of cross-linking.[22] The extent of cross-linking is critical and can depend on the sensitivity of the protein. Too much cross-linking may mask epitopes and too little cross-linking may lead to incomplete fixation.[22]

b. The extent of sonication should also be monitored since chromatin fragments that are too large may not be readily immunoprecipitated. Set aside a 50-μl aliquot of the lysate before immunoprecipitating for determining the average size of sheared DNA. Add 200 μl 1% (w/v) sodium dodecyl sulfate (SDS)/1× TE

[21] V. Orlando and R. Paro, *Cell* **75,** 1187 (1993).

[22] V. Orlando, *TIBS* **25,** 99 (2000).

(10 mM Tris-HCl, pH 7.5, 1 mM EDTA) to this total cell lysate and then treat this sample in parallel with immunoprecipitates through steps 11–13. This total lysate DNA can then be analyzed by gel electrophoresis to determine the extent of shearing. A 2% agarose gel or 6% polyacrylamide gel is recommended for this analysis. If DNA fragments larger than 2 kb are observed, then increase the number of sonication pulses.

c. The amount of antibody used for immunoprecipitation is another critical parameter. Preliminary immunoprecipitation experiments should be performed to determine the appropriate amount of antibody to be used for specific purification of the protein of interest. For the mouse monoclonal anti-hemagglutinin (HA) antibody 12CA5 (BabCo, Richmond, CA), a 1 : 200 dilution of the serum works well for immunoprecipitation of most proteins tagged with three copies of the hemagglutinin epitope. To ensure that the cross-linking is not rendering the protein refractory to immunoprecipitation, 10% of material eluted from the beads in step 10 can be analyzed by SDS–PAGE and Western blotting. The material should be boiled in sample buffer for 20 min before SDS–PAGE.

Labeling Probe DNA

Chromatin immunoprecipitation yields a very small amount of DNA (about 1–10 ng), so all of the precipitated DNA must be used in the labeling steps that follow. Because so little DNA is immunoprecipitated, a robust PCR amplification is required both before labeling and during the labeling step, which is different from previous methods of microarray probe labeling. DNA immunoprecipitated from the tagged and untagged strains are labeled with different fluorophores (Cy3, Cy5). Use similar amounts of DNA for each of these labeling reactions. Labeling involves three stages of PCR, random nanomers with fixed sequence linkers and fluorophore-conjugated dCTP. The labeling procedure is adapted from that used in Iyer *et al.*[12] Other probe preparation protocols are available at http://cmgm.stanford.edu/pbrown.

1. Immunopurified DNA is brought to a final volume of 8 μl in 1× TE and 1 μl 10× T7 DNA polymerase buffer (USB:), and 1 μl 50 μM Primer 1 (5′-GTTTCCCAGTCACGATCNNNNNNNNN-3′) is added to it. The DNA is denatured in a thermalcycler at 94° for 2 min and then cooled to 8°, at which point 5 μl of reaction mix 1 [1× T7 DNA polymerase buffer, 0.15 mM dNTPs, 0.015 M DTT, 0.75 μg BSA, 4U T7 DNA polymerase (USB, Cleveland, OH)] is added to the tube. The reaction mixed is maintained at 8° for 2 min and then ramped to 37° over 8 min. The DNA is allowed to polymerize for 8 min at 37° and then is denatured at 94° for 2 min. The reaction tubes are cooled to 8° and 1 μl of reaction mix 2 [0.4U T7 DNA polymerase in dilution buffer (USB)] is added. The reaction is held at 8° for 2 min, ramped to 37°, and then held at 37° again for 8 min. The products are diluted with 35 μl 1× TE.

2. The diluted products of the first stage of DNA synthesis are then used as a template in the second stage of PCR amplification. Each 100 μl reaction contains 15 μl of diluted product from step 1, 10 μl 10× PCR buffer (Qiagen), 2.5 μl 10 m*M* dNTPs, 2 μl 25 m*M* $MgCl_2$, 2.5 μl 50 μM Primer 2 (5′-GTTTCCCAGTCA CGATC-3′), 1 μl 5U/μl *Taq* polymerase (Qiagen), and 67 μl water. Thermal cycling conditions are as follows: 92° for 30 sec, 40° for 30 sec, 50° for 30 sec, 72° for 1 min for 25 cycles.

3. A fraction of the PCR products from step 2 is used as template in the third stage of PCR amplification. For a 50 μl reaction, the PCR conditions are as follows: 15 μl PCR product from step 2, 5 μl 10× PCR buffer (Qiagen), 0.5 μl F-33× dNTP stock solution, 1 μl 50 μM Primer 2, 1 μl 25 m*M* $MgCl_2$, 0.5 μl 5 U/μl *Taq* polymerase (Qiagen), and 24 μl water. The F-33× dNTP stock solution contains 3.3 μl 100 m*M* dATP, 3.3, μl 100 m*M* dGTP, 3.3 μl 100 m*M* dTTP, 1.7 μl 100 m*M* dCTP, and 28.4 μl 1× TE. After these reaction components are mixed, add 1 μl of Cy3- or Cy5-conjugated dCTP (Amersham, Piscataway, NJ) to the 50 μl reaction. Thermal cycling conditions are identical to those in step 2.

4. The labeled DNA should be purified immediately following the labeling reaction. Cy3- and Cy5-labeled DNA probes are combined and purified using Qiagen PCR purification system according to the manufacturer's instructions. The DNA is eluted from the spin column with 25 μl elution buffer and incubated at room temperature for 1 min before the final centrifugation step. The eluted probe mixture should be visibly purple in color.

Troubleshooting. If the purified probe mix is colorless or only one of the Cy dyes (pink or blue) is visible, the Cy-conjugated nucleotide probably has to be replaced. Some batches of Cy dye do not incorporate well. When repeating step 3 above, use the same template generated by PCR step 2, but a different vial of Cy-dCTP with a different batch number. If problems persist, a protocol involving aminoallyl coupling can be followed in lieu of steps 3 and 4, which has been adapted from a protocol available through http://cmgm.stanford.edu.

3. All PCR reaction components are identical to those in step 3 above except in place of F-33× dNTPs, 0.5 μl of a 50× dNTP mix using a 4 : 1 ratio of aminoallyl-dUTP (Sigma, St. Louis, MO) to dTTP is used and no Cy-conjugated nucleotides are added. The 50× dNTP mix contains 10 μl each of 100 m*M* dATP, dCTP, dGTP, 2 μl 100 m*M* dTTP, and 8 μl 100 m*M* aminoallyl-dUTP. Aminoallyl-dUTP is dissolved in 0.04 1 *N* NaOH (1 mg/18 μl), so that the final pH is roughly 7.

4. All traces of Tris must be removed from the PCR products so that the free amine groups do not interfere with the coupling reaction below (step 5), so the products are purified and washed with neutralized water. (*Note:* The pH of the water must be approximately 7 in order for the DNA to adhere to the purification filter.) Dilute the PCR reactions with 450 μl water and then transfer the diluted products

to Microcon-30 filters (Amicon, Danvers, MA). Spin at 14,000 rpm for 7 min at room temperature and discard flow-through. Wash filters twice with 450 μl neutralized water and repeat centrifugation. Concentrate volume to 10 μl with one to two 1 min centrifugations at 14,000 rpm. Collect sample by inverting the filter tube into a fresh microfuge tube. Samples can be stored at -20° indefinitely.

5. Before coupling the monofunctional NHS-ester Cy3 (Amersham) and Cy5 (Amersham) to the aminoallyl-incorporated DNA, add 0.5 μl 1 *M* sodium bicarbonate, pH 9. The Cy dyes are supplied as a dry pellet. Dissolve the pellet in 20 μl DMSO and divide into ten 2-μl aliquots that are then dried under vacuum and stored at 4°. (*Note:* Do not store the dyes in water or DMSO as they are rapidly hydrolyzed in water.) Add the 10.5-μl reaction to the desiccated dye pellet and incubate the coupling reaction for 1 hr at room temperature in the dark. Mix by tapping the reaction tube approximately every 15 min.

6. Quench the coupling with 4.5 μl 4 *M* hydroxylamine and incubate for 15 min in the dark. Combine the Cy3 and Cy5 reactions and purify with the Qiagen PCR purification kit according to manufacturer specifications. Elute the labeled DNA with 25 μl elution buffer and incubate at room temperature for 1 min before the final centrifugation step. Again, the probe mixture should be visibly purple in color.

Another potential explanation for faint or colorless labeled DNA is that the concentration of template DNA used in the labeling step is too low. This problem may be corrected in one of three ways: first, try to immunoprecipitate more DNA by starting with a larger culture of yeast cells; second, increase the number of amplification cycles in steps 2 and 3 of the labeling procedure up to 28 cycles, which is still within the linear range; and third, purify and concentrate the reaction products of steps 1 or 2 of the labeling procedure for use in the steps that follow.

Hybridization

The following hybridization protocol yields the best results for arrays printed on CMT-GAPS coated slides with DMSO (as described above). Other hybridization protocols are available for slides coated with polylysine or slides printed with 3× SSC through various Web sites (http://www.cmt.corning.com, http://sequence.aecom.yu.edu, http://www.microarrays.org).

1. The microarray slide is first prehybridized by adding 50 μl prehybridization buffer [25% (v/v) formamide, 5× SSC, 0.1% (w/v) SDS, 1% (w/v) bovine serum albumin (BSA)] to the printed side of the array and then covering with a coverslip that has been rinsed with water and then with ethanol and air-dried. The slide is placed into a HybChamber (Genemachines, San Carlos, CA) according to manufacturer's instructions and incubated at 42° for 1 hr. The slide is then removed, rinsed in water, and dried by centrifugation at 500 rpm for 5 min at room temperature.

2. An equal volume of 2× hybridization buffer (50% formamide, 10× SSC, 0.2% SDS) is added to the purified probe; the hybridization mix is subsequently boiled for 5 min and then centrifuged for 2 min at 14,000 rpm at room temperature. The probe should not be placed on ice. The hybridization mix is then pipetted onto the prehybridized array and covered with a coverslip that has again been rinsed with water and ethanol. The slide is incubated at 42° in a HybChamber for 12–16 hr.

3. When washing, separate wash containers should be used for slides hybridized with different probes and slides should not be allowed to dry between washes. The coverslip is first removed by immersing in 42° 2× SSC/0.1% SDS until the coverslip moves freely away from the slide. Slide is then washed in 2× SSC/0.1% SDS for 5 min at 42° and then in 0.1× SSC/0.1% SDS for 10 min at room temperature. The array is washed 5 times in 0.1× SSC for 1 min each and then rinsed in water and dried by centrifugation at 500 rpm for 5 min at room temperature. The slide is then ready for scanning.

Troubleshooting. If, on scanning, the hybridization signals are generally weak (low fluorescent intensities), the signal may be increased by altering several parameters in the labeling procedure (see above). Nonspecific hybridization may be a concern if significantly high signal is detected for spots containing no DNA or heterologous DNA. The specificity may be increased by prehybridizing for longer periods of time. Alternatively, the hybridization temperature can be increased.

Scanning and Analysis

The slide is scanned to determine the Cy3 and Cy5 fluorescence intensity for each array element. The ratio of signal from the fluorophore labeling the transcription factor-associated DNA (usually Cy5) to the signal from the fluorophore labeling the background DNA (usually Cy3) represents the extent of binding of the transcription factor to the specific genomic fragment. The background is mock immunoprecipitated DNA so that genomic fragments that are nonspecifically precipitated will not be deemed binding targets. Background subtracted data is normalized and scaled. Analysis of data from 2–3 replicate experiments should be performed to identify those intergenic regions that are consistently and significantly enriched by chromatin immunoprecipitation and are therefore candidate binding targets of the transcription factor of interest. The ratio quantities of individual intergenic regions can vary significantly between duplicate experiments; therefore the results determined are simply binary, fragments that are bound vs those that are not bound. The affinity of a transcription factor for its target DNA cannot be determined with the ChIP-chip assay, as the ratio quantities determined may be a function of sensitivity of certain region of chromatin to fixation or immunoprecipitation as well as promoter context. The analysis steps below are adapted from the methods used in Iyer *et al.*[12]

1. Create a list of array elements and their corresponding address on the microarray from the list of intergenic regions in their 96-well format supplied by Research Genetics. The 96-well address (plate number, well designation) should be stepwise deconvoluted to the 384-well address (plate number, well designation) and then to the microarray address (block number, column number, row number). A deconvolution program, such as CloneTracker (BioDiscovery, Los Angeles, CA), can be used or a program can be created in Microsoft Excel for this purpose. Add an appropriate header to the array list and convert the list to a tab-limited text file format so that it can be read by the scanner software.

2. Open Genepix Pro version 3.0 software (Axon Instruments, Foster City, CA) and scan the slide at the appropriate photomultiplier tube (PMT) voltage for both the 532 nm (Cy3 excitation wavelength) and 635 nm (Cy5 excitation wavelength) lasers using a Genepix 4000A scanner (Axon Instruments). (*Note:* Cy3-dCTP generally incorporates less efficiently than Cy5-dCTP. Compensate for this effect by increasing the PMT voltage of the 532 nm laser during scanning.) The optimal PMT voltage for both channels is one at which a maximum number of array elements are visible, without saturation. White pixels represent saturation.

3. Load the array list generated in step 1. Align the blocks to the spot features on the scanned image. The Genepix software will define all of the array elements or intergenic regions. The software will also compute the fluorescence intensity of each array element in both channels and calculate the ratio of flourescence intensities for the red channel (Cy5) over the green channel (Cy3) when the *Analyze* button is pressed. The raw data and subsequent computations are sent to the Results tab spreadsheet.

4. Filter data by flagging spots with obvious defects and spots with total fluoresence intensities that is below an empirically determined threshold for both channels.

5. Genepix Pro 3.0 does not normalize the generated data, but it does compute a normalization factor for each of the ratio quantities. Once the analysis of the image has been completed, the results can be saved as a tab-delimited text file, reopened, and normalized in Microsoft Excel. To normalize, the *ratio of medians* (pixel-by-pixel ratio) for each intergenic region is simply divided by the normalization factor given in the data header for the *ratio of medians*. Other programs are available for normalization of microarray data such as GeneSpring (Silicon Genetics, San Carlos, CA).

6. The data must also be scaled before they can be compared with duplicate ChIP-chip experiments or contrasted with ChIP-chip experiments using different conditions or for different transcription factors. Scaling can also be performed in Microsoft Excel by calculating the PERCENT RANK or percentile ranking (a statistical function) of the normalized *ratio of medians* for each intergenic region. When computing the PERCENT RANK, the array or range of data is zero to

the maximum quantity within the normalized *ratio of medians* column for that experiment and "x" is the normalized ratio for each intergenic region.

7. In Microsoft Excel, combine the scaled data with the processed data from at least one other identical experiment to identify intergenic regions that are consistently enriched in immunoprecipitates with the transcription factor of interest. Find the average percentile rank for each intergenic region and then sort the average percentile ranks in ascending order. Use the histogram analysis tool to bin the ranks. Make 100 bins from 0.01 to 1.0 and then chart the distribution of the ranks using the column graph type option from the Chart Wizard. The distribution should be bimodal with candidate gene targets represented by the high percentile ranks that do not fall within the normal Gaussian distribution.

Troubleshooting. If the bimodal distribution is not observed, the problem may be corrected by adjusting the bins with the Histogram analysis tool. Some features of the distribution may have been obscured during the binning process. Create more bins with smaller intervals and look at the distribution of percentile ranks again. If restructuring the data fails, then add more data from a replicate experiment. If a bimodal distribution is still not observed after adding data, then examine the data to see if the observed enrichments are inconsistent. If the problem is inconsistency, it may be beneficial to see if the chromatin immunoprecipitation is working with a PCR assay (see above). If, however, consistent enrichments are observed, then it could be that the analysis is too rigorous. Candidate binding targets can then be identified on the basis of their consistent enrichment, especially if the potential binding targets flank genes that are functionally related or have similar transcriptional profiles.

Conclusions and Future Directions

ChIP-chip can be used to identify the genome-wide binding sites of transcription factors in yeast. When this technique is coupled to genome-wide expression analysis, the contribution of a particular transcription factor to the expression of each of its target genes can be determined. Target gene data can then be clustered by their functional class and by their expression profiles. Not only will these approaches describe the transcriptional role of known and unknown transcription factors, but because transcription factors frequently regulate the expression of other transcription factors, these techniques should advance description of transcriptional cascades, generating a temporal map of the transcriptional circuitry of the yeast cell.

Identifying all the genomic binding targets of a transcription factor will also be useful for elucidating the consensus binding sequence for uncharacterized transcription factors and refining the consensus sequences for previously studied DNA-binding proteins. The genomic segments bound by a factor can be scanned for common sequence motifs. There are many programs available for this type of analysis, including Consensus,[23] MEME,[24] and Gibbs' sampling.[25,26] It will also be interesting to see how promoter sequence motifs relate to functionally distinct targets or target clusters with similar expression profiles.

The ChIP-chip technique can be expanded to all DNA and chromatin interacting proteins. Using a microarray that includes all genomic elements, both ORFs and intergenic regions, the binding profiles of chromatin structural proteins, DNA silencing factors, and proteins involved in recombination can be examined. The technique can also be applied to proteins that do not directly interact with DNA. The chemical cross-linker formaldehyde fixes protein–protein interactions as well as protein–DNA interactions, so DNA indirectly associated with a protein can be immunopurified. The binding of signaling molecules important for the function or activation of a transcription factor can thus also be mapped on the genome.

When the binding profiles are examined for all of these proteins under different environmental conditions and at different points in the cell cycle, we will gain a better understanding of transcriptional regulation and chromosome dynamics in yeast. The hope is, of course, to apply the ChIP-chip method to higher eukaryotes as their genomic sequences become available. Generating transcription factor binding profiles for higher eukaryotes will bring us closer to deciphering the complexities of gene regulation in these organisms.

Acknowledgments

Past and present members of the M. Snyder, P. Brown, and D. Botstein laboratories have contributed to the development of these protocols. We are especially grateful to V. R. Iyer and C. S. Scafe for pioneering work to develop the yeast intergenic microarray. Thanks also to J. Rinn for critical comments of this manuscript. C. Horak is supported by a predoctoral fellowship from the Howard Hughes Medical Institute.

[23] G. D. Stormo and G. W. Hartzell, *Proc. Natl. Acad. Sci. U.S.A.* **86,** 1183 (1989).

[24] T. L. Bailey and C. Elkan, *in* "Proceedings of the Second International Conference on Intelligent Systems for Molecular Biology, Menlo Park, California." American Association for Artificial Intelligence, 1994.

[25] A. F. Neuwald, J. S. Liu, and C. E. Lawrence, *Protein Sci.* **4,** 1618 (1995).

[26] C. E. Lawrence, S. F. Anschul, M. S. Bogouski, J. S. Liu, A. F. Neuwald, and J. C. Wooten, *Science* **262,** 208 (1993).

[27] Computational Approaches to Identifying Transcription Factor Binding Sites in Yeast Genome

By HAO LI

Introduction

Cells respond to environmental or genetic perturbations by modifying their transcriptional programs. The identification of transcription factor binding sites is important for identifying possible regulators and their transcriptional targets. Traditionally the problem has been tackled by experimental approaches such as promoter bashing which could be very laborious. However, the availability of the complete genome sequence of yeast and DNA microarray technology has changed the situation dramatically. DNA microarrays are now used routinely to identify potential transcriptional targets on a genomic scale. From the complete genome sequence, the regulatory regions (upstream noncoding regions) of the target genes can be extracted. A number of computational algorithms have been developed which are capable of identifying putative transcription factor binding sites by combining sequence information with genome-wide mRNA expression data. In this chapter I will describe the basic ideas behind some of the algorithms and direct the reader to the relevant Web sites for online access.

In the yeast genome, there are an estimated 500–600 transcription factors. The targets of these factors can range from a few to a few hundred genes. Typically the upstream region of a gene includes the binding sites of different transcription factors which control the expression of the gene under different conditions. Thus, on a genomic scale, transcriptional regulation has a huge combinatorial complexity and the identification of regulatory sites is a very challenging problem. Several approaches have been developed to tackle this problem. One commonly used approach has been to delineate as sharply as possible a group of coregulated genes and search for common sequence patterns in their upstream regulatory regions. The computational algorithms range from finding overrepresented substrings or regular expression patterns to multiple local sequence alignment. The prerequisite for this class of algorithms is a cleanly defined subset of genes that may share a few common motifs. I will describe a few representatives in this category. An alternative approach is to delineate combinatorial motifs from a large collection of regulatory sequences without the need for defining coregulated groups. I will describe two algorithms in this category that we have developed. One algorithm is based on a mathematical model of probabilistic segmentation, a generalization of segmentation models used in statistical language processing. Another algorithm identifies sequence patterns in the promoter regions of genes that strongly correlate with the genome-wide gene expression data.

0076-6879/02 $35.00

TABLE I
URLs OF WEB SITES WHICH PROVIDE ONLINE ACCESS TO DATABASES AND COMPUTATIONAL ALGORITHMS DISCUSSED IN THIS CHAPTER

URL	Name
http://transfac.gbf.de/TRANSFAC/	TRANSFAC
http://cgsigma.cshl.org/jian	SCPD, Sequence Retrieval
http://embnet.cifn.unam.mx/rsa-tools/	Oligonucleotide analysis, Consensus Gibbs Sampler, Sequence Retrieval
http://ural.wustl.edu/~jhc1/consensus	Consensus
http://bayesweb.wadsworth.org/gibbs/gibbs.html	Gibbs Sampler
http://meme.sdsc.edu/meme/website/	MEME
http://mobydick.ucsf.edu/~haoli/tools	Mobydick, REDUCE
http://bussemaker.bio.columbia.edu/reduce	REDUCE
http://bioprospector.stanford.edu/	BioProspector

Preparing Sequence Data

One class of algorithms that I will describe needs to define a group of coregulated genes that may share similar regulatory motifs. A commonly used method is to cluster genes based on their similar expression profiles. The expression profile of a gene can be characterized by its expression levels at multiple time points (such as different time during cell cycle or sporulation) or across different experiments (e.g., responses to various environmental stresses and different mutant strains). Genes can also be grouped based on their biochemical functions. In cases where there are not enough data points for cluster analysis, genes may be grouped based on similar response to certain conditions (e.g., genes up- or down-regulated in a mutant strain compared to the wild type, or when a transcription factor is overexpressed).

It is known that transcription factor binding sites in the yeast genome are typically positioned within ~600 bp upstream from the translation start site (ATG). Thus, to identify putative transcription factor binding sites, it is often sufficient to search for a region a few hundred base pairs upstream of ATG (rarely beyond 1000 bp). The position of each gene is annotated in the Stanford Genome Database (for information provided by SGD, see elsewhere in this volume[1]). Using the genome annotation and the complete genome sequences, one can extract the sequences in the upstream noncoding regions of the genes of interest. There are also a number of Web sites which provide online sequence retrieval tools (see Table I). The user can input a list of open reading frame (ORF) names and get upstream noncoding sequences returned.

[1] L. Issel-Tarver, K. R. Christie, K. Dolinski, R. Andrada, R. Balakrishana, C. A. Ball, G. Binkley, S. Dong, S. S. Duright, D. G. Fisk, M. Harris, M. Scheeder, A. Sethuraman, K. Tse, S. Weng, D. Botstein, and J. M. Cherry, *Methods Enzymol.* **350,** [19], 2002 (this volume).

Knowledge-Based Approach to Identifying Binding Sites

Considerable knowledge of yeast transcription factors and their binding sites has accumulated over years of experimental studies. This knowledge can be used to identify putative binding sites of characterized transcription factors. One popular resource is the transcription factor database (TRANSFAC) where binding sites from literatures are archived.[1a] A more specialized database for yeast is the Promoter Database of *Saccharomyces cerevisiae* (SCPD), where transcription factor binding sites for yeast are collected[2] (see Table I for Web links to TRANSFAC and SCPD). The current version of SCPD has ~700 binding sites representing ~100 factors. For yeast, the current public version of TRANSFAC has about half as many sites. There are a number of search tools available in SCPD and TRANSFAC. The user can search for the known target genes and binding sites of a given transcription factor, or search for experimentally mapped binding sites in the regulatory region of a given gene. SCPD also allows the user to input a specific motif and search for known binding sites that contain the motif with a specified number of mismatches. This could be useful for linking motifs identified by *de novo* methods (see the next section) to a known factor.

Both SCPD and TRANSFAC provide online tools for searching putative binding sites of characterized transcription factors based on their binding site profiles. There are a number of transcription factors (about 30) for which many examples of binding sites are known. For these factors, different binding sites can be aligned to derive consensus sequences and quantitative profiles—position-specific probability matrices. A consensus sequence consists of A, C, G, T and the degenerate IUPAC symbols. For example, the consensus sequence for the MSE site recognized by Ndt80 is CRCAAAW, where R stands for A or G and W stands for A or T. A position-specific probability matrix is more quantitative than a consensus sequence. It specifies the probability of occurrence of the four nucleotide bases at each single base position of the binding site (see the next section for a more detailed description). These matrices can be used to derive a normalized score between 0 and 1 (where 1 is the highest possible score) to potential sites in the promoter sequences that the user inputs. Subsequences matching preferred bases at all the positions of the site will get a high score. The user can specify a cutoff value for the score between 0 and 1 depending on the tolerance on false positives. A lower cutoff yields more predictions (thus higher sensitivity) and a higher false positive rate.

Unfortunately, examples in these databases cover only a small fraction of the transcription factors in the genome. The SCPD database has ~700 binding sites, which covers at most a few percent of all possible regulatory sites in the yeast

[1a] E. Wingender, X. Chen, R. Hehl, H. Karas, I. Liebich, V. Matys, T. Meinhardt, M. Prüβ, I. Reuter, and F. Schacherer, *Nucleic Acid Res.* **28,** 316 (2000).

[2] J. Zhu and M. Q. Zhang, *Bioinformatics* **15,** 607 (1999).

genome. Thus, in most cases it is necessary to employ computational algorithms that can identify putative sites *de novo*.

Identifying Binding Sites *de Novo*

A number of computational algorithms have been developed that can identify putative binding sites *de novo*. In the past few years, they have been successfully applied to analyzing DNA microarray data. Instead of trying to give a complete survey, I will describe a few examples that represent different approaches. These algorithms are all available online and the corresponding Web sites are listed in Table I.

Algorithm Based on Frequency Count of Oligonucleotides

The method developed by van Helden, Andre, and Collado-Vides is a representative of a class of algorithms that search for overrepresented substrings or regular expression patterns in the upstream regions of a group of coregulated genes.[3] The algorithm searches for oligonucleotides that are overrepresented based on their expected frequencies by exhaustively enumerating all possible oligonucleotides up to a certain length (typically 6 to 8). In the default setting, the expected frequencies are taken from the frequencies observed in all the noncoding regions in the genome. For a given oligonucleotide w, the algorithm counts the observed occurrences $n(w)$ in the dataset and calculates the expected occurrences based on the expected frequency of w and the size of the data. To find out if $n(w)$ is significantly more than expected, the algorithm calculates a P value (occurrence probability) associated with w, the chance probability of observing occurrences equal to or larger than $n(w)$, based on a binomial background distribution, where the probability for the binomial distribution is given by the expected frequency of w and the number of trials is given by the number of possible positions in the dataset where w can occur. A highly overrepresented oligonucleotide will have a small P value. The statistical significance (occurrence significance) is given by $-\log_{10}(PN_{\text{oligo}})$, where N_{oligo} is the number of oligonucleotides screened. An oligonucleotide is considered to be significantly overrepresented in the dataset only if its occurrence significance is larger than 0, i.e., its P value is smaller than $1/N_{\text{oligo}}$. This threshold for P value is set so that only $\sim$1 false positives is expected. The algorithm has a good background calibration as it uses all the upstream regions in the genome to calculate background frequencies. Given a group of genes with a binding site sufficiently overrepresented, the algorithm is capable of identifying the conserved core of the site. Since the algorithm enumerates all oligonucleotides, the P value cutoff has to be set very low to control false positives; thus the algorithm may miss

[3] J. van Helden, B. Andre, and J. Collado-Vides, *J. Mol. Biol.* **281,** 827 (1998); *Yeast* **16,** 177 (2000).

motifs that are not highly overrepresented in the dataset. It is also likely to miss long and fuzzy motifs. In case significant motifs can be detected, the algorithm usually outputs a large number of overlapping oligonucleotides that need to be assembled.

Algorithms Based on Local Multiple Sequence Alignment

This class of algorithms searches for common sequence patterns in the regulatory regions of a group of coregulated genes by finding subsequences that give statistically significant multiple alignments. Three popular algorithms in this category are Consensus, Gibbs Sampler, and MEME. These algorithms all use a position-specific probability matrix to describe the motif (the motif model). A position-specific probability matrix specifies the probability of occurrence of the four nucleotides at each single base position of the motif. Such a matrix reflects the base preferences of the corresponding transcription factor. Shown in Fig. 1 is a pedagogical example of several binding sites aligned and the resulting position-specific probability matrix. The matrix elements are given by the normalized counts of the four bases at different positions in the alignment. For simplicity I ignored pseudo counts usually added for Bayesian inference. In addition to the motif model, there is also a model to describe the background. Typically the background is modeled as uncorrelated random sequences with appropriate single base frequencies. The specificity of the motif depends on the sharpness of the alignment, relative to the

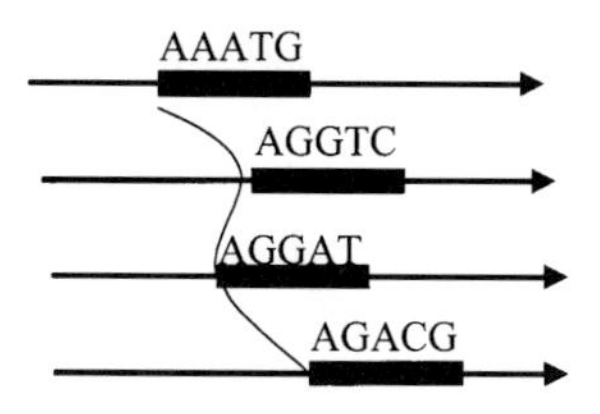

position specific frequency matrix $f_{i\sigma}$

	1	2	3	4	5
A	1.00	0.25	0.50	0.25	0.00
C	0.00	0.00	0.00	0.25	0.25
G	0.00	0.75	0.50	0.00	0.50
T	0.00	0.00	0.00	0.50	0.25

FIG. 1. An example of several subsequences aligned (*top*) and the position-specific probability matrix derived from the alignment (*bottom*). The columns correspond to different positions and the rows correspond to different bases. The matrix element at position i and base σ gives probability of occurrence of base σ at the ith position in the motif. For example, base A occurs with probability 1 at the first position.

background, as quantified by the information content. For example, in the alignment in Fig. 1, position 1 is specified to be base A with probability 1. Assuming that the background frequencies are the same for all the four bases, specifying A out of 4 equal possibilities needs 2 bits of information; thus the information content at position 1 is 2 bits. The information content of the motif is given by adding the information contents of all the positions in the motif together:

$$\text{Information} = \sum_{i,\sigma} f_{i\sigma} \log_2\left[f_{i\sigma}/f_\sigma^0\right]$$

where $f_{i\sigma}$ gives the probability of occurrence of base σ at position i in the motif, and f_σ^0 gives the background probability of base σ. This information content is related to the log likelihood ratio—the logarithm of the ratio of the probability of occurrence of the motif given the motif model vs that given the background model. Since the positions of the motif in the sequences are unknown, essentially all the algorithms try to find a local alignment that maximizes the information content of the resulting position-specific probability matrix. To be more precise, they try to find the alignment with information content least expected by chance. The algorithms differ by the mathematical approaches to finding the optimal alignment. These algorithms are capable of identifying transcription factor binding sites from the upstream sequences of a group of coregulated genes if a significant fraction of these sequences share a similar site. They can identify fuzzy motifs that do not have a conserved core repeated in different promoter sequences.

Consensus. Developed by Hertz, Stormo, and others.[4] Because the number of possible local alignments grows exponentially with the number of sequences, it is impractical to find the best alignment by enumerating all possibilities. Consensus and its relative Wconsensus use a heuristic approach to find the best alignment. For Consensus, the user needs to input the length of the motif being sought. The algorithm starts by aligning pairs of subsequences of the specified length. A user-specified number of top pairwise alignment matrices (those with highest information contents) are saved. The saved matrices are then aligned with a third subsequence not already contained in the pairwise alignment to generate new matrices given by the alignments of three subsequences. The top matrices from this cycle will be saved and aligned with a fourth subsequence, so on and so forth. The algorithm calculates a P value—the probability by chance that an alignment of the same number of random sequences with the same length will have information content equal to or higher than the observed value, and the expected frequency (P value multiplied by the number of possible alignments). An alignment is considered to be significant only if the expected frequency is much less than 1. The algorithm outputs a few top alignments (the user can specify the number), the corresponding

[4] G. Z. Hertz and G. D. Stormo, *Bioinformatics* **15,** 563 (1999); G. D. Stormo and G. W. Hartzell, *Proc. Natl. Acad. Sci. U.S.A.* **86,** 1183 (1989).

matrices, their information contents, *P* values, and expected frequencies. Wconsensus automatically determines the length of the motif by comparing the information contents for motifs of different lengths. Extending a motif will inevitably lead to higher information content even if the flanking region is random because the base frequencies in an aligned column of random bases will still be different from the background frequencies due to finite size fluctuation. Wconsensus compares motifs of different lengths after subtracting a term proportional to the length to adjust the finite size effect. The user can specify how many standard deviations (s value) in addition to the mean per base to subtract from the information content in order to compare different lengths. A large s value will prefer short motifs while small s prefers long motifs. Typically s is chosen to be between 1 and 2. In cases where the motif is very sharp, the result is insensitive to the choice of s.

Gibbs Sampler. Developed by Lawrence, Liu, Newald, and others.[5] Instead of trying to find the optimal alignment, this algorithm samples different alignments probabilistically using a Gibb sampling strategy.[6] The algorithm initializes with positions of subsequences picked at random. It then updates the positions one at a time while keeping the rest fixed. The probability of picking a new position is proportional to the likelihood ratio of its being the start of the motif to its being a general background position, with the likelihood ratio calculated from the motif model based on the alignment of the rest of the positions currently held fixed. The algorithm cycles through all positions until an equilibrium distribution is reached. The basic idea is that once some binding sites are found by chance, the resulting position-specific probability matrix will give other binding sites higher probability to be sampled, and thus will recruit more binding sites into the alignment. More binding sites in the alignment will in turn give stronger preference for other binding sites to be sampled. When a significant motif is found, the algorithm converges to a probability distribution where the optimal alignment has a large statistical weight, and thus is sampled most of the time. By sampling alignments probabilistically, the algorithm can avoid local minimum traps often encountered by deterministic algorithms. Gibbs Sampler outputs the optimal alignment and the alignment of subsequences sampled more than a certain percentage of the time (typically 50%). For each alignment, the algorithm gives the position-specific probability matrix, the information content at each position, and the maximum *a posteriori* (MAP) score. The algorithm does not give an expected number of occurrences for the observed MAP score; thus there is no direct assessment of the statistical significance. A significant motif is typically associated with a MAP

[5] C. E. Lawrence, S. F. Altschul, M. S. Boguski, J. S. Liu, A. F. Neuwald, J. C. Wootton, *Science* **262,** 208 (1993); J. S. Liu, A. F. Neuwald, and C. E. Lawrence, *J. Am. Statist. Assoc.* **90,** 1156 (1999).

[6] J. S. Liu, "Monte Carlo Strategies in Scientific Computing," Springer Series in Statistics. Springer, New York, 2001.

score much larger than 0. A more powerful version of Gibbs Sampler called Bio-Prospector has been developed recently.[6a]

MEME. Developed by Baily and Elkan.[7] The algorithm models the sequences as a mixture of background and motifs and uses an expectation maximization (EM) algorithm to find the motif models and the mixing parameters that maximize the likelihood of observing the sequence data. The density of a motif is determined automatically from the corresponding mixing parameter. The algorithm outputs the sequence alignment, the position-specific probability matrix for the motif, a graphical representation of the information content at each position, the log likelihood ratio of the alignment (the ratio of the likelihood of the alignment given the motif model vs that given the background model), and an E value, which gives the statistical significance of the alignment. The E value is the expected number of motifs with the given log likelihood ratio (or higher), and with the same width and number of occurrences, that one would find in a similar sized set of random sequences. To have a statistically significant motif, the E value should be much less than 1.

Although the above three algorithms use different mathematical approaches, their spirits are similar and very often their performances are also similar. I suggest that the user try different algorithms and compare the results. All the algorithms have a number of parameters that can be used to fine-tune the performance. The user should refer to the corresponding references and documentation for a better understanding of the effects of the parameters.

Algorithms for Finding Combinatorial Motifs

Algorithms based on frequency contrast of oligonucleotides or multiple local sequence alignment need to have a well-defined group of coregulated genes that share a few common motifs. However, because of the combinatorial nature of transcriptional regulation, it is quite typical that genes responding similarly to a specific environmental or genetic perturbation may be regulated by many different transcription factors. For example, genes repressed by the general repressor Tup1 can be divided into different subsets controlled by various regulators working together with Tup1.[8] Thus different genes induced by Tup1 deletion will have the binding sites of different factors. On the other hand, genes sharing a common motif may not respond similarly to all conditions, and thus may not be clustered together. Several approaches have been developed to tackle the problem of

[6a] X. Liu, J. S. Liu, and D. L. Brutlag, *Proc. Pac. Symp. Biocomput.,* p. 127. World Scientific Press, 2001.

[7] T. L. Bailey and C. Elkan, Proceedings of Second International Conference on Intelligent Systems for Molecular Biology, p. 28. AAAI Press, Menlo Park, CA; T. L. Bailey and C. Elkan, Technical Report CS 93-302, Dept. of Computer Science, University of California, San Diego, August 1993.

[8] M. Wahi, K. Komachi, and A. D. Johnson, *Cold Spring Harbor Symp. Quant. Biol.* **63,** 447 (1998).

identifying combinatorial motifs without the need for clustering genes. I will describe two such algorithms, Mobydick and REDUCE, developed by Bussemaker, Siggia, and myself. Both programs are capable of identifying multiple regulatory motifs simultaneously from a large collection of promoter sequences. Mobydick uses a statistical model of segmentation to delineate biologically meaningful "words" from upstream noncoding sequences that are likely to be regulatory elements. REDUCE uses genome-wide gene expression data (such as those measured by DNA microarrays) to extract regulatory elements responsive to the specific experimental condition where the expression data are obtained.

Mobydick. This algorithm is based on a mathematical model of probabilistic segmentation and the maximum likelihood approach.[9] The algorithm explores an interesting analogy between regulatory sequences and language: regarding the binding site of a transcription factor as a "word," the regulatory regions of genes can then be viewed as sentences made by concatenations of words and background sequences. A word can be used in the regulatory regions of many genes (regulon), and the regulatory region of a gene can have multiple words (combinatorial control). When formulated this way, the problem of identifying regulatory elements is similar to delineating words from a scrambled English text, created by replacing all word separators with strings of random letters. Mobydick solves the problem by modeling the sequences as concatenations of words drawn from a probabilistic dictionary that specifies words and their associated probabilities. To infer the dictionary from the observed sequences, Mobydick segments the sequences probabilistically into a collection of plausible words and determines their probabilities by maximizing the likelihood of observing the input sequences. The dictionary is constructed iteratively by starting with single letters and gradually building words from their fragments. The algorithm has been tested on the scrambled novel *Moby-Dick* (hence the name for the algorithm) as well as the noncoding regions of $\sim$6000 genes in the yeast genome.[9] The advantage of Mobydick is that it can analyze a large collection of sequence data (typically on the order of 10^6 bps) and identify multiple motifs (hundreds of words) in parallel. The algorithm is sensitive and can identify motifs that only occur in a small subset of input sequences.

Mobydick can be applied to analyzing the noncoding sequences of the whole genome as well as the regulatory regions of a subset of genes of interest. For analyzing genome-wide gene expression data, the user can first define a subset of genes, for example, all genes derepressed in Tup1 deletion mutant compared to the wild-type strain, or all genes up-regulated during sporulation. These genes do not need to be coregulated at different time points or across different experiments and they can be controlled by many different transcription factors. Mobydick can then

[9] H. J. Bussemaker, H. Li, and E. D. Siggia, *Proc. Natl. Acad. Sci. U.S.A.* **97,** 10096 (2000); Proceedings of Eighth International Conference on Intelligent Systems for Molecular Biology, p. 67. AAAI Press, Menlo Park, CA, 2000.

be used to decompose the upstream noncoding regions of these genes into a dictionary of plausible motifs. Some of these will be general motifs (for example, TATA box) which are not specific to the experiment. To find out the motifs that are responsive to a specific condition, for example, motifs responsible for turning on genes during sporulation, one can screen for words that are overrepresented in the subset of the genes based on their frequencies in the whole genome noncoding regions. Although this word screening step is similar to the oligonucleotide frequency analysis algorithm, the whole approach is quite different as words themselves are first identified without external frequency calibration. Thus overrepresented words are less likely to be false positives due to compositional fluctuations in the promoter regions of the subset of genes analyzed. In addition, the number of words is typically much smaller than the number of all possible oligonucleotides of comparable length; thus the statistical noise is much lower, giving rise to higher sensitivity.

Both the Mobydick algorithm and the word screening program are available online (see Table I). Mobydick takes general sequences a user inputs and returns a list of words with their associated statistical properties, such as probability, statistical significance, and quality factor (see the manual for more detailed explanation). To run the word screening program, the user needs to input reference sequences to calibrate background frequencies. For yeast promoter analysis, we have developed an interface which combines sequence retrieval, Mobydick, and the word screening programs together. The user just needs to input a list of ORF (open reading frame) names. The program automatically extracts upstream noncoding sequences for the input ORFs, runs Mobydick to generate a list of words, and screens the words by using the 5′ noncoding sequences in the whole genome as reference sequences. The algorithm returns a list of words rank ordered by their overrepresentation with other statistical properties. Mobydick allows the user to specify a couple of parameters. One is the number of expected false positives. This parameter sets the statistical criterion for adding a new word to the dictionary during the iterative dictionary construction process.

REDUCE. The name stands for "Regulatory Element Detection Using Correlation with Expression." This program searches for sequence patterns in the promoter regions of genes that significantly correlate with the genome-wide mRNA expression data.[10] The basic assumption used by the REDUCE program is that if a motif is responsible for regulating mRNA expression under a specific condition, the expression level of a gene should correlate with the occurrences of the motif in its promoter region. Although having the motif may not be sufficient for a gene to respond, on the whole genome scale, one still can observe a correlation as long as a reasonable fraction of the genes having the motif respond to the condition. REDUCE does not need to cluster genes or select up- or down-regulated genes by a threshold cutoff. It takes as input the fold changes of the mRNA levels for

[10] H. J. Bussemaker, H. Li, and E. D. Siggia, *Nat. Genet.* **27,** 167 (2001).

all the genes in the genome and outputs a set of motifs that significantly correlate with the expression. The algorithm searches for motifs by enumerating all oligonucleotides up to a certain length (typically 6–8) as well as all dimer pattern of the form w1N· · ·Nw2, where w1 and w2 are short oligonucleotides (typically of lengths 3 to 5), N matches any base, and the number of N's typically vary from 0 to 50. For a given oligonucleotide or dimer w, the algorithm counts $N_g(w)$, the number of occurrences of w in the upstream noncoding region of gene g, and calculates the Pearson correlation coefficient $r(w)$ between $N_g(w)$ and $\log(E_g)$, the log of the ratio of gene g in the experiment, for all the genes g in the genome. The program computes a P value—the probability by chance to have a Pearson correlation coefficient equal to or larger than $r(w)$ under the null hypothesis that w has no correlation with the expression. The statistical significance of a motif is given by $-\log_{10}(PN_{\mathrm{m}})$, where N_{m} is the number of motifs screened. Only those motifs with statistical significance larger than 0 will be kept. This threshold is set so that only $\sim$1 false positives is expected.

REDUCE uses an iterative scheme to find independent motifs. Typically if a motif correlates with the gene expression, many strings overlapping with the motif will also correlate with the expression. For example, the MCB box ACGCGT correlates with cell cycle expression data; various strings that overlap with this motif (such as ACGCG or ACGCGTT) also correlate with the expression. REDUCE removes the redundancy by iteration. Once N motifs are found, the algorithm uses them to do a linear fit of the expression data, finds the residual of the fit (the part of expression data unexplained by the N motifs), searches for the motif that has strongest correlation with the residual data, and add it to the motif list if the correlation is statistically significant. The algorithm iterates through fitting and screening steps until no more motifs can be found. At the end, REDUCE outputs a list of distinct motifs with their statistical significance and fitting coefficients. Motifs responsible for the induction of gene expression have positive fitting coefficients while those responsible for depression of gene expression have negative coefficients. We have used REDUCE to analyze a number of microarray data and found that the program typically gives quite specific predictions with a low false positive rate.

I have described several computational algorithms that can be used to identify statistically significant patterns as potential transcription factor binding sites. The reader should be cautioned that statistical significance does not equal biological significance. Statistical significance is always calculated based on a background model. For example, algorithms based on multiple local sequence alignment typically assume a random background described by single base frequencies (although more complicated background models can be incorporated). However, it is not unusual for DNA sequences to have various correlations, e.g., simple repeats. Such patterns may be identified as statistically significant, even though they are unlikely to be transcription factor binding sites. Similarly other algorithms may find statistical significant patterns that are not regulatory elements. To evaluate the biological relevance of a motif, one may look at the positional distribution of the motif with respect to ATG in the whole genome and examine the functions of genes having the motif in their promoter regions. In many known cases, regulatory motifs are distributed in preferred positional windows, and genes pulled out by these motifs tend to be enriched in certain groups that are functionally related. One may also check SCPD to see if the motif matches experimentally determined binding sites in the database. Ultimately, the identified putative regulatory elements have to be tested by experiments.

The availability of the complete genome sequences and DNA microarray technology will make it more and more popular to use the "bottom up" approach to deciphering regulatory networks. In this approach the transcriptional targets of a regulatory pathway are first identified by whole-genome transcriptional profiling. Computational methods can then be used to identify putative regulatory motifs from these target genes. These motifs can then be tested (e.g., by using artificial promoter constructs) and the transcription factors that bind to these motifs can be screened by (e.g., using the yeast one hybrid library). Such a combination of computational and experimental approaches will drastically speed up the discovery process in the new genomic era.

Section IV

Proteomics

[28] Array-Based Methods for Identifying Protein–Protein and Protein–Nucleic Acid Interactions

By JOSEPH F. GERA, TONY R. HAZBUN, and STANLEY FIELDS

Introduction

Genomic studies to date have focused largely on measuring changes in transcription, and only recently have begun to assay protein function on a genome-wide basis (for example, Refs. 1–3). One approach to assigning function to an uncharacterized protein is the identification of its interacting protein and/or nucleic acid partners. In many instances, this procedure can enable a prediction of a putative function to be made. Genetic assays have been developed in yeast to identify candidate protein–protein (two-hybrid),[4] DNA–protein (one-hybrid),[5,6] and RNA–protein (three-hybrid)[7] interactions. These screens can be configured such that an ordered array of yeast transformants, each expressing a unique open reading frame (ORF), is searched rapidly to uncover potential interactors.[3] For protein–protein interactions, these associations can lead to the construction of large-scale interaction maps reflecting the networks that underlie such processes as signal transduction, replication, transcription, and RNA processing.[8,9]

The approach we have taken has been to develop a high-throughput array-based method to conduct two-hybrid screens in yeast.[3] This method can also be adapted to screen for DNA–protein or RNA–protein interactions using the same array. The two-hybrid assay is based on reconstituting a functional transcription factor composed of two fusion proteins (Fig. 1A). An interaction between a protein fused to the Gal4 DNA-binding domain (DBD) and a protein fused to the Gal4 activation domain (AD) leads to transcriptional activation of a reporter gene containing Gal4 upstream activating sequences. The one- and three-hybrid assays rely on the same general principle of transcriptional activation of a reporter gene to detect

[1] M. Martzen, S. McCraith, S. Spinelli, F. Torres, S. Fields, E. Grayhack, and E. Phizicky, *Science* **286,** 1153 (1999).

[2] E. Marcotte, M. Pellegrini, M. Thompson, T. Yeates, and D. Eisenberg, *Nature* **402,** 83 (1999).

[3] P. Uetz, L. Giot, G. Cagney, T. Mansfield, R. Judson, J. Knight, D. Lockshon, V. Narayan, M. Srinivasan, P. Pochart, A. Qureshi-Emili, Y. Li, B. Godwin, D. Conover, T. Kalbfleisch, G. Vijayadamodar, M. Yang, M. Johnston, S. Fields, and J. Rothberg, *Nature* **403,** 623 (2000).

[4] S. Fields and O. Song, *Nature* **340,** 245 (1989).

[5] M. Wang and R. Reed, *Nature* **364,** 121 (1993).

[6] J. Li and I. Herskowitz, *Science* **262,** 1870 (1993).

[7] D. SenGupta, B. Zhang, B. Kraemer, P. Pochart, S. Fields, and M. Wickens, *Proc. Natl. Acad. Sci. U.S.A.* **93,** 8496 (1996).

[8] B. Schwikowski, P. Uetz, and S. Fields, *Nat. Biotechnol.* **18,** 1257 (2000).

[9] C. Tucker, J. Gera, and P. Uetz, *Trends Cell Biol.* **11,** 102 (2001).

METHODS IN ENZYMOLOGY, VOL. 350
0076-6879/02 $35.00

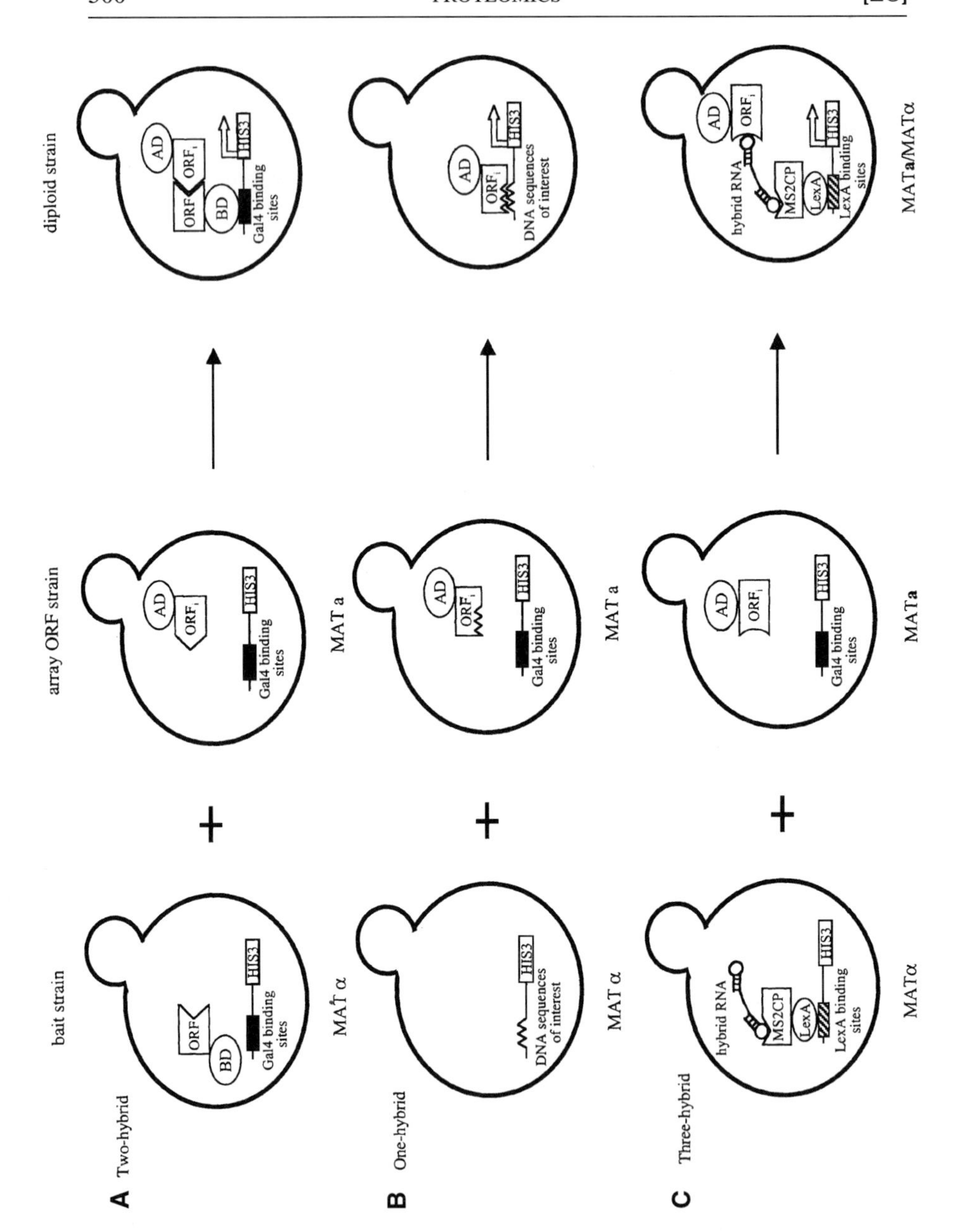
A Two-hybrid
bait strain
array ORF strain
diploid strain
ORF
BD
AD
ORF_i
Gal4 binding sites
HIS3
MAT α
MAT a
B One-hybrid
DNA sequences of interest
C Three-hybrid
hybrid RNA
MS2CP
LexA
LexA binding sites
MATα
MATa
MATa/MATα

interactions.[10] In the one-hybrid assay[5,6] (Fig. 1B), a DNA sequence of interest is inserted upstream of a reporter gene that contains no upstream activating sequence, and an interaction between this inserted sequence and a protein fused to the Gal4 AD leads to transcriptional activation. In this manner, it is possible to identify DNA-binding proteins that interact with a particular sequence. In the three-hybrid assay[7] (Fig. 1C), the localization of the Gal4 AD to a reporter gene is accomplished by the interactions of two protein fusions with a hybrid RNA molecule. In this method, there is one protein consisting of a DNA-binding protein (LexA repressor) fused to a characterized RNA-binding protein (bacteriophage MS2 coat protein), another protein consisting of the Gal4 AD fused to a second RNA-binding protein, and a hybrid RNA composed of MS2 coat protein binding sites fused to an RNA sequence of interest. Interactions between the hybrid RNA sequence and both the MS2 coat protein and the other RNA-binding protein lead to formation of a ternary complex and result in reporter gene activation. All three of these assays lead to the identification of putative interactions, which can be verified by other approaches.

The array we have constructed to conduct two-hybrid screens is a spatially ordered set of separated transformants, each potentially expressing one of the ~6000 open reading frames (ORFs) in *Saccharomyces cerevisiae* as a fusion to the Gal4 AD.[3] To screen the array, we use a mating strategy whereby a strain expressing a protein as a fusion to the Gal4 DBD (the "bait" protein) is mated to each of the array transformants.[11] The diploid positives that arise in these screens (potential interacting proteins) are easily identifiable based on their location within the array. Additionally, each element serves as a reference for adjacent elements such that several thousand assays can be compared for each bait screened. A particularly useful

[10] P. Bartel and S. Fields (eds.), "The Yeast Two-Hybrid System." Oxford University Press, Oxford, UK, 1997.

[11] G. Cagney, P. Uetz, and S. Fields, *Methods Enzymol.* **328,** 3 (2000).

FIG. 1. Mating strategy employed for one-, two-, and three-hybrid screens. (A) For two-hybrid screens, a Gal4 DBD protein fusion (bait) is expressed in an α strain containing a Gal4-dependent *HIS3* reporter gene (black boxes). This strain is mated to each of the ~6000 array strains expressing an individual Gal4AD-ORF fusion and also containing a Gal4-dependent *HIS3* reporter gene. A protein–protein interaction leads to transcriptional activation of the *HIS3* reporters present in the diploids. (B) For one-hybrid screens, a target DNA sequence (jagged line) is inserted in a reporter gene such as *HIS3* and the resultant strain mated to the array strains. A DNA–protein interaction is identified by reporter gene expression when an Gal4AD-ORF fusion binds to the target sequence. (C) In the three-hybrid assay, an α strain is generated that expresses a hybrid RNA containing MS2 coat protein binding sites fused to the RNA sequence of interest. This strain also expresses a fusion between the *Escherichia coli* LexA repressor and the MS2 bacteriophage coat protein and contains a *HIS3* reporter regulated by LexA binding sites (striped boxes). This strain is mated to the 6000 array strains. Interactions between the RNA sequence of interest and an Gal4AD-ORF fusion results in reporter activation.

feature of this comparison is that specific positives that appear in screens of many unrelated proteins can be identified and eliminated as false positives. Also, since each potential interacting partner exists only at one location within the array, the problem of identifying many positives corresponding to a single interaction, which can occur in library searches, is eliminated. The array can also be made to high density to facilitate the screening of many proteins. Our emphasis in this chapter is to describe the construction, maintenance, and use of these large-scale arrays in screening experiments.

Construction of Protein Array for *Saccharomyces cerevisiae*

An array can be constructed that contains nearly all of the predicted ORFs contained within the *S. cerevisiae* genome sequence, expressed as fusions to the Gal4 AD in an **a** mating type strain. An example of such an array of ~6000 transformants, spotted onto solid YPD media, is shown in Fig. 2. Generation of these tranformants requires the cloning of each of the full-length ORFs into

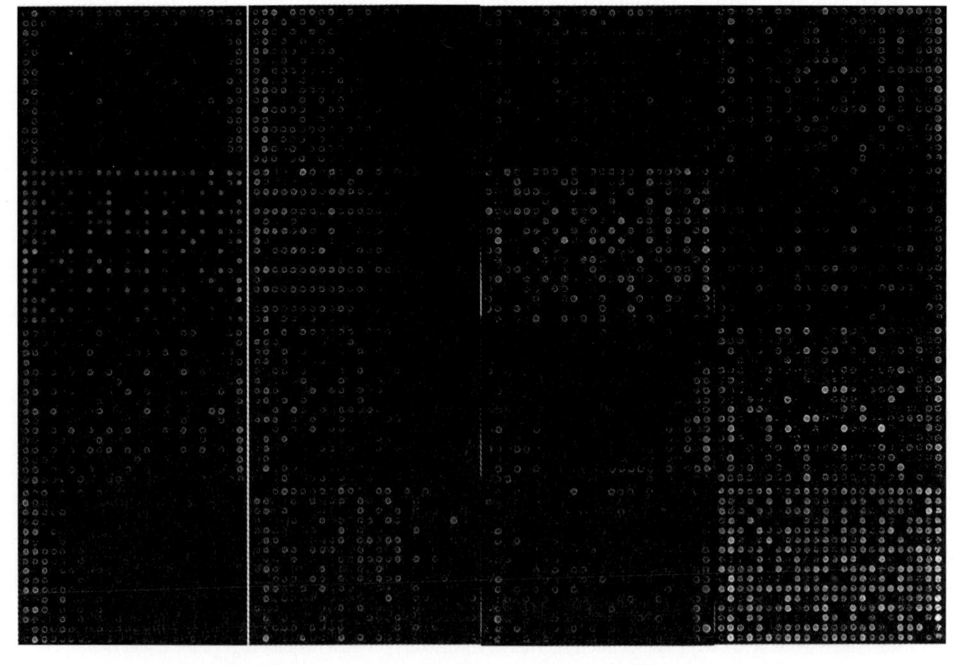

FIG. 2. Array of 6144 yeast transformants expressing each yeast ORF fused to the Gal4 activation domain. Sixteen OmniPlates each containing 384 colonies have been grown on YPD media. Spaces within the array that are empty correspond to gene products that are toxic when overexpressed in PJ69-4a, or the transformation has failed for other reasons.

the activation domain plasmid pOAD by homologous recombination.[12–14] Each of the ORFs can be amplified using gene-specific primers, and then amplified again to add additional vector-specific flanking sequences required for efficient recombinational cloning (Fig. 3).[15] The initial amplification is made possible by a set of primers (available from Invitrogen, Inc., Carlsbad, CA) containing 20 to 30 bases of complementarity to the gene of interest and a common tail of 20 bases of sequence homologous to vector sequences present in both the two-hybrid vectors pOAD and pOBD2.[3,15] Following this amplification, each ORF is reamplified with a single set of primers to increase the length of homologous vector sequence present at the ends of the first-round PCR products. These second-round primers, of 70 bases each in our case, contain sequence complementary to the common 20 base ends of the first-round amplification products and 50 bases of additional vector sequence. In total, each ORF will be generated with 70 base pairs of flanking vector sequence on each end (Fig. 3). Efficient recombinational cloning in yeast typically can be done with as few as 30 to 40 bases of homology between the insert and vector.[14] Primer sequences and typical amplification conditions are shown below.

Amplification of ORFs

Initial Gene-Specific Amplification Primers with Common 20-mer Ends

*Pvu*II *Nco*I

Forward primer: 5′-AA TTC CAG CTGACC ACC ATG XXX$_{20\text{-}30}$-3′

Reverse primer: 5′-GA TCC CCG GGA ATT GCC ATG XXX XXX$_{20\text{-}30}$-3′

Secondary Amplification Primers (70-mer)

Forward primer: 5′-C TAT CTA TTC GAT GAT GAA GAT ACC CCA CCA AAC CCA AAA AAA GAG ATC GAA TTC CAG CTG ACC ACC ATG-3′

Reverse primer: 5′-C TTG CGG GGT TTT TCA GTA TCT ACG ATT CAT AGA TCT CTG CAG GTC GAC GGA TCC CCG GGA ATT GCC ATG-3′

An X represents gene specific sequences within the forward and reverse first-round primers. The first triplet (XXX) in the reverse primer represents the reverse complement of one of the three possible stop codons. ORFs can be amplified as follows.

[12] T. L. Orr-Weaver, J. Szostack, and R. Rothstein, *Proc. Natl. Acad. Sci. U.S.A.* **78,** 6354 (1981).

[13] H. Ma, S. Kunes, P. Schatz, and D. Botstein, *Gene* **58,** 201 (1987).

[14] K. Oldenburg, K. Vo, S. Michaelis, and C. Paddon, *Nucleic Acids Res.* **25,** 451 (1997).

[15] J. Hudson, E. Dawson, K. Rushing, C. Jackson, D. Lockshon, D. Conover, C. Lanciault, J. Harris, S. Simmons, R. Rothstein, and S. Fields, *Genome Res.* **7,** 1169 (1997).

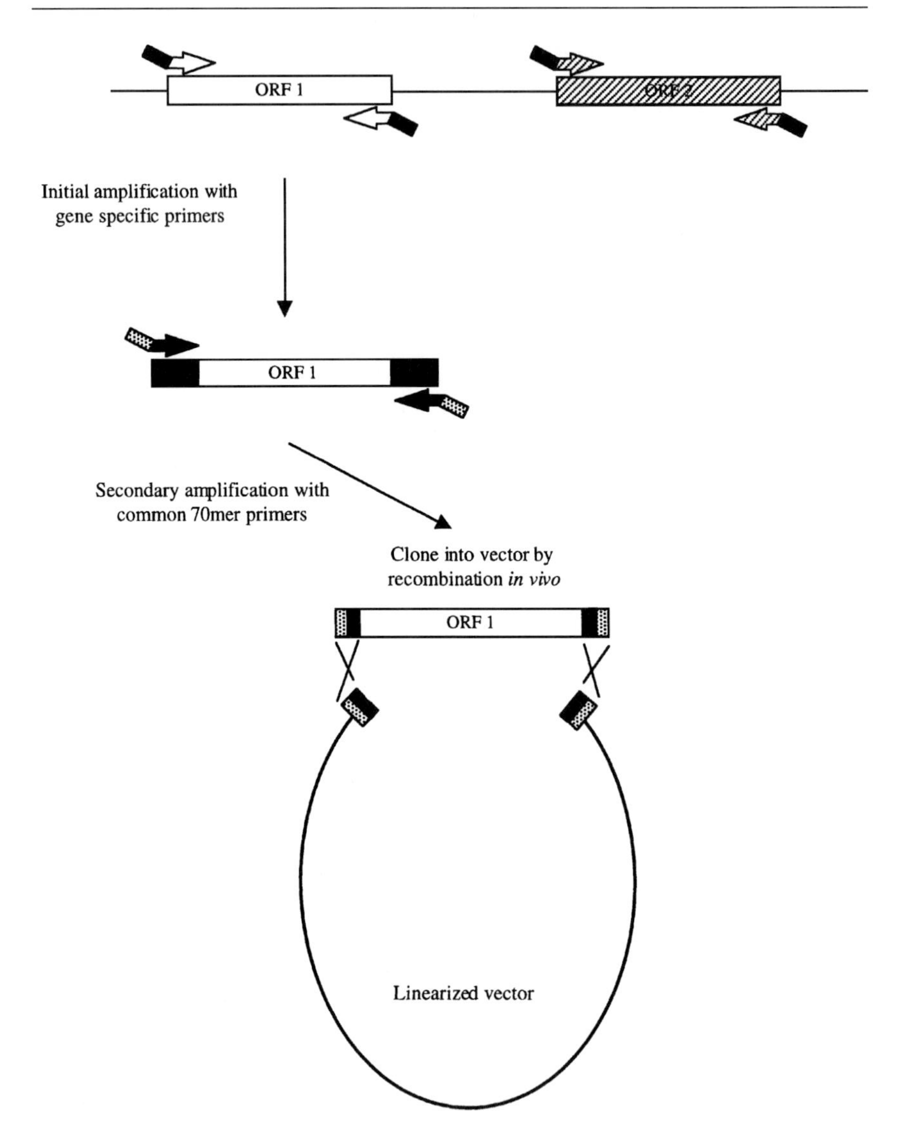

FIG. 3. Schematic diagram showing the cloning strategy employed. Each ORF in the genome is amplified using a set of gene-specific primers tailed with 20 nucleotides of vector sequence (open arrows). These PCR products are then amplified again with a single set of primers hybridizing to the 20 bp of common sequence present at the ends of each ORF. The secondary amplification primers are tailed with 50 nucleotides of additional vector sequence (closed arrows). Each ORF following the secondary amplification contains 70 bp of flanking vector sequence. The ORFs are then cotransformed with linearized vector DNA into an appropriate strain to generate in-frame fusions to either the Gal4 DBD or AD.

Step 1. Primers should be PAGE purified. A standard initial amplification reaction contains 30 ng of template genomic DNA, 1.5 m*M* MgCl, 5 units of *Taq* polymerase, 0.003 units of *Pfu* polymerase, and 20 pmol of each gene-specific primer.

Step 2. Cycling conditions are: one PCR cycle at 95° for 3 min, followed by 36 cycles of 50° for 45 sec, 72° for 210 sec, 95° for 60 sec, and a final cycle of 72° for 8 min.

Step 3. For the secondary amplification, we typically use only 5 ng of the initial amplification product, 0.6 units of *Taq* polymerase, and 25 pmol of each common 70-mer primer.[15] Some ORFs can be difficult to amplify for a variety of reasons. In many instances, troublesome amplifications have required altering conditions to eliminate primer dimerization by employing a "hot start" cycling procedure.

Step 4. Cycling conditions are 94° for 45 sec, followed by 35 two-temperature cycles of 94° (15 sec) and 72° (2–12 min, depending on ORF size).

Each amplification product should be confirmed by analyzing its size.

Cloning by Homologous Recombination

The *S. cerevisiae* ORFs can be cloned by recombination into the plasmids pOAD and pOBD2 (Fig. 4). The vector used to generate the hybrid RNA for use in the three-hybrid assay is pIII/MS2-2 (Fig. 4). For a one-hybrid assay, any convenient vector, such as pHISi-1 (Clontech, Palo Alto, CA), can be used that contains a suitable yeast marker. It should be noted, however, that the reporter construct containing a target element in the plasmid pHISi-1 must be integrated into a yeast strain that can then be mated to the array strains. Linearized plasmid DNA is cotransformed along with the amplification product into yeast, which recombines the two molecules to form a closed circular plasmid in which the ORF is joined in-frame to the activation or DNA-binding domain of Gal4. pOAD contains a *LEU2* marker and pOBD2 a *TRP1* marker, while the hybrid–RNA plasmid contains a *URA3* marker. Plasmids can be introduced into the following strains. The genotypes are shown below:

PJ69-4a[16] *MAT***a** *trp1-901, leu2-3,112, ura3-52, his3-200, gal4Δ gal80Δ LYS2::GAL1-HIS3, GAL2-ADE2, met2::GAL7-lacZ*

R40coatK *MATα ura3-52, leu2-3, 112, his3Δ200, trp1Δ1, ade2, LYS2::(lexA op)-HIS3, ura3::(lexA-op)-lacZ, LexA-MS2 coat protein (TRP1), gal4::Kan*

[16] P. James, J. Halladay, and E.A. Craig, *Genetics* **144,** 1425 (1996).

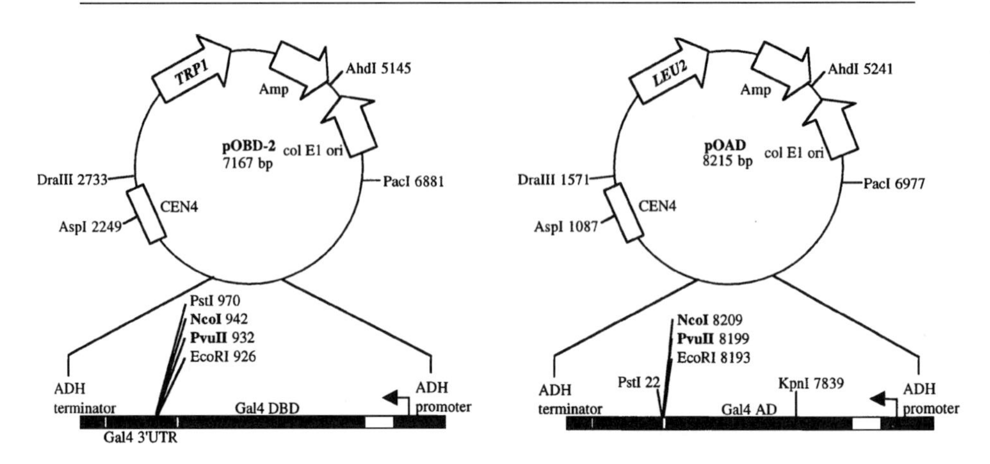

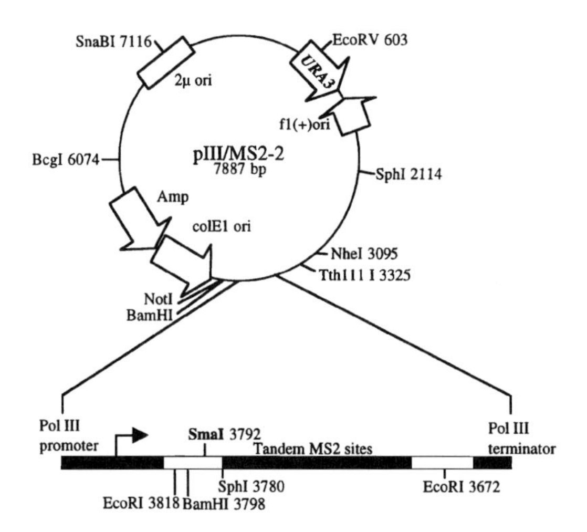

FIG. 4. Yeast shuttle vectors used in screening experiments with the Gal4 activation domain–ORF protein fusion array. Particular unique restriction sites are shown and sites where sequences have been inserted to generate appropriate fusions are in boldface type.

pOBD2 plasmids can be introduced into PJ69-4α, which is identical to PJ69-4a but of the opposite mating type. Large-scale transformation of the amplified ORFs into the two-hybrid vectors can be carried out as follows.

Prepare the following reagents:

1. Carrier DNA solution is prepared by dissolving 7.75 mg/ml of salmon sperm DNA (Sigma, St. Louis, MO) in water followed by a 15 min autoclave cycle.
2. 45% PEG solution contains 45.6 g polyethylene glycol (PEG-3350) (Sigma),

6.1 ml 2 *M* lithium acetate, 1.14 ml 1 *M* Tris-HCl pH 7.5, and 232 μl of 0.5 *M* EDTA dissolved in water.

Step 1. Prepare a 50-ml overnight culture of the host strain (PJ69-4a or PJ69-4α) and grow at 30°. Allow cells to grow for a minimum of 15 hr ($OD_{600} = 0.7$).

Step 2. Harvest the cells by centrifugation, remove supernatant and resuspend in 2 ml of 100 m*M* lithium acetate.

Step 3. Boil 580 μl of the carrier DNA solution for 5 min and place on ice.

Step 4. Add 20.73 ml of the 45% PEG solution, 580 μl of boiled carrier DNA, 200 ng of linearized plasmid DNA, and 2.62 ml of DMSO.

Step 5. Add the 2 ml of cells in lithium acetate to the solution in step 4 and vortex.

Step 6. Pour the mixture into a sterile trough and pipette 245 μl into each well of a 96-well microtiter plate.

Step 7. Add 3 μl of insert DNA to each well, seal with plastic, and vortex for 4 min.

Step 8. Incubate for 30 min at 42°.

Step 9. Centrifuge the microtiter plate to pellet the cells and remove the supernatant from each well using an 8-channel micropipetter.

Step 10. Add 200 μl of water to each well and resuspend the cells.

Step 11. Transfer each of the cells from the wells to individual 35-mm plates containing selective media appropriate for the plasmid used in the transformations and incubate at 30° for 2 to 3 days.

Spotting of Array Elements

From the transformants that grow on selective media, colonies can be picked individually or pooled to constitute an individual element within an array. Errors in primer synthesis, amplification, and recombinational cloning were determined to result in approximately a 10–15% failure rate in generating the correct ORF-AD cloning during the construction of our array.[3] In our experiments, we pooled two transformants in order to increase the likelihood that the desired ORF might be expressed within an array element. Transformants from each 35-mm plate were grown up individually in 96-well microtiter plates in SD–Leu media at 30° and the two cultures were then combined and aliquots spotted onto solid media contained in single-well OmniTrays (Nalge Nunc, Rochester, NY). We typically store the spotted array on SD-Leu media for up to 4 months from which a working copy can be generated onto YEPD to be used in screening procedures.

To save a permanent copy of each transformant, transfer cells to 96-well plates containing YEPD media and grow for 2 days at 30°. Add glycerol to each of the 96-well liquid cultures and freeze at –70°.

Validation of Array

If possible, confirming the expression of each of the in-frame fusions by Western analysis would be ideal. Since this may prove impractical in most situations, sequencing the initial base pairs of the ORF and the junction sequences flanking it provides an indication that the correct fusion gene has been generated. One possible means of determining expression might be to introduce a selectable marker in frame with the ORF that allows a simple phenotypic selection. However, this approach would not rule out the possibility of deleterious point mutations or deletions and insertions occurring in the ORF that are not readily detected in such a strategy. It also results in a more complex hybrid protein in which the ORF has additional protein sequences at both its N and C termini.

Screening Procedures for One-, Two-, and Three-Hybrid Analysis

A mating strategy is used to conduct screens with protein, DNA, or RNA baits. The array contains haploid cells of **a** mating type potentially expressing each of the Gal4AD-ORF fusions. A given bait is expressed in a strain of α mating type, the two strains are mated, and the diploids selected by plating on medium that requires the combination of plasmid markers on the appropriate plasmids required for each type of screen. The diploids are then transferred to medium selective for both the plasmid markers as well as for reporter gene expression. For example, in two-hybrid screens, a bait (pOBD2-ORF fusion) is expressed from a plasmid containing a *TRP1* marker in strain PJ69-4α. This strain is then mated to each of the array strains which potentially express an Gal4AD-ORF fusion from a plasmid containing a *LEU2* marker. The diploids are selected on SD-Leu, -Trp media. These diploid cells are then transferred onto media selective for reporter gene activity (SD-Leu, -Trp, -His). In the case of two-hybrid screens, the *HIS3* reporter gene is under control of Gal4 DNA-binding sites (PJ69-4a and PJ69-4α). The two-hybrid strains also contain a Gal4-based *ADE2* reporter, and more stringent screens can be done utilizing this reporter. However, use of this reporter runs the risk of not detecting weaker interactions, and so we have chosen to use only the *HIS3* reporters in our experiments. We also add a competitive inhibitor of His3 to the media when selecting for reporter activity. 3-Aminotriazole (3-AT) reduces the background of growth due to *HIS3* reporter gene activity in the absence of a protein–protein or protein–nucleic acid interaction.

For one-hybrid screens, a strain must be constructed that contains the *cis*-acting element of interest proximal to a suitable reporter. This strain must also be of the opposite mating type to that of the array strains and carry a disrupted *GAL4* locus if the *HIS3* gene is used as the reporter. Disruption of the *GAL4* gene is necessary so that the Gal4-dependent *HIS3* reporter present in the strains of the array is not globally activated. In this type of screen, the α strain expressing a *HIS3* reporter, for example, would be mated to the array strains and the diploids selected on SD-His, -Leu. The diploids are then transferred to media lacking histidine and leucine and containing 3-AT.

A three-hybrid screen can be conducted in the strain R40coatK by expressing an RNA bait from plasmid pIIIMS2-2 which contains a *URA3* marker. This strain contains a LexA-MS2 coat protein fusion gene that has been stably integrated in the genome along with the *TRP1* gene. The bait-expressing strain is mated to the array strains and the diploids selected for on SD-Trp, -Ura, -Leu. The diploids are then transferred to SD-Trp, -Ura, -Leu, -His and containing 3-AT to detect the expression of the LexA operator-based *HIS3* reporter present in the R40coatK strain.

Before carrying out one-, two-, or three-hybrid screens, it is advisable to predetermine the reporter gene activity with a given bait prior to conducting a screen. In some cases, screens with various proteins, DNAs, or RNAs can be hampered by strong reporter activation alone of the DBD protein fusion (two-hybrid screen), DNA sequence (one-hybrid screen), or hybrid RNA molecule (three-hybrid screen). In fact, titrating the haploid strain expressing either the DBD fusion, new reporter gene, or hybrid RNA fusion bait on a series of 3-AT concentrations prior to the screen is advisable to determine the amount needed to reduce the background of growth in the absence of histidine. Typical concentrations of 3-AT to use for this titration are 5, 10, 25 and 50 m*M* (although we have carried out screens at up to 150 m*M* 3-AT). Alternatively, screens with background reporter activity can be carried out using the more stringent *ADE2* reporter gene. It has also been observed that some DBD protein fusions can inhibit mating efficiencies, which makes screening with these baits more problematic. To date, an RNA bait that inhibits the mating step has not been encountered, although we have done many fewer RNA searches against the array compared to the number of protein baits screened.

If possible, it is advisable to carry out as many of the procedures of constructing and assaying the array as is feasible by robotic methods in order to decrease the possibility of introducing errors and to increase throughput. These unfortunately are not trivial issues because of the sheer numbers of colonies involved. We have used a Biomek 2000 (Fig. 5A) from Beckman Coulter, Inc., Fullerton, CA, as the robotic workstation to spot colonies from liquid media on to solid media, to create 384 colony plates from 96 colony ones, and to transfer colonies in each step of the screening protocol. These transfers are carried out using high-density replicating tools with 96 or 384 pins (Fig. 5B). Although manual transfer of colonies can be carried out using a similar manual pinning device (such as manufactured by Nalge Nunc), this procedure is tedious, more likely to result in errors, and not as accurate for reproducibly pinning colonies directly on top of others, as is required for mating.

The working surface of a Biomek 2000 has capacity for 12 tools or plates, which are configured according to the task and graphically displayed in a software program from Beckman Coulter, Inc. (Fig. 6). Typically, for two-hybrid screens, the

A

B

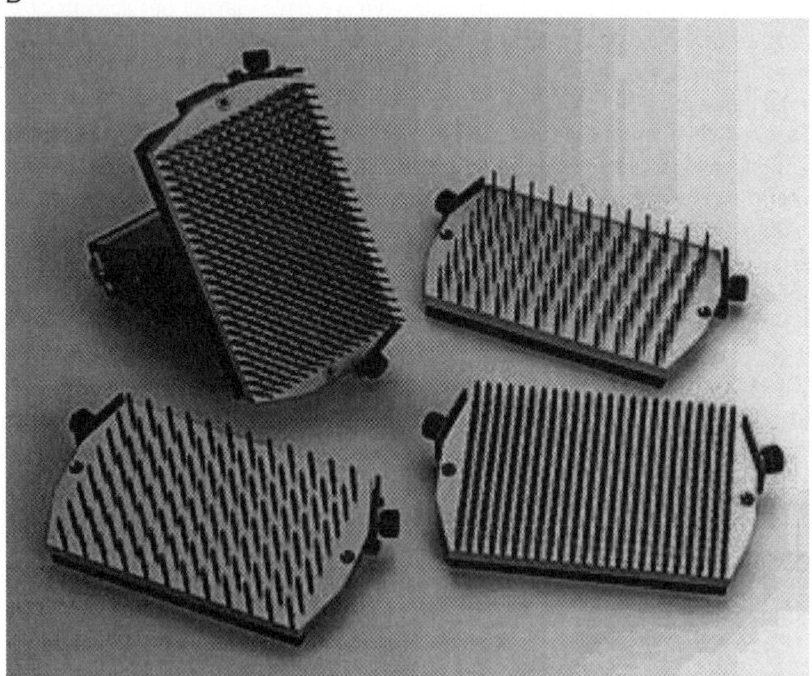

FIG. 5. (A) The Biomek 2000 workstation. (B) High-density replica pinning tools for the Biomek 2000. [Photos compliments of Beckman Coulter, Inc.]

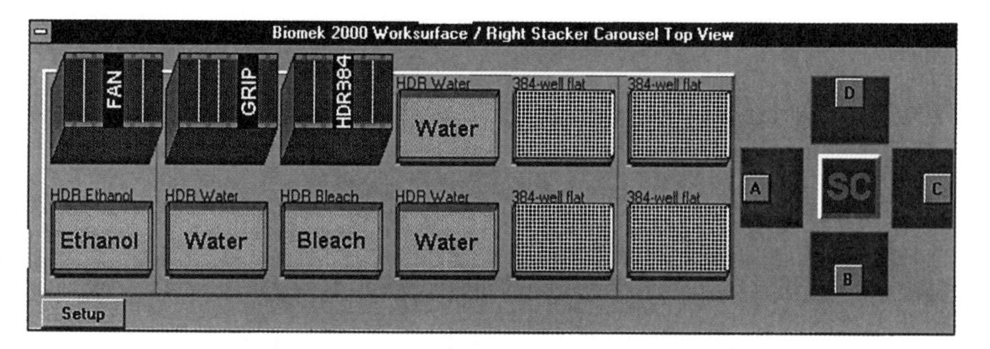

FIG. 6. A Bioworks software window displaying the Biomek 2000 work surface and right stacker carousel (SC), with tools, reservoirs, and plates positioned appropriately for an array screen. The sterilization cycle involves sequentially dipping the pins into water, bleach, water, and ethanol, followed by drying with the fan. The pins are then prewetted with sterile water before continuing the pinning cycle. [Photo compliments of Beckman Coulter, Inc.]

surface would contain a fan for drying the pins after a sterilization cycle, a pinning tool, a gripping tool, reservoirs of ethanol, water, and bleach for sterilization and one of water for wetting the pins, and source plates and destination plates that will carry the array. The Biomek can be configured with a stacker carousel that has a total capacity of 80 plates in four stackers of 20. Up to 20 bait proteins can be screened in a week, a number that can be scaled up by increasing either the density of the array or the number of workstations. Below is our standard protocol for screening the array in a two-hybrid format. Differences in the protocol when screening for one- and three-hybrid interactions are noted.

Prepare the following reagents:

Single-well OmniPlates containing either YEPD plus adenine or appropriate selective medium

Step 1. Grow an overnight 6-ml culture in YEPD of the strain expressing the bait plasmid. Strains that express baits that inhibit mating or reduce growth can be grown in three- to fivefold larger volumes and can be concentrated by centrifugation to achieve higher density cultures.

Step 2. Pour the culture on top of a sterile Omniplate containing solid YEPD medium. Transfer the bait expressing strain by pinning to 16 Omniplates containing solid YEPD medium. Make sure all plates are sufficiently dry such that colonies do not run together. Allow the spotted yeast to dry for 20 to 30 min.

Step 3. Pin the colonies from the ORF array directly on top of the dried bait yeast using a pinning device (Biomek workstation, Beckman, 384 pinning device). Alternatively, colonies from the array can first be pinned onto YEPD, followed by transfer of the bait-expressing strain onto the pinned array.

Step 4. Incubate for 1 to 2 days at 30° to allow mating of the two strains.

Step 5. Transfer the colonies to Omniplates containing SD −Leu, −Trp selective medium. In the case of a one-hybrid screen, transfer the colonies to SD −Leu, −His. For a three-hybrid screen, transfer the colonies to SD −Leu,−Trp,−Ura medium.

Step 6. Incubate for 2 days at 30° to allow the diploid cells to grow.

Step 7. Transfer the colonies to Omniplates containing SD −Leu, −Trp, −His selective medium supplemented with an appropriate amount of 3-AT. For nonactivating DBD-ORF strains we typically use 3 m*M* 3-AT. In the case of a one-hybrid assay transfer colonies to SD −Leu, −His and containing a predetermined amount of 3-AT to inhibit background growth. For a three-hybrid screen, plate onto SD −Leu, −Trp, −Ura, −His selective medium containing 5 m*M* 3-AT.

Step 8. Incubate for 1 to 3 weeks at 30°.

Colonies that grow above background are considered positive interactions. During the course of the two-hybrid experiments, we found that many of the positives that arise during these screens are not reproducible. We typically conduct two screens with a given bait protein and consider only positives that arise in both screens as potential interactors. These interactions can then be confirmed using other techniques. However, positives that appear in many screens are classified as false positives.

In the methods described here, we have used yeast ORFs, in which introns are a rarity. This has allowed the direct amplification from genomic sequences in order to clone each of the predicted ORFs. To adapt these procedures to higher eukaryotic systems will require a source of full-length cDNAs. Advances in 5′ end sequence selection and isolation of full-length cDNAs should aid in the adaptation of the assays described here to other systems.[17]

Acknowledgments

This work was supported by NIH Grants GM54415 and RR11823 and by a grant from the Merck Genome Research Institute. J.F.G. and T.R.H. are associates and S.F. is an investigator of the Howard Hughes Medical Institute.

[17] S. Bashiardes and M. Lovett, *Curr. Opin. Chem. Biol.* **5,** 15 (2001).

[29] Building Protein–Protein Networks by Two-Hybrid Mating Strategy

By MICHELINE FROMONT-RACINE, JEAN-CHRISTOPHE RAIN, and PIERRE LEGRAIN

Introduction

The availability of the complete genomic sequences of many different organisms makes the characterization of the biological role of the genes, most of which encode proteins, an important challenge. The function of proteins can often be elucidated by the network of interactions in which the protein is involved. Therefore, an efficient approach for elucidating the function of unknown proteins is to use a large-scale two-hybrid assay to produce protein interaction maps.

The basic concept of the yeast two-hybrid system is to detect the interaction between two proteins via transcriptional activation of a reporter gene.[1] A classical eukaryotic transcription activator contains a domain that specifically binds to DNA sequences (the binding domain, BD) and a domain that recruits the transcription machinery (the activation domain, AD). In the two-hybrid system, these two domains are distinct polypeptides, each fused to a polypeptide, X and Y, respectively. The basis of the assay is that transcription will occur only if X and Y interact. The yeast two-hybrid system can detect interactions between two known proteins or polypeptides (the "matrix approach") and can also search for unknown partners (prey) of a given protein (bait) (the whole library screening approach) (for review, see Ref. 2). Briefly, the "matrix approach" uses a collection of predefined open reading frames (ORFs) as both bait and prey for interaction assays. Combinations of bait and prey can be assessed individually or after pooling cells expressing different bait or prey proteins. It was first used to explore interactions among *Drosophila* cell cycle factors.[3] Large-scale approaches have been published for the vaccinia virus[4] and for the yeast proteome.[5–7] The whole-library screening

[1] S. Fields and O. Song, *Nature* **340,** 245 (1989).

[2] M. Vidal and P. Legrain, *Nucleic Acids Res.* **27,** 919 (1999).

[3] R. L. Finley, Jr. and R. Brent, *Proc. Natl. Acad. Sci. U.S.A.* **91,** 12980 (1994).

[4] S. McCraith, T. Holtzman, B. Moss, and S. Fields, *Proc. Natl. Acad. Sci. U.S.A.* **97,** 4879 (2000).

[5] T. Ito, K. Tashiro, S. Muta, R. Ozawa, T. Chiba, M. Nishizawa, K. Yamamoto, S. Kuhara, and Y. Sakaki, *Proc. Natl. Acad. Sci. U.S.A.* **97,** 1143 (2000).

[6] T. Ito, T. Chiba, R. Ozawa, M. Yoshida, M. Hattori, and Y. Sakaki, *Proc. Natl. Acad. Sci. U.S.A.* **98,** 4569 (2001).

[7] P. Uetz, L. Giot, G. Cagney, T. A. Mansfield, R. S. Judson, J. R. Knight, D. Lockshon, V. Narayan, M. Srinivasan, P. Pochart, A. Qureshi-Emili, Y. Li, B. Godwin, D. Conover, T. Kalbfleisch, G. Vijayadamodar, M. J. Yang, M. Johnston, S. Fields, and J. M. Rothberg, *Nature* **403,** 623 (2000).

METHODS IN ENZYMOLOGY, VOL. 350
0076-6879/02 $35.00

approach was first applied to determine protein networks for the T7 phage proteome which contains 55 proteins.[8] In 1997, we used a version of the two-hybrid system to screen the complete yeast genome selectively, and to identify a limited set of proteins that potentially interact with a given protein.[9] In this preliminary study, several proteins known to be involved in pre-mRNA splicing were used as bait in the first round of two-hybrid screening. Based on the information gained from the selected sets of prey fragments, a new series of bait proteins was used for a second round of screens. This approach of multicycle two-hybrid screens led to the characterization of a network of interactions. Moreover, starting from known proteins, this approach identified new factors and suggested novel functional links between biological pathways. This study was followed by others,[10–13] which emphasized that the main advantage of this approach is the characterization of fragments of proteins that encode interacting domains. In addition, when used on a large scale, this technique generates a large amount of experimental data on interactions. Therefore statistical analyses can be performed, giving rise to a predictive score for every single interaction.[13]

Nevertheless, the yeast two-hybrid system is an indirect genetic assay. Intrinsic limitations of the yeast two-hybrid system include reliance on complex transcriptional activation of reporter genes. Incorrect folding, inappropriate subcellular localization (bait and prey proteins must interact in a nuclear environment), or degradation of chimeric proteins and absence of certain types of posttranslational modifications in yeast could lead to false negatives. Other properties of the assay may lead to the selection of false positives. For example, bait proteins might have a predisposition to activate the transcription of reporter genes above the threshold level and some prey proteins or fragments might be selected in a two-hybrid assay in combination with a wide variety of bait proteins. Confirmation of interactions should be obtained via various independent genetic, biochemical or functional assays, such as copurification in complexes, colocalization, or demonstration of a functional association.

[8] P. L. Bartel, J. A. Roecklein, D. SenGupta, and S. Fields, *Nat. Genet.* **12,** 72 (1996).

[9] M. Fromont-Racine, J. C. Rain, and P. Legrain, *Nat. Genet.* **16,** 277 (1997).

[10] A. Flores, J. F. Briand, O. Gadal, J. C. Andrau, L. Rubbi, V. Van Mullem, C. Boschiero, M. Goussot, C. Marck, C. Carles, P. Thuriaux, A. Sentenac, and M. Werner, *Proc. Natl. Acad. Sci. U.S.A.* **96,** 7815 (1999).

[11] M. Flajolet, G. Rotondo, L. Daviet, F. Bergametti, G. Inchauspé, p. Tiollais, C. Transy, and P. Legrain, *Gene* **242,** 369 (2000).

[12] M. Fromont-Racine, A. E. Mayes, A. Brunet-Simon, J.-C. Rain, A. Colley, I. Dix, L. Decourty, N. Joly, F. Ricard, J. D. Beggs, and P. Legrain, *Yeast Comparative Functional Genomics* **17,** 95 (2000).

[13] J. C. Rain, L. Selig, H. De Reuse, V. Battaglia, C. Reverdy, S. Simon, G. Lenzen, F. Petel, J. Wojcik, V. Schachter, Y. Chemama, A. Labigne, and P. Legrain, *Nature* **409,** 211 (2001).

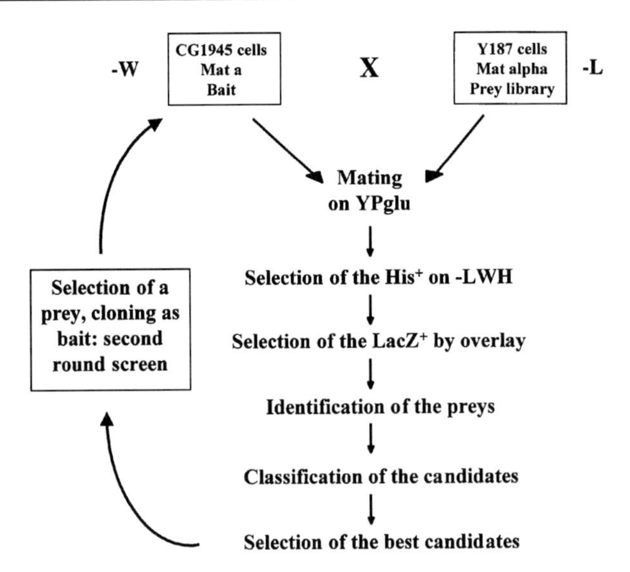

FIG. 1. Schematic representation of multiple round two-hybrid screens of a genomic library by a mating strategy.

General Strategy

To take full advantage of the two-hybrid genomic library screening technology, a reproducible experimental protocol that can screen the entire library must be performed. For this purpose, we have developed an efficient mating strategy (see Fig. 1).[9] In brief, we use a CG1945 strain transformed with the bait plasmid and a Y187 strain transformed with a yeast genomic DNA library (FRYL library). Thirteen million Y187 colonies have been transformed with the library and collected, pooled, and stored in aliquots at -80°. This procedure allows us to perform highly reproductive screens using the same genomic library for hundreds of screens. Moreover, the mating procedure allows us to use selective plates for direct selection because the two fusion proteins are already produced by the parental cells. No replica plating is required. Prey plasmids from all colonies are sequenced at the Gal4 domain junction and prey fragments are classified according to their heuristic value. The classification of interacting proteins found in a two-hybrid screen allows us to select prey proteins to be used as bait for second-round screens.

Materials and Methods

Strains

Y187 (*MAT*α *gal4*Δ *gal80*Δ *ade2-101, his3-200, leu2-3,-112, trp1-901, ura3-52, URA3 :: GAL1*$_{UAS}$ *GAL1*$_{TATA}$*LacZ, met*)

CG1945 (*MAT***a,** *gal4-542, gal80-538, ade2-101, his3-200, leu2-3,-112, trp1-901, ura3-52, lys2-801, URA3 :: $GAL4_{17mers\,(x3)}$ $Cyc1_{TATA}LacZ$ LYS2:: $GAL1_{UAS}GAL1_{TATA}HIS3$, CYH^R*)

L40ΔG (*MAT***a,** *ade2, trp1-901, leu2-3,-112, lys2-801am, his3Δ200, LYS2 :: $(lexAop)_4$-HIS3, URA3 :: $(lexAop)_8$-lacZ,Δgal4 :: $KANA^R$*)

Bait Plasmids

Bait plasmids are usually derived from the pAS2ΔΔ vector. This vector is derived from the pAS2 vector (Clontech, Palo Alto, CA) by deletion of the CYH2 cassette, the expression of which is toxic in yeast, and the HA epitope, which increases the rate of false positives.[9] pAS2ΔΔ derived plasmids are transformed into the CG1945 strain. Alternatively, bait plasmids are derived from the pBTM116 vector, which encodes a LexA fusion protein. These plasmids are transformed into the L40ΔG strain containing *His3* and *LacZ* reporter genes under the transcriptional control of LexA binding sites.[12]

Transfer of Genomic Library into Yeast Cells

Large-scale analysis of a genome by the two-hybrid approach requires a well-represented library. We have constructed a yeast genomic DNA library in pACTIIst.[9] Random fragments are generated by shearing the genomic DNA by sonication. We follow a similar procedure to that described by Elledge and colleagues.[14] A linker is added to the genomic DNA fragments before cloning. This prevents the cloning of multiple fragments in the same plasmid and the self-ligation of the vector. Given that a genomic DNA library is composed of random fragments, all the genomic regions should be present with more or less the same frequency. This can be assessed by randomly sequencing selected clones from the library and estimating the frequency of each genomic region potentially encoding an interacting domain. We have recovered 5×10^6 independent *Escherichia coli* clones (75% of which contain a genomic insert); these have been pooled to constitute the FRYL library. Given the size of the yeast genome (16×10^6 bp), a fusion event to the Gal4 activating domain occurs statistically once every four bases.

The Y187 yeast strain is transformed with the FRYL library DNA according to standard procedures. Typically, several hundred plates (90 mm diameter), each containing between 10,000 and 30,000 colonies, are used. We aim to collect three times more yeast colonies than the initial number of *E. coli* clones. This results in 95% probability of recovering the original clones. One severe limitation at this step is the occurrence of contaminant colonies (either bacteria or fungi) that will then be propagated. We add tetracycline (10 μg/ml) to the medium to avoid most bacterial

[14] S. J. Elledge, J. T. Mulligan, S. W. Ramer, M. Spottswood, and R. W. Davis, *Proc. Natl. Acad. Sci. U.S.A.* **88,** 1731 (1991).

contamination. Thirteen million transformed yeast colonies have been collected in rich medium, pooled, and stored in glycerol [40 g of 87% glycerol per 100 ml of the cell mixture] at −80°. Each aliquot contains at least 20 cells from each colony.

Mating Experiment

Two strains are preferentially used: Y187, which harbors a sensitive *lacZ* reporter gene for Gal4 binding domain bait proteins, and CG1945, which has a nonleaky *HIS3* reporter gene that enables the selection of positive diploids in the absence of 3-aminotriazole. The Y187 strain is transformed with the genomic library (see above) and CG1945 cells harbor GAL4 bait plasmids. For LexA bait plasmids, the strain L40ΔG is used. It is generated by replacing the entire GAL4 coding sequence in L40 with the kanamycin-resistance cassette.[12] Thus, Gal4 activation domain-derived libraries transformed into Y187 are used for two-hybrid screens with LexA bait proteins. Diploid cells generated by mating L40ΔG cells with Y187 cells contain two *lacZ* genes. The one from the L40ΔG strain is controlled by LexA binding sites and functions as a reporter gene for LexA baits, and the one from Y187 cells is controlled by Gal4 binding sites and is inactive in these diploid cells.

Methods

Day 1: Grow a preculture of CG1945 or L40ΔG cells carrying the bait plasmid at 30° in 20 ml medium lacking tryptophan (−W medium) with vigorous agitation.

Day 2: Inoculate 150 ml −W medium at OD_{600} 0.006 (may depend on bait plasmid and on local growth conditions). Grow overnight at 30° with vigorous agitation.

Day 3: The OD_{600} of the culture of CG1945 or L40ΔG cells carrying the bait plasmid should be around 1. For convenience, the protocol is given as an example for a vial containing 4×10^8 viable Y187 cells transformed with the yeast DNA library. For the mating step, twice as many bait cells as library cells are used.

Eighty OD_{600} units of bait culture (roughly 8×10^8 cells) are put into a 250-ml flask. (Do not leave the CG1945 cells without agitation because cells will aggregate.) Thaw a vial containing the library slowly on ice. Add the contents of the vial to 20 ml YPGlu. Let the cells recover at 30° for 10 min accompanied with gentle agitation. Add the Y187 cell suspension to the bait culture. Mix the library/bait cells by hand. Centrifuge the cells at 5000*g* for 3 min at room temperature. Resuspend the pellet with 2 ml of YPGlu medium. Spread 400 μl of the diploid mixture onto YPGlu plates. Incubate plates at 30° for 4 hr 30 min. To get a good mating efficiency cells must be collected at a density of 4.5×10^6 cells per cm^2.

Cells are collected from the plates with 3 ml of –LWH medium per plate. Then, plates are washed with an additional 3 ml of –LWH medium. This step is repeated with all plates. Approximately 30 ml of cell suspension is obtained. Shake the cell suspension and distribute cells in 400 μl aliquots over 75 –LWH plates. Incubate plates at 30° for 3 days. Typically, for unknown bait, the optimal selection conditions must first be determined by plating mated cells samples onto –LWH plates without or with 3-aminotriazole at various concentrations (ranging from 1 to 50 m*M*). Depending on the characteristics of the bait (e.g., its tendency to autoactivate reporter genes) it might be necessary to adapt selective pressure by adding a given concentration of 3-aminotriazole to the selective medium. To determine the mating efficiency, the number of diploids is estimated. For this purpose, an aliquot of cell suspension is diluted by three 10-fold dilution steps. Fifty μl of the 1000-fold dilution are spread onto −L and −LW plates and 50 μl of the 10,000-fold dilution are spread onto −W plates.

Day 5: Estimation of the number of diploids. Colonies corresponding to parental cells (−L and −W medium) or diploid cells resulting from the mating (−LW medium) are counted. The total number of diploid cells assayed for interaction in the screen is then calculated (−LW colonies $\times$ 20 $\times$ 30 $\times$ 1000). Three times more diploid cells than the number of original yeast colonies present in the library are usually assayed to cover 95% of the prey library.

β-Galactosidase Detection by X-Gal Overlay

To identify the $LacZ^+$ colonies, an X-Gal overlay assay is performed directly on the selective medium −LWH plates after scoring the number of His^+ colonies. This procedure is less sensitive than the filter assay but is faster and permits a better recovery of cells because it does not require breaking cells open in liquid nitrogen.

For every plate 10 ml overlay mixture is necessary: 5 ml 0.5 *M* phosphate buffer pH 7.5; 4.7 ml 1% (w/v) agar; 200 μl 2% (w/v) X-Gal in dimethylformamide (DMF); 100 μl 10% (w/v) sodium dodecyl sulfate (SDS).

A batch of overlay mix is prepared immediately before use by mixing 175 ml 0.5 *M* phosphate buffer, 165 ml 1% agar, and 3.5 ml 10% SDS. As DMF is toxic, 7 ml of 2% X-Gal is added under a chemical hood. The overlay mix is incubated in a water bath at 50° and 10 ml is poured onto every plate. The plates are incubated cell side up at 30°. The blue colonies are checked after 30 min, 1 hr, 2 hr, 4 hr, and 6 hr of incubation. The positive colonies are streaked on –LWH plates using a sterile toothpick (by picking through the agar overlay into the colony and collecting some cells). Viable cells can still be collected from positive colonies after 24 hr incubation time with the mix overlay but recovery from small colonies is sometimes poor.

Several points should be noted. (1) What counts as a "real" positive (i.e., blue color) can differ from one screen to another. It might be related to the nature of the bait fusion, or to its expression level. (2) Another factor is the sensitization of yeast cells to X-Gal. In a given screen, blue colonies might show up after 1 hr, whereas under slightly different conditions, it may take 6 hr. The intensity of the blue color also varies from deep dark to almost gray-light blue. Sometimes a blue halo is visible around the colonies or the color is restricted to the center of the colony. It is important to compare the color of putative positives under the same conditions for one experiment (colonies on the same plate or on plates from the same screen). (3) It is convenient to keep track of all this information (size of colonies, intensity of blue color, length of incubation to get colonies turning blue).

Stock Solutions

0.5 *M* Na_2HPO_4 pH 7.5 (71 g Na_2HPO_4 per 1 liter + 4 ml orthophosphoric acid), stored at 50° in 175 ml aliquots (not filtered or autoclaved)
1% Bacto-agar (1.65 g/165 ml), autoclaved and stored at 50°
2% X-Gal in DMF stored at −20°

Identification of Prey

PCR on Yeast Colonies. Genomic inserts are amplified from library plasmids by PCR on colonies, using oligonucleotides that are complementary to pACTIIst vector sequences.[12] This is an efficient procedure for the identification of sequences cloned into this plasmid. However, it is not a standardized protocol: it varies from strain to strain and depends on experimental conditions (number of cells, *Taq* polymerase source, etc). It should be optimized for specific local conditions.

Remove one colony with a toothpick. Resuspend the cells in 10 μl freshly prepared 0.02 *M* NaOH at room temperature. Incubate for 5 min at 100°. For a large series, vortex the tubes to counter cell sedimentation before incubation at 100°. Transfer to ice and briefly centrifuge the tube to spin down the drops of condensation on the cap. Resuspend the pellet by vortexing and add 2 μl of cell extract to the PCR mix.

For one reaction use 24.6 μl H_2O; 3.2 μl 10× PCR buffer; 0.7 μl 10 m*M* dNTP; 0.5 μl 20 μM oligo5′ ABS1 (5′-GCGTTTGGAATCACTACAGG-3′); 0.5 μl 20 μM oligo3′ ABS2 (5′-CACGATGCACAGTTGAAGTG-3′); 0.5 μl *Taq* polymerase). PCR cycling conditions: 31 cycles of 94° 30 sec; 55° 1 min 30 sec; 72° 3 min. Check the quality, quantity, and length of the PCR fragment on 1% agarose gel (load 3 μl). The length of the cloned fragment is the estimated length of the PCR fragment minus 300 bases corresponding to the amplified flanking vector sequences.

Sequence Analysis. PCR fragments can be sequenced without the purification step. However, an enzymatic pretreatment must be performed (according to

Amersham, Buckinghamshire, England) to degrade the oligonucleotides and the nucleotides present in the PCR product.

It is convenient to use a thermal cycler (block of 96 wells) and 0.2 ml 8 tube strips. Add 6 μl of PCR amplification product (do not vortex the tube to avoid taking the cellular extract at the bottom of the tube) to 1 μl of *Exo*I (10 units) and 1 μl of shrimp alkaline phosphatase (1 unit). Incubate at 37° for 15 min. Inactivate the enzymes by heating at 80° for 15 min. Chill on ice.

Add 4 μl of 0.8 pmole/μl primer JC90 5′-CGATGATGAAGATACCCCA CCAAA-3′ and 8 μl of Terminator Ready Reaction Mix (PerkinElmer, Norwalk, CT) to each tube. Place the 8 tube strips in the thermal cycler programmed for 10 sec at 96°; 5 sec at 50°; 4 min at 60°. Repeat for 25 cycles. Precipitate at room temperature with 75 μl of 70% (v/v) ethanol in 0.5 m*M* $MgCl_2$. Centrifuge at 3000*g* for 15 min at room temperature. Wash pellet with 200 μl of 70% ethanol. Resuspend the pellet in 5 μl of loading buffer [deionized formamide : dextran (5 : 1)]. The dextran blue solution is dissolved in 50 mg/ml in 25 m*M* EDTA (pH 8). Heat the samples at 95° for 3 min. Load 2 μl of the samples on a 6% acrylamide gel. [Dissolve 25 g urea with 19 ml of H_2O, 5 ml of 10× TBE, 7.5 ml of 40% acrylamide (Bio-Rad, Hercules, CA). Add 250 μl of 10% (w/v) APS and 23 μl of TEMED]. Run gel on ABI 373 machine according to manufacturer's instructions (PerkinElmer). Slight modifications of the procedure are required to run gels on an ABI 377 machine.

Classification of Preys

For each clone, a sequence file is created by the PerkinElmer sequencer. The file is sent to the DOGEL software developed at the Pasteur Institute by Nicolas Joly. This software recognizes and eliminates the 20 bases (after the *Bam*HI restriction site) from the linker added to the genomic DNA fragments to make the FRYL library. It analyzes mutations within the linker (including frameshifts).

It searches the chromosomal coordinates (chromosome number, coding strand or noncoding strand, position) of the beginning of the insert, by reading 50 nucleotides with the SGD blast program (SGD; http://genome-www.stanford.edu). It identifies the ORF and defines the exact location of the beginning of the insert in relation to the initiation codon. Biological information on the ORF can be extracted from the YPD database (YPD; http://www.proteome.com).

Given the size of the library, which contains a fusion point once every four bases, one expects to find a given ORF several times as independent clones selected in a complete screen of such a library. However, the probability of selecting a given fusion depends on the length of the interacting domain and on the position of the interacting domain on the coding sequence. These parameters predict that every candidate falls into one of the following categories of different heuristic values. A1 ORFs (the best one) are those for which at least two distinct overlapping fusions are found. A2, A3, and A4 ORFs are found only as a single fusion, even though the same fusion is found several times. A2 ORFs are fusions starting close to the initiation codon of a yeast ORF, within 150 bases of the in-frame stop codon located upstream of this ORF. A3 ORFs are fusions containing coding inserts of more than 1000 bases (significantly longer than the mean size of fragments in the library). A4 are the other candidates. B fusions express nonbiological peptides, i.e., antisense or intergenic regions. Out-of-frame fusions might be selected in the process. They are not discarded as they may encode genuine yeast polypeptides resulting from a frameshifting event.[9]

Applications

Building Network: Assignment of Biological Functions and Identification of New Complexes

We have carried out over 150 exhaustive genomic screens using yeast proteins. This two-hybrid database was used to assign potential functions to previously uncharacterized proteins, which are now embedded into a specific network of interactions. Several criteria were applied when drawing local protein–protein interaction maps. The first criterion takes into account the different categories of distinct heuristic values (see above). The second one reflects the connectivity of proteins and specificity of interactions. Proteins connected to many proteins with no known functional relations between them are considered to be probable "sticky" proteins and are labeled as such. Potential functions were further confirmed or invalidated by biochemical or functional assays. Several examples are briefly described below, illustrating the strengths and limitations or the approach.

Two-hybrid screens performed on the eight yeast Lsm proteins revealed that they have multiple connections with both splicing factors and the proteins involved in the degradation of cytoplasmic mRNA. This suggested a role for Lsm complexes in two distinct RNA processes.[12] These two-hybrid results have been confirmed by functional assays.[15,16]

Based on our criteria for two-hybrid assays, Ykl155c was chosen for further functional analysis because this protein was shown to interact with the Prp11 splicing factor[9] and with several nucleocytoplasmic transport factors (when used as a bait protein; M.F.R., P.L., and J.-C. Rain, unpublished data, 1998). To determine whether Ykl155c is part of a complex involved in splicing or mRNA transport we purified a biochemical complex by using the TAP strategy.[17] Unexpectedly, the biochemical complex associated with Ykl155c was the small ribosomal subunit of the mitochondria.[18] This illustrates the limitations of the approach. Indeed, the

[15] A. E. Mayes, L. Verdone, P. Legrain, and J. D. Beggs, *EMBO J.* **18,** 4321 (1999).

[16] S. Tharun, W. He, A. E. Mayes, P. Lennertz, J. D. Beggs, and R. Parker, *Nature* **404,** 515 (2000).

[17] G. Rigaut, A. Shevchenko, B. Rutz, M. Wilm, M. Mann, and B. Séraphin, *Nat. Biotechnol.* **17,** 1030 (1999).

[18] C. Saveanu, M. Fromont-Racine, A. Harington, F. Ricard, A. Namane, and A. Jacquier, *J. Biol. Chem.* **276,** 15861 (2001).

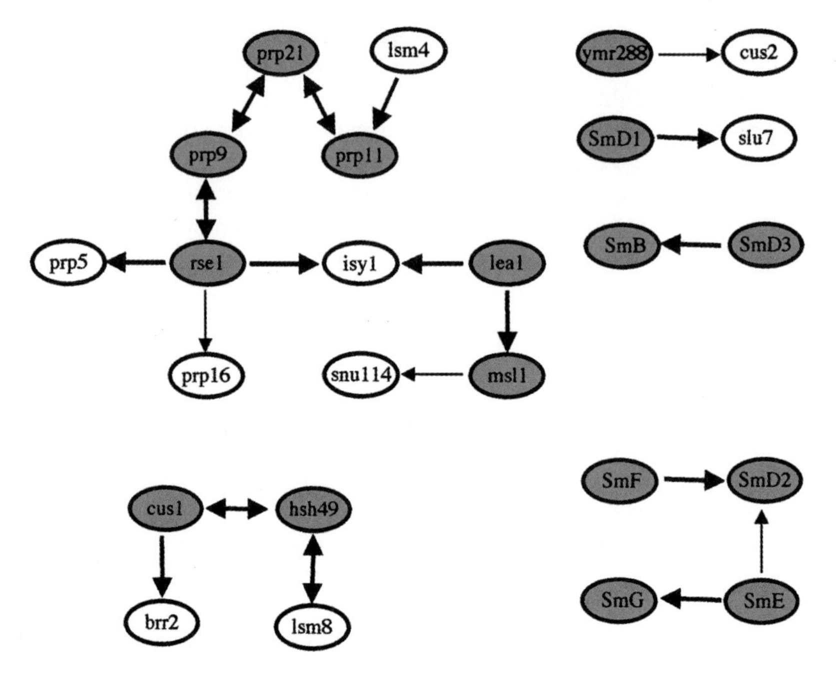

FIG. 2. A yeast U2snRNP protein interaction map. All the interactions selected in two-hybrid screens using U2snRNP proteins as bait and selecting prey proteins known to be involved in pre-mRNA splicing are indicated. The arrows show the prey proteins found with bait. The double arrow indicates that the two reciprocal experiments were performed with success. The gray ovals represent all the proteins from the U2snRNP complex that were purified as a biochemical complex. Bold arrows correspond to interactions with high heuristic values.

selected two-hybrid interactions were specific but not physiologically meaningful and probably occurred because of the artificial nuclear localization of Ykl155c.[18]

All proteins of the U2snRNP particle, which is involved in the spliceosome formation, were used as bait proteins. Figure 2 compares two-hybrid data with the biochemical identification of the U2 snRNP components.[19] Two-hybrid assays showed that most of the proteins identified via complex purification (in gray) were found associated (represented by arrows). For example, we identified a network of interactions between Prp9, Prp11, Prp21, Rse1, Lea1, and Msl1. It is noteworthy that the two-hybrid network published in 1997 includes two novel splicing factors, Rse1 (yml049c) and Hsh49 (yor319w).[9] Moreover, the two-hybrid data reveal additional interactions between previously described splicing factors (Fig. 2).

A model of the RNA polymerase III preinitiation complex has been proposed based on large-scale exhaustive two-hybrid screening and genetic studies.[10]

[19] F. Caspary, A. Shevchenko, M. Wilm, and B. Séraphin, *EMBO J.* **18,** 3463 (1999).

Domains of interaction were defined for many components of the complex, filling gaps between 3D structures of monomers and the functional definition of the active complex.

Selected Fragments in Two-Hybrid Screens Defining Interacting Domains That Are Consistent with Genetic and Structural Data

The screening of randomly generated fragments allows us to determine which domains interact. The common sequence shared by the selected overlapping prey fragments experimentally defines the smallest selected docking site of the bait, thus allowing the precise mapping of a functional domain. In an exhaustive screen performed with the Kap104 as bait to understand the nuclear import mediated by this karyopherin,[20] we selected a highly represented candidate, the Nab2 protein, which was previously identified as a substrate of Kap104. Six independent fusions were selected in the screen. Comparison of the various fusions led to the definition of a small minimally interacting domain that was independently identified through tedious deletion mutant experiments.[21] This result emphasizes one of the advantages of screening an exhaustive two-hybrid library.

Finally, an exhaustive proteome-wide approach for building the protein interaction map of *Helicobacter pylori* has been completed.[13] This map links half of the proteins of the organism in a comprehensive network of protein–protein interactions. This study identified complexes that have been characterized in other organisms, such as in *E. coli*. In several cases, interacting fragments can be mapped on 3D structures of proteins and assigned to a functional domain.

Concluding Remarks

Intrinsic limitations of the yeast two-hybrid system include its reliance on complex transcriptional activation of reporter genes. False negatives (i.e., interactions that are not found by two-hybrid screening) occur because of incorrect folding, inappropriate subcellular localization, degradation of chimeric proteins, or the absence of certain types of posttranslational modifications in yeast. False positives result from the activation of the transcription of reporter genes by bait proteins alone or in combination with some prey proteins or fragments that have the tendency to stick to a wide variety of bait proteins. These are key issues that should be kept in mind when performing large-scale protein–protein interaction screens. To enable the analysis of these large data sets, a new type of databases was created: the PimRider, which is a tool that allows the visualization of protein–protein interaction maps with direct access to experimental raw data (as an example see

[20] M. C. Siomi, M. Fromont, J. C. Rain, L. Wan, F. Wang, P. Legrain, and G. Dreyfuss, *Mol. Cell. Biol.* **18,** 4141 (1998).

[21] D. C. Lee and J. D. Aitchison, *J. Biol. Chem.* **274,** 29031 (1999).

pim.hybrigenics.com for the protein interaction map of *H. pylori*[13]). Biological validation and/or integration of data from other sources will help in predicting the biological relevance of these interactions. Protein interaction maps also represent a new and potentially rich source of data for the prediction of functions by bioinformatic techniques. "Guilty-by-association" methods annotate proteins based on the annotations of their interacting partners. It should be emphasized that these bioinformatics algorithms are still very difficult to use because they rely on both high-quality and complete data. Moreover, such techniques suffer from a lack of independent validation methods. For example, Ynr053c and other proteins have been predicted to have functions related to RNA splicing,[22] and these functions seem to be consistent with other predictions based on different bioinformatic tools. In fact, the only experimental data supporting this annotation comes from two-hybrid screens which have several limitations (see above).[9] Independent experimental data have revealed that Ynr053c is indeed involved in RNA metabolism but not in splicing.[23] Large-scale protein interaction mapping corresponds to technology-driven experiments rather than hypothesis-driven experiments. These approaches are still in their early stages but they will undoubtedly be valuable tools for protein function assignment in the future. Related bioinformatic tools are also in the early stages of development and need to be optimized before becoming useful predictive tools. Finally, functional annotations cannot replace primary experimental sources of information that should be also accessible through functional databases available on the Web.

Acknowledgments

This technology was developed at the Pasteur Institute in the frame of the European TAPIR network (Biotech 95007). This work was financed in part by the Centre National de Recherche Scientifique and the GIP-HMR. We thank Alaim Jacquier, Laurence Decourty, and Caroline Clerte for critical reading of the manuscript.

[22] B. Schwikowski, P. Uetz, and S. Fields, *Nat. Biotechnol.* **18,** 1257 (2000).

[23] C. Saveanu, D. Bienvenu, A. Namane, P. E. Gleizes, N. Gas, A. Jacquier, and M. Fromont-Racine, *EMBO J.* **20,** 6475 (2001).

[30] Integrated Version of Reverse Two-Hybrid System for the Postproteomic Era

By HIDEKI ENDOH, SYLVIE VINCENT, YVES JACOB, ELÉONORE RÉAL, ALBERTHA J. M. WALHOUT, and MARC VIDAL

Introduction

Functional genomic databases provide information for many of the genes predicted from genome and expressed sequence tags (ESTs) sequencing projects. These databases should gradually relieve molecular biologists from the tedious and repetitive steps involved in conventional gene cloning and basic protein characterization. This in turn might help extend studies primarily focused on the function of proteins toward more precisely deciphering the function of *interactions* between proteins. Here we describe an improved version of the yeast reverse two-hybrid system to generate and characterize interaction-defective alleles (IDAs) that can be used to study the function of physical interactions between proteins.

Biological Atlas of Proteome

Comprehensive interactome mapping projects using large-scale yeast two-hybrid systems are underway for *Saccharomyces cerevisiae* and *Caenorhabditis elegans*.[1–3] Similar projects should emerge soon for other model organisms and for humans. Typically two-hybrid interactome maps consist of large lists of potential binary interactions made available on the World Wide Web. In parallel to the development of interactome maps, other functional mapping projects such as expression profiling, large-scale phenotypic analyses, and systematic protein localization studies have also been developed.[4–6] These maps should provide additional layers of information on various aspects of protein function (Fig. 1).

A potential caveat of using this information is that most functional genomic approaches used to generate maps can give rise to false negative and false positive information. Hence, the relatively high-throughput settings can preclude the finding of already known information. For example, protein–protein interactions already described in the literature might not be found in large-scale interactome mapping efforts. On the other hand, the artificial nature of the assays used in functional genomic mapping is such that one should be cautious about the biological

[1] J. F. Gera, T. R. Hazbun, and S. Fields, *Methods Enzymol.* **350,** [28], 2002 (this volume).

[2] M. Fromont-Racine, J.-C. Rain, and P. Legrain, *Methods Enzymol.* **350,** [29], 2002 (this volume).

[3] A. J. M. Walhout and M. Vidal, *Nat. Rev. Mol. Cell. Biol.* **2,** 55 (2001).

[4] D. J. Lockhart and E. A. Winzeler, *Nature* **405,** 827 (2000).

[5] A. Kumar and M. Snyder, *Nat. Rev. Genet.* **2,** 302 (2001).

[6] P. W. Sternberg, *Cell* **105,** 173 (2001).

0076-6879/02 $35.00

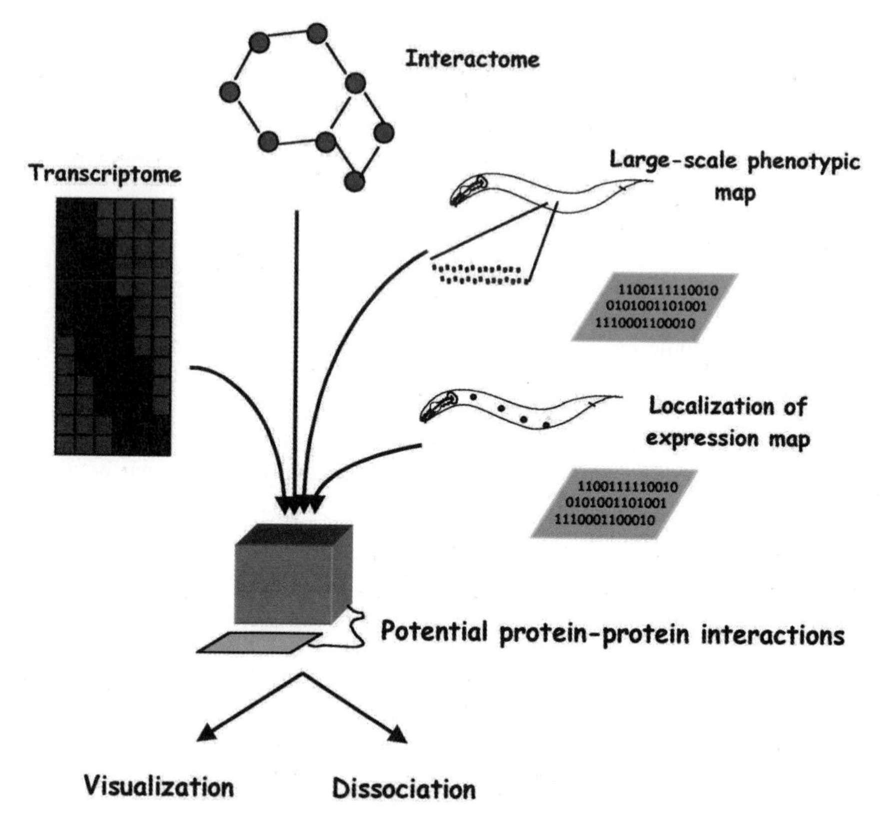

FIG. 1. Integration of functional genomic maps (see text) is expected to give rise to meaningful hypotheses of protein–protein interactions. An example is shown here for functional genomics in *C. elegans*. These hypotheses can be validated and characterized using visualization methods and genetic strategies. Although very important, the techniques related to visualizing interactions *in vivo,* such as fluorescence resonance energy transfer (FRET), will not be discussed here. Descriptions of these techniques can be found elsewhere [M. Chalfie, Y. Tu, G. Euskirchen, W. W. Ward, and D. C. Prasher, *Science* **263,** 802 (1994); D. A. Zacharias, G. S. Baird, and R. Y. Tsien, *Curr. Opin. Neurobiol.* **10,** 416 (2000)]. The characterization of loss-of-interaction phenotypes should give clues to the function of protein interactions.

relevance of the information. For example, it is assumed that a number of interactions listed in interactome maps might not correspond to any physiological binding. However, it has been proposed that, despite the caveats inherent to functional genomic strategies, the progressive integration of different functional maps into a "biological atlas of the proteome" should help in formulating hypotheses with increasing accuracy[7] (Fig. 1[7a,7b]).

[7] M. Vidal, *Cell* **104,** 333 (2001).

[7a] M. Chalfie, Y. Tu, G. Euskirchen, W. W. Ward, and D. C. Prasher, *Science* **263,** 802 (1994).

[7b] D. A. Zacharias, G. S. Baird, and R. Y. Tsien, *Curr. Opin. Neurobiol.* **10,** 416 (2000).

Indeed it is imaginable that, in the near future, potentially interesting protein–protein interactions, such as those that link seemingly unrelated biological processes, will be primarily found *in silico* by consulting interactome map databases, rather than *in vivo* or *in vitro* by using laboratory based experimental strategies. Furthermore, upon finding such potential interactions, it should become increasingly possible to "surf" the Internet for additional supporting evidence. For example, the genes encoding potentially interacting proteins might already have been found to belong to identical expression clusters from microarray analyses. In addition, their respective knockouts might have been shown to confer identical or opposite phenotypes. Finally, their products might have been found to localize in similar tissues, or cells, or subcellular compartments. Thus by overlapping data from various functional maps, it should be possible to select, among all interactions available in an interactome map, a subset that might help understanding a biological process of interest. Consequently biological questions might evolve from a primary focus on finding the function of proteins to deciphering the function of interactions between proteins.

From Function of Proteins to Function of Interactions between Proteins: Need for Interaction-Defective Alleles (IDAs)

Most proteins function in complex regulatory networks or biological machines. In addition, many proteins mediate multiple functions, and splicing variants or posttranslationally modified polypeptides can even mediate opposite functions. These facts sometimes obscure the interpretation of the gross alterations created by null mutations. Hence, in order to represent complex networks, proteins (and RNA molecules) should not necessarily be considered as the main operational units of signal transduction pathways and molecular machines. Their actions, i.e., the physical interactions they mediate and/or the enzymatic reactions they catalyze, should also be considered as operational units. This notion has already been embraced by several computational biologists.[8] Thus models that optimally describe biosynthetic and signal transduction pathways use relationships of the type "protein 1—action(s)—protein 2," with descriptors for each of these three operational units, rather than simply "protein 1—protein 2"[8] (Fig. 2). The inclusion of molecular actions as units of biological models allows a greater flexibility when describing complex relationships between proteins.[8] For example, two or more actions can be represented separately for a single protein: "X—interacts with—protein Y" and "protein X—phosphorylates—protein Z." In this context, the genetic entities that can be considered for mutagenesis and phenotypic analysis are not only the proteins themselves but also their sites of action.

[8] J. van Helden, A. Naim, R. Mancuso, M. Eldridge, L. Wernisch, D. Gilbert, and S. J. Wodak, *Biol. Chem.* **381,** 921 (2000).

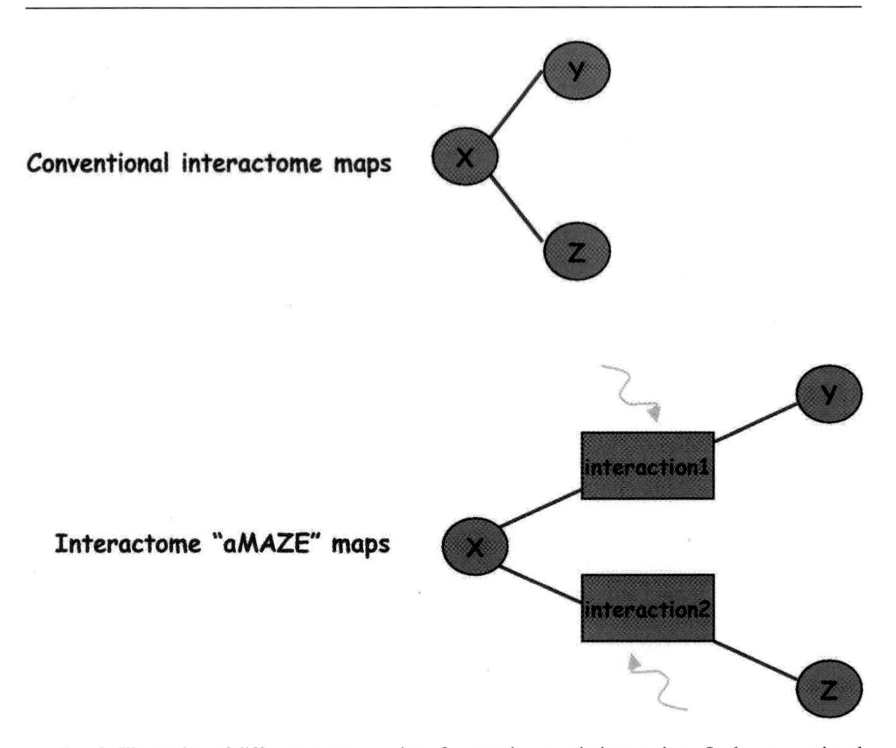

FIG. 2. Illustration of different representations for protein–protein interactions. In the conventional representation, X, Y, and Z are the operational units and are linked according to two-hybrid binary interactions. In the aMAZE representation (see text), X, Y, Z and both X/Y and X/Z interactions are all operational units. The function of X/Y and X/Z can be addressed experimental using the strategy described here.

For each potentially interesting interaction selected from an interactome map and supported by functional information contained in other databases, any combination of the following three basic questions can be experimentally addressed: (1) Does the interaction really take place *in vivo*? (2) What is its role in the relevant biochemical, cellular, or organismal context? (3) Where (in which tissue, cell or subcellular compartment) and when (at which stage of development or in response to what signal) does the interaction take place?

Attempts to address these questions have so far mainly relied on biochemical methods. Coimmunoprecipitation experiments from relatively complex extracts are often used to support the notion that proteins interact physically. However, biochemical experiments usually fail to unequivocally demonstrate *in vivo* direct interactions between partners. Indeed, it can be relatively cumbersome to verify that the observed physical interactions did not take place during or after the generation

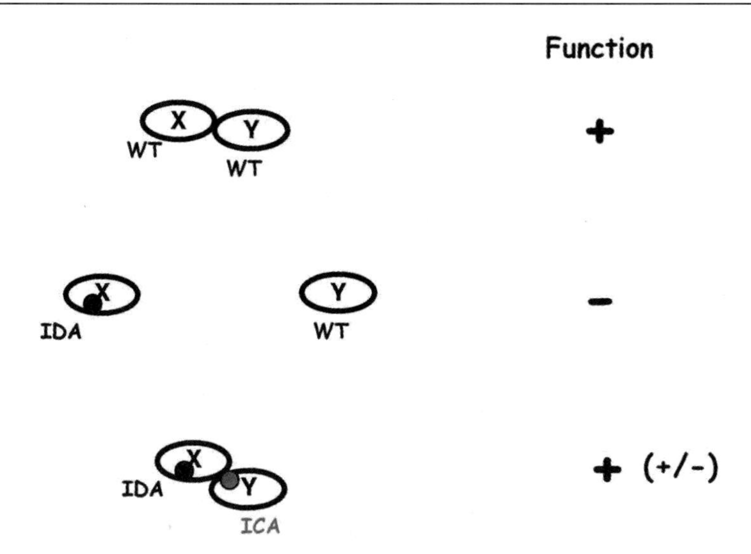

FIG. 3. Interaction-defective alleles (IDAs) can be compared to wild-type for their ability to mediate a particular function. According to the key–lock model, IDAs, together with interaction-compensatory alleles (ICAs), can be used to formally demonstrate an interaction *in vivo*.

of the extracts.[9] Biochemical methods can also be very labor intensive. Finally, biochemical methods are not well suited to study the role(s) of physical protein–protein interactions in their biological context.

One of the genetic ways to address the validity and the role of a particular protein–protein interaction is to prevent or eliminate this interaction *in vivo* while maintaining other actions mediated by both partners, i.e., physical interactions and enzymatic activities. Under these conditions, a function can be inferred for the interaction of interest by observing the resulting "loss-of-interaction" phenotypes (Fig. 3). Interaction-defective alleles (IDAs) and interaction–dissociator molecules, such as peptides (interaction–dissociator peptides or IDPs) or chemical compounds (interaction–dissociator compounds or IDCs) can be used to prevent or eliminate protein–protein interactions *in vivo* (for review, see Refs. 10 and 11) [see also Appendix for standardized descriptive systems for IDAs and interaction-compensatory alleles (ICAs)]. Although these approaches are relatively simple conceptually, they have remained difficult to implement in practice because the generation of specific IDAs, IDPs, and IDCs can be cumbersome, especially in the absence of detailed structural information on the proteins involved. Large-scale

[9] E. Harlow, P. Whyte, B. J. Franza, and C. Schley, *Mol. Cell. Biol.* **6,** 1579 (1986).
[10] M. Vidal and H. Endoh, *Trends Biotechnol.* **17,** 374 (1999).
[11] M. Vidal and P. Legrain, *Nucleic Acids Res.* **27,** 919 (1999).

approaches to systematically mutagenize the sites of action of proteins have not been extensively developed yet. Particularly the generation and characterization of specific alleles that affect physical interactions between proteins (IDAs) are still relatively complicated and thus not used systematically. Below we describe a genetic system referred to as the reverse two-hybrid system that can be used to conveniently isolate IDAs.

Reverse Two-Hybrid System

The reverse two-hybrid system was developed to circumvent the difficulties inherent in finding IDAs, IDPs, and IDCs. In short, this system allows a genetic selection against wild-type interactions that have been reconstituted in the yeast two-hybrid system.[11] With this system, relatively rare IDAs and IDCs provide a selective advantage and hence can be readily isolated from complex libraries of alleles or chemical compounds, respectively.

The yeast two-hybrid system allows the detection of an interaction between two proteins, X and Y, through the functional reconstitution of a transcription factor.[12] X and Y are fused to a DNA-binding domain (DB) and an activation domain (AD), respectively, resulting in DB-X and AD-Y fusion proteins. The most commonly used DBs derive from yeast Gal4p or bacterial LexA.[13] However, other DBs are compatible with the system as well. Interaction between DB-X and AD-Y activates the transcription of reporter gene(s) containing DB binding sites. In the forward two-hybrid system, reporter gene expression confers either a measurable amount of enzymatic activity or a selective growth advantage on defined media.[13] In the reverse two-hybrid system, DB-X/AD-Y wild-type interactions activate the expression of a counterselectable gene and thus confer a sensitivity phenotype on appropriate media.[14–16]

IDAs can be genetically selected from complex libraries of alleles generated randomly by using either mutagenic PCR (polymerase chain reaction) or conventional mutagenesis, without any knowledge of the three-dimensional structure of the two partners.[15,17,18] Alternative binding methods such as glutathion-S-transferase (GST) pull-down and/or coimmunoprecipitation are then used to make sure that the IDAs are genuine and specific.[15,17] Ideally IDAs should contain a single amino acid substitution that minimally affects full-length expression,

[12] S. Fields and O. Song, *Nature* **340,** 245 (1989).

[13] I. G. Serebriiskii, V. Khazak, and E. A. Golemis, *Biotechniques* **30,** 634 (2001).

[14] M. Vidal, R. Brachmann, A. Fattaey, E. Harlow, and J. D. Boeke, *Proc. Natl. Acad. Sci. U.S.A.* **93,** 10315 (1996).

[15] H. M. Shih, P. S. Goldman, A. J. DeMaggio, S. M. Hollenberg, R. H. Goodman, and M. F. Hoekstra, *Proc. Natl. Acad. Sci. U.S.A.* **93,** 13896 (1996).

[16] C. A. Leanna and M. Hannink, *Nucleic Acids Res.* **24,** 3341 (1996).

[17] M. Vidal, P. Braun, E. Chen, J. D. Boeke, and E. Harlow, *Proc. Natl. Acad. Sci. U.S.A.* **93,** 10321 (1996).

[18] C. Inouye, N. Dhillon, T. Durfee, P. C. Zambryski, and J. Thorner, *Genetics* **147,** 479 (1997).

stability, folding, interaction with other partners, and enzymatic activities if relevant.[19] In addition, once selected in the reverse two-hybrid system, IDAs should be conveniently transferable to many expression vectors with which their activity can be tested in biologically relevant systems.

Experimental Design to Validate Interactions and Study Their Function

The experimental design described here is based on the "key–lock" model. This model was originally formulated from two basic observations using phage or yeast genetics.[20–23] First, X/Y interactions can be disrupted by amino acid substitutions in X that confer a conditional phenotype (referred to here as IDAs). Second, in many cases one can isolate compensatory amino acid substitutions in Y (referred to here as interaction-compensatory alleles or ICAs) that restore the interaction (X^{ida}/Y^{ica}).

X^{ida} and Y^{ica} pairs of alleles can be used to validate and characterize the role of X/Y protein-protein interactions *in vivo* (Fig. 3). Such experiments were mostly developed in yeast since appropriate IDAs and ICAs can be genetically selected from large numbers of mutants. In a few cases, key–lock experiments were also described for multicellular organisms.[24,25] However, the fact that the IDAs and ICAs had to be generated by site-directed mutagenesis based on the three-dimensional structure of the interacting domains prevented a wide development of the system.

The reverse and forward two-hybrid systems bypass this limitation by providing a genetic selection, first for IDAs in the reverse configuration, and then for ICAs in the forward configuration. Hence, strategies used so far only for simple model organisms can now be extended to multicellular organisms.

Examples of Published IDAs and IDA/ICA Pairs

As a way to exemplify the uses of the reverse and forward two-hybrid system protocols described below and to illustrate the success of the method, we summarize below a few examples available in the literature. Even though conventional genetic screens can be conveniently developed when characterizing yeast protein interactions, several investigators have decided to select IDAs in the context of the reverse two-hybrid system. The phenotypes described in the literature are indicative that IDAs selected through the reverse two-hybrid system can address questions related to specific roles of the protein–protein interactions.

[19] T. Yasugi, M. Vidal, H. Sakai, P. M. Howley, and J. D. Benson, *J. Virol.* **71,** 5942 (1997).
[20] J. Jarvik and D. Botstein, *Proc. Natl. Acad. Sci. U.S.A.* **72,** 2738 (1975).
[21] A. E. Adams, D. Botstein, and D. G. Drubin, *Science* **243,** 231 (1989).
[22] A. E. Adams and D. Botstein, *Genetics* **121,** 675 (1989).
[23] P. Novick, B. C. Osmond, and D. Botstein, *Genetics* **121,** 659 (1989).
[24] B. Amati, M. W. Brooks, N. Levy, T. D. Littlewood, G. I. Evan, and H. Land, *Cell* **72,** 233 (1993).
[25] S. M. Kaech, C. W. Whitfield, and S. K. Kim, *Cell* **94,** 761 (1998).

To dissect the role of protein–protein interactions involved in the yeast mating response pathway, Ste5p IDAs that fail to bind Ste7p but not Ste11p, and others that fail to bind Ste11p but not Ste7p, were selected using a dual reverse two-hybrid system.[18] A double mutation allele of Ste5p failed to interact with Ste7p but exhibited wild-type interactions with Ste11p and also Fus3p and Ste4p as tested by both two-hybrid and coimmunoprecipitation experiments. Similarly, two single mutation alleles of Ste5p failed to interact with Ste11p but exhibited wild-type interaction with Ste7p and also Fus3p and Ste4p as tested by both two-hybrid and coimmunoprecipitation experiments. The fact that each one of these specific alleles fail to rescue the mating deficiency of a *ste5*Δ mutant demonstrated that both Ste5p/Ste7p and Ste5p/Ste11p interactions are important for the mating response pathway.

One of the most complete utilizations of the reverse two-hybrid system was described in the context of two proteins involved in the genome replication of the tobacco etch potyvirus (TEV).[26] Several reports had described a potential interaction between the proteinase domain of the VPg-proteinase protein Nia and the Nib RNA-dependent RNA polymerase (see also Appendix). However, the validity and the biological role of that interaction had remained unclear. Thermosensitive Nia IDAs were selected using first a forward two-hybrid selection at 20° (permissive temperature) and subsequently a reverse two-hybrid selection at 30° (restrictive temperature). The structural integrity of various Nia IDAs was tested biochemically by comparing their autoproteolytic activity to that of the wild-type Nia protein. Two IDAs contained single amino acid substitutions, tested negative in their ability to bind to Nib upon retransformation in yeast cells, and retained proteolytic activity at both temperatures. One of these alleles exhibited a *ts* (temperature sensitive) genome amplification phenotype when tested back *in vivo* in the context of a TEV recombinant strain. Seventeen Nib ICAs were selected from a forward two-hybrid selection performed at restrictive temperature using this particular IDA as bait. The two ICAs obtained in this forward two-hybrid selection partially restored the loss-of-function phenotype conferred by the starting IDA, demonstrating that the NiA/Nib interaction is essential for TEV genome amplification.

Another optimal implementation of the reverse two-hybrid system was described in the context of GATA-1 and FOG-1, two mouse proteins essential for erythroid and megakaryocyte development.[27] A GATA-1/FOG-1 interaction had been detected in the two-hybrid system and was supported by subsequent biochemical experiments. However, the validity and the *in vivo* role of that interaction in differentiation had not been formally demonstrated. First GATA-1 IDAs were selected using a reverse two-hybrid selection and a fusion to β-galactosidase to eliminate truncations (see below). The structural integrity of several potentially

[26] J. A. Daros, M. C. Schaad, and J. C. Carrington, *J. Virol.* **73,** 8732 (1999).
[27] J. D. Crispino, M. B. Lodish, J. P. MacKay, and S. H. Orkin, *Mol. Cell* **3,** 219 (1999).

interesting GATA-1 IDAs was tested biochemically by measuring their DNA binding activity. Three IDAs tested wild-type for their DNA binding activity. Compared to wild-type GATA-1, all three IDAs failed to rescue the terminal erythroid maturation of GATA-1 knockout G1E cells. Three FOG-1 ICAs were selected from a forward two-hybrid selection using one of the functionally defective GATA-1 IDAs as bait. One of the resulting FOG-1 ICAs contained a single amino acid substitution. Interestingly, this ICA restored both interaction and function with the corresponding GATA-1 IDA but not with another one. This level of specificity of the key–lock model demonstrates that the GATA-1/FOG-1 interaction is essential for erythroid differentiation.

In some applications of the reverse two-hybrid system, IDAs and ICAs were isolated simply as a way to characterize the structure of the corresponding domains. HBO1 was found to interact with both MCM2 and ORC1 in the forward two-hybrid system, *in vitro* in GST pull-down experiments, and *in vivo* on expression from a recombinant adenovirus.[28] To further characterize the interaction domains, MCM2 IDAs affected in their interaction with HBO1 were generated using a reverse two-hybrid selection.[28] A single amino acid change (L222P), obtained from a mutant containing three substitutions, conferred a strong interaction-deficient phenotype. As for many IDAs, it is always possible that the spurious presence of a proline at position 222 in this allele, and not the absence of a leucine at that position, causes the lack of interaction. To test this notion, a substitution to alanine at the same position was generated by site-directed mutagenesis and confirmed to be interaction-defective as well. HBO ICAs were then selected using a forward two-hybrid selection with the MCM2 IDA as bait. The finding that a single amino substitution in HBO1 (I380T) can restore the interaction suggests that the MCM2/HBO1 interaction is direct. The role of the MCM/HBO1 interaction can now be tested in a biological assay using the MCM2 IDA and HBO1 ICA pair of alleles.

Formally ICAs should always be used as ultimate controls of experiments performed with IDAs. However, the characterization of IDAs alone can also strongly suggest the existence of potential interactions. A few examples of interesting IDAs are described here to further exemplify the use of the reverse two-hybrid system.

To validate the potential interactions obtained from a forward two-hybrid selection between human papillomavirus type 16 (HPV16) E1 protein (E1) and hUBC9 and 16E1-BP, E1 IDAs alleles were selected using a reverse two-hybrid system.[19] Among these alleles, one was specifically affected for hUBC9 interaction, i.e., exhibited nearly wild-type interaction with 16E1-BP, E2, and E1. This allele also exhibited ATPase activity similar to that of wild-type E1. The fact that this IDA is significantly affected in its ability to support HPV16 genome amplification suggests strongly that E1/hUBC9 interaction is important for that function.

[28] T. W. Burke, J. G. Cook, M. Asano, and J. R. Nevins, *J. Biol. Chem.* **276,** 15397 (2001).

A two-hybrid interaction between Bim, a Bcl-2 family member with proapoptotic activity, and LC8, the cytoplasmic dynein light chain, suggested a functional link between apoptosis and the dynein motor complex.[29] To substantiate this possibility, Bim IDAs affected in their interactions with LC8 were selected using a reverse two-hybrid system. Two IDAs contained single amino acid substitutions and were completely altered in their ability to interact with LC8 but retained their ability to interact with Bcl-2. The inability of these two Bim IDAs to associate with dynein complex supports the hypothesis.

It should be noted that IDAs have also been used as dominant negative alleles[30] and as negative controls of dominant negative contructs.[31]

Protocols

In the reverse two-hybrid system described here, Gal4p DB is used together with a *URA3* reporter gene containing 10 Gal4p binding sites (*SPAL10::URA3*) (see below).[14] In this context, DB-X/AD-Y interaction confers a 5-fluoroorotic acid (5FOA) sensitivity phenotype (Foa^S). Events that dissociate or prevent X/Y interaction and provide a Foa^R phenotype can thus be genetically selected from complex libraries. This was demonstrated using known mutations, peptides, and compounds that are known to affect known protein interactions.[10,14] Variations on the reverse two-hybrid system theme have been described in detail elsewhere and include the use of other counterselectable markers such as *CYH2* or Tet^R-*HIS3* as well as other DBs such as LexA.[15,16,18]

In practice, it is difficult to address the basic challenges in isolating IDAs as described above (e.g., selection of single amino acid changes and downstream cloning into expression vectors). Variations and improved versions of the reverse two-hybrid system have been developed with these issues in mind. Full-length expression, stability, and folding can be addressed by using a conveniently detectable C-terminal Tag to the protein to be mutagenized. The system described here uses the green fluorescent protein (GFP).[32,33] Thus full-length IDAs can be detected by screening among Foa^R colonies for those that retain fluorescence (Fig. 4). The selection of IDAs that retain the ability to interact with another interactor is possible using the dual reverse two-hybrid system (see Ref. 13). However, for proteins that interact with multiple partners, it is preferable to use GFP and isolate a large collection of alleles that can then be classified according to their specificity of interaction for each of the partners.[32] Finally, we have integrated the Gateway

[29] H. Puthalakath, D. C. Huang, L. A. O'Reilly, S. M. King, and A. Strasser, *Mol. Cell.* **3,** 287 (1999).
[30] U. Schaeper, N. H. Gehring, K. P. Fuchs, M. Sachs, B. Kempkes, and W. Birchmeier, *J. Cell. Biol.* **149,** 1419 (2000).
[31] C. F. Chen, P. L. Chen, Q. Zhong, Z. D. Sharp, and W. H. Lee, *J. Biol. Chem.* **274,** 32931 (1999).
[32] H. Endoh, A. J. M. Walhout, and M. Vidal, *Methods Enzymol.* **328,** 74 (2000).
[33] H. Puthalakath, A. Strasser, and D. C. S. Huang, *Biotechniques* **30,** 984 (2001).

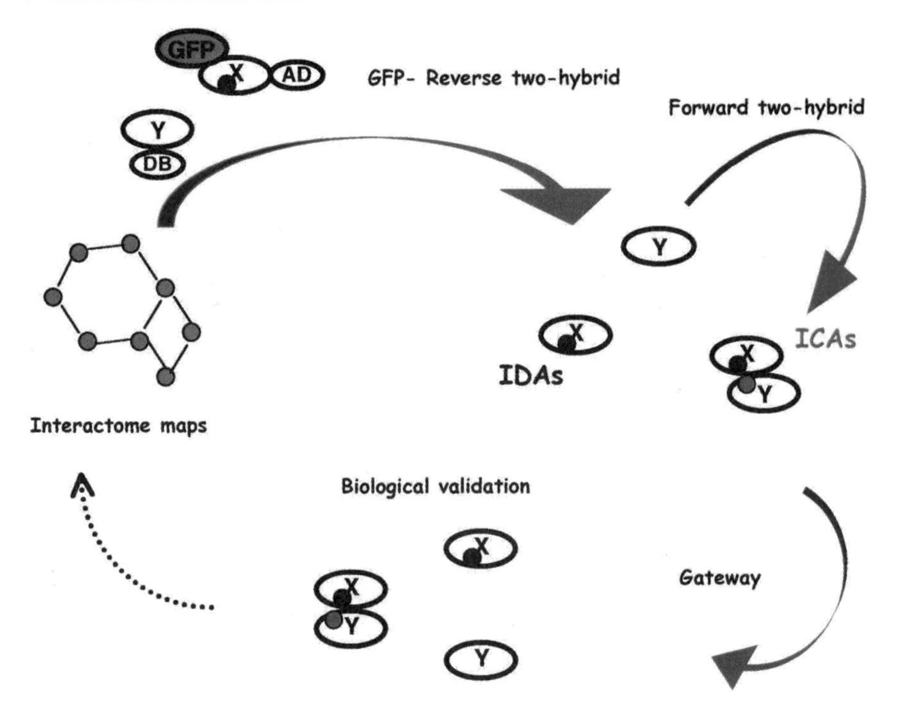

FIG. 4. Protein–protein interactions found in interactome maps can be studied genetically using the strategy shown here. The GFP reverse two-hybrid system is used as a genetic selection for full-length interaction-defective alleles (IDAs). IDAs that score as functionally defective are then used as baits to select for interaction-compensatory alleles (ICAs) using the forward two-hybrid system. Subsequently IDAs and ICAs are cloned into various expression vectors using the Gateway cloning system. Biological validation is performed by comparing the phenotypes conferred by the following pairs of alleles: X/Y, X^{ida}/Y, X^{ida}/Y^{ica}.

technology[34–36] together with the reverse two-hybrid system to allow rapid transfer of IDAs to various expression vectors (see below).

The experimental design is as follows (Fig. 4):

1. The functional assay. A biological assay is generated such that exogenously expressed wild-type X and X IDAs can be compared for their ability to mediate a biological function. For yeast, *Caenorhabditis elegans, Drosophila,* mouse, and other model organisms, X knockout mutants can be used for this analysis.

[34] J. L. Hartley, F. T. Temple, and M. A. Brasch, *Genome Res.* **10,** 1788 (2000).

[35] A. J. M. Walhout, R. Sordella, X. Lu, J. L. Hartley, G. F. Temple, M. A. Brasch, N. Thierry-Mieg, and M. Vidal, *Science* **287,** 116 (2000).

[36] A. J. M. Walhout, G. F. Temple, M. A. Brasch, J. L. Hartley, M. A. Lorson, S. van den Heuvel, and M. Vidal, *Methods Enzymol.* **328,** 575 (2000).

Biochemical settings can also be used. In this case X depletion experiments can be performed with α-X antibodies and subsequently, purified X and X IDAs can be compared by adding them back into the reaction.

2. *Gateway-GFP-reverse two-hybrid system.* DB-X-GFP/AD-Y or DB-X/AD-Y-GFP interactions are reconstituted in the reverse two-hybrid system and the Foa^S and Gfp^+ phenotypes are confirmed. X or Y IDAs are selected as Foa^R GFP^+ colonies from a complex library of randomly generated alleles. X or Y IDAs are then PCR amplified from Foa^R colonies using Gateway-compatible primers (see below), cloned into the Gateway resource vector, and sequenced.[34–36] IDAs corresponding to single amino acid changes can be transferred to various Destination vectors for two-hybrid tests, biochemical characterization, and complementation analysis.

3. *Two-hybrid retests and biochemical characterization.* IDAs should be retested to confirm their loss-of-interaction phenotypes. When possible, they should also be tested against other interacting partners and the most specific IDAs should be retained. Specific IDAs are those that remain able to interact with as many other partners as possible. Finally, for proteins that exhibit a detectable enzymatic activity, IDAs that retain such activity should be chosen preferentially.

4. *Interaction-compensatory alleles (ICAs).* ICAs in partner Y can be selected using the forward two-hybrid system with a particular X IDA as bait and a library of randomly mutated Y alleles. ICAs should be Gateway cloned in the same subset of vectors as the IDAs (see above). ICAs and ICAs can then be treated in parallel.

5. *Biological tests.* X IDAs are first compared to their wild-type X counterpart for their ability to mediate a predefined biological function (see *1*). Functionally defective IDAs are then expressed together with their cognate ICA. X^{ida}/Y^{ida} reconstitution of the biological function validates the X/Y protein–protein interaction.

Example

To demonstrate that our GFP/reverse/Gateway two-hybrid system can be used to rapidly select and characterize missense IDAs, the FKBP12/RIC interaction[37] was chosen as a test case, mainly because the three-dimensional structure of this heterodimer is available.[38] After PCR mutagenesis, reverse two-hybrid selection, and Gfp screening, alleles were obtained that exhibit negative two-hybrid readouts (Fig. 5A) as well as a Gfp^+ phenotype (Fig. 5B). In all cases, the expression levels of the IDAs were similar to wild-type FKBP12 (Fig. 5C). The nucleotide sequence of the FKBP12-IDAs confirmed that single amino acid substitutions were selected using the GFP/reverse system (data not shown). Importantly, although the

[37] T. Wang, P. K. Donahoe, and A. S. Zervos, *Science* **265,** 674 (1994).

[38] M. Huse, Y. G. Chen, J. Massague, and J. Kuriyan, *Cell* **96,** 425 (1999).

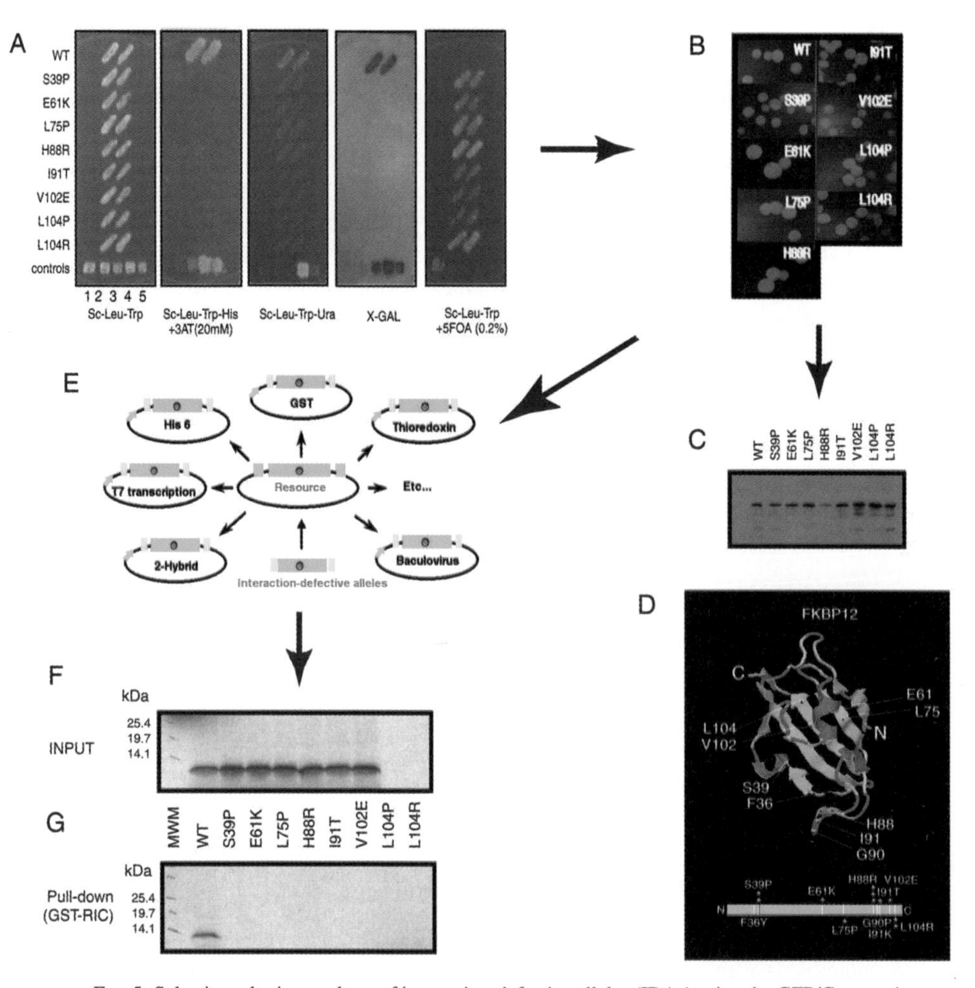

FIG. 5. Selection, cloning, and use of interaction-defective alleles (IDAs) using the GFP/Gateway/Reverse two-hybrid system. (A) Yeast phenotypes. Eight FKBP12/RIC IDAs were selected as Foa^R/Gfp^+ colonies and tested for 4 two-hybrid phenotypes (see Ref. 40). Compared to the wild-type interaction, the IDAs confer a $3AT^S$, Ura^-, β-Gal^- and Foa^R phenotype. The five controls (named 1–5) at the bottom [M. Vidal, *in* "The Yeast Two-Hybrid System" (P. Bartels and S. Fields, eds.), p. 109. Oxford University Press, New York, 1997] provide an indication of the quality of the plates and incubation conditions. (B) The fluorescence conferred by each IDA is unchanged compared to that of cells expressing wild-type interactions. (C) Western-blot analysis using anti-Gal4AD antibodies (see Ref. 32) demonstrates that the IDAs encode full-length proteins expressed at wild-type levels. (D) Amino acid changes of IDAs generated by the GFP/Gateway/Reverse two-hybrid system cluster into a cognate FKBP12 protein interaction pocket. (E) The Gateway recombinational cloning system (see Refs. 34–36) can be used to transfer IDAs from the GFP/Reverse two-hybrid plasmids simultaneously to various Destination vectors. (F) Protein expression from a Destination vector in which wild-type FKBP12 or corresponding IDAs have been cloned using Gateway. (G) FKBP12 IDAs fail to bind to RIC under conditions in which wild-type FKBP12/RIC interaction can be readily detected.

substitutions are distributed over the primary sequence of the encoded protein, they cluster into the interacting domain of the three-dimensional structure (Fig. 5D).

With the Gateway recombinational cloning technology,[34–36] it should be possible to manipulate hundreds of IDAs simultaneously in a highly reliable, standardized and automatable manner (Fig. 5E).[39] To illustrate the use of Gateway in the validation of potential interactions, we generated "Entry" clones containing FKBP12 IDAs and transferred these into a Destination vector for *in vitro* transcription and translation. The proteins expressed from this system (Fig. 5F) were tested for their ability to interact with RIC fused to glutathione *S*-transferase (GST-RIC). As expected, all FKBP12 IDAs were altered in their ability to interact with RIC *in vitro* (Fig. 5G). Using Gateway, these alleles can now be transferred into any other Destination vector for additional functional and structural characterization (Fig. 5E).

Strains

MaV103 (*MAT***a**, *his3-200, ade2–101, trp1-901, leu2-3,112, gal4*Δ, *gal80*Δ, *SPAL10::URA3, GAL1::HIS3@LYS2, GAL1::lacZ, can1*^{R}, *cyh2*^{R}) and MaV203 (*MAT*α, same genotype) can be used for reverse and forward two-hybrid selections. They carry three reporter genes each regulated by different promoters that contain Gal4p binding.[40]

The *GAL1::HIS3* reporter gene allows for positive selection. By increasing the expression levels of *GAL1::HIS3,* DB-X/AD-Y interaction increases the resistance to 3-amino-1,2,4-triazole (3-AT), a competitive inhibitor of the His3p enzyme. 3-AT concentrations can be titrated depending on the level of self-activation conferred by DB-X. In most cases 10 m*M* 3-AT is optimal. However, for self-activating DB-X fusions, increased 3-AT concentrations can be used up to 100 m*M*. The *GAL1::lacZ* reporter gene allows a quantifiable readout. Increased expression of β-galactosidase in response to DB-X/AD-Y interactions confers a blue phenotype in the presence of the chromogenic substrate X-Gal. The *SPAL10::URA3* reporter gene allows for both positive and negative selection. Elevated levels of *SPAL10::URA3* expression confers growth on media lacking uracil. Conversely, since the Ura3p enzyme encoded by the *URA3* gene can catalyze the conversion of 5-FOA into a toxic product, DB-X/AD-Y interactions confer a Foa^S phenotype.

MaV203 (*MAT*α, same genotype as MaV103) can also be used for reverse and forward two-hybrid selections.

[39] J. Reboul, P. Vaglio, N. Tzellas, N. Thierry-Mieg, T. Moore, C. Jackson, T. Shin-i, Y. Kohara, D. Thierry-Mieg, J. Thierry-Mieg, H. Lee, J. Hitti, L. Doucette-Stamm, J. L. Hartley, G. F. Temple, M. A. Brasch, J. Vandenhaute, P. E. Lamesch, D. E. Hill, and M. Vidal, *Nat. Genet.* **27,** 332 (2001).

[40] M. Vidal, *in* "The Yeast Two-Hybrid System" (P. Bartels and S. Fields, eds.), p. 109. Oxford University Press, New York, 1997.

Two-Hybrid Vectors

For most DB-X/AD-Y potential interactions, it should be possible to recover interaction-defective alleles of X (X*), or Y (Y*), that cannot interact with Y or X, respectively. We designed a GFP reverse two-hybrid system that allows the use of DB-X-GFP or AD-Y-GFP fusion proteins, together with AD-Y or DB-X, respectively.[32]

Multicopy plasmids (2 μm) are used because they confer relatively high expression levels of the protein to be characterized, which allows convenient detection of the GFP activity of Foa^R yeast colonies (Gfp$^+$). Multicopy plasmids used in the conventional two-hybrid system were adapted to the GFP reverse two-hybrid system by cloning a sequence encoding one of the GFP derivatives into their multiple cloning site (MCS). Since the Gap-repair method is used to introduce PCR-generated libraries into these plasmids (see below), it is important to eliminate the frequent transformants that arise from plasmid recircularization. This is possible because the GFP encoding sequence is cloned in a different frame than that of DB or AD. Thus upon recircularization of the plasmid, GFP is not expressed and the corresponding transformants can be eliminated (Foa^R Gfp$^-$).

The AD-GFP plasmid (pACT2-viGFP) is derived from pACT2 (Clontech, Palo Alto, CA) and encodes the GFP mutant derivative Phe64Leu, Ser65Thr, Val68Ile ("viGFP"). The DB-GFP plasmid (pAS2-EGFP) is derived from pAS2-1 (Clontech) and encodes another GFP derivative, "EGFP."

Primers for Yeast Gap Repair

Primers for the amplification of DNA fragments to be mutagenized should be designed with 40-nucleotide tags identical to the coding sequences of Gal4pDB or Gal4pAD (forward primers) and EGFP or viGFP (reverse primers).[32] PCR products tailed with such tags can be inserted in the relevant two-hybrid plasmids by Gap repair. The primer sequences are as follows:

Forward primers used to generate DB-X fusions: 5′AGAGAGTAGTAAC AAAGGTCAAAGACAGTTGACTGTATCGNNNNNNNNN3′

The underlined sequence corresponds to Gal4pDB and the N's correspond to 20–25 nucleotides (nt) of X N terminus-encoding sequence (melting temperature (T_m): ~60°).

Forward primers used to generate AD-Y fusions: 5′ ATTCGATGATGAAGAT ACCCCACCAAACCCAAAAAAAGAGNNNNNNNNN3′

The underlined sequence corresponds to Gal4pAD and the N's correspond to 20–25 nt of Y N terminus-encoding sequence (T_m: ~60°).

Reverse primers to generate X-EGFP fusions: 5′GCACCACCCCGGTGAAC AGCTCCTCGCCCTTGCTCACCATNNNNNNNNN3′

The underlined sequence corresponds to EGFP and the N's correspond to 20–25 nt of X C terminus-encoding sequence (T_m: $\sim 60 \pm 4°$). The stop codon is omitted from the sequence.

Reverse primers to generate Y-viGFP fusions: 5′ TTGGGACAACTCCAGTG AAAAGTTCTTCTCCTTTACCCATNNNNNNNN3′

The underlined sequence corresponds to viGFP and the N's correspond to 20–25 nt of Y C terminus-encoding sequence (T_m: $\sim 60 \pm 4°$). The stop codon is omitted from the sequence.

PCR Conditions

Concentrations of manganese and individual nucleotides should be titrated to obtain a mutation rate of about 10^{-3}. The optimal conditions will depend on the length and sequence of the fragment to be mutagenized. There are two standard protocols.[32]

For sequences to be mutagenized equal or shorter than 1 kb:

1. For each PCR reaction prepare a 100 μl mixture containing 100 ng DNA template, 0.3 μM primers, 50 μM each dNTP, 50 mM KCl, 10 mM Tris-Cl pH 9.0, 0.1% Triton X-100, 1.5 mM $MgCl_2$, 1 mg/ml BSA (bovine serum albumin), and 2.5 units of *Taq* DNA polymerase.
2. First perform a set of 10 elongation cycles:
   Step 1. 94°, 1 min
   Step 2. 45°, 1 min
   Step 3. 72°, 2 min
3. Add 100 μM $MnCl_2$ to the reaction.
4. Perform another set of 30 amplification cycles as follows:
   Step 1. 94°, 1 min
   Step 2. 45°, 1 min
   Step 3. 72°, 2 min

Because the misincorporation rate of *Taq* polymerase is sufficient to generate approximately one mutation in sequences longer than 1 kb, conventional PCR conditions should be used to prevent the occurrence of multiple mutations per molecule.

5. For each PCR reaction prepare a 100 μl mixture containing 100 ng DNA template, 0.3 μM primers, 50 μM each dNTP, 50 mM KCl, 10 mM Tris-Cl pH 8.3, 0.1% Triton X-100, 2.5 mM $MgCl_2$, 1 mg/ml BSA, and 2.5 units of *Taq* DNA polymerase.
6. Perform a set of 35 elongation cycles:
   Step 1. 94°, 1 min
   Step 2. 45°, 1 min
   Step 3. 72°, 2 min

Yeast Transformations, Reverse Two-Hybrid Selections, and Fluorescence Microscopy

The pACT2-viGFP plasmid harbors the *LEU2* selectable marker. Thus, the recipient cells of mutagenized AD-Y-GFP PCR products should contain unmutagenized DB-X plasmids that harbor *TRP1*. For example, pGBT9 (Clontech) or pMAB20 (a centromeric DB vector based on pPC97) can be used for this purpose. Linearization of pACT2-viGFP plasmid can be performed with *Bam*HI since the corresponding site is unique on the plasmid and located between the coding region of Gal4pAD and viGFP.

The pAS2-EGFP plasmid harbors the *TRP1* selectable marker. Thus, the recipient cells of mutagenized DB-X-GFP PCR products should contain unmutagenized AD-Y plasmids that harbor *LEU2*. For example, pGAD10, pGAD424, or pACT2 (Clontech) can be used for this purpose. Linearization of pAS2-EGFP plasmid can be performed with *Sal*I since the corresponding site is unique on the plasmid and located between the coding region of Gal4pDB and EGFP.

The transformation protocol is as follows[32]:

1. Approximately 100 ng of linearized pAS2-EGFP or pACT2-viGFP plasmid should be mixed with 100 ng PCR product and transformed into MaV103 cells carrying the relevant AD-Y or DB-X plasmid, respectively.
2. Yeast transformation protocols that are optimal for this yeast strain have been described.[39]
3. Controls for Gap repair should include transformation with a reference circular plasmid, no DNA, and circularized vector alone.
4. After transformation, the cells are plated on Synthetic complete plate (Sc) lacking leucine and tryptophan (Sc-L-T). Transformants are incubated at 30° for 2 days. The Sc-L-T plates are then replica plated onto Sc-L-T + 5-FOA. Most experiments can be performed at 5-FOA concentrations between 0.15 and 0.2%. The 5-FOA plates are then immediately replica cleaned and incubated for 24 hr at 30°. Foa^R yeast colonies can then be scored for GFP activity by fluorescence microscopy.
5. Light and fluorescence microscopy can be used to detect Foa^R Gfp^+ colonies directly on plates containing 2% agar. By switching back and forth from light to fluorescence, Gfp^+ colonies can conveniently be discriminated from Gfp^- colonies. We currently use a Leica MZ8 microscope (KRAMER Scientific, Burlington, MA).

PCR Recovery of IDAs Using Gateway Primers

The Gateway technology[34–36] was originally developed to allow convenient cloning of large numbers of ORFs predicted from genome and EST sequencing projects into many different vectors. For example Gateway is currently used to clone the (nearly) complete set of ORFs, or "ORFeome," of *C. elegans*.[39] In short, the system is based on phage lambda integration (BP reaction) into and excision (LR reaction) from the *Escherichia coli* genome. The naturally occurring *cis*-acting sites have been modified into pairs of derivatives (attB to B1 and B2, attP to P1 and P2, attL to L1 and L2, attR to R1 and R2). PCR products generated with B1 and B2 tails can recombine into the Entry Gateway vector through P1 and P2. The resulting Entry clones contain inserts flanked by L1 and L2. These inserts can be transferred to Destination vectors through recombination with R1 and R2 sites.

In the reverse two-hybrid implementation of Gateway, IDAs are PCR amplified from the Foa^R Gfp^+ yeast colonies using B1 and B2 primers. The resulting B1-IDA-B2 PCR products are then Gateway-cloned into the P1-P2 Entry vector and the resulting L1-IDA-L2 Entry clones can then be transferred into R1-R2 Destination vectors for expression in different settings. Each cloning step can be performed on 96-well plates using automation devices.

The forward and reverse primers used to PCR amplify IDAs contain B1 and B2, respectively:

B1-AD forward primer: 5′G GGG ACA AGT TTG TAC AAA AAA GCA GGC T *CC CCA CCA AAC CCA AAA AAA GAG* 3′

The underlined sequence corresponds to the B1 recombination site and the italicized sequence corresponds to AD.

B1-DB forward primer: 5′G GGG ACA AGT TTG TAC AAA AAA GCA GGC T *GT CAA AGA CAG TTG ACT GTA TCG* 3′

The underlined sequence corresponds to the B1 recombination site and the italicized sequence corresponds to DB.

EGFP-B2 primer: 5′GGG GAC CAC TTT GTA CAA GAA AGC TGG GT*C AGC TCC TCG CCC TTG CTC ACC AT* 3′

The underlined sequence corresponds to the B2 recombination site and the italicized sequence corresponds to EGFP.

viGFP-B2 primer: 5′GGG GAC CAC TTT GTA CAA GAA AGC TGG GT*G AAA AGT TCT TCT CCT TTA CCC AT* 3′

The underlined sequence corresponds to the B2 recombination site and the italicized sequence corresponds to viGFP.

The conditions for yeast colony PCR reactions are identical to those described above for long DNA fragments at the exception that the template DNA is prepared as follows. A volume of approximately 3 μl of yeast cells is resuspended in 10 μl of 0.02 *N* NaOH and incubated for 30 sec in a microwave at maximum power.

Gateway Cloning of B1-IDA-B2 PCR Products into Entry[36]

Fresh B1-IDA-B2 PCR products should be used, preferably generated the same day. Minipreparation quality vector DNA is sufficient for Gateway reactions. The controls for the Gateway reaction should be: (1) vector without PCR product, (2) PCR product without vector, (3) no BP clonase, and (4) no DNA in the transformation. As a positive control we used the B1-*let-60*-B2 PCR product which works consistently.

1. Take the BP clonase from the −80° freezer and place on dry ice.
2. Carry out the reactions on ice.
3. Combine in sterile tubes: 125 ng Donor vector DNA (pDNR-gent), 5 μl 5× BP reaction buffer, 5 μl PCR product, and add TE to a final volume of 16 μl.
4. Thaw the BP clonase on ice; vortex briefly.
5. Add 4 μl BP clonase to each reaction and mix by pipetting.
6. Incubate 1 hr at 25°.
7. Add 2 μl proteinase K (1 mg/ml); mix by pipetting.
8. Incubate 10 min at 37°.
9. Transform 2 μl of each reaction into a 100 μl aliquot of competent DH5 alpha cells, plate the whole transformation reaction on LB-gent plates (15 μg/ml), and incubate overnight at 37°. The number of colonies obtained varies between 10 and >1000.
10. IDAs should be sequenced from the Entry clones.

Destination Cloning of IDAs into Two-Hybrid and Other Expression Vectors[36]

In addition to the two-hybrid vectors described below, many other Destination vectors are available or currently being developed. These include a baculovirus vector (for expression in insect cells), a GFP vector (for protein localization studies), tagging vectors such as His-tag (for expression in bacteria), a T7 transcription vector (for *in vitro* transcription) and a CMV-neo vector (for expression in mammalian cells) (Fig. 5E).

1. Linearize the DB- and AD-Destination vectors (minipreparation quality) using *Sma*I. Verify digestion on an agarose gel by comparing to the undigested DNA.
2. Take the LR clonase from the −80° freezer and place on dry ice.
3. Carry out the reactions on ice.
4. Combine in sterile tubes: 100 ng Entry clone, 300 ng linearized Destination vector, 5 μl 5× LR buffer and TE such that the final volume is 14 μl.
5. Thaw the LR clonase on ice.
6. Add 6 μl LR clonase to each reaction and mix by pipetting.
7. Incubate 1 hr at 25°.
8. Add 2 μl proteinase K (1 mg/ml), mix by pipetting.
9. Incubate 10 min at 37°.

10. Transform 2 μl of each reaction in DH5 alpha cells as described above, plate the whole transformation reaction on LB-amp (AD-dest) or LB-kan (DB-dest) plates (both 50 μg/ml), and incubate overnight at 37°.

Selection for ICAs

ICAs can be selected in the interactor of a X IDA using forward two-hybrid selections such as 3-AT in response to elevated levels of the *GAL1::HIS3* reporter gene. Libraries of randomly generated alleles can be generated as described above for IDAs. Particular caution is needed if libraries are generated as DB-Y fusions since spurious self-activators can be readily obtained this way.

Summary

Although extremely useful, biochemical methods are time-consuming and fail to unequivocally demonstrate protein–protein interactions. Genetic methods can be used to address this problem. The protocols described here allow the rapid generation and manipulation of interaction-defective alleles that can be used to study protein–protein interactions genetically.

Appendix

Suggestions for Standardized Descriptive System of IDAs and ICAs

Classical genetics benefited from the development of standardized descriptive systems. In *S. cerevisiae* and *C. elegans,* for example, the creation of a commonly used standard system for naming genes and proteins, as well as alleles and phenotypes, greatly facilitated the exchange of information between laboratories. No such system exists to specifically describe genetically altered protein–protein interactions and hence the description of IDAs and ICAs in the literature is somewhat confusing. Thus we propose to add to current nomenclatures a few standardized descriptors for IDAs and ICAs:

1. For IDAs in protein X: X[*idaAApositionAA**(*interactorY*$^-$*; interactorZ*$^+$)] where X is the standard name of the protein according to the nomenclature of the organism. If the allele was first identified in conventional genetic screens and subsequently found to be IDA, its classical allele name can be indicated according to the nomenclature for each model organism; *ida* refers to interaction-defective allele; *AApositionAA** refers to the position of the modified amino acid and the nature of the change from wild-type (AA) to mutant AA*. Multiple mutations can be indicated and separated by commas *AApositionAA**, *AApositionAA**, ... ; *interactorY* and *interactorZ* refer to the name of known interactors of X according to the nomenclature of the organism; (*interactorY*$^-$*; interactorZ*$^+$) refers to the

experimentally defined interaction pattern of the IDA. Here the X IDA fails to interact with interactor X but successfully interacts with interactor Z. For ts and cs interaction phenotypes, (*interactorY*^{ts}) or (*interactorZ*^{cs}) can be used.

2. For ICAs in protein Y: Y[*icaAApositionAA**(*interactorX*$^{AApositionAA*+}$; *interactorX*$^{-}$; *interactorZ*$^{+}$)] where Y is the standard name of the protein according to the nomenclature of the organism. If the allele was first identified in conventional genetic screens and subsequently found to be ICA, its classical allele name can be indicated according to the nomenclature for each model organism; *ica* refers to interaction-compensatory allele; *AApositionAA* refers to the position of the modified amino acid and the nature of the change from wild-type (AA) to mutant AA*. Multiple mutations can be indicated and separated by commas *AApositionAA**, *AApositionAA; interactorX* and *interactorZ* refer to the name of Y interactors according to the nomenclature of the organism; (*interactorX*$^{AApositionAA*+}$; *interactorX*$^{-}$; *interactorZ*$^{+}$) refers to the known interaction pattern of the Y ICA. In this example, the ICA restores interaction with the mutated version of interactor X containing the AApositionAA* mutation. It also fails to interact with the wild-type version of interactor X and maintains its interaction with interactor Z.

It should be noted that, as in conventional genetic descriptions of genotypes, unknown parameters cannot be indicated, but their potential existence should be kept in mind.

IDA Nomenclature Applied to Nia/Nib Interaction

We describe below an example of how the IDA nomenclature could be used in the context of Nia and Nib, the two proteins involved in the genome replication of the tobacco etch potyvirus (TEV)[26] (see above). The two Nia IDAs can be referred to as Nia[*idaQ384P*(*Nib*^{ts})] and Nia[*idaN393D*(*Nib*^{ts})] while the two Nib ICAs selected using Nia[*idaN393D*(*Nib*^{ts})] as a bait can be referred to as Nib[*icaI94T*(*Nia*$^{N393D+}$)] and Nib[*icaC380R*(*Nia*$^{N393D+}$)].

Acknowledgments

We thank Katie Bremberg for assistance in preparing this manuscript. This work was supported by Grants 5R01HG01715-02 (NHGRI), P01CA80111-02 (NCI), 7 R33 CA81658-02 (NCI), 232 (MGRI) awarded to M.V., and a Medical Foundation grant to S.V.

[31] Biochemical Genomics Approach to Map Activities to Genes

By ERIC M. PHIZICKY, MARK R. MARTZEN, STEPHEN M. MCCRAITH, SHERRY L. SPINELLI, FENG XING, NEIL P. SHULL, CERI VAN SLYKE, REBECCA K. MONTAGNE, FRANCY M. TORRES, STANLEY FIELDS, and ELIZABETH J. GRAYHACK

Introduction

This chapter describes a genomic method to map biochemical activities to their corresponding genes. Understanding the biochemical activity of a gene product is a critical step toward deducing the mechanism by which a gene exerts its function, in some cases providing the first clue to that function. It also often leads to the identification of new protein motifs, molecular understanding of regulation, clues to the nature of some drug targets, and insight into the evolution of protein function, as well as a better understanding of metabolism and enzymatic catalysis. Identification of the gene responsible for some biochemical activities has resulted in some surprising links between two apparently unrelated processes, such as the link between *tyr* tRNA synthetase and cytokines produced during apoptosis.[1]

Previously, assignment of biochemical activities to genes was restricted to two difficult and time-consuming approaches, in the absence of clues from genetic analysis. The most common approach is purification of a protein, followed by sequence analysis and gene isolation, a process which can take as long as several years for difficult proteins. An alternative approach is expression cloning, which has been widely used for proteins and activities that are assayable from the outside of the cell or from the cell membrane.[2–5] The three requirements for use of expression cloning are a rich source of the activity from which to produce a cDNA library, an organism or cell line in which the activity is absent to serve as a host for the cDNA library, and an assay that can be applied either to many individual transformants or pools of transformants. These requirements make expression cloning an *ad hoc* method which is both time consuming and limited by sensitivity when used for nonsurface enzymatic activities.[6–10] The biochemical genomics approach we have

[1] K. Wakasugi and P. Schimmel, *Science* **284,** 147 (1999).
[2] B. Seed, *Curr. Opin. Biotechnol.* **6,** 567 (1995).
[3] M. F. Romero, Y. Kanai, H. Gunshin, and M. A. Hediger, *Methods Enzymol.* **296,** 17 (1998).
[4] H. Simonsen and H. F. Lodish, *Trends Pharmacol. Sci.* **15,** 437 (1994).
[5] M. Fukuda, M. F. Bierhuizen, and J. Nakayama, *Glycobiology* **6,** 683 (1996).
[6] G. H. Jones and D. A. Hopwood, *J. Biol. Chem.* **259,** 14151 (1984).
[7] J. W. Tobias and A. Varshavsky, *J. Biol. Chem.* **266,** 12021 (1991).
[8] K. A. White, S. Lin, R. J. Cotter, and C. R. Raetz, *J. Biol. Chem.* **274,** 31391 (1999).

0076-6879/02 $35.00

Array of yeast transformants, each producing a distinct GST-ORF fusion.

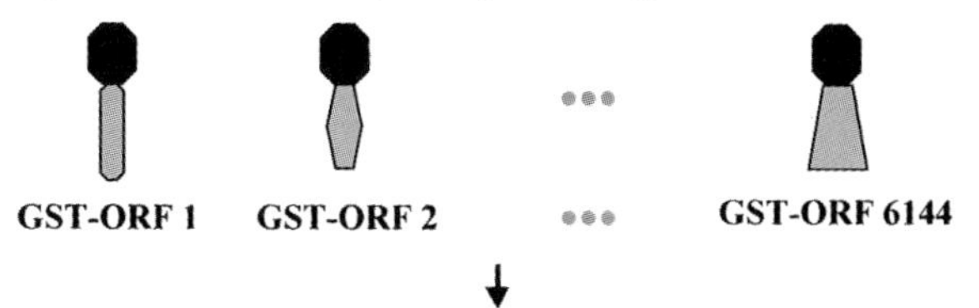

↓

Purify GST-ORF fusion proteins from 64 pools of 96 transformants each.

↓

Screen the pools for a biochemical activity.

↓

Subdivide a positive pool to identify protein and gene of interest.

FIG. 1. Overall scheme of the biochemical genomics approach to map activities to genes.

developed to identify genes associated with an activity has several advantages over these methods: it is rapid, systematic, sensitive, and general.

Overall Scheme for Gene Identification

To identify genes associated with activities, we first constructed a genomic ordered array of 6144 *Saccharomyces cerevisiae* strains, each expressing a unique yeast ORF (open reading frame) fused at its N terminus to the C terminus of GST (glutathione *S*-transferase) protein, under control of the P_{CUP1} promoter[11,12] (see Fig. 1). This collection is stored on 64-microtiter plates, each with 96 wells.

We then purified 64 pools of GST–ORF proteins, each derived from the strains on a microtiter plate (Fig. 1). To obtain these pools, we first cocultured mixtures of the 96 strains, induced expression of the GST-ORFs, and harvested cells. Then crude lysates were made, GST–ORFs were purified by a one-step batch affinity purification, and samples were dialyzed for storage.

To identify genes associated with activities, the purified GST-ORF pools are assayed for activity. Then active pools are deconvoluted by growth and analysis of subpools of strains to identify the source strain and ORF responsible for activity. We have used this approach to identify genes associated with a number of different activities (Refs. 11 and 12).

[9] S. J. Li and M. Hochstrasser, *Nature* **398,** 246 (1999).

[10] J. Li-Hawkins, E. G. Lund, A. D. Bronson, and D. W. Russell, *J. Biol. Chem.* **275,** 16543 (2000).

[11] M. R. Martzen, S. M. McCraith, S. L. Spinelli, F. M. Torres, S. Fields, E. J. Grayhack, and E. M. Phizicky, *Science* **286,** 1153 (1999).

[12] E. J. Grayhack and E. M. Phizicky, *Curr. Opin. Chem. Biol.* **5,** 34 (2001).

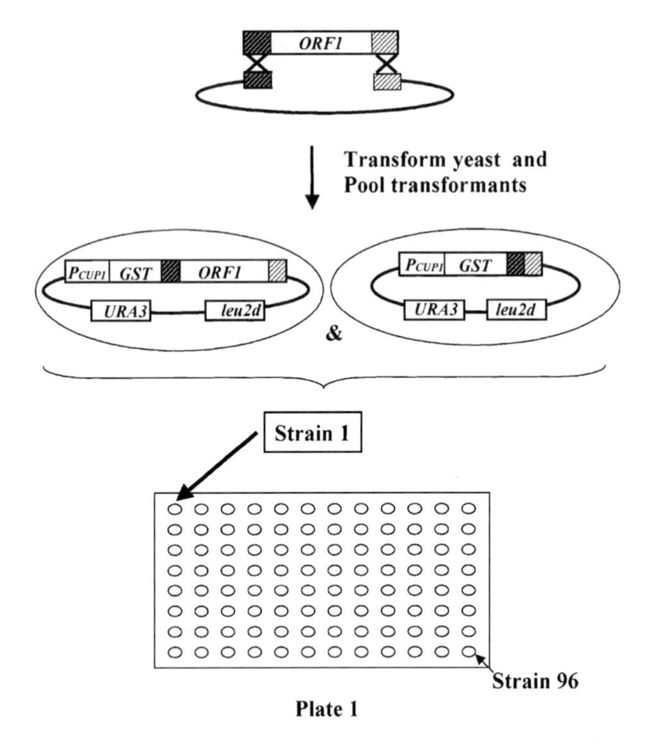

FIG. 2. Construction of the strain array. Each strain in the library was constructed by recombination-mediated cotransformation of one of the PCR amplified ORFs in the previously constructed genomic collection [J. R. Hudson, Jr., E. P. Dawson, K. L. Rushing, C. H. Jackson, D. Lockshon, D. Conover, C. Lanciault, J. R. Harris, S. J. Simmons, R. Rothstein, and S. Fields, *Genome Res.* **7,** 1169 (1997)], together with a linearized vector with the same 5′ and 3′ ends as in the ORF set. As indicated, each strain is a pool of all transformants from each reaction mixture.

Construction and Analysis of Libraries

Composition of Strain Library

Each strain in the library was constructed by cotransformation of linearized vector with the appropriate PCR-amplified ORF DNA from a previously constructed genomic collection of ORF DNAs.[13] Since all the ORFs have common 5′ and 3′ ends that match the ends of the linearized vector, this results in efficient recombination-mediated transformation of the plasmid and ORF DNA, to generate an in-frame fusion of GST with the ORF [14,15] (Fig. 2). Each transformation

[13] J. R. Hudson, Jr., E. P. Dawson, K. L. Rushing, C. H. Jackson, D. Lockshon, D. Conover, C. Lanciault, J. R. Harris, S. J. Simmons, R. Rothstein, and S. Fields, *Genome Res.* **7,** 1169 (1997).

[14] H. Ma, S. Kunes, P. J. Schatz, and D. Botstein, *Gene* **58,** 201 (1987).

[15] K. R. Oldenburg, K. T. Vo, S. Michaelis, and C. Paddon, *Nucleic Acids Res.* **25,** 451 (1997).

yields four products: vector containing the desired ORF insert; vector containing the ORF insert with deleterious PCR mistakes; resealed or uncut vector; and vector containing primer dimers which arose during the second PCR amplification step of the original collection of ORFs. When we constructed the GST-ORF library, we pooled and saved all the transformants arising from each transformation reaction; thus each strain has all four classes of transformants, and for that reason every time we use the strain we use a patch of colonies. This population-based approach avoids the risk of using a single colony with a plasmid ORF insert bearing a deleterious PCR error, or with a plasmid lacking an ORF insert, and ensures that as many ORFs as possible are represented in functional form in the library.

Representation of ORFs in Library Strains

In an analysis of a random set of strains, we found that most strains contain GST-ORF fusions and that the frequency of ORF inserts in plasmids is high. We examined the ORF insertion frequency in 63 random strains from plate 29 (which contains ORFs of 1000–1200 bp) and 4 strains from plate 25. We found that nearly every strain (66/67) contains plasmids with inserts. Of these 66, all but 5 had inserts in 50% or more of the plasmids examined, and the insert frequency in the remaining 5 strains was 22–43% of the plasmids. Of the 530 plasmids examined overall, 75% had insert; thus for such strains signal detection is 75% of the maximum it can be (see below).

We also found that the frequency of inserts in plasmids from strains with kinase ORFs is much lower than that from random strains. Thus, of 18 strains with ORFs encoding kinases, only 37% of the plasmids examined had inserts (56/150 examined). In particular, 3 strains had 0 or 1 insert out of 12, and 3 strains had no inserts out of 4 or 5 examined, although other kinase ORFs were present at high frequencies in plasmids. The contrast is striking between randomly analyzed strains and those with kinase ORFs. The most likely interpretation is that the P_{CUP1} promoter allows leaky expression of ORFs under noninducing conditions; this would be deleterious to growth for strains containing certain kinase ORFs, causing them to be selected against in our population.

Purification of Soluble GST–ORF Proteins

Strains and Plasmids

The host strain for the array is EJ 758 [*MAT***a** *his3-Δ200, leu2-3,112, ura3-52, pep4::HIS3*], a derivative of JHRY-20-2Ca. This strain has behaved well as a source of extracts for purification of proteins.[16–18] Each strain is transformed with the 2μ

[16] S. M. McCraith and E. M. Phizicky, *Mol. Cell. Biol.* **10,** 1049 (1990).

[17] G. M. Culver, S. A. Consaul, K. T. Tycowski, W. Filipowicz, and E. M. Phizicky, *J. Biol. Chem.* **269,** 24928 (1994).

[18] G. M. Culver, S. M. McCraith, S. A. Consaul, D. R. Stanford, and E. M. Phizicky, *J. Biol. Chem.* **272,** 13203 (1997).

URA3 leu2d plasmid vector pYEX4T-1 + rec, which was derived from pYEX4T-1 (Invitrogen Corp., Carlsbad, CA) by insertion of the recombination domain to allow recombination-based insertion of the ORFs during transformation.

Materials

Glutathione Sepharose-4B is from Amersham/Pharmacia (Piscataway, NJ); glutathione is from Sigma (St. Louis, MO); glass beads are from Braun Biotech (Bethlehem, PA); cell scrapers are from Becton Dickinson and Company, Franklin Lakes, NJ.

Buffers

Extraction Buffer

50 m*M* Tris-HCl (pH 7.5)
1 m*M* EDTA
4 m*M* $MgCl_2$
5 m*M* Dithiothreitol (DTT)
10% (v/v) Glycerol
1 *M* NaCl
2.5 μg/ml Leupeptin (Add just prior to use from a 1 mg/ml solution)
2.5 μg/ml Pepstatin (Add just prior to use from a 1 mg/ml solution)

Wash Buffer. For No Salt Wash buffer; omit NaCl.

50 m*M* Tris-HCl (pH 7.5)
4 m*M* $MgCl_2$
1 m*M* DTT
10% Glycerol
0.5 *M* NaCl

Elution Buffer. (Make up right before use.)

Wash Buffer	29.7 ml
20 m*M* NaOH	0.3 ml of 2 *M* stock
25 m*M* Glutathione	0.23 g

Mix Wash Buffer and NaOH first, then add glutathione.

Dialysis Buffer

20 m*M* Tris-HCl (pH 7.5)
2 m*M* EDTA
4 m*M* $MgCl_2$
1 m*M* DTT
55 m*M* NaCl
50% (v/v) Glycerol

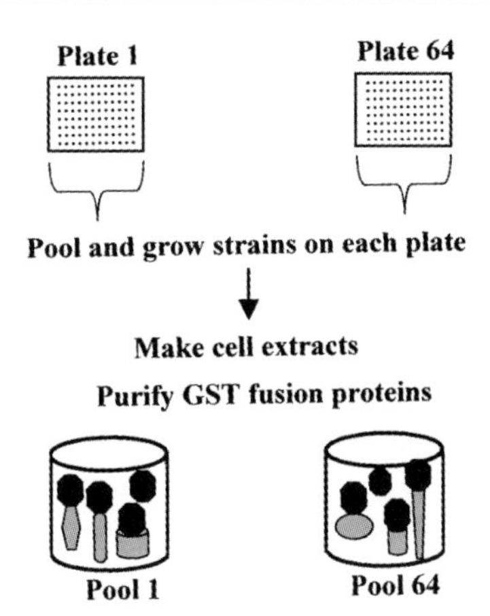

FIG. 3. Scheme for preparing pools of purified GST–ORF proteins.

Equilibration of Glutathione Sepharose Resin

Use 100 μl glutathione (GSH) Sepharose resin for each 2 ml extract. Before starting, the resin is prewashed five times to equilibrate it with buffer and to remove free glutathione, by adding 1 ml Wash Buffer, sedimenting the resin for 25–40 sec at low speed (in either a microfuge or a clinical centrifuge), and removing the buffer with a Pasteur pipette.

Procedure

The overall scheme to obtain the pools of purified GST-ORF proteins is illustrated in Fig. 3.

1. Growth of Strains. To obtain the set of strains from a microtiter plate, the microtiter plates are first thawed, and then cells are spotted on plates. Generally, we find that the cells have settled to the bottom of the wells, so we first mix the material in the individual wells using a multichannel pipettor. Then the cells are transferred to a large (150 mm) SD-Ura plate using either a 96-well pinner or a multichannel pipettor (2 μl/well). This results in a dense patch of growth after 1 to 2 days at 30°. We always use a mixture of cells from any well or plate, since (as described above) each strain in the collection is a mixture of transformants expressing not only the desired GST-ORF, but also empty vector, vector with primer dimers, and vector with the ORF containing PCR (polymerase chain reaction)-generated errors.

The 96 patched strains are transferred from the SD-Ura plate to liquid SD-Ura medium at $OD_{600} = 0.2$ for growth overnight. To do this, cells are scraped from the plate into 5 ml YPD, yielding a thick slurry of OD $\sim$ 50, and an appropriate aliquot is serially diluted into SD-Ura at the correct OD. The rest of the slurry from the plate is diluted to OD of 10, supplemented with dimethyl sulfoxide (DMSO) to 7%, aliquoted, and saved at −70°.

After growth overnight at 30°, the cells are inoculated into 50 ml of SD −Ura −Leu medium at OD_{600} = 0.2 and grown overnight (to amplify the plasmid, which has an inefficiently expressed *leu2-d* marker). Then cells are reinoculated into 500 ml SD −Ura −Leu medium at OD_{600} = 0.2, grown to OD_{600} = 0.8, and supplemented with 0.5 m*M* copper sulfate for 2 hr to induce transcription of the GST–ORF from the P_{CUP1} promoter before harvest.[19] The cells are harvested by centrifugation in a GS3 rotor tube, washed with cold water in two Falcon tubes (Becton Dickinson and Company, Franklin Lakes, NJ), and frozen at −70°. One tube (from 250 ml culture) is used for the protein purification.

As an alternative to growing 50 ml culture overnight, we frequently dilute the 50 ml SD −Leu −Ura culture into the large culture the night before so it will be ready the next morning for copper induction and subsequent steps (assuming a 2.5-hr generation time).

2. Cell Lysis and Extract Formation. Frozen cell pellets from 250 ml culture are resuspended in 1 ml Extraction Buffer, and leupeptin and pepstatin are added. The mixture is then transferred to a 2-ml bead beater vial (BioSpec Products, Inc., Bartlesville, OK), and glass beads (0.45–0.5 mm) are added to the top of the tube (usually about 1 ml).[16] Cells are lysed by shearing with glass beads at 4° for 200 sec in a Biospec mini bead beater, using 10 cycles of shearing for 20 sec, followed by 1 min of cooling in an ice–water bath. In the absence of the glass bead beater, adequate extracts can be made by vortexing with glass beads. Other alternatives include use of a French press or standard lysis methods following enzymatic removal of the cell wall.

After lysis, the cell extract is removed from the beads. We puncture the bottom of the tube with a hot (flamed) 25-gauge needle and gently force out the liquid with low-pressure air, while retaining the beads in the tube. The beads are then washed with 0.5 ml Extraction Buffer which is also forced through the tube with air. Finally, 2 μl of 1 *M* phenylmethylsulfonyl fluoride (PMSF) is added, and debris is removed by centrifugation for 10 min at 4° in a microfuge. The supernatant (crude extract) is transferred to a new microfuge tube and a sample is used to assess lysis by measuring protein concentration. Crude extracts prepared like this generally contain about 20–30 mg/ml protein, as measured by Bradford assay.[20] The rest of the extract is used for purification of GST-ORF proteins, as described below (see Fig. 3).

3. GST–ORF Purification. GST-ORF fusion proteins are purified by a single step affinity purification procedure on a glutathione-Sepharose column, essentially as described.[21,22]

[19] I. G. Macreadie, O. Horaitis, A. J. Verkuylen, and K. W. Savin, *Gene* **104,** 107 (1991).
[20] M. M. Bradford, *Anal. Biochem.* **72,** 248 (1976).
[21] D. B. Smith and K. S. Johnson, *Gene* **67,** 31 (1988).
[22] J. R. Nelson, C. W. Lawrence, and D. C. Hinkle, *Science* **272,** 1646 (1996).

Addition of resin. The crude extract is diluted with an equal volume of No Salt Wash Buffer to bring the final salt concentration to 0.5 *M* NaCl. Then 100 μl of preequilibrated glutathione-Sepharose resin is added per 2 ml of extract (as measured after addition of No Salt Wash Buffer) and the tube is mixed gently in the cold room for 3 hr. We use a nutator for this step, but a roller or rocker would be as good. We chose this modest salt concentration because it is high enough to prevent most nonspecific protein–ligand or protein–protein interactions, while still preserving required functional interactions.

Removal of unbound protein. To remove unbound protein, the mixture is centrifuged for 20–30 sec at low speed in the microfuge (near lowest setting 1–2), and the supernatant, containing unbound protein, is removed using a Pipetman. This is followed by two wash steps in which 1 ml Wash Buffer is added to the resin, followed by mixing for 10–20 min, centrifugation and removal of buffer.

Elution of bound protein. Bound protein is eluted from the resin with Elution Buffer, which contains 25 m*M* glutathione to compete with glutathione on the resin. To do this, 2 ml Elution Buffer is added to the resin, followed by 40 min or more of mixing, and then by low speed centrifugation. This supernatant, which contains the purified proteins, is then dialyzed overnight against Dialysis Buffer and stored at -20°.

Yield. The usual yield from a 250-ml culture of cells is 0.5 ml of protein at about 250 μg/ml (range: 150 to 350 μg/ml).

4. Quality of Protein Array. We have done three different types of analysis of the protein array. First, we have examined the purity of the proteins. As judged by SDS–PAGE and silver staining, the GST–ORF proteins are highly purified and uniformly free of most background proteins usually present in crude extracts (Fig. 4). Typically, each preparation contains three prominent bands. One band corresponds

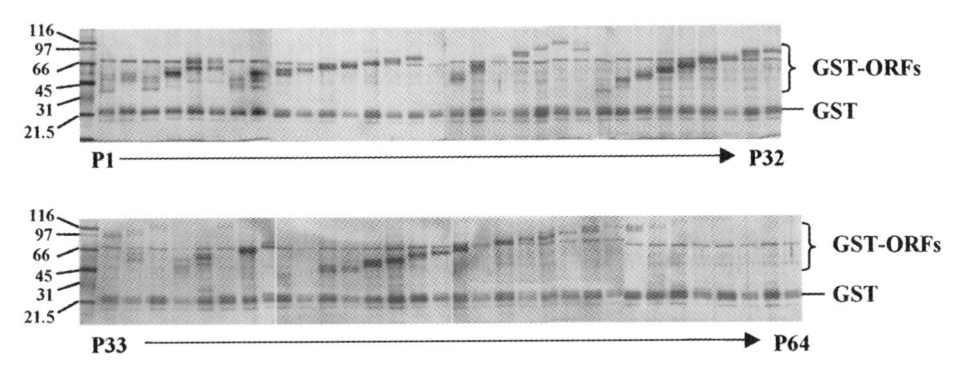

FIG. 4. Analysis of pools of purified GST–ORF proteins. Pools of GST–ORFs, prepared as described in the text, were analyzed by SDS–PAGE. A 10-μl aliquot was resolved on a 10% gel, followed by silver staining of the polypeptides. Migration of molecular weight markers is shown in the first lane of each row and indicated at the left.

in size to the cluster of GST–ORF proteins predicted from the particular microtiter plate. The GST–ORFs from each plate form a cluster because the PCR-amplified ORFs used to make the library were originally grouped according to size to facilitate multiwell PCR amplification. The second prominent band corresponds to GST, and the third (much fainter) band is from a single contaminant of mass ~70 kDa. For the majority of the pools, there is as much or more of the GST–ORF proteins as there is of the GST protein. Thus for this majority of preparations, the final yield is on the order of 125 μg/ml of GST–ORFs. Since each pool is derived from 96 strains, about one-third of which are expected to express GST–ORFs that are membrane proteins (which are not extracted under these conditions), each GST–ORF in these pools is likely present at ~0.9 μg/ml (~20 n*M*), assuming equal representation of all the soluble proteins. However, we note that there is notably less of the GST–ORF proteins in 11 of the pools (pools 16, 35, 38, 42, and 58–64). Since for all but two of these pools, the encoded proteins are predicted to be larger than 90 kDa, we conclude that larger ORFs are either underexpressed or underrepresented in the library. Nonetheless, two of the activities we have uncovered were found in these pools of underrepresented ORFs—one activity in pool 35 and one in pool 58.

Second, there is minimal contamination by biochemical activities from wild-type proteins in the extract. As described below, removal of such background proteins is critical for success of the procedure, since most of the activities are already present in crude extracts and must be removed in order to see the signal from the GST–ORF. We have found that most activities we have assayed are not detectable at all in pools other than the correct ones, under any tested conditions. However, for a couple of activities, the background signal becomes visible when large amounts of protein are incubated for extended periods of time; for example, cyclic phosphodiesterase activity in the positive plate pool is only 10-fold greater than the background in negative pools. As a further test of the removal of background, we have looked at alcohol dehydrogenase, an activity that is present in large amounts in crude extracts and unlikely to be overproduced from the P_{CUP1} promoter. We find that more than 99% of the alcohol dehydrogenase activity is removed by the GST–ORF purification procedure, and although a substantial amount is left as background, we can still see the increase in signal from two of the pools that ought to contain alcohol dehydrogenase activity.

Third, the collection is reasonably complete for simple soluble enzymes. This conclusion is based on the fact that we have found most of the enzymatic activities for which we have looked: 17 of about 20. Thus, the representation of the protein pools is excellent from an enzymatic standpoint.

Identification of Genes Associated with Biochemical Activity

Procedure

1. Assay GST–ORF Pools for Activity. Identification of GST–ORF pool(s) containing a particular activity requires an assay with slightly different features than normally used in traditional purifications. To begin a purification, one needs an assay that is both sensitive and specific in crude extracts, since this is the starting point for the purification. Development of such an assay can be difficult because of numerous competing proteins in the extract that degrade or alter the substrate, the product, or the proteins, and considerable effort can be devoted to finding conditions to prevent such losses (different extracts, salt conditions, competitors and inhibitors of various side reactions, etc.). Furthermore, the assay has to be sufficiently specific that only the desired activity is detected; otherwise, one might invest a lot of time and effort in purification of the wrong protein.

Use of the biochemical genomics approach to identify activities requires only a sensitive assay. Because the pools of proteins are already largely free of bulk contaminants, there is little need for measures to inhibit side reactions. Specificity is also not as important as in extracts, since it requires so little effort to identify pools with activities. Thus, if two or more activities appear, it is simple enough to impose a more selective assay at that time, or to first deconvolute the activities to identify the relevant genes, and then sort through the various possibilities later (subunits of the same protein; different proteins with the same biochemical activity, but different *in vivo* substrates, etc.).

The requirement for a sensitive assay is crucial because the purified GST–ORFs are present at relatively low concentrations, of the order of 20 nM (0.9 μg/ml) of each protein. This concentration of individual GST–ORFs is about fivefold less than that of a typical modestly represented protein in a crude extract (assuming a standard crude extract of 40 mg/ml, and a protein that normally requires a 10,000-fold purification), and correspondingly less for underrepresented GST–ORFs. Nonetheless, this amount of protein is usually sufficient to detect proteins that catalyze metabolic reactions, particularly when there is a convenient labeled substrate for ease of detection, and efficient methods such as polyacrylamide gel electrophoresis (PAGE), thin layer chromatography (TLC), high-performance liquid chromatography (HPLC), or trichloroacetic acid (TCA) precipitation to separate products from reactants. One major advantage of the purified GST–ORF pools is that assay sensitivity can be significantly increased by use of more protein from the pools and longer incubation times. A striking example is illustrated by assay of cytochrome *c* methyltransferase, Ctm1.[23] Increasing the amount of protein in the assay from 2 μl to 4 μl and the incubation time from 1 hr at 37° to 8 hr at 30° results in a signal peak that is 28-fold higher, while the background remains constant (our unpublished results). This extended incubation time will almost never work in crude extracts. We have obtained similar results for several other activities. The low concentration of GST–ORFs is a particular problem when assaying binding, since the detection of such activity generally requires protein concentration within range of the binding constant (using labeled ligand). We have circumvented this

[23] B. Polevoda, M. R. Martzen, B. Das, E. M. Phizicky, and F. Sherman, *J. Biol. Chem.* **275,** 20508 (2000).

problem for EMSA assays of DNA binding by using labeled DNA near the K_d in the presence of excess nonspecific competitor (our unpublished results). We note that higher concentrations of individual GST–ORF proteins can be attained by concentrating the protein, by starting with larger number of cells to begin purification, or by use of smaller pool sizes (see below).

Signal detection using the GST–ORF pools is affected by several other factors, most of which are fixed by the system we used for strain construction, cell growth, and protein purification, as well as by the pooling approach. These factors are conveniently summarized by the expression below:

$$\text{Signal} = \frac{k\ (\textit{protein/cell})\ (\textit{yield purified protein})\ (f_{\textit{ORF inserts}})\ (f_{\textit{correct PCRs}})}{(\text{background protein})\ (\text{number of ORFs in pool})}$$

a. Protein/cell. To produce as much GST-ORF fusion protein per cell as possible, we used two strategies. First, we chose the strong Cu^{2+}/-inducible P_{CUP1} promoter for expression of large amounts of protein.[19] Second, we maximized gene dosage by use of a 2μ plasmid with both a *URA3* marker for normal selection and a *leu2-d* marker for increased amplification of the plasmid on medium lacking leucine. Indeed, for three gene products that we have assayed (Ctm1, Smm1, and Tpt1), expression is increased more than 300-fold in the crude extracts containing the GST–ORF, relative to that observed in control extracts. One drawback with this promoter is that there are difficulties with keeping expression low during nonselected growth; this appears to have impaired retention of ORFs (like kinases) whose expression can be deleterious to cell growth.

b. Yield of purified protein and background protein. We used the GST affinity tag because anecdotal evidence suggests that it rarely interferes with activity and because purification with this tag is efficient and can be performed at high salt concentration to prevent nonspecific interactions. The result is extremely pure protein preparations (low background protein), but the yields of GST–ORF are not as high as we would like (10–20%, as measured by activity). Improving this yield by a factor of 5 would correspondingly increase signal strength.

c. Fraction of plasmids with ORF insertions ($f_{\textit{ORF inserts}}$). The effect of insert frequency is relatively minor. As described above, our analysis of the frequency of ORF inserts in the plasmids of 67 strains showed that 75% of the plasmids examined contained inserts. Thus, $f_{\textit{ORF inserts}} = 0.75$ in these strains. As can be seen from the relationship above, this representation of inserts reduces the maximum possible signal strength for detection of activities by only 25%. Since almost all the activities we have detected in GST–ORF pools are severalfold above background levels [and some such as Tpt1[11] and cytochrome *c* methyltransferase (see below) are 25- to 100-fold above background], the 25% reduction in maximum signal strength is quantitatively minor. For those strains (such as kinase strains) where the ORFs or proteins are underrepresented, this problem is correspondingly more important.

d. Fraction of plasmids with ORF insertions that are biochemically active ($f_{correct\ PCRs}$). Using a population of cells from each strain also ensures that the effect of incorrect PCR products is small. The PCR products used to construct this library were made with a mixture of *Taq* and *Pfu* polymerases. For 20 effective cycles of amplification of 1000 bases with *Taq,* 16% of the DNAs will have a single mutation.[24] Since at most only about 20% of the random changes in a protein sequence are likely to be deleterious, this implies a 3.2% loss of functional proteins in each amplification step. Thus, $f_{correct\ PCRs}$ for the two steps of amplification is 0.94 and results in a negligible effect on signal detection.

e. Number of ORFs in pool (pool size). Clearly, the number of ORFs that are represented in each pool is a major factor in signal detection, since the activity from any single ORF is reduced directly according to its representation in the pool. We initally chose mixtures of 96 strains for combined growth and analysis of GST–ORF proteins, after reconstruction tests with two tRNA splicing activities demonstrated that activity was easily detected in mixtures of purified GST–ORFs of this complexity.[11] However, for activities that are more problematic, one could easily use smaller pool sizes. In the extreme, the pool size would be 1, which would entail assay of every individual purified protein, but would enhance the signal 96-fold from that which we now observe.

2. Deconvolute Positive Pools to Identify Source Strain and GST–ORF. Pools that have activity are deconvoluted by preparing and analyzing subpools of GST-ORFs from the strains on the corresponding microtiter plate. To do this, we usually prepare 8 subpools from the rows of strains on the plate and 12 subpools from the columns of strains, and assay each subpool (see Fig. 2). For activities due to single subunit proteins, the signal will increase in the subpools because the number of different ORFs is smaller. (For activities that require more than one component, the activity may stay the same in the subpools.) After we have identified the appropriate row and column fraction, we test the individual implicated strain.

3. Isolate and Sequence Plasmid DNA. Although the identity of the gene corresponding to the activity should be apparent from its position in the plate and its identification in the data base, we always sequence part of the plasmid ORF to confirm it. To do this, we first isolate the plasmid, by transformation of *Escherichia coli* with a preparation of DNA from a yeast overnight culture. Of the 20 or so clones we have checked in this way, the only mistakes we have found are in the orientation of plates 27 and 51, both of which are rotated by 180 degrees so that position A1 is really H12, etc.

4. Confirm that Activity Copurifies with GST–ORF. Finally the isolated DNA is used to transform the host yeast strain, and the individual GST–ORF is purified from a single transformant to confirm that the activity copurifies with it. This step guards against the (unlikely) possibility that the source of DNA used to make the

[24] J. Cline, J. C. Braman, and H. H. Hogrefe, *Nucleic Acids Res.* **24,** 3546 (1996).

library was cross-contaminated with some fraction of another ORF that was itself responsible for activity. We have not found such a case of contamination.

5. Determine If Identified GST–ORF Is Sufficient for Catalytic Activity. The GST–ORF associated with an activity is either the active polypeptide, part of a multisubunit enzyme in which several proteins are required for activity, or an associated submit that copurifies with the active polypeptide. To determine if the GST–ORF is active, two experiments are done. First, the amount of activity in extracts from cells that overexpress the GST–ORF is compared to that in control extracts. If the GST–ORF extracts overproduce activity, then the ORF is the limiting component in the assay and is likely the polypeptide responsible for activity. Second, the ORF [tagged with hexahistidine (His_6) at its amino terminus] is expressed and purified from *E. coli,* and then assayed for activity. If the preparation from *E. coli* cells is active, the ORF is almost certainly the catalytic polypeptide (otherwise it would have to fortuitously bind and copurify with a heterologous catalytic subunit from *E. coli*).

Requirements and Strengths of This Approach

The biochemical genomics approach described here has four minimal requirements for success. First, the GST–ORF has to be present in the gene and protein pools. As discussed above, the vast majority of ORFs are represented in the strain library; furthermore, a large fraction of the soluble enzymatic activities we have looked for are found in the pools, indicating that most such proteins are present. Second, the ORF must be functional as an N-terminal GST–ORF fusion. This is true of a large number of fusions, based on the success of large scale two-hybid screens and on anecdotal evidence. A notable exception may come from the class of membrane and secreted proteins since the native N terminus may be important for correct localization of such proteins. Another exception may arise in cases where dimerization of GST interferes with function of the Protein. Third, the GST–ORF has to be purified with other components required for activity. As discussed below, the one-step affinity purification maximizes the likelihood that this will occur. Fourth, there must be a reasonably sensitive activity assay with which to identify active pools.

There are five particular strengths of this approach. First, deconvolution to a single gene is rapid and easy. Starting from the 64 pools of purified GST–ORFs (which themselves take only 3 to 4 weeks to prepare), it takes only about an additional 2 weeks to implicate a gene associated with a particular activity. Because of this, it is possible to deconvolute several positives in parallel if several different proteins with the same activity are detected in different pools. This is turn eases requirements for a highly specific assay. Second, the method is sensitive. Because the GST–ORF pools are highly purified from other cellular proteins before assay, activities can often be assayed for hours without interference from contaminants;

this can result in a high signal-to-noise ratio. Third, the method is completely general and can be used for any activity normally present in wild-type cell extracts. This is possible because the affinity purification of GST–ORFs efficiently removes background proteins, even those that are present at high levels in extracts. This is in contrast to expression cloning, which relies on prior identification of a host lacking the desired activity, rather than on purification of the expressed proteins. Fourth, the method can be used to detect activities that require more than one component, provided than the GST–ORF binds the other component(s) tightly enough to survive the single-step purification that is employed. Since GST pull-down experiments often work, and this is a homologous system, the normal binding partner of a particular GST–ORF is likely to be present in the cell and to be copurified in pools containing the GST–ORF. Although the activity from a GST–ORF complex is lower than that from a single subunit GST–ORF (because other ORFs in the complex are not overproduced), it is still often sufficiently above background to be detected. Fifth, the method is comprised of reasonably stable components. The strains are easily stored and can be used repeatedly (subject only to losses arising from propagation of mixed cultures). Moreover, because the pools of GST–ORFs are reasonably stable during storage, a single set of pools can be used repeatedly for assay of new activities. Indeed, we used the original pools for up to 15 months after their purification to identify activities.

Acknowledgments

This work was supported by grants from the Merck Genome Research Institute (Grant #196) and NIH (HG-02311).

[32] Use of Two-Dimensional Gels in Yeast Proteomics

By ANDERS BLOMBERG

1. Introduction

The word proteomics was invented in the mid-1990s to describe the analysis of the complete protein complement encoded by a genome—the proteome (with the analogy to all genes in a genome). Thus, it should be distinguished from protein chemistry/analysis in general and should be used exclusively for attempted genome-wide analysis, be it global analysis of expression changes, global analysis of modification changes, or global analysis at some of the other levels of complexity encountered at the protein level. Proteomics was initially technically synonymous with the use of quantitative two-dimensional polyacrylamide gel electrophoresis

0076-6879/02 $35.00

(2D-PAGE) and subsequent protein identification by microsequencing or mass spectrometry. More recently proteomics has also come to include other genome-wide technologies at the protein level, e.g., global analysis of protein–protein interactions, where the two-hybrid system in yeast (see elsewhere in this volume[1]) has set the standard for high-throughput analysis.

2D-PAGE analysis was in use long before the word proteomics was invented.[1a] This analysis is based on the separation of proteins according to their isoelectric point (p*I*) in the first-dimensional separation (*x* axis) and their molecular weight (M_r) in second-dimensional separation (*y* axis) (Fig. 1). Differences in p*I* values of different proteins can be analyzed by isoelectric focusing. In this technology proteins migrate in an electric field in a pH gradient, until they experience the pH where their net charge is zero (a prerequisite to migration in an electric field is charge), where the protein stops and focuses. This results in a banding pattern, and for a total cell lysate approximately 100 bands can be distinguished. The isoelectric focusing strip (or gel rod) is then applied on top of a large-format slab gel, molded in place, and proteins separated by sieving in the presence of the detergent sodium dodecyl sulfate (SDS). Analysis of a total cell lysate by only this SDS-based technology yields about 100 distinct bands. However, by applying the combination of the two technologies in tandem, the analysis of a total yeast cell lysate will generate at least 1500 information units as small or large protein spots (Fig. 2). Thus, proteins which exhibit the same p*I* value, and therefore migrate to identical positions in the first dimension, will most often be separated in the second dimension because of molecular weight differences (Fig. 1). This overall design of the 2D-PAGE technology was invented in the mid-1970s, so 2D analysis has sometimes been described as an older technology. However, the past few years have seen some drastic refinements in 2D-PAGE-based proteomics technology,[2] which are mainly linked to (*i*) the improved 2D pattern standardization and extension by the use of immobilized pH gradients in the first-dimensional separation[3] and (*ii*) the improved downstream identification of resolved proteins by mass spectrometry.[4]

The 2D technology was initially applied to specific studies in yeast research, e.g., the molecular physiology of heat adaptation in yeast,[5] as well as the analysis of industrial strains.[6] In addition, various research groups started in the 1980s to aim for a complete understanding of the resolved protein pattern by utilizing a number of ingenious identification strategies.[7] Our own interest and experience in the

[1] M. Fromont-Racine, J.-C. Rain, and P. Legrain, *Methods Enzymol.* **350,** [29], 2002 (this volume).

[1a] P. H. O'Farrell, *J. Biol. Chem.* **250,** 4007 (1975).

[2] S. J. Fey and P. M. Larsen, *Curr. Opin. Chem. Biol.* **5,** 26 (2001).

[3] A. Görg, C. Obermaier, G. Boguth, A. Harder, B. Scheibe, R. Wildgruber, and W. Weiss, *Electrophoresis* **21,** 1037 (2000).

[4] A. Pandey and M. Mann, *Nature* **405,** 837 (2000).

[5] M. J. Miller, N.-H. Xuong, and E. P. Geiduschek, *J. Bacteriol.* **151,** 311 (1982).

[6] M. Brousse, N. Bataillé, and H. Boucherie, *Appl. Environ. Microbiol.* **50,** 951 (1985).

[7] A. Blomberg, *Yeast,* in press, 2002.

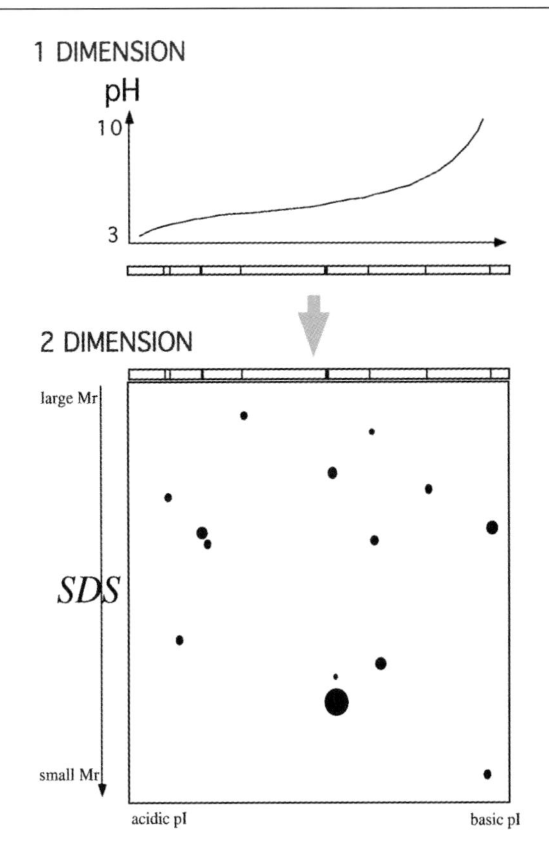

FIG. 1. The general principles underlining separation of proteins by two-dimensional polyacrylamide gel electrophoresis (2D-PAGE). The first-dimensional separation in this example is based on the use of nonlinear pH gradient in the range 3–10. Most commercial strips, however, are of the linear type. The use of a nonlinear pH gradient has provided us with a well-resolved yeast protein pattern over the whole pH range (see Fig. 2).

technology comes from studies on the adaptive physiology of cellular dehydration of yeast, and we have thus mainly focused on the identification of proteins that displayed interesting behavior under the stress conditions studied.[8] At present, several groups have annotated 2D pattern databases on the Internet and these can be used by others as a valuable source of information (see Section 9). Soon there will probably be databases presenting 2D patterns where every single protein spot has been identified and for which protein expression and modification changes under many conditions can be found. The list of experimental studies in the yeast field that have made use of proteomics based on 2D electrophoretic separation of

[8] J. Norbeck and A. Blomberg, *Yeast* **16,** 121 (2000).

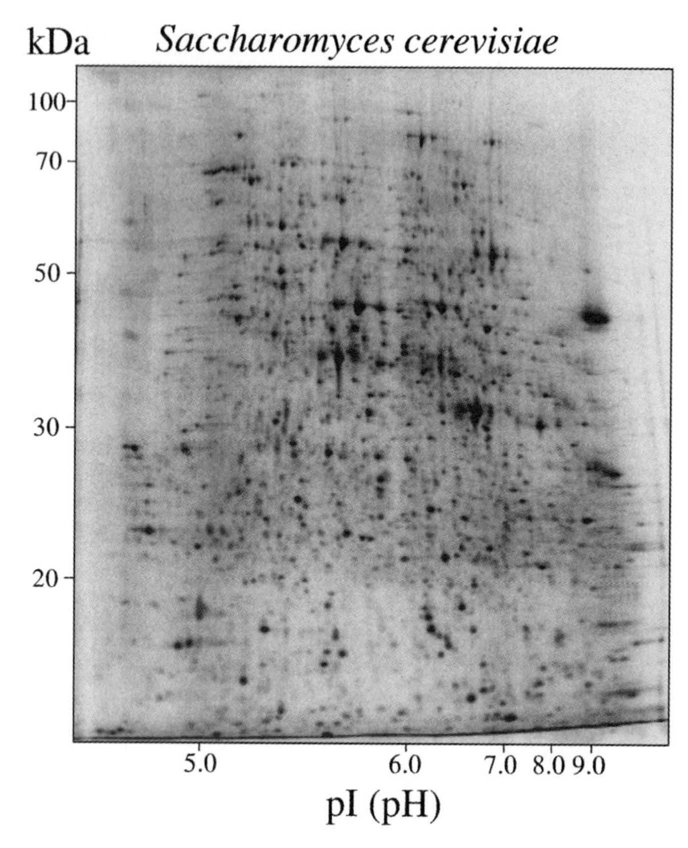

FIG. 2. Two-dimensional gel image of exponentially growing *Saccharomyces cerevisiae* in 1 *M* NaCl minimal YNB medium with glucose as carbon and energy source and ammonium as nitrogen source. Cells were labeled with [^{35}S]methionine for 30 min. Proteins were separated in a nonlinear pH gradient 3–10 in the first dimension and in a 12.5% acrylamide SDS gel in the second dimension. Visualization of proteins by phosphorimager technology.

proteins is long and diverse.[7] Most of these analyses have used 2D in the black-box approach: look for altered protein expression, identify the responders, and then pursue other types of analysis. However, there are also groups that have adopted the technology more systematically for a number of years and proven its utility in quite a number of applications. In addition to expression analysis, the technology is well suited for global analysis of protein modifications, particularly for N-terminal acetylation.[9]

[9] B. Polevoda, J. Norbeck, H. Takakura, A. Blomberg, and F. Sherman, *EMBO J.* **18,** 6155 (1999).

TABLE I
DIFFERENCES IN USE OF STRAINS, GROWTH CONDITIONS, AND LABELING PROCEDURES[a]

Parameter	Nawrocki *et al.* (1998) (14)	Hodges *et al.* (1999) (32)	Sanchez *et al.* (1996) (33)	Perrot *et al.* (1999d) (16)	Gygi *et al.* (1999) (34)	Norbeck and Blomberg (1997) (17)
Strain	W303-1B	W303-1A	X2180	S288c	YPH499	SKQ2n
Medium	YNB	YNB	?	YNB	YPD	YNB
Carbon source	2% Glucose	2% Glucose	?	2% Glucose	2% Glucose	2% Glucose
Nitrogen source	NH_4	?	?	NH_4	Various rich growth medium	NH_4
pH control	Succinate; pH 5.8	?	?	Succinate; pH 5.8	No	No
Amino acid plus nucleotide		All amino acids besides Met added	?	None	None	None
Temperature	22°	30°	?	23°	30°	30°
Shaker (rpm)	?	Roller ?	?	Rotary shaker	?	Rotary shaker 110 rpm
Growth vessel	100 ml E-flask		?	500 ml E-flask	?	100 ml E-flask
Growth volume	10 ml	?	?	50 ml	?	10 ml
Recording growth	OD 600 nm	?	?	OD	?	OD 610 nm
Growth phase	OD = 1.0	5×10^6 cells/ml	post diauxii?	OD = 1 2.4×10^7 cells/ml	Early log phase	OD = 0.5 5×10^6 cells/ml
Labeling volume	1 ml	1 ml	No label—silver staining	2 ml	No label—silver staining	10 ml
Labeling vessel	?	?	—	?	—	100 ml E-flask
Compound labeling	[^{35}S]Methionine	[^{35}S]Methionine	—	[^{35}S]Methionine	—	[^{35}S]Methionine
Labeling time	30 min	10–15 min	—	30 min	—	30 min

[a] Differences between laboratories for which 2D image-based protein information has been presented in publications or on the Internet. Only those conditions for which annotations on images are available have been included in the list.

2. Growth and Labeling of Cells

The 2D pattern generated (Fig. 2) should ideally be viewed as an array in which individual proteins could be identified by their position. However, both qualitative and quantitative aspects of the 2D pattern generated are tightly linked to the particular yeast strain under study as well as the physiological state of the culture at harvest and/or labeling. When comparisons of 2D patterns between laboratories are made, differences in strains or growth conditions are usually mistakenly overlooked, and for some of the 2D data in current proteomics databases or publications, these factors are even poorly defined (Table I). For an entity such as the yeast proteome with its extreme plasticity, this is prone to cause misunderstanding and problems in the interpretation of data.

2.1. Yeast Strains

A survey of yeast 2D studies published to date reveals that a number of different laboratory strains have been in use (Table I). It is recommended that the same strain be used if proper comparisons are to be made between positional similarities in the different studies. This statement is particularly relevant when industrial strains[10] or wild-type isolates[11] are analyzed in which the resulting differences in the 2D patterns become extensively more complex (as a result of point mutations, gene duplications, modification differences, and drastic expression changes). However, strain standardization applies even more to the analysis and databasing of expression profiles of particular genes/proteins, where minor changes in the genetic background can have profound implications on the response pattern of particular genes/protein. From the data in Fig. 3[8] it is clear that the salt-dependent increase in Tdh1p production is closely linked to level of activity of protein kinase A (PKA). A high PKA activity (as for the *bcy1*Δ strain) results in constitutively low levels of Tdh1p while the tpk^{wl} strain with low protein kinase A activity displays salt-regulated and high Tdh1p expression.[8] This is just one example of many. The strain issue is a long-standing dispute in the yeast field and it can only be hoped for that some of the complete strain collections for deletions mutants[12] will set the standard for a more uniform use of yeast strains among yeast researchers in the future. This general comment is of course not restricted to 2D-based yeast studies; however, strain differences are highly apparent in the context of proteome analysis.[12a]

2.2. Growth Media and Growth Conditions

Growth media and growth conditions used also differ between laboratories (Table I). Defined minimal medium should be used for standardization [we currently use YNB medium consisting of 2% (w/v) glucose, 0.14% (w/v) Yeast Nitrogen Base without amino acids (YNB; Difco Detroit, MI), 0.5% (w/v) ammonium sulfate, and 1% (w/v) succinic acid, supplemented with 20 mg/liter uracil, 20 mg/liter tryptophan, 20 mg/liter histidine, and 100 mg/liter leucine depending on the mutant requirement, pH 5.8]. It is advantageous to use buffered minimal media, since this will make the growth and labeling procedure more robust, and thus more reproducible. The type of buffering agent to be used can be discussed at length, and a thorough treatment of the subject is outside the scope of this chapter. The most commonly used substance, succinate, is by no means neutral in cell physiology, since it is an intermediate in metabolism (citric acid cycle) and also a weak acid (it can thus penetrate the membrane in its undissociated form, dissociate

[10] R. Joubert, P. Brignon, C. Lehmann, C. Monribot, F. Gendre, and H. Boucherie, *Yeast* **16,** 511 (2000).

[11] T. Andlid, L. Blomberg, L. Gustafsson, and A. Blomberg, *Syst. Appl. Microbiol.* **22,** 145 (1999).

[12] C. B. Brachmann, A. Davies, G. J. Cost, E. Caputo, J. Li, P. Hieter, and J. D. Boeke, *Yeast* **14,** 115 (1998).

[12a] A. Rogowska-Wrzesinska, P. M. Larsen, A. Blomberg, A. Görg, P. Roepstorff, J. Norbeck, and S. J. Fey, *Comp. Function. Genomics* **2,** 207 (2001).

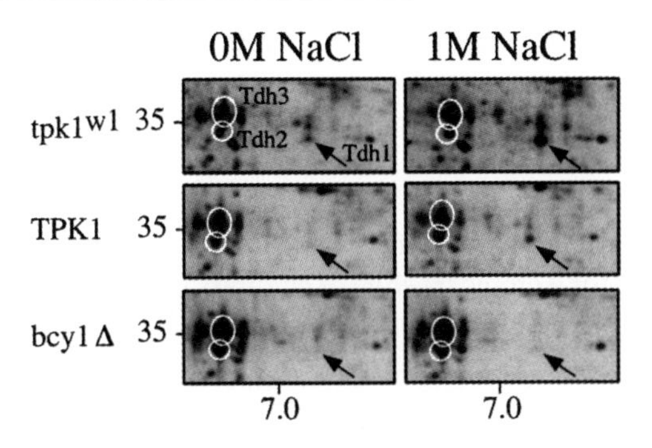

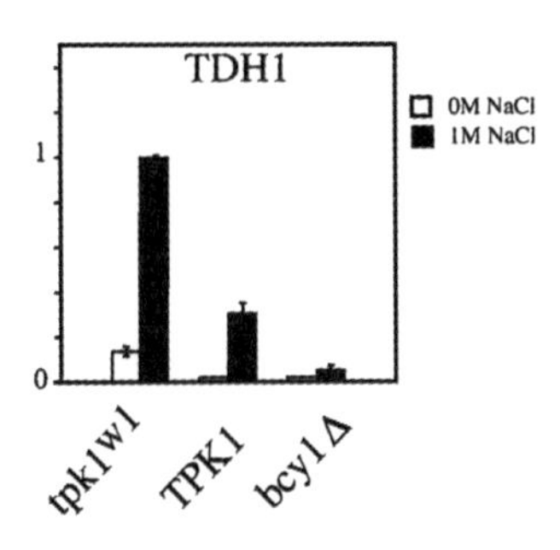

FIG. 3. One example of a protein expression change as a functon of growth conditions (0 *M* and 1 *M* NaCl) and the strain under study (different mutants of protein kinase A; tpk 1w1—a mutated form of TPK1 that exhibits constitutive low activity of protein kinase A; TPK1—the wild-type allele of TPK1; bcy1Δ—a deletion mutant of the regulatory subunit that results in constitutive high activity of protein kinase A). *Upper:* A small section of the whole 2D gel pattern is displayed where some of the identified proteins in that region are indicated (the TDH genes encodes the different isoforms of the glycolytic protein glyceraldehyde-3-phosphate dehydrogenase). The arrow marks the position of the salt regulated form Tdh1p. *Lower:* The Tdh1 protein was quantified by image analysis (PDQUEST). The relative amounts at the different conditions are indicated (1 corresponds to 4515 ppm); open bars, 0 *M* NaCl medium and closed bars, 1 *M* NaCl medium [J. Norbeck and A. Blomberg, *Yeast* **16,** 121 (2000)].

inside the cell, and thus lower the internal pH). In any case, the type of buffer will certainly influence the resulting 2D pattern (Norbeck and Blomberg, unpublished data). In addition, a minimal medium such as YNB will produce a 2D pattern rather different from the pattern generated by growing cells in rich YPD medium. The most apparent difference will, of course, be encountered for proteins involved in anabolic processes because their expression will be repressed by a rich complete medium such as YPD. However, changes in expression of other proteins not directly linked to anabolism will also be displayed (Norbeck, Warringer, Larsson, Karlsson, and Blomberg, unpublished data).

Other parameters, e.g., the growth vessel and the growth temperature, as well as the incubators used (Table I), have an effect on the physiology of yeast and thus the 2D pattern. It should also be stressed that if exponentially growing cells are to be analyzed, growth must be properly recorded during the whole period of the experiment to ensure exponential growth.

2.3. *Isotopic Labeling of Yeast*

One of the great advantages of working with yeast is that the usually rapid growth rate is ideal for isotopic labeling of cultures. The following is the procedure that we have adopted for labeling exponentially growing yeast cells with [^{35}S] methionine:

The optical density (OD) of the overnight preculture is measured. An appropriate volume is taken as inoculum for the experimental culture (10 ml medium in 100 ml E-flask) in order to yield a cell density of 1×10^6 cell/ml (in our laboratory corresponding to an OD_{610} of 0.07). At an OD value of 0.35–0.7, add 10 μl of [^{35}S]methionine solution (total amount applied will be around 150 μCi) and label for 30 min or longer. One method is to always label for a fixed length of time. Another is to label for a fixed proportion of the generation time since the rate of growth might be highly different under different growth conditions (e.g., different carbon sources) or different strains (e.g., mutants that display impaired growth). The latter approach is suggested to provide a better base for comparison between strains/conditions.

Stop labeling by placing the culture on ice and add 300 μl of cycloheximide stock solution (0.14 g/10 ml = 50 m*M*) to stop further protein synthesis. The addition of cycloheximide has not been shown to be essential and should be regarded as optional. Transfer to a 15-ml Falcon tube, spin cells down (3000 rpm, 2 min, 4°), and wash once in 1 ml of ice-cold isotonic solution. Transfer the 1-ml suspension to a microcentrifuge tube for a final short spin (10 sec) at 13,000 rpm, 4°. Freeze the cell pellet at −80°.

3. Protein Extraction Procedures for 2D-PAGE

There are a number of methods for the preparation of protein extract from yeast cells to be used in 2D analysis.[13] The main technical details to consider in the design and selection of the procedure are to address problems related to proteolytic degradation, efficient solubilization of most proteins, and removal of interfering substances such as nucleic acids. The procedure described below appears to address most of these issues; however, other procedures which are based on sonication in immobilized pH gradient (IPG) rehydration buffer[14] or the disruption of dry yeast powder[15] appear equally efficient. The sonication procedure

[13] A. Harder, R. Wildgruber, A. Nawrocki, S. J. Fey, P. M. Larsen, and A. Görg, *Electrophoresis* **20,** 826 (1999).

[14] A. Nawrocki, M. R. Larsen, A. V. Podtelejnikov, O. N. Jensen, M. Mann, P. Roepstorff, A. Görg, S. J. Fey, and P. Mose Larsen, *Electrophoresis* **19,** 1024 (1998).

[15] H. Boucherie, G. Dujardin, M. Kermorgant, C. Monribot, P. Slonimski, and M. Perrot, *Yeast* **11,** 601 (1995).

is especially useful for preparative analysis (more than 500 μg protein applied to each gel) because this minimizes the amount of SDS being loaded (SDS in high amounts will interfere with the first-dimensional separation).

3.1. Protein Extraction–Cell Disruption with Glass Beads

Sample Buffer I

SDS	0.3 g
β-Mercaptoethanol	5.0 ml
Tris-HCl	0.444 g
Tris base	0.266 g
Milli-Q water (Millipore Corporation, Bedford, MA)	to 10.0 ml

Sample Buffer II

Tris base 1.5 *M* (1.817 g/10 ml)	80 μl
Tris-HCl 1.5 *M* (2.36 g/10 ml)	1585 μl
$MgCl_2$ 1.0 *M*	250 μl
Milli-Q water	3000 μl (to 5 ml)
DNAs I (Worthington Biochemical Corporation, Lakewood, NJ)	5 mg
RNAs A (Worthington)	1.25 mg

Resuspend the cell pellet (approximately 5×10^7 cells) in 160 μl of ice-cold Milli-Q water and add 0.2 g of acid-washed glass beads (0.5 mm in diameter). To minimize the risk of proteolytic degradation it is advisable to include a protease inhibitor cocktail (e.g., Complete; Roche Diagnostic, Bromma, Sweden). However, rapid processing and proper cooling appear to provide enough precautions since no difference in 2D pattern is observed in our hands with or without protease inhibitors. Vortex on an ordinary lab-bench vortex for 30 sec at 2000 rpm, 20°, four times with intermittent placement on ice for >1 min.

Add 20 μl of sample buffer I and mix using a vortex. Punch a hole in the lid (or secure the lid with a cap) and place in water bath or a heating block set at 95° for 5 min. This heating step is to ensure efficient solubilization of proteins and at the same time denaturation of any proteases. Place the tube on ice and add 20 μl of sample buffer II; mix by vortex, 2000 rpm, 5 sec, 20°. Sample buffer II contains efficient nucleases which should degrade any nucleic acids from the samples that might otherwise disturb the isoelectric focusing. After 15 min incubation on ice spin the extract in a cold centrifuge at 13,000 rpm for 10 min. Transfer supernatant to a new microcentrifuge tube. A maximum of 100 μl of this supernatant can be applied to the IPG rehydration buffer for the strip rehydration and simultaneous loading of sample.

4. Extract Pre-2D Analysis

Prior to 2D-PAGE analysis the amount of protein and the specific radioactivity incorporated into proteins have to be determined. This is to ensure uniform handling and adequate standardization. We use 2×10^6 dpm (disintegrations per minute) per gel for radioactive samples and 50 μg protein/gel for nonradioactive samples that are to be visualized by silver staining.

4.1. *Protein Determination with TCA Precipitation*

The amount of protein can be determined using any of a number of commercially available protein kits. The analysis must remove contaminating substances. A safe procedure is to eliminate interfering substances such as SDS by trichloroacetic acid (TCA) precipitation prior to analysis (this is included as the first step in some commercial protein kits). If the procedure described above for extract preparation is followed we usually get a protein concentration in the extract of 4–5 μg/μl. Because this is in a total volume of about 120 μl (after aspiration from the glass beads) the total amount from 5×10^7 cells (from 10 ml medium volume) is about 500 μg. This amount of protein is also sufficient for most preparative purposes.

4.2. *Determination of Incorporated [^{35}S]Methionine*

The level of incorporation can be determined in the following way: Dilute 5 μl of protein extract 100-fold by the addition of 495 μl of Milli-Q water. Fifteen μl of bovine serum albumin (BSA) stock solution (10 μg/μl) is then added to 10 μl of the diluted sample in a microcentrifuge tube. Add 1 ml of an ice-cold solution of 10% (w/v) TCA containing 1% (w/v) casein hydrolyzate to the microcentrifuge tube; centrifuge at 13,000 rpm for 10 min at 4°. Aspire off the supernatant and repeat the last step with the ice-cold TCA/casein hydolyzate and spin down at 13,000 rpm for 10 min at 4°. Add 1 ml of scintillation liquid (Ready-Safe, Beckman Coulter Inc., Fullerton, CA), vortex at 2000 rpm, 15 min, 20°, and place tube in a scintillation vial. Measure radioactivity in a scintillation counter. We normally get about 1×10^5 dpm/μl when labeling actively growing cells in basal YNB medium, of which usually 2×10^6 dpm is applied per gel.

5. Two-Dimensional Polyacrylamide Gel Electrophoresis (2D-PAGE)

5.1. *General Comments about Technology*

There are several detailed procedures for separation of proteins based on two-dimensional polyacrylamide gel electrophoresis. The main technical difference between the procedures is found in the first-dimensional separation. Historically carrier ampholytes were used for the isoelectric focusing. Carrier ampholytes are small organic molecules that orient themselves in an electric field to produce a pH gradient. Carrier ampholytes are still used in some laboratories, and this is particularly apparent in the yeast field since the data in most annotated yeast proteome databases are derived from this technology[16] (see Section 9). This is a good technique for generating well-resolved proteins in a reproducible way. Although we used this technology for many years, we have replaced it with the Immobilines technology[17] (Amersham Bioscience Inc., Uppsala, Sweden) for the following reasons. (*i*) It is not possible with carrier ampholytes to separate by isoelectric focusing proteins with a p*I* greater than 6.5, because the pH gradient above that value collapses with focusing time[18] (thus ampholytes cannot generate information about the p*I* of proteins in that range). (*ii*) In the region close to pH 6.5 the reproducibility of the pattern is not very good and many gels displayed poor resolution and streaking. (*iii*) Reproducibility is affected because ampholyte batches vary in performance. (*iv*) The Immobiline strips are easier to handle and provide good gels. (*v*) The Immobiline technology enables flexible zoom-in on certain areas in the 2D pattern, (e.g., p*I* 5.5–6.5) which provides increased resolution of complex protein constellations.[19] The zoom-in gels are particularly important when it comes to the preparative separation of proteins for subsequent identification. A thorough discussion of the benefits of Immobilines is reviewed elsewhere.[3]

All chemicals used should be of high quality. Most companies can provide electrophoresis-grade chemicals. However, companies change their production lines occasionally which can make the quality variable over time. In an interlaboratory comparison between three European laboratories there were quite a number of different brands of chemicals used; however, all were of the highest quality.[20] In that comparison all the brands seemed to perform equally well. Some companies provide chemicals that have been tested in their quality control laboratories for 2D applications.

5.2. *Our Procedure*

We use the following apparatus for our 2D separation: In the first dimension a horizontal electrophoresis system with a power supply (Multiphore II; Amersham Bioscience Inc.) is used in which the running tray is placed on a cooler coupled to a standard high performance temperature-controlled water bath for proper temperature control to 20° during the run. The second dimension is a vertical electrophoresis system that can hold 5 gels per tank that are fully submerged in the buffer and temperature-controlled to 20° by a Peltier cooler

[16] M. Perrot, F. Sagliocco, T. Mini, C. Monribot, U. Scheider, A. Schevchenko, M. Mann, P. Jenö, and H. Boucherie, *Electrophoresis* **20,** 2280 (1999).

[17] J. Norbeck and A. Blomberg, *Yeast* **13,** 529 (1997).

[18] P. Z. O'Farrell, H. M. Goodman, and P. H. O'Farrell, *Cell* **12,** 1133 (1977).

[19] R. Wildgruber, A. Harber, C. Obermaier, G. Boguth, W. Weiss, S. J. Fey, P. M. Larsen, and A. Görg, *Electrophoresis* **21,** 2610 (2000).

[20] A. Blomberg, L. Blomberg, J. Norbeck, S. Fey, P. Mose Larsen, P. Roepstorff, H. Degand, M. Boutry, A. Posch, and A. Görg, *Electrophoresis* **16,** 1935 (1995).

(Investigator; Genomic Solutions Inc., Ann Arbor, MI). This vertical system takes 22 × 22 cm gels. The size of the gels appears to matter in that the larger the gels, the better the resolution. Currently we run 18-cm strips in the first dimension which can easily be applied to the second dimension gels. The longer 24-cm IPG strips that have recently been marketed should also fit in this 22 × 22-cm format. However, the use of larger second dimension gels will cause more practical problems in handling.

5.2.1. Separation According to Isoelectric Point (pI): First Dimension

Selection of strips. The immobilized pH gradient (IPG) strips are usually 3 mm in width and the gel is cast on a plastic film to a thickness of about 0.5 mm. The IPG strips we most often use have a nonlinear gradient and cover pH 3–10 and are 18 cm long (Amersham Bioscience Inc.). Most IPG strips sold are linear; however, for the analysis of yeast samples we have found that the nonlinear gradient provides a pattern with a good spread of proteins. For other types of cell material however, some other pH gradients might be more appropriate. Currently this gradient is the only nonlinear IPG strip on the market. A large number of linear strips are commercially available in different lengths and the pH interval encompassed varies. The widest can range from pH 3 to 12[21] while the narrowest are no wider than one pH unit: pH 4–5, 5–6, etc.[19] Strips, at least the linear ones, can also be homemade, which of course provides flexibility and makes the procedure inexpensive, and at least some groups cast their IPG strips routinely.[3] The casting of the nonlinear strips with a certain gradient profile is, however, somewhat problematic. The main arguments for the purchase of IPG strips are the quality control provided by the manufacturer, the possibility for easier standardization between laboratories, and the ability to use nonlinear pH gradients. The wide pH gradients are used as a "world map" of the proteome; the intention is to separate most proteins well enough to be able to determine general trends of changes. If this first analysis indicates that there are regions of the gel where expression or modification changes appear to take place in areas that are too crowded, the high-resolution zoom-in strips can be used in that particular pH interval. It is also clear that some spots will be composed of more than one protein, which can be revealed by these zoom-in gels.[19,22] Optimally, multiple narrow-range zoom-in gels should be used to maximize the number of resolved proteins. However, this issue is really a cost/benefit problem because the use of a large number of gels per sample will not only improve resolution, but also slow down dramatically the analysis of samples. In the area of functional genomics, where high-throughput analysis is the rule, the regular use of multiple gels will most likely hamper the speed of many experimental studies.

[21] A. Görg, C. Obermaier, G. Boguth, and W. Weiss, *Electrophoresis* **20,** 712 (1999).

[22] K. C. Parker, J. I. Garrels, W. Hines, E. M. Butler, A. H. McKee, D. Patterson, and S. Martin, *Electrophoresis* **19,** 1920 (1998).

Rehydration. Before use the dry IPG strips are rehydrated overnight at room temperature in an IPG-rehydration solution. One commonly used rehydration solution contains 8 *M* urea, 1% (v/v) Nonidet p-40 (NP-40), 13 m*M* dithiothreitol (DTT), and 1% (v/v) Pharmalyte (pH range 3–10) (Amersham Bioscience Inc.) and a trace of bromphenol blue, ≈0.01% (w/v). Alternatively, the chaotrophic agent urea is supplemented with the stronger chaotroph thiourea; a commonly used mixture is 7 *M* urea and 2 *M* thiourea. The addition of thiourea has been shown to improve the solubility of some proteins and the recovery of proteins especially at the preparative scale.[13] We now routinely use this urea/thiourea mixture for all analysis. The complete rehydration solution can be prepared beforehand and stored frozen for long periods of time.

Sample application. Samples can be applied to the first-dimensional separation in different ways (Fig. 4). In the original set-up, a plastic cup is placed on top of the reswollen and mounted strip and then the sample (in a volume of about 25 μl) is applied to the cup (Fig. 4A). The proteins will then diffuse into the gel during the run and start to migrate in the electrical field. This procedure has been used

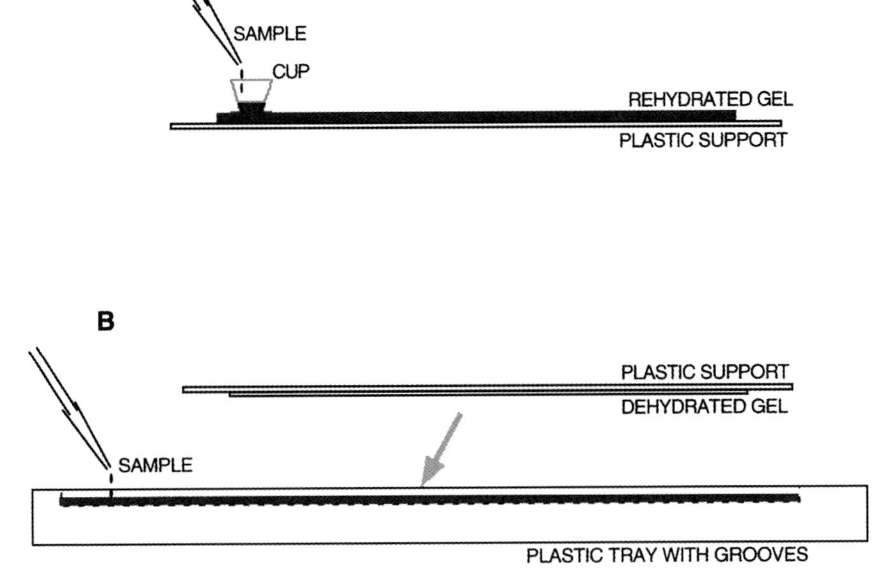

FIG. 4. Different strategies for application of sample to the first-dimensional IPG strip. (A) The original method of sample application to a cup that rests on top of the rehydrated gel (rehydration solution contains bromphenol blue, which makes the gel blue). (B) The in-gel rehydration procedure which is used in our laboratory. The dehydrated gel is placed with gel downward onto 500 μl rehydration solution to which the sample has been applied. Rehydration is usually overnight.

extensively and works well for most proteins. However, for some proteins there appears to be a position-dependent effect so that an anodic or cathodic application will influence the result. An alternative procedure to cup-loading that has a number of advantages is the application of the samples into the IPG rehydration solution (Fig. 4B).[23] The proteins are then transported via suction into the gel together with the rehydration solution. In this procedure the sample (maximum 100 μl) is dissolved in rehydration buffer to a final volume of 500 μl. This mix is heated at 37° for 30 min and spun at 13,000 rpm for 10 min at 20°. Make sure not to heat samples at temperatures higher than 37° because this will cause carbamylation of proteins leading to charge heterogeneity. Pipette the entire volume into the respective position in the rehydration tray. This device is a plastic block with grooves that allows the different strips to reswell individually. Remove the protective cover from the IPG strip and place it carefully, gel-side down, in the rehydration tray. Overlay with 2 ml of silicone oil and leave the strips overnight. The silicone oil protects against evaporation which can be a severe problem in these highly concentrated solutions of urea. If evaporation occurs, urea crystals will form.

The amount of protein applied to the gel depends on the strain and growth conditions, the question to be studied, the procedure for visualization, and whether analytical or preparative gels are to be run. Applying too much protein will hamper resolution. It can easily be seen that at the preparative scale (more than 500 μg applied per gel) using the wide gradient (pH 3–10), some of the dominant spots will cover less dominant neighbors. Thus, for the analytical scale using less protein means better results. In general, not more than 50 μg should be used. This is particularly relevant for isotopically labeled protein extract where high specific activities provide the option of applying very low protein amounts.

Running conditions. The rehydrated IPG strips (Fig. 4B) are placed in a running tray and properly cooled. The cooling, or more precisely, adequate temperature control, is absolutely essential because the p*I* of a protein is dependent on the pK_a values of the individual amino acids in the protein which in turn is dependent on the temperature. Thus, the p*I* values of different proteins will vary with temperature; running the same extract under identical conditions except for the temperature will inevitably result in altered 2D patterns.

Apply a stripe of silicone oil (approximately 5 ml; Amersham Bioscience Inc.) on the cooling plate and place the running tray on top. Then apply a stripe of the same silicone oil (approximately 10 ml) in the running tray and position the gel aligner in which the individual gel strips will be positioned. Pick the rehydrated gels out of the rehydration cassette and place them gel-side down on a wet filter paper (5 sec) and then gel-side up on a moist filter paper (5 sec) before placing them in their respective positions in the gel aligner. This drying step is important since excess water will cause protein streakings in the final 2D patterns because of

[23] T. Rabilloud, C. Valette, and J. J. Lawrence, *Electrophoresis* **15,** 1552 (1994).

electroosmotic effects. It is essential that the acidic end (usually pointed) face the anode end of the tray and that the gel side be up! When all gels are in place, wet two electrode strips with Milli-Q water and dry them well with a paper tissue so that they are only moist. Place the electrode strips across the ends of the gels, place the electrodes in position, and press them down firmly but without breaking the gels. Cover the gels with silicone oil and put on the lid. Some laboratories run gels without the silicone oil on top. This is not a problem provided that the chamber is kept moist to prevent evaporation and subsequent urea crystallization.

Check the program of the power supply. A number of different programs can be used so long as the program provides good resolution with little streaking and the pattern is fixed on prolonged incubation. We have adopted the following conditions for a wide range of samples (read yeast strains and growth conditions): A linear voltage gradient that goes from 0 V to 500 V in 5 hr; 5 hr of constant voltage at 500 V; a linear gradient going from 500 V to 3500 V in 9.5 hr; 3500 V maintained for another 2.5 hr. If necessary the 3500 V can be maintained for longer periods of time, mainly for the running of preparative gels.

Finishing the 1D run. The strips are unmounted and the silicone oil is drawn off the strips with a filter paper. The strips can now be either directly put into the equilibration solution (see below) or put in the freezer after being wrapped in plastic film. We have not seen any change in the 2D pattern after storage for long periods of time at either $-20°$ or $-85°$.

5.2.2. Separation According to Molecular Weight (M_r): Second Dimension

Casting gels. Gels are cast between glass plates with plastic spacers (usually 1 mm wide). It is important that the glass plates be clean; glass surfaces that face the gel should be cleaned with ethanol. This is particularly important if gels are to be silver stained because this staining procedure easily picks up dirt or dust. In addition, an untreated surface is more prone during the casting to catch air bubbles that will later hamper proper resolution of proteins. Set up the casting box and place the glass sandwiches in the box. The usual design of these boxes is that the acrylamide solution is let in from underneath the box.

The second-dimension gels are usually made up of 10 or 12.5% acrylamide with a bisacrylamide concentration of about 0.6%. Most companies sell good quality acrylamide. There is also a substituted acrylamide brand called Duracryl (Genomic Solutions Inc.) which performs well. One advantage of these Duracryl gels is that they do not tear apart as easily during the handling, e.g., silver staining.[24] Another way to improve the tensile strength of the gels is to decrease the bisacrylamide concentration.[14] Changing the concentration of bisacrylamide will alter the mobility of proteins and thus alter the 2D pattern obtained. The gels and the running buffer should also contain 0.1% SDS and the buffer is usually 0.6 *M* Tris.

[24] W. F. Patton, M. F. Lopez, P. Barry, and W. M. Skea, *BioTechniques* **12,** 580 (1992).

For 10 gels take:

	10% T (12.5% T)
Duracryl (0.65% bisacrylamide)	359.0 (448) ml
Tris Slab Gel Buffer (see below)	267.0 (267) ml
Tris base	(122 g/mol) 130.8 g
Tris-HCl	(159 g/mol) 66.3 g
Milli-Q water	to 1000.0 ml
Milli-Q water	442.0 (352) ml
Degas 10 min	
SDS (w/v)	11.0 ml
TEMED	541.0 μl
APS 10%	2706.0 μl

Cast gels up to approximately 5 mm below the top of the glass plates and overlay each gel with 600 μl of water-saturated isobutanol. One can cast gels at room temperature or in the cold room; however, the latter takes longer to polymerize (at room temperature it takes about 2 hr). Wash away isobutanol with deionized water when polymerization is complete (if gels are to be left overnight, overlay with 200 ml 4× diluted slab gel buffer with 2 ml of 10% SDS added).

Gel running. Set up the gels in the running containers. The commercially available containers usually hold between 5 and 12 gels and have slightly different designs. The running temperature (the cooling) must be well controlled so that separation can be performed under constant temperature. This is important because the mobility of proteins differs with temperature and thus changes in temperature will alter the 2D pattern. Fill the lower tank with running buffer (25 m*M* Tris-base, 0.1% w/v SDS, and 192 m*M* glycine).

Thaw the first-dimension strips and put them in petri dishes (or glass tubes with cork screw tops) for the equilibration. The equilibration step is set to prepare the proteins for the size separation by treatment with SDS. The first-dimension IPG strips are equilibrated for 30 min on a rotary shaker (we use the same shaker as the one we use later for the washing steps in silver staining) in 20 ml of equilibration solution containing 3% (w/v) SDS, 50 m*M* DTT, 0.3 *M* Tris-base, 0.075 *M* Tris-HCl, and roughly 0.01% (w/v) bromphenol blue. After equilibration stand the gels on edge and drop dry on a moist filter paper.

1D gel equilibration buffer:

Tris-base	3.66 g
Tris-HCl	1.19 g
SDS	3.0 g
DTT	0.77 g
Bromphenol blue	A few grains
Milli-Q	To 100 ml

The first-dimension gels are now ready to be placed on top of the slab gels. Prepare 50 ml of 0.5% agarose in 4× diluted slab gel buffer with 0.1% SDS and warm to

65° (agarose is easily dissolved by heating briefly in a microwave oven). Coat the first SDS gel with hot agarose. Dip the gel quickly in 4× diluted slab gel buffer and place it on the second dimension gel. Press it down gently onto the polyacrylamide surface, so that no air bubbles are trapped between the strip and the gel surface.

When all of the gels are in place, fill the upper tank with 2000 ml of 2× running buffer. Gels should be fully submerged in running buffer for efficient cooling. Set the temperature control to 20°.

Setting up the power supply. Electrophoresis in the second dimension is performed at a limiting power of 16,000 mW (maximum voltage 500 V) per gel for about 5 hr until the dye front reaches 0.5 cm from the bottom of the gel. Alternatively set the power to 1000 mW/gel for an overnight run.

Finishing the second dimension. Remove the glass plate sandwich from the tank and take the glass plates apart. If samples were isotopically labeled, put the gel on a filter paper and transfer to a gel dryer. The gel surface has to be covered by Saran wrap or some other kind of plastic film during the drying process. There are a number of devices for drying gels on the market. The vacuum in the dryer should be easily applied and reliably maintained during the run. If the vacuum is released suddenly, the whole gel usually cracks into small pieces and the only option is to start from the beginning again.

If gels are to be stained, carefully remove the gels from the glass and place them in a tray containing 500 ml of the appropriate solution (see below).

6. Visualization of Proteins

A number of different procedures can be applied for the visualization of separated proteins, and the visualization method used will strongly influence the outcome of the analysis. For example, when proteins do not contain methionine (e.g., Tpilp and Hsp26p), the most frequent amino acid used for isotopic labeling, qualitative differences are seen when autoradiographic images are compared to images produced by some of the staining procedures. In addition, because proteins can contain very different amounts of methionine, proper stoichimetric comparison of different proteins can only be performed after methionine normalization.[17]

6.1. Autoradiography

The use of films that are sensitive for radioactive emissions had been the method of choice for visualization of isotopically labeled proteins. Photographic films are in some ways ideal for visualization of 2D-resolved proteins because of their superior spatial resolution, in which the silver grains in the film set the effective resolution to about 5 μm. However, the major drawback with the use of films is poor quantitative resolution: the quantitative window is usually not greater than about 2 orders of magnitude. Since the variation in signals generated from proteins resolved in a normal cell lysate is substantially greater (about 50,000-fold difference), global quantitative 2D analysis with films is problematic. The solution to this problem in the past was to utilize the option of multiple exposures of gels for different durations and to include calibration strips to correct for the nonlinear film response.[25,26] This procedure enabled quantitative merging of the different exposures to generate a composite image containing valid quantitative information for the full dynamic range. Even though this procedure is rather straightforward, the advent of the phosphorimager technology, in which image plates with a dynamic range of up to 5 orders of magnitude are used, has drastically improved quantitative analysis and simplified the quantification procedure. Moreover, the response curve for the phosphorimager plates is linear over a great span of intensities, enabling precise quantification even without the inclusion of a calibration strip. The major drawback with the phosphorimager technology is its relatively poor spatial resolution: it provides no better than about 200 μm in effective resolution (defined as the distance between lines where separation is achieved at half-maximum amplitude).[27] Even if most modern phosphorimager scanners provide a scanning pixel resolution of 25 μm or better, this should not be confused with the overall effective resolution, which is not that good. The rather poor spatial resolution of the phosphorimager is one of the reasons why large-format 2D gels are to be preferred (on the order of 22 cm $\times$ 22 cm) because they will increase the physical distance between individual protein spots. The spot sizes will also be greater with these large gels, which ensures that even minor spots are represented by a reasonable number of pixels. The whole technology appears robust, and the expensive image plates, if carefully handled, can last for years.

6.2. Silver Staining

In those cases where cells cannot be isotopically labeled, silver staining is typically the visualization method of choice. An example where this is relevant for yeast (and where isotopic handling is not appropriate) occurs when large-scale industrial processes need to be monitored. Withdrawing small amounts of samples from these huge industrial tanks in order to isotopically label cells in the milliliter scale *in situ* is not recommended because this will inevitably introduce expression artifacts. Isotopic labeling should also be avoided for yeast samples with low metabolic activity (starved cells or cells subjected to severe stress conditions) that display low levels of incorporation of amino acids.

Silver staining has long been in use and there are many protocols available that detect proteins in the low nanogram per spot range. With the use of silver

[25] J. I. Garrels, *J. Biol. Chem.* **264,** 5269 (1989).

[26] A. Blomberg, *J. Bacteriol.* **177,** 3563 (1995).

[27] R. F. Johnston, S. C. Pickett, and D. L. Barker, *Electrophoresis* **11,** 355 (1990).

nitrate-based silver staining it is possible to achieve quantitative results by developing gels for a defined length of time (which should not exceed 10 min).[28,29] Furthermore, we have adopted the use of a type of acrylamide solution with increased performance in silver staining[24] (Duracryl; Genomic Solutions Inc.). Silver staining intensity is strongly dependent on the protein and some proteins do not even stain with silver. It follows that the intensities of different protein spots on the same gel should not be compared, and that quantitative analysis should be restricted to measuring expression changes for one particular protein relative to itself. Another problem in quantitative analysis is that some of the silver staining procedures generate spots with different colors. Even though this might provide a nice (almost beautiful) 2D image, the scanning and subsequent image analysis gets more complicated. This is the main reason for our choice of silver staining protocol,[30] which is not the most sensitive, but which is robust, is mostly protein nonspecific and provides an almost identical blackish coloring of spots.

All steps in the staining procedure are performed at room temperature (roughly 23°) in plastic photographic trays and incubation is on a rotary shaker set to about 50 rpm. Gel cassettes are dismounted and gels placed in 500 ml of fixation solution [50% (v/v) ethanol plus 10% (v/v) acetic acid] for at least 2 hr (this step can be prolonged to overnight). Gels are subsequently washed three times in 500 ml of Milli-Q water for 20 min, transferred to a 500-ml DTT solution (0.013 g/2500 ml) for 30 min, which is supposed to sensitize the SH groups of the proteins, and thereafter transferred to 500 ml silver nitrate solution (5.1 g/2500 ml) and incubated for 30 min. In the following steps, timing is very important: First, wash the gels in 500 ml Milli-Q water for 30 sec. Pour off the water and add 200 ml of sodium bicarbonate solution (86.8 g/2500 ml) to which formaldehyde (1.25 ml/2500 ml) has been added and incubate until a yellow color appears (usually takes 10–20 sec). Pour the yellow solution off as quickly as possible and add 500 ml of the same sodium bicarbonate/formaldehyde solution. Watch the gels during development and stop development when protein spots are clearly visible and background is not too high, usually 5–6 min. Under no circumstances should development last more than 10 min. The development is stopped by pouring off the development solution and adding 500 ml of acetic acid solution (75 ml concentrated acetic acid/2500 ml). Incubate for 10–30 min.

The gels are now ready to be scanned and subsequently wrapped in plastic for storage. To properly quantify proteins, a calibration strip (1cm × 22 cm; protein

[28] W. F. Patton, *J. Chromatogr.* **698,** 55 (1995).

[29] L. V. Rodriguez, D. M. Gersten, L. S. Ramagli, and D. A. Johnston, *Electrophoresis* **14,** 628 (1993).

[30] J. H. Morrisey, *Anal. Biochem.* **117,** 307 (1981).

steps 1 cm × 1 cm) with 12 stepwise, roughly 2-fold increases in concentrations of bovine serum albumin (range 10–5000 ng/cm^2) is added in the first washing step, simultaneously stained, and used in the computer-aided image analysis. This analysis converts optical values into protein density values and thus normalizes gels and compensates for the nonlinear behavior of the staining close to the background and at higher protein concentrations. Gels can be scanned with a white light scanner and images (whole gel and calibration strip images) analyzed with imaging software such as PDQUEST (Bio-Rad, Hercules, CA)[25] where calibration of staining response (with the calibration strip) can be performed. It should be noted that the silver staining procedure with all the different solutions and incubations/washing steps is more prone to experimental variability than isotopic labeling and autoradiography.

6.3. *Coomassie Staining*

Colloidal stains for protein visualization have long been in use: The most common is the Coomassie stain which has proven its utility over many years. It is used mainly when preparative gels are run and spots need to be cut out for further analysis by mass spectrometry (see below).

Put gel in stain without prior fixation. Stain with 0.2% (w/v) of Coomassie Brilliant Blue G-250 in 20% (v/v) methanol and 0.5% (v/v) acetic acid for 30 min. Five hundred ml staining solution should be used for large-format 2D gels (22 cm × 22 cm). Destain in 3× 500 ml overnight with 30% (v/v) methanol. The destaining solution is usually changed during the first day and once (or more than once depending on the level of destaining) during the second day.

6.4. *Fluorescence*

Besides silver staining, highly sensitive detection of nonisotopically labeled proteins can be obtained using fluorescent compounds. A method for postelectrophoresis staining based on the dye SYPRO Ruby provides a good dynamic range, high sensitivity, and a linear response. However, procedures for pre-electrophoresis staining can also be applied (3). The major drawback of the latter method is charge modifications of proteins that accompany the use of some of the dyes thus leading to 2D pattern alterations.

Incubate gels in 250 ml of 40% methanol and 7% acetic acid for 30 min at room temperature and with gentle shaking. Incubate in 250 ml of SYPRO Ruby (concentrated solution; Molecular Probes Inc., Leiden, The Netherlands) overnight (at least 15 hr at room temperature with gentle shaking). The gels cannot be overstained. The stain is very expensive but it can also be reused at least once. Gels are washed in 250 ml 10% methanol and 7% acetic acid for 30 min to 2 hr (time is not so critical). Before scanning the gels can be washed in water for 60 min (to decrease the contamination of the scanner).

7. Image Analysis

7.1. Determining Quantitative Expression Changes

The incentive for separating and detecting proteins through the procedures described above is to provide quantitative data on global changes in protein expression or modification (Fig. 3). Some kind of software tool for image analysis of the obtained 2D patterns is usually applied. However, the imaging software will typically not discover information that is not revealed to the naked eye. Thus, image analysis should not be implemented as a means of improving poor 2D data but should be used (i) to get an objective measure of the relative amount of the individual proteins that can be statistically treated and (ii) to manage the task of handling the quantitative information from thousands of proteins. However, if precise quantitative values are needed for only a small number of proteins there is no need to use image analysis. A simple procedure where spots are cut out from the gels with a scalpel, placed in individual scintillation vials with a normal scintillation cocktail, and subsequently counted in a scintillation counter works well and provides relative quantitative values.

However, if the global response is to be analyzed to indicate the total number of spots that have changed under certain conditions, computer-aided image analysis should be applied. There are a number of image analysis software packages available having different prices and features. Most of the more powerful software packages have similar capabilities. The user interface differs but this is mostly a matter of personal preference. One of the substantial differences is the way the different software packages identify objects; some identify and quantify objects as Gaussian volumes while others will identify the borders of the spot and then simply sum the values of all pixels within those borders. The quantitative outcome of either procedure tends to be similar.

In practice the following steps are undertaken in the image analysis:

Image smoothing. The scanners will generate some electronic noise which should be reduced before the identification of objects in the image. If a smoothing algorithm is not applied the spot detection algorithms usually have problems in identifying the right objects or they find too many objects.

Background subtraction. The background can vary in different images either globally over the whole or a great portion of the gel or locally around specific spots. The background subtraction algorithms usually are of the rolling disk type (a wheel radius of about 50 pixels has been used in our laboratory with good results) or an average background value is taken from pixels adjacent to a spot.

Spot identification and spot quantification. The peak of the spots is identified and the volume determined by either a Gaussian volume fitting or pixel summation. The problem with the former is nonideal Gaussian shapes of some protein spots which make fitting nonoptimal; the latter approach suffers from difficulties in splitting overlapping protein spots.

Manual editing. Spots in the front or at the very top of the gels should be manually erased. For some of the more dominant spots in some gels multiple spot centers will have been positioned and these need to be manually corrected into one single spot center.

Gel matching. When the different gels in an experimental set have been processed, they should be matched to each other to allow quantitative comparison. The different imaging software packages contain automatic procedures for matching and usually only a couple of landmarks (starting points for the automatic procedures) have to be given. However, our experience is that these automatic procedures perform well only for the more well-resolved and dominant spots[17] and do not manage very well with clusters of spots or spots close to the background. Thus, a great deal of manual intervention is still necessary at this final step, which makes the analysis slow and somewhat subjective. A better use of training sets for automation of this procedure can be expected in the future. This is essential if true large-scale proteomics utilizing 2D gels is to be implemented.

Statistical evaluation. Individual quantification of resolved proteins is usually performed by normalization of the total amount of radioactivity in all quantified spots in the gel. This accounts for any inconsistency in the transfer of proteins into the first dimension or from the first to the second dimension. This will also handle any pipetting errors made while determining radioactivity or loading samples to the first-dimensional gels. The normalized data can now be analyzed for statistical changes. Most 2D image analysis software packages have internal tools for statistical evaluation, which usually include some kind of Student's *t*-test. One should keep in mind that one of the requirements of the Student's *t*-test is that the variance be independent of the average value of the analyzed entity. This is not the case for 2D generated data where low values (spots close to the background) will have low absolute standard deviation while the more dominant spots will have higher absolute variability. A standard way in statistics to deal with these differences in variance is to log transform the data before the Student's *t*-test analysis is performed. This can usually be performed by the software packages.

Selection of proteins that display statistically significant changes can now be performed. However, in this kind of analysis there is always a risk of getting spurious hits just because the variation in the compared samples was low. One can then introduce some other filters, one of which is a threshold selection for at least twofold changes. This value is of course only chosen *ad hoc,* but would have the rationale of providing a selection threshold roughly fivefold higher than the average standard deviation. The standard deviation varies between different proteins, but on average (if one sample is analyzed a few times) is about 20%.[17,20] If one adds the biological/experimental variation the overall standard deviation can be estimated to be about 30–35%, which is why one has to judge minor expression changes with caution.

7.2. Determining M_r and pI Scale

The pattern generated not only indicates changes in expression, but also provides information about the apparent M_r and pI of the resolved proteins. For this purpose a number of calibration proteins have to be selected. Calibration can be performed with the addition of standard proteins with known M_r and pI values. However, because we have a good idea of the identity of quite a number of the more dominant proteins in the yeast pattern these could be used equally well as standards. This approach has been tested and a number of proteins that appear to be well suited for this purpose (adhering to a general trend in M_r or pI) have been selected[17] (Table II). If other proteins are used as standards slightly different experimental values will be estimated for all the proteins in the pattern. Thus, when values are indicated for experimentally determined M_r and pI the proteins used for calibration should be provided together with the theoretical values used (Table II).

8. Protein Identification

Proteins resolved in a 2D pattern can be identified by a number of means.[7] In the past a stringent, but not so fast, technology was N-terminal sequencing of the

TABLE II
PROTEINS USED AS M_r AND pI STANDARDS IN 2D ANALYSIS FOR LABORATORY STRAINS OF *S. cerevisiae*[a]

Gene name	ORF	Protein description	Number of Met	Theoretical pI/M_r	Used for pI and/or M_r calibration
EFB1	YAL003W	Elongation factor 1 β	3	4.30/22.7	pI
ENO1	YGR254W	Enolase 1	5	6.17/46.7	pI and M_r
GPM1	YKL152C	Phosphoglycerate mutase	1	8.86/27.5	pI
IPP1	YBR011C	Inorganic pyrophosphatase	2	5.36/32.2	pI and M_r
LYS9	YNR050C	Saccharopine dehydrogenase ($NADP^+$)	9	5.10/4 8.9	pI
MET17	YLR303W	O-Acetylhomoserine sulfhydrylase	1	5.98/48.5	pI
SSB1	YDL229W	Heat shock homolog ssb1	9	5.32/66.5	M_r
SHM2	YLR058C	Serine hydroxymethyltransferase	11	6.98/52.2	pI
TDH1	YJL052W	Glyceraldehyde-3-phosphate dehydrogenase 1	6	8.32/35.6	pI
TDH3	YGR192C	Glyceraldehyde-3-phosphate dehydrogenase 3	6	6.5/35.6	pI
YST1	YGR214W	Nucleic acid-binding protein	1	4.65/27.9	pI

[a] The information about the location of these proteins in the 2D pattern generated by our procedure can be found at http://yeast-2dpage.gmm.gu.se and in Ref. 17. Values for the theoretically calculated values for M_r and pI as well as information about the number of methionines per protein are taken from Swissprot.

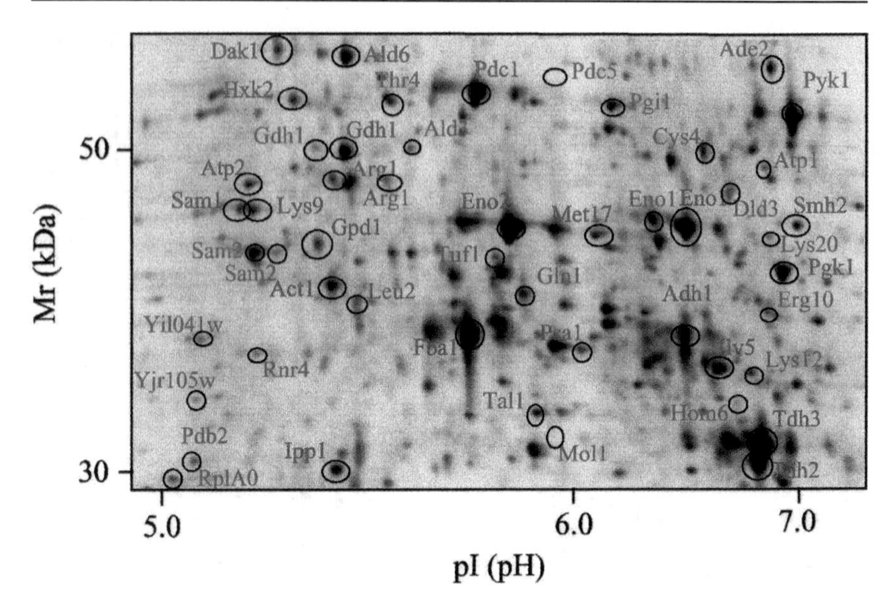

FIG. 5. A central portion of a nonlinear pH gradient 3–10 2D gel with identified proteins encircled and with gene/protein names indicated. Same cells as in Fig. 3. Proteins identified by microsequencing are indicated in red and those identified by mass spectrometry are indicated in blue.

full-length protein or of trypsin-generated peptides.[31] Even if this technology provides good data it has in most laboratories been replaced by identification via matrix-assisted desorption ionization–mass spectrometry (MALDI-MS). This MALDI-MS technology is currently being developed into a truly high-throughput method which promises to enable future development of local 2D databases where most, if not all, proteins resolved are identified. Here I will only provide a description of a procedure that we have adopted for the MALDI-MS analysis of protein spots from 2D gels (Fig. 5) which in our hands has worked well. The procedure is mainly designed to be used for digestion of protein visualized with the Coomassie blue stain. It can also be adapted for silver-stained gels but subsequent cleaning with ZipTips (Millipore Corporation) should be applied.

Protein trypsination. This procedure should be carried out in a clean, dust-free part of the laboratory. Dust particles are one of the main problems in MS analysis

[31] J. Norbeck and A. Blomberg, *Electrophoresis* **16,** 149 (1995).

[32] P. E. Hodges, A. H. McKee, B. P. Davis, W. E. Payne, and J. I. Garrels, *Nucleic Acids Res.* **27,** 69 (1999).

[33] J.-C. Sanchez, O. Golaz, S. Frutiger, D. Schaller, R. Appel, A. Bairoch, G. Hughes, and D. Hochstrasser, *Electrophoresis* **17,** 556 (1996).

[34] S. P. Gygi, Y. Rochon, B. R. Franza, and R. Aebersold, *Mol. Cell. Biol.* **19,** 1720 (1999).

because keratins from dust or clothes often appear as contaminants. A clean fume hood works well and general good laboratory practice appears sufficient. However, when small amounts of protein material are to be handled, it is advisable that a dedicated low dust particle area and extreme care should be used. Cut out the whole spot (for a not so dominant spot) or the central part of spot (if a dominant spot) with an ethanol sterilized/cleaned scalpel. A small square (about 5–25 mm^2) is placed in a microcentrifuge tube (preferably silanized microcentrifuge tubes should be used, especially when a small amount of material is to be analyzed). A three-step washing regime is then carried out: (i) Wash gel piece for 15 min in 1 ml of 10% acetonitrile at room temperature. Remove fluid. (ii) Wash gel piece for 15 min in 1 ml of 50% acetonitrile at room temperature. Remove fluid. (iii) Wash gel piece for 15 min in 1 ml of 100% acetonitrile at room temperature. Remove fluid. Dry the gel piece in the SpeedVac, 30 min. The gel piece will now be totally dried out and appear as a tiny dust particle. Be careful when opening and closing the lid on the tube so that the particle does not "fly out." Add 10 μl trypsin solution (12.5ng/μl in 50 m*M* ammonium hydrogen carbonate, 5 m*M* calcium chloride). The trypsin solution should not be kept for a period of extended time in the freezer. Rehydrate on ice, 60 min. Aspirate off the surplus trypsin solution with a pipette and incubate at 37° overnight. Add 2–8 μl of 10% (v/v) TFA in 75% acetonitrile (the volume is

TABLE III

CURRENTLY AVAILABLE YEAST PROTEOMICS DATABASES OVER THE INTERNET

YPM: Yeast Protein Map
URL: http://www.ibgc.u-bordeaux2.fr/YPM/
Location: Bordeaux, France
Technology: Carrier ampholytes in the first dimension
Annotations: 410 identified spots corresponding to 282 different genes
Expression information: Identified proteins under heat shock (23° to 36°)

SWISS-2DPAGE
URL: http://www.expasy.ch/ch2d/publi/yeast.html
Location: Geneva, Switzerland
Technology: Immobiline gels nonlinear pH 3–10 in first dimension
Annotations: 98 identified spots corresponding to 61 different genes
Expression information: No values for protein expression reported

Yeast-2DPAGE
URL: http://yeast-2dpage.gmm.gu.se
Location: Göteborg, Sweden
Technology: Immobiline gels nonlinear pH 3–10 in first dimension
Annotations: 136 identified spots corresponding to 92 different genes
Expression information: Expression changes for all identified proteins under a number of different growth conditions; possible to cluster proteins in relation to their pattern of expression

added in relation to the gel area; about 5μl is added to a 10 mm^2 gel piece). This addition elutes the generated peptides. Let the tubes stand at room temperature for at least 30 min to allow for elution of the peptides before applying to MALDI target. The tube (with gel piece and solution) can also be frozen at -20° or -80° for later MS analysis.

Application to MALDI Target. Of the elution solution, 0.5 μl is removed from the tube containing the gel piece and mixed on the MALDI target with 0.5 μl of matrix solution (usually α-cyano-4-hydroxycinnamic acid [10 mg in 1 ml of water/acetonitrile (1 : 1)]. Crystals form on the MS target by evaporation and the sample is ready to be loaded into the MALDI-MS for analysis.

9. Access to Databases

There are currently a number of laboratories that maintain yeast proteomics databases (Table III). These databases for yeast are, like most 2D-related proteomics databases for other organisms, mainly concerned with presenting annotations for resolved proteins in their 2D-PAGE pattern. However, the YPM database contains information about the protein expression change during a shift from 23° to 36°, and YEAST-2DPAGE has launched the presentation of expression values for all the identified proteins for different experimental samples (different growth conditions or genetic backgrounds). Standardization in 2D analysis has always been a concern because the position of individual proteins in the pattern is sensitive to minor deviations in running conditions between laboratories. This is a major drawback in building extensive proteomics databases. Some of the databases are based on the carrier ampholyte technology for the first-dimensional separation, whereas others have adapted the Immobilines. Transfer of data between these two systems is unfortunately difficult for anything but the most abundant proteins.[14,17] The Immobiline technology appears to be one step toward the goal of interlaboratory 2D pattern standardization.[20]

Acknowledgments

This work was supported by the Swedish Research Council.

Author Index

Numbers in parentheses are footnote reference numbers and indicate that an author's work is referred to although the name is not cited in the text.

C

D

E

F

G

H

K

M

N

O

P

Q

R

S

T

W

Subject Index

A

B

C

D

E

F

G

H

I

K

N

O

P

R

S

T

U

Y

ISBN 0-12-182253-2